Volume 7

Mechanisms of Inorganic and Organometallic Reactions

Volume 7

Mechanisms of Inorganic and Organometallic Reactions

Edited by

M. V. Twigg

Imperial Chemical Industries P.L.C.
Billingham, Cleveland, United Kingdom

SPRINGER SCIENCE+BUSINESS MEDIA, LLC

The Library of Congress has cataloged this serial title as follows:

Mechanisms of inorganic and organometallic reactions. — Vol. 1- — New York: Plenum Press, c1983–

    v.: ill.; 24 cm.
    Editor: M. V. Twigg.
    ISSN 0740-8900 = Mechanisms of inorganic and organometallic reactions.

    1. Chemical reactions — Collected works. 2. Chemistry, Inorganic — Collected works. 3. Organometallic chemistry — Collected works. 4. Organometallic compounds — Collected works. I. Twigg, M. V.
QD501.M425   541.3'9'.05 — dc19   87-648073
                                                   AACR 2   MARC-S
Library of Congress   [8706]

ISBN 978-1-4613-6650-8   ISBN 978-1-4615-3710-6 (eBook)
DOI 10.1007/978-1-4615-3710-6

© 1991 Springer Science+Business Media New York
Originally published by Plenum Press, New York in 1991
Softcover reprint of the hardcover 1st edition 1991

Contributors

M. Bochmann   School of Chemical Sciences, University of East Anglia, University Plain, Norwich NR4 7TJ, U.K.

J. Burgess   Chemistry Department, The University, Leicester LE1 7RH, U.K.

E. C. Constable   University Chemical Laboratory, Lensfield Road, Cambridge CB2 1EW, U.K.

R. J. Cross   Chemistry Department, University of Glasgow, Glasgow, Scotland G12 8QQ, U.K.

R. van Eldik   Institut für Anorganische Chemie, Universität Witten/Herdecke, Stockumer Strasse 10, 5810 Witten-Annen, Germany

J. F. Endicott   Chemistry Department, Wayne State University, Detroit, Michigan, 48202

R. W. Hay   Chemistry Department, University of St. Andrews, St. Andrews, Scotland KY16 9ST, U.K.

D. A. House   Chemistry Department, University of Canterbury, Christchurch 1, New Zealand

L. A. P. Kane-Maguire   Chemistry Department, Wollongong University, Wollongong, NSW 2500, Australia

S. F. Lincoln   Department of Inorganic and Organic Chemistry, University of Adelaide, Adelaide, South Australia 5001, Australia

D. H. Macartney   Department of Chemistry, Queen's University, Kingston, Ontario K7L 3N6, Canada

A. McAuley   Chemistry Department, University of Victoria, Victoria, British Columbia V8W 3P6, Canada

K. G. Orrell   Chemistry Department, University of Exeter, Stocker Road, Exeter, Devon EX4 4QD, U.K.

R. D. Pike   Chemistry Department, Brown University, Providence, Rhode Island, 02912

A. J. Poë   Chemistry Department, University of Toronto, Toronto, Ontario M5S 1A1, Canada

K. Schneider   Institut für Anorganische Chemie, Universität Witten/Herdecke, Stockumer Strasse 10, 5810 Witten-Annen, Germany

G. Steadman   Chemistry Department, University of Swansea, Singleton Park, Swansea, Wales SA2 8PP, U.K.

T. Whitcombe   Chemistry Department, University of Victoria, Victoria, British Columbia V8W 3P6, Canada

Preface

The objective of *Mechanisms of Inorganic and Organometallic Reactions* is to provide an ongoing critical review of the literature concerned with the mechanisms of reactions of inorganic and organometallic compounds. The main focus is on reactions in solution, although solid state and gas phase studies are included where they provide relevant mechanistic insight.

Each volume covers an eighteen month literature period, and this, the seventh volume in the series, deals with papers published during July 1988 through December 1989. Where appropriate, there are references to earlier work, and also to specific sections in previous volumes. Coverage continues to span the whole area as comprehensively as possible in each volume, and although it is impossible for it to be absolutely complete, every effort is made to include all the important published work that is relevant to the elucidation of reaction mechanisms.

Numerical data are reported in the units used by the original authors, and they are only converted to common units when making comparisons. The basic format of earlier volumes is retained to facilitate tracing progress over several years in a particular topic; this can now be done for more than a decade worth of research.

In the last volume, ligand reactivity of both coordination and organometallic compounds were brought together in Chapter 12, and, in response to numerous positive comments from readers, this arrangement has been maintained. There have been some similar suggestions about oscillating reactions, and this topic may have a separate section in the next volume.

As always, thanks are due to all of the contributors for the tremendous effort that goes into collecting references and producing the chapters, and to the publishers for help and support during the production and distribution of the series.

The support and encouragement of many readers who send reprints or preprints of their publications are also very much appreciated. General comments and suggestions from readers are always most welcome.

M. V. Twigg

Contents

Chapter 3.   *Metal–Ligand Redox Reactions*

A. McAuley and T. W. Whitcombe

Part II. Substitution and Related Reactions

Chapter 4. Reactions of Compounds of the Nonmetallic Elements

G. Steadman

Chapter 5. Substitution Reactions of Inert-Metal Complexes— Coordination Numbers 4 and 5

R. J. Cross

Chapter 6.   Substitution Reactions of Inert-Metal Complexes—Coordination Numbers 6 and Above: Chromium

D. A. House

Chapter 7.   Substitution Reactions of Inert-Metal Complexes—Coordination Numbers 6 and Above: Cobalt

R. W. Hay

Chapter 8. Substitution Reactions of Inert-Metal Complexes— Coordination Numbers 6 and Above: Other Inert Centers

J. Burgess

Chapter 9. Substitution Reactions of Labile Metal Complexes

S. F. Lincoln

Part III. Reactions of Organometallic Compounds

Chapter 10. Substitution and Insertion Reactions

A. J. Poë

Chapter 11. Metal-Alkyl and Metal-Hydride Bond Formation and Fission; Oxidative Addition and Reductive Elimination

R. D. Pike

Chapter 12. Reactivity of Coordinated Ligands

R. W. Hay, E. C. Constable, and L. A. P. Kane-Maguire

Chapter 13. Rearrangements, Intramolecular Exchanges, and Isomerizations of Organometallic Compounds

K. G. Orrell

Part I
Electron Transfer Reactions

Chapter 1

Electron Transfer: General and Theoretical

1.1. Overview and General Aspects of Reactions in Fluid Media

The great majority of studies of electron transfer problems continue to be structured around the separate treatment of electronic and nuclear coordinates, as exemplified in Eq. (1), where the κ_i (i = el, nu) are electronic and nuclear transmission

$$k_{el} = \kappa_{el}\kappa_{nu}\nu_{nu} \tag{1}$$

coefficients, respectively, and ν_{nu} is an effective nuclear frequency. This general approach provides a convenient basis for the organization of material in this chapter.

Of considerable general interest is a four-volume work *Photoinduced Electron Transfer* edited by Fox and Chanon.[1-4] Many of the individual chapters in this work deal with important aspects of electron transfer and have little or no specific photochemical bias. The first volume of this series, *Part A. Conceptual Basis*,[1] contains several chapters on theoretical aspects of electron transfer, as is noted in the appropriate sections below. A monograph has been published which deals with tunneling processes, especially tunneling in electron transfer reactions.[5] Purcell and Blaive[6] have presented a general description of classical and quantum theories of electron transfer. The classical theory presented in this Chapter is largely that of Marcus.[7] The quantum approach presented includes a discussion of the basis of the Fermi golden rule, and the authors generate a golden-rule-based analytical expression for the rate constant which occupies six lines of text. This expression is relatively neatly factored into solvent and molecular vibrational contributions, and an electronic factor. Mikkelsen and Ratner[8-10] have been developing a time-dependent density operator approach for a dynamical theory of electron transfer reactions. This treatment differs from the more common use of time-independent functions, to approximate Franck–Condon factors in golden-rule-based theories, in that electronic–vibrational coupling is included for each matrix element and not introduced as a perturbational correction of the stationary state solutions of the wave equation.

Reimers and Hush[11] have published the first part of a general treatment of electron and energy transfer in bridged donor–acceptor systems. This approach generalizes electronic relaxation theories whose origin was in the theory of non-radiative excited state relaxation, and its key element is the strength of the coupling, l_s, between the acceptor and the solvent medium. It is deduced that golden-rule-based treatments are only appropriate when: (a) l_s is large compared to the electronic coupling, H_{RP}, between reactant and product states; and (b) l_s^2 is large compared to the square of the energy difference between initial and final states (reaction driving force). Brunschwig and Sutin[12] have presented a detailed semiclassical analysis of electron transfer reactions involving unstable conformational intermediates. In the normal free-energy regime (i.e., when the free-energy difference between reactants and products, $\Delta G°$, is smaller in magnitude than the vertical reorganizational energy, λ_{reorg}) intermediate pathways were shown to be important only when the λ^I_{reorg}, for electron transfer to form the unstable intermediate (I), is smaller than the reorganizational energy required to form the ultimate products. The authors also discussed the intervention of intermediates in the inverted region ($|\Delta G°| > \lambda_{reorg}$), and the conditions which prevail in order that conformational intermediates can "gate" electron transfer rates.

1.2. Electronic Coupling (κ_{el})

There has been a great deal of theoretical work on the problem of electronic contributions to electron transfer reactions, and some of these are reviewed in the recent monograph on tunneling.[5] Most of the recent theoretical studies have ranged through the often interrelated problems of the distance dependence of electron transfer,[5,9,13–16] the modulation of the effects of poor electronic coupling by vibrational motions,[9,10,17,18] the magnetic[19] or electric[20] field perturbations of the electronic coupling, the combinations of factors that contribute to electron transfer decay rates in frozen solutions,[21] and the relationship between electron transfer rates and electronic absorption (or emission) spectra[22] or nonradiative excited state relaxation.[23] The work discussed in this section deals with studies of electronic coupling in the "normal" free-energy domain ($|\Delta G°| < \lambda_{reorg}$). Other theoretical studies of the nonadiabatic regime are included in later sections on the "inverted" free-energy domain and on solvent effects.

Several theoretical treatments have shown that ν_{nu} in Eq. (1) should be an inverse function of the solvent relaxation time (usually the longitudinal relaxation time, τ_L; e.g., see Refs. 16, 18, 24, and 25). This becomes especially important when λ_{reorg} is mostly (or entirely) a function of solvent contributions. The significance of the inferred contributions of τ_L to ν_{nu}, in regard to material in this section, is that the condition $\kappa_{el} < 1$ implies $\nu_{el} < 2\nu_{nu}$, so that variations in τ_L should be inversely correlated with rate constant variations only for adiabatic reactions, while nonadiabatic reactions should be nearly independent of τ_L. Weaver and co-workers have capitalized on this property in their studies of metallocene-metallocinium self-exchange reactions.[26-29] Thus, the $[M(Cp)_2]^{+/0}$ self-exchange reactions which are very nearly "diffusion limited" in the gas phase[30] are several orders of magnitude slower in solution, indicating that the solvent component dominates λ_{reorg}. Furthermore, the $[Co(Cp)_2]^{+/0}$ self-exchange rate constant (k_{exch}^{Co}) is three times larger than k_{exch}^{Fe} for $[Fe(Cp)_2]^{+/0}$ in the gas phase, even though the M—Cp bond length changes are larger for the cobalt couple.[30] In constrast, the $k_{exch}^{Co} : k_{exch}^{Fe}$ ratios vary from about 1:1 to 20:1 in the solution phase, depending on the solvent, and with a much larger solvent dependence found for k_{exch}^{Co} than for k_{exch}^{Fe}.[29] Weaver and co-workers have inferred that the $[Co(Cp)_2]^{+/0}$ self-exchange reaction is adiabatic, while the $[Fe(Cp)_2]^{+/0}$ self-exchange is non-adiabatic, and they conclude that these differences in adiabaticity arise from the differences in donor–acceptor orbital overlap (the redox orbitals were inferred to be significantly ligand-centered in the former and largely metal-centered in the latter couple[28,29]). This inference is also consistent with the relative intensities of intervalence transitions in the corresponding bis-metallocene cations[28]; however, the comparison of the electronic coupling matrix elements of the thermally and photoinduced electron transfer events is complicated by the necessary differences in the perturbational Hamiltonians for these different processes.

The question of the adiabacity of the $[Co(NH_3)_6]^{3+/2+}$ self-exchange reaction has continued to be discussed. This self-exchange reaction is formally spin-allowed, but since it is a three-electron process it is orbitally forbidden. Newton has used an *ab-initio*, SCF approach to treat the $(^1A_{1g})[Co(NH_3)_6]^{3+}$-$(^4T_{1g})[Co(NH_3)_6]^{2+}$ self-exchange, and he has found that spin–orbit coupling ($^1T_{1g} \leftrightarrow {}^3T_{1g}$ and $^2E_g \leftrightarrow {}^4T_{1g}$) in the transition state results in a nearly allowed electron transfer with $\kappa_{el} \sim 10^{-2}$.[31] Geselowitz[32] has invoked a donor–acceptor "exchange" coupling argument to justify the proposals that the $[Co(NH_3)_6]^{3+/2+}$ self-exchange reaction is not spin-forbidden and that it is probably adiabatic. The exchange formalism has been employed by many workers as a factor contributing to the electronic matrix element, H_{RP}, which describes the coupling of the reactant and product potential-energy surfaces in the neighborhood of their formal crossing (for examples see Refs. 33 and 34). The Geselowitz application of this formalism seems to confuse a contribution to H_{RP} with the actual process of electron transfer.

1.2.1. The Distance Dependence of Electron Transfer Rates

As usual, most of the work on the contributions of electronic factors to electron transfer rates has dealt with the transfer of electrons between widely separated donors and acceptors. Most of the literature in the time period surveyed has dealt with elegant experimental studies of synthetic, modified metalloenzyme or of photosynthetic donor–acceptor systems.

The theoretical aspects of "long-range electron transfer" have been reviewed in a recent monograph.[5] Mikkelsen and Ratner have used their density matrix approach to examine the effects of a molecular species (H_2O) placed between the benzene radical anion donor and the benzene molecule acceptor and of the relative orientation of the donor and acceptor.[9,10] Electron transfer rates: (a) decreased rapidly with distance; (b) were dependent on the relative orientation of donor and acceptor (their C_6 planes parallel) most strongly at relatively small separation distances; and (c) were increased when a water molecule was inserted into the space between the planes of well separated, parallel donor and acceptor species. The facilitating effect of the intervening water molecule did not involve any increase in electron density on H_2O. Mikkelsen *et al.*[10] considered the influence of fluctuations in solvent polarization on the amplitude of the electronic wave functions far from the nuclei of origin, and they found that such fluctuations can alter the magnitude of the electronic transmission coefficient when the donor and acceptor are appreciably separated. These authors infer that this coupling between solvent motions and electronic wavefunctions can lead to a free-energy dependence of κ_{el}. Finally, Mukamel and co-workers[13,35] have considered the limit in which electronic coupling is small enough that electron transfer couples with solvent fluctuations, and they develop the theoretical analogy between the superexchange and electron hopping mechanisms for electron transfer with coherent and incoherent mechanisms of nonlinear spectroscopy. It is proposed that the nature of the electronic coupling could be best determined from spectroscopic measurements.[13,35]

Padden-Row, Verhoeven and co-workers have continued their studies of elegant molecules in which donors and acceptors are separated by relatively rigid, norbornyl and related linkers.[36-39] A theme of much of this work has been to establish the importance of "through bond" donor–acceptor coupling as a mechanism for long-range electron transfer. Recent studies involve electron transfer quenching of fluorescence from the electronically excited bis-methoxynaphthyl acceptor by the dicyanoethyl donor.[36,37] These authors have used time-resolved microwave conductivity measurements to establish the dipole moments of the product species which results from the quenching process, and they have inferred interchromophoric electronic coupling (H_{RP}) in related model compounds from perturbations of optical absorptions[39] or from measurements of ionization energies and electron affinities and calculated π and π^* orbital splitting energies.[36,38] Among the interesting results of this work are the observations that the electron transfer quenching rates in molecules (1) and (2) are different and solvent-dependent although the donor–acceptor, center-to-center distances are about the same.[36]

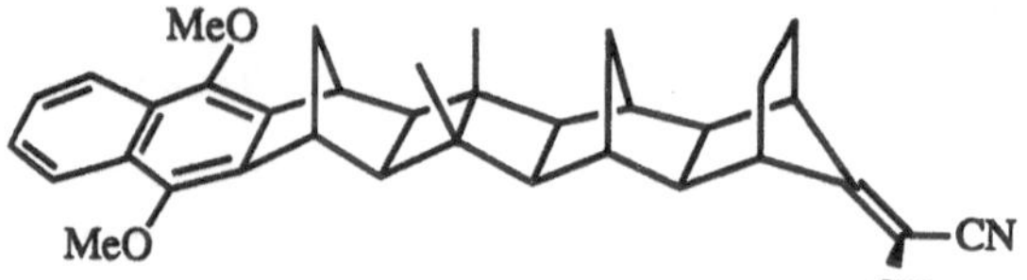

(1)

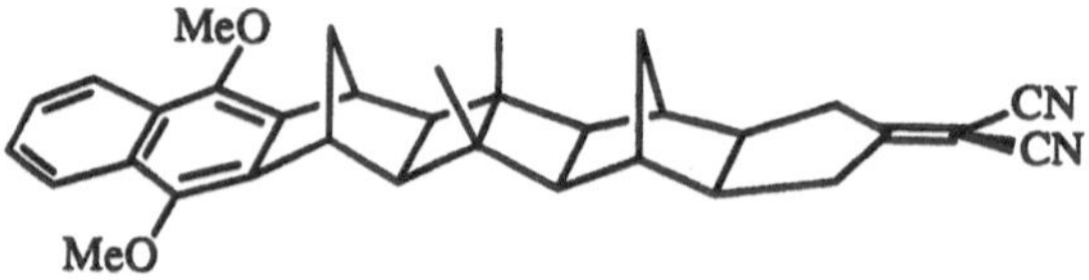

(2)

Table 1.1. Solvent Dependence of the Electron Transfer Quenching of the Donor Fluorescence in Covalently Linked Benznaphthyl-Dicyanoethyl Molecules

Solvent	D_{op} [a]	D_s [a]	$(1 - D_s^{-1})$ [b]	$k_1 \times 10^{-10}$ [c] (s^{-1})	$k_2 \times 10^{-10}$ [d] (s^{-1})	k_1/k_2
Dibutyl ether	1.958			4.7	1.1	4.1
Benzene	2.244	2.27	0.56	5.2	2.61	2.0
Diethyl ether	1.828	4.22	0.76	4.7	1.32	3.6
Ethyl acetate	1.883	6.02	0.83	4.5	1.48	3.0
Tetrahydrofuran	1.971	7.4	0.86	6.7	2.07	3.2
Acetonitrile	1.807	36.2	0.97	3.9	0.59	5.1

[a] Data from Refs. 52 and 53.
[b] Form of solvent dependence of $\Delta G°$ for charge separation based on the Born equation (e.g., see Ref. 43).
[c] Data from Refs. 54 and 55.
[d] Data from Ref. 36.

The authors interpret this in terms of a hyperconjugative component of the through-bond coupling and its solvational perturbation. That the ratio of $k_1:k_2$ is approximately correlated with the static dielectric constant of the solvent (see Table 1.1) could arise from a more mundane dependence of the electron transfer relaxation rate constant on κ_{nu}. Thus, when Eq. (2) holds[6,7], the nuclear contribution to the rate constant ratio becomes the relationship in Eq. (3). One does not

$$RT \ln \kappa_{nu} = (\lambda_{reorg}/4)(1 + \Delta G°/\lambda_{reorg})^2 \qquad (2)$$

$$-RT \ln k_1/k_2 = [\lambda_{reorg}(1) - \lambda_{reorg}(2)]/4$$
$$+ (\Delta G°)^2/4\lambda_{reorg}(1) - (\Delta G°)^2/4\lambda_{reorg}(2) \qquad (3)$$

expect these terms to cancel since the solvational component, λ_{solv}, of λ_{reorg} is a function of solvent polarizability ($e^2[D_{op}^{-1} - D_s^{-1}]$ in the classical limit) and molecular shape (the classical shape function for spherical molecules is $[(2r_A)^{-1} + (2rD)^{-1} - (r_{DA})^{-1}]$)[6,7,40-42] and these molecules differ in shape. Depending on the details of the molecular systems either the reorganizational or the free-energy-dependent terms might dominate Eq. (3). Contributions from either term would lead to a medium dependence of the rate ratio: the reorganizational term as noted above, and the free-energy term through the work required to separate charge in the dielectric medium.[43] Additional solvent-dependent variations, parallel in k_1 and k_2, can be attributed to ν_{nu} (see material above and in Section 1.4 below). Thus much of the variation observed in k_1 and k_2 can be attributed to $\kappa_{nu}\nu_{nu}$ rather than to κ_{el}. How much is not clear since the pertinent reorganizational and free-energy parameters are not known accurately enough to permit evaluation.

    Gray and Malmström have published a short review of studies of long-range electron transfer involving metalloproteins and modified metalloproteins[44] that includes a brief discussion of work of Gray's group on the distance-dependent rate constants of myoglobin in which the heme group was replaced by zinc mesoporphyrin (ZnP), and $[Ru(NH_3)_5]^{2+}$ was attached to one of four "surface histidines" of the protein. Electron transfer was induced by flash photolytic excitation of ZnP. The authors do not report any attempt to evaluate the distance-dependent contributions of κ_{nu}. As noted in the preceding paragraph, the measurement of decay rates cannot unequivocally be used to estimate κ_{el}, unless κ_{nu} is accurately assessed. The dependence of metalloprotein electron transfer rates on κ_{nu} has also been reviewed.[41,44,45] Durham, Millett and co-workers have reported[46] a method of attaching $[Ru(bpy)_3]^{2+}$ (one bpy a dicarboxybipyridine) to the lysine amino groups in cytochrome c. This permits the study of a large number of isomeric species in which photoinduced electron transfer $([*Ru(bpy)_3]^{2+} \rightarrow Fe^{III}(heme))$ and thermally induced recombinations $(Fe^{II}(heme) \rightarrow [Ru(bpy)_3]^{2+})$ can be examined.[46,47] These reactions were estimated to fall in the regime where $|\Delta G°| \sim \lambda_{reorg}$ and the variations in κ_{nu} were assumed to be small. These authors also examined the plausible range of lysine side-chain orientations, in order to set limits on the possible ranges of donor–acceptor distances.[47] Even in such seemingly rigid molecules, small changes in orientation can result in an appreciable range of donor–acceptor distances. Meade et al.[48] have examined the effects of driving-force variations on electron transfer rates of ZnP–cytochrome c containing various Ru(II)-am(m)ines substituted at histidine-33. The observations were fitted to Eq. (1) with κ_{nu} based on Eq. (2), and the authors inferred very small values of $\kappa_{el}\nu_{nu} \sim 4.4 \times 10^6$ and $2.9 \times 10^6\,s^{-1}$ for the charge separation and recombination steps, respectively. These values can be compared to $\kappa_{el}\nu_{nu} \sim 10^{12}$–$10^{13}\,s^{-1}$ expected for adiabatic electron transfer processes.[41]

Johnson et al.[49] have reported that the distance dependence of the rate constants (k^+) for charge transfer relaxation between a biphenyl-radical cation and a naphthyl group, linked with steroidal spacers, was very similar to that (k^-) found previously for analogous biphenyl radical anion-spacer-naphthyl molecules.[50] The authors note that the portion of the distance dependence originating from κ_{el} is difficult to unscramble from the distance dependence of κ_{nu}. Closs et al. have also noted that the product of these charge transfer rate constants (k^+k^-), roughly correlates with intramolecular triplet–triplet energy transfer rate constants in the closely related (4-benzophenyl)-(steroidal spacer)-(2-naphthyl) molecules[51] despite the difficulty in evaluating the different contributions expected from κ_{nu} in electron and energy transfer reactions.

1.2.2. Electric and Magnetic Field Effects on Electronic Coupling and Related Problems of Photoinduced Electron Transfer

External magnetic or electric fields can be used to alter the electronic matrix element for a nonadiabatic process either: (a) by changing the relative energies of the reaction transition state and the perturbing excited state when superexchange coupling contributes significantly to H_{RP}; or (b) by means of a change in the selection rules for coupling to the perturbing state owing to the lowered symmetry in the presence of an external field. In addition, the energy differences between reactants and products and between reactants and the transition state will generally be different in and out of an external field, and these variations will lead to different values of κ_{nu}.

There has been a continuing controversy about the interpretation of field effects on the primary step in photosynthesis.[56-64] This step consists of the transfer of an electron over an appreciable distance from a photoexcited bacteriochlorophyl dimer (B_D) to a bacteriopheophytin (Ph). The recent discussion has centered on the role of an accessory bacteriochlorophyll (B_A) which is spatially intermediate between the donor and acceptor. The models most discussed can be simplified as: (a) a sequence of steps, or electron "hops," (B_D^*, B_A, Ph) $\rightarrow (B_D^+, B_A^-, Ph) \rightarrow (B_D^+, B_A^-, Ph^-)$; or a single-step process, $(B_D^*, B_A, Ph) \rightarrow$ (B_D^+, B_A, Ph^-), in which the electronic coupling of the donor and acceptor is enhanced through superexchange couplings with the accessory bacteriochlorophyl. The superexchange pathway has been advocated by Jortner and co-workers.[58-60] Marcus has objected to this model, arguing that calculated superexchange coupling terms are too small to account for the observed rates, and that matrix elements required for this model are not consistent with the observed singlet–triplet energy difference of the (B_D^+, B_A, Ph^-) radical pair.[56] Marcus[56,57] further argues that a hopping model is consistent with observations if the intermediate and product potential-energy surfaces are so strongly coupled that the (B_D^+, B_A^-, Ph) state does not have a measurable lifetime (a "nonadiabatic/adiabatic" mechanism), and he proposes that the magnitude of the influence of an external electric field on the primary electron transfer reaction rate could distinguish between these mechanisms.[56] Boxer and co-workers[61,62] have given evidence that the electric-field effect on the fluorescence yield is much smaller than predicted by the superexchange model. Scherer and Fischer[63] have treated the six pigments of the *Rps. viridis* reaction center quantum mechanically, in the π-electron approximation, as a hexamer, and their results are consistent with the "nonadiabatic/adiabatic" mechanism. Lösche *et al.*[64] have discussed recent Stark spectroscopic studies of the bacteriochlorophyll reaction center, and they have presented a critical review of attempts to infer the nature of charge transfer excited state coupling to the optical absorption. These authors tentatively infer that the superexchange mechanism appears to be inconsistent with the observations, but that local field corrections introduce enough uncertainty to preclude a definitive interpretation. Bixon *et al.*[65] have presented a superexchange coupling argument which accounts for the reaction center fluorescence polarization data of Boxer and co-workers,[61,66] and which also accounts for the directionality of electron transfer in the photosynthetic reaction center. This model attributes most of the primary electron transfer behavior in the reaction center to superexchange coupling, but it does allow for some contributions of thermally activated electron transfer to the accessory bacteriochlorophyll. Reimers and Hush[11] have inferred that an accessory bacteriochlorophyll would have to be nearly at resonance with electronically excited bacteriochlorophyll donor in order to significantly impact the donor-to-acceptor electron transfer rate through "superexchange" coupling.

The X-ray structural studies of the photosystem II reaction center of *Rhodopseudomonas viridis* and related aspects of plant membrane structure have been reviewed.[66] Photoinduced charge separation in synthetic, porphyrin-based molecules which mimic aspects of the natural photosystems have also been reviwed.[67]

1.3. The Free-Energy Dependence of Electron Transfer Reactions: The "Inverted Region" Problem

The role of the solvent in dictating the behavior of electron transfer rate constants when $|\Delta G^\circ| > \lambda_{reorg}$ has continued to be vigorously discussed. The Reimers and Hush approach emphasizes the coupling between products and the solvent,[11] and this implies different kinds of systems might couple to the solvent better than do others in this regime. Thus, the form of the variation of κ_{el} with ΔG° may be system-dependent. Kakitani, Mataga and co-workers[21,68,69] have continued to explore their hypothesis[70] that much of the difficulty in observing inverted region behavior has originated from the larger force constants required to describe near solvent motions around ions than around neutral molecules. Much of their recent work has involved Monte Carlo simulations of solvent motion accompanying charge separation and charge recombination processes. In a closely related study, Carter and Hynes[71] have used a molecular dynamics approach to demonstrate a difference in effective solvational force constants for charged and neutral solutes, and they also infer a significant dependence of the free-energy dependence of the rate constant on these solvational differences. The asymmetries in the free-energy dependencies of electron transfer rate constants that have been inferred by Kakitani and Mataga and by Carter and Hynes have been shown by Tachiya[72,73] to be the result of the way in which those authors defined the reaction coordinate. With the reaction coordinate properly defined, the rate constant has a symmetrical dependence on ΔG°,[69,72,73] but the dependence may be appreciably flattened near the regime where $|\Delta G^\circ| = \lambda_{reorg}$.[69] Marcus[22] has published a very readable paper which develops the analogy between charge transfer (CT) absorption or emission spectra and the free-energy dependence of electron transfer rate constants. He points out that the observation of absorption maxima in the CT spectra is equivalent to the observation of the inverted region for thermally activated electron transfer processes, and that the CT absorption bandwidths and absorption maxima are functions of the same reorganizational parameters which determine electron transfer rates. As a consequence, a comparison of charge transfer spectra is equivalent to the comparison of the free-energy dependencies of electron transfer rate constants, and a comparison of the spectra of photoinduced charge separation, charge recombination, and charge shift processes indicates that there is no significant peculiarity in band shape which can be attributed to potential-energy surface asymmetries induced by differences in solvational force constants.[22]

Yang and Cukier[74] have used a density matrix approach to examine the influence of solvent dynamics on electron transfer rates in the inverted region. They found nonadiabatic rate constants in the highly exoergic regime correspond to surface crossing over an extended potential-energy region (i.e., beyond the formal surface crossing region) owing to nuclear tunneling and, since the tunneling frequency depends on the barrier width, the rates in this regime should be a function of solvent relaxation time.

Gould, Farid and co-workers[75-77] have continued to report on their studies of organic radical ion recombination reactions. These authors generate radical ion pairs by means of electron transfer quenching of electronically excited donors. They infer electron transfer recombinations rate constants, within the resulting collision complexes, based on spectroscopically determined yields (complemented by photoacoustic detection methods in Ref. 75) of the quenching products and on the assumption that the rate of diffusional escape from the primary solvent cage is independent of the nature of donor or acceptor. This technique provides a convenient means for generating electron transfer reagents with very large driving forces. The most recent contribution[75] demonstrates there is very little difference in the inverted region behavior of charge recombination and charge shift reactions (although small differences in λ_{solv} were inferred). This group has also compared the inverted region recombination rates of ion pairs formed from excited state electron transfer quenching reactions to those formed from direct excitation of charge transfer complexes.[76] It was presumed that these different strategies would generate solvent separated and contact radical ion pairs, respectively. The parameters extracted from the fit of the inferred electron transfer rate constants to the low-temperature limiting form of an expression proposed by Kestner et al.[78] (i.e., an expression applicable when there is one active high-frequency molecular vibrational mode contributing to the molecular component of λ_{reorg}, and when the vibrational quanta of this mode are much greater than $k_B T$) were consistent with this presumption: they found the "electronic coupling" parameter to be larger and the solvent reorganizational energy to be smaller in the contact radial ion pair reactions. They also found that there were no qualitative departures from expectation based on simple classical or semiclassical models. This group has also found that the recombination rates were slowed the most by perdeuteration of methyl substituents (only aromatic cationic species were deuterated). The resulting k_H/k_D ratios tended to increase (from 1.1 to 1.8) with $|\Delta G°|$, qualitatively consistent with expectation for inverted region behavior. Such an isotope effect indicates that vibrational quantization is an important factor in this region and could arise from the effects of changes in zero point energy, nuclear tunneling, and/or the density of acceptor states. The observed isotope effects are only likely to be consistent with C—H stretching frequencies, if the small value of $\lambda_{vibn} \sim 0.3$ eV and the assumption of a single quantized acceptor mode are both correct; however, it is not clear how the single-mode argument can be correct.

In a related study, Levin et al.[79] have examined the relaxation of triplet exciplexes through electron transfer pathways. The magnetic field dependence of the radical ion pair recombination rate constants indicated that these reactions were predominantly of triplet radical ion pairs recombining through electron transfer to regenerate the singlet ground states. The inferred rate constants exhibited inverted region behavior, a classical (6, 7) dependence on the dielectric medium, and they were fitted to the single quantum mode equations of Kestner *et al.*[78]; a value of $\lambda_{vibn} \sim 0.7$ eV was employed, and a very small medium-dependent (attributed to variations in geminate radical separation) electronic coupling matrix element (<1 cm^{-1}) was inferred.

Elliott, Kelley and co-workers[80] have reported that the electron transfer quenching rate constants of electronically excited polypyridyl-ruthenium(II) complexes by covalently linked, diquat acceptors decreased as the reaction driving force increased. Meyer and co-workers[81] have reported some very clever studies in which the charge transfer excited states of substituted [(4,4'-X$_2$bpy)Re(CO)$_3$(Py-D)]$^+$ complexes are quenched by electron transfer from the covalently linked donor to Re(II). This results in a radical anion pby$^-$-to-radical cation D$^+$ pair held together by coordination to Re(I). This radical ion recombination rate was found to be about two orders of magnitude slower than the electron transfer CT-excited state quenching step, and well within the inverted region, so that increases in driving force (achieved by means of variations in X) resulted in decreases in the electron transfer recombination rate constants, k_{elr}. These authors[81] compared the energy-gap dependencies of k_{elr} to those of the related [(4,4'-X$_2$bpy)Re(CO)$_3$(py-L)]$^+$ (L an innocent functional group) triplet state relaxation rate constants, k_{tnr}. They found the respective slopes of plots of ln k_{elr} and ln k_{tnr} *vs* $\Delta E_{1/2}$ (difference in electrochemical potentials) or E_{em} to be -3.4 and -7.3 (eV)$^{-1}$. Since these are both formally charge recombination processes, the contrasting dependencies on driving force seem surprising. These authors discuss this behavior using a theoretical formalism which has been developed[82] to describe electronic excited state relaxation of molecules in the limit that there is a single high-frequency (usual OH, NH, or CH) nuclear motion which couples to the relaxation process. This approach implies low-frequency solvent modes do not play a role in these electron transfer processes in ambient fluid solutions. It seems more likely that the larger charge separation in the k_{elr} process leads to a larger value of λ_{solv} than is required for the k_{enr} process, so that the electron transfer recombinations do not fall as deeply in the inverted region as do the CT excited state relaxations. An additional contributing factor could arise from the appreciable change of charge density at Re in the CT excited state relaxation process, since this could result in some low-frequency vibrational modes contributing significantly to λ_{vibn} (see the discussion in Section 1.5), and the low-frequency modes may couple better to the solvent leading to more efficient dissipation of the excess energy of reaction and more classical behavior, as suggested by the arguments of Reimers and Hush,[11] for k_{enr} than for k_{elr}.

1.4. The Effects of Solvent Dynamics

Calef[24] and Maroncelli *et al.*[83] have published very readable reviews, the former summarizing the basic theory of coupling between electron transfer and solvent relaxation times, while the latter emphasizes current directions in modeling non-Debye solvents (i.e., solvents with multiple, or nonexponential relaxation times). Bagchi[25] has written a broader, more technical review. Much recent work has involved the "mean spherical approximation" (MSA) in which an ionic (or dipolar) solute and the dipolar solvent are treated as hard spheres[17,83-87] and by means of which some molecular properties can be attributed to solvent molecules in the first solvation shell. The MSA model works well for aprotic solvents, but not so well for protic solvents,[83,86] and the inadequacies of the MSA model in protic solvents have been attributed to the disruptive (or "structure breaking") effects of solute solvation on solvent hydrogen bonding networks.[86] Computer simulations have been employed to compare aspects of the different models.[83,87] A theoretical treatment of Debye solvents which have multiple relaxation times has appeared[88] and a very general approach[89] has related the effect of solvation dynamics on electron transfer rate constants to parameters from nonlinear optical measurements. A damped quantum oscillator model has been used to describe the solvent coupling to the reactants' progress along the electron transfer reaction coordinate.[89,90] This model indicates that solvent dynamics will only be important in the inverted regime,[89] and the theoretical behavior of such systems indicates frictional effects could be a significant factor even at very low temperatures.[90] Helman[91] has used a stochastic approach to obtain relations describing the coupling of electron transfer rates to the dielectric relaxation times of a homogeneous dielectric medium.

Kosower and co-workers[92] have found the photoinduced, barrierless charge separation processes of substituted polyaromatics to be controlled by solvent relaxation behavior over a large temperature range in alcohol solvents. Heitele and Michel-Beyerle[93] reported on the complex solvent- and temperature-dependent electron transfer fluorescence quenching in some covalently linked aromatic donor–acceptor compounds in viscous solvents. These authors have attempted a critical comparison between current theoretical models and their experimental results, and the limitations of current theoretical models are discussed.

Eisenthal and co-workers[94] and Gaudel *et al.*[95] have published preliminary reports of their fs flash photolysis studies of recombination processes following the two-photon ionization of water. Both studies report that the "geminate," $H_3O^+-e_{aq}^-$ (or $OH-e_{aq}^-$) recombination processes occur during the first 100 ps following excitation, and that the distance over which electron and cation motions are correlated is relatively large, ≥ 11 Å. In principle, such studies could provide sensitive insights into the effects of solvent dynamics on charge recombination rates. Mataga and co-workers[96] have reported the preliminary results of photo-induced charge separation rates for a CT complex (tetracyanobenzene–toluene) in acetonitrile using fs and ps flash photolysis. Hirata and Mataga have used ps flash photolysis techniques to examine electron–cation geminate recombination in nonpolar solvents.[97] Magata and co-workers[98] have examined the rates of the photoinduced charge separation and electron transfer recombination processes in covalently linked porphyrin complexes. The approximately 30-fold faster charge separation was attributed to solvational differences of the ions and neutral molecules.

1.5. Metal-to-Metal and Ligand-to-Ligand Charge Transfer ("Intervalence" Transfer)

Piepho[99] has treated intervalence transitions using a relatively simple three-center molecular orbital model in which vibronic coupling is introduced as a pseudo-Jahn–Teller perturbation. This could be a valuable approach, and it can probably be generalized to a great variety of problems. In this report,[99] Piepho uses the approach to calculate intervalence transition band shapes for a range of vibronic couplings, and she discusses the vibronic criteria for valence trapped and delocalized systems.

Zhang et al.[100] have published a quantum-mechanical treatment of absorption line shapes in bridged, mixed valent dimers. This treatment allows vibronic perturbations to mix the electronic potential-energy surfaces in complexes in which the exchanging electron is completely or partially delocalized. When this three-center treatment is applied to the Creutz–Taube ion, it is consistent with complete delocalization of the odd electron and with very little mixing of the three-center electronic states in this complex.

Katriel and Ratner[101] have developed a simple model to describe the effects of selective solvation on intervalence absorption spectra (e.g., see Ref. 102). Lewis and Obeng[103] have reported that the ionic-strength-dependent intervalence absorption maximum (E_{op}) is also dependent on the reduction potential of the cationic oxidant. This observation is attributed to an effect of ionic aggregation. Jozefiak et al.[104] have reported infrared evidence that mixed valence, conjugated semiquinones, prepared electrochemically in DMF, are valence trapped.

Doorn and Hupp have used preresonance Raman spectra in an analysis of the vibronic components which contribute to the intervalence absorption maximum of $[(CN)_5Ru^{II}\text{-}CN\text{-}Ru^{III}(NH_3)_5]$[105] and to the MLCT absorption maximum of $[(bpy)Ru(NH_3)_4]^{2+}$.[106] These authors employ the time-dependent scattering approach of Heller[107] to obtain the nuclear displacements of several vibrational modes coupled to the electronic transitions. They find in each case that several vibrational modes, spanning a wide range of frequencies, do contribute significantly to the photoinduced electron transfer processes. Hopkins and co-workers[108,109] have used a two-color, ps Raman technique to investigate interligand electron transfer in Ru(II)–*tris*-polypyridyl complexes, and they find vibrational relaxation of the electronically excited mole ule occurs within about 30 ps of excitation, after which interligand equilibration occurs more slowly than 5×10^{-7} s.

Chapter 2
Redox Reactions between Two Metal Complexes

2.1. Introduction

The diversification of the investigations of the kinetics and mechanisms of electron transfer reactions between metal complexes is continuing. An increased interest in the reactions of heavier transition metals and the emergence of a clearer understanding of the kinetics and energetics of electron exchange and electron transfer in nonaqueous solution is observed. The format of this chapter remains much the same as in previous volumes, with sections covering metal aqua and oxo ions, metal ion complexes, and metalloproteins. The rate constants and activation parameters are presented in four tables, with the data for electron transfer reactions between metal complexes in Table 2.1, directly measured electron self-exchange rate data in Table 2.2, and the data for intramolecular and intermolecular reactions involving metalloproteins in Tables 2.3 and 2.4, respectively. The discussion in each section of the text and the data in the tables are arranged, for the most part, in the order of increasing atomic number of the central metal ion in the reductant.

2.2. Reactions of Metal Aqua and Oxo Ions

2.2.1. Titanium

The outer-sphere reductions of $[Ru_2(O_2CCH_3)_4Cl_2]^-$ by the Ti^{3+} aqua ion[1] and by $[Ti(hedta)(H_2O)]^{(2)}$ are dominated by pathways involving their respective conjugate bases, with the difference observed in the rate constants attributed to the relative redox potentials of the reductants. The reductive quenching of the ^2E

excited state of a series of tris(polypyridine) chromium(III) complexes by Ti(III) involves both Ti^{3+} and TiOH^{2+} pathways.[3]

2.2.2. Vanadium and Chromium

The pH dependences of the kinetics of the Cr^{2+} and V^{2+} reductions of several isomers of a tetranuclear cobalt(III) ammine complex (1) containing bridging

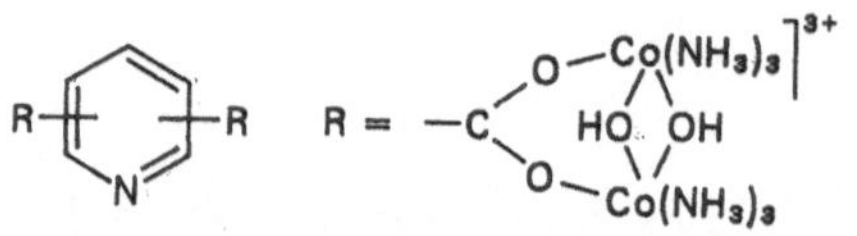

(1)

pyridinedicarboxylate ligands have been investigated.[4] The reductions of the 2,4-, 2,5-, and 3,5-dicarboxylate isomers by Cr^{2+} proceed by an inner-sphere mechanism when the pyridine nitrogen is deprotonated and by an outer-sphere pathway upon protonation. Steric hindrance at the pyridine nitrogen of the 2,6-dicarboxylate isomer dictates an outer-sphere mechanism for the acid-independent reductions by Cr^{2+} and V^{2+}. The Cr^{2+} and V^{2+} reductions of pentacyano(N-heterocycle)cobalt(III) complexes occur by different mechanisms.[5] In the Cr^{2+} reactions an inner-sphere mechanism, with a rate-determining electron transfer from Cr^{2+} to the bridging ligand, is proposed on the basis of a comparison of the rate constants to those for the Cr^{2+} reductions of analogous [Co(NH$_3$)$_5$L]$^{3+}$ species. A "dead-end" cyano-bridged binuclear complex is formed between V^{2+} and the oxidant, with the electron transfer proceeding by an outer-sphere reaction between the monomeric reactants.

The inner-sphere reductions of [Co(NH$_3$)$_5$(SCONHR)]$^{2+}$ and [Co(NH$_3$)$_5$(OCSNHR)]$^{2+}$ by Cr^{2+} involve attack at the remote oxygen and sulfur atoms, respectively, with a subsequent isomerization of the O-bonded chromium(III) product in the former reaction.[6] The unusually rapid reactions of the S-bonded cobalt(III) complexes are attributed to a structural *trans*-effect on the Co—N bond length, reducing the reorganization energy needed to form the transition state. A kinetic study of the Cr^{2+} reduction of [Co(NH$_3$)$_5$(pyruvate)]$^{2+}$ reveals that the rate of reduction is dependent on the nature of pyruvate ligand, with the keto form about 400 times as reactive as the hydrated form.[7] An inner-sphere mechanism has be postulated for the Cr^{2+} reduction of [Co(NH$_3$)$_5$(pyridine N-oxide)]$^{3+}$ on the basis of the rate and activation parameters.[8] The outer-sphere Cr^{2+} reduction of [Co(sepulchrate)]$^{3+}$ is catalyzed by halide ions, with the ion-pair formation constants for [Co(sep), X]$^{2+}$ estimated to be 5.5, 2.3, and 1.7 M^{-1} for Cl$^-$, Br$^-$, and I$^-$, respectively.[9]

An application of the Marcus relationship to the rate constants for the oxidations of [VO(TCDA)] (TCDA = 1,4,7-triazacyclononane-N,N'-diacetate) and [VO(DOCDA)] (DOCDA = 1-oxa-4,7-diazacyclononane-N,N'-diacetate) by [Ni([9]aneN$_3$)$_2$]$^{3+}$ yields self-exchange rate constants of 0.75 and 0.54 M^{-1} s^{-1} for the $[VO(TCDA)]^{+/0}$ and $[VO(DOCDA)]^{+/0}$ couples, respectively.[10] The oxidations of pentaaquaorganochromium(III) complexes by $[Ni(cyclam)]^{3+}$ and $[Ni([9]aneN_3)_2]^{3+}$ yield short-lived CrR^{3+} species, which decompose by homolysis or intramolecular electron transfer reactions.[11] The oxidation of Cr^{3+} by Mn^{3+} in sulfuric acid solution involves the rate-determining oxidation to Cr(IV), followed by rapid oxidations to yield Cr(VI).[12]

The reductions of the superoxochromium(III) ion, CrO_2^{2+}, by $[Ru(NH_3)_6]^{2+}$, $[Co(sep)]^{2+}$, and $[V(H_2O)_6]^{2+}$ proceed by an outer-sphere mechanism.[13] An inner-sphere mechanism is postulated for the reductions of CrO_2^{2+} by $[Fe(H_2O)_6]^{2+}$, $[Co([14]aneN_4)(H_2O)_2]^{2+}$, and $[Co([15]aneN_4)(H_2O)_2]^{2+}$ on the basis of observed μ-peroxo intermediates. The kinetics of the reduction of a peroxochromium(IV) complex, $[Cr(dien)(O_2)_2]$, with VO^{2+}, $[Fe(CN)_6]^{4-}$, and Ti^{3+} have been studied.[14] An inner-sphere reduction of only the Cr(IV) center is observed with VO^{2+}, while Ti^{3+} and $[Fe(CN)_6]^{4-}$ reduce both Cr(IV) and peroxo functions, in inner- and outer-sphere processes, respectively.

2.2.3. Iron

The effects of micellar medium on the kinetics of the outer-sphere reduction of $[IrCl_6]^{2-}$ by Fe^{2+} have been studied in the presence of hexadecyltrimethylammonium chloride and sulfate.[15] The experimental data are accounted for in terms of a two-pseudophase model. The oxidation of $[Mo(H_2O)_6]^{3+}$ by $[Fe(H_2O)_6]^{3+}$ in p-toluenesulfonic acid proceeds through Mo(V) monomer and dimer, $[Mo_2O_4]^{2+}$, species.[16] The mechanism of the reduction of MoO_2Cl_2 by Fe^{2+} in 8 M HCl is not known but believed to be a chloro-bridged inner-sphere process.

2.2.4. Molybdenum and Tungsten

The kinetics of the oxidations of a cuboidal Mo(III) ion, $[Mo_4S_4(H_2O)_{12}]^{4+}$, by $[Co(NH_3)_5X]^{n+}$ (X = halide, H_2O, and $O_2CCH_3^-$) and $[Co(bpy)_3]^{3+}$ have been studied, and on the basis of log–log correlations with the corresponding reductions by V^{2+} and $[Ru(NH_3)_6]^{2+}$, an outer-sphere mechanism is postulated, with the exceptions of X = $O_2CCH_3^-$ and perhaps F^-.[17] The application of the Marcus relationship to the rate constants for the reductions of $[Mo_4S_4(H_2O)_{12}]^{5+}$ by Cr^{2+} and V^{2+} yielded a $[Mo_4S_4]^{4+/5+}$ self-exchange rate constant of 760 M^{-1} s^{-1}.[18] From the kinetic studies of several outer-sphere cross reactions of the $[Mo_4S_4(edta)_2]^{4-/3-}$ and $[Mo_4S_4(edta)_2]^{3-/2-}$ couples it was observed that replacing the coordinated waters by $edta^{4-}$ ligands increases the electron self-exchange rate constants to 1.5×10^7 and 7.7×10^5 M^{-1} s^{-1}, respectively.[19] The acid dependence observed in the kinetics of the oxidation of $[Mo_4S_4(edta)_2]^{3-}$ by VO_2^+ is attributed to a facilitation of the electron transfer through protonation of the oxidant.[20]

The reduction of $[IrCl_6]^{2-}$ by the W(IV) aquo ion, $[W_3O_4(H_2O)_9]^{4+}$, proceeds by an outer-sphere pathway involving the conjugate base of $W_3O_4^{4+}$ ($K_a = 0.24$ M), with a rate constant that is $\approx 10^6$ greater than for the $Mo_3O_4^{4+}$ analog.[21] Inner-sphere mechanisms, involving the formation of hydroxyl radical intermediates, have been proposed for the reductions of $[MoO(OH)(O_2)_2]^-$ and $[WO(OH)(O_2)_2]^-$ by Fe^{2+} and Eu^{2+} in acidic aqueous solution.[22]

2.3. Reactions of Metal Ion Complexes

2.3.1. Chromium

The electron self-exchange rate constant for the $[Cr(CNdipp)_6]^{0/+}$ couple (CNdipp = 2,6-diisopropylphenyl isocyanide) in CD_2Cl_2 has been measured between -89 and $+22\,°C$ using ^1H NMR line-broadening techniques, with an extrapolated value of $1.8 \times 10^8\,M^{-1}\,s^{-1}$ determined for 25 °C.[23] The kinetics of the outer-sphere oxidations of tris(polypyridine)chromium(II) complexes by a series of tris(chelate)cobalt(III) species have been studied in aqueous solution.[24] The cross-reaction rate constants obey the Marcus relationship, with the exception of $[Co(bpy)_3]^{3+}$ and $[Co(phen)_3]^{3+}$, for which mild nonadiabaticity ($\kappa_{12} = 0.13$) was observed.

The quenching of $[^*Cr(bpy)_3]^{3+}$ by substituted ferrocenium ions proceeds by an energy transfer mechanism, while the diffusion-controlled quenching by the corresponding ferrocenes involves both energy and electron transfer pathways. The $[Cr(bpy)_3]^{2+}$ and ferrocenium products of electron-transfer quenching undergo very rapid back electron-transfer reactions.[25] Reductive quenching and back electron transfer reactions have also been studied between a series of excited state substituted tris(polypyridine)chromium(III) complexes and $[Co([14]andN_4)$-$(H_2O)_2]^{2+}$.[26] A Marcus analysis of the back electron transfer data indicates that the reactions are adiabatic.

2.3.2. Manganese

The rate constants and activation parameters (including ΔV^*) for electron self-exchange in the $[Mn(CNC(CH)_3)_6]^{+/2+}$ and $[Mn(CNC_6H_{11})_6]^{+/2+}$ couples have been determined by ^{55}Mn NMR line broadening in several pure and binary organic solvent systems.[27] The values of ΔV^* cover a range of about $12\,cm^3\,mol^{-1}$ (-9 to $-21\,cm^3\,mol^{-1}$) with no simple correlation with solvent parameters observed. A self-exchange rate constant of $0.7 \pm 0.4\,M^{-1}\,s^{-1}$ has been calculated for the $[Mn(edta)(H_2O)]^{0/-}$ and $[Mn(cdta)(H_2O)]^{0/-}$ couples from the application of the Marcus relationship to outer-sphere cross-reactions with a variety of metal complexes in aqueous solution.[28] Deviations from the correlation were observed for the nonadiabatic reactions with osmium tris(polypyridine) complexes.

The rate constants for electron transfer between a manganese carbonyl cation and anion, and for the disproportionation of the radical products have been determined from cyclic voltammograms in THF[29,30]; see Eq. (1). The electron transfer from the anion to the cation is slower for the *trans* isomer ($k_{et} = 35\,M^{-1}\,s^{-1}$, $k_d = 5.3\,M^{-1}\,s^{-1}$)[29] than for the *cis* isomer ($k_{et} = 200\,M^{-1}\,s^{-1}$, $k_d = 20\,M^{-1}\,s^{-1}$),[30]

$[Mn(CO)_2(dppe)_2]^+ + [Mn(CO)_2(dppe)(\eta^1\text{-dppe})]^-$

$$\underset{k_d}{\overset{k_{et}}{\rightleftharpoons}} 2[Mn(CO)_2(dppe)(\eta^1\text{-dppe})] \quad (1)$$

owing to the 150 mV difference in the reduction potential of the isomeric $[Mn(CO)_2(dppe)_2]^+$ cation. The reversible three-electron transfer between two manganese porphyrins, $[(TTP)Mn\equiv N]$ and $[Mn(OEP)]$, is mediated by a nitrogen atom transfer.[31]

2.3.3. Iron, Ruthenium, and Osmium

The study of the pressure effects on the rate of electron exchange in the $[Fe(phen)_3]^{2+/3+}$ couple in aqueous and acetonitrile media yielded $\Delta V^+ = -2.2$ and $-5.9 \text{ cm}^3 \text{ mol}^{-1}$, respectively.[32] A volume of activation of $-7.0 \pm 2.0 \text{ cm}^3 \text{ mol}^{-1}$ has been determined for the electron exchange reaction of the ferrocene/ferrocenium couple in acetonitrile.[33] The effects of temperature and pressure ($\Delta V^+ = -3.0 \pm 0.2 \text{ cm}^3 \text{ mol}^{-1}$) of the ruthenocene/bromoruthene electron exchange rate constant in acetonitrile have been studied.[34] The self-exchange rate constants for the (carboxymethyl)cobaltocene/cobaltocenium and (hydroxymethyl)ferrocene/ferrocenium couples have been measured in aqueous and nonaqueous solvents, and used with other metallocene data to evaluate the electron matrix coupling element.[35] The rate constants determined for the oxidation of $[Fe((CH_3)_n phen)_3]^{2+}$ complexes ($n = 4, 6$) by $[Fe(DMF)_m(CH_3CN)_{6-m}]^{3+}$ species in acetonitrile have been correlated with the driving forces for the respective reactions.[36]

The outer-sphere electron transfer reaction between $[Fe(CN)_6]^{4-}$ and $[Co(NH_3)_5H_2O]^{3+}$ has been studied as a function of temperature and pressure in a variety of water–glycerol solvent mixtures.[37] While ΔV^+ remains essentially constant at $28 \text{ cm}^3 \text{ mol}^{-1}$, both ΔH^+ and ΔS^+ increase with an increase in the viscosity of the solvent. The intramolecular electron transfer processes in $(NH_3)_5CoLFe(CN)_5$ complexes, where L is 3,3'-dimethyl-4,4'-bipyridine, 4,4'-bipyridylacetylene, 2,7-diazapyrene, and 3,8-phenathroline, have been studied.[38] The activation free energies display an inverse dependence on the Fe–Co distance and, when corrected for solvent reorganization energies, are relatively constant at $14.0 \pm 0.5 \text{ kcal mol}^{-1}$.

The electron self-exchange rate constants for several Fe(II)/Fe(III) porphyrin couples have been measured by ^1H NMR line-broadening techniques in 5:1 acetone/water at $-20 \,^\circ C$.[39] The relative rate constants for the $[Fe(P)(1\text{-MeIm})_2]$ couples, P = octaethylporphyrin $\approx$ chlorin $<$ isobacteriochlorin, have been attributed to differences in outer-sphere reorganization, related to the steric bulk. The rate-determining step in the metallopophyrin-catalyzed reductions of dioxygen by substituted ferrocenes is the electron transfer between the ferrocene and the metalloporphyrin (M = Fe, Co, and Mn).[40] The Marcus relationship provides a rationale of the relative magnitudes of the rate constants in terms of the reaction driving forces and the self-exchange rate constants of the metalloporphyrins. Long-range electron transfer from photoexcited $[Ru(phen)_3]^{2+}$ to tris(polypyridine) complexes of cobalt(III), rhodium(III), and chromium(III) is mediated by DNA.[41] The DNA polymer provides an efficient intervening medium for donor–acceptor coupling, with surface-bound complexes displaying greater rate enhancements than intercalatively-bound species.

The electron self-exchange rate constants for the $[Ru([9]aneN_3)_2]^{2+/3+}$ ($5 \times 10^4 \ M^{-1} s^{-1}$)[42] and $[Fe([9]aneN_3)_2]^{2+/3+}$ ($4.6 \times 10^3 \ M^{-1} s^{-1}$)[43] couples have been measured by NMR line-broadening techniques and Marcus cross-reaction calculations, respectively. The autooxidation of the encapsulated ruthenium(II) complex, $[Ru(sar)]^{2+}$ (sar = 3,6,10,13,16,19-hexaazabicyclo[6.6.6]eicosane), involves a reversible disproportionation reaction between $[Ru(sar)]^{2+}$ and the deprotonated $[Ru(sar(-H))]^{2+}$ ion.[44] The comproportionation reaction of $[(bpy)_2Ru(O)(OH)]^{2+}$ and $[(bpy)_2Ru(OH)(OH_2)]^{2+}$ yields the ruthenium(IV) complex $[(bpy)_2Ru(O)(OH_2)]^{2+}$.[45]

The reductive quenching of excited state ruthenium(II) polypyridyl complexes by metal cyano complexes, $[M(CN)_8]^{4-}$ (M = Mo and W) and $[M(CN)_6]^{4-}$ (M = Fe, Ru, and Os), has been investigated.[46] While the cage escape efficiencies were considerably higher for the octacyanometallates than for the hexacyanometallates, there was little observed dependence on the relative thermodynamic driving forces. Enantioselectivity has been observed in the electron transfer reactions between $[Co(acac)_3]$ and two chiral ruthenium photosensitizers, $[Ru((-)-mncb)_3]^{2+}$ [(-)-mncb = 4,4'-((1R,2S,5R)-(-)-menthylcarboxy-2,2'-bipyridine][47] with $k_\Delta/k_\Lambda =$ 1.32 and $[Ru(S(-)-PhEtbpy)_3]^{2+}$ ($k_\Delta/k_\Lambda = 1.54$) or $[Ru(R(+)-PhEtbpy)_3]^{2+}$ [PhEtbpy = 4,4'-bis(C(O)HNCH(CH_3)C_6H_5)-2,2'-bipyridine].[48]

2.3.4. Cobalt and Rhodium

The electron self-exchange rate constants have been determined for the low-spin d^7/d^6 $[Co([9]aneS_3)_2]^{2+/3+}$ ($1.6 \times 10^5 \ M^{-1} s^{-1}$) and $[Co([9]aneN_3)-([9]aneS_3)]^{2+/3+}$ ($4.2 \times 10^4 \ M^{-1} s^{-1}$) couples, the former by ^1H NMR line-broadening techniques and the latter from the application of the Marcus relationship to the cross-reaction with $[Ru(NH_3)_6]^{2+}$.[42] The self-exchange rate constants derived for the high-spin d^7/low-spin d^6 $[Co(taptacn)]^{2+/3+}$ couple [taptacn = 1,4,7-tris(3-aminopropyl)1,4,7-triazacyclononane] from the reactions of $[Co(taptacn)]^{3+}$ with $[Co(sep)]^{2+}$ and $[Co([9]aneN_3)]_2^{2+}$ are in excellent agreement with the value ($0.04 \ M^{-1} s^{-1}$) determined directly from the reaction of $(+)$-$[Co(taptacn)]^{3+}$ with $[Co(taptacn)]^{2+}$.[49]

The rate constants for electron transfer reactions between stereo- and diastereoisomeric pairs of various $[Co\{(NH_3)_2,Me_3sar\}]^{5+/4+}$ ions ($\{(NH_3)_2,Me_3sar\} =$ 4,12,17- or 4,11,17-trimethyl-3,6,10,13,16,19-hexaazabicyclo[6.6.6]eicosane-1,8,-diamine) have been measured polarimetrically.[50] The surprisingly large differences in the observed rate constants for similar species are linked to the effects of lel_3 vs. ob_3 configurations on ligand field strengths and reduction potentials. The rate and activation parameters for the electron transfer reactions between several cobalt chlathrochelate complexes have been studied as a function of added electrolyte in a number of nonaqueous solvents.[51] The electron exchange reactions of a tetrahedral cobalt complex, $[Co(1-nor)]^{n+}$ (1-nor = 1-norbornyl) in the unusual oxidation states of Co(IV)/Co(III) and Co(V)/Co(IV) have been investigated.[52] Self-exchange rate constants of 1.3×10^6 and 2.0×10^6 M^{-1} s^{-1} have been determined at 40 °C for the $[Co(1\text{-nor})]^{0/-}$ (THF-d_8) and $[Co(1\text{-nor})]^{+/0}$ (CD_2Cl_2) couples, respectively, using ^1H and ^{59}Co NMR line-broadening measurements.

The kinetic data for a series of outer-sphere electron transfer reactions between the $[Rh_2(O_2CCH_3)_4(CH_3CN)_2]^{0/+}$ couple and nickel tetraaza macrocycles and iron and ruthenium tris(polypyridine) complexes in acetonitrile have been correlated in terms of the Marcus relationship, yielding a $[Rh_2]^{0/+}$ electron exchange rate constant of $3.0 \pm 1.7 \times 10^5$ M^{-1} s^{-1}.[53] A somewhat smaller value of $5.3 \pm 1.3 \times 10^4$ M^{-1} s^{-1} was determined directly for the couple, using ^1H NMR line-broadening techniques. The reductions of the $17e$ radical $[Rh(dmgH)_2PPh_3]$ by a series of $[XCo(dmgH)_2PPh_3]$ and $[CO(NH_3)_5X]^{2+}$ complexes (where X^- is a halide or pseudohalide ion) proceed by an inner-sphere mechanism, with the "normal" reactivity order, $Br^- > Cl^- > F^-$, observed.[54]

2.3.5. Nickel, Palladium, and Platinum

The self-exchange rate constants for several Ni(I)/Ni(II) couples, $[Ni(Me_6[14]aneN_4)]^{+/2+}$ ($k_{11} = 1.0 \times 10^4$ M^{-1} s^{-1}), $[Ni(Me_6[14]dieneN_4)]^{+/2+}$ (2.3×10^6), $[Ni(Me_2pyo[14]trieneN_4)]^{+/2+}$ (4.7×10^7), and $[Ni(Me_6[14]\text{-}1,4,8,11\text{-}tetraeneN_4)]^{+/2+}$ (1.5×10^8), have been determined from cross-reactions of the Ni(I) species with a series of outer-sphere oxidants.[55] The oxidants of sterically crowded N-methylated nickel(I) tetraaza macrocyclic complexes, $[Ni(Me_{10}[14]aneN_4)]^+$ and $[Ni(Me_4[14]aneN_4)]^+$, with cobalt(III) amine complexes proceed by an outer-sphere mechanism.[56]

A self-exchange rate constant of 48 M^{-1} s^{-1} was determined for nickel bis(dipeptido) complex couple, $[Ni(H_{-1}Aib_2)_2]^{2-/-}$, from the cross-reaction of $[Ni(H_{-1}Aib_2)_2]^-$ with $[Cu(H_{-3}Aib_3a)]^-$.[57] A novel ^{61}Ni EPR line-broadening technique has been used to directly determine the self-exchange rate constant for the $[Ni(H_{-2}Aib_3)]^{-/0}$ couple, 450 M^{-1} s^{-1}.[58] Kinetic studies have been carried out on the reductions of bis(6-amino-3-methyl-4-azahex-3-en-2-one oximato)nickel(IV), $[Ni(L^3)_2]^{2+}$, and 15-amino-3-methyl-4,7,10,13-tetraazapentadec-3-en-2-one oximato)nickel(III), $[Ni(L^1)]^{2+}$ by $[Co(phen)_3]^{2+}$ and the oxidation of $[Ni(HL^1)]^{2+}$ by $[Ni([9]aneN_3)_2]^{3+}$ in aqueous solution.[59] Marcus analysis of the cross-reaction kinetic data have led to calculations of self-exchange rate constants for the $[Ni(L^3)_2]^{+/2+}$ ($k_{11} = 7 \times 10^4$ M^{-1} s^{-1}), $[Ni(L^3)_2]^{0/+}$ (8×10^3), $[Ni(L^3)(HL^3)]^{+/2+}$ (5×10^2), $[Ni(L^1)]^{+/2+}$ (6×10^2), and $[Ni(HL^1)]^{2+/3+}$ (5.5×10^2) couples.

Marcus analyses of the cross-reactions of $[Pd([9]aneN_3)_2]^{2+}$ and $[Pd([9]aneN_3)_2]^{3+}$ with Ni(III) and Co(II) polyaza macrocyclic complexes have yielded self-exchange rate constants of 3×10^{-6} and 7×10^2 M^{-1} s^{-1} for the Pd(II)/Pd(III) and Pd(III)/Pd(IV) couples, respectively.[60] In the former couple, a large inner-sphere reorganization is involved in the conversion of the square planar d^8 Pd(II) to an octahedral d^7 Pd(III) species by axial coordination of free nitrogen donor atoms. In the latter couple both species have an octahedral configuration. The mechanism for the oxidation of $[Pd([9]aneN_3)_2]^{2+}$ by $[Ni([9]aneN_3)_2]^{3+}$

and $[Ni(Me[9]aneN_3)_2]^{3+}$ is comprised of pathways involving $[Pd(L)_2]^{2+}$ (k_1), $[Pd(L)(LH)]^{3+}$ (k_2), and $[Pd(LH)_2]^{4+}$ (k_3), with $k_1 > k_2 > k_3$. The rate constants for the oxidation of a series of $[Pt(NH_3)_2(NHR)_2]^{2+}$ complexes by $[IrCl_6]^{2-}$ may be correlated with the steric hindrance of the equatorial ligands on Pt(II) and with the thermodynamic driving forces of the reactions.[61]

2.3.6. Copper and Silver

The rate constants for the cross-reactions of Cu(I) tetraaza macrocycles with a series of outer-sphere oxidants have been used to determine the electron self-exchange rate constants of several Cu(I)/Cu(II) couples: $[Cu(Me_6[14]dieneN_4)]^{+/2+}$ ($k_{11} = 23$ $M^{-1} s^{-1}$), $[Cu(Me_4[14]-1,3,8,10-$ tetraeneN$_4$)]$^{+/2+}$ (4.8×10^2), $[Cu(Me_2pyo[14]trieneN_4)]^{+/2+}$ (2.7×10^3), and $[Cu(Me_2pyo[14]eneN_4)]^{+/2+}$ (~ 1).[55] The self-exchange rate constants for $[Ag(Me_6[14]aneN_4)]^{+/2+}$ (2.0×10^{-2}), $[Ag([14]aneN_4)]^{+/2+}$ (3.0×10^{-2}), and $[Ag([14]aneN_4)]^{2+/3+}$ (2.0×10^9) have been determined in a similar manner.[55]

The electron self-exchange rates for the [1,7-bis(5-methylimidazol-4-yl)-2,6-dithiaheptane]copper(I)/(II) couple in DMSO have been determined by ^{1}H NMR line broadening as a function of temperature, with $k_{11} = 4 \times 10^3$ $M^{-1} s^{-1}$ at 28 °C.[62] An electron self-exchange rate constant of 1.3×10^4 $M^{-1} s^{-1}$ has been measured for the $[Cu((imidH)_2DAP)]^{+/2+}$ couple {(imidH)$_2$DAP = 2,6-bis[1-((2-imidazol-4-ylethyl)imino)ethyl]pyridine} in acetonitrile.[63] The self-exchange rate constants for the $[Cu(dmp)_2]^{+/2+}$ couple (dmp = 2,9-dimethyl-1,10-phenanthroline) have been determined in aqueous, acetone, and acetonitrile media.[64] The volume of activations were found to be −3.4 and −7.8 cm^3 mol^{-1} in acetonitrile and acetone, respectively.

Further evidence for conformational intermediates in electron transfer reactions of Cu(I)/Cu(II) tetrazamacrocycle couples has been provided by cyclic voltammetic studies at low temperatures and high scan rates.[65] The oxidation of bis(dimethylglyoxime)copper(II) by $[Cu(H_{-2}DGEN)]^+$ (DGEN = diglycylethylenediamine) proceeds with pathways involving both $[Cu(dmgH)_2(H_2O)]$ ($pK_a = 10.9$) and its conjugate base $[Cu(dmgH)(dmg)-(H_2O)]^-$.[66] The application of the Marcus relationship to the rate constants for the reductions of a series of copper(III) imine–oxime complexes by $[Ni([14]aneN_4)]^{2+}$ yields a Cu(II)/Cu(III) self-exchange rate constant of about 5×10^5 $M^{-1} s^{-1}$.[67]

2.3.7. Technetium and Rhenium

The kinetics of the oxidations of trans-$[Tc(dppe)X_2]$ (X = Cl$^-$, Br$^-$) by Co(III) amine complexes have been investigated in pure and mixed DMF and CH$_3$CN solvent systems, with an observed increase in the rate constant with DMF concentrations.[68] The rate constants for the reductions of three metal carbonyl dimers $[M_2, M = Mn(CO)_5, CpMo(CO)_3,$ and $Co(CO)_4]$ by $[Re(CO)_5]^-$ in THF (measured using infrared stopped-flow spectroscopy) do not parallel the reduction

Table 2.1. Electron Transfer Reactions between Metal Ion Complexes at 25 °C.

Reaction	Medium I (M)	k $(M^{-1}\,s^{-1})$	$\Delta H^{\ddagger}$ $(kcal\,mol^{-1})$	$\Delta S^{\ddagger}$ $(cal\,K^{-1}\,mol^{-1})$	Ref.
Titanium(III)					
$[Ru_2(O_2CCH_3)_4Cl_2]^- + Ti^{3+}$	1.0	7.5×10^{1} a			1
$[Ru_2(O_2CCH_3)_4Cl_2]^- + TiOH^{2+}$	1.0	1.1×10^{3} a			1
$[Ru_2(O_2CCH_3)_4Cl_2]^- + [Ti(hedta)(H_2O)]$	1.0	1.6×10^{4}			2
$[Ru(NH_3)_5pyr]^{3+} + [Ti(hedta)(H_2O)]$	1.0	2.0×10^{5} b			2
$[*Cr(bpy)_3]^{3+} + Ti^{3+}$	1.00	1.9×10^{7}			3
$[*Cr(bpy)_3]^{3+} + TiOH^{2+}$	1.00	6.0×10^{8}			3
$[*Cr(4,4'-Me_2bpy)_3]^{3+} + Ti^{3+}$	1.00	2.6×10^{6}			3
$[*Cr(4,4'-Me_2bpy)_3]^{3+} + TiOH^{2+}$	1.00	1.0×10^{8}			3
$[*Cr(phen)_3]^{3+} + Ti^{3+}$	1.00	1.6×10^{7}			3
$[*Cr(phen)_3]^{3+} + TiOH^{2+}$	1.00	4.6×10^{8}			3
$[*Cr(5-Clphen)_3]^{3+} + Ti^{3+}$	1.00	3.7×10^{7}			3
$[*Cr(5-Clphen)_3]^{3+} + TiOH^{2+}$	1.00	9.7×10^{8}			3
$[*Cr(5-Mephen)_3]^{3+} + Ti^{3+}$	1.00	1.4×10^{7}			3
$[*Cr(5-Mephen)_3]^{3+} + TiOH^{2+}$	1.00	4.7×10^{8}			3
$[*Cr(5,6-Me_2phen)_3]^{3+} + Ti^{3+}$	1.00	1.2×10^{7}			3
$[*Cr(5,6-Me_2phen)_3]^{3+} + TiOH^{2+}$	1.00	3.0×10^{8}			3
$[*Cr(4,7-Me_2phen)_3]^{3+} + Ti^{3+}$	1.00	2.8×10^{6}			3
$[*Cr(4,7-Me_2phen)_3]^{3+} + TiOH^{2+}$	1.00	8.1×10^{7}			3
Vanadium(II)					
$[\{Co_2(\mu\text{-}OH)_2(NH_3)_6\}_2(\mu\text{-}2,6\text{-}CO_2pyr)]^{6+} + V^{2+}$	1.0	1.9×10^{-1}			4
$[\{Co_2(\mu\text{-}OH)_2(NH_3)_6\}_2(\mu\text{-}3,5\text{-}CO_3pyr)]^{6+} + V^{2+}$	1.0	1.7×10^{-1}			4
$[Co(CN)_5(pyazine)]^{2-} + V^{2+}$	1.0	1.3×10^{1}			5
$[Co(CN)_5(pyridine)]^{2-} + V^{2+}$	1.0	2.3			5
$[Co(CN)_5(isonicotinamide)]^{2-} + V^{2+}$	1.0	1.4×10^{1}			5
$[CrO_2(H_2O)_5]^{2+} + V^{2+}$	0.10	2.28×10^{5}			13
$[Mo_4S_4(H_2O)_{12}]^{5+} + V^{2+}$	2.0	2.45×10^{4}	2.3	−30.8	18

(continued)

Table 2.1. (continued)

Reaction	Medium I (M)	k (M^{-1} s^{-1})	ΔH^+ (kcal mol^{-1})	ΔS^+ (cal K^{-1} mol^{-1})	Ref.
Vanadium(IV)					
[Ni([9]aneN$_3$)$_2$]$^{3+}$ + [VO(TCDA)]	0.56	6.85×10^2			10
[Ni([9]aneN$_3$)$_2$]$^{3+}$ + [VO(DOCDA)]	0.56	1.4×10^1			10
Chromium(II)					
[{Co$_2$(μ-OH)$_2$(NH$_3$)$_6$}$_2$(μ-2,6-CO$_2$pyr)]$^{6+}$ + Cr^{2+}	1.00	5×10^{-3}			4
[Co$_2$(μ-OH)$_2$(NH$_3$)$_6$}$_2$(μ-3,5-CO$_2$pyrH)]$^{7+}$ + Cr^{2+}	1.00	6.9×10^{-3}			4
[Co$_2$(μ-OH)$_2$(NH$_3$)$_6$}$_2$(μ-3,5-CO$_2$pyr)]$^{6+}$ + Cr^{2+}	1.00	1.34×10^{-2}			4
[Co$_2$(μ-OH)$_2$(NH$_3$)$_6$}$_2$(μ-2,4-CO$_2$pyrH)]$^{7+}$ + Cr^{2+}	1.00	3.9×10^{-2}			4
[Co$_2$(μ-OH)$_2$(NH$_3$)$_6$}$_2$(μ-2,4-CO$_2$pyr)]$^{6+}$ + Cr^{2+}	1.00	1.44			4
[Co$_2$(μ-OH)$_2$(NH$_3$)$_6$}$_2$(μ-2,5-CO$_2$pyrH)]$^{7+}$ + Cr^{2+}	1.00	7.7×10^{-2}			4
[Co$_2$(μ-OH)$_2$(NH$_3$)$_6$}$_2$(μ-2,5-CO$_2$pyr)]$^{6+}$ + Cr^{2+}	1.00	3.93×10^{-1}			4
[Co(CN)$_5$(pyrazine)]$^{2-}$ + Cr^{2+}	1.00	1.5×10^3			5
[Co(CN)$_5$(4-cyanopyridine)]$^{2-}$ + Cr^{2+}	1.00	2.8×10^1			5
[Co(CN)$_5$(isonicotinamide)]$^{2-}$ + Cr^{2+}	1.00	1.9×10^1			5
[Co(CN)$_5$(pyridine)]$^{2-}$ + Cr^{2+}	1.00	1.2			5
[Co(NH$_3$)$_5$(SCONH(CH$_3$))]$^{2+}$ + Cr^{2+}	1.0	6.5×10^4			6
[Co(NH$_3$)$_5$(SCONH(C$_6$H$_5$))]$^{2+}$ + Cr^{2+}	1.0	3.5×10^4			6
[Co(NH$_3$)$_5$(SCONH(CH$_2$C$_6$H$_5$))]$^{2+}$ + Cr^{2+}	1.0	4.0×10^4			6
[Co(NH$_3$)$_5$(SCONH(4-CN-C$_6$H$_4$))]$^{2+}$ + Cr^{2+}	1.0	2.2×10^4			6
[Co(NH$_3$)$_5$(OCSNHCH$_3$)]$^{2+}$ + Cr^{2+}	1.0	6.8×10^1			6
[Co(NH$_3$)$_5$(OCSNH(CH$_2$C$_6$H$_5$)]$^{2+}$ + Cr^{2+}	1.0	5.5×10^1			6
[Co(NH$_3$)$_5$(OCSNH(C$_6$H$_5$)]$^{2+}$ + Cr^{2+}	1.0	6.4×10^{-1}			6
[Co(NH$_3$)$_5$(O$_2$CC(OH)$_2$CH$_3$)]$^{2+}$ + Cr^{2+}	1.0	2.55×10^1			7
[Co(NH$_3$)$_5$(pyro)]$^{3+}$ + Cr^{2+}	1.0	3.71×10^{-1}	3.8	−48	8
[Co(sep)]$^{3+}$ + Cr^{2+}	0.5	8.7×10^{-4} (k_0)	9	−43	9
	0.5	2.5×10^{-2} (k_X)	11	−28	9
[Mo$_4$S$_4$(H$_2$O)$_{12}$]$^{5+}$ + Cr^{2+}	2.0	6.1×10^2	4.0	−32.3	18
[Co(en)$_3$]$^{3+}$ + [Cr(5-Mephen)$_3$]$^{2+}$	0.15	1.6×10^4 [a]			24

Reaction	Conditions	k	Ref.
$[Co(en)_3]^{3+} + [Cr(4,4'\text{-}Me_2bpy)_3]^{2+}$	0.15	6.6×10^4 [a]	24
$[Co(en)_3]^{3+} + [Cr(4,7\text{-}Me_2phen)_3]^{2+}$	0.15	1.1×10^5 [a]	24
$[Co(bpy)_3]^{3+} + [Cr(5\text{-}Clphen)_3]^{2+}$	0.15	1.2×10^8 [a]	24
$[Co(bpy)_3]^{3+} + [Cr(bpy)_3]^{2+}$	0.15	1.1×10^8 [a]	24
$[Co(bpy)_3]^{3+} + [Cr(phen)_3]^{2+}$	0.15	2.0×10^8 [a]	24
$[Co(bpy)_3]^{3+} + [Cr(4,4'\text{-}Me_2bpy)_3]^{2+}$	0.15	2.8×10^8 [a]	24
$[Co(bpy)_3]^{3+} + [Cr(4,7\text{-}Me_2phen)_2]^{2+}$	0.15	3.5×10^8 [a]	24
$[Co(phen)_3]^{3+} + [Cr(5\text{-}Clphen)_3]^{2+}$	0.15	1.6×10^8 [a]	24
$[Co(phen)_3]^{3+} + [Cr(bpy)_3]^{2+}$	0.15	1.3×10^8 [a]	24
$[Co(phen)_3]^{3+} + [Cr(phen)_3]^{3+}$	0.15	2.0×10^8 [a]	24
$[Co(phen)_3]^{3+} + [Cr(4,4'\text{-}Me_2bpy)_3]^{2+}$	0.15	3.6×10^8 [a]	24
$[Co(phen)_3]^{3+} + [Cr(4,7\text{-}Me_2phen)_3]^{2+}$	0.15	4.1×10^8 [a]	24
$[Fe(C_5H_4COOH)_2]^+ + [Cr(bpy)_3]^{2+}$	0.050 (70% CH_3CN)	8.9×10^9	25
$[Fe(C_5H_5)(C_5H_4CHO)]^+ + [Cr(bpy)_3]^{2+}$	0.050 (70% CH_3CN)	8.8×10^9	25
$[Fe(C_5H_5)(C_5H_4COOH)]^+ + [Cr(bpy)_3]^{2+}$	0.050 (70% CH_3CN)	6.2×10^9	25
$[Fe(C_5H_5)(C_5H_4OH)]^+ + [Cr(bpy)_3]^{2+}$	0.050 (70% CH_3CN)	6.0×10^9	25
$[Fe(C_5H_5)(C_5H_4\text{-}nBu)]^+ + [Cr(bpy)_3]^{2+}$	0.050 (70% CH_3CN)	5.3×10^9	25
$[Fe(C_5H_4CH_3)_2]^+ + [Cr(bpy)_3]^{2+}$	0.050 (70% CH_3CN)	4.5×10^9	25
$[Fe(C_5H_5)(C_5H_4CH_2NMe_2)]^+ + [Cr(pby)_3]^{2+}$	0.050 (70% CH_3CN)	3.6×10^9	25
$[Fe(C_5H_5)_2]^+ + [Cr(bpy)_3]^{2+}$	0.050 (70% CH_3CN)	3.5×10^9	25
$[Co([14]aneN_4)]^{3+} + [Cr(5\text{-}Clphen)_3]^{2+}$	1.0	1.70×10^7	26
$[Co([14]aneN_4)]^{3+} + [Cr(bpy)_3]^{2+}$	1.0	4.4×10^7	26
$[Co([14]aneN_4)]^{3+} + [Cr(phen)_3]^{2+}$	1.0	1.3×10^8	26
$[Co([14]aneN_4)]^{3+} + [Cr(5,6\text{-}Me_2phen)_3]^{2+}$	1.0	1.58×10^8	26
$[Co([14]aneN_4)]^{3+} + [Cr(5\text{-}Mephen)_3]^{2+}$	1.0	1.05×10^8	26

(continued)

Table 2.1. (continued)

Reaction	Medium I (M)	k ($M^{-1}s^{-1}$)	$\Delta H^{\ddagger}$ (kcal mol^{-1})	$\Delta S^{\ddagger}$ (cal K^{-1} mol^{-1})	Ref.
[Co([14]aneN₄)]³⁺ + [Cr(4,4'-Me₂bpy)₃]²⁺	1.0	≈7 × 10⁸			26
[Co([14]aneN₄)]³⁺ + [Cr(4,7-Me₂phen)₃]²⁺	1.0	≈7 × 10⁸			26
[Ag([14]aneN₄)]²⁺ + [Cr(bpy)₃]²⁺	0.03	1.3 × 10⁸			55
[Ag(bpy)₃]²⁺ + [Cr(pby)₃]²⁺	0.1	3.7 × 10⁸			55
Chromium(III)					
[Ni([14]aneN₄)]³⁺ + [Cr(H₂O)₅(CH₂Ph)]²⁺	1.0	1.88 × 10⁴ c			11
[Ni([14]aneN₄)]³⁺ + [Cr(H₂O)₅(CH(CH₃)OC₂H₅)]²⁺	1.0	3.44 × 10³ c			11
[Ni([14]aneN₄)]³⁺ + [Cr(H₂O)₅(CH₂OCH₃)]²⁺	1.0	2.45 × 10³ c			11
[Ni([14]aneN₄)]³⁺ + [Cr(H₂O)₅(CH(CH₃)₂)]²⁺	1.0	2.20 × 10² c			11
[Ni([14]aneN₄)]³⁺ + [Cr(H₂O)₅(CH₂CH₃)]²⁺	1.0	1.30 × 10¹ c			11
[Ni([9]aneN₃)₂]³⁺ + [Cr(H₂O)₅(CH₂Ph)]²⁺	1.0	1.31 × 10⁴ c			11
[Ni([9]aneN₃)₂]³⁺ + [Cr(H₂O)₅(CH₂(CH₃)₂)]²⁺	1.0	9.5 × 10¹ c			11
Manganese(II)					
[Os(bpy)₃]³⁺ + [Mn(edta)]²⁻	0.10	1.25 × 10⁴	11.5	−1	28
[Os(bpy)₃]³⁺ + [Mn(cdta)]²⁻	0.10	2.03 × 10⁴	8.7	−10	28
[Os(phen)₃]³⁺ + [Mn(edta)]²⁻	0.10	1.44 × 10⁴			28
[Os(phen)₃]³⁺ + [Mn(cdta)]²⁻	0.10	4.18 × 10⁴	11.1	0	28
[Os(5-Clphen)₃]³⁺ + [Mn(edta)]²⁻	0.10	9.69 × 10⁴	10.0	−2	28
[Os(5-Clphen)₃]³⁺ + [Mn(cdta)]²⁻	0.10	1.32 × 10⁵			28
[IrCl₆]²⁻ + [Mn(edta)]²⁻	0.10	2.96 × 10³	8.6	−14	28
[IrCl₆]²⁻ + [Mn(cdta)]²⁻	0.10	4.61 × 10²	9.1	−16	28
[Ni([9]aneN₃)₂]³⁺ + [Mn(edta)]²⁻	0.10	1.34 × 10⁴	13.8	−7	28
[Ni([9]aneN₃)₂]³⁺ + [Mn(cdta)]²⁻	0.10	2.69 × 10⁴	10.8	−2	28
[Ni(Hdiox)]²⁺ + [Mn(edta)]²⁻	0.10	7.0 × 10³			28
[Ni(Hdiox)]²⁺ + [Mn(cdta)]²⁻	0.10	4.43 × 10⁴			28
[Ni(Hdiox)]⁺ + [Mn(edta)]²⁻	0.10	1.29 × 10⁵			28
[Ni(diox)] + [Mn(cdta)]²⁻	0.10	1.2 × 10⁵			28

Reaction				Ref.
[Ni(diox)] + [Mn(edta)]²⁻	0.10	4.3×10^5		28
[(TTP)MnN] + [Mn(OEP)]	THF	$4.2 \times 10^{3\,d}$		31
Iron(II)				
[CrO₂(H₂O)₅]²⁺ + Fe²⁺	0.10	4.5×10^3		13
MoO₂Cl₂ + Fe²⁺	8.0	$3.6 \times 10^{3\,d}$		16
MoO(O₂)₂ + Fe²⁺	1.0	2.8×10^4	19	22
[MoO(OH)(O₂)₂]⁻ + Fe²⁺	1.0	5.6×10^4		22
[Co(edta)]⁻ + [Fe(edta)]²⁻	1.0	4.4		19
[Co([9]aneS₃)₂]³⁺ + [Fe([9]aneN₃)₂]²⁺	0.05	3.3×10^6		43
[Co(NH₃)₅(OH₂)]³⁺ + [Fe(CN)₆]⁴⁻	1.0	$1.27 \times 10^1 \text{ s}^{-1}$	24.4	37
[*Ru(4,4'-CO₂bpy)₃]⁴⁻ + [Fe(CN)₆]⁴⁻	0.03	$2.0 \times 10^{7\,e}$		46
*Ru(bpy)₂(4,4'-CO₂bpy) + [Fe(CN)₆]⁴⁻	0.03	$1.1 \times 10^{10\,e}$		46
[*Ru(bpy)₃]²⁺ + [Fe(CN)₆]⁴⁻	0.03	$6.4 \times 10^{9\,e}$		46
*Ru(bpy)₂(Mebpy-Mebpy)Ru(bpy)₂]⁴⁺ + [Fe(CN)₆]⁴⁻	0.03	$9.9 \times 10^{9\,e}$		46
[Co(NH₃)₅(μ-DMBP)]³⁺ + [Fe(CN)₅OH₂]³⁻	0.10	$2.3 \times 10^{-3} \text{ s}^{-1}$		38
[Co(NH₃)₅(DMBP)]³⁺ + [Fe(CN)₅]³⁻	0.10	1.9×10^2		38
[Co(NH₃)₅(μ-BPA)]³⁺ + [Fe(CN)₅]³⁻	0.10	$1.7 \times 10^{-3} \text{ s}^{-1}$		38
[Co(NH₃)₅(BPA)]³⁺ + [Fe(CN)₅OH₂]³⁻	0.10	3.8×10^2		38
[Co(NH₃)₅(μ-BPBD)]³⁺ + [Fe(CN)₅]³⁻	0.10	$6.9 \times 10^{-4} \text{ s}^{-1}$		38
[Co(NH₃)₅(BPBD)]³⁺ + [Fe(CN)₅OH₂]³⁻	0.10	2.9×10^2		38
[Co(NH₃)₅(μ-PHEN)]³⁺ + [Fe(CN)₅OH₂]³⁻	0.10	$4.2 \times 10^{-3} \text{ s}^{-1}$		38
[Co(NM₃)₅(PHEN)]³⁺ + [Fe(CN)₅OH₂]³⁻	0.10	1.1×10^2		38
[Co(NH₃)₅(μ-DAP)]³⁺ + [Fe(CN)₅OH₂]³⁻	0.10	$9.3 \times 10^{-3} \text{ s}^{-1}$		38
[Co(NH₃)₅(DAP)]³⁺ + [Fe(CN)₅OH₂]³⁻	0.10	1.5×10^2		38
[Co(TPP)]⁺ + [Fe(C₅H₅)₂]	0.02(CH₃CN)	2.1×10^4		40
[Fe(TPP)]⁺ + [Fe(C₅H₅)₂]	0.02(CH₃CN)	1.5×10^6		40
[Co(TPP)]⁺ + [Fe(C₅H₄CH₃)₂]	0.02(CH₃CN)	4.6×10^4		40
[Fe(TPP)]⁺ + [Fe(C₅H₄CH₃)₂]	0.02(CH₃CN)	6.0×10^6		40
[Mn(TPP)]⁺ + [Fe(C₅H₄CH₃)₂]	0.02(CH₃CN)	1.3×10^2		40
[Co(TPP)]⁺ + [Fe(C₅(CH₃)₅)₂]	0.02(CH₃CN)	6.0×10^5		40
[Mn(TPP)]⁺ + [Fe(C₅(CH₃)₅)₂]	0.02(CH₃CN)	7.0×10^4		40
[Co(TIM)]³⁺ + [Fe(C₅(CH₃)₅)₂]	0.01(CH₃CN)	1.2×10^4		40

(*continued*)

Table 2.1. (continued)

Reaction	Medium I (M)	k (M^{-1} s^{-1})	$\Delta H^{\ddagger}$ (kcal mol^{-1})	$\Delta S^{\ddagger}$ (cal K^{-1} mol^{-1})	Ref.
[Co(TIM)]$^{3+}$ + [Fe(C$_5$H$_4$CH$_3$)$_2$]	0.01(CH$_3$CN)	5			40
[Co(TIM)]$^{3+}$ + [Fe(C$_5$H$_5$)(C$_5$H$_4$-nBu)]	0.01(CH$_3$CN)	3.3			40
[Co(TIM)]$^{3+}$ + [Fe(C$_5$H$_5$)(C$_5$H$_4$-nAmyl)]	0.01(CH$_3$CN)	3.0			40
[Co(TIM)]$^{3+}$ + [Fe(C$_5$H$_5$)$_2$]	0.01(CH$_3$CN)	1.1			40
[Fe(DMF)$_3$(CH$_3$CN)$_3$]$^{3+}$ + [Fe(Me$_6$phen)$_3$]$^{2+}$	CH$_3$CN	3.87 × 10^4			36
[Fe(DMF)$_3$(CH$_3$CN)$_3$]$^{3+}$ + [Fe(Me$_4$phen)$_3$]$^{2+}$	CH$_3$CN	1.91 × 10^4			36
[Fe(DMF)$_4$(CH$_3$CN)$_2$]$^{3+}$ + [Fe(Me$_6$phen)$_3$]$^{2+}$	CH$_3$CN	5.95 × 10^3			36
[Fe(DMF)$_4$(CH$_3$CN)$_2$]$^{3+}$ + [Fe(Me$_4$phen)$_3$]$^{2+}$	CH$_3$CN	2.39 × 10^2			36
[Fe(Me$_4$phen)$_3$]$^{3+}$ + [Fe(DMF)$_4$(CH$_3$CN)$_2$]$^{2+}$	CH$_3$CN	2.80 × 10^4			36
[Fe(Me$_4$phen)$_3$]$^{3+}$ + [Fe(DMF)$_4$(CH$_3$CN)$_2$]$^{2+}$	CH$_3$CN	1.05 × 10^4			36
Cobalt(II)					
[*Cr(5-Clphen)$_3$]$^{3+}$ + [Co([14]aneN$_4$)]$^{2+}$	1.0	5.2 × 10^8			26
[*Cr(bpy)$_3$]$^{3+}$ + [Co([14]aneN$_4$)]$^{2+}$	1.0	1.8 × 10^8			26
[*Cr(phen)$_3$]$^{3+}$ + [Co([14]aneN$_4$)]$^{2+}$	1.0	3.3 × 10^8			26
[*Cr(5,6-Me$_2$phen)$_3$]$^{3+}$ + [Co([14]aneN$_4$)]$^{2+}$	1.0	2.55 × 10^8			26
[*Cr(5-Mephen)$_3$]$^{3+}$ + [Co([14]aneN$_4$)]$^{2+}$	1.0	4.04 × 10^8			26
[*Cr(4,4'-Me$_2$bpy)$_3$]$^{3+}$ + [Co([14]aneN$_4$)]$^{2+}$	1.0	7.0 × 10^7			26
[*Cr(4,7-Me$_2$phen)$_3$]$^{3+}$ + [Co([14]aneN$_4$)]$^{2+}$	1.0	1.28 × 10^8			26
[Mn(cdta)]$^-$ + [Co(5-Clphen)$_3$]$^{2+}$	0.10	1.48 × 10^3	8.1		28
[Co(nox)$_3$(BF)$_2$]$^+$ + [Co(dpg)$_3$(BPh)$_2$]	0.10(CH$_3$CN)	6.0 × 10^3		−14	51
[CrO$_2$(H$_2$O)$_5$]$^{2+}$ + [Co([15]aneN$_4$)]$^{2+}$	0.10	6.2 × 10^5			13
[CrO$_2$(H$_2$O)$_5$]$^{2+}$ + [Co(sep)]$^{2+}$	0.10	8.5 × 10^5			13
[Pd([9]aneN$_3$)$_2$]$^{3+}$ + [Co(sep)]$^{2+}$	0.2	5.6 × 10^2			60
[Pd([9]aneN$_3$)$_2$]$^{3+}$ + [Co([9]aneN$_3$)$_2$]$^{2+}$	0.2	1.17 × 10^2			60
[Co(taptacn)]$^{3+}$ + [Co(sep)]$^{2+}$	0.1	1.03 × 10^2			49

$[CrO_2(H_2O)_5]^{2+} + [Co([15]aneN_4)]^{2+}$	0.10	6.2×10^5	13
$[CrO_2(H_2O)_5]^{2+} + [Co(sep)]^{2+}$	0.10	8.5×10^5	13
$[Pd([9]aneN_3)_2]^{3+} + [Co(sep)]^{2+}$	0.2	5.6×10^2	60
$[Pd([9]aneN_3)_2]^{3+} + [Co([9]aneN_3)_2]^{2+}$	0.2	1.17×10^2	60
$[Co(taptacn)]^{3+} + [Co(sep)]^{2+}$	0.1	1.03×10^2	49
$[Co(taptacn)]^{3+} + [Co([9]aneN_3)_2]^{2+}$	0.1	6.8×10^1	49
Ni(amine-oxime)]$^{2+}$ + $[Co(phen)_3]^{2+}$	0.10	1.65×10^3	59
$[Ni(mono-oxime)_2]^{2+} + [Co(phen)_2]^{2+}$	0.10	1.97×10^5	59
$[mer-\Lambda-lel_3-Co\{(NH_3)_2Me_3sar\}]^{5+}$ $+ [fac-\Delta-ob_3-Co\{(NH_3)_2Me_3sar\}]^{4+}$	0.2	5.4×10^1	50
$[mer-\Delta-lel_3-Co\{(NH_3)_2Me_3sar\}]^{5+}$ $+ [fac-\Lambda-ob_3-Co\{(NH_3)_2Me_3sar\}]^{4+}$	0.2	4.5×10^1	50
$[fac-\Lambda-lel_3-Co\{(NH_3)_2Me_3sar\}]^{5+}$ $+ [fac-\Delta-ob_3-Co\{(NH_3)_2Me_3sar\}]^{4+}$	0.2	4.0×10^1	50
$[fac-\Delta-lel_3-Co\{(NH_3)_2Me_3sar\}]^{5+}$ $+ [fac-\Lambda-ob_3-Co\{(NH_3)_2Me_3sar\}]^{4+}$	0.2	3.2×10^1	50
$[mer-\Lambda-lel_3-Co\{(NH_3)_2Me_3sar\}]^{5+}$ $+ [mer-\Delta-ob_3-Co\{(NH_3)_2Me_3sar\}]^{4+}$	0.2	1.7×10^1	50
$[mer-\Delta-lel_3-Co\{(NH_3)_2Me_3sar\}]^{5+}$ $+ [mer-\Lambda-ob_3-Co\{(NH_3)_2Me_3sar\}]^{4+}$	0.2	1.4×10^1	50
$[fac-\Lambda-lel_3-Co\{(NH_3)_2Me_3sar\}]^{5+}$ $+ [mer-\Delta-ob_3-Co\{(NH_3)_2Me_3sar\}]^{4+}$	0.2	1.3×10^1	50
$[fac-\Delta-lel_3-Co\{(NH_3)_2Me_3sar\}]^{5+}$ $+ [mer-\Lambda-ob_3-Co\{(NH_3)_2Me_3sar\}]^{4+}$	0.2	1.0×10^1	50
Nickel(I)			
$[Ru(NH_3)_6]^{3+} + [Ni(Me_6[14]aneN_4)]^+$	0.01	6.9×10^8	55
$[Ru(NH_3)_6]^{3+} + [Ni(Me_6[14]dieneN_4)]^+$	0.01	4.5×10^8	55
$[Cr(bpy)_3]^{3+} + [Ni(Me_6[14]dieneN_4)]^+$	0.01	3.1×10^8	55
$[Co(en)_3]^{3+} + [Ni(Me_6[14]dieneN_4)]^+$	0.01	1.1×10^6	55
$[Co(Me_6[14]-1,4,8,11-tetraeneN_4)]^{3+}$ $+ [Ni(Me_6[14]dieneN_4)]^+$	0.01	4.0×10^7	55

(continued)

Table 2.1. (continued)

Reaction	Medium I (M)	k ($M^{-1}\,s^{-1}$)	$\Delta H^{\ddagger}$ (kcal mol^{-1})	$\Delta S^{\ddagger}$ (cal K^{-1} mol^{-1})	Ref.
[Ni(Me$_2$pyo[14]trieneN$_4$]$^{2+}$ + [Ni(Me$_6$[14]dieneN$_4$)]$^+$	0.01	2.1×10^7			55
[Cu(Me$_2$pyo[14]trieneN$_4$]$^{2+}$ + [Ni(Me$_6$[14]dieneN$_4$)]$^+$	0.01	1.0×10^8			55
[Ru(NH$_3$)$_6$]$^{3+}$ + [Ni(Me$_4$[14]-1,4,8,11-tetraeneN$_4$)]$^+$	0.01	3.0×10^8			55
[Cr(bpy)$_3$]$^{3+}$ + [Ni(Me$_4$[14]-1,4,8,11-tetraeneN$_4$)]$^+$	0.01	8.0×10^8			55
[Ru(NH$_3$)$_6$]$^{3+}$ + [Ni(Me$_2$pyo[14]trieneN$_4$)]$^+$	0.01	2.3×10^8			55
[Cr(bpy)$_3$]$^{3+}$ + [Ni(Me$_2$pyo[14]trieneN$_4$)]$^+$	0.01	5.1×10^8			55
[Ag[14]aneN$_4$]$^{2+}$ + [Ni(Me$_6$[14]dieneN$_4$)]$^+$	0.03	5.7×10^8			55
[Co(NH$_3$)$_6$]$^{3+}$ + [Ni(*meso*-Me$_{10}$[14]aneN$_4$)]$^+$	0.1	6.32×10^3			56
[Co(en)$_3$]$^{3+}$ + [Ni(*meso*-Me$_{10}$[14]aneN$_4$)]$^+$	0.1	2.37×10^3			56
[Co(sep)]$^{3+}$ + [Ni(*meso*-Me$_{10}$[14]aneN$_4$)]$^+$	0.1	9.59×10^3			56
[Co(NH$_3$)$_5$F]$^{2+}$ + [Ni(*meso*-Me$_{10}$[14]aneN$_4$)]$^+$	0.1	6.0×10^3			56
[Co(NH$_3$)$_6$]$^{3+}$ + [Ni(Me$_4$[14]aneN$_4$)]$^+$	0.1	3.0×10^4			56
[Co(en)$_3$]$^{3+}$ + [Ni(Me$_4$[14]aneN$_4$)]$^+$	0.1	9.0×10^3			56
[Co(sep)]$^{3+}$ + [Ni(Me$_4$[14]aneN$_4$)]$^+$	0.1	7.2×10^4			56
Nickel(II)					
Ni(H$_{-2}$Aib$_3$)(pyr)(H$_2$O) + [Ni(H$_{-2}$Aib$_3$)]$^-$	0.1	1.03×10^2 ^{a}			58
[Cu(Me$_2$dioxH)]$^{2+}$ + [Ni([14]aneN$_4$)]$^{2+}$	0.50	3.67×10^5 [H$^+$]			67
[Cu(MeEtdioxH)]$^{2+}$ + [Ni([14]aneN$_4$)]$^{2+}$	0.50	4.47×10^5 [H$^+$]			67
[Cu(PhMedioxH)]$^{2+}$ + [Ni([14]aneN$_4$)]$^{2+}$	0.50	3.67×10^5 [H$^+$]			67
[Cu(Pr^{i}MedioxH)]$^{2+}$ + [Ni([14]aneN$_4$)]$^{2+}$	0.50	3.67×10^5 [H$^+$]			67
[Ni([9]aneN$_3$)$_2$]$^{3+}$ + [Ni(amine–oxime)]$^{2+}$	0.10	4.84×10^5			59
[Fe(5,6-Me$_2$phen)$_3$]$^{3+}$ + [Ni([14]aneN$_4$)]$^{2+}$	0.52(CH$_3$CN)	2.28×10^5			53
[Fe(5,6-Me$_2$phen)$_3$]$^{3+}$ + [Ni(Me$_2$[14]dieneN$_4$)]$^{2+}$	0.052(CH$_3$CN)	2.55×10^5			53
[Rh$_2$(O$_2$CCH$_3$)$_4$(CH$_3$CN)$_2$]$^+$ + [Ni([14]aneN$_4$)]$^{2+}$	0.10(CH$_3$CN)	7.3×10^5			53
[Rh$_2$(O$_2$CCH$_3$)$_4$(CH$_3$CN)$_2$]$^+$ + [Ni(Me$_2$[14]dieneN$_4$)]$^{2+}$	0.10(CH$_3$CN)	8.1×10^5			53
Copper(I)					
[Ru(NH$_3$)$_6$]$^{3+}$ + [Cu(Me$_6$[14]dieneN$_4$)]$^+$	0.01	7.2×10^4			55

Reaction			Ref.
$[Cr(bpy)_3]^{3+} + [Cu(Me_6[14]dieneN_4)]^+$	0.01	3.7×10^6	55
$[Co(bpy)_3]^{3+} + [Cu(Me_6[14]\text{-}1,3,8,10\text{-}tetraeneN_4)]^+$	0.01	5.6×10^6	55
$[Ru(NH_3)_6]^{3+} + [Cu(Me_4[14]\text{-}1,3,8,10\text{-}tetraeneN_4)]^+$	0.01	1.2×10^5	55
$[Co(en)_3]^{3+} + [Cu(Me_6[14]\text{-}1,3,8,10\text{-}tetraeneN_4)]^+$	0.01	3.3	55
$[Ru(NH_3)_6]^{3+} + [Cu(Me_2pyo[14]trieneN_4)]^+$	0.01	1.5×10^6	55
$[Co(bpy)_3]^{3+} + [Cu(Me_2pyo[14]trieneN_4)]^+$	0.01	2.5×10^7	55
$[Rh(bpy)_3]^{3+} + [Cu(Me_2pyo[14]trieneN_4)]^+$	0.7	1.3×10^7	55
Copper(II)			
$[Ni(H_{-1}Aib_2)_2]^- + [Cu(H_{-3}Aib_3a)]^-$	0.1	8.3×10^2	57
$[Cu(H_{-2}DGEN)]^+ + [Cu(dmgH)_2(OH_2)]$	0.10	1.95×10^4	66
$[Cu(H_{-2}DGEN)]^+ + [Cu(dmgH)(dmg)(OH_2)]^-$	0.10	$\sim 8 \times 10^8$	66
Molybdenum(III)			
$[Fe(H_2O)_6]^{3+} + [Mo(H_2O)_6]^{3+}$	1.0	1.30×10^3	16
$[Co(NH_3)_5(O_2CCH_3)]^{2+} + [Mo_4S_4(H_2O)_{12}]^{4+}$	2.0	2.6×10^{-1}	17
$[Co(NH_3)_5(H_2O)]^{3+} + [Mo_4S_4(H_2O)_{12}]^{4+}$	2.0	4.0×10^{-1}	17
$[Co(NH_3)_5F]^{2+} + [Mo_4S_4(H_2O)_{12}]^{4+}$	2.0	3.9×10^{-1}	17
$[Co(NH_3)_5Cl]^{2+} + [Mo_4S_4(H_2O)_{12}]^{4+}$	2.0	8.1	17
$[Co(NH_3)_5Br]^{2+} + [Mo_4S_4(H_2O)_{12}]^{4+}$	2.0	2.89×10^1	17
$[Co(NH_3)_5I]^{2+} + [Mo_4S_4(H_2O)_{12}]^{4+}$	2.0	2.08×10^2	17
$[Co(bpy)_3]^{3+} + [Mo_4S_4(H_2O)_{12}]^{4+}$	2.0	4.60×10^2	17
$[Fe(edta)]^- + [Mo_4S_4(edta)_2]^{4-}$	1.0	2.4×10^6	19
$[Co(edta)]^- + [Mo_4S_4(edta)_2]^{4-}$	1.0	5.4×10^3	19
$[Co(dipic)]^- + [Mo_4S_4(edta)_2]^{3-}$	1.0	1.78×10^1	19
$[VO_2(H_2O)_4]^+ + [Mo_4S_4(edta)_2]^{4-}$	1.00	8.3×10^1	20
		$4.6 \times 10^3 \,[H^+]$	20
Molybdenum(IV) and (V)			
$[*Ru(4,4'\text{-}CO_2\text{-}bpy)_3]^{4-} + [Mo(CN)_8]^{4-}$	0.03	$1.6 \times 10^{8\,e}$	46
$[*Ru(bpy)_2(4,4'\text{-}CO_2bpy)] + [Mo(CN)_8]^{4-}$	0.03	$2.9 \times 10^{9\,e}$	46
$[*Ru(bpy)_3]^{2+} + [Mo(CN)_8]^{4-}$	0.03	$1.6 \times 10^{9\,e}$	46
$[*Ru(bpy)_2\{Mebpy\text{–}Mebpy)Ru(bpy)_2\}]^{4+} + [Mo(CN)_8]^{4-}$	0.03	$1.2 \times 10^{9\,e}$	46
$[Fe(H_2O)_6]^{3+} + [Mo_2O_4]^{2+}$	1.0	1.06	16

(continued)

Table 2.1. (continued)

Reaction	Medium I (M)	k $(M^{-1}\,s^{-1})$	$\Delta H^{\ddagger}$ $(kcal\,mol^{-1})$	$\Delta S^{\ddagger}$ $(cal\,K^{-1}\,mol^{-1})$	Ref.
Technetium(II)					
$[Co(en)_2(S(CH_2C_6H_4CH_3)CH_2CH_2NH_2)]^{3+}$ + trans-$[Tc(dppe)_2Cl_2]$	0.10$[CH_3CN]$	3.0×10^4			68
$[Co(en)_2(S(CH_2C_6H_4CH_3)CH_2CH_2NH_2)]^{3+}$ + trans-$[Tc(dppe)_2Cl_2]$	0.10(DMF)	3.7×10^1			68
$[Co(en)_2(S(CH_2C_6H_4CH_3)CH_2CH_2NH_2)]^{3+}$ + trans-$[Tc(dppe)_2Br_2]$	0.10(CH_3CN)	1.2×10^4			68
Ruthenium(II) and (III)					
$[CrO_2(H_2O)_5]^{2+}$ + $[Ru(NH_3)_6]^{2+}$	0.10	9.5×10^5			13
$[Co(NH_3)_5F]^{2+}$ + $[Ru(NH_3)_6]^{2+}$	2.0	1.1			17
$[Co([9]aneN_3)([9]aneS_3)]^{3+}$ + $[Ru(NH_3)_6]^{2+}$	0.1	1.3×10^4			43
$[Ru(sar-H)]^{3+}$ + $[Ru(sar)]^{2+}$	0.1	6.2×10^3			44
cis-$[Ru(bpy)_2(O)(OH)]^{2+}$ + cis-$[Ru(bpy)_2(OH_2)]^{3+}$	0.1	4.8×10^4 ^{f}			45
$[Ru(C_6H_5)_2Br]^{2+}$ + $[Ru(C_6H_5)_2]$	CD_3CN	1.6×10^3 ^{d}	8.0	−16.3	34
$[Cu(2,9\text{-}Me_2phen)_2]^{2+}$ + $[Ru(hfac)]^{-}$	0.04(CH_3CN)	2.86×10^5 ^{g}	8.6	−4.2	64
rac-$[Co(phen)_3]^{3+}$ + $[*\Lambda\text{-}Ru(phen)_3]^{2+}$	0.05	1.4×10^9			41
rac-$[Co(phen)_3]^{3+}$ + $[*\Delta\text{-}Ru(phen)_3]^{2+}$	0.05	1.4×10^9			41
rac-$[Co(phen)_3]^{3+}$ + $[*\Delta\text{-}Ru(phen)_3]^{2+}$ (DNA surface bound)	0.05	1.7×10^{10}			41
rac-$[Co(phen)_3]^{3+}$ + $[*\Lambda\text{-}Ru(phen)_3]^{2+}$ (DNA surface bound)	0.05	3.9×10^{10}			41
rac-$[Co(phen)_3]^{3+}$ + $[*\Delta\text{-}Ru(phen)_3]^{2+}$ (DNA intercalated)	0.05	9.0×10^{10}			41
rac-$[Co(phen)_3]^{3+}$ + $[*\Lambda\text{-}Ru(phen)_3]^{2+}$ (DNA intercalated)	0.05	1.6×10^{10}			41
rac-$[Co(bpy)_3]^{3+}$ + $[*\Delta\text{-}Ru(phen)_3]^{2+}$	0.05	2.0×10^9			41
rac-$[Co(bpy)_3]^{3+}$ + $[*\Lambda\text{-}Ru(phen)_3]^{2+}$	0.05	2.0×10^9			41
rac-$[Co(bpy)_3]^{3+}$ + $[*\Delta\text{-}Ru(phen)_3]^{2+}$ (DNA surface bound)	0.05	2.4×10^{10}			41

Reaction			
rac-[Co(bpy)$_3$]$^{3+}$ + [*Λ-Ru(phen)$_3$]$^{2+}$ (DNA surface bound)	0.05	4.7×10^{10}	41
rac-[Co(bpy)$_3$]$^{3+}$ + [*Δ-Ru(phen)$_3$]$^{2+}$ (DNA intercalated)	0.05	1.2×10^{10}	41
rac-[Co(bpy)$_3$]$^{3+}$ + [*Λ-Ru(phen)$_3$]$^{2+}$ (DNA intercalated)	0.05	1.7×10^{10}	41
[NiMe$_6$[14]dieneN$_4$)]$^{3+}$ + [Ru(bpy)$_3$]$^{2+}$	0.10(CH$_3$CN)	1.68×10^4	53
[Ni(rac-Me$_6$[14]aneN$_4$)]$^{3+}$ + [Ru(bpy)$_3$]$^{2+}$	0.10(CH$_3$CN)	8.78×10^3	53
[Ni(meso-Me$_6$[14]aneN$_4$)]$^{3+}$ + [Ru(bpy)$_3$]$^{2+}$	0.10(CH$_3$CN)	1.22×10^4	53
[Rh$_2$(O$_2$CCH$_3$)$_4$(CH$_3$CN)$_2$]$^+$ + [Ru(4,4'-Ph$_2$bpy)$_3$]$^{2+}$	0.10(CH$_3$CN)	2.0×10^6	53
Rhodium(II)			
[Ag([14]aneN$_4$)]$^{2+}$ + [Rh(bpy)$_3$]$^{2+}$	0.04	1.3×10^8	55
[Ni(rac-Me$_6$[14]aneN$_4$)]$^{3+}$ + [Rh$_2$(O$_2$CCH$_3$)$_4$(CH$_3$CN)$_2$]	0.10(CH$_3$CN)	1.50×10^4	53
[Ni(meso-Me$_6$[14]aneN$_4$)]$^{3+}$ + [Rh$_2$(O$_2$CCH$_3$)$_4$(CH$_3$CN)$_2$]	0.10(CH$_3$CN)	2.18×10^4	53
[NiMe$_6$[14]dieneN$_4$)]$^{3+}$ + [Rh$_2$(O$_2$CCH$_3$)$_4$(CH$_3$CN)$_2$]	0.10(CH$_3$CN)	1.63×10^4	53
[Fe(5-Clphen)$_3$]$^{3+}$ + [Rh$_2$(O$_2$CCH$_3$)$_4$(CH$_3$CN)$_2$]	0.05(CH$_3$CN)	3.3×10^6	53
Co(dmgH)$_2$PPh$_3$Cl + Rh(dmgH)$_2$PPh$_2$	95% ethanol	1.3×10^8	54
Co(dmgH)$_2$PPh$_3$Br + Rh(dmgH)$_2$PPh$_2$	95% ethanol	1.8×10^8	54
[Co(NH$_3$)$_5$F]$^{2+}$ + Rh(dmgH)$_2$PPh$_2$	95% ethanol	2.9×10^6	54
[Co(NH$_3$)$_5$Cl]$^{2+}$ + Rh(dmgH)$_2$PPh$_2$	95% ethanol	2.3×10^8	54
[Co(NH$_3$)$_5$N$_3$]$^{2+}$ + Rh(dmgH)$_2$PPh$_2$	95% ethanol	2.6×10^8	54
[Co(NH$_3$)$_5$NCS]$^{2+}$ + Rh(dmgH)$_2$PPh$_2$	95% ethanol	2.6×10^8	54
[Co(NH$_3$)$_5$Br]$^{2+}$ + Rh(dmgH)$_2$PPh$_2$	95% ethanol	3.0×10^8	54
[Co(NH$_3$)$_5$OH$_2$]$^{2+}$ + Rh(dmgH)$_2$PPh$_2$	95% ethanol	$<7 \times 10^5$	54
[Co(NH$_3$)$_5$CN]$^{2+}$ + Rh(dmgH)$_2$PPh$_2$	95% ethanol	$<3 \times 10^6$	54
[Co(NH$_3$)$_6$]$^{3+}$ + Rh(dmgH)$_2$PPh$_2$	95% ethanol	$<1 \times 10^6$	54
Palladium(II)			
[Ni([9]aneN$_3$)$_2$]$^{3+}$ + [Pd([9]aneN$_3$)$_2$]$^{2+}$	1.0	2.06×10^3	60
[Ni([9]aneN$_3$)$_2$]$^{3+}$ + [Pd([9]aneN$_3$)([9]aneN$_2$(NH)]$^{2+}$	1.0	$\approx 1.2 \times 10^2$	60

(continued)

Table 2.1. (continued)

Reaction	Medium I (M)	k ($M^{-1}\,s^{-1}$)	$\Delta H^{\ddagger}$ (kcal mol^{-1})	$\Delta S^{\ddagger}$ (cal K^{-1} mol^{-1})	Ref.
$[Ni([9]aneN_3)_2]^{3+} + [Pd([9]aneN_2(NH))_2]^{2+}$	1.0	$\approx 5 \times 10^1$			60
$[Ni(Me[9]aneN_3)_2]^{3+} + [Pd([9]aneN_3)_2]^{2+}$	1.0	1.8×10^3			60
$[Ni(Me[9]aneN_3)_2]^{3+} + [Pd([9]aneN_3)([9]aneN_2(NH))]^{2+}$	1.0	$\approx 9.6 \times 10^1$			60
$[Ni(Me[9]aneN_3)_2]^{3+} [Pd([9]aneN_2(NH))_2]^{2+}$	1.0	$\approx 3 \times 10^1$			60
Palladium(III)					
$[Ni([9]aneN_3)_2]^{3+} + [Pd([9]aneN_2(NH))_2]^{5+}$	1.0	5.7×10^3			60
$Co^{3+} + [Pd([9]aneN_2(NH))_2]^{5+}$	1.0	7.7×10^2			60
$CoOH^{2+} + [Pd([9]aneN_2(NH))_2]^{5+}$	1.0	3.1×10^5			60
Tungsten(IV)					
$[IrCl_6]^{2-} + [W_3O_4(H_2O)_8OH]^{3+}$	2.0	1.06×10^6			21
$[*Ru(4,4'-CO_2bpy)_3]^{4-} + [W(CN)_8]^{4-}$	0.03	2.0×10^8 e			46
$[*Ru(bpy)_2(4,4'-CO_2bpy)] + [W(CN)_8]^{4-}$	0.03	1.7×10^{10} e			46
$[*Ru(bpy)_3]^{2+} + [W(CN)_8]^{4-}$	0.03	1.1×10^{10} e			46
$[*Ru(bpy)_2(Mebpy\text{-}Mebpy)\text{-}Ru(bpy)_2]^{4+} + [W(CN)_8]^{4-}$	0.03	2.0×10^{10} e			46
Rhenium(-I)					
$[Co_2(CO)_8] + [Re(CO)_5]^-$	THF	1.3×10^1			69
$[Mn_2(CO)_{10}] + [Re(CO)_5]^-$	THF	1.4			69
$[Cp_2Mo_2(CO)_6] + [Re(CO)_5]^-$	THF	2.0×10^1			69
Osmium(II)					
$[Mn(edta)]^- + [Os(5,6\text{-}Me_2phen)_3]^{2+}$	0.10	1.54×10^4			28
$[*Ru(4,4'-CO_2bpy)_3]^{4-} + [Os(CN)_6]^{4-}$	0.03	3.3×10^6 e			46
$*Ru(bpy)_2(4,4'-CO_2bpy) + [Os(CN)_6]^{4-}$	0.03	4.1×10^9 e			46
$[*Ru(bpy)_3]^{2+} + [Os(CN)_6]^{4-}$	0.03	3.0×10^9 e			46
$[*Ru(bpy)_2(Mebpy\text{-}Mebpy)Ru(bpy)_2]^{4+} + [Os(CN)_6]^{4-}$	0.03	2.1×10^9 e			46

Platinum(II)

Reaction					
$[IrCl_6]^{2-} + cis\text{-}[Pt(NH_3)_2(NH_2Me)_2]^{2+}$	1.0	$7.30 \times 10^3[Cl^-]$[d]	1.4	−36	61
$[IrCl_6]^{2-} + cis\text{-}[Pt(NH_3)_2(NH_2Et)_2]^{2+}$	1.0	$3.22 \times 10^3[Cl^-]$[d]	1.9	−36	61
$[IrCl_6]^{2-} + cis\text{-}[Pt(NH_3)_2(NH_2Pr^n)_2]^{2+}$	1.0	$3.19 \times 10^3[Cl^-]$[d]	2.3	−35	61
$[IrCl_6]^{2-} + cis\text{-}[Pt(NH_3)_2(NH_2Pr^i)_2]^{2+}$	1.0	$2.54 \times 10^2[Cl^-]$[d]	4.3	−33	61
$[IrCl_6]^{2-} + cis\text{-}[Pt(NH_3)_2(NH_2CH_2Bu^t)_2]^{2+}$	1.0	$1.46 \times 10^3[Cl^-]$[d]	2.7	−35	61
$[IrCl_6]^{2-} + cis\text{-}[Pt(NH_3)_2(NH_2CH_2Ph)_2]^{2+}$	1.0	$5.38 \times 10^2[Cl^-]$[d]	3.4	−35	61
$[IrCl_6]^{2-} + cis\text{-}[Pt(NH_3)_2(NH_2CH_2Ph)_2]^{2+}$	1.0	$3.11 \times 10^2[Cl^-]$[d]	4.5	−32	61
$[IrCl_6]^{2-} + cis\text{-}[Pt(NH_3)_2(NH_2C_5H_9)_2]^{2+}$	1.0	$1.56 \times 10^2[Cl^-]$[d]	4.6	−33	61
$[IrCl_6]^{2-} + cis\text{-}[Pt(NH_3)_2(NH_2C_6H_{11})_2]^{2+}$	1.0	$7.85 \times 10^1[Cl^-]$[d]	5.2	−32	61
$[IrCl_6]^{2-} + cis\text{-}[Pt(NH_3)_2(pyr)_2]^{2+}$	1.0	$1.0 \times 10^2[Cl^-]$[d]			61
$[IrCl_6]^{2-} + cis\text{-}[Pt(NH_3)_2(3\text{-}CNpyr)_2]^{2+}$	1.0	$1.05 \times 10^4[Cl^-]$[d]	1.6	−35	61
$[IrCl_6]^{2-} + cis\text{-}[Pt(NH_3)_2(meen)_2]^{2+}$	1.0	$1.12 \times 10^3[Cl^-]$[d]	3.7	−32	61
$[IrCl_6]^{2-} + cis\text{-}[Pt(NH_3)_2(dmen)_2]^{2+}$	1.0	$4.13 \times 10^1[Cl^-]$[d]	6.1	−30	61

Silver(II)

Reaction					
$[Ni(Me_6[14]dieneN_4)Cl_2]^- + [Ag[14]aneN_4]^{2+}$	0.1	4.8×10^8			55

Europium(II)

Reaction					
$[MoO(O_2)_2] + Eu^{2+}$	0.1	1.9×10^3			22
$[MoO(OH)(O_2)_2]^- + Eu^{2+}$	0.1	3.0×10^3			22
$[WO(OH)(O_2)_2]^- + Eu^{2+}$	0.1	2.8×10^4			22

[a] 23 °C. [b] 4.5 °C. [c] 23.4 °C. [d] 20 °C. [e] 22–23 °C. [f] 22 °C. [g] 20.1 °C, $kg\ mol^{-1}\ s^{-1}$.

potential of the dimers, leading to the postulation of an inner-sphere mechanism involving attack of the anion on a dimer carbonyl carbon.[69]

2.3.8. Ytterbium

Proton NMR line-broadening techniques have been used to measure the electron self-exchange rate constants for the bis(t-butyl[8]annulene)ytter-bate(II)/(III) couple in THF (19.3 °C) in the presence ($k_{11} = 1 \times 10^7 \, M^{-1} \, s^{-1}$) and absence of solvating diglyme ($k_{11} = 5 \times 10^7 \, M^{-1} \, s^{-1}$).[70] Mechanistic schemes involving potassium ion association and diglyme coordination have been postulated to account for the observed rate and activation parameters.

2.4. Reactions with Metalloproteins

2.4.1. Introduction

The kinetic investigations of electron transfer reactions of metalloproteins continue to appear at a rapid pace. The labeling of protein surface amino acids by covalently attached redox-active metal complexes has been extended from common use of histidine to include glutamic acid and lysine residues. The current status of the study of long-range electron transfer in multisite metalloproteins, including modified proteins, has been reviewed.[71] The use of metal-labeled blue copper proteins in the investigation of their structure–reactivity relationships has also been reviewed.[72] The Proceedings of the 4th International Conference on Bioinorganic Chemistry (1989) have been published in a special issue of the *Journal of Inorganic Biochemistry*.[73]

2.4.2. Copper Proteins

The kinetics of electron transfer reactions between spinach plastocyanin and $[Fe(CN)_6]^{3-}$, $[Co(phen)_3]^{3+}$, and Fe(II) cytochrome c have been studied as a function of ionic strength.[74] Applications of the equations of Van Leeuwen support the proposal of two sites of electron transfer, with $[Co(phen)_3]^{3+}$ binding near residues 42–45 and the interaction of $[Fe(CN)_6]^{3-}$ at a hydrophobic region near the copper ion. Pulse radiolysis has been employed to measure the rates of electron transfer from Ru(II) to Cu(II) in plastocyanins from *Anabaena variabilis* and *Scenedesmus obliquus* which have been modified at His-59 by $[Ru(NH_3)_5]^{2+}$.[75] The small intramolecular rates (<0.082 and $<0.26 \, s^{-1}$, respectively) over a donor–acceptor distance of ~12 Å indicate that electron transfer from the His-59 site to the Cu center is not a preferred pathway. A more favorable route, via the acidic (residues 42–44) patch (≈ 14 Å to Cu), is supported by the rate of $>5 \times 10^3 \, s^{-1}$ for the reduction of PCu(II) by unattached $[Ru(NH_3)_5im]^{2+}$. The intramolecular electron transfer from Fe(II) in horse cytochrome c to Cu(II) in French bean plastocyanin (~12 Å from heme edge to Cys-84 S), in a carbodiimide cross-linked covalent complex, proceeds with a rate of $1.05 \times 10^3 \, s^{-1}$.[76] The presence of the

Table 2.2. Electron Self-Exchange Reactions of Transition Metal Complex Couples at 25 °C

Reaction	Medium I (M)	k $(M^{-1}\,s^{-1})$	$\Delta H^{\ddagger}$ $(kcal\,mol^{-1})$	$\Delta S^{\ddagger}$ $(cal\,K^{-1}\,mol^{-1})$	Ref.
$[Cr(CNdipp)_6]^{+/0}$	CD_2Cl_2	1.8×10^8	1.5	-15.6	23
$[Mn(CNC(CH_3)_3)_6]^{2+/+}$	$0.1(C_6H_5CN)$	5.3×10^4 [a]	5.0	-19	27
	$0.1(CH_3CH_2OH)$	3.3×10^4 [a]	5.7	-17	27
	$0.1((CH_3)_2CO)$	3.3×10^4 [a]	4.3	-22	27
	$0.1((CH_3O)_3PO)$	5.4×10^4 [a]	5.1	-18	27
	$0.1((CH_3CH_2)_2CO)$	2.9×10^4 [a]	4.3	-22	27
	$0.3(CH_2Cl_2)$	6.5×10^4 [a]	4.0	-21	27
$[Fe(phen)_3]^{3+/2+}$	$0.4(D_2O)$	$1.31 \times 10^7\ m^{-1}\,s^{-1}$ [b]	0.4	-2.4	32
	$0.3(CD_3CN)$	$8.0 \times 10^6\ m^{-1}\,s^{-1}$ [c]	2.6	-17.3	32
$[Fe(C_5H_5)_2]^{+/0}$	$0.01(CD_3CN)$	8.2×10^6	5.3	-9.0	33
$[Fe(C_5H_5)(C_5H_4CH_2OH)]^{+/0}$	$0.01(CD_3CN)$	5.5×10^6			35
	$0.01(propionitrile)$	$\sim 9 \times 10^6$			35
	$0.01((CD_3)_2CO)$	7.2×10^6			35
	$0.01(D_2O)$	1.4×10^7			35
	$0.01(CD_3NO_2)$	$\sim 1.4 \times 10^7$			35
	$0.01(C_6D_5CN)$	1.8×10^7			35
	$0.01(C_6D_5NO_2)$	1.4×10^7			35
$[Fe(OEC)(1\text{-}MeIm)]^{+/0}$	$5:1\ acetone/H_2O$	2.8×10^7 [d]			39
$[Fe(OEP)(1\text{-}MeIm)]^{+/0}$	$5:1\ acetone/H_2O$	4.0×10^7 [d]			39
$[Fe(OEiBC)(1\text{-}MeIm)]^{+/0}$	$5:1\ acetone/H_2O$	1.7×10^8 [d]			39
$[Co(C_5H_4COOH)_2]^{+/0}$	$0.01(CD_3CN)$	2.0×10^8			35
	$0.01((CD_3)_2CO)$	1.2×10^8			35
	$0.01(D_2O)$	$\sim 3.0 \times 10^8$			35
	$0.01(CD_3NO_2)$	2.4×10^8			35
	$0.01(C_6D_5CN)$	8×10^7			35
	$0.01(C_6D_5NO_2)$	$\sim 4.5 \times 10^8$			35
	$0.01(TMU)$	4.9×10^8			35
	$0.01(PC)$	1.4×10^8			35
	$0.01(CD_3OD)$	$\sim 1.5 \times 10^8$			35

(continued)

Table 2.2. (continued)

Reaction	Medium I (M)	k $(M^{-1}\,s^{-1})$	$\Delta H^{\ddagger}$ (kcal mol^{-1})	$\Delta S^{\ddagger}$ (cal K^{-1} mol^{-1})	Ref.
[Co(1-nor)$_4$]$^{0/-}$	THF	7.9×10^5 [e]	12.0	8	52
[Co(1-nor)$_4$]$^{+/0}$	C$_5$D$_5$N	3.8×10^5 [f]	8.5	-6	52
[Co(1-nor)$_4$]$^{+/0}$	CD$_2$Cl$_2$	2.0×10^6 [f]			52
[Co(taptacn)]$^{3+/3+}$	0.1	4.2×10^{-2}			49
[Co((9)aneS$_3$)$_2$]$^{3+/2+}$	0.2	1.6×10^5	8.3	-7	43
[fac-lel_3-Co((NH$_3$)$_2$Me$_3$sar]]$^{5+/4+}$	0.2	3.1×10^{-2}			50
[mer-lel_3-Co((NH$_3$)$_2$Me$_3$sar]]$^{5+/4+}$	0.2	3.3×10^{-2}			50
[fac-ob_3-Co((NH$_3$)$_2$Me$_3$sar]]$^{5+/4+}$	0.2	9.7×10^{-1}			50
[Ni(H$_{-2}$Aib$_3$)]$^{0/-}$	0.1	4.5×10^2 [g]			58
[Cu(bidhp)]$^{2+/+}$	DMSO	4×10^3 [h]	9.1	-11.8	62
[Cu((imidH)$_2$DAP)]$^{2+/+}$	0.0223	1.31×10^4			63
[Cu(2,9-Me$_2$phen)$_2$]$^{2+/+}$	0.002	2.0×10^5 m^{-1} s^{-1}	5.7	-15	64
	0.1(CD$_3$CN)	4.9×10^3 m^{-1} s^{-1}	7.1	-18	64
	0.1((CD$_3$)$_2$CO)	3.0×10^3 m^{-1} s^{-1}	7.0	-19	64
[Ru((9)aneN$_3$)$_2$]$^{3+/2+}$	0.1	5×10^4 [i]			42
[Rh$_2$(O$_2$CCH$_3$)$_4$(CH$_3$CN)$_2$]$^{+/0}$	0.10(CD$_3$CN)	5.4×10^4			53
[Yb(t-Bu[8]annulene)$_2$]$^{2-/-}$	THF-d$_8$	5×10^7 [j]	11.3	15	70
[Yb(t-Bu[8]annulene)$_2$]$^{2-/-}$	THF/diglyme	1×10^7 [j]	8.8	3	70

[a] 0 °C. [b] 3 °C. [c] 4 °C. [d] -20 °C. [e] 30 °C. [f] 40 °C. [g] 24 °C. [h] 28 °C. [i] 23 °C. [j] 19.3 °C.

plastocyanin inhibits the oxidation of the cytochrome c(II) by $[Fe(CN)_6]^{3-}$ and promotes oxidation by $[Fe(C_5H_5)_2]^+$ as a result of electrostatic and steric effects. *Rhus* stellacyanin has been modified with two $[Ru(NH_3)_5]^{2+}$ units at neighboring His-32 and His-100 residues. The electron transfer to the Cu(II) center, over a distance of $\sim$18 Å, proceeds with a rate of 0.05 s^{-1}.[77]

Investigations of the effects of coordinating anions on the kinetics of the redox reactions of Cu(II)/(I) superoxide dismutase with the $[Fe(CN)_6]^{4-/3-}$ couple indicate that the Cu(II) geometry of the anion adduct [unreactive square-planar (with CN$^-$ and N$_3^-$) vs. reactive five-coordinate (with OCN$^-$ and SCN$^-$)] affects its ability to accept an electron from $[Fe(CN)_6]^{4-}$.[78]

2.4.3. Hemoglobin and Myoglobin

The long-range electron transfer reactions in ruthenium-modified myoglobin, in which the labile heme unit has been replaced by various metalloporphyrins, have been reviewed.[79] The reductions of the $[Ru(NH_3)_5]^{3+}$ moiety, attached at His-48, by Pd- and Pt-substituted hemes in myoglobin proceed at rates of $9 \cdot 1 \times 10^3$ and 1.2×10^4 s^{-1}, respectively.[80] The difference in rates for electron transfer between Fe^{2+}(heme) and Mg or Zn(porphyrin)$^+$ centers in $[\alpha(Fe(II)P),\beta(M^{+\cdot}P)]$ hemoglobin hybrids indicates a direct process as opposed to the involvement of a conformational "gate."[81] Using $[Co(NH_3)_5Cl]^{2+}$ to quench the ^{3}Zn* state, a rate constant of 2.4×10^3 s^{-1} has been measured for back electron transfer within $[\alpha(Zn^{+\cdot}P)\beta(Fe(III)CN)]$.[82]

The rate constants for the reduction of BrCN-modified metmyoglobin, possessing a pentacoordinate geometry, by several metal complexes have been measured and compared to the values for the analogous hexacoordinate native metmyoglobin.[83] The calculated self-exchange rate constant for the BrCN-modified couple ($\sim$10^4 M^{-1} s^{-1}) is considerably larger than for the native myoglobin couple ($\sim$1 M^{-1} s^{-1}), and is attributed to the geometry changes (water dissociation) at the iron center upon reduction.

2.4.4. Cytochromes

The intramolecular electron transfer between Ru(II) and Fe(III) in *P. stutzeri* cyt c_{551} modified at His-47 by $[Ru(NH_3)_5]^{2+}$ proceeds over a distance of 7.9 Å at a rate of 13 s^{-1}.[84] A rate of 1.7 s^{-1} has been measured for the Ru(II) to Fe(III) electron transfer within a mutant yeast cytochrome c modified at His-62 with $[Ru(NH_3)_5]^{2+}$.[85] The dependence of the rate constants for the Ru(II) to ZnP$^{+\cdot}$ and ZnP* to Ru(III) electron transfer processes in a $[Ru(NH_3)_4L(His-33)]$ (L = pyridine or isonicotinamide) modified zinc-substituted cytochrome c (11.7 Å separation) on the driving forces is in agreement with the semiclassical electron transfer model.[86] The pH dependences (or lack thereof) of the rate constants measured for the electron transfer reactions between various metal complexes and horse (unmodified and His-33 diethyl pyrocarbonate modified), tuna (no His-33), and *Candida krusei* cytochromes c have been attributed to the presence of an accessible His-33 residue.[87] A rate of 2.8 s^{-1} has been measured for electron transfer from

Table 2.3. Intramolecular Electron Transfer Reactions Involving Metalloproteins at 25 °C

Oxidant	Reductant	Medium I (M)	k (s^{-1})	$\Delta H^{\ddagger}$ (kcal mol^{-1})	$\Delta S^{\ddagger}$ (cal K^{-1} mol^{-1})	Ref.
Ru(NH$_3$)$_5$(His-59)	A. variabilis PCu(I)	0.31	$<8.2 \times 10^{-2}$ [a]			75
Ru(NH$_3$)$_5$(His-59)	S. obliquus PCu(I)	0.31	$<2.6 \times 10^{-1}$ [a]			75
S. obliquus PCu(II)	[Ru(NH$_3$)$_5$(imidazole)]$^{2+}$	0.31	$>5 \times 10^{3}$ [b]			75
P. aeruginosa PCu(II)	Ru(NH$_3$)$_4$(H$_2$O)(His-83)	0.31	2.5			75
PCu(II)	Fe cyt c(II)	0.001	1.05×10^{3}			76
Rhus StCu(II)	{Ru(NH$_3$)$_5$}$_2$(His-32,100)	0.10	5×10^{-2} [c]			77
Ru(NH$_3$)$_5$(His-48)	Mb(Pt*)	0.1	1.2×10^{4} [d]			80
Hb(βZnP)$^+$	Hb(αFeP)	0.01	3.5×10^{2} [e]			81
Hb(αFeP)	Hb(β^3ZnP)	0.01	1.12×10^{2} [e]			81
Hb(βMgP)$^+$	Hb(αFeP)	0.01	1.55×10^{2} [e]			81
Hb(αFeP)	Hb(β^3MgP)	0.01	4.7×10^{1} [e]			81
Hb(αZnP)$^+$	Hb(βFeP)	0.01	2.4×10^{3} [f]			82
P. stutzeri cyt c_{551}(III)	Ru(NH$_3$)$_5$(His-47)	0.31	1.3×10^{1} [g]			84
yeast(N62H) cyt c(III)	Ru(NH$_3$)$_5$(His-62)	0.1	1.7			85
Zn cyt c$^+$	Ru(NH$_3$)$_4$(isn)(His-33)	0.1	2.0×10^{5} [h]	<0.5	-35	86
Zn cyt c$^+$	Ru(NH$_3$)$_4$(pyr)(His-33)	0.1	3.5×10^{5} [h]	<0.5	-34	86
Ru(NH$_3$)$_4$(isn)(His-33)	^{3}Zn cyt c	0.1	3.3×10^{6} [h]	2.2	-22	86
Ru(NH$_3$)$_4$(pyr)(His-33)	^{3}Zn cyt c	0.1	2.9×10^{6} [h]	<0.5	-30	86
horse Fe cyt c(III)	Co(II)(diAMsar)(Glu)	0.1	2.8			88
horse Fe cyt c(III)	*Ru(bpy)$_2$(dcbpy)(Lys-13)	0.1	1.6×10^{7}			90
horse Fe cyt c(III)	*Ru(bpy)$_2$(dcbpy)(Lys-72)	0.1	1.4×10^{7}			90
horse Fe cyt c(III)	*Ru(bpy)$_2$(dcbpy)(Lys-25)	0.1	1.0×10^{6}			90
horse Fe cyt c(III)	*Ru(bpy)$_2$(dcbpy)(Lys-27)	0.1	2.0×10^{7}			90
horse Fe cyt c(III)	*Ru(bpy)$_2$(dcbpy)(Lys-7)	0.1	3×10^{5}			90
Ru(bpy)$_2$(dcbpy)(Lys-13)	horse Fe cyt c(II)	0.1	2.6×10^{7}			90
Ru(bpy)$_2$(dcbpy)(Lys-72)	horse Fe cyt c(II)	0.1	2.4×10^{7}			90
Ru(bpy)$_2$(dcbpy)(Lys-25)	horse Fe cyt c(III)	0.1	1.5×10^{6}			90
Ru(bpy)$_2$(dcbpy)(Lys-27)	horse Fe cyt c(II)	0.1	3.0×10^{7}			90
Ru(bpy)$_2$(dcbpy)(Lys-7)	horse Fe cyt c (II)	0.1	6×10^{5}			90

cyt c oxidase Fe_a(III)	cyt c oxidase Cu_A(I)	0.05	1.8×10^4	96
cyt c oxidase Cu_A(II)	cyt c oxidase Fe_a(II)	0.15	1.7×10^4 [i]	97
CcP(IV,R·+)	horse Fe cyt c(II)	0.01	4.5×10^2	100
CcP(III,R·+)	horse Fe cyt c(II)	0.01	5	100
CcP(IV,R·+)	horse Fe cyt c(II)	0.114	4.5×10^3	100
CcP(IV,R·+)(MI-His-181)	horse Fe cyt c(II)	0.114	3.45×10^3	101
CcP(IV,R·+)(MI-Gly-181)	horse Fe cyt c(II)	0.114	1.85×10^3	101
Fe cyt c(III)	liposomal Fe cyt c_1(II)	0.125	1.2×10^1 [j]	103

[a] 20°C. [b] 8.7°C. [c] 18°C. [d] 22-24°C. [e] 5°C. [f] 28°C. [g] 19°C. [h] 22°C. [i] 25.5°C. [j] 2°C.

a Co(II) macrocyclic cage, [Co(diAMsar)]$^{2+}$, covalently attached at Glu-66 or Glu-69, to Fe(III) of cytochrome c (14–15 Å).[88] An inverse dependence of the increased luminescence decay rates of cytochrome c derivatives, attached at several lysine residues with [Ru(bpy)$_2$(4,4'-CO$_2$bpy)], on the Ru(II) to heme-edge distance suggested a contribution from *Ru(II) to Fe(III) electron transfer.[89] The rate constants for the thermal back-reactions between Fe(II) and the Ru(III)-labeled lysines were fastest at the Lys-13, -27, and -72 positions surrounding the heme crevice ($1 - 3 \times 10^7$ s^{-1}), with significantly smaller rate constants (10^5–10^6 s^{-1}) for electron transfer to sites further removed from the heme group.[90]

Rate data, obtained from the literature, on the reactivities of CDNP-lysine modified cytochromes c toward strongly and weakly binding redox partners has been analyzed in terms of the relative contributions of the dipole moments of CDNP-cytochrome c and the electrostatic potential at the heme edge.[91] The ionic strength dependences of the rate constants for the reduction of various cytochromes with [Co(phen)$_3$]$^{3+}$ and [Fe(CN)$_6$]$^{3-}$ have been used to calculate the sign and magnitude of the surface charges at the sites of electron transfer.[92] The results indicate the presence of more than one site for electron transfer on c-type cytochromes, with electrostatic interactions controlling the site preference of redox partners. The rate constants for electron transfer between ferrocytochrome c and [M(CN)$_6$]$^{3-}$ (M = Fe, Ru, and Os), generated by photoexcitation, have been measured as a function of ionic strength.[93,94] The inverse ionic strength dependences have been related to the complex-protein binding constants, and the trend in the rate constants (Ru > Os > Fe) is consistent with the relative thermodynamic driving forces. The ionic strength and solvent (aqueous ethanol) dependences of the stereoselectivity observed in the oxidation of ferrocytochrome c by tris(acetylacetonato)cobalt(II) have been related to the hydrophobic–hydrophillic interactions at the heme edge.[95]

The intramolecular electron transfer from the Cu$_A$(I) center to Fe(III) in heme a of cytochrome c oxidase has a rate of 1.8×10^4 s^{-1}, with an activation energy of 2.8 kcal mol^{-1}.[96] The apparent rate of reverse electron transfer, from the Fe(II)$_a$ to Cu$_A$(II), has been measured at 1.7×10^4 s^{-1} in the CO-inhibited form of cytochrome c oxidase.[97] The binding rate constants for the oxidation of native[98] and mutant[99] cytochromes c by cytochrome c oxidase have been measured. Replacement of Phe-82 on cytochrome c by other residues affects both the protein–protein binding constants and electron transfer rates through rearrangements in the vicinity of the heme. The kinetics of the sequential reductions by cytochrome c of cytochrome c peroxidase compounds I [Fe(IV),R$^{\cdot+}$] and II [Fe(III), R$^{\cdot+}$] have been investigated.[100] A small change in the rate constant for the cytochrome c reduction of cytochrome c peroxidase [Fe(IV), R$^{\cdot}$] is observed when His-181 is replaced by a glycine residue, indicating the absence of a role of the imidazole side-chain in the electron transfer pathway.[101] A rate constant of 6.5×10^7 M^{-1} s^{-1} has been measured for electron transfer between *Desulfovibria gigas* cytochrome c$_3$(II) and its physiological partner *D. gigas* hydrogenase.[102] The ionic strength dependence of the rate constants for the electron transfer from liposomal cytochrome c_1 to cytochrome c has been related to a polyelectrolytic microenvironment about membrane-bound cytochrome c_1.[103]

Table 2.4. Electron Transfer Reactions involving Metalloproteins at 25 °C

Reaction	Medium I (M)	k (M^{-1} s^{-1})	$\Delta H^{\ddagger}$ (kcal mol^{-1})	$\Delta S^{\ddagger}$ (cal K^{-1} mol^{-1})	Ref.
[Fe(CN)$_6$]$^{3-}$ + spinach PCu(I)	0.10	7.0×10^4			74
[Co(phen)$_3$]$^{3+}$ + spinach PCu(I)	0.105	1.77×10^3			74
spinach PCu(II) + horse cyt c(II)	0.105	3.2×10^6			74
spinach PCu(II) + horse cyt c(II)(CDNP-Lys-13)	0.10	2.7×10^6			74
spinach PCu(II) + horse cyt c(II)(CDNP-Lys-72)	0.10	2.0×10^6			74
horse cyt c(III)(CDNP-Lys-13) + [Fe(dtpa)]$^{3-}$	0.11	2.4×10^4			74
[Co(phen)$_3$]$^{3+}$ + {spinach PCu(I)Co(NH$_3$)$_5$$^{3+}$}	0.10	6.0×10^2			74
{A. variabilis PCu(II)(His-59)Ru(NH$_3$)$_5$$^{3+}$} + {A. variabilis PCu(II)(His-59)Ru(NH$_3$)$_5$$^{2+}$}	0.31	1.2×10^5 [a]			75
{S. obliquus PCu(II)(His-59)Ru(NH$_3$)$_5$$^{3+}$} + {S. obliquus PCu(II)(His-59)Ru(NH$_3$)$_5$$^{2+}$}	0.31	3.3×10^5 [a]			75
SODCu(II) + [Fe(CN)$_6$]$^{4-}$	0.15	7.2 (pH 7.7)			78
[Fe(CN)$_6$]$^{3-}$ + SODCu(I)	0.15	3.9×10^1 (pH 7.7)			78
metMb(H$_2$O) + [Ru(NH$_3$)$_6$]$^{2+}$	0.2	3.0×10^2	8.6	−18	83
(BrCN-modified)metMb + [Ru(NH$_3$)$_6$]$^{2+}$	0.2	1.8×10^5	5.6	−16	83
DPmetMb(H$_2$O) + [Ru(NH$_3$)$_6$]$^{2+}$	0.2	1.9×10^2			83
DADPmetMb(H$_2$O) + [Ru(NH$_3$)$_6$]$^{2+}$	0.1	1.0×10^3			83
DPDPmetMb(H$_2$O) + [Ru(NH$_3$)$_6$]$^{2+}$	0.1	2.0×10^3			83
metMb(H$_2$O) + [Ru(edta)(H$_2$O)]$^{2-}$	0.2	2.4×10^3	8.7	−14	83
(BrCN-modified)metMb + [Ru(edta)(H$_2$O)]$^{2-}$	0.2	3.8×10^5	5.7	−14	83
metMb(H$_2$O) + [Co(sep)]$^{2+}$	0.1	1.1×10^3	8.0	−18	83
(BrCN-modified)metMb + [Co(sep)]$^{2+}$	0.1	9.6×10^2	9.6	−13	83
[Fe(CN)$_6$]$^{3-}$ + horse cyt c(II)/Pc(I)	0.085	2.1×10^6			76
[Fe(C$_5$H$_5$)$_2$]$^{+}$ + horse cyt c(II)/Pc(I)	0.085	9.0×10^6			76
[Co(phen)$_3$]$^{3+}$ + horse cyt c(II)	0.10	1.8×10^3 (pH 8.0); 1.1×10^3 (pH 5.0)			87
[Co(phen)$_3$]$^{3+}$ + horse cyt c(II)(DEPC-His-33)	0.10	1.8×10^3			87

(continued)

Table 2.4. (continued)

Reaction	Medium I (M)	k $(M^{-1}\,s^{-1})$	$\Delta H^{\ddagger}$ (kcal mol^{-1})	$\Delta S^{\ddagger}$ (cal K^{-1} mol^{-1})	Ref.
$[Co(terpy)_2]^{3+}$ + horse cyt c(II)	0.10	9.2×10^2 (pH 8.0)			87
		6.3×10^2 (pH 5.0)			87
$[Co(terpy)_2]^{3+}$ + horse cyt c(II)(DEPC–His-33)	0.10	9.0×10^2			87
$[Fe(CN)_6]^{3-}$ + horse cyt c(II)	0.10	9.0×10^2			87
$[Co(dipic)_2]^-$ + horse cyt c(II)	0.10	1.0×10^4			87
horse cyt c(III) + $[Co(terpy)_2]^{2+}$	0.10	1.0×10^3			87
horse cyt c(III) + $[Ru(NH_3)_5py]^{2+}$	0.10	4.1×10^3			87
horse cyt c(III) + $[Co(sep)]^{2+}$	0.10	3.4×10^5			87
$[Co(phen)_3]^{3+}$ + tuna cyt c(II)	0.10	1.5×10^3			87
$[Co(phen)_3]^{3+}$ + C. krusei cyt c(II)	0.10	2.8×10^3 (pH 8.0)			87
		2.4×10^3 (pH 5.0)			87
$[Co(phen)_3]^{3+}$ + C. krusei cyt c(II)(DEPC–His-33)	0.10	2.8×10^3			87
$[Co(phen)_3]^{3+}$ + Pseudomonas c-551(II)	∞	3.2×10^4 ^{b}			92
$[Co(phen)_3]^{3+}$ + Azobacter c_5(II)	∞	4.9×10^4 ^{b}			92
$[Co(phen)_3]^{3+}$ + Paracoccus c_2(II)	∞	3.5×10^4 ^{b}			92
$[Co(phen)_3]^{3+}$ + horse cyt c(II)	∞	7.0×10^3 ^{b}			92
$[Fe(CN)_6]^{3-}$ + Pseudomonas c-551(II)	∞	3.1×10^5 ^{b}			92
$[Fe(CN)_6]^{3-}$ + Azobacter c_5(II)	∞	2.0×10^5 ^{b}			92
$[Fe(CN)_6]^{3-}$ + Paracoccus c_2(II)	∞	2.0×10^6 ^{b}			92
$[Fe(CN)_6]^{3-}$ + horse cyt c(II)	∞	1.2×10^6 ^{b}			92
$[Fe(CN)_6]^{3-}$ + horse cyt c(II)	0.110	8.5×10^6 ^{b}			94
$[Os(CN)_6]^{3-}$ + horse cyt c(II)	0.110	3.6×10^8 ^{b}			94
$[Ru(CN)_6]^{3-}$ + horse cyt c(II)	0.110	1.2×10^9 ^{b}			94
cyt c oxidase + cyt c(II)(Phe-82)	0.049	1.82×10^8 ^{b}			99
cyt c oxidase + cyt c(II)(Tyr-82)	0.049	1.37×10^8 ^{b}			99
cyt c oxidase + cyt c(II)(Gly-82)	0.049	7.27×10^7 ^{b}			99
cyt c oxidase + cyt c(II)(Ser-82)	0.049	1.24×10^7 ^{b}			99
cyt c oxidase + cyt c(II)(Leu-82)	0.049	1.12×10^8 ^{b}			99
cyt c oxidase + cyt c(II)(Ile-82)	0.049	4.36×10^7 ^{b}			99

Reaction			
D. gigas hydrogenase + *D. gigas* cyt c_3(II)	0.05	6.5×10^7	102
HiPIP$_{ox}$ + HiPIP$_{red}$	0.1	1.7×10^4	104
HiPIP(Fe$_4$Se$_4$)$_{ox}$ + HiPIP(Fe$_4$Se$_4$)$_{red}$	0.1	7×10^4	104
[Co(phen)$_3$]$^{3+}$ + HiPIP$_{red}$	0.10	3.04×10^3 (pH 7.0)	105
[Co(phen)$_3$]$^{3+}$ + HiPIP$_{red}$(DEPC-His-42)	0.10	6.8×10^3	105
[Fe(CN)$_6$]$^{3-}$ + HiPIP$_{red}$	0.10	2.09×10^3 (pH 7.0)	105
[Fe(CN)$_6$]$^{3-}$ + HiPIP$_{red}$(DEPC-His-42)	0.10	2.9×10^3	105
HiPIP$_{ox}$ + [Co(phen)$_3$]$^{2+}$	0.10	1.45×10^3 (pH 7.0)	105
HiPIP$_{ox}$(DEPC-His-42) + [Co(phen)$_3$]$^{2+}$	0.10	1.1×10^3	105
HiPIP$_{ox}$ + [Fe(CN)$_6$]$^{4-}$	0.10	2.07×10^2 (pH 7.0)	105
HiPIP$_{ox}$(DEPC-His-42) + [Fe(CN)$_6$]$^{4-}$	0.10	1.0×10^2	105

[a] 20 °C. [b] 23.5 °C.

2.4.5. Iron-Sulfur Proteins

Electron self-exchange rate constants of the native and [Fe$_4$Se$_4$] reconstituted HIPIP$_{red/ox}$ couples have been determined to be 1.7×10^4 and $7 \times 10^4 \, M^{-1} \, s^{-1}$, respectively, from ^{1}H NMR T_1 measurements.[104] Diethyl pyrocarbonate modification of His-42 of HIPIP$_{red}$ and HIPIP$_{ox}$ has indicated that the pH rate dependences of the redox reactions of HIPIP$_{red/ox}$ with the [Co(phen)$_3$]$^{3+/2+}$ and [Fe(CN)$_6$]$^{3-/4-}$ couples are related to the proton equilibrium of His-42 (pK_a = 7.0), and the resulting changes in the redox potential.[105]

Chapter 3
Metal–Ligand Redox Reactions

3.1. Introduction

This chapter reviews the literature published during the period July 1988 to December 1989. In keeping with previous practice, some selectivity has been exercised. Also, the chapter includes redox reactions between metallic and nonmetallic species, regardless of whether or not the nonmetallic component is coordinated to the metal center during electron transfer. The format is such that reactions have been grouped according to their central or reactive nonmetallic element. Some topics have received scant coverage, particularly if they are dealt with elsewhere in this volume. Thus, oxidative addition/reductive elimination reactions are not within the scope of this section.

A number of review articles have recently appeared on metal–ligand redox reactions. These include a comprehensive compilation of rate constants for a wide variety of inorganic radicals produced by radiolysis and photolysis.[1] Details are provided on reactions with both inorganic and organic compounds. The kinetics and mechanism of the catalyzed free radical decomposition of peroxymonosulphate by various metals has been reviewed.[2] The reaction types by which cobalt dioxygen are converted irreversibly to inert complexes have been discussed.[3] The role of metals in oxygen radical chemistry, particularly in Fenton-type reactions, has been examined.[4] A review of dirhodium complexes has appeared with respect to their catalytic properties in reduction reactions of olefins and ketones.[5] An interesting discussion has been published on studies of the mechanism of photographic development from the viewpoint of modern theories of electron transfer dynamics.[6] The paper also includes recent results on attempts at developing a pure AgBr emulsion by phenylhydrazine and its derivatives. A review of pyrophosphito-bridged diplatinum complexes presents the reaction kinetics of free radicals and oxidative addition reactions with a variety of substrates.[7] The sites of oxidation and reduction involving metallic and nonmetallic porphyrin complexes has been reviewed.[8] Applictions of $[Fe(CN)_5L]^{3-}$ in analytical and bioinorganic chemistry, particularly with respect to electron transfer, have been discussed.[9] In addition, some related articles have appeared addressing complexes of paramagnetic metals with paramagnetic ligands,[10] 17- and 19-electron organometallic radicals and their reactions with ligands,[11] and the chemistry of metallo- and organometallic radicals.[12] An article addressing the chemical mechanism for electron transfer and bridge reducibility has appeared.[13] Discussion is focused upon one limiting mechanism for inner-sphere electron transfer which involves radical formation in the bridging ligand.

3.2. Oxygen, Peroxide, and Other Oxygen Compounds

3.2.1. Dioxygen

Time-resolved Raman spectroscopy has been utilized to detect $\nu(Fe\text{-}O)$ in an early intermediate in the reduction of O_2 by cytochrome oxidase.[14] Consideration of other data on dioxygen adducts indicates that this species is most likely the cytochrome $a_3^{2+}\text{-}O_2$ complex. The $\nu(Fe^{2+}\text{-}O_2)$ frequency is elevated relative to that observed for other systems from which it may be concluded that at 10 μs into the oxidase reaction the O=O bond is weakened. The ligand, HAPH (1) (R = H, R' = 4-imidazolyl), has been been employed as a model for the metal binding site of Bleomycin (BLM).[15] [Fe(II)(HAPH)] and [Fe(BLM)] were subjected to oxidation by O_2 in the presence of a spin trapping agent. Detection of the trapped adduct allows an estimation to be made of the relative efficiencies for the conversion of O_2 (53%) and H_2O_2 (63%) to HO' by [Fe(II)(HAPH)] compared to [Fe(BLM)]. A similar study using a modification of (1) has been reported.[16]

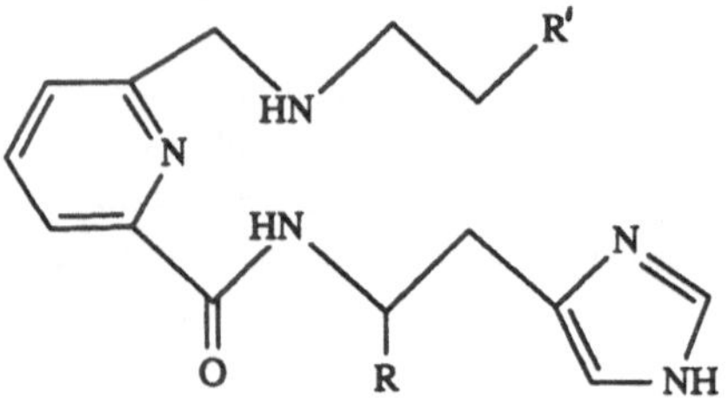

(1)

Oxygenation of $[Fe(DTBC)_3]^{3-}$ [$DTBCH_2$ = 3,5-bis(*t*-butyl)catechol] leads to changes in both the UV/visible and ESR spectra.[17] These changes suggest the reversible binding of O_2 in the complex and the formation of a semiquinonato-Fe(III)-(O_2^-) complex. It should be noted that this implies ligand loss to form an inner-sphere complex or strong ion pairing interactions to form an outer-sphere complex. The reduction of O_2 by ferrocene derivatives is catalyzed by metalloporphyrins or metallomacrocycles in acetonitrile in the presence of $HClO_4$.[18] The strong inner-sphere nature of electron transfer reactions involving metalloporphyrins plays an essential role in the catalytic reduction of dioxygen. Air oxidation (O_2) of cuboidal mixed-valence molybdenum/sulfido clusters leads to a variety of interconversions.[19] For example, upon complexing $[Mo_4S_4]^{5+}$ by NCS^-, air oxidation gives $[Mo_4S_4]^{6+}$ overnight. Hydrolysis of $[MoLBr_3]$ (L = 1,4,7-trimethyl-1,4,7-triazacyclononane) solution affords $[L_2Mo_2(\mu\text{-}OH)(\mu\text{-}OAc)_2](PF_6)_3$, which may be deprotonated to $[L_2Mo_2(\mu\text{-}O)(\mu\text{-}OAc)_1]^{2+}$.[20] Air oxidation of the latter complex yields $[L_2Mo(III)Mo(IV)(\mu\text{-}O)(\mu\text{-}OAc)_2](ClO_4)_3$. $[Mo(O)(CN)_5]^{3-}$ reacts readily with O_2 under phase transfer conditions to yield $[Mo(O)(O_2)(CN)_4]^{2-}$.[21] The plausible implications that such a process would have on the catalytic oxidation on metal oxide surfaces are discussed. Mn(III) porphyrins undergo facile one-electron oxidation in alkaline media producing a manganese(IV) μ-oxo dimer, $[PMn(IV)\text{-}O\text{-}Mn(IV)P]^{2+}$, according to the best evidence available.[22] The same species is also produced by the reaction of Mn(II) porphyrin complexes and O_2. $[([14]aneN_4)MnO]_2(ClO_4)_3$ is prepared as a mixed-valence complex.[23] Detailed studies of the formation of this complex indicate that the bridging oxygen comes from solvent water and that the bridge formation occurs with both Mn(III) ions. Oxidation of $[Ru(III)(Saloph)Cl_2]^+$ and $[Ru(III)(Saloph)XCl]$ (SalophH$_2$ = bis(salicylidene)-o-phenylenediamine; X = imidazole, 2-methylimidazole) by O_2 gives $[Ru(V)(Saloph)(O)Cl]$ and $[Ru(V)(Saloph)(O)X]^+$, respectively.[24] The oxo complexes were also prepared by reaction with PhIO or H_2O_2. The kinetics of the oxygenation reaction are presented. Further studies suggest that the reaction proceeds through an intermediate superoxo species, $[Ru(IV)(O_2)(Saloph)(Im)]$, before dissociating to form Ru(V)(O).[25] Reactions with $[Ru(Salen)(Im)Cl]$ yield similar products. $[Ru_3O(pfb)_6(Et_6O)_3](pfb)$ (pfb = perfluorobutyric acid) has been shown to catalytically oxidize a variety of olefins using molecular oxygen as the tertiary oxidant.[26] Mechanistic studies implicate a free radical chain pathway. The autooxidation of $[Ru(sar)]^{2+}$ (sar = sarcophogine) exhibits a pronounced pH dependence in the range 2.5 to 11.[27] Under acidic conditions (pH < 2.5), the reaction is first order in both $[Ru(sar)^{2+}]$ and $[O_2]$ with a pH-independent rate constant of $1.4 \pm 0.1 \, M^{-1} \, s^{-1}$, while under basic conditions (pH > 11), the pH-independent rate constant is much larger, $33.4 \pm 1.0 \, M^{-1} \, s^{-1}$. A mechanism is presented to account for the observed reaction rates.

The reversible reaction of $[Co_2(\mu\text{-}OH)L]^{3+}$ (L = 7,19,30-trioxa-4,10,16,22,27,33-hexaazabicyclo-[11,11,11]-pentatriacontane) with O_2 to form $[Co_2(\mu\text{-}OH)(O_2)L]^{3+}$ and the subsequent autooxidation of this complex in acid to $[Co_2(OH)L]^{5+}$ have been examined by UV/visible spectroscopy.[28] The autooxidation rate constant at 95 °C is only $3.7 \times 10^{-5} \, s^{-1}$. Similarly, the decomposition of cobalt dioxygen complexes with the ligand 3,9,17,23,29,30-hexaaza-6,20-dioxatricyclo-[23.3.1.1^{11,15}]triaconta-1(28),11,13,15(30),25(29),26-hexaene (BIS-BAMP) have been explored at 50 °C between pH 9 and 12.[29] Under these conditions, the dioxygen complexes undergo irreversible metal-centered autooxidation to form Co(III) complexes. Co(II)(1,5,8,12-tetraazadibenzo[3,4:9,10]cyclotetradecane-1,11-diene) may be oxidized by air to a 1:1 (superoxo) and 2:1 Co:O_2 derivative.[30]

Work has continued on the oxidation of dicopper(I) complexes by O_2 to give peroxo-dicopper(II) and hydroperoxo-dicopper(II) species.[31] These complexes utilize a phenolic dinucleating ligand possessing two tridentate sites, providing the copper nucleii with a constrained ligand environment. Further reports outlining kinetic data for two synthetic systems in which dinuclear copper(I) complexes exhibit reversible oxygen binding behavior and a crystallographic examination of the oxidized products have appeared.[32,33] The Cu(I) complexes of several Schiff-base-type ligands have been examined and their reaction with dioxygen probed.[34] The data available suggest that regeneration of Cu(I) from the adduct is not through a pathway involving the simple release of oxygen. $[Cu_4(py)_4Cl_4]$ is oxidized to $[Cu_4(py)_4Cl_4O_2]$ by molecular oxygen in nitrobenzene with a second-order rate law.[35] Insertion of O_2 into the halo/copper core is the proposed rate-determining step. The reaction of O_2 with dicopper(I) complexes of 20-membered tetraaza macrocyclic ligands proceeds via an intermediate with spectra characteristic of a mixed-valence species.[36] The air oxidation rate of Cu_2O has been measured at 120–170 °C.[37]

The reductive quenching of the ^2E excited states of $[CrL_3]^{3+}$ (L = bipyridine, phenanthroline, and derivatives) generates $[CrL_3]^{2+}$ which was subsequently used to study the reduction of O_2.[38] The rate constants exhibit a linear dependence on free energy, as predicted by Marcus theory and, from the electrostatically corrected Marcus cross-relation, a value of $2 \pm 1\ M^{-1}\ s^{-1}$ was obtained for k_{22}, the self-exchange rate, of the O_2/O_2^- couple. A suggestion is made that k_{22} lies in the range 1–$10\ M^{-1}\ s^{-1}$. The tetrakis(μ-2-anilinopyridinato)dirhodium(II,III) cation, $[(ap)_4Rh(II)Rh(III)]^+$, may be electrochemically reduced to $[(ap)_4Rh_2(II)]$.[39] This complex rapidly binds molecular oxygen to form $[(ap)_4Rh(II)Rh(III)(O_2^-)]$ as detected by ESR spectroscopy. The oxygenated product is capable of reacting with CH_2Cl_2 yielding CH_2O, Cl^-, and the original $[(ap)_4Rh_2(II)]$ species, which may react with another O_2. The photolytically generated bis(dimethylglyoximato)-(triphenylphosphine)rhodium(II) radical reacts rapidly with O_2 ($k = 1.4 \times 10^9\ M^{-1}\ s^{-1}$), with the first step being adduct formation.[40] This reaction ultimately results in the formation of hydrogen peroxide.

Ligand oxidation may be affected by molecular oxygen in the nickel complex of N,N-bis(2-mercaptoethyl)-2-methylthioethylamine) forming a sulfinato complex.[41] O_2^- has been detected by ESR in dry Nafion perfluorinated membranes neutralized by Ti^{3+} with the most likely source of the radical being electron transfer from Ti^{3+}.[42] The interaction of Pt(II) with creatinine generates a monomeric species which undergoes a slow redox process forming blue to green paramagnetic species.[43] The reaction is sensitive to the presence of oxygen and to the nature and acidity of the solvent. Yellow syn-$[L_2W_2O_4]I_2$ and its purple $anti$-isomer (L = 1,4,7-triazacyclononane) may be generated from Zn-dust-reduced aqueous solutions of LWO_3 or $[L_2W_2O_5](Br_3)_2$ by oxidation with molecular oxygen.[44]

3.2.2. Hydrogen Peroxide

Reactions of [Fe(II)(NTA)] and [Fe(II)(EDDA)] (NTA = nitrilotriacetate; EDDA = ethylenediamine-N,N-diacetate) with H_2O_2 at neutral pH and in the presence of formate leads to a transient reactive intermediate.[45] The kinetic results and product distribution are interpreted by a proposed mechanism. The decomposition of H_2O_2 by hydroxonitrilotri(methylenephosphonato)iron(III) has been examined in nitrate media.[46] At 26 °C, the specific rate constant for the reactions (1) and (2) were estimated as $k_1 = 11$ and $k_2 = 100\ M^{-1}\,s^{-1}$. The kinetic data

$$[Fe(NTMP)(OH)]^{4-} + HO_2^- \xrightarrow{k_1} [Fe(NTMP)(OH)]^{5-} + O_2^- + H^+ \qquad (1)$$

$$[Fe(NTMP)(OH)]^{4-} + O_2^- \xrightarrow{k_2} [Fe(NTMP)(OH)]^{5-} + O_2 \qquad (2)$$

support a mechanism in which electron transfer is the rate-determining step. The Fe(II) complex of *trans*-1,2-aminocyclohexane-N,N,N',N'-tetraacetic acid reacts with H_2O_2 at neutral pH to yield an ill-characterized reactive intermediate.[47] This reactive species is formed in a second-order reaction, first order in both species, with a bimolecular rate constant of $1.26 \pm 0.19 \times 10^3\ M^{-1}\,s^{-1}$. Excess H_2O_2 leads to a chain mechanism which mimics hydroperoxidase activity. Catalysis of H_2O_2 disproportionation and lumomagneson (H_2L) oxidation by Fe(II) ions has been shown to proceed via the sequential formation of OH^- and O_2^- radicals.[48] However, at high ratios of H_2L : Fe(II)(>5), a reversal in the sequence of radical initiation is observed. The catalytic effects of Fe^{2+} on the oxidation of indigocarmine by hydrogen peroxide has been demonstrated to be a sensitive method for the detection of trace quantities of iron.[49] The reaction of $[Cr(IV)(dien)(O_2)_2]$ with $[Fe(CN)_6]^{4-}$ and Ti(III) results in the reduction of both the Cr(IV) center and the peroxo ligands.[50] The unbuffered oxidation of hydrogen sulfite ions by H_2O_2, which is autocatalytic in H^+, may be linked to the reaction between $[Fe(CN)_6]^{4-}$ and H_2O_2, which consumes H^+, to produce an oscillatory reaction.[51] Similarly, oscillatory behavior may be observed in the reaction of $[Fe(CN)_6]^{4-}$ ion with H_2O_2.[52] The stability of this reaction is dependent upon the presence of light which, under certain conditions, will flip the reaction from a stable to an oscillating state, or *vice versa*. The catalytic decomposition of H_2O_2 by Fe^{2+}, Fe^{3+}, Cu^{2+}, and Co^{2+} adsorbed on neutral α-Al_2O_3 has been examined.[53] Oxidation of secondary $C-H$ bonds of saturated hydrocarbons by H_2O_2 in the presence of copper or iron perchlorates in pyridine solution yields the ketones and alcohols expected without the formation of alkyl radicals.[54]

Dowex 50W resin charged with an ethylamine–Cu(II) complex ion was utilized as a potential active catalyst for the decomposition of H_2O_2 in aqueous medium.[55] From the kinetic data, it may be observed that the rate constant is inversely related to the degree of cross-linking within the resin. The Cu-catalyzed redox reaction $2H_2O_2 + N_2H_4 \rightarrow 4H_2O + N_2$ is marked by an induction period the length of which varies inversely with the catalyst concentration.[56] The effects of a variety of physical techniques and chemical additives upon the reaction kinetics have been monitored. Further studies indicates that the mechanism involves hydrazyl radicals and diimides, autocatalysis, and solvent cage effects.[57] The $CuSO_4$-catalyzed reaction of H_2O_2 with SCN^- ion is reported to be the first example of a homogeneous, liquid-phase, halogen-free system that oscillates under batch conditions.[58] A detailed(!) mechanism is proposed involving 30 reactions and 26 variables, which simulates very well the observed oscillations in absorbance and O_2 evolution and two of the three bistability points. The kinetics of the Cu^{2+}-catalyzed decomposition of H_2O_2 proceeds by the formation of a superoxide–Cu(I) complex.[59] A mechanism involving the formation of this complex along with OH, HO_2, and O_2^- radicals simulates the observed kinetic data. The H_2O_2 hydroxylation of a dicopper complex (2) to the phenoxo and hydroxo bridged product (3) shown in Eq. (3) follows the rate law (4), with $k = 0.4\ M^{-1}\ s^{-1}$ and $K = 4500\ M^{-1}$.[60]

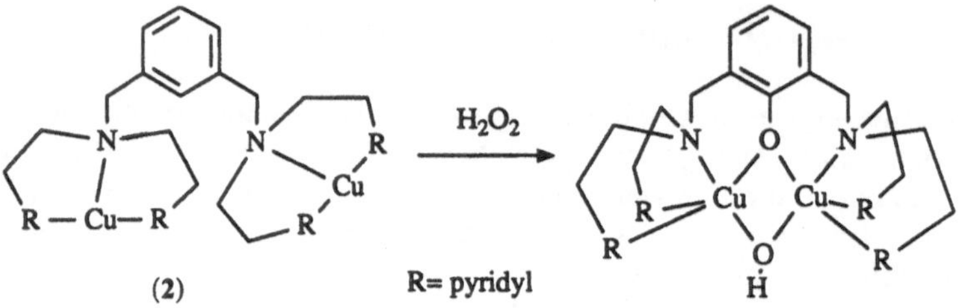

$$\text{(3)}$$

(2)   R= pyridyl

(3)

$$\text{rate} = k[Cu_2(L-H)^{4+}]^2[H_2O_2]/([H^+](1 + K[Cu_2(L-H)^{4+}])) \qquad (4)$$

The reaction is postulated to proceed through a hydroperoxide intermediate, $[Cu_2(L-H)(OOH)]^{3+}$. Experimental studies into Fenton and "Fenton-like" reactions point out that the detailed mechanism involves the formation of a low-valent cation.[61] However, while $[(H_2O)_5Cr^{2+}O_2H^-]$ decomposes directly to $[Cr(H_2O)_6]^{3+}$ and OH in the presence of $>0.1\ M$ EtOH, 2-PrOH, and 2-BuOH, $[(H_2O)_mCu^+O_2H^-]$ reacts directly with the alcohols. In the absence of organic substrate, the latter complex does yield $[Cu(H_2O)_n]^{2+}$ and OH.

H_2O_2 oxidation of $[Cr(O_2)_2(en)(H_2O)]$ and $[Cr(O_2)_2(dien)]$ proceeds through the formation of a diperoxo intermediate of the form, $[Cr(O)(O_2)_2L]$ (L = en or dien).[62] A mechanism which involves the entry of free peroxide into the coordination sphere of the metal ion prior to the redox process is favored based upon the kinetic data. The proposed mechanism for the reaction of $[Co(bpy)_3]^+$ with H_2O_2 is $[Co(bpy)_3]^+ \rightarrow [Co(bpy)_2]^+ + bpy$, $[Co(bpy)_2]^+ + H_2O_2 \rightarrow [Co(bpy)_2(H_2O_2)] \rightarrow$ oxidation products.[63] The limiting step is presumably the formation of the inner-sphere complex prior to electron transfer and, in excess H_2O_2, a value of $15\ s^{-1}$ is obtained. The reaction is inhibited by free bipyridine. An inner-sphere mechanism is also apparently involved in the reaction of square-planar nickel macrocyclic complexes with H_2O_2 or t-butyl peroxide.[64] Evidence is provided by the much slower reaction of Ni(decamethylcyclam) than the corresponding but sterically less restricted complex, Ni(tetramethylcylam). The oxo diperoxo complexes of Mo(VI) and W(VI) are obtained in the presence of H_2O_2.[65] These metal complexes act as oxygen transfer agents, oxidizing $[(en)_2Co(SCH_2CH_2NH_2)]^{2+}$ first to $[(en)_2Co(S\{O\}CH_2CH_2NH_2)]^{2+}$ and then, more slowly, to $[(en)_2Co(S\{O\}_2CH_2CH_2NH_2)]$. The metal catalyzed reaction is much more rapid than the corresponding reactions with H_2O_2 alone. The rate constants for the formation of $[MoO(OH)(O_2)_2]^-$ and $[WO(OH)(O_2)_2]^-$ are $8.6 \pm 0.8\ M^{-1}\ s^{-1}$ and $3.0 \pm 0.4\ M^{-1}\ s^{-1}$, respectively at pH 7.[66] Subsequent reduction of these complexes by Fe(II) proceeds through a ligand-bridged electron transfer with an intermediate, tentatively formulated as $[M(VI)-(O_2^{2-})-Fe(III)]$, being detected. The kinetics of the reduction with a number of other substrates has also been reported. In aqueous media, [ethylenebis(biguanide)]silver(III) oxidizes H_2O_2 to O_2 while being reduced to Ag(I) and free ligand.[67] The kinetics demonstrated an inverse hydrogen ion dependence which is attributed to the deprotonation of H_2O_2 facilitated by its axial coordination to the complex ion. The probable mechanism and plausible role of any Ag(II) intermediate are discussed. The kinetics of the formation of μ-peroxo bis(ethylenediaminetetraacetate)ruthenate(IV) by the oxidation of the $[Ru(III)(EDTA)Cl]^-$ ion with H_2O_2 are first order with respect to both the metal complex and H_2O_2.[68] The OsO_4-catalyzed decomposition of H_2O_2 is induced by the nucleophilic attack of HO_2^- ion on aqueous OsO_4.[69] The initial peroxide adduct of the catalyst undergoes rapid intramolecular electron transfer. Platinum greens may be synthesized by the careful addition of H_2O_2, under Ar, to aqueous mixtures of cis-$[Pt(NH_3)_2I_2]$ and pyrimidine derivatives.[70] Singlet oxygen is implicated in the mechanism of oxidation. The metal catalyzed epoxidation of olefins has been examined for platinum complexes[71] and manganese porphyrins.[72] In both cases, effective catalytic reaction procedures were developed. The redox behavior of Np(V) has been investigated in solutions containing H_2O_2 in 2-6 M HNO_3.[73] A reaction mechanism involving HO and HO_2 radicals is proposed. A kinetic method has been presented for the determination of U(VI), based upon its catalytic action in the decomposition of H_2O_2 in alkaline media.[74]

3.2.3. Alkyl Hydroperoxides

The reaction of $[Fe(OPPh_3)_4](ClO_4)_2$ with oxidants such as HOOH, t-BuOOH, m-ClC$_6$H$_4$C(O)OOH, $PhI(OAc)_2$, $Bu_4N(IO_4)$, O_3, or NaOCl in anhydrous MeCN gave $[(Ph_3PO)_4FeOOFe(OPPh_3)_4 \cdot 2H_2O](ClO_4)_4$.[75] Mechanistic pathways for the formation of the dimeric species are proposed. The reaction of $[W(VI)OS_3]^{2-}$ with t-BuOOH, $S_2O_8^{2-}$, I_2, or PhSSPh yields $[W(V)_2O_2(\mu-S)_2(S_2)_2]^{2-}$ which reverts to $[W(VI)OS_3]^{2-}$ on treatment with S_x^{2-} via an induced electron transfer.[76] The dimeric species also undergoes simple substitution and ligand centered reactions. The kinetics of hydroperoxide, ROOH [R = H, t-butyl, $Ph(CH_3)_2C$] reduction by Cu(I) O,O-dialkyl dithiophosphates has been examined by UV/vis spectroscopic methods.[77] Rate constants for the reaction were determined and led to a suggested mechanism. The oxidation by ROOH [R = t-butyl, $Ph(CH_3)_2C$] of a water-soluble, dimeric Fe(III) porphyrin with no μ-oxo linkage is proposed to proceed through a solvent caged $[(por)Fe(IV)(X)(OH)-t-BuO^{\cdot}]$ intermediate.[78] Analysis of the product distribution and use of a spin trapping agent led to the proposed mechanistic scheme. Further studies of the oxidation of the same iron porphyrin system with t-BuOOH and other aklyl hydroperoxides as well as m-chloroperbenzoic acid suggest that the mechanism involves homolytic cleavage of the O—O bond.[79] The oxidation of $[(H_2O)_2Co(dmgBF_2)_2]$ by HOOH and t-BuOOH has been reported.[80] Ruthenium catalyzes the oxidation of tertiary amines by alkyl hydroperoxides.[81] The reaction yields the corresponding α-(t-butyl-dioxy)alkyl-amines from which the iminium ion intermediates may be generated by the addition of acid. *cis*-$[Ru(II)(L)_2(OH_2)_2]^{2+}$ complexes (L = substituted 2,2'-bipyridines or 1,10-phenanthrolines) catalyze the oxidation of saturated hydrocarbons to alcohols by *t*-BuOOH.[82] The value of the kinetic isotope effect and the tertiary to secondary C−H bond reactivity suggest a mechanism involving hydrogen atom abstraction by a radical species. Similarly, Pt(II) catalyzes the oxidation of olefins to either ketones or epoxides by *t*-BuOOH or $KHSO_5$.[83] Alkane epoxidation may also be obtained using clay supported Mn(III)–Schiff base complexes as catalysts with *t*-BuOOH as oxidant.[84]

The combination of $[FeL(ClO_4)]$ $[H_2L$ = tetrakis(2,6-dichlorophenyl)-porphine] with *m*-chloroperbenzoic acid, O_3, or C_6F_5IO results in the formation of a ligand oxidized complex, $[Fe(II)(L^{\dot+})O]^+$.[85] Reduction of this species by 1*e* equivalent leads to the formation of $[(L)Fe(II)(O)]$ which is spectroscopically similar to compound II of Horse radish peroxidase. A high-valent chloroperbenzoic acid to the monofluoroiron(III) porphyrin at −78 °C.[86] This oxidized iron complex is capable of olefin epoxidation at this low temperature. A Ferrylporphyrin cation radical may be generated by the addition of *m*-chloroperbenzoic acid in aprotic solvents.[87] High-valent(tetraarylporphinato)manganese complexes may be generated via the low temperature reaction of the Mn(III) complexes with *m*-chloroperbenzoic acid, OCl_2, or Cl.[88] The reversible redox behavior of these systems is demonstrated by their reaction with cyclohexane or I^-. Similar studies on the oxidation of chloro(5,10,15,20-tetramesitylporphyrinato)manganese(III) with *m*-chloroperbenzoic acid and tetramethylammonium hydroxide have been reported.[89] The reaction of the oxidized manganese complex with olefins has also been examined. Oxidation of $[Ru(II)(OEP)(CO)]$ (OEP = octaethylporphyrin) by *m*-chloroperbenzoic acid in MeOH yields $[Ru(VI)(OEP)(O)_2]$.[90] In alcohols, the alkene epoxidation reaction gives $[Ru(IV)-(OEP)O(ROH)]$, which dimerizes to $[Ru(IV)(OEP)(OH)]_2O$ in noncoordinating solvents. The oxidation of $[Os(P)(CO)(EtOH)]$ (P = porphyrin dianion) by *m*-chloroperbenzoic acid in CH_2Cl_2 to $[Os(P)(O)_2]$ proceeds through the formation of a stable Os(VI) intermediate.[91] However, in contrast to the other oxidized metalloporphyrin complexes discussed above, these species are ineffective in the epoxidation of olefins. The mechanism of oxidation of $[Fe(III)TMP(X)]$ (TMP = tetramesitylporphyrin) by peroxyacids has been examined.[92] In methylene chloride at low temperature, the oxoiron(IV) porphyrin cation radical $[Fe(IV)TMP^+(O)]$ is obtained. The kinetics of oxidation of $[Fe(terpy)_2]^{2+}$ by O_3 have been studied in aqueous acidic solution.[93] The results indicate a 1*e* transfer reaction forming O_3^- which decomposes to form the OH⋅ radical. The OH⋅ radical may be scavenged by Cl^- resulting in $Cl_2^{\dot-}$ radical which causes additional oxidation of the $[Fe(terpy)_2]^{2+}$.

A number of studies have been reported in which a metal atom catalyzes the oxidation of alkenes by PhIO or its derivatives. Three studies have employed iron or manganese model systems for cytochrome P-450 in the oxidation of alkenes.[94-96] $[Ru(III)$ $(salen)(X)(Y)]^{n-}$ catalyze the oxidation of alkenes by PhIO at room temperature.[97] Similarly, ruthenium(II) complexes of bidentate phosphines may be utilized to promote the epoxidation of alkenes in the presence of iodosylbenzene and in low coordinating media.[98] $[Ni(cyclam)]^{2+}$ promotes the oxidation of cyclohexene and various aryl-substituted alkenes with PhIO as the terminal oxidant.[99] Also, the addition of the tin Lewis acids, $Ph_3SnO_2C_8F_{15}$ and $Ph_3SnO_3SC_6F_{13}$, dependent upon the species present, either increases or reduces the radical character in hydrocarbon oxidation reactions using PhIO and manganese porphyrins, [Mn(TPP)OAc] and [Mn(TyP)OAc], by substitution of the ⁻OAc with the perfluoro anion.[100]

3.3. Nitrogen Compounds and Oxyanions

3.3.1. Hydrazine, Azides, Hydroxylamines, and Derivatives

The kinetics of the reduction of slurried MnO_2 samples by N_2H_4 have been examined.[101] A faster rate at lower pH implies acid catalysis. Oxidation of hydrazine by permanganate in aqueous acidic solutions has been reported.[102] The dynamics of the reaction suggest it as a possible oscillatory system. Hexachloroosmate(IV) reacts with hydrazine producing a mixture of $[Os(NH_3)_5(N_2)]Cl_2$ and $[Os(NH_3)_4(N_2)_2]Cl_2$.[103] However, the addition of $N_2H_4 \cdot H_2O$ to $(NH_4)_2[OsCl_6]$ yields the nitrido-bridged complex $[(NH_3)_5OsNOs(NH_3)_5]Cl_5 \cdot H_2O$ in 90% yield.

Biphasic kinetics are observed in the oxidation of hydrazoic acid by bis(2,2'-bipyridine)manganese(III) ions.[104] Analysis of the kinetics suggests that both $[Mn(bpy)_2N_3]^{2+}$ and $[Mn(bpy)_2N_3]^+$ are involved in the oxidation process producing free radicals, N_3. The diazido salt cis-$[Co(en)_2(N_3)_2]^+$ is reduced by Fe(II) in acidic solution.[105] The kinetics may be explained by an inner-sphere mechanism utilizing the azide ions as bridges.

Iron is found to catalyze the reduction of peroxide-bound chromium(IV) by hydroxylamine in acetate buffered media.[106] The proposed catalytic sequence involves recycling of dissolved iron between the +2 and +3 states, the rapid oxidation of Fe(II) by Cr(IV) ($k = 3 \times 10^3 \ M^{-1} s^{-1}$), a rate-determining one-electron oxidation of Fe(III)-bound NH_3OH^+ ion by Cr(IV), and the rapid reaction of Fe(III)(ONNH$_2$) with excess NH_3OH^+. The kinetics of oxidation of $HONH_2 \cdot HCl$ by chloramine-B in HCl medium have been reported.[107]

3.3.2. Oxynitrogen Compounds

The dirhenium(II) complexes, $[Re_2X_3(\mu\text{-dppm})_2(NCR)_2]PF_6$ (X = Cl, Br; R = Me, Et, i-Pr, t-Bu, Ph), reacts with nitriles to give the green, paramagnetic complexes $[Re_2X_3(\mu\text{-HN}_2C_2R_2)(\mu\text{-dppm})_2(NCR)](PF_6)$ which may be subsequently oxidized to the red, diamagnetic salts $[Re_2X_3(\mu\text{-HN}_2C_2R_2)(\mu\text{-dppm})$-$(NCR)](PF_6)_2$ by $NO^+PF_6^-$ in dichloromethane.[108] No kinetic data were provided. Similarly, $[Re_2H_5(PMe_3)_6]PF_6$ may be readily oxidized to purple $[Re_2H_5(PMe_3)_6]$-$(PF_6)_2$ by $NO^+PF_9^-$.[109] Ag may be oxidized to the solvated Ag(I) cation by the nitrosonium cation in acetonitrile.[110] The Ag(I) cation is readily used in synthesis of complexes with pyridine or P(OMe)$_3$. Reaction of nitrosoarenes with the [(cyclohexadienyl)Fe(CO)$_3$]$^+$ cation yields nitroxides of the type

$[(OC)_3Fe(C_6H_7)(Ar)NO]$.[111] In some cases, dispropostionation occurs and a mechanism is postulated for the observed reactions.

$H_2N_2O_2$ may be oxidized by ethylenediaminetetraacetatothallium(III) initially yielding 1 mol of HNO_2 and $\frac{1}{2}$ mol of N_2O gas.[112] Subsequently, HNO_2 is oxidized by Tl(III). In H_2SO_4 solutions, one mole of $H_2N_2O_2$ reacts with one mole of Ce(IV) producing NO_3^-, N_2O, and N_2.[113] The active oxidant and substrate were postulated to be $CeSO_4^{2+}$ and $HN_2O_2^-$. Oxidation of $N_2O_2^{2-}$ by $[Fe(CN)_6]^{3-}$ in alkaline solution is proposed to occur by two paths.[114] The intermediates are NO_2 and peroxonitrite while N_2O, N_2, and NO_3^- are observed as the final products.

The reaction between $[Fe(phen)_3]^{3+}$ and HNO_2 was examined[115] as an extension of the investigation of the reversible reaction (5). The kinetic results were compared to those obtained previously and comply with a two-step mechanism

$$2\,[Fe(phen)_3]^{2+} + NO_3^- + 3H^+ \rightleftharpoons 2\,[Fe(phen)_3]^{3+} + HNO_2 + H_2O \qquad (5)$$

for the forward reaction. NO, HONO, and NO_2^- react with [Fe(II)(edta)] to give $[Fe(II)(edta)(NO)]$.[116] The suggested mechanism involves the rate-determining formation of NO^+ and the reduction of HONO to NO by Fe(II)(edta) which account for the parallel reaction paths. The electrocatalytic reduction of NO_2^- to NH_3 may be achieved by "simple" coordination compounds of both iron and ruthenium.[117] Nitrite may be oxidized to nitrate by chlorine dioxide in aqueous media.[118] The reaction is biphasic with the slower generation of Cl^- and NO_3^-.

The kinetics of oxidation of $[Mo^{IV}O(L—NS_2(DMF)]$ $[L—NS_2 = 2,6\text{-bis}(2,2\text{-diphenyl-2-mercaptoethyl})\text{-pyridine(2-)}]$ by NO_3^- to yield $[Mo^{VI}O_2(L—NS_2)]$ have been reported.[119] The reaction is characterized by saturation kinetics in which nitrate reversibly forms a substrate–Mo(IV) complex. The reaction of $[Co^{III}L_2(NH_3)_2]Cl\cdot H_2O$ (L = salicylideneamine) with 60% HNO_3 leads to the formation of the ligand radical species $[Co^{III}(L')_2(NH_3)_2](Cl)(NO_3)\cdot 2H_2O$ (HL' = 5-nitososalicylideneamine) and the biradical $[Co^{II}(L')_2(NH_3)_2]$ $Cl(NO_3)\cdot 2H_2O$.[120] Trinuclear, mixed Pt_2Pd-1-methyluracil and -1-methylthymine complexes may be oxidized to an average metal oxidation state of +2.33 by nitric acid.[121] Electronic delocalization leads to the strong blue color observed for these complexes. The enzyme-catalyzed reduction of NO_3^- and NO_2^- to ammonia has been reported.[122] The reaction employs the two enzymes, nitrate reductase and nitrite reductase, as catalysts and the photogenerated N,N'-dimethyl-4,4'-bipyridinium cation as an electron carrier.

3.3.3. Amines and Nitriles

The reduction of Fe(III) porphyrin hydroxides by the heterocyclic aromatic amines, pyridine, 1-methylimidazole, and derivatives takes place in toluene to give the bis-amine Fe(II) porphyrin complexes.[123] The preliminary data suggest that the initial step is formation of a bis-amine Fe(III) complex. Electrochemical oxidation of $[Os(II)(terpy)(bpy)(NH_3)](PF_6)_2$ in the presence of Et_2NH or morpholine yielded bound nitrosoamine complexes, $[Os(II)(terpy)(bpy)(N(O)NR_2)]^{2+}$.[124] The kinetics and stoichiometry of the reaction between

$[Cu(dien)(OH)]^+$ and $[Fe(CN)_6]^{3-}$ in alkaline media suggest a mechanism in which there is initial, fast formation of a doubly deprotonated Cu(II)–diamido complex.[125] Subsequent slow internal redox occurs regnerating the amine species. Cu(II) has also been observed to oxidize a variety of tertiary amine buffers.[126] The results are compared extensively to previous reports.[127] The luminescence quenching of excited $[^*Ru(bpy)_3]^{2+}$ by a variety of aromatic amines (i.e., 1-naphthylamine) in several organic solvents has been examined.[128] The activation parameters for individual amines are solvent-independent, and these results have been analyzed within the framework of Marcus theory. Similarly, the quenching of excited *cis*-$[Ru(II)(phen)_2(CN)_2]$ by aromatic amines in acetonitrile has been reported.[129] $[M(cbqo)_2]$ [M = Cu, Ni; Hcbqo = 4-chloro-1,2-benzoquinone-2-oxime) may be reduced by $NCO^-/MeOH$ giving a semiquinone complex.[130] The reaction between $[Fe^{II}(\eta^5\text{-}C_5Me_4H)_2]$ and a variety of cyano acceptors (i.e., tetracyanoethylene, hexacyanobutadiene) results in the formation of a 1:1 charge transfer complex, $[Fe^{III}(C_5Me_4H)_2]^+[A]^{\pm}$.[131] Tungsten polyoxometalates oxidize organic substrates photocatalytically, such as tetracyanoethylene, on the picosecond time scale.[132]

3.4. Sulfur Compounds and Oxyanions

3.4.1. Peroxodisulfate and Peroxomonosulfate

The oxidation of Mn(II) by $S_2O_8^{2-}$ to Mn(VII) in H_3PO_4 medium proceeds via stable Mn(IV) species.[133] The activation parameters and a mechanism consistent with experimental observations are provided. The rate constant for oxidation of $[Mo(CN)_8]^{2-}$ by $S_2O_8^{2-}$ exhibits a dependence upon the salt concentration.[134] The kinetic data confirm that ion pairs play a key role in determining reactivities in these systems. The Tc(III/IV) dimer, $[(tcta)Tc(\mu\text{-}O_2)Tc(tcta)]^{3-}$ (tcta = 1,4,7-triacetato-1,4,7-triazacyclononane), may be oxidized by $S_2O_8^{2-}$ to give $[(tcta)Tc(\mu\text{-}O_2)Tc(tcta)]^{2-}$ via a reversible electron transfer.[135] A Ru(IV) complex is formed when $Na_2S_2O_8$ is used to oxidize $[Ru(II)(hedta)(H_2O)]$ [H_3hedta = N-(2-hydroxyethyl)ethylenediaminetriacetic acid].[136] Analysis supports the formation of a bridging Ru—O—Ru structure. The oxidation of the dinuclear $[\mu\text{-}1,2\text{-}$bis(4-pyridyl)ethane]pentaamine ruthenium(III) pentacyanoferrate(II) **(4)** by $S_2O_8^{2-}$ has been investigated.[137] Oxidation is postulated to proceed with **(5)** as the active species.

$(NH_3)_5Ru^{III}N$ ⟨⟩⟨⟩ $NFe^{II}(CN)_5$

(4)

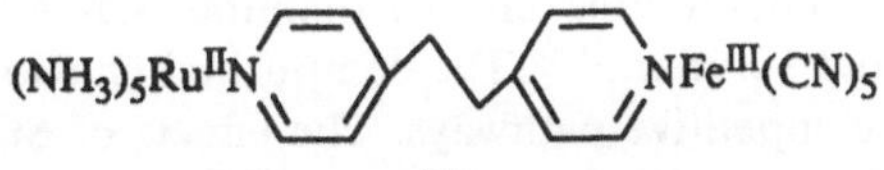

$(NH_3)_5Ru^{II}N$ ⟨⟩⟨⟩ $NFe^{III}(CN)_5$

(5)

As(III) may be oxidized to As(V) by $S_2O_8^{2-}$ in aqueous alkaline solutions.[138] At pH $\geq$ 12.0, the observed rate constant is $1.6 \pm 0.3 \times 10^{-2}\ M^{-1}\,s^{-1}$, at an ionic strength of 0.1 M. Sustained oscillations in redox potential, pH, and the concentration of dissolved O_2 are reported in the Cu(II)-catalyzed reaction betwen $K_2S_2O_8$ and $Na_2S_2O_3$ in a stirred tank reactor.[139] A free radical mechanism involving Cu(I), Cu(II), and the radicals SO_4^- and $S_2O_3^-$ may be used to account for the dynamic behavior of this sytem. The kinetics and mechanism of the oxidation of thiosulfate coordinated to cobalt(III) by peroxymonosulfate have been reported.[140] The reaction proceeds via two consecutive nucleophilic additions of the terminal peroxy oxygen atom to the coordinated $S_2O_3^{2-}$.

3.4.2. Sulfur Dioxide and Sulfite Ions

The reactions of SO_2 with $[Cr(NN)_3]^{3+}$ (NN = bpy, phen, and derivatives) when subjected to visible light (laser pulse) are threefold.[141] Quenching yields $[Cr(NN)_3]^{2+}$ which undergoes back electron transfer. The predominant reaction is electron transfer between SO_2 and $[Cr(NN)_3]^{2+}$ yielding the transient SO_2^-. The rate constants obtained may be utilized in a Marcus cross-correlation relation to calculate a self-exchange rate of between $1 \times 10^4\ M^{-1}\,s^{-1}$ and $18 \times 10^4\ M^{-1}\,s^{-1}$ for SO_2/SO_2^- in 1.0 M H_2SO_4. The reaction $[V(O_2)_3]^-$ and $SO_{2(g)}$ is biphasic going through a yellow, ESR inactive solution to yield an ESR active blue-green solution.[142] The final product has been identified as $[VO(SO_4)(H_2O)_3]^{2-}$ with the sulfate ion monodentate. Intermediate oxidation states are realized when $SO_{2(g)}$ is bubbled through solutions of $[Au(\mu\text{-}(CH_2)_2PPh_2)]_2$ and $[Au_2Pt(\mu\text{-}C_1S-CH_2PPH_2S)_4]$.[143] The oxidation state of the metal ions is reflected in the metal–metal separation. Reaction of SO_2 with $[Fe^{II}(TPP)]$ and $[Fe^{III}(TPP)]^+$ (TPP = tetraphenylporphyrin) results in the formation of adducts and sulfate species.[144]

The kinetics of SO_3^{2-} oxidation by $[PtCl_6]^{2-}$ have been studied over a wide range of experimental conditions.[145] On the basis of the experimental evidence, a 2e reduction mechanism was proposed. The reduction of Cu(II) by SO_3^{2-} proceeds via $Cu_2^{II}SO_3^{2+}$ and $Cu_2^{II}SO_3OH^+$ intermediates.[146] The rate law is consistent with consecutive first-order reactions leading to a mechanism involving complexation followed by a rate-determining electron transfer. Inner-sphere electron transfer is also assumed for the electron transfer between Fe(III) and SO_3^{2-}.[147] The reduction of $[Ag(OH)_4]^-$ by SO_3^{2-} to give Ag(I) and SO_4^{2-} in strong base obeys a second-order rate law with $k = 9.1 \pm 0.1 \times 10^2\ M^{-2}\,s^{-1}$.[148] Two kinetically indistinguishable pathways are postulated. SO_3^{2-} oxidation by Ag(III) is $\sim 10^4$ faster than that of HPO_3^{2-}, while no reaction is observed with NO_2^-. The oxidation of SO_3^{2-} by V^{5+} (it is unlikely that this species exists as V^{5+}) was studied in $HClO_4$ medium.[149] The observed rate constant was not sensitive to the solution pH, but increased with increasing ionic strength. An inner-sphere mechanism is proposed for the reduction of *trans*-cyclohexane-1,2-diamine-N,N,N',N'-tetraacetatomanganese(III) in aqueous solution.[150] The composite behavior implies the involvement of a series of competitive pathways. The effects of SO_3^{2-} on the Belousov-Zhabotinskii oscillating reaction have been studied and compared with the effects on the nonreactive Cl^- ion.[151] Pulsed-accelerated-flow spectroscopy has been used to measure pseudo-first-order rate constants for the reactions of excess SO_3^{2-} with HOCl and OCl^- in carbonate buffered medium.[152] Rate constants of $7.6 \times 10^8 \ M^{-1} s^{-1}$ and $2.3 \times 10^4 \ M^{-1} s^{-1}$, respectively, were obtained. The oxidation of dissolved SO_2 by HSO_5^- has been examined.[153] A mechanism has been proposed in which the rate-determining step involves the acid-catalyzed decomposition of a peroxide–sulfite intermediate to $S_2O_7^{2-}$ and ultimately, H_2SO_4.

3.4.3. Sulfoxides

Mono-, bis-, and tris-sulfito complexes of Fe(III), produced from the reaction of Fe(III)$_{aq}$ and S(IV) oxides, undergo two successive redox reactions and produce Fe(II), SO_4^{2-}, and $S_2O_6^{2-}$.[154] The mechanistic aspects of the reported data are discussed in reference to the suggested complex formation mechanism and results previously reported. Rate constants for the oxidation of $S_2O_3^{2-}$ by $[PtCl_6]^{2-}$ follow the expected second-order rate law, first order in both $S_2O_3^{2-}$ and Pt(IV).[155] While the influence of pH is small, the rate does depend on the nature and concentration of the added cation in the order; $Cs^+ > Rb^+ > K^+ > Na^+ > Li^+$. The reduction of Cu(II) to Cu(I) by $S_2O_3^{2-}$ induces the subsequent reduction of $[Co(C_2O_4)_3]^{2-}$.[156] The rate constant for the reaction between Cu(I) and $[Co(C_2O_4)_3]^{2-}$ is discussed in terms of the Marcus theory. Sodium dithionite reacts with N-alkylhemin to yield the σ-alkyliron(III) complex.[157] This ligand conversion may also be achieved by electrochemical reduction.

The oxidation of the substitution-inert Ru(II) complex ion, $[Ru(bpy)_3]^{2+}$, by SO_4F^- obeys a bimolecular rate law with a 1:1 stoichiometry.[158] At 13.3 °C, corrected to finite dilution, $k_2 = 1346 \pm 19 \ M^{-1} s^{-1}$. SO_4F^- and XeF_2 were also utilized in the oxidation of Pu(III) to Pu(IV) in 1 M $HClO_4$.[159] In both reactions, two moles of Pu(III) are oxidized per mole of oxidant, with no evidence of Pu(V) or Pu(VI) being formed. At 2 °C, the reactions have bimolecular rate constants of $120 \pm 9 \ M^{-1} s^{-1}$ (SO_4F^-) and $1160 \pm 70 \ M^{-1} s^{-1}$ (XeF_2). The reduction kinetics of $[Fe(EDTA)]^-$ by $HOCH_2CH_2SO_2Na$ have been examined by a polarographic method.[160] The reaction proceeds by a postulated "associative" mechanism shown in Eqs. (6)–(9). $[Co(en)_2L](CF_3SO_3)_2$ (HL = hydroxyl-L-proline) is oxidized by

$$HOCH_2SO_2^- \rightarrow HOCH_2OSO^- \tag{6}$$

$$[Fe(EDTA)]^- + HOCH_2OSO^- \rightarrow SO_2^- + [Fe(EDTA)]^{2-} + HCHO + H^+ \tag{7}$$

$$[Fe(EDTA)]^- + SO_2^- \rightarrow Fe(EDTA)^{2-} + SO_2 \tag{8}$$

$$SO_2 + HCHO + OH^- \rightarrow HOCH_2SO_3^- \tag{9}$$

$SOCl_2$ to $[Co(en)L']Cl_2$ (HL' = hydroxydihydropyrrolecarboxylic acid).[161] A study of the redox kinetics and reaction mechanisms of some sulfur oxyanions has appeared.[162] Whole organisms of *T. tepidarius* oxidize thiosulfate by a mechanistic system quite different from that previously described for *T. versutus*.[163]

3.4.4. Alkyl Sulfur Compounds

Thiols RSH (RSH = MeSH, cysteine and derivatives, glutathione, penicillamine, and N-acetylpenicillamine) react with nitroprusside ($[Fe(CN)_5(NO)]^{2-}$) to give the pentacoordinate Fe(I) complex $[Fe(CN)_4(NO)]^{2-}$.[164] The fate of this complex is dependent upon reaction conditions and the presence of oxygen. With an excess of ligand, substitution can occur at $[Fe(CN)_4(NO)]^{2-}$ to give $[Fe(NO)_2(SR)_2]$ and subsequently $[Fe_2(SMe)_2(NO)_4]$ (when R = Me) or $[Fe_4S_3(NO)_7]^-$ (when R = H). Chloroperoxidase-catalyze oxidation of a variety of sulfides (R—S—R'; R = Me, Et; R' = Ph, p-tolyl, o-tolyl, p-ClC_6H_4, p-$O_2NC_6H_4$, p-$MeOC_6H_4$, o-$MeOC_6H_4$, 2-naphthyl, benzyl, n-Bu) affords the corresponding sulfoxides having the (R) absolute configuration in enantiomeric excesses of up to 92%.[165] t-BuOOH was utilized as the sacrificial oxidizing agent. The reaction of N,N-bis(2-mercaptoethyl)-2-(methylthio)ethylamine with anhydrous Ni(OAc)$_2$ yields the dimeric product (6).[166] Oxidation of (6) by I_2 results in a four-electron transfer and the formation of a distorted square pyramidal Ni(II) complex with a coordinated disulfide ligand, (7). Presumably, the RSSR' moiety is formed through a metal-assisted oxidative reaction. Treatment of the dimer

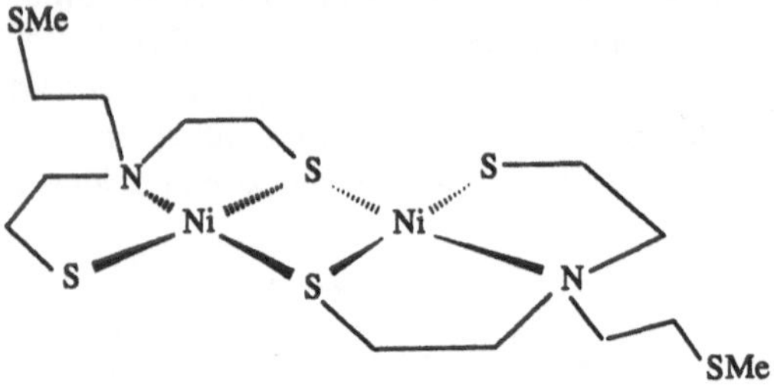

(6) 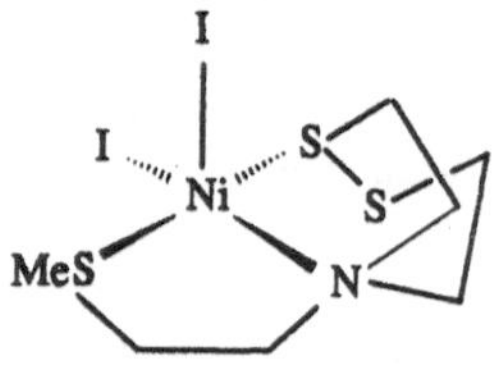   (7)

$[Ru(OEP)]_2$ (OEP = octaethylporphyrin^{2-}) with n-decyl methyl sufide results in the low-spin complex $[Ru(OEP)L_2]$ where L is the sulfide.[167] This species catalyzes the autooxidation of the sulfide, possibly via an outer-sphere process, and the *in situ* generation of H_2O_2. The CO_2-containing complex $[Ni(Pcy_3)_2(CO_2)]$ (cy = cyclohexyl) reacts with RSH (R = H, Ph), in toluene and under nitrogen, to give $[Ni(Pcy_3)_2(CO)_2]$, H_2O, and $[Ni(Pcy_3)_2(SR)_2]$.[168] The process involves the protonation/reduction of coordinated CO_2. The role of the —SH group was investigated.

The reaction of Ni(acac)$_2$ with pyridine-2,6-dimethylthiol (H$_2$pdmt) affords $[Ni(pdmt)]_2$.[169] This product is cleaved by thiolates to yield $[Ni(pdmt)(SR)]^-$ (R = Ph, Et) which may subsequently undergo irreversible ligand-based oxidation to give RSSR. The metallobleomycin analog $[Cu(PMA)]ClO_4 \cdot 0.5H_2O \cdot 0.5MeOH$ (PMA = 8) may be reduced to a Cu(I) species by dithiothreitol.[170] Subsequent oxygenation produces OH$^{\cdot}$ and induces double-strand breaks in DNA.

The kinetics and mechanism have been elucidated for the reduction of cobalt(II) 4,4',4'',4'''-tetrasulfophthalocyanine by 2-mercaptoethanol.[171] A rate constant of 228 ± 3.8 s^{-1} was obtained for the "rate-limiting" electron transfer in the formation of the 2-hydroxyethyl disulfide. The reduction of Cr(VI) by glutathione has been examined by visible spectroscopy ($\lambda_{max} = 650$ nm)[172] and by

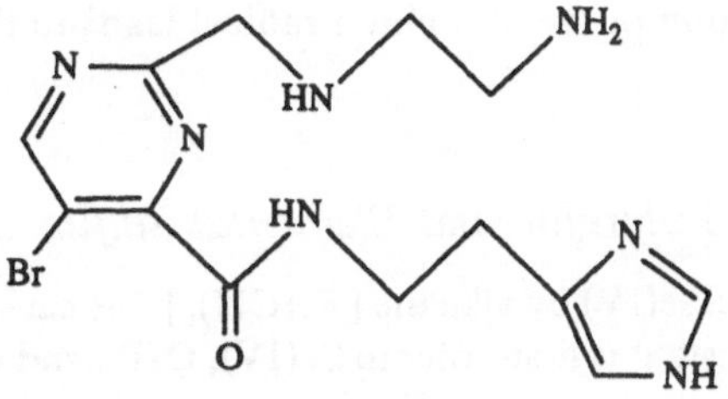

(8)

ESR spectroscopy.[173] In both cases, an intermediate Cr(V) species was detected. In the latter report, comparison was drawn with reactions where cysteine was employed as the reductant. The decomposition of manganic cysteinates follows a reaction rate law, expressed by the disappearance of the Mn(III) and Mn(IV) complexes, that is, second-order with respect to the absorbant complex species and zero-order in the reducing substrate.[174] The oxidation of thiourea by permanganate ions in aqueous acidic media was found to be autocatalytic.[175] The presence of Mn^{2+} thiourea complex ions were detected and a mechanism was proposed to account for the stoichiometry of the oxidation process as well as the product distribution. $[Ag(OH)_4]^-$ is rapidly reduced by thiourea in accordance with the three-term rate law (10), where $k_1 = 1.08 \text{ s}^{-1}$, $k_2 = 1.46 \times 10^3 \ M^{-1}\text{s}^{-1}$, and $k_3 = 2.02 \times 10^3 \ M^{-2}\text{s}^{-1}$.[176] The k_1 pathway occurs via the rate-determining

$$\text{rate} = \{k_1 + (k_2 + k_3[\text{OH}^-])[\text{thiourea}]\}[\text{Ag(III)}] \tag{10}$$

aquation of $[Ag(OH)_4]^-$, while the other two pathways involve axial attack of thiourea on Ag(III). The Cr(II) reduction of several S- and O-bonded (thiocarbanato)pentaamine cobalt(II) complexes have been reported.[177] The reduction occurs via remote attack at either the O or S atoms in the S- or O-bonded systems, with the rate constants obtained for the S-bonded system being a factor of $\sim 10^3$ faster. This enhanced rate is attributed to a structural *trans* effect. The reaction between $[\text{Cu(tet b)}]^{2+}$ (tet b = rac-5,7,7,12,14,14-hexamethyl-1,4,8,11-tetraazacyclotetradecane) and the mercaptoacetate dianion in alkaline methanol yields the unusual persulfidoacetate complex, $[(\text{tet-b})\text{Cu(II)}(\text{SSCH}_2\text{CO}_2)]$.[178] The proposed mechanism features the oxidation of the coordinated mercaptoacetate to afford a transient Cu(I)–thiyl radical that is rapidly attacked by a second mercaptoacetate to give Cu(I)–disulfide anion radical intermediate. Iron dramatically catalyzes the reduction of $[\text{Cr(IV)(dien)(O}_2)_2]$ by mercaptocarboxylic acids.[179] The rate law accommodates a rapid formation of an Fe(III)-thiol complex ($K > 10^{5.6} \ M^{-1}$), which undergoes electron transfer to Cr(IV) ($k = 1.9 \times 10^3 \ M^{-1}\text{s}^{-1}$) yielding an RS˙ radical (which dimerizes) and a Cr(III)peroxo complex. Further experimental evidence supports this mechanistic scheme. The kinetics of the oxidation of 2-mercaptopyrimidine by a variety of silver macrocycles follows a second-order rate law.[180] A rate–pH profile characteristic of an acid–base equilibrium for the thiol was also observed. The oxidation of $CuClO_4$ by thiuram disulfide (tds) proceeds by a rate-limiting inner-sphere electron tranfer in the $[\text{Cu(tds)}]^+$ complex.[181]

Subsequent dissociation of the tds$^{\pm}$ anion radical leads to the formation of Cu(I) dithiocarbanate.

3.4.5. Selenium, Tellurium, and Elemental Sulfur

The oxidation of Se(IV) by alkaline $[Fe(CN)_6]^{3-}$ is catalyzed by Os(VIII).[182] The reaction rate constant is first-order in Se(IV), OH^-, and Os(VIII) but independent of $[Fe(CN)_6]^{3-}$ concentration. The complexation of Au^+ by Se_4^{2-} results in unexpected products due to the remarkable interplay of electrons transferred from Au^+ to Se_x^{2-}.[183] No kinetic details are provided. Tellurium(IV) is oxidized by manganese(III) in sulfuric acid.[184] The reaction order for Te(IV) is unity while for Mn(III) it is between 1 and 2 and, from the kinetic data, a mechanism is proposed wherein both Mn^{3+} and $MnOH^{2+}$ are the reactive species. In aprotic solvents, hydroxide reacts with elemental sulfur to give the trisulfur anion radical, $S_3^{\pm}$.[185] The results are consistent with those obtained from electrochemical investigation.

3.5. Halogens, Halides, and Halogen Oxyanions

3.5.1. Halogens

The contribution of the electron transfer process with iodine to the cleavage of the cobalt–carbon in alkylcobalt(III) complexes has been evaluated.[186] By comparison with electron transfer reactions with ferrocene and from the product distribution studies expected from electrophilic reactions, iodine is considered to act as an oxidizing agent. Similarly, electron transfer appears to be the rate-limiting step in the oxidation of six-coordinate organocobalt(III) complexes by iodine.[187] The transient formation of organocobalt(IV) is consistent with ESR spectroscopic observations. It should be noted, however, that detection of this intermediate argues against the authors implication of initial electron transfer as the rate-limiting step since this species is formed after the original oxidation and hence, if the mechanism is correct, should not be present in any significant concentration.

The nickel hexa-aza-macrocyclic complex (**9**) may be oxidized non-stoichiometrically by iodine producing a black powder.[188] This material demonstrates spectroscopic features anticipated for a partially oxidized nickel complex.

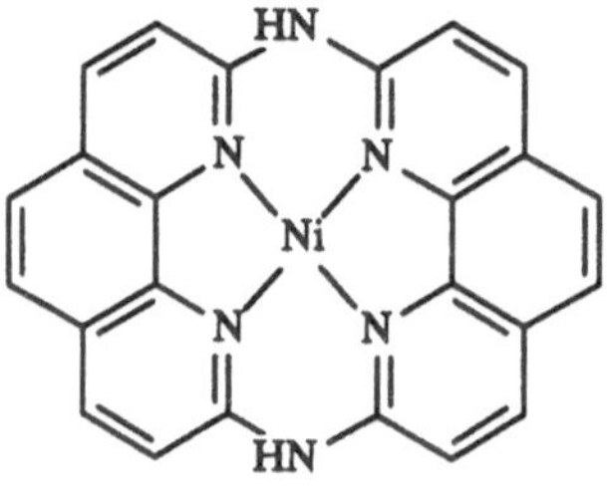

(**9**)

The reaction of the potentially tripodal N,N-bis(2-mercapto-ethyl)-2-(methylthio-ethylamine with anhydrous $Ni(OAc)_2$ yields a dimeric complex.[189] Chemical oxidation with iodine in a $2e$ process yields metal oxidized products which exist as two monomeric nickel species. One of these complexes is believed to have been derived from the formation of thiyl radicals. Both the reactions of solid $[CuL_2]$ (HL = acetylacetone, trifluoroacetylacetone, benzoylacetone) with iodine vapor and those of $[CuL_2]$ solutions with iodine solutions in weakly polar solvents have been interpreted as proceeding through an oxidative mechanism.[190]

3.5.2. Halides

The kinetics of Cl^- oxidation by $[Ni(bpy)_3]^{3+}$ have been investigated in aqueous media.[191] The reaction mechanism was interpreted as involving outer-sphere complexation of Cl^- and two parallel redox pathways. The oxidation of Cl^- to Cl_2 by Ce^{4+} with $RuO_2 \cdot xH_2O$ as catalyst has been examined.[192] The results were inconsistent with an electrochemical model in which the catalyst particles act as microelectrodes and the current density due to Cl_2 evolution is proprtional to $[Cl_2]^2$. Oxidation of Br^- by $Mn^{III}(bpy)_2$ in aqueous perchlorate is postulated to proceed through $Mn(bpy)_2Br_{aq}^{2+}$, the decomposition of which is the rate-controlling step.[193] The dependence of the rate constant upon acid concentration also led to the conclusion that $Mn(bpy)(bpyH^+)Br_{aq}^{3+}$ was probably part of the mechanistic pathway. 6 M HBr was utilized in the reduction of TcO_4^- via a two-step process: $Tc(VII) \rightarrow Tc(V) \rightarrow Tc(IV)$.[194] Both reactions were studied and the spectral characteristics of the Tc(V) intermediate species were found to be similar to the Rb^+ and Cs^+ salts of $[TcOBr_5]^{2-}$. The EDTA catalyzed oxidation of I^- by $HCrO_4^-$ obeys first-order kinetics with respect to Cr(VI) and H^+ but fractional order with respect to EDTA.[195] The order in I^- changes from zero to unity leading to the proposal of a termolecular complex as part of the mechanistic pathway. The higher oxidation states (III, IV, V) of iridium in aqueous acidic solution and their interconversion by various chemical oxidants and reductants have been examined.[196] Iodide has been utilized as a reductant converting the Ir(V) to blue-purple Ir(IV) and then to the pale yellow Ir(III). The kinetics of oxidation of N-substituted urea–titanium(III) complexes by I_3^- has been examined in ethanol and acetonitrile.[197] The mechanism proposed involves an inner-sphere redox reaction of a bonded I_3^-. The redox reaction between $RR'I^+$ ($R = C_3F_7, C_6F_{13}$; $R' = Ph$) and ferrocene was found to be first order in both reactants with rate constants of $6.4 \pm 0.6 \times 10^2\ M^{-1}\,s^{-1}(R = C_3F_7)$ and $1.4 \pm 0.15\ M^{-1}\,s^{-1}$ ($R = C_6F_{13}$).[198]

The Ce(III)-catalyzed Br(V) oxidation of oxalic acid in the presence of $HClO_4/H_2SO_4$ and Hg(II) exhibits second order behavior in [Br(V)] and acid but fractional order between 0.4 and 0.0 in Ce(III).[199] However, owing to the nature of the reagents, a complex mechanistic scheme is not unexpected! The oxidation of the thiocyanate and iodide ions by hydrogen atoms in acidic solutions has been examined by pulse radiolysis.[200] The reaction between chlorine(III) and iodide was investigated by stopped-flow techniques at low pH.[201] The mechanism adopted leads to a rate law involving a second-order term in H^+ which might have consequences in understanding the nonlinear behvaior of this reaction. The rapid reactions between HOCl and I⁻ and NCl₃ and I⁻ have been measured by pulsed-accelerated-flow spectroscopy.[202] The effects of chloride ion on the oxidation of Pt(II) complexes by hexachloroiridate(IV) have been investigated.[203] The kinetic acceleration of the Cr(II) reduction of $[Co(sep)]^{3+}$ by halide ions is postulated to arise from a favorable entropic contribution to the activation energy through hydrogen-bonding interactions.[204]

3.5.3. Oxyhalogen Compounds

The reaction kinetics for the oxidation of hexacyanoferrate(II) by bromate led to a proposed mechanism involving $[Fe(CN)_5(H_2O)]^{3-}$ as the active species.[205] Numerical simulations with the mechanism agree with the experimentally obtained data. However, the reaction of BrO_3^-, SO_3^{2-}, and $[Fe(CN)_6]^{4-}$ in a continuous stirred tank reactor exhibit sustained oscillations.[206] The observed oscillations can be simulated by a nine-step mechanism. The relationship between this system and the mixed Landolf oscillator has been discussed. The BrO_3^- oxidation of $[Fe(phen)_3]^{2+}$ and Fe_{aq}^{2+} have been examined in sulfuric acid.[207] The rate laws for both reactions generally can be described as $-d[Fe(II)]/6dt = d[Br^-]/dt = k[Fe(II)][BrO_3^-]$ with the results suggesting that the BrO_3^-/Fe_{aq}^{2+} reaction proceeds by an inner-sphere mechanism while the $BrO_3^-/[(Fe(phen)_3]^{2+}$ reaction proceeds by an outer-sphere mechanism. Aquation and dissociation of $[Fe(bpy)_3]^{2+}$ occurs concurrently with oxidation by BrO_3^-.[208] The oxidation is autocatalytic through $HBrO_2$ and an empirical rate law of the form $d[Fe(bpy)_3^{3+}]/dt = k[BrO_3^-]^2[H_2SO_4]^{(1-2)}$ was obtained. The oxidation of $[Cr(III)(H_2O)(N-(2-$ hydroxyethyl)ethylenediamine-N,N,N'-triacetate] and [Cr(III)(nitrilotriacetate)- $(H_2O)_2]$ by IO_4^- are both proposed to proceed through inner-sphere mechanisms with the hydroxo ligand of the Cr(III) complex bridging the reactants.[209,210] A bridged dimer of molybdenum(V) is readily oxidized to Mo(VI) by a number of oxyhalogen species under acidic conditions.[211] Reactions of BrO_3^-, $HBrO_2$, $HClO_2$, and HOCl with excess reductant yield the corresponding halide ions while reduction of ClO_3^- produces Cl⁻ even when the oxidant is in large excess. The iron complex of deuteroporphyrin IX undergoes oxidation by hypochlorite ion.[212] The intermediate state generated is spectroscopically identical to that obtained by reaction with other O atom oxidants such as iodobenzene, peroxide, peroxybenzoic acid, and chorite ions. Similarly, reaction of hypochlorite with [(Me₁₂TPP)Mn(III) (Cl)] [Me₁₂TPP = 5,10,15,20-tetrakis(2,4,6-trimethylphenyl)porphyrin] results in oxidation and, presumably, oxygen transfer.[213] The resulting "activated" manganese species subsequently reacts with alkenes to produce epoxides. A kinetic chemiluminescence method for the determination of V(V) based on its catalytic effect on the ClO_3^-/I^- reaction with the luminol/I₂ reaction as monitor has been reported.[214] V(V) was also determined by its catalytic effects upon the oxidation of chromotropic acid by BrO_3^- in acetate buffered media.[215]

Also, addition of HOBr can eliminate the induction period for oscillations in a Belousov–Zhabotinskii system.[216] However, in a solution saturated with air, the initial addition of HOBr causes a train of only a few oscillations. Oscillating reactions have been examined very extensively in recent issues of the *Journal of*

Physical Chemistry[217] and the *Journal of Chemical Education.*[218] Interested readers are directed toward these issues, and also Chapter 4.

3.6. Phosphorus, Arsenic, and Oxycompounds

3.6.1. Phosphorus Oxyanions

The kinetics of oxidation of the $H_2PO_2^-$ ion to $H_2PO_3^-$ by $[Ag(OH)_4]^-$ have been examined in strongly basic solution using stopped-flow spectrophotometry.[219] The stoichiometry is consistent with a $2e$ change, which is equivalent to an O atom transfer as has been observed with other oxidizing species. The proposed mechanism includes the formation of an intermediate, involving an axial ligation of the $H_2PO_2^-$ ion on $[Ag(OH)_4]^-$ which undergoes both internal and OH^--assisted redox. Similarly, an internal electron transfer in the complex $[(NH_3)_5CoO_2PH_2]^{2+}$ leads to the quantitative formation of Co(II) and a 1:1 mixture of $H_2PO_2^-$ and HPO_3^{2-}.[220] The reaction proceeds through a Co(I) species which reduces the remaining Co(III) ions, thus accounting for the release of $H_2PO_2^-$. With a second Co(III) oxidant present, yields of $H_2PO_2^-$ are reduced accordingly The oxidation of $H_2PO_2^-$ by $[Cu(TeO_6)_2]^{9-}$ has been studied[221] in alkaline media under pseudo-first-order conditions ($H_2PO_2^-$ in excess) and was found to be first order in $H_2PO_2^-$ but independent of OH^- over the range 0.4×10^{-2}–3.6×10^{-2} mol l^{-1}. Two reaction schemes accounting for the data were proposed. However, the large negative charge would likely lead to ion pairing, which was not considered by the authors. The kinetics of Os(VIII)-catalyzed oxidation of HPO_3^{2-} by diperiodoargentate(III) have been investigated.[222] Curiously, the reaction rate is independent of the concentration of the silver species while being first order in both Os(VIII) and $H_2PO_3^-$, and dependent upon the OH^- concentration. $Mn(II)_{(aq)}$ is oxidized to $Mn(III)_{(aq)}$ when treated with excess peroxodiphosphate at pH 4.0–5.3.[223] Further acidification inhibits the reaction rate. Coordinated phosphates have been observed to undergo intramolecular attack with the reaction rate presumably enhanced by the metal center (M = Co, Ir).[224]

3.6.2. Phosphines and Arsines

The mechanisms of the oxidation of phosphines and arsines by chromium(VI) have been examined both in solution and on a diatomite support.[225] Kinetic parameters are presented for both supported and solution reactions. A ruthenium complex of 1,4,8,11-tetramethyl-1,4,8,11-tetraazacyclotetradecane has been utilized to oxidize triphenylphosphine in acetonitrile.[226] Although a limited temperature range was utilized, a $\Delta H^{\ddagger}$ value of 8.7 ± 0.8 kcal mol^{-1} and a $\Delta S^{\ddagger}$ value of -20 ± 2 cal K^{-1} mol^{-1} were calculated. The secondary phosphine oxides, $HP(O)R_x$ (R = n-butyl, isobutyl, cyclohexyl) and 9H-9-phosphabicyclononane-9-oxide, react with cobaltocene to yield dihydrogen and cobalt(I) compounds.[227] With the less bulky phosphorus ligands at elevated temperatures trinuclear cobalt(III, II) complexes may be obtained. Arsenious acid may be utilized to catalyze the oxygen atom exchange reaction between CrO_4^{2-} ions and H_2O.[228] The rate of the redox reaction between CrO_4^{2-} and As(III) is considerably slower than the observed oxygen transfer rate leading to the proposed mechanism.

3.7. Inorganic Radicals

A study[229] of the superoxide anion radical, $O_2^{\cdot-}$, reacting as an oxidant toward several amine complexes of iron(II) and ruthenium(II), suggests that the mechanism proceeds through a hydrogen abstraction step from the amine to yield the perhydroxyl radical, $HO_2^{\cdot}$. The metal center acts to stabilize the ligand radical intermediate formed. The reactions of superoxide ion with $[Mn^{II}(TPP)(py)]$ and $[Mn^{III}TPP]^+$ have been studied[230] and evidence obtained for the formation of two complexes depending upon superoxide ion concentration. ESR spectroscopy suggests that one of these complexes is a mixed-valence $Mn^{III/IV}$ oxo-bridged dimer. Similarly, ESR spectroscopy has been utilized in determining a mechanism for the interaction of the binuclear complex (10) with $O_2^{\cdot-}$.[231] The light-sensitive reduction of both monomeric and dimeric Co di- or tetrasulfophthalocyanines has been followed spectrophotometrically.[232] The mechanism involves formation of an outer-sphere cobalt(II)-ligand-hydroxo complex which yields $OH^{\cdot}$ radicals.

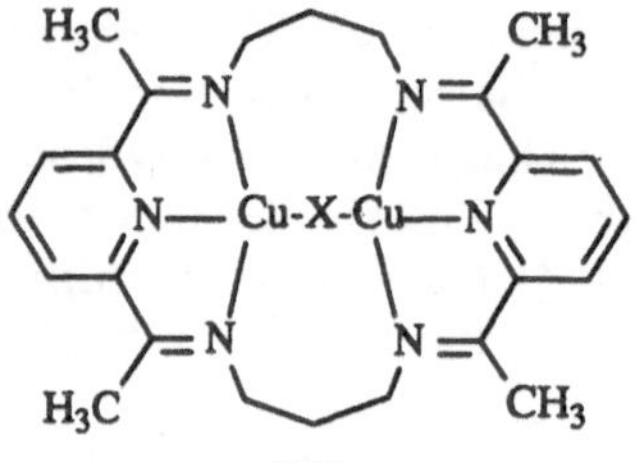

(10)

Reaction with O_2 produces $O_2^{\cdot-}$. A series of rigidly structured Ni(II) macrocyclic and acyclic complexes has been shown to react rapidly with hydroxyl radicals and aquated electrons.[233] Addition of e_{aq} generates a Ni(I) transient except when the ligand incorporates a nitro group. Reaction with OH leads to a longer-lived Ni(III) species. $OH^{\cdot}$ radicals have also been shown to react with Pu(III) with a specific rate of oxidation of $1.83 \pm 0.78 \times 10^9\ M^{-1}\ s^{-1}$ at pH 4.0.[234] Comparison has been made with oxidation rates of U(III) and Am(II). The reactivity of Pd(II) to $OH^{\cdot}$, $NO_3^{\cdot}$, and $SO_4^{\cdot-}$ has been examined in acidic media by pulse radiolysis.[235] An "intrasphere" mechanism of electron transfer is postulated. Hydroxyl radicals have also been utilized in the oxidation of $[PtCl_4]^{2-}$ to $[PtCl_4OH]^{2-}$, while $[PtCl_6]^{2-}$ is reduced by e_{aq} or H to $[PtCl_6]^{3-}$.[236] The Pt(III) complexes subsequently undergo reaction with H_2O to produce $[PtCl_4(OH)_2]^{3-}$ and H^+. From these studies, the early stages of the reductive elimination/oxidative addition mechanism for Pt(II) and Pt(IV) have been elucidated. Five different routes involving pulse radiolysis and laser photolysis have been employed in the synthesis of Pt(III) species.[237] The mechanisms for the formation of the Pt(III) complex ions by these processes are discussed. Similarly, a variety of As(IV) species have been observed by the pulse radiolysis technique in reactions of aqueous arsenious acid, arsenite, and arsenate solutions.[238]

CO_2^- has been utilized to generate a disulfide radical anion in a synthetically altered human α-haemoglobin chain.[239] The electron of the disulfide radical anion then transfers to the protein heme group at a nominal distance of approximately 12 Å with an intramolecular rate constant of 188 ± 23 s^{-1}, at room temperature and pH 7. Similarly, CO_2^- reduces the [Fe(CN)$_6$]$^{3-}$ modified [ferricytochrome C$_{551}$ Ru(III)] with a rate constant of 1.9×10^9 M^{-1} s^{-1} to give the stable Fe^{II}Ru^{III} and metastable Fe^{III}Ru^{II} products.[240] The metastable complex subsequently undergoes an internal electron transfer reaction, over a distance of approximately 7.9 Å, with a rate constant of 13 s^{-1} to yield the Fe^{II}Ru^{III} species. The reactions of ferrous poly(amino carboxylate) complexes with CO_2^- have been studied at neutral pH in aqueous media.[241] The transient complexes formed are in equilibrium with the ferrous poly(amino carboxylate) complex and CO_2^- free radical. CO_2^- radicals generated by pulse radiolysis have been employed in the reduction of [(NH$_3$)$_5$Co(mbpy)]$^{4+}$ (mbpy = 1-methyl-4,4'-bipyridinium).[242] The spectral characteristics of the reduced species resemble the free ligand radical, mbpy. The Br$_2^-$ radical reacts with Fe(II) through an inner-sphere mechanism in MeOH/Et$_2$O mixtures with a rate constant of $5.9 \pm 0.5 \times 10^7$ M^{-1} s^{-1} at 298 K.[243] The rate constant is significantly larger than that obtained in aqueous media owing to the longer lifetime of the (Br$_2^-$ $\cdots$ Fe^{2+}) adduct. The relationship has been examined between differences in redox potential and the reactivities of inorganic radicals, generated by pulse radiolysis, with various water soluble ferrocene derivatives.[244]

The rhodium(II) radical [Rh(dmgH)$_2$PPh$_3$], formed by laser photolysis of the corresponding dimer, is oxidized by a range of cobalt(II) complexes which have coordinated halides.[245] Electron transfer is believed to occur through a ligand-mediated pathway and the reaction rate is dependent upon the choice of halide. In an examination of the Ti(III)-induced cyclization of epoxyolefins, a mechanism has been invoked which involves a metal-centered radical.[246]

3.8. Ascorbic Acid, Quinols, Catechols, and Diols

3.8.1. Ascorbic Acid

The reaction of Fe(III) with ascorbic acid has been reinvestigated.[247] In the present study, Fe(ClO$_4$)$_3$ and not FeCl$_3$ was used as the source of Fe(III). The absence of chloride ion decreases the rate constants obtained by a factor of 30 compared to previous results.[248] It is suggested that in the previous work, [Fe(H$_2$O)$_5$Cl]$^{3+}$ was the species reacting with the reductant. Similarly, the effects of chloride ion on the reaction between Cu(II) and L-ascorbic acid have been measured.[249] The observed kinetics with the coordinating ligand present are fundamentally different from those in the presence of a noncoordinating counterion such as perchlorate. Ascorbic acid also assists the complexation of Cu(II) by macrocyclic ligands.[250] A mechanism has been proposed with a rate-limiting reduction of Cu(II) by ascorbic acid while a subsequent re-oxidation of the Cu(I) ion by dissolved oxygen presumably facilitates formation of the macrocyclic complex. $[Cr^{IV}(dien)(O_2)_2]$ reacts slowly with ascorbic acid but the reaction is catalyzed by Fe(II) or Fe(III).[251] Both the chromium center and the peroxo ligands are reduced, with the Cr(IV) product containing bound ascorbate groups. Similarly Cu(II), V(IV), and V(V) have been found[252] to catalyze the reduction of $[Cr^{IV}(dien)(O_2)_2]$ by ascorbic acid, although the proposed reaction mechanism is not the same in each instance.

The oxidation of organic substrates by a ruthenium(III) analog of the Udenfriend system, involving Ru(III)–EDTA–ascorbic acid and O_2 has been reported.[253] The catalysis of the ascorbic acid oxidation by H_2O_2 has been shown to be promoted by a hexakis(trichloroacetato)oxocobaltdiiron(1+) cluster.[254] Ascorbic acid has also been utilized in the reductive dissolution of α-Fe_2O_3.[255] The oxidation/reduction mechanism for the ascorbic/dehydroascorbic acid system at the dropping mercury electrode has been reinvestigated.[256] Ascorbate has also found application as a test for trace metal ion concentration in buffer solutions.[257] The first-order rate constant for the loss of ascorbate in an air-saturated, catalytic-metal-free solution is less than $6 \times 10^{-7} \, s^{-1}$ at pH 7.0. Calibration graphs for ascorbate autooxidation in the presence of trace quantities of Cu and Fe were found to be linear.

3.8.2. Aromatic Diols and Diones

The nickel(IV) oxime, bis(6-amino-3-methyl-4-azahexa-3-ene-2-one oxime)nickel(IV), and the nickel(III) oxime, (15-amino-3-methyl-4,7,10,13-tetraazapentadeca-3-ene-2-one oxime)nickel(III), complexes react with hydroquinone.[258] Proton-related equilibria for both the nickel complexes and the hydroquinone could be elucidated from the kinetic details. For reactions with the complexes and the hydroquinone could be elucidated from the kinetic details. For reactions with the Ni(III) complex there is evidence of an inner-sphere process. The [NiL(TCCat)] complex (TCCatH$_2$ = tetrachlorocatechol; L = 2,4,4-trimethyl-1,5,9-triazacyclododec-1-ene) forms a 1:1 adduct with tetrachloro-1,2-benzoquinone.[259] Spectroscopic evidence suggests that this compound can be described formally as a quinone adduct of a Ni(I)–semiquinone moiety arising from inner-sphere ligand oxidation. The crystallographically determined structure (11) is shown below. Several copper complexes of a vareity of semiquinones also exist in solution in equilibrium with the corresponding catechol complexes.[260]

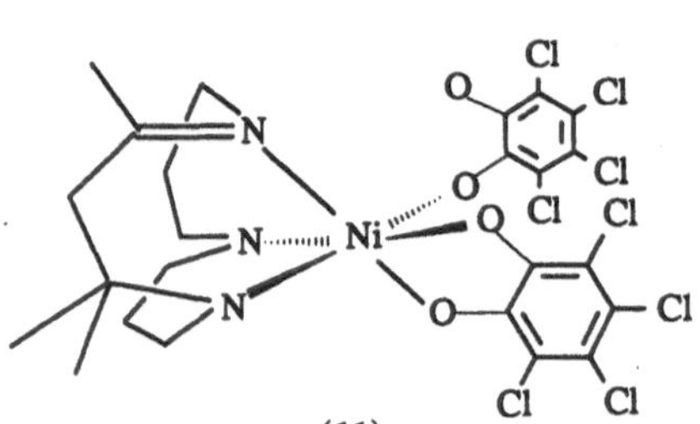

(11)

With the Ru dimers, $[\{Ru(bpy)_2\}_2(\mu\text{-L})]^{3+}$, ESR data indicate that these binuclear semiquinone complexes are near the borderline between anion radical complexes and metal-centered mixed-valence species.[261] The thermal reaction between $[W(CO)_6]$ and tetrachloro-1,2-benzoquinone in toluene produces the tris(quinone) tungsten complex.[262] Features of the molecule indicate that the quinones are coordinated as catecholate ligands and the tungsten ions are in a formal +6 oxidation state. Quinones trapped within a polymeric electrode film can act as electron sinks and/or sources.[263] I_3^-/I^- and $[Fe(CN)_6]^{3-/4-}$ have been utilized as charge-release mediators.

The Cu(II) complexes the tripodal ligand, N,N-bis(3,5-dimethylpyrazol-1-ylmethyl)-1-hydroxy-2-aminoethane (bpmhe), have been examined for their reactivity toward the oxidation of a catechol to quinone.[264] The rate constants obtained were dependent upon the choice of the fifth coordinating ligand, the azido complex being seen to have the fastest rate. Two papers addressing the substitution and oxidation kinetics in catechol–iron(III) systems have been published.[265,266] The mechanism postulated involves the coordination of the catechol to a Fe(III) but electron transfer to a second Fe(III) ion. Several schemes are proposed and discussed in terms of the generated ion of the experimentally observed rate law. 1,8-Dihydroxyanthraquinone forms complexes with Fe(III) in 1:1–1:3 metal/ligand ratios.[267] Reduction of these complexes results in the release of a metal-free semiquinone radical, and may serve as a model for the actions of anthracyclines in biological systems. Long chain poly-enes may be viewed as "molecular wires" to facilitate electron transfer.[268] For two Zn-porphyrin-polyene-anthraquinone molecules possessing either 5 or 9 all *trans* double bonds, electron transfer, subsequent to photoexcitation, is very rapid despite the long distances involved (25 Å and 35 Å, respectively).

Np(VI) reduction by Kojic acid (5-hydroxy-2-hydroxymethyl-4-pyrone) has been studied.[269] The rate constant and activation parameters, at $\mu = 1.0\ M$ and $T = 25\,^{\circ}\text{C}$, are $k = 1.63 \pm 0.08\ \text{s}^{-1}$, $\Delta H^{\ddagger} = 82.7 \pm 3.4\ \text{kJ mol}^{-1}$ and $\Delta S^{\ddagger} = 34 \pm 12\ \text{J K}^{-1}\,\text{mol}^{-1}$. A mechanistic scheme was proposed to account for these data. A variety of oscillations has been observed in the Co, Br^--catalyzed O_2 oxidation of cyclohexanone.[270] A kinetic model based on the formation of Co^{3+} is proposed. Benzophenone ketyl has been employed as an electron transfer initiator in the selective catalytic ligation of $[\{\mu\text{-}(CF_3)_2C_2\}Co_2(CO)_6L]$.[271] Rapid incorporation of labeled CO may be observed.

3.8.3. Aromatic and Aliphatic Alcohols

The heterobimetallic catalyst, $[Os(N)R_2(CrO)_4]^-$ $(R = CH_3,\ CH_2SiMe_3)$, readily oxidizes a wide variety of alcohols to the corresponding ketones in the presence of molecular oxygen.[272] Enhanced reaction rates were observed upon addition of $Cu(acetate)_2$ to the reaction medium. *Trans*-$[Ru(NO_2)(PMe_3)_2(trpy)]^{2+}$ may be generated electrochemically or by chemical oxidation.[273] This complex reacts with $PhCH_2OH$ in H_2O $(k_2 = 1.3\ M^{-1}\,\text{s}^{-1}$ at 5 °C) or o-dichlorobenzene $(k_2 = 2.9 \times 10^{-4}\ M^{-1}\,\text{s}^{-1}$ at 22 °C), producing PhCHO. Similarly, oxidation of BuOH in H_2O $(k = 2.5 \times 10^{-1}\ M^{-1}\,\text{s}^{-1}$ at 5 °C) produces the expected aldehyde.

Both reactions proceed through e transfer with no other change in the ruthenium complex. In contrast,[274] [Au^{IV}(O)(py)(bpy)$_2$]$^{2+}$ and [Ru^{III}(OH)(py)(bpy)$_2$]$^{2+}$ oxidize phenol and alkylated phenols to the corresponding quinones via a detectable intermediate. Oxygen labeling studies show that quantitative oxygen transfer occurs. Ferrocenium ion reacts via a photoinduced pathway with benzylic alcohol, benzhydrol, and 1-propanol.[275] Reaction products are dependent upon the choice of alcohol. The Ni(III) macrocycles, [Ni(cyclam)]$^{3+}$ and [Ni([9]aneN$_3$)$_2$]$^{3+}$, react with pentaaquoorganochromium(III) complexes to give Cr^{3+} and olefins or alcohols, depending upon the structure of the organo group.[276]

$$CrR^{2+} + NiL^{3+} \rightarrow CrR^{3+} + NiL^{2+} \tag{11}$$

$$CrR^{3+} \rightarrow Cr^{3+} + R^{\cdot} \tag{12}$$

$$R^{\cdot} + NiL^{3+} \rightarrow ROH \text{ or } R-H + NiL^{2+} \tag{13}$$

The reaction proceeds through a short-lived CrR^{3+} complex which may decompose by intramolcular e^- transfer. The results have been compared to the one-electron oxidation of organocobaloximes.

3.9. Carboxylic Acids, Carboxylates, Carbon Dioxide, and Carbon Monoxide

3.9.1. Carboxylic Acids and Carboxylates

The oxidation of HCO$_2^-$ is induced by the reaction of the Cu(I) complex of phenanthroline with H$_2$O$_2$.[277] The reaction has been studied by radiolysis of the Cu(II) complex and by mixing high concentrations of the Cu(I) species with H$_2$O$_2$ and HCO$_2^-$. It is proposed that the peroxo complex, [(phen)$_2$CuH$_2$O$_2$]$^+$, oxidized HCO$_2^-$ with a rate constant of $3 \times 10^7 \, M^{-1} \, s^{-1}$. Cationic rhodium complexes catalyze the reduction of pyridine coenzymes by sodium formate very efficiently.[278] The proposed mechanism proceeds by reaction of the formate ion with the rhodium complex which in turn reacts with the pyridine coenzyme.

Ag(II) reacts rapidly with [Co(NH$_3$)$_5$(OCHO)](ClO$_4$)$_2$ in acidic conditions.[279] Product distribution data reflect the variety of paths on which the silver ion may attack the coordinated formate ligand, although ultimately a Co(NH$_3$)$_5^{3+}$ moiety and CO$_2$ are obtained. Ag(I) may also catalyze the oxidation of formic acid by the unusual Ag(III) complex [ethylene bis(biguanidine)]silver(III) in perchloric acid media.[280] Evidence is provided for the Ag$^+$ ion acting as the oxidant for HCO$_2$H with the complex regenerating the Ag$^+$ ion from Ag^0. The same bis(oxalato^{2-})chromate(V) intermediate is postulated in the oxidation of oxalic acid by either Cr(VI) or *trans*-bis[3-ethyl-2-hydroxy-butanoato-(2-)]oxochromate(V).[281] The results implicate ligand exchange as a prelude to the reduction of Cr(V) by organic substrates, necessitating the reassessment of the previously postulated mechanism.[282] Oxalate may also be used to quench reductively ^2E [CrL$_3$]$^{3+}$ (L = 2,2'-bipyridine, 1,10-phenanthroline and derivatives).[283] The mechanism involves ion pairing followed by the rate-limiting electron transfer.

Rate constants vary from $85.4 \times 10^5 \, M^{-1} \, s^{-1}$ for the 5-chloro-1,10-phenanthroline complex to $0.21 \times 10^5 \, M^{-1} \, s^{-1}$ for the 4,4'-dimethyl-2,2'-bipyridine species. Photoexcited $[Ru(bpz)_3]^{2+}$ may be reductively quenched by $Na_2C_2O_4$ with a rate constant of $6.6 \times 10^7 \, M^{-1} \, s^{-1}$.[284] The resulting $[Ru(bpz)_3]^+$ ion may be coupled to the ITO electrode providing a photocurrent. Ce(IV) induces electron transfer in pentaamine oxaloacetato cobalt(III) perchlorate yielding CO_2 and Co(II) quantitatively.[285] The reaction of $[Ag(OH)_4]^-$ with 4-t-butyl-phenolate (L^-) yields a biphasic absorbance change, with each phase exhibiting pseudo-first-order kinetics.[286] Both reactions represent $1e$ transfers from L^- to the silver ion yielding Ag(II) as an intermediate. Each process is thought to proceed by a substitution-controlled mechanism.

3.9.2. Carbon Dioxide and Carbon Monoxide

Carbon dioxide rapidly forms an adduct with CoL^+ ($L = 5,7,7,$ 12,14,14-hexamethyl-1,4,8,11-tetraazacyclotetra-4,11-diene) in acetonitrile.[287] This adduct reacts to give CoL^{2+}, CO, and $NaHCO_3$ with a rate expression: $-d[CoL(CO_2)^+]/dt = k_{app}[CoL(CO_2)^+]^2$ ($k_{app} = 1.0 \pm 0.1 \times 10^{-3} \, M^{-1} \, s^{-1}$), implying a cooperative reaction involving two cobalt centers. The *cis*-$[M(bpy)_2(CF_3SO_3)_2]^+$ complexes [M = Rh(III), Ir(III)] act as electrocatalysts for the reduction of CO_2.[288] Up to 80% current efficiency may be accounted for by the production of formate. The formate proton is obtained from the supporting electrolyte. TaN powders may be oxidized by CO_2/CO mixtures under extreme (!) conditions [850–990 °C; $0.1 \leq P(CO_2)/P(CO) \leq 9$].[289] A study of reaction medium effects on CO oxidation in the presence of Pd(II), Ag(I), and Cu(I, II) complexes revealed that acetate ions can both inhibit and enhance the observed reaction rate.[290] The oxidation of CO to a bidentate carbonate ligand by $[(CH_3)Ir(P(p\text{-tolyl})_3)_2(CO)(O_2)]$ has been reported.[291] The $PtCl_4^{2-}/CuCl_2$ catalyzed oxidation of CO proceeds in ^{18}O-enriched H_2O to incorporate the labeled oxygen atom in 75% of the CO_2 produced.[292] Redox reactions between a number of metal carbonyl cations and anions have been examined.[293] Mechanically, the reactions are equivalent to a $2e$ change with, formally, CO^{2+} being transferred.

3.10. Alkyl Halides

The reaction of Ni(I) complexes with alkyl halides has been examined. The octaethylisobacteriochlorin complex reacts with CH_3I to produce the neutral Ni(II) complex, CH_4, and I^-.[294] Evidence is presented for the intermediacy of an alkyl-Ni(III) species. Ni(I)(tetra-N-methylcyclam) reacts with 1,5-dihaloalkanes yielding cyclopentane.[295] Comparison of the data for these reactions suggests a common inner-sphere electron transfer. With other disubstituted alkanes, the reaction leads to the formation of alkenes.[296] No transient organonickel species are observed. The reaction between electrochemically generated $[(TPP)Co(I)]$ (TPP = tetraphenylporphyrin) and 17 different alkyl halides has been monitored by cyclic voltammetry.[297] An s_N2 reaction mechanism was postulated for all species except CH_3I, where the reaction involved features of an outer-sphere electron transfer. The reaction of metal-coordinated allylic alkenes $[(C_5R_5)M(CH_2CHCH_2Cl)(CO)_n]$ (R = H, Me; M = Rh, Co, Mn) to give cationic metal allyls is thought to proceed through a rate-limiting initial electron transfer reaction.[298] Bis(dimethylglyoximato)cobalt(II) reacts with $NCCH_2I$ producing an alkyl-type radical.[299] Two papers exploring various other aspects of alkyl halide oxidation/reduction have also appeared.[300,301]

3.11. Organic Radicals

The reaction of cobalt(II) tetrasulfophthalocyanine, $[Co(II)(tspc)]^{4-}$, with $\cdot CH_3$, $\cdot CH_2CH_2OH$, $\cdot CH(CH_3)CH_2OH$, $\cdot CH(CH_3)CH(CH_3)OH$, and $\cdot CH_2C(CH_3)_2OH$ free radicals has been studied.[302] Results indicate the initial formation of a $[Co(II)(tspc-R)]^{4-}$ species where the exact nature of the interaction of $\cdot R$ and tspc is not clear. There follows a subsequent internal redox formation of $[(tspc)Co(III)-R]^{4-}$ via a first-order process. Subsequent decomposition produces methane for $\cdot R$ or alkenes for $\cdot ROH$. Interaction of methyl radical with $[Co(II)(nta)(H_2O)_2]^-$ [nta = $N(CH_2CO_2)_3^{3-}$] yields $[(nta)(H_2O)Co(III)-CH_3]^-$ ($K_1 = 2.7 \pm 0.5 \times 10^6 \, M^{-1}$ and $k_{-1} = 60 \pm 10 \, s^{-1}$.[303] This reaction is followed by

$$O_2 + \cdot CH_3 \rightarrow \cdot O_2CH_3 \quad (k_3 = 3 \times 10^9 \, M^{-1} \, s^{-1})$$

and

$$\cdot O_2CH_3 + [Co(II)(nta)(H_2O)_2]^- \rightarrow [(nta)(H_2O)Co(III)(O_2CH_3)]$$
$$(k_4 = 1.0 \pm 0.1 \times 10^8 \, M^{-1} \, s^{-1})$$

The 1-hydroxy-1-methylethyl radical has been utilized as a reducing agent for a series of $[(NH_3)_5Co(X-Im)]^{3+}$ complexes (X = H, 4-CH_3, 1-CH_3, 2-CH_3, and 4-Cl).[304] The rate constants obtained were found to correlate to the lowest-energy $d-d$ absorption band of these complexes. Similarly, the $\cdot C(CH_3)_2OH$ radical was used in a study of Co(III) amine complexes involving bi- and multi-dentate ligands.[305] However, the previously observed correlation between the lowest-energy $d-d$ absorption band for a variety of Co(III) amine complexes and reaction rates appears to break down for $[Co(sep)]^{3+}$, where a much faster rate than anticipated is observed. $[RCo([14]aneN_4)(H_2O)]^{2+}$ (R = CH_3, C_2H_5, 1-C_3H_7, CH_2Cl, CH_2Br, CH_2OCH_3) undergoes photolysis of the carbon–cobalt bond providing a convenient source of carbon-centered radicals.[306] The cobalt(II) macrocycles, $[LCo(II)]$ (L = C-meso-Me_6[14]aneN_4, Me_6[14]-4,11-dieneN_4, corrin, and Me_4[14]-1,3,8,10-tetraene), interact with primary and substituted alkyl radicals.[307] The reactions result in Co—C bond formation, with the rate constant dependent upon the macrocyclic ligand but almost independent of the radical species. The energy transfer from ^2E $[Cr(bpy)_3]^{3+}$ to $[(H_2O)Co([14]aneN_4)R]^{2+}$ is followed by two competitive reactions, namely, relaxation of $[^*(H_2O)Co([14]aneN_4)R]^{2+}$ to the ground state and homolytic cleavage of the Co—R bond.[308] The reactivity pattern is not typical of outer-sphere electron transfer, as evidenced by comparative reactions with $[Ru(bpy)_3]^{3+}$. Various organocobaloxime species containing thioalkyl groups undergo homolytic cleavage during photolysis to give radical reactions involving the thioester group.[309]

The reaction of methyl radicals with $[Ni(cyclam)]^{2+}$ proceeds via $[(cyclam)-(H_2O)Ni(III)-CH_3]^{2+}$ ($K = 1.1 \times 10^7\ M^{-1}$) to ultimately yield C_2H_6.[310] In the presence of O_2, the intermediate Ni(III) species decomposes by homolysis with the subsequent reactions:

$$O_2 + {}^{\cdot}CH_3 \rightarrow {}^{\cdot}O_2CH_3$$

and

$${}^{\cdot}O_2CH_3 + [Ni(cyclam)]^{2+} \rightarrow [(cyclam)(H_2O)Ni(III)-O_2CH_3]^{2+}$$

Fe(V) has been generated via pulse radiolytic reduction of aqueous ferrate(VI) by 2-PrOH radicals.[311] The protonated forms of Fe(V) are unstable and decompose by a second-order process to H_2O_2 and Fe^{3+} ions. Pulse radiolysis studies of the reaction between carbon-centered radicals and (diethylenetriaminepentaace-tato)iron(III) lead, eventually, to the production of the Fe(II) complex.[312] Cu(I) phenanthroline is oxidized by ${}^{\cdot}CH_2CH_2OH$ and ${}^{\cdot}CH_2C(CH_3)_2OH$ free radicals giving a Cu(II) species and the corresponding alkenes.[313] The data imply the formation of a transient Cu—C σ-bond and a mechanism accounting for this species is provided. The reduction of Ce(IV) complexed to dicarboxymethyl $[{}^{\cdot}CH(COOH)_2]$ has been investigated by kinetic ESR spectroscopy in aqueous $HClO_4$ media.[314] Also, $[(\eta^4\text{-COT})RhCp]$ undergoes an irreversible one-electron oxidation to a transient radical cation, which undergoes rearrangement ultimately leading to a dimerized organic ligand bridging two RhCp moieties.[315]

Part II
Substitution and Related Reactions

Chapter 4

Reaction of Compounds of the Nonmetallic Elements

4.1. Boron

The pyrolysis of the boranes is a topic of continuing interest. Lipscomb *et al.*[1] have published calculations that cast doubt on the long-held view that in the pyrolysis of diborane the decomposition of B_3H_9, an adduct of BH_3 and B_2H_6, is a rate-determining stage. The same group have also published calculations that are consistent with the idea that a BH_5 species is an intermediate in the aqueous hydrolysis of the borohydride anion.[2] Greenwood *et al.*[3] suggest reactions (1)–(3)

$$B_6H_{12} \rightarrow B_5H_9 + BH_3 \quad \text{slow} \tag{1}$$

$$2BH_3 \rightleftharpoons B_2H_6 \tag{2}$$

$$B_5H_{11} \rightarrow B_4H_8 + BH_3 \tag{3}$$

for the pyrolysis of B_6H_{12} and B_5H_{11}. Calculations on the diamond–square–diamond isomerization mechanism in $C_2B_5H_7$ have been reported.[4] However, for the thermal rearrangements of the icosahedral carboranes the extended triangle rotation mechanism is held to rationalize the data better.[5] Barrier heights for bridge hydrogen scrambling in B_6H_{10} and CB_5H_7 have been calculated.[6] Slow deuterium–protium exchange for B_2H_6 in FSO_3D/SbF_5 at low temperature probably involves a $B_2H_7^+$ species, $[H_2B(\mu H)_2BH_2D]^+$.[7] The reaction of *nido*-2,3-$R_2C_2B_4H_5^-$ with organic halides R'X gives B(4)-substituted products together with B(6)-substituted enantiomers, probably[8] via a bridged intermediate $R_2C_2B_4H_5$-μ-(4, 5)-R'.

Studies of the interaction of the interaction of (1) with (2) have been made using NMR. Several intermediates are involved.[9] The insertion of MO or MN bonds in $M(OMe)_3$ and $M(NMe_2)_3$, where M = P, As, or Sb, into the triple

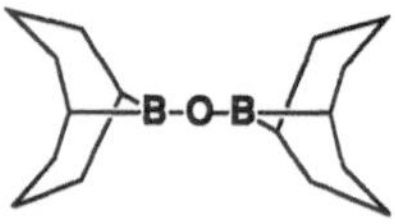

(1)    (2)

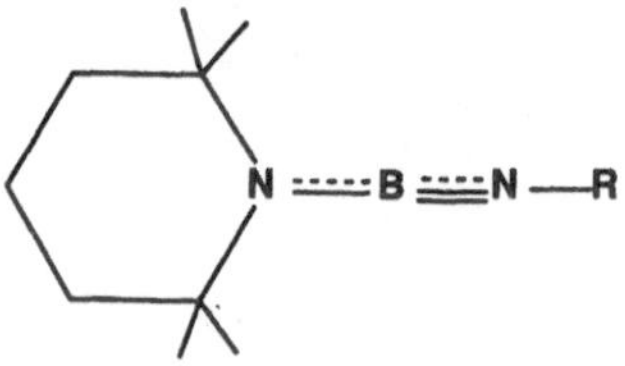

(3)

boron–nitrogen bond in (3) depends upon the Lewis acidity of MX_3 (P < As < Sb) and also upon steric factors.[10] The exchange reactions of $B_3N_3Me_5Cl$ with $CH_2=CHMgCl$ have been studied.[11] The fluorination of BCl_3, BBr_3, B_2Cl_4, B_2Br_4, B_4Cl_4, and $1\text{-}BCl_2B_5H_8$ with $CFCl_3$, $CFBr_3$, and $Hg(CF_3)_2$ have been studied by ^{11}B and ^{19}F NMR. The results for B_4Cl_4 suggest attack on a face of the tetrahedron, rather than at a B—Cl bond.[12] Roberts has published accounts of several ESR studies of the reaction of boron-containing radicals.[13-15] The flash photolysis of sodium metaborate + potassium persulfate results in the formation of $B(OH)_4^-$. The pK_a is 10.75 and $E°$ is +1.4 V. The reaction with amines and phenols appears to involve the conjugate base as the active species.[16]

4.2. Carbon

A study of the kinetics of the reaction of CO_2 with various amines in aqueous solution yields a rate law given in Eq. (4). The first term is suggested to involve a

$$k_{obs} = k_1[R_2NH]^2 + k_2[R_2NH][H_2O] \tag{4}$$

termolecular process with one of the amine molecules functioning as a nucleophile and the other as a base.[17] The reaction of nitrite ion with arylisocyanates in organic solvents prodeeds[18] by the mechanism in Eq. (5)-(7). The solvolysis of

$$ArN=C=O + NO_2^- \rightarrow Ar\bar{N}\text{—}CO\text{—}O\text{—}NO \tag{5}$$

$$Ar\bar{N}\text{—}CO\text{—}O\text{—}NO \rightarrow ArN=N\text{—}O^- + CO_2 \tag{6}$$

$$ArN=N\text{—}O^- + ArNCO \rightarrow ArNH\text{—}N=N\text{—}Ar + CO_2 \tag{7}$$

thiocyanate by trifluoracetic acid has been investigated, and probably involves $NH_2C(S)OOCCF_3$ as an intermediate.[19] Parallel routes for decomposition to NH_4^+, OCS, and $(CF_3CO)_2O$ are postulated. Kinetic data on the NH_4SCN to $(NH_2)_2CS$ isomerization reaction in melts has been reported.[20] The kinetics of detritiation of a range of carbon acids such as Cl_3CH, $PhC\equiv CH$, and p-$NO_2C_6H_4CH_2CN$ by OH^-/H_2O, $OMe^-/MeOH$, and $OEt^-/EtOH$ has been investigated.[21]

4.3. Silicon

Some mechanistic material is included in an annual review of silicon chemistry.[22] A novel mechanism for the dehydrogenative polymerization of silanes

Scheme 1

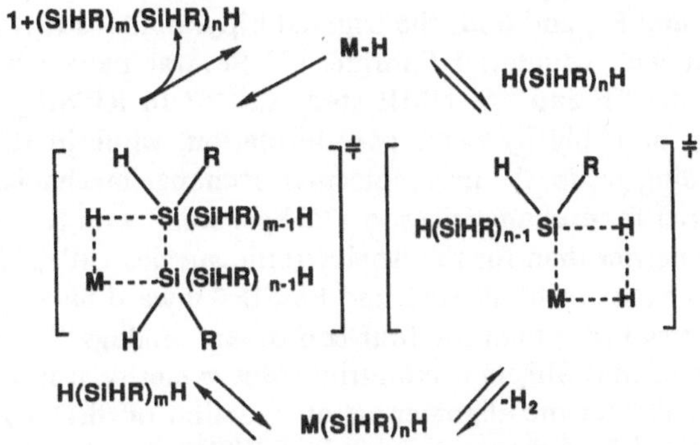

to polysilanes, catalyzed by zirconocene and hafnocene, has been described[23] (Scheme 1) and Corey has proposed a mechanism for the dehydrogenative coupling of dihydrosilanes to disilanes and short-chain oligomers.[24] Wurtz-type reactions involving $Na/C_6H_5CH_3$ and $MePrSiCl_2$ reacting to form organopolysilanes are catalyzed[25] by the use of 18-crown-6 as an additive. The mechanism of the insertion of sulfur into Si—Si bonds in the reactions of cyclic $(Me_2Si)_6$ and linear $(Me_2Si)_x$ involves the polysilane acting as a nucleophile, with elimination of Me_2SiS.[26]

The role of five-coordinate intermediates in substitution at silicon remains a topic of continuing attention. Corriu proposes such intermediates in the exchange reactions of Si—C and Si—H bonds in RR^1SiH_2 and $RSiH_3$ in the presence of $LiAlH_4$ with NaH and KH catalysts,[27] and in the racemization of optically active 1-NpPhMeSiH(D) with KH, $LiAlH_4$, and $LiAlD_4$ in tetrahydrofuran.[28] He has surveyed and provided a mechanistic interpretation of the nucleophilic activation reactions of silylenolethers and esters, again involving the rapid formation of a hypervalent silicon species.[29] *Ab-initio* and MNDO calculations on a wide variety of pentacoordinate silicon anions have been carried out; 86 out of 91 are predicted to be stable to the loss of an anion.[30] The MNDO results are generally consistent with the results of Corriu and co-workers; where there are discrepancies, as in the case of compounds containing fluorine, the difficulties are usually removed by the use of *ab-initio* calculations. It can be useful to study the structure of stable compounds containing five-coordinate silicon, particularly by NMR methods; Corriu reports[31] a study on 2-(dimethylaminomethyl)phenyl-1-naphthylsilane, while Dixon *et al.* report on $[R_3Si(CN)_2]^-$ species.[32] Evidence for a five-coordinate structure (4) with M = Si, Sn has been published[33] and other studies also propose

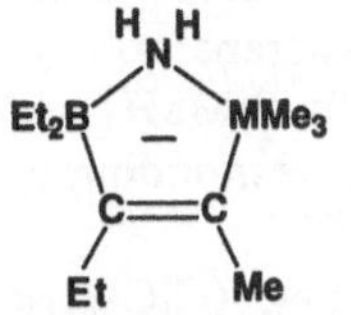

(4)

such species.[34,35] A number of papers deal with five-coordinate species in silicon–fluorine systems. Streitweiser has carried out a theoretical study of the $S_N2(Si)$ reaction of SiH_4 and F^-, and finds the trigonal bipyramidal structure more stable with apical than with equatorial fluorines.[36] Several papers by Holmes and co-workers describe ^{19}F and ^{29}Si NMR studies.[37-39] In $RR'SiF_3^-$ intramolecular exchange occurs, probably by Berry pseudorotation, while in $RSiF_4^-$, if R is a sterically demanding group, the intramolecular exchange mechanism is effectively stopped; the barrier for exchange rises to 12.8 kcal mol^{-1} with $R = 2,4,6$-Bu^{t}C$_6$H$_2$. This is a greater barrier than for the isoelectronic species $ClPF_4$, Me_2NPF_4, and $Pr_2^iNPF_4$. The reactivity of $MePhSiF_3^-$ and $Ph_3SiF_2^-$ toward nucleophilic substitution is 10–100 times faster than for four-coordinate analogs.[40] Following work on the hydrolysis of Mes_2SiF_2 in acetonitrile (Mes = mesitylgroup), a mechanism has been suggested[40] for the oligomerization reaction of $Si(OH)_4$. The effect of substitution on the ^{19}F exchange barrier in $RR'SiF_3^-$ has been studied, taking particular care to avoid problems due to traces of moisture.[41] Six-coordinate species are also of interest in the mechanism of reaction of silicon compounds. Corriu has produced NMR evidence[42] for nondissociative permutational isomerization in such species, and there is a claim for a six-coordinate intermediate in the reactions of salts of bis(1,2-benzenediolato)allyl silicates with aromatic aldehydes.[43]

Roberts has extended his ESR studies on silicon-based radials, investigating the reaction of $R_3Si^•$ and $(RO)_3Si^•$ with CO_2, the reaction of $R_3Si^•$ with $R^1R^2N \rightleftharpoons BH_2$, and the halogen abstraction reactions of $(RO)_3Si^•$ with allyl chloride and allyl bromide.[44-46] The UV irradiation of poly-$(R_2Si)_n^-$ in solution is known to produce initial persistent radical species. These have now been identified[47] as $-SiR_2-SiR^•-SiR_2^-$. Silylenes, R_2Si:, can be stabilized by bulky groups. Ultrasonic irradiation of $Cl_2SiBu_2^i + Li + Et_3SiH$ can readily generate Bu_2^iSi:, whose reactivity can then be investigated.[48] The authors doubt if this produces a truly "free" silylene. Photolysis of $Bu_2^iSi(N_3)_2$ in a matrix also produces Bu_2^iSi:.[49] Silylenes can be produced by the laser flash photolysis[50] of cyclic tetrasilanes according to Eq. (8). Photolysis of cyclic $(Ph_2SiCH(CH_2Bu^t))_2$ produced

$$(R^1R^2Si)_4 \xrightarrow{h\nu} R^1R^2Si: + (R^1R^2Si)_3 \qquad (8)$$

a ring-opened singlet 1,4-biradical.[51] The far-UV photolysis of (5) has been studied mechanistically, and an $Si=C$ species is thought[52] to be involved as an intermediate in the ring opening and cleavage reaction. Intermediates in the reaction of species of the type $>Si=C<$ and $>Si=N-$ with nonenolizable aldehydes and ketones have been studied.[53] Silylenium ions in $Ph_3Si^+ClO_4^-$ and $Me_3Si^+ClO_4^-$ in sulfalone have been studied by chlorine NMR; the results show a rapid exchange between ion pairs and free ions, and it is claimed that the percent of free silylenium ions can be calculated.[54] Fourier-transform cyclotron resonance spectroscopy has been used to derive H^- affinities for $SiMeH_2^+$, $SiMe_2H^+$, and $SiMe_3^+$. The silylenium ions are more stable than the corresponding carbocations with H^- as a base, but are less stable with F^- as base.[55]

The reaction of chlorosilanes, $(ClCH_2)R_2SiCl$, with phenols and amines involve kinetically controlled attack on the silicon.[56] The kinetics of reaction of

(5)   (6)

ArO$^-$ in a range of solvents follows the sequence Ph$_2$MeSiCl > Me$_3$SiCl > Bu^{t}Ph$_2$SiCl > Pr$^i_3$SiCl. Ion pairing effects can be important.[57] The mechanism of silylation of silanols by chlorosilanes in the presence of nucleophiles has been reported.[58] Competition betwen attack at silicon and at a proton in the reaction of (6) in tertiary butanol (LG = leaving group) has been discussed.[59]

Relaxation studies by ^{29}Si NMR of solutions of "silica" in aqueous alkali metal hydroxides show a time-dependent line broadening almost entirely due to Si/Si exchange processes involving Si(OH)$_4$. Intramolecular cyclization is faster than intermolecular polymerization. Past disagreement about the origin of temperature-dependent line broadening is attributed to failure to consider the effect of pH on the leaching of line-broadening contaminents from glass sample tubes.[60] High-field ^{17}O NMR spectra of aqueous silicate solutions show overlapping spectral regions due to terminal and bridging oxygen. Preliminary kinetic measurements indicate a much faster rate of exchange of water with terminal as compared to bridging oxygen.[61] Studies by ^{29}Si NMR of the dynamics of silicon exchange between silicate anions in alkali yield rates of monomer/dimer and monomer/dimer/cyclic trimer exchange. The rate is a function of the alkali metal cation size and ion pairs are involved.[62] It is claimed that the base-catalyzed cleavage of RMeSi(OH)$_2$ involves the unimolecular dissociation of RMeSi(O)$_2^{2-}$ at high [OH$^-$] into R$^-$ and Me(O)Si=O$^-$, while for RSi(OH)$_3$ it appears[63] that a *meta*-silicate is formed as shown in reaction (9). The kinetics of reaction of

$$R(OH)Si(O)_2^- \rightarrow R^- + HOSiO_2^- \qquad (9)$$

Me$_3$Si(OSiMe$_2$)$_4$OH with Me$_3$SiOSiMe$_2$OCOCH$_3$ involve base activation of the Si—OH group; bases that are strongly nucleophilic may attack the silicon bonded to the acetate ligand to form a five-coordinate species.[64] Ester exchange occurs between Et$_3$SiOBu^n and Bu^{s}OH, catalyzed by IBr and ICl; a four-centered transition state is favored.[65] *Trans*-esterification in Si(OEt)$_4$ + BuOH has also been investigated.[66] Cleavage of the Si—O bond by HCl in dioxane has been studied; reaction shows an induction period due to the autocatalytic formation of H$_2$O as a reaction product.[67] Silicon–oxygen bond cleavage catalyzed by BF$_3$ (as the dioxane:BF$_3$ reagent) shows complex kinetics, that can be followed by PMR. The reaction order with respect to [BF$_3$] is 4.7! The mechanism shown in Eq. (10)–(12)

$$(R_3Si)_2OBF_3 \rightarrow R_3SiF + R_3SiOBF_2 \qquad (10)$$

$$R_3SiOBF_2 + BF_3 \cdot solvent \rightleftharpoons R_3SiOBF_2 \cdot BF_3 \qquad (11)$$

$$R_3SiOBF_2 \cdot BF_3 \rightarrow R_3SiF + (BF_2)_2O \qquad (12)$$

is suggested.[68] The reaction of (CH$_3$O)$_3$SiH to form (CH$_3$O)$_4$Si and SiH$_4$ is catalyzed by hydrotalcite-type minerals with interlayer chloride and methoxide ions.[69] A kinetic investigation of the reaction of Ph_3SiH with Ph_3SiOH catalyzed by $Me_3SiO^-Na^+$ is reported to be zero order in $[Ph_3SiOH]$; a multistep mechasism is postulated.[70] The possibility of intramolecular effects in the hydrolysis of compounds containing $SiMe_2OPh$ and a range of phosphorus-containing groups, linked by a carbon chain, has been investigated. Reactions are both acid and base catalyzed, but no evidence was found for P=O acting as a nucleophile and attacking

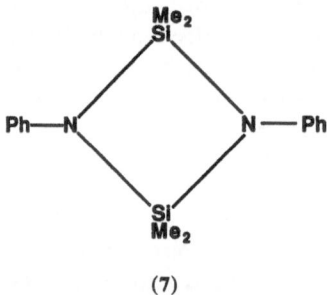

(7)

the silicon center.[71] An interesting study of the base- and acid-catalyzed hydrolysis of cyclodisilazanes, (7), has been reported by Perry. The mechanism of the base-catalyzed process in tetrahydrofuran is shown in Eqs. (13) and (14) where the

$$PhN(\mu SiMe_2)_2NPh + OH^- \rightarrow HOSiMe_2NPhSiMe_2NHPh \qquad (13)$$

$$HOSiMe_2NPhSiMe_2NHPh \rightarrow (PhNHSiMe_2)_2O \qquad (14)$$

intermediate silanol undergoes proton loss from the hydroxyl group, followed by intramolecular attack of the $-O^-$ upon the second silicon. The species $(PhNHSiMe_2)_2O$ undergoes a slow further hydrolysis to aniline and dimethylsiloxane. The acid-catalyzed hydrolysis is very rapid and proceeds directly to aniline and a dimethylsiloxy compound. For the cyclodisilazane $PhN(\mu SiPh_2)_2NPh$, catalysis by HBF_4 is particularly rapid and pathways involving the formation of silicon–fluorine compounds such as $FPh_2SiNPhSiPh_2NHPh$ are suggested.[72]

4.4. Germanium

A common feature of many papers on the mechanisms of reactions of germanium compounds is the involvement of GeR_2: species. There is spectroscopic evidence for Me_2Ge: in the photolysis of $Me_2Ge(SePh)_2$.[73] The thermal decomposition of (8) produces Me_2Ge: ; its reaction with styrene has been studied.[74] Laser flash photolysis of $[Ph_nMe_{3-n}Ge]_2$ results in the formation of germyl radicals and germylenes.[75] Photolysis of cyclic $(GeAr_2)_3$ produces $Ar_2Ge=GeAr_2$, which can react with CH_2N_2 to produce cyclic $(Ar_2Ge)_2CH_2$. This in turn can photolyze to form Ar_2Ge:[76]. Photolysis of hexamesitylcyclotrigermane yields $Mes_2Ge=GeMes_2$ and Mes_2Ge:[77]. This is also formed in the reaction of (9) with diazomethane.[78] Addition occurs across the C=Ge bond to form a three-membered ring species, which on photolysis yields Mes_2Ge: . Insertion reactions of free Me_2Ge: with C—Cl and C—Br bands of alkynyl and propargyl halides

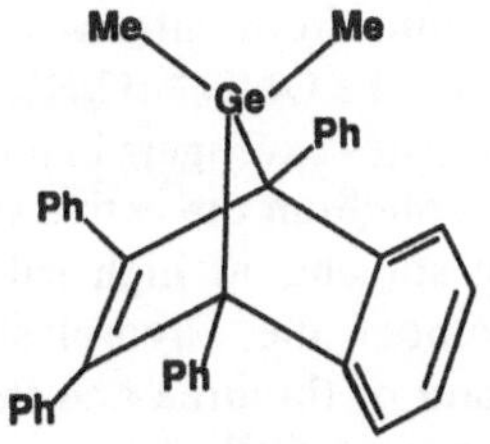

(8)

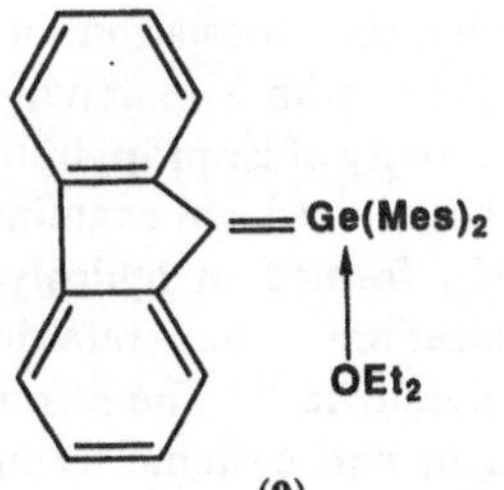

(9)

occur[79] by a concentrated mechanism. Transient species $Me_2Ge=CH_2$, $Me_2Ge=S$, Me_2GeCH_2S, and $Me_2GeCH_2GeMe_2S$ are probably involved in the thermolytic and photolytic reactions of $Me_2GeCH_2SCH_2$, $Me_2GeCH_2SSCH_2$, and $Me_2GeCH_2SO_2CH_2$.[80] A different intermediate $[GeR_3]^+$, R = Me, Ph, can be formed in the reaction of R_3GeH with trityl perchlorate.[81] The reaction of 2,2′-bipyridine with closo-1-Ge(II)-2,3-$(SiMe_2)_2$-2·3-$C_2B_4H_4$ in benzene involves the slow formation of an adduct, which is followed by a rapid process in which all boron atoms become equivalent on the NMR timescale.[82]

4.5. Nitrogen

A careful and detailed study of the solubility and decomposition kinetics of nitrous acid in aqueous solution has appeared.[83] This is part of an excellent program of work on atmospheric chemistry, and provides valuable quantitative data, particularly in terms of solubility parameters. Stanbury et al.[84] have bubbled NO_2 in a stream of N_2 into alkaline aqueous solutions of $Na^{15}NO_2$, and determined the amount of $^{15}NO_3^-$. This is a difficult experiment to do well, and has produced a value of 580 $M^{-1}s^{-1}$ for the self-exchange rate constant for NO_2 and NO_2^-. The authors estimate an uncertainty factor of three, but even with this possible error the value is about four orders of magnitude greater than that calculated by use of the Marcus relationship. A value of the solvent isotope effect $k_D/k_H = 2.55 \pm 0.28$ has been estimated for the formation of NO^+ from HNO_2.[85] The reaction of nitrous acid with [Fe(II)edta] has been studied as a reaction of potential interest for the absorption of NO_X from flue gases. The kinetics follow rate law (15) to produce [Fe(II)(edta)NO]. The first term represents a rate-determining formation

$$-d[Fe(II)edta]/dt = k_a[HNO_2]^2 + k_b[Fe(II)edta][HNO_2] \qquad (15)$$

of N_2O_3, a well-known process, although the authors write it in an unfamiliar guise. The second term (k_b) is interpreted as reactions (16) and (17), with the latter process being very fast.[86]

$$HNO_2 + [Fe(II)edta] \rightarrow [Fe(III)edta] + NO^· + OH^- \qquad (16)$$

$$NO^· + [Fe(II)edta] \rightarrow [Fe(II)edta(NO)] \qquad (17)$$

Interest in the mechanism of action of the alkylnitrites as electrophilic nitrosating agents has developed in recent years. Williams has produced evidence[87] for a rate-limiting formation of NO^+ from alkynitrites and from nitrous acid in acetonitrile, with a reactivity order $HNO_2 > Bu^iONO > Pr^iONO > C_5H_{11}^iONO$. The reactivity of isopropylnitrite used with a range of nitrite scavengers in aqueous isopropanol has been examined. In this partly aqueous medium the active species is HNO_2 formed in hydrolytic equilibrium with the solvent; at high substrate concentrations the rate-determining stage becomes the hydrolysis of isopropylnitrite.[88] The nitrosation of acetylacetone and of thiourea and thioglycolic acid and cysteine by alkylnitrites have been reported.[89-91] Nitrosation at sulfur centers continues to arouse interest. The nitrosation of thiomorpholine involves an initial S-nitrosation followed by a sulfur to nitrogen shift of the nitroso group.[92] Further quantitative studies of the reactivity of the nitrosothiosulfate anion, $ONSSO_3^-$, have been reported; the value found for the formation constant, $2 \times 10^6 \, M^{-2}$, is somewhat lower than the value originally proposed.[93] The influence of kinetic factors on the products of the decomposition of the S-nitrosoalkylthioureas has been studied.[94] Other papers report on nitrosation of 2,4,6-trimethylphenylurea[95] (which leads to a value of $0.15 \, M^{-1}$ for the equilibrium constant $[>C=O-NO^+]/([>C=O][NO^+])$), and the nitrosation of malononitrile.[96] NMR evidence for the existence of a protonated N-nitrosamine, $[Me_2NHNO]^+$, has been found; the K_a value is *estimated* to be 10^{12}-$10^{13} \, M$.[97] The promotion of iodination reactions by $[NO]^+[BF_4]^-$ is attributed[98] to the reactions (18) and (19). Nitrosation by *N*-methyl-*N*-nitrosotoluene-*p*-sulfonamide

$$I_2 + NO^+BF_4^- \rightarrow I^+BF_4^- + INO \qquad (18)$$

$$2INO \rightarrow I_2 + 2NO \qquad (19)$$

has been studied,[99] and there is a report on the base hydrolysis and transnitrosation reactions of $RCON(NO)Me$.[100] Crampton has looked[101] at the nitrosation of hexamethylenetetraamine to give the partially ring-opened compound (10), and

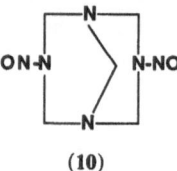

(10)

Williams has studied the nitrosation of 2,3-diaminonaphthalene, which results in triazole formation.[102] Zollinger[103] has suggested that in the acid-catalyzed diazotisation of $ArNH_3^+$ and deamination of NH_3OH^+, which are thought to involve initial nitrosation at the aromatic ring and the hydroxyl group, respectively, the loss of an $N-H$ proton precedes the transfer of the nitroso group to the nitrogen, rather than occurring as a synchronous process. This is consistent with the views of Bourke.[104] Another, more exclusively inorganic system is the reaction of nitrous acid with hyponitrous acid. Hughes has shown[105] by an isotopic and product study that in addition to the reactions (20)-(22) there is an additional process

$$H_2N_2O_2 + HNO_2 \rightarrow HNO_3 + N_2 + H_2O \qquad (20)$$

$$3HNO_2 \rightarrow 2NO + HNO_3 + H_2O \qquad (21)$$

$$H_2N_2O_2 \rightarrow N_2O + H_2O \tag{22}$$

$$0.55\ ^{15}NO + 0.55H_2\ ^{14}N_2O_2 \rightarrow 0.29\ ^{14}NO + 0.63\ ^{14}N^{15}NO + 0.1\ H\ ^{15}NO_3 \tag{23}$$

(23). Hughes has also reported a study of the kinetics of reaction up to 11.6 M perchloric acid. A very sharp maximum in the rate at 7 M acid suggests that HNO_2 and not NO^+ may be the active species.[106] Studies with ^{15}N and ^{18}O have also been used to investigate denitrification processes using cell-free extracts.[107]

The autocatalytic oxidation reactions of nitric acid have been studied by Bazsa *et al.* They have examined the oxidation of formaldehyde, and suggest[108] a rate-determining step involving $H_2C(OH)_2$ and N_2O_4. A much more complex system is the oxidation of bromide. The reaction has been studied in both forward and backward directions, and has been found to exhibit bistability when studied in a continuous-flow stirred tank reactor. A detailed mechanism has been proposed [see Eqs. (25)–(31)] together with values for the rate constants. The reverse reaction, the oxidation of nitrous acid by bromine, gives rate law (24), a much simpler

$$v = k[Br_2]^{1/2}[HNO_2]h^{-0.5} \tag{24}$$

$$H^+ + HNO_2 + NO_3^- \rightleftharpoons 2NO_2 + H_2O \tag{25}$$

$$2NO_2 + Br^- + H^+ \rightleftharpoons BrNO_2 + HNO_2 \tag{26}$$

$$HNO_2 + H^+ + Br^- \rightleftharpoons ONBr + H_2O \tag{27}$$

$$ONBr + NO_2 \rightarrow BrNO_2 + NO \tag{28}$$

$$BrNO_2 + H_2O \rightleftharpoons HNO_2 + HOBr \tag{29}$$

$$Br^- + HOBr + H^+ \rightleftharpoons Br_2 + H_2O \tag{30}$$

$$2HNO_2 \rightleftharpoons NO + NO_2 + H_2O \tag{31}$$

expression than was found by earlier workers.[109] In a system of such complexity this mechanism may well be modified in the future. Another complex system, the reaction of NO_2^- with ClO_2^-, has been investigated by Stanbury. The rate constant for the electron transfer reaction between these two species is some three orders of magnitude less than that predicted by use of the Marcus cross-relationship, implying a strong overlap mechanism.[110] Another nitrogen/halogen system of interest is the reaction of INO_3 and $BrNO_3$ with phenols in acetonitrile or chloroform. A mixture of halogenated and nitrated products is obtained.[111] When a reagent that complexes halogens is present there is no reaction observed, while in the presence of pyridine or Et_3N only halogenated products are observed. The electrochemical reduction of NO_2Cl in sulfalone at a platinum electrode leads to a rate constant for the reaction (32), which is close to the value for the gas phase reaction of NO and NO_2Cl.[112]

$$NO_{elec} + NO_2Cl \rightarrow NOCl + NO_2 \tag{32}$$

There have been a number of papers on nitration processes. Species formally corresponding to the nitric–acidium ion $H_2NO_3^+$ have been reported[113] in mass-spectrometric studies; $(H_2O)NO_2^+$ is considered to be more stable than $(HO)_2NO^+$. In recent years the possibility of a radical pathway in aromatic nitration has attracted a lot of attention. It had been suggested that the radical pathway might be negligible for the nitration of naphthalene, but Ridd has found evidence for a small but significant amount of reaction via an electron transfer process.[114] Kinetic studies of nitration with reactive and unreactive substrates have been reported.[115,116] A mild nitrating agent is CF_3COONO_2.[117]

The reduction of nitrate and nitrite species to ammonia is of continuing interest. The selective reduction by a $(Bu_4N)[MoFe_3S_4(SPh)_3(O_2C_6Cl_4)]_2$ modified glassy carbon electrode can occur[118] under electrolysis at -1.25 V (vs. SCE) at pH 10, while at -1.00 V nitrite is reduced to N_2. Hyponitrite is an intermediate in the reduction of NO_2^- by $[Fe_4S_r(SPh)_4]^{2-}$ in acetonitrile.[119] The mechanism of the electrochemical reduction of NO_2^- to NH_3 using $[Ru(Hedta)(H_2O)]$ and $[Fe(Hedta)H_2O]^-$ has been studied,[120] as has the reduction of NO_2^- and NO to NH_3 using iron(II)-substituted polyoxotungstates.[121] An addition complex, written as $HNO_2 \cdot Fe(II) \cdot XW_{11}O_{39}$ (X = Si, Ge, As, P), is involved. Holm has produced a model system for the molybdenum mediated reduction of NO_3^- to NO_2^-, an analog of nitratereductase.[122]

This section ends with a number unrelated pieces of work. Isotopic labeling studies have produced evidence for a transient cyclic form of N_3^- when a N_2 matrix is bombarded with 5 keV neon atoms or ions. Calculations indicate that the triplet cyclic azide ion is about 43 kcal above the singlet ground state of linear N_3^-.[123] Calculations have been made to determine the barrier height for the conversion of NH_2-ONO to NH_2NO_2.[124] The kinetics of oxidation of $NH_2NHCSNHNH_2$ by chloramine B and dichloramine B have been studied.[125] An analytical study of the reaction of hydrazine and $2,4-(NO_2)_2C_6H_3F$ has followed the release of fluoride ions.[126] The kinetics of the well-known Berthelot reaction for the colorimetric determination of ammonia have also been reported.[127] Finally, we note that the earlier work of Candlin and Wilkins on the hydrolysis of hydroxysulfamic acid to hydroxylamine and bisulfate ion has been extended recently.[128] The equilibrium between $(NSCl)_3$ and 3 NSCl has been studied by ^{14}N NMR. The forward and back reactions are described as "very kinetically hindered."[129]

4.6. Phosphorus

Metaphosphate species are often postulated to be found in reactions of phosphates. A high-pressure mass-spectrometric study of clustering of D_2O and PO_3^- suggests that the first two water molecules form simple adducts, but that addition of a third involves isomerization to a dihydrate of $D_2PO_4^-$.[130] Metaphosphate is suggested[131] as an intermediate in the reactions of phosphorus compounds containing P—OH bonds in molten urea. Isocyanic acid may be involved, as shown in Eqs. (33)–(36). Metaphosphate is not formed as an intermediate in the reactions

$$(NH_2)_2CO \rightarrow NH_3 + HNCO \qquad (33)$$

$$HNCO + H_2PO_4^- \rightarrow NH_2CO \cdot OPO_3H^- \qquad (34)$$

$$NH_2CO \cdot OPO_3H^- \rightarrow PO_3^- + NH_3 + CO_2 \qquad (35)$$

$$PO_3^- + H_2PO_4^- \rightarrow H_2P_2O_7^{2-} \qquad (36)$$

of phosphorylated pyridine anions, though the conditions under which this might occur are discussed.[132] Radical-based dephosphorylation can involve[133] a monomeric metaphosphate intermediate; see Eq. (37).

$$RCH_2P(O)(OR^1)O^{\cdot} \rightarrow RCH_2^{\cdot} + (R^1O)PO_2 \qquad (37)$$

Measurements of the volume of activation, $\Delta V^{\ddagger}$, indicate a dissociative mechanism for the reaction of $2,4\text{-}(NO_2)_2C_6H_3OP(S)(O)_2^{2-}$ in aqueous solution to form a free, monomeric thiometaphosphate ion PSO_2^-.[134] Nitrophenol esters of phosphoric acids are favorite substrates for hydrolytic studies. Metal complexes catalyze the hydrolysis of $2,4\text{-}(NO_2)_2C_6H_3OP(O)(OEt)(Me)$,[135] and $[Cu(bipy)]^{2+}$ accelerates the hydrolysis of $p\text{-}NO_2C_6H_4OP(OEt)_2O^-$.[136] In the latter case ^{18}O studies show P—O bond cleavage, and a mechanism involving metal binding to the leaving group is proposed. The hydrolysis of 2-(1,10-phenantholyl)phosphate is affected by metal ions which coordinate at the nitrogen atoms.[137] Metal–oxygen interaction retards metaphosphate formation and enhances nucleophilic attack at phosphorus. Evidence for the planar metathiophosphate species $Bu^iP(S)(O)$ has been obtained from the product of the thermal and photochemical decomposition of (11).[138] Iodosobenzoate, $[2,4\text{-}(CO_2)(R)C_6H_3IO]^-$, catalyzes the hydrolysis of $p\text{-}NO_2C_6H_4OP(O)(R^1)(R^2)$.[139] Micellar media have been used to study the cleavage of $p\text{-}NO_2C_6H_4(PhO)_2PO$ by (12).[140] The displacement reactions of the

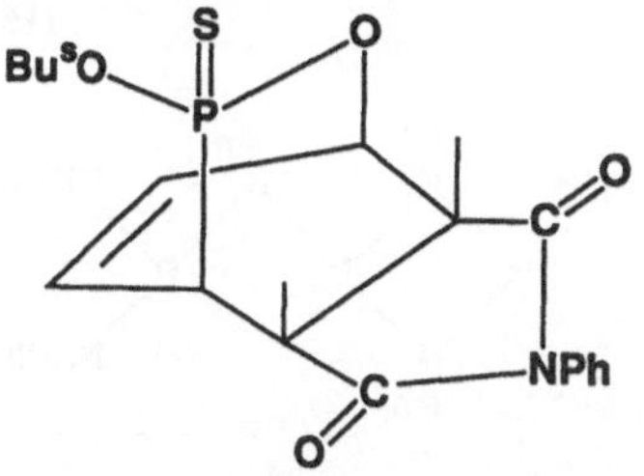

(11) 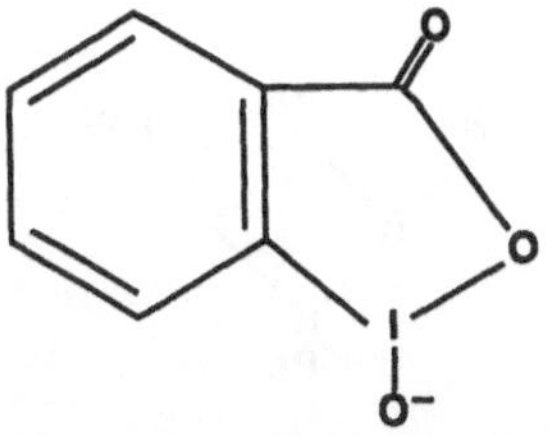   (12)

nucleophilic ethoxide ion with $R^1P(O)(OR^2)(SMe)$ and $R^1P(O)(OR^2)(OR^3)$ show a slight preference for the displacement of MeS^- over MeO^- for $R^1 = Ph$ and $R^2 = Me$.[141] Barriers to pseudorotation in monocyclic oxyphosphoranes have been measured[142] by ^{13}C NMR. Experiments in ^{18}O-enriched water throw light on the mechanism of the desulfurization of $[O_2P(S)OCHPhCH_2OP(S)O_2]^{4-}$. The products include some $P_2O_7^{4-}$, and 45% of the tracer is found in the bridging oxygen. This shows that loss of sulfur must occur at one center, followed by a torsional rotation of the resultant ^{18}O-labeled phosphate, followed by capture at the other phosphorothioate center.[143] Hypophosphorous acid, H_3PO_2, is best known as a reducing agent. The reaction of $H_2PO_2^-$ with $[Ag(III)(OH)_4]^-$ involves a 1:1 complex, with oxygen transfer from silver to phosphorus.[144] The other well-known reducing oxyacid of phosphorus H_3PO_3 reduces Cu(III) in $[Cu(III)\text{-}(TeO_6)_2]^{9-}$, the active species being HPO_3^{2-}. An inner-sphere pathway is suggested.[145]

The reaction of one of the classic phosphonitrilic halides (phosphazanes), $P_3N_3Cl_6$, with $AlMe_3$ has been studied by GC and ^{31}P NMR. Reaction proceeds by parallel geminal and nongeminal pathways, as well as by a ring-opening mechanism.[146] The reaction of geminal $P_3N_3Ph_2Cl_4$ with aliphatic bifunctional alcohols, amines, and aminoalcohols has been studied.[147] The species $(ArNP(O)Cl)_3$ where Ar is the phenyl group, and its *para*-methyl and *para*-methoxy derivatives, are relatively inert to nucleophilic attack.[148] However, reaction with EtO⁻ and OH⁻ gives products that suggest a transannular attack of nitrogen upon phosphorus, following the sequence (13) → (14) → (15). A review of the chemistry of polyalkyl/arylphosphazanes and their precursors contains some mechanistic material.[149] The thermolysis of $PhN(P(NHPh)_2)_2$ produces (16) via the intermediate (17).[150] Intermediates $1,2\text{-}C_6H_4(NH)_2PNEt_2$ and (18) can be identified by ³¹P

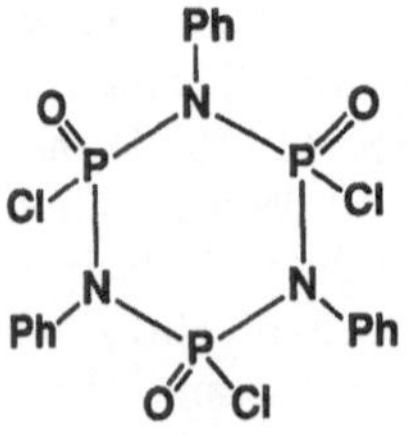

(13) 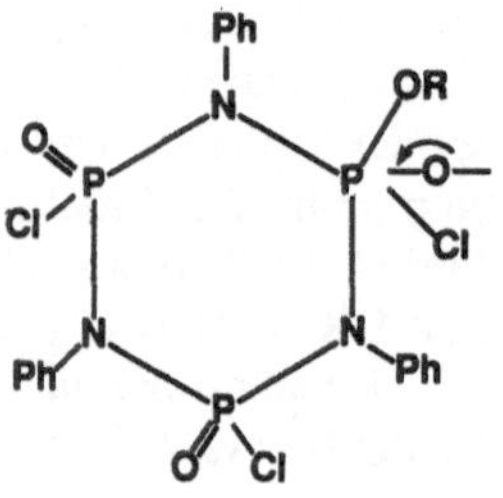　　　　　　　　　　(14) 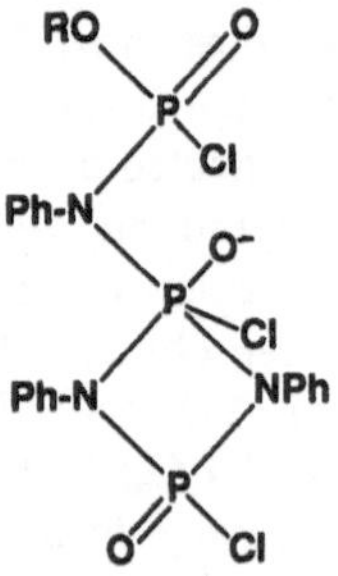　　　　　　　　　　(15)

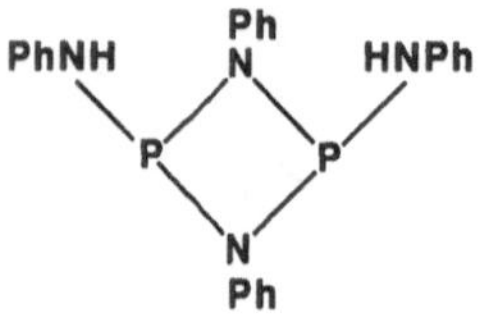

(16) 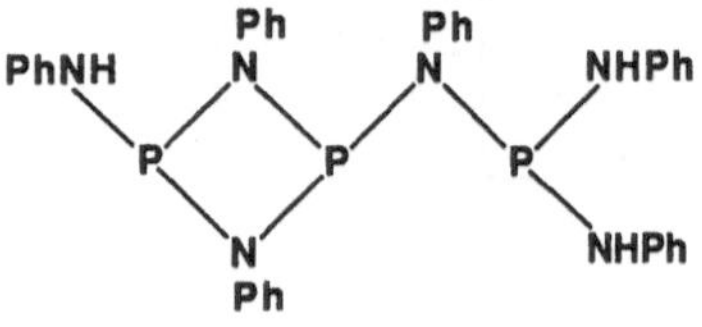　　　　　　　　　　　　　　　　(17)

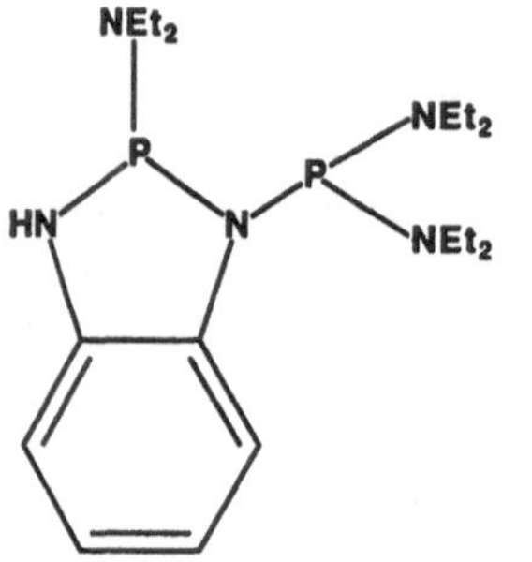

(18)

NMR in the reaction of $1,2\text{-}(NH_2)C_6H_4$ with $P(NEt_2)_3$ to form linear $C_6H_4N_2[P(NEt_2)_2]_2PNEt_2$.[151] Further reactions of this product have been described and discussed.

A mechanistic scheme for the reactions of $PhPCl_2$ with amines to produce a range of phosphazanes has been given.[152] A mechanism[153] for the exclusive

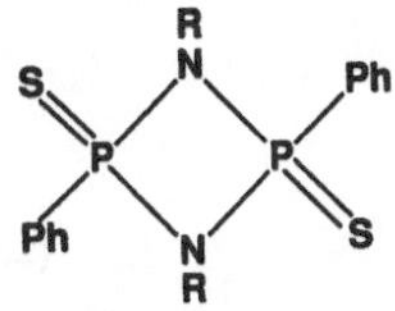

(19)

formation of the *trans* dimer (19) from $PhP(S)Cl_2$ and RNH_2 involves reactions (38)-(40).

$$PhP(S)Cl_2 + RNH_2 \rightarrow PhP(S)(NHR)Cl \qquad (38)$$

$$PhP(S)(NHR)Cl + base \rightarrow PhP(S)(NR) \qquad (39)$$

$$2PhP(S)(NR) \rightarrow (19) \qquad (40)$$

The reactions of compounds with P—P bonds have been investigated by several groups. Mills *et al.* have examined the insertion of a CF_3P unit into a P—P bond in reactions of type (41). This appears to involve a general type of mechanism

$$R_2PPR_2 + (1/n)(CF_3P)_n \rightarrow R_2PP(CF_3)PR_2 \qquad (41)$$

with a four-centered intermediate. Thus for the reaction of R_2PPR_2 with cyclic $(PCF_3)_4$ an addition process leads to formation of (20), which can cleave as shown in Eq. (42). This linear species can be further attacked by R_2PPR_2 to form (21),

$$(20) \rightarrow R_2P(CF_3P)_4PR_2 \qquad (42)$$

which can cleave to $R_2P(PCF_3)_3PR_2 + R_2P(PCF_3)PR_2$. The five phosphorus species can be further attacked, etc. When a P—H bond is involved, as in an acyclic molecule such as $CF_3(H)PP(H)CF_3$, then this may be involved in the formation of a four-membered ring as in (22) which can lead to the formation of Me_2PH. The reaction of Me_2PH with cyclic $(CF_3P)_4$ produces a mixture of products, thought to derive from (23). The exchange/scrambling reactions of $(CF_3)_2PP(CF_3)_2$ with a range of pnicogen-pinogen, P—H and P—Cl compounds have been studied.[154-156] The term pnicogen refers to an element in the N to Bi subgroup, and a dipnicogen contains a bond such as P—P or P—As. In contrast to this a silyl-substituted species such as $(Me_3Si)_3CP=PSiMe_3$ decomposes to (24) without any evidence for a cyclotetraphosphorous species.[157] Ring opening of (25) with BuLi to form (26) occurs by a conrotatory process.[158] The scrambling of P—SP and P—SC bonds between P_4S_{10} and $(RS)_3PS$ has been studied.[159]

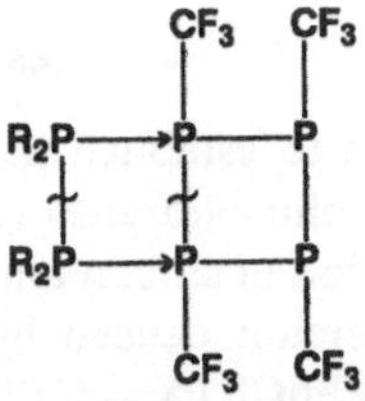

(20)

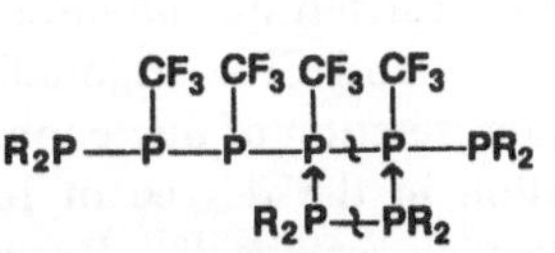

(21)

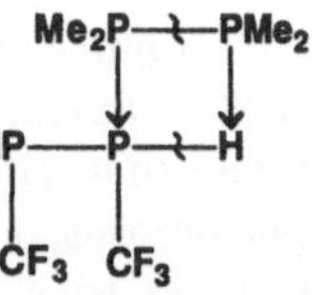

(22)

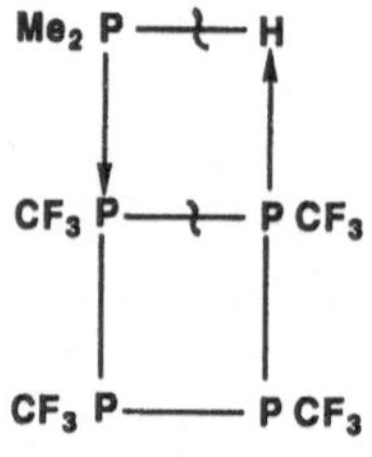

Me$_2$ P $\longrightarrow$ H

CF$_3$ P $\longrightarrow$ P CF$_3$

CF$_3$ P $\longrightarrow$ P CF$_3$

(23)

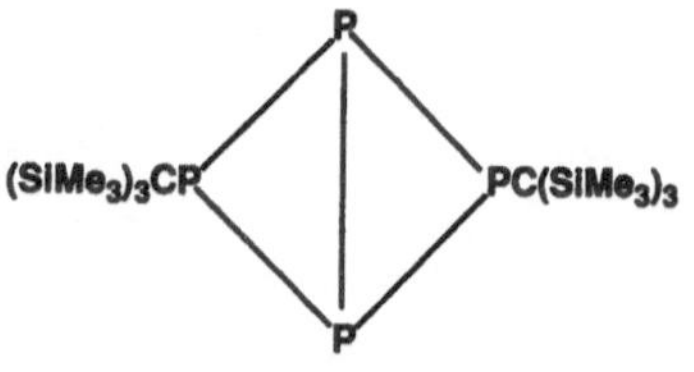

(SiMe$_3$)$_3$CP — P — PC(SiMe$_3$)$_3$

(24)

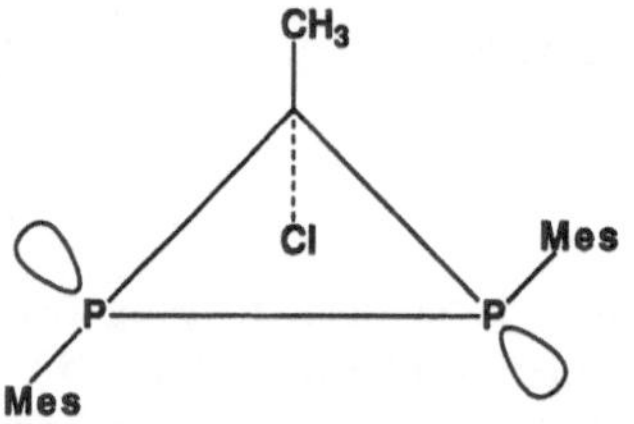

CH$_3$

Cl

P   P

Mes   Mes

(25)

Mes   CH$_3$

P=C

Mes

P

Bu

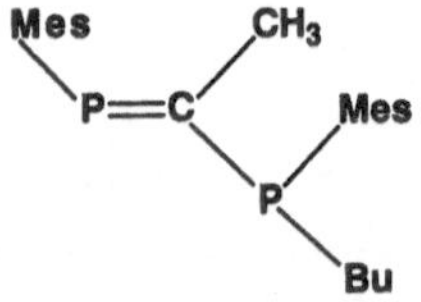

(26)

The Wittig reaction involving phosphonium ylides R$_3$P=CXY has been reviewed, and a great deal of mechanistic material summarized.[160] The Wittig reaction of conjugated ylides has been found to be governed by kinetic rather than equilibrium factors.[161] Methylenephosphenium ions, $>\overset{+}{P}=C<$, have been postulated as transients, but never isolated. The species $(Pr^i_2N)_2\overset{+}{P}=C(SiMe_3)_2$ has been isolated, and its structure determined.[162] The cleavage of phosphinoacetic acids in toluene at 90 °C proceeds by a mechanism shown in Eqs. (43) and (44) involving an ylid that decomposes to $R^1R^2PCH_2R + CO_2$ by a cyclic transition state.[163] The

$$R^1R^2PCH_2COOH \rightarrow R^1R^2\overset{+}{P}(H)CH_2CO_2^- \qquad (43)$$

$$R^1R^2\overset{+}{P}(H)CH_2CO_2^- \rightarrow R^1R^2P(H)=CH_2 \qquad (44)$$

mechanisms of isomerization of P_7^{3-} and related species such as P_6As^{3-} and $As_3S_4^+$ have been considered.[164] Finally, we mention some work on phosphorous–halogen systems. The reaction of $[PRCl_5]^-$ with excess AgCN to form $[PEt(CN)_5]^-$, $[P(C_6F_5)Cl(CN)_4]^-$, and $[P(CCl_3)Cl_2(CN)_3]^-$ has been studied by ^{31}P NMR and several intermediates detected.[165] A reaction mechanism for the formation of PH_3F_2 from P_2H_4 and HF has been suggested[166] and is outlined in Eqs. (45) and (46).

$$P_2H_4 + H^+ \rightarrow PH_2PH_3^+ \xrightarrow{F^-} F\overset{-}{P}H_2\overset{+}{P}H_3 \qquad (45)$$

$$F\overset{-}{P}H_2\overset{+}{P}H_3 \rightarrow [PH_2F] \xrightarrow{HF} PH_3F_2 \qquad (46)$$

Holmes and Holmes have carried out *ab-initio* calculations to establish the relative apicophilicities of H, F, and Cl in trigonalbipyramidal phosphoranes to establish a self-consistent interpretation of nucleophilic substitution at silicon and phosphorus, and the variation in the degree of retention/inversion caused by changes in entering and leaving groups.[167] Reactions of XYPhCl (X = OEt, Y = Ph and XY = OCH$_2$CH$_2$O) with *p*-CH$_3$C$_6$H$_4$C(O)N(CH$_3$)OH can form P(III) intermediates, which can decompose by homolysis of the N—O bond. CIDNIP evidence confirms a radical cage process.[168]

4.7. Arsenic

The reaction of As_4 and O_3 in an argon matrix irradiated with red light produces AsO and As_4O_x, $x = 1$–5. Under these conditions, As_4 is more reactive than P_4.[169] Codeposition of AsH_3 and O_3 in argon matrices gives $AsH_3 \cdot O_3$. Red light photolysis involves the formation of intermediate, possibly HAsO.[170] The reaction of $Ph_3\overset{+}{P}(CN)(As(NPr_2^i)_2)$ and $Cl_2AsNPr_2^i$ involves the formation of intermediate (27), which decomposes to form (28).[171] The spiro arsorane (29) undergoes base hydrolysis by an associative mechanism involving a six-coordinate arsenic species.[172]

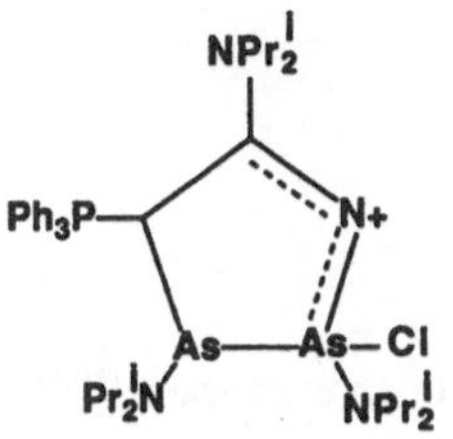

(27) 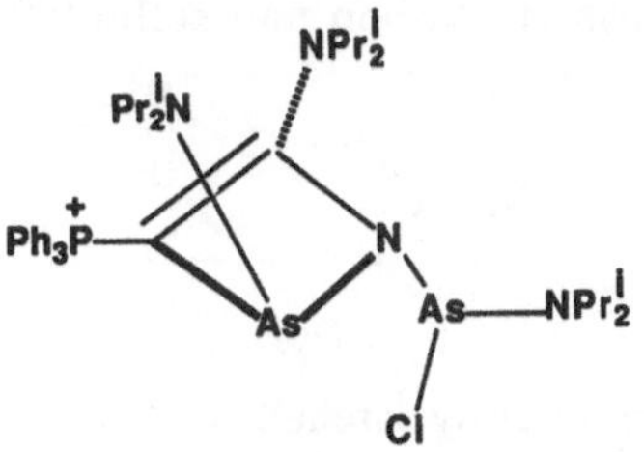   (28) 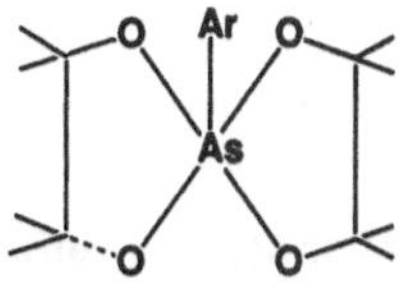   (29)

A nucleophilic addition involving the formation of a —CN—As— bridged complex occurs[173] in the oxidation of As(III) to As(V) by $[Mn(CN)_6]^{4-}$. A similar oxidation in aqueous alkali by $S_2O_8^{2-}$ involves only one active arsenic species, $[H_2AsO_3]^-$. No evidence for the formation of As(IV) was observed, and a two-electron transfer mechanism is favored.[174] Arsenic(IV) is thought to be involved in the pulse radiolysis of As(III), leading to the formation of the species $As(OH)_4^{\cdot}$, $[HOAsO_2]^-$, $[OAs(OH)_3]^-$, and $[AsO_3]^{2-}$. The kinetics of the dehydration interconversion reactions have been studied. As(IV) functions as both a powerful oxidant and a powerful reductant.[175]

4.8. Oxygen

A value for the self-exchange rate constant for O_2/O_2^- has been deduced by Espenson *et al.*[176] The reaction of $O_2(^1\Delta)$ with I^- probably involves the formation of HOOI, followed by a fast reaction with I^- to yield $I_3^- + H_2O_2$.[177] Intermediates in the reaction of $^1O_2$ with Et_2S can be trapped efficiently with trimethylphosphite.[178] The kinetics of the reaction of $(ArO)_3P$ with $(PhO)_3P \cdot O_3$ to form $^1O_2$ and $(ArO)_3PO$ show a spontaneous and an $(ArO)_3P$-catalyzed pathway.[179] A complex mechanism is involved in the Cu(II)-catalyzed decomposition of hydrogen peroxide in the range pH 11 to 12. A superoxide Cu(I) complex is involved.[180]

4.9. Sulfur

The reaction of S_8 with OH^- in aprotic solvents forms $S_3^{\cdot-}$ as the major product, and initial rate measurements follow a simple second-order rate law.[181] A possible scheme is shown in Eqs. (47) and (48). A thermal method has been used to obtain

$$S_8 + OH^- \rightarrow [HOS_7S^-] \rightarrow HOS_4S^{\cdot} + S_3^{\cdot-} \tag{47}$$

$$HOS_4S^{\cdot} + OH^- \rightarrow S_3^{\cdot-} + HOSSOH \rightarrow 0.25S_8 + H_2O_2 \tag{48}$$

kinetic parameters for the reaction between S_8 and $P(OR)_3$ in carbondisulfide solution.[182] The reaction of S_8 in liquid NH_3 has been followed by UV measurements. An intermediate, S_3N^-, a precursor to S_4N^-, is observed; S_7N^- was not observed. A complex mechanism is proposed.[183,184] Exchange of ^{15}N between different sites in (30) can be followed by NMR. Reaction proceeds by simultaneous 1,3-nitrogen shift pathways via carbon and sulfur.[185] The studies of metal ion

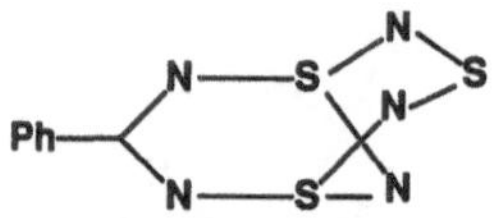

(30)

hydrolyses of sulfur compounds by Satchell and Satchell have been extended[186,187] to $PhRC(SEt)_2/Ag^+$ and $PhSSPh/Hg^{2+}$. In the latter case, the reaction pathway involves the formation of $PhSOH$, $PhSO_2H$, and finally $PhSO_2SPh$. The kinetics of oxidation of R_2S by $NaOCl$, with a phase transfer catalyst, have been examined.[188] The cationic species Me_3S^+ (and also Me_3Se^+, $Me_3Te^+CF_3SO_3^-$) in superacid solution act as electrophilic methylating agents.[189] The reaction of R_2SO and Me_3SiX produces Me_2SiX^+, where X is a halogen, which can act as either a sulfur- or halogen-centered electrophile.[190]

The oxidation of SO_2 has been reviewed.[191] Mass-spectrometric evidence for H_2SO_3 and $H_2SO_3^+$ has been obtained,[192] and there is also evidence for $H_3SO_3^+$ (thought to be a trihydroxy species).[193] The oxidation of bisulfite by HSO_5^- is thought to form $H_2S_2O_8^{2-}$ that converts to $S_2O_7^{2-}$, which is finally hydrolyzed to $2HSO_4^-$.[194] the oxidation of S(IV) by Fe(III)[195] and by V(V),[196] and the photoassisted autoxidation of SO_2 in the presence of colloidal α-Fe_2O_3 have been studied.[197] The reaction of SO_2 with dry $(CH_3)_3NO$ produces $(CH_3)_2(H)NCH_2SO_3$. This species is oxidized by aqueous I_2, and the kinetics have been measured.[198]

Nucleophilic substitution at sulfonylhalides is a topic of continuing interest. Studies have appeared on the kinetics of hydrolysis of $ClO_2SAr-B-ArSO_2Cl$, where B is a bridging group, and on the hydrolysis of substituted thiophene-2-sulfonyl halides.[199,200] A SET mechanism is suggested for the reaction of p-$MeC_6H_4SO_2Cl$ with $R^1R^2N^-Li^+$, and the presence of a species with unpaired electrons has been detected by ESR.[201] The same technique has been used to investigate the reactions of a variety of aliphatic and aromatic sulfonyl radicals $RSO_2^{\cdot}$.[202]

4.10. Selenium and Tellurium

The reactions of Se_2Ph_2 and Te_2Ph_2 with Bi_2Ph_4 produce Ph_2BiEPh (E = Se, Te), which can react with CH_2N_2 to form Ph_2BiCH_2EPh.[203] The kinetics of exchange between RSeH and RSeSeR, and between thiols and disulfides in D_2O have been studied by NMR.[204] Reaction is slow for sulfur, and occurs by an S_N2 mechanism, but is fast for selenium. It is concluded that RSe^- is a better nucleophile, and better leaving group than RS^-. The reaction of $(Me_3Si)_3C^-$ with $(Me_3Si)_3CSeSeC(SiMe_3)_3$ or with the corresponding ditellurides does not lead to the expected substitution products but to a range of products thought to arise from a single electron transfer mechanism (SET) involving formation of $(Me_3Si)_3CSe^·$ and $(Me_3Si)_3CTe^·$, which are trapped by the solvent.[205] Nucleophilic substitution reactions with $2\text{-}NO_2C_6H_4CH_2XCN$ and XCN^-, X = S, Se, Te, in acetonitrile proceed via nucleophilic attack at the CH_2 group. The order of nucleophilic reactivity is NCTe $\gg$ NCSe > NCS while the order of effectiveness as leaving group is $NCTe^- \geq NCSe^- \geq NCS^-$. Data on the 3-$NO_2$ isomer is also reported.[206] A mechanism is proposed for the reaction of $SeCl_2/SeOCl_2$ and *p*-tosylhydrazones of aldehydes and ketones.[207] Solubility considerations are important in determining the products of the reaction between Se_2Cl_2, SCl_2, and KI.[208] An elegant study of the solvolysis of $ArTeCl_3$ in $MeOH/C_6H_6$ or $MeOH/dioxane$ solvents is interpreted[209] by the mechanism in Eqs (49)–(51).

$$ArTeCl_3 + H^+ \rightleftharpoons ArTeCl_2(ClH)^+ \qquad (49)$$

$$ArTeCl_2(ClH)^+ \rightarrow ArTeCl_2^+ + HCl \qquad (50)$$

$$ArTeCl_2^+ + MeOH \rightarrow products \qquad (51)$$

The reaction of $PhTeCl_3$ with $[S_2CNEt_2]^-$, $[S_2P(OEt)_2]^-$, and $[S_2COEt]^-$ has been studied by variable-temperature NMR. A variety of mixed ligand species are observed, and the solvent dependence of the intramolecular exchange processes observed.[210] The same group has used ^{125}Te NMR to study the intramolecular exchange of monodentate and bidentate $S_2P(OEt)_2^-$ ligands in $Te[OTe(C_8H_8)\text{-}(S_2P(OEt)_2)]_6$, a species containing both Te(IV) and Te(VI).[211] They have also investigated reaction (52), where dtc = dithiocarbamate and tds = thiouramidisulfide. Intermolecular dtc exchange occurred more rapidly with Te(II)

$$Te(dtc)_4 \rightleftharpoons Te(dtc)_2 + tds \qquad (52)$$

than with Te(IV).[212] The reactions of tellurium fluorides have been studied by Janzen *et al.* The *mer*-isomer of $[TePh_3F_2(OH)]$ with *cis* fluorine atoms undergoes slow intermolecular fluorine exchange by the mechanism in Scheme 2. Addition

Scheme 2

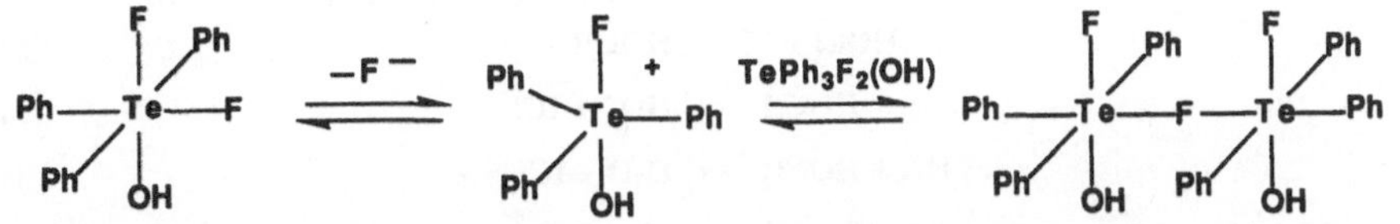

of excess fluoride ion suppresses the exchange by reversing the formation of the five-coordinate intermediate. The isomers *cis-* and *trans-* F_2TePh_3Cl can be synthesized stereoselectively via five-coordinate cations such as Ph_3TeFCl^+ and $Ph_3TeF_2^+$ as the three phenyl groups adopt a *mer-*arrangement. The *trans* compound isomerizes to the more stable *cis* form, and the *cis* isomer undergoes stereoselective fluorine exchange by mechanisms involving tellurium–fluorine bond cleavage to form five-coordinate species, which can then bridge with the fluorine of a six-coordinate complex.[213,214]

4.11. Halogens, Krypton, and Xenon

4.11.1. Flourine

A review has appeared on the use of $F_2(g)$ as a reagent for selective fluorination.[215] Acetylhypofluorite and $F_2(g)$ are used as electrophilic fluorinating agents, the former being the more selective.[216] Gillespie has published H_0 values for solutions of SbF_5 in anhydrous HF, and also data for solutions of H_2O, KF, HSO_3F, PF_5, NbF_6, and AsF_5 in HF.[217] The kinetics of oxidation of $[Ru(bipy)_3]^{2+}$ by SO_4F^- have been reported. Carbon dioxide is one of the products, so some oxidative degradation of the ligands must occur.[218]

4.11.2. Chlorine

Solutions of HCl or Cl_2 in $HNO_3 + H_2SO_4$ can both chlorinate and nitrate deactivated aromatic compounds. A new species, $NO_2Cl_2^+$, may be a possible electrophilic chlorinating agent.[219] The N-chlorination of Me_2NH and Et_2NH and the kinetics of chlorine transfer from NH_2Cl to CH_3NH_2 has been described.[220,221] The N-halogenoamines are very selective chlorinating and brominating agents.[222] The chlorination of electron-rich aromatics is suggested to involve an arenium ion mechanism, with transfer of Cl^+ from R_2NHCl^+ to the aromatic substrate.[223,224] Such electron-rich aromatics are very susceptible to one-electron oxidations, and chlorination can also proceed by an electron transfer chain reaction. A number of papers have appeared on the kinetics of oxidation reactions of N-halogeno compounds and substrates include SCN^-,[225,266] $NH_2NHCSNH_2$,[227] and NH_3OH^+.[228] Two related studies have been mentioned[125,127] in the section on nitrogen. A pulsed accelerated flow study of the oxidation of I^- by HOCl has been reported by Margerum, with the mechanism involving reactions (53)–(56).[229] The reaction of NCl_3 with I^- exhibits saturation

$$HOCl + I^- \rightleftharpoons HOClI^- \tag{53}$$

$$HOClI^- \rightarrow HO^- + ICl \tag{54}$$

$$HA + HOClI^- \rightarrow H_2O + ICl + A^- \tag{55}$$

$$ICl + 2I^- \rightarrow I_3^- + Cl^- \tag{56}$$

kinetics due to formation of NCl_3I^-, which undergoes decomposition to $NHCl_2 +$ $ICl + OH^-$. The oxidation of SO_3^{2-} by HOCl is another rapid process studied by the pulsed accelerated flow method. Reaction involves formation of $HOClSO_3^{2-}$, which decomposes to $ClSO_3^-$ and is finally hydrolyzed to $Cl^- + SO_4^{2-}$.[230] The ClO_2^-/I^- reaction has been investigated under acidic conditions, and rate law (57) observed.[231] The term that is second order in $[H^+]$ may be helpful in understanding

$$0.5d[I_2 + I_3^-]/dt = (k_1[H^+] + k_2[H^+]^2)[Cl(III)][I^-]/([H^+] + K_a) \qquad (57)$$

the chlorite–iodide oscillator. An elegant study of the oxidation of Cl^- to Cl_2 by Ce^{4+}, catalyzed by RuO_2, has appeared. The results can be interpreted in terms of an electrochemical mechanistic model, regarding the catalyst particles as microelectrodes.[232] ESR studies of $Cl^{\cdot}$ generated in aqueous solution by reaction of Cl^- with $SO_4^{\cdot-}$ and $H_2PO_4^{\cdot}$ have been reported.[233] The kinetics of the nanosecond laser flash-photolysis of ClO_2^- in Cl_3CF involves the formation of $ClO^{\cdot}$ and Cl_2O_3 as intermediates.[234]

4.11.3. Bromine

Once again a good deal of bromine chemistry is covered in the section on oscillatory reactions. The species Br_5^- may be involved in olefin brominations at high concentrations of bromine.[235] The kinetics and equilibria of reaction (58) have been studied, and rate constants for (59) and (60) evaluated.[236] The mechanism of oxidtion of ferrocyanide by bromate involves $[Fe(CN)_5H_2O]^{3-}$ as an intermediate.[237] The uncatalyzed and the vanadium(V) catalyzed reaction of bromate with methylene blue have been reported.[238]

$$5HOBr \rightleftharpoons 2Br_2 + BrO_3^- + H^+ + 2H_2O \qquad (58)$$

$$2HOBr \rightarrow Br^- + HBrO_2 + H^+ \qquad (59)$$

$$Br^- + BrO_3^- + 2H^+ \rightarrow HOBr + HBrO_2 \qquad (60)$$

4.11.4. Iodine

Electrophilic iodination by $ICN/AlCl_3$ has been investigated; $ICNAlCl_3$ is suggested as the reagent.[239] The pulsed accelerated flow technique has been used to study the hydrolysis of iodine monochloride in neutral and basic solution. It is proposed that iodine monochloride in aqueous solution is a mixture of ICl and a hydrated species H_2OICl (18% and 82%). The equilibrium constant $[IClaq][Cl^-]/[ICl_2^-]$ was determined as $0.13\ M$ and $[HOI][Cl^-]^2[H^+]/[ICl_2^-]$ as $1.06 \times 10^{-6}\ M^3$. Of the two species ICl is the more reactive, and can react at the encounter rate. The addition of Cl^- suppresses the rate of hydrolysis due to the formation of ICl_2^-, while OH^- accelerates hydrolysis. Hydrolysis catalyzed by buffer, and species such as $[(HO)_2P(O)O{-}I{-}Cl]^-$ may be involved.[240] The kinetics of oxidation of SCN^- by ICl and by I_2 in aqueous acid is reported to involve the formation of a complex between HNCS and ICl leading to the formation of INCS and ISCN.[241] A much more complete study shows that there are diffrent stoichiometries above and below pH 4, and two mechanisms involving Eqs. (61), (62) and (63), (64) are postulated. The conversion to H_2SO_4, the final sulfur product, proceeds via HO_2SCN.[242]

$$I_2 + SCN^- \rightleftharpoons I_2SCN^- \rightleftharpoons ISCN + I^- \tag{61}$$

$$ISCN + H_2O \rightarrow HOSCN + H^+ + I^- \tag{62}$$

$$I_2 + H_2O \rightleftharpoons HOI + H^+ + I^- \tag{63}$$

$$HOI + SCN^- \rightarrow HOSCN + I^- \tag{64}$$

A kinetic study of the reaction of HOI and 4-[2-(methylamino)propyl]phenol has been carried out as part of a study on an oscillating system.[243]

4.11.5. Krypton and Xenon

The krypton cation salt $[HC{\equiv}N{-}Kr{-}F]^+[AsF_6]^-$ in BrF_5 is stable over several hours around -55 to $-58\,°C$, but on warming decomposes to NF_4^+, CF_4, CF_3H, and Kr.[244] The reaction of $B(C_6F_5)_3$ with XeF_2 in dichloromethane gives nucleophilic displacement of fluorine, with the formation of $[C_6F_5Xe]^+[C_6F_5BF_3]^-$. The $[C_6F_5Xe]^+$ can act as an electrophile, transferring the $[C_6F_5]^+$ group. In acetonitrile decomposition occurs by a radical reaction with the solvent.[245] The reaction of XeF_2 with sulfur(II) compounds such as Me_2S produces an electrophilic fluorinating agent.[246] A stopped-flow study of the oxidation of Pu(III) to Pu(IV) by XeF_2, and by SO_4F^- is interpreted in terms of one-electron reaction steps involving formation of XeF^- and $SO_4^{\pm}$.[247] One of the most interesting compounds of xenon is H_4XeO_6. Stopped-flow measurements have given reliable zero-time absorbances from which pK_a values of 4.29 and 10.81 for $[H_3XeO_6]^-$ and $[H_2XeO_6]^{2-}$, respectively, have been deduced. However, the kinetics of the subsequent decomposition reactions were complex and irreproducible.[248]

4.12. Oscillating Reactions

In this section the whole field of exotic dynamics is considered; this term includes not merely oscillating reactions but also oligo-oscillatory reactions, multiple steady states, spatial phenomena such as travelling reaction waves, and chaotic systems. All of these have common roots in autocatalytic processes. This area has continued to expand, and there is a case for treatment in future volumes by a specialist reviewer. An entry into the literature can be gained from a recent series of articles in a chemical education journal,[249] and in a festschrift issue in honor of Professor R. M. Noyes.[250] Other useful sources are a volume of conference proceedings,[251] and a volume of lecture preprints of a 1989 conference.[252] The present summary is concerned with the chemical rather than the mathematical aspects of the topic.

Chemical systems often show oscillatory behavior only over a narrow range of conditions, and one reason for the popularity of the classic Belousev–Zhabotinskii reaction (BZ) and related systems is that they show exotic behaviour over a wide range of conditions. Evidence has been obtained[253] for an additional negative feedback loop in the BZ reaction involving the malonyl radical, and ^1H NMR has been used to study the organic component in a BZ reaction using methylmalonic acid.[254] Shifts in the H_2O peak have been related to changes in BrO_3^- and H^+ concentrations. Mass spectrometry has been used to monitor CO_2 evolution and O_2 consumption.[255] The effect of electric and magnetic fields has been studied.[256] Single or periodic pulse perturbation effects on the BZ reaction,[257] and the influence of delay time in a BZ system in a CSTR when the concentration of the input is controlled by the concentration of a reaction species have been studied.[258] The BZ reaction has been studied using $CH_3COCH_2COCH_3$ as substrate in a pair of CSTR coupled by mass flow between them.[259] Ferroin and Ce^{4+} are often used together in BZ systems. A study of the kinetics of reaction between Ce^{4+} and ferroin shows that $[Fe(II)(bipy)_3^{2+}]$ decreases to a minimum, then increases to a maximum before finally decreasing. A mechanism involving an iron(IV) species is proposed.[260] The effect of added Ag^+ has been a topic of controversy. Further work has appeared,[261] and the dispute has been resolved.[262] The problem concerned the value to be assigned $(10^9-10^4\ M^{-1}\ s^{-1})$ to the rate constant for the formation of AgBr(s) from Ag^+ and Br^-. The actual formation of small oligomers of silver bromide is very rapid, and in such a system bromide can still react with $HBrO_2$ in the BZ reaction. The actual precipitation of AgBr(s) is a much slower process. The effect of O_2,[263] of added salts and organic solvents,[264] and added micelles have been examined.[265] Temporal oscillation in a frozen BZ system containing Mn^{2+} at -10 to $-15\ °C$ continues, as indicated by the EPR signals due to manganese(II).[266] Quenching effects of added Ce^{4+}, $HBrO_2$, $HBrO$, Br_2, and Br^- have been studied.[267] In a system containing ascorbic acid, acetone, and cyclohexylamine two types of oscillation are observed, separated by a time pause. The first regime is attributed to ascorbic acid/acetone as the organic substrate, the pause to oxidation of ascorbic acid to oxalic acid, and the second regime to oxalic acid/cyclohexanone as the organic substrate.[268] A review article on bromate oscillators has been published,[269] and oscillations in the $PhNH_2/KBrO_3/H_2SO_4$ system,[270] the gallic acid/$KBrO_3$ system,[271] and the $BrO_3^-/SO_3^{2-}/[Fe(CN)_6]^{4-}$ system have been investigated.[272] In the latter case a nine-step mechanism accounts for the sustained pH oscillation observed in a continuously stirred tank reactor, CSTR. The system is usefully viewed as related to the $IO_3^-/SO_3^{2-}/[Fe(CN)_6]^{4-}$ oscillator. The effect of external perturbations, a periodic variation in the value of $[BrO_3^-]$ in the supply to a CSTR for the $Mn^{2+}/BrO_3^-/$citric acid system, has been studied.[273] Under some conditions chaotic oscillations are observed. For the $Mn^{2+}/BrO_3^-/H_2PO_2^-$ system oscillations can be observed if the gaseous products are driven off by a stream of N_2 for certain ranges of gas flow rates. In the absence of gaseous purging no oscillations occur.[274] Oscillations in the $BrO_3^-/Br^-/Ce^{3+}$ system in a CSTR and the $BrO_3^-/Ce^{4+}/(COOH)_2$ systems under batch conditions have been simulated.[275,276]

The system used in the Landolt "clock reaction," IO_3^-/SO_3^{2-}, when treated with $[Fe(CN)_6]^{4-}$, shows oscillations in a CSTR.[277] A modification of the mechanism proposed previously resolves a number of problems. A slightly different system, $IO_3^-/HSO_3^-/S_2O_3^{2-}$ produces a small number of high-amplitude pH oscillations (up to 2 units of pH) over a narrow range of conditions in a closed system.[278]

The oxidation of SCN^- by IO_3^- in acid conditions gives oligo-oscillatory behavior (concentration vs. time curves with several maxima/minima).[279] The oscillatory behavior of another iodate reaction, the Bray–Liebhasky reaction, $H_2O_2/IO_3^-/H^+$, and the CSTR behavior of the $IO_3^-/H_3AsO_3/H^+$ reaction have been studied.[280,281] Bistability has been observed. The $IO_4^-/S_2O_3^{2-}$ system in buffered solution shows a variable stoichiometry. In excess periodate the reaction products are $IO_3^- + SO_4^{2-}$, while in excess thiosulfate $I^- + S_4O_6^{2-}$ are formed. In a CSTR experiment, conditions for pH and redox potential oscillations can be found.[282] The IO_4^- oxidation of NH_3OH^+ in a CSTR can also show sustained oscillations, although over a narrow range of conditions.[283]

Hydrogen peroxide is a common reagent in these studies. Light-sensitive oscillations in the H_2O_2 oxidation of $[Fe(CN)_6]^{4-}$ have been observed,[284] and in a systematically designed pH oscillator, the $H_2O_2/SO_3^{2-}/[Fe(CN)_6]^{4-}$ system, variations of up to 3 units of pH have been reported.[285] Both of these reactions were studied in a CSTR. Phenylmalonic acid, PhMA, can serve as substrate in a Briggs–Rauscher-type oscillator $H_2O_2/IO_3^-/PhMA/Mn^{2+}/H^+$.[286] Oscillating systems involving permanganate have received more attention recently, with phosphate species being added to stabilize Mn(III) and Mn(IV) intermediate oxidation states. In the $MnO_4^-/H_2O_2/H_3PO_4$ system oscillations in a CSTR have been reported; three types of bistability and tristability have been observed.[287] The phosphate stabilizes colloidal MnO_2. Orbán and Epstein have designed a minimal permangante oscillation in a CSTR, using the Guyard reaction (65). Nucleation and surface effects are important.[288] Oscillations and bistability occur in

$$2MnO_4^- + 3Mn^{2+} + 2H_2O \rightarrow 5MnO_2 + 4H^+ \qquad (65)$$

$MnO_4^-/NO_2^-/HCOOH/MeOH$.[289] A useful background study is the oxidation of Mn(II) to Mn(III) by MnO_4^- in the presence of $C_2O_4^{2-}$ or $P_2O_7^{4-}$. A binuclear Mn(II)speces is involved, bridged with either oxalate or phosphate.[290] The $MnO_4^-/N_2H_5^+$ reaction may be worth further study. Reaction is autocatalytic, and there is evidence[291] for a rate-law term involving $[MnO_4^-]^2$. Copper is another element that has been found to be involved in oscillations. Oscillations in chemiluminescence occur[292] in the CSTR reaction of $luminol/H_2O_2/SCN^-/Cu^{2+}$, and oscillations in redox potential, pH, and $[O_2]$ can be observed[293] in the Cu(II)-catalyzed reaction of $S_2O_3^{2-}$ and $S_2O_8^{2-}$. Copper(II) also catalyzes the H_2O_2/SCN^- reaction, and in alkaline conditions oscillation can be observed under batch conditions.[294]

An interesting system is the methylene blue catalyzed oxidation of HS^- by O_2 in a CSTR. Oscillation occurs over a narrow range of flow rates, and a mechanism has been proposed.[295,296] Oscillations have also been found in the cobalt/bromide catalyzed oxidation of cyclohexanone by O_2,[297] in the electrochemical oxidation of HCHO and $HCOOH/HCOO^-$,[298] and in the electrodissolution of copper in acidic chloride solution,[299] and the anodic dissolution of nickel in aqueous sulfuric acid.[300] Damped oscillations are predicted for reactions in stirred diaphragm cells.[301]

A rather difficult area of oscillating reaction chemistry is that encompassed under the heading "chaos." Irreproducibility in the ClO_2^-/I^- reaction observed by

Epstein has been attributed to fluctuations in solution after they have been well mixed.[302] Sagués and Sancho[303] have produced a theoretical approach to the problem. Field and Györgi suggest that the deterministic behavior (chaos) apparent in CSTR studies of the BZ reaction may be due to coupling of reactions in different areas of the CSTR with slightly different concentrations, due to imperfect mixing.[304] Such a system would behave as independent coupled oscillators. A hydrodynamic oscillating system has been used to study the coupling between a pair of chemical oscillators.[305] A method for organizing data and qualitative modeling of complex dynamic systems has been proposed and applied to the $ClO_2^-/S_2O_3^{2-}$ system.[306] Oscillation and chaotic behavior are observed in a model of the peroxidase-catalyzed oxidation of NADH.[307]

We finally turn to the traveling waves or traveling reaction fronts that are sometimes observed in systems involving autocatalytic reactions. The most striking advance is in a paper by Nagypál et al., where a general method for the design of propagating reaction fronts has been described. Attention is focused on acid-catalyzed reactions of oxy-anions where protons are consumed by the oxidant and released by the reductant. By considering the ratio of protons released to "electrons consumed" by the oxidant (and vice versa for the reductant), Nagypál has found twenty new systems in which a traveling reaction front can be observed. Reaction is initiated by acidification at one spot in an initially homogeneous reaction mixture, and the movement of the reaction front, involving changes in pH, is followed visually by use of a suitable acid–base indicator.[308] Another significant paper concerns traveling waves in the reaction of ferroin with acidic solutions of nitrate, studied by Póta et al.[309] These authors have developed an analytical treatment of the traveling wave problem, and have produced a formula for the thickness of the wave front. This looks to be a valuable development. A study of the reverse reaction between nitrous acid and ferroin has been reported.[310] A video camera has been used to study the buildup of trigger waves in a BZ system, and concentration profiles have been measured for the Ce^{4+}-catalyzed BZ reaction over a wide range of conditions.[311,312] Trigger waves in the BrO_3^-/Ce^{4+}/malonic acid reaction have been compared[313] with those in the system catalyzed by ferroin and $[Ru(bipy)_3]^{2+}$. Spatial patterns have been observed for a BZ reaction where the ferroin catalyst is immobilized on the surface of cation exchange resin beads.[314] Propagating waves across the surface of the beads can be observed. Winfree has reported studies of vortexes of chemical activity in BZ reactions,[315] and of the long-term motion of the remarkable scrolling of activity in BZ systems.[316] Calculations have been made of moving reaction fronts in the acetylcholinesterase reaction (the hydrolysis of acetylcholine to choline and acetic acid).[317] The theory of Liesegang patterns formed by diffusion control processes has been discussed.[318]

Chapter 5

Substitution Reactions of Inert-Metal Complexes— Coordination Numbers 4 and 5

5.1. Introduction

Among reviews of general interest are two very useful accounts by van Eldik and co-workers on the application of high-pressure methods to kinetics and mechanisms.[1,2] Reference 1 contains a valuable compilation of activation and reaction volumes of inorganic reactions, and considers their interpretation for various reaction types. Reference 2 dwells more on the experimental details and design of equipment, but also contains illustrations from reactions of square-planar compounds. A review by Blandamer and Burgess on a thermochemical approach to solvation of transition metal complexes is also of relevance to this section,[3] and a report by Burgess and Pelizzetti on medium effects points out the importance of solvation in reactions of square-planar compounds, emphasizing the role of leaving-group solvation.[4] Second-sphere coordination of cyclodextrins to square-planar complexes has been surveyed by Stoddart and co-workers.[5] The potential importance of such interactions is illustrated by a report that the complex ion $[Pt(bipy)(NH_3)_2]^{2+}$ can form a bipy–crown ether interaction which then protects it from some photochemically induced reactions.[6]

The many works related to platinum-containing anticancer drugs include a review of the binding of *Cisplatin* to DNA.[7] Also of importance in this field are two communications[8,9] concerning the known preference of Pt(II) to bond to N-7 of guanine and its derivatives. The affinity of platinum for particular sites is enhanced by the presence of an adjacent guanine. The active species of *Cisplatin* are *cis*-$[PtCl_2(NH_3)_2]$ itself and the mono-aquo cation with which it equilibrates, but not the bis-aquo derivative.[9] Both of these bind to GMP to form a chloro-GMP intermediate, which reacts further with more substrate or GMP. Equations (1)–(5)

$$PtCl_2 + H_2O \rightleftharpoons PtCl(OH_2)^+ + Cl^- \tag{1}$$

$$PtCl_2 + GMP \rightarrow PtCl(GMP)^+ + Cl^- \tag{2}$$

$$PtCl(OH_2)^+ + GMP \rightarrow PtCl(GMP)^+ \tag{3}$$

$$PtCl(GMP)^+ + PtCl_2 \rightarrow Pt_2Cl_2(GMP)^{2+} + Cl^- \tag{4}$$

$$PtCl(GMP)^+ + GMP \rightarrow Pt(GMP)_2^{2+} + Cl^- \tag{5}$$

summarize this. ^{14}N NMR spectroscopy has been successfully employed to detect NH$_3$ release from *Cisplatin* with biological molecules,[10] and ^{1}H, ^{13}C, ^{15}N, and ^{195}Pt NMR observations have been used to examine the reaction of *cis*-[Pt(OH$_2$)$_2$(NH$_3$)$_2$]$^{2+}$ with amino acids.[11]

Several square-planar molecules of d^4 Cr(II) have been reported, along with a survey of all the known compounds of this type.[12] UV/visible spectroscopy of cobalt(II) halides in 4-methylpyridine at various temperatures and pressures reveals equilibria between tetrahedral and octahedral structures, the latter favored by high pressure and low temperature.[13] Most 4-coordinate cobalt(I) complexes are tetrahedral, however, and [Co(PMe$_3$)$_4$]BPh$_4$ is isostructural with its nickel analog.[14] Oxidative addition reactions of tetrahedral [CoBr(PMe$_3$)$_3$] have been described.[15] The rhodium(I) complex [Rh(PMe$_2$Ph)$_4$]BF$_4$ is also distorted toward tetrahedral geometry, with transoid P—Rh—P angles of about 150°, though it remains diamagnetic.[16] The origin of the distortion is the need to interleave the twelve organic substituents. Meanwhile, solid-state NMR measurements (^{13}C and ^{31}P) on a variety of bisphosphine complexes of palladium and platinum have revealed that most of them are distorted somewhat from square planes in this phase.[17] A report of an X-ray crystal structure determination of *cis*-[PtCl$_2$(PCy$_3$)$_2$] served as a vehicle to survey P—Pt—P bond angles in a number of platinum complexes of this sterically demanding ligand. The ligand substituents tend to enmesh to reduce the effective cone angle subtended at the metal atom.[18]

Another area of structural concern, where the results are often of direct relevance to mechanistic studies, is that of ambident ligands. The mode of attachment (and controlling influences) of thiocyanate and dmso in particular have often been touched upon in this section. Structural analysis of a series of complexes [Pd(CNS)$_2$(biL)] (biL are various diphosphines) show that the generally accepted premise that increasing steric hindrance in *cis*-bisthiocyanato complexes tends to exclusively promote *N*-bonding is wrong.[19] *N,S* bonding predominates in such complexes, suggesting a conjunctive response in which this configuration minimizes interaction with each other and with the other ligands. A complicating feature is that the solid-state structures may not retain their integrity in solution. Indeed the crystallizing solvent can have an effect on the product isolated.[20] The dmso in [MCl(dmso)(biL)]$^+$ (M is Pd or Pt; biL is bipy, phen, or en) is *O*-bonded in the palladium bipy and phen compounds, and *S*-bonded in the others, in the solid state (X-ray and IR measurements).[21] Proton NMR investigations in CD$_3$NO$_2$ solution reveal that the *O*-bonded ligands equilibrate to a mixture of *O*- and

S-bonded isomers. Interconversion between the two isomers is slow on the NMR timescale in the absence of free dmso, a finding which casts doubt on some previous suggestions that dmso might enter coordination as an oxygen-bonded ligand and then isomerize.

5.2. Associative Ligand Exchange at Square-Planar Platinum(II)

The cation $[Pt(dmso)_4]^{2+}$ contains two O-bonded and two S-bonded ligands, an arrangement which is retained in CD_3NO_2 solution. Variable-temperature and variable-pressure ^1H NMR spectroscopy reveals that, as expected from *trans*-effect values, the O-bonded ligands undergo fast exchange with free dmso at a lower temperature than do the S-bonded groups.[22] The independent processes are both first order in substrate and dmso. Activation parameters for the exchange of one O-bonded ligand are $\Delta H^+ = 32.8 \pm 0.2\,kJ\,mol^{-1}$, $\Delta S^+ = -62.0 \pm 0.7\,J\,K^{-1}\,mol^{-1}$, and $\Delta V^+ = -2.5 \pm 0.3\,cm^3\,mol^{-1}$ (at 264.4 K), and the respective values for the exchange of the S-bonded group are $47 \pm 4\,kJ\,mol^{-1}$, $-74 \pm 10\,J\,K^{-1}\,mol^{-1}$, and $-5 \pm 3\,cm^3\,mol^{-1}$ (at 360 K). These data, in common with the data from the few other solvent exchange reactions examined for Pt(II) and Pd(II), clearly support an associative activation process, but thus far there are too few values of ΔV^+ to differentiate between an I_a or limiting A mechanism for any of them.

Water exchange with $[Pt(OH_2)_4]^{2+}$ and *trans*-$[PtCl_2(OH_2)_2]$ has been studied using ^{18}O tracers and a novel quenching technique which employs Cl_2 oxidation to produce substitution-inert Pt(IV) complexes.[23] The ^{195}Pt chemical shifts of the products are sensitive to the oxygen isotopes, allowing analysis of the products. The water exchange at the chloro complex is 30 times slower than that of the tetra-aquo complex, an effect similar to that previously reported for chloride anation reactions of these two substrates. Variable-temperature measurements on the $[Pt(OH_2)_4]^{2+}$ system gave ΔH^+ as $92 \pm 2\,kJ\,mol^{-1}$ and ΔS^+ as $0 \pm 5\,J\,K^{-1}\,mol^{-1}$.

More comparisons of ligand nucleophilicity have been reported. (See also Section 5.8.) The rates of chloride displacement from *trans*-$[PtCl_2(py)_2]$, $K[PtCl_3(dmso)]$, and $[PtCl(dien)]ClO_4$ by substituted thioureas in methanol at 30 °C revealed that the electron-donating properties of the incoming nitrogen donor had no significant effect on the ligand nucleophilicity.[24] There was a marked steric effect, however, the anionic complex being the most sensitive to incoming ligand bulkiness. Rates of chloride displacement from $[PtCl(triL)]^-$ (triL = pyridine-2,6-dicarboxylate) by H_2O, OH^-, Br^-, I^-, SCN^-, $SeCN^-$, CN^-, and SO_3^{2-} in water at 25 °C have been determined, and deviations from a linear plot of $\log k_2^0$ vs. n_{Pt}^0 of points relating to neutral and dianionic nucleophiles discussed.[25] The reactions follow the usual two-term rate law. The neutral ligand, water, was more reactive than expected from comparisons with its reactions with neutral substrates, and the dianionic nucleophile, SO_3^{2-}, was less reactive.

The kinetic behavior of the anticancer drug *Carboplatin*, $[Pt(NH_3)_2(biL)]$ (biL = 1,1-cyclobutanedicarboxylate), has been examined.[26] In the absence of acid and other nucleophiles, the complex is quite inert. With addition of chloride, the dicarboxylate ligand is displaced according to a first-order dependence on [Cl⁻]. After initial displacement of one carboxylate, the dicarboxylate ring readily closes again, but this reverse step can be prevented in the presence of acid, and the process then resembles the successive displacement of two monodentate carboxylates.

A study of some phosphineplatinum complexes by ^{31}P 2D exchange NMR spectroscopy has shown it to be a useful technique in detecting multiple exchange processes and in elucidating reaction mechanisms.[27] Where isotopomers are possible (e.g., ^{195}Pt–^{31}P, ^{194}Pt–^{31}P or ^{117,119}Sn–^{31}P, ^{118}Sn–^{31}P) the method differentiated between intermolecular and intramolecular processes. For example, the 2D exchange ^{31}P NMR spectrum of complex (1) revealed exchange between P^1 and

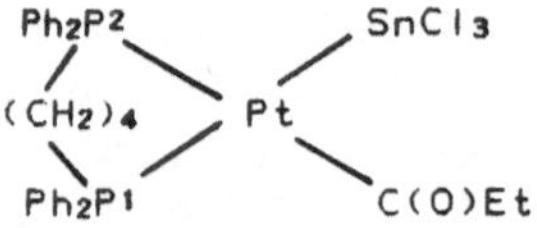

(1)

P^2, but no exchange between ^{195}Pt–P and ^{194}Pt–P, eliminating phosphine exchange as the mechanism. Intermolecular transfer of SnCl$_2$ was implicated in the process, and Scheme 1, involving a five-coordinate SnCl$_2$ adduct of a type reported in Volume 6 of this series, accounted for all of the observations. The five-coordinate species was required to be fluxional, presumably by the Berry or turnstile route.

Scheme 1

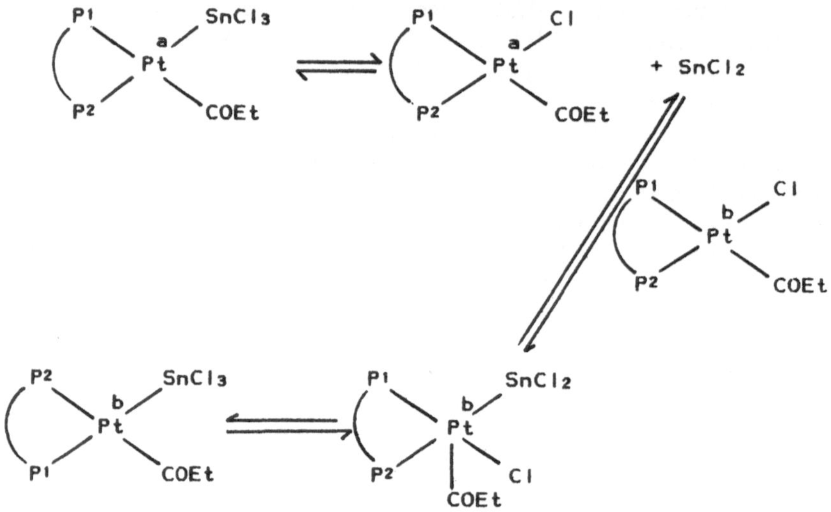

Dynamic NMR spectroscopic methods have been used to study more inversions of coordinated sulfur. The rate of S-inversion of compound (2) varies little with the nature of the solvent, in keeping with a process involving no Pt—S bond breaking.[28] Two separate sulfide inversions were detected in compound (3). The

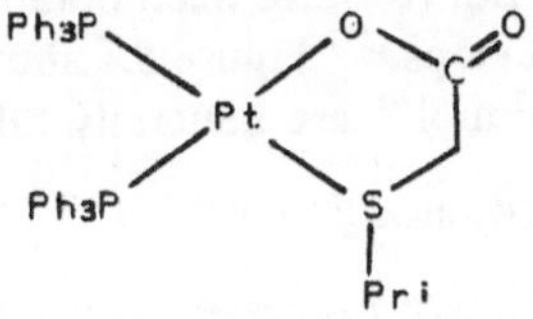

$$(2) \qquad\qquad (3)$$

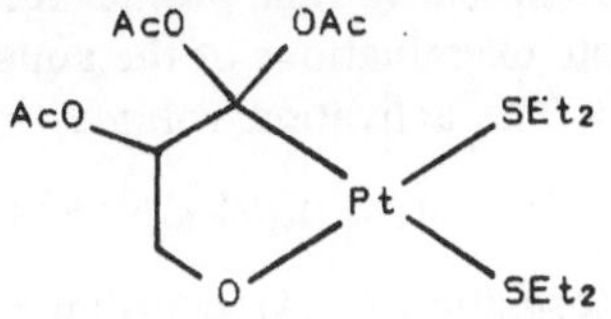

sulfur *trans* to oxygen inverts by an intramolecular mechanism, but that *trans* to carbon reveals no Pt—H or Pt—C coupling, and appears to invert by intermolecular ligand exchange.[29] Addition of free Et_2S accelerates this latter process. The crystal structures of the *meso* and *dl* isomers of $[PdBr_2(PhTe(CH_2)_3TePh)]$ have been examined, though examples of Te inversion remain rare.[30]

5.3. Associative Ligand Exchange at Square-Planar Palladium(II)

Kinetic studies on palladium complexes of dien, and of its more bulky N-substituted forms, continue to furnish valuable mechanistic information. Iodide anation reactions of $[Pd(OH_2)(R_5\text{-dien})]^{2+}$ (R = H, Me, or Et) at pH values between 7 and 12 throw light on previously reported deviations in kinetic data above pH 9.[31] The value of k_{obs} increases with increasing $[I^-]$, and decreases with increasing steric bulk of R_5-dien and increasing pH. For pH values up to 9, plots of k_{obs} against $[I^-]$ are linear and pass through the origin, and can be accounted for by the reactions of Scheme 2 where $[Pd(OH)(R_5\text{-dien})]^+$ is kinetically inert.

Scheme 2

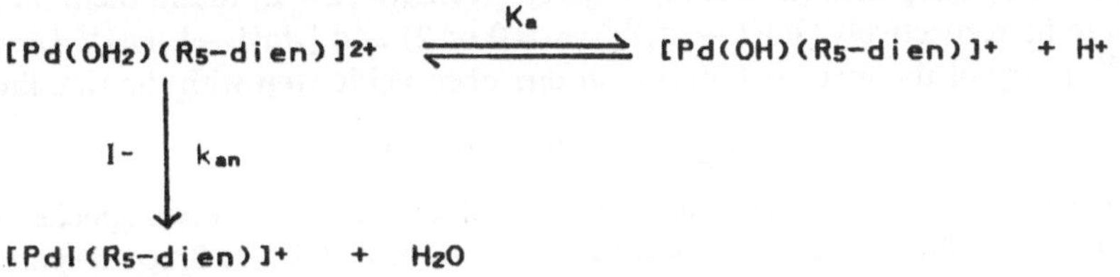

$$[Pd(OH_2)(R_5\text{-dien})]^{2+} \underset{}{\overset{K_a}{\rightleftharpoons}} [Pd(OH)(R_5\text{-dien})]^+ \;+\; H^+$$

$$I^- \downarrow k_{an}$$

$$[PdI(R_5\text{-dien})]^+ \;+\; H_2O$$

Above pH 9, the concentration of the aquo ion is very small and the reaction consequently very slow, but significant deviations in the kinetics are observed, resulting in a positive intercept. This enhanced lability is suggestive of conjugate base formation by deprotonation of dien, but since it is also observed in the

Me$_5$-dien and Et$_5$-dien complexes this explanation cannot hold. The explanation presented is that, under these conditions, the reverse aquation of the product [PdI(R$_5$-dien)]$^+$ becomes important, significantly boosting the small concentrations of the aquo ion available for the anation reaction.

Complete volume profiles for the reactions of Eq. (6) have been obtained by separate examinations of the aquation and anation steps.[32] Figure 5.1 shows the results. The activation volumes of -10 to -12 cm^3 mol^{-1} are generally taken to

$$[PdX(R_5\text{-dien})]^{(2-n)+} + H_2O \rightleftharpoons [Pd(OH_2)(R_5\text{-dien})]^{2+} + X^{n-} \qquad (6)$$

infer formation of five-coordinate species by associative processes, and many of these reactions conform to this interpretation (although a contribution to ΔV^+ from a geometry change of a five-coordinate intermediate may be involved). Values for the Et$_5$-dien iodide and oxalate systems are somewhat exceptional, indicating a rather unique situation for these large ionic leaving groups at the most sterically hindered complex.

Activation volumes for water exchange at [Pd(OH$_2$)(R$_5$-dien)]$^{2+}$ (R = Me, Et) have been obtained, along with those for substitution of water by thiourea and its methyl-substituted forms.[33] The negative values of ΔV^+ are again consistent with associative activation, and the degree of steric congestion does not affect this. Finally, solvolysis of [Pd(py)(Me$_5$-dien)]$^{2+}$ has been followed as a function of temperature and pressure in the solvents H$_2$O, MeOH, EtOH, dmso, dmf, and MeCN.[34] No charge changes were involved, so it was argued that possible contributions from electrostriction would probably not be large. The activation entropy values were all negative, and values for ΔV^+ varied between 0 and -6 cm^3 mol^{-1}, leading to the conclusion that associative substitution dominated throughout.

The natures of palladium aquo cations in solution have again come under scrutiny. A study of [Pd(OH$_2$)$_2$(biL)]$^{2+}$ (biL is bipy or phen) lead to the conclusion that it is virtually nonexistent in aqueous solution at normal pH.[35] Its acidic nature ensures that a proton is lost, probably resulting in hydroxy-bridged dimers which were isolated as [(biL)Pd(μ-OH)$_2$Pd(biL)](ClO$_4$)$_2$. A separate study of the reactions of the resultant conjugate bases in solution (with biL as bipy or en) with bidentate nucleophiles en, tmeda, or glycylglycinate (L—L) found the final products to be respectively [Pd(L—L)$_2$]$^{n+}$ (n = 0 or 2) and [Pd(L—L)en]$^{n+}$ (n = 1 or 2).[36] Entry of the first L—L occurs in one observable step with the rate law (7).

$$k_{obs} = (a + b/[OH^-])[L\text{—}L] \qquad (7)$$

This law was accounted for in terms of a mechanism in which two species in fast protonic equilibrium lead to the product by reactions with L—L. Since the participation of hydroxy-bridged dimers and oligomers did not satisfactorily account for the observed solution experimental data, these authors proposed an equilibrium between mononuclear monohydroxy and dihydroxy species (Scheme 3).

The reaction of [Pd(OH)(par)] (par is 4-(2-pyridylazo)resorcinol) with cyanide, followed by conventional stopped-flow methods, is first order in both species when cyanide is in large excess but zero order in [CN$^-$] at very low cyanide concentrations.[37] The final product is the ion [Pd(CN)$_4$]$^{2-}$. The reverse reaction

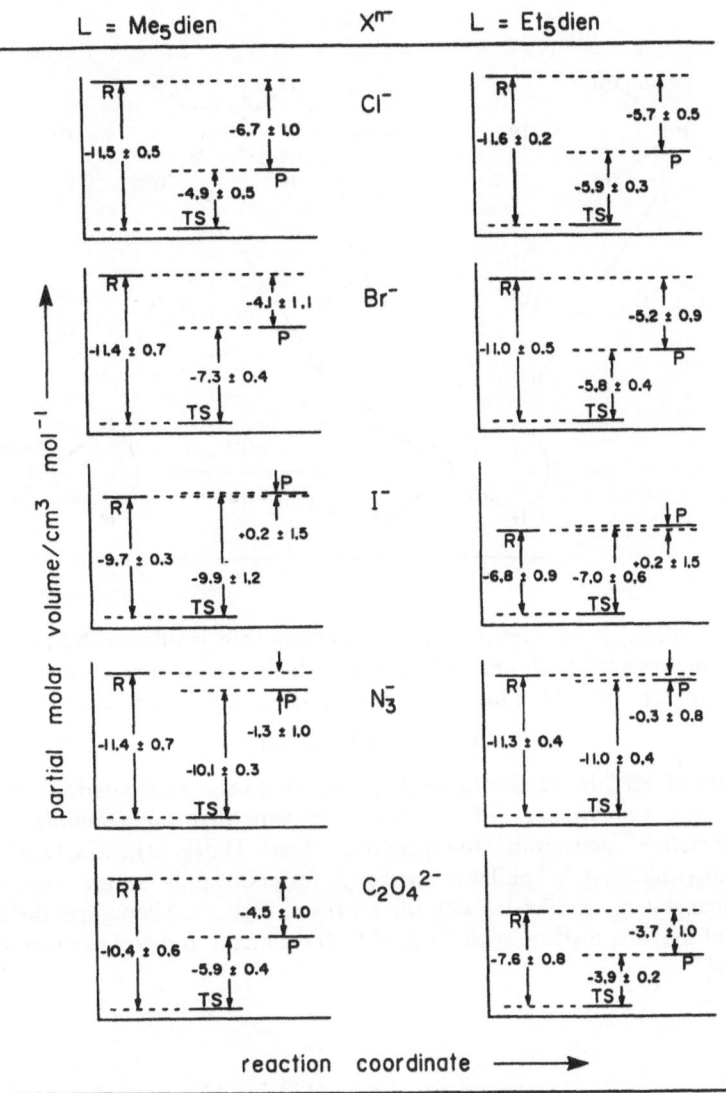

reaction coordinate ⟶

R = reactants; TS = transition state; P = products
Data taken from Table I and II

Figure 5.1. Volume profiles for the reactions $[PdX(R_5\text{-dien})]^+ + H_2O \rightleftharpoons [Pd(OH_2)(R_5\text{-dien})]^{2+} + X^{n-}$ (Ref. 32). Reprinted with permission from *Inorganic Chemistry.* Copyright 1989 American Chemical Society.

between this ion and par is also first order in each reagent, and exhibits an inverse first-order dependence on $[CN^-]$. Slow loss of the hydroxy group is proposed to account for the observed zero order in CN^- of the forward reaction, though the nature of that step is not discussed. Formation of $[PdCl_4]^{2-}$ from the reaction of

Scheme 3

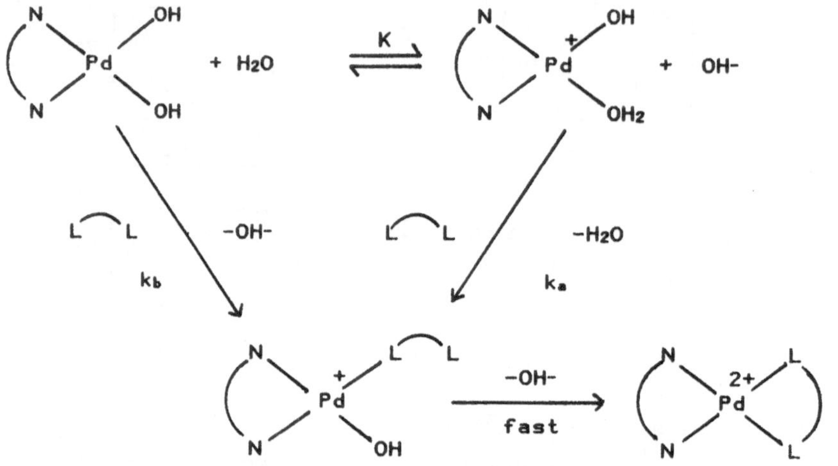

Cl^- with $[Pd(ox)_2]^{2-}$ or $[Pd(mal)_2]^{2-}$ in aqueous acid is affected by protonation of the complexes.[38] Both proceed by two well-separated consecutive steps, and conform to rate law (8). Finally, kinetics and equilibria for reactions of aquo

$$k_{obs} = k_0 + k_2[H^+][Cl^-] \qquad (8)$$

complexes of Pd^{2+} in water with acetonitrile have been elucidated.[39] Rate data and entropy and volumes of activation indicate that the reactions proceed by associative activation to produce first $[Pd(H_2O)_3(MeCN)]^{2+}$ then $[Pd(H_2O)_2(MeCN)_2]^{2+}$. The latter species is formed rapidly as the *trans* isomer, which then slowly equilibrates with the *cis* form. Figure 5.2 compares the volume profiles of the first reaction with those of H_2O exchange and replacement of water by dmso.[39]

5.4. Ligand Exchange at Platinum(II) by Dissociative Processes

Rate constants for the ligand exchange reactions of *cis*-$[PtR_2L_2]$ (R = Me, Ph; L = Me$_2$S, dmso) with free L have been determined in chloroform or benzene by ^{1}H NMR magnetization transfer experiments, and as a function of pressure.[40] Values of $\Delta V^{\ne}$ of about +5 cm^3 mol^{-1} provide more evidence that these reactions proceed by dissociative activation, supporting previous findings for the same systems (see Volume 6). The combination of the Pt—C bonds, with their high *trans* influence, and the readily lost ligands dmso or Me$_2$S, is believed to be responsible for the change from the associative behavior which dominates ligand exchange reactions in related square-planar species.

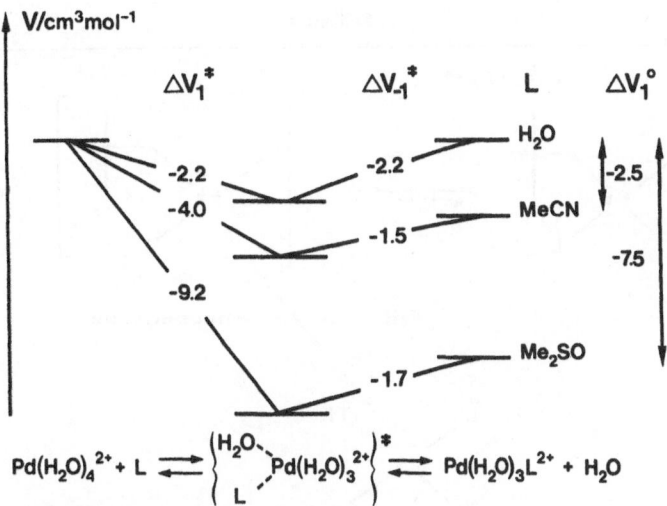

Figure 5.2. Volume profiles for the formations and dissociations of monocomplexes between aqueous Pd^{2+} and ligands, L (Ref. 39). Reprinted with permission from *Inorganic Chemistry*. Copyright 1988 American Chemical Society.

The complexes *cis*-[PtPh$_2$L$_2$] (L are thioethers) lose L in solution according to Eq. (9). Replacement kinetics for the substitution of L by bipy in CH$_2$Cl$_2$ follows rate law (10), identical to those found for related reactions for which dissociative

$$2cis\text{-}[PtPh_2L_2] \rightleftharpoons [Pt_2Ph_4(\mu\text{-}L)_2] + 2L \tag{9}$$

$$k_{obs} = a[bipy]/(b[L] + [bipy]) + c[bipy] \tag{10}$$

activation was proposed. It thus seems plausible that a dissociative mechanism operates in these cases also. Reactivity decreases linearly with increasing σ-donor ability of the ligands, L, but bulky substituents accelerate the reaction.[41]

A different system, this time involving the readily-lost ligand trifluoromethanesulfonate (tf), has also been proposed to react via a dissociative mechanism.[42] The complex *trans*-[PtR(tf)(PMe$_3$)$_2$] (R is neopentyl) reacts with benzene at elevated temperatures to form *trans*-[PtPh(tf)(PMe$_3$)$_2$] and neopentane. The reaction rate is decreased when [Bu$_4$N]tf is added to the solution, but increased when [Bu$_4$]BF$_4$ is added, consistent with the operation of a tf dissociation step. The loss of tf is rapid and reversible (Scheme 4), with the rate-limiting step involving either oxidative addition of benzene/reductive elimination of neopentane, or electrophilic attack at benzene (the authors could not rule out either possibility).

Despite steadily accumulating evidence over the years that some reactions of some platinum complexes proceed by ligand dissociation to three-coordinate T-shaped 14-electron intermediates, and notwithstanding the persuasive activation volume data reported above, the inability of kinetic data alone to rule out the intervention of solvento intermediates as alternative pathways has always left room for doubt. Thus a paper on the reversible loss of olefin from *trans*-[PtCl$_2$(ol)L]

Scheme 4

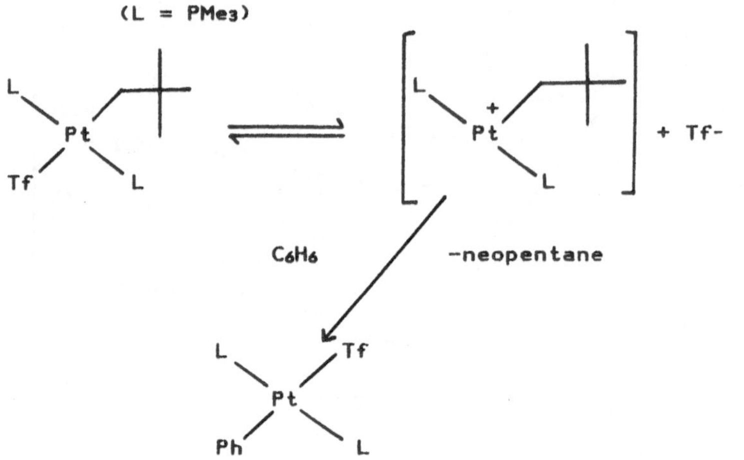

(ol is 4-methyl-1-pentene; L is pyridine-1-oxide) followed by NMR spectroscopy in $CDBr_3$ found evidence for a dissociative process, but the authors interpreted their results in terms of solvent reversibly displacing the olefin.[43] Interpretations of some kinetic data reported previously as supporting dissociative activation (e.g., see Volumes 5 and 6) have hinged on the very poor coordinating ability of the solvents used—typically benzene, chloroform, or dichloromethane. It is therefore worth keeping in mind that solvents of this nature are not necessarily innocent; for example, CH_2Cl_2 coordination to platinum has been known for many years,[44] and this year sees a report of CH_2Cl_2 acting as a bidentate ligand.[45] Clearly caution is still necessary before interpreting unsupported kinetic results in terms of *D* reactions.

5.5. Ligand Exchange at Nickel

Replacement reactions of the bromide of *trans*-$[NiBrRL_2]$ (R is C_6F_5, C_6Cl_5, mesityl, or C_6Cl_4H; L is $PMePh_2$, $PEtPh_2$, or PPh_3) by other anions have been followed in acetone solution.[46] The conventional two-term rate laws of associative reactions were found, and the rates were considerably slowed by steric hindrance, either from *ortho*-substituents on R or from the bulk of L. Similar investigations on related compounds (R is C_6Cl_5 or C_6Br_5; L is $PMePh_2$, $PEtPh_2$, or $PPrPh_2$), in which the coordinated bromide was replaced by I^-, NO_2^-, or SCN^- in EtOH, established that solvolysis and direct replacement were able to compete in the pentachlorophenyl compounds, but with R = C_6Br_5 the solvolysis pathway dominated.[47]

The reactions of four square-planar nickel complexes $[Ni(R\text{-}sal)_2]$ [R-sal is (N-alkyl)salicylaldiminate: R = $(CH_2)_2Ph$, $(CH_2)_3Ph$, $(CH_2)_4Ph$, and $(CH_2)_2$-

C_6H_4-4-OH)] with N,N'-disalicylidene ethylenediamine (H_2-salen) have been followed by stopped-flow methods in acetone. Only one step was kinetically observable, giving the second-order rate law (11). Since the values of k did not vary greatly with the alkyl chain length of R, it was concluded that no contribution from aromatic "ring stacking" [possible sterically only when R = $(CH_2)_3Ph$ or $(CH_2)_4Ph$] was made to the kinetics.[48] A similar work involves the complexes with R = $CH(CH_2OH)CH(OH)Ph$, $CHMeCH(OH)Ph$, and CH_2CH_2Ph.[49] Here, the diversity of geometrical arrangements possible at nickel presents a complication. Only the last complex is diamagnetic and square planar: the dihydroxy ligand complex is octahedral and paramagnetic in acetone, and the UV/visible spectrum of the monohydroxy ligand complex shows the characteristics of both modes of coordination. Rate law (11) is again adhered to for the reactions with H_2-salen.

$$\text{rate} = k[\text{H}_2\text{-salen}][\text{Ni}(\text{R-sal})_2] \qquad (11)$$

The planar complex reacts much faster than the other two, with a negative ΔS^+ suggesting associative activation. The substitution of R-sal by H_2-salen for the other two complexes appears to be dominated by rapid planar/octahedral equilibria in solution, and the octahedral form is kinetically inert. The rate differences span four orders of magnitude in acetone, but are reduced by a factor of 30 when MeOH is employed as solvent. This is probably caused by coordination of the more polar methanol, leading to more stable octahedral adducts. Interestingly, the observed rate constant for the reaction of the planar complex is greater in methanol than it is in acetone. The likely explanation is a contribution from an associative solvolysis pathway in the more polar solvent.

A UV/vis spectroscopic investigation of bromonickel complexes in dmso emphasizes the variable geometry attainable by Ni(II).[50] Increasing bromide concentrations convert $[\text{Ni}(\text{dmso})_6]^{2+}$ to octahedral $[\text{NiBr}(\text{dmso})_5]^+$ and $[\text{NiBr}_2(\text{dmso})_4]$ first, but the next bromide causes a geometry change to tetrahedral $[\text{NiBr}_3(\text{dmso})]^-$ then $[\text{NiBr}_4]^{2-}$. The equilibrium position between the last two species reverses as the temperature is raised, probably due to the lowering dielectric constant of the medium destabilizing the doubly charged $[\text{NiBr}_4]^{2-}$. Addition of en to planar bis(N,N-diethyldithiocarbamato)nickel results in an octahedral complex (Scheme 5).[51] Thermodynamic parameters in several solvents are given.

The nickel(II) macrocyclic ligand complex (4) is diamagnetic and planar as the perchlorate, but paramagnetic and octahedral with additional ligands such as

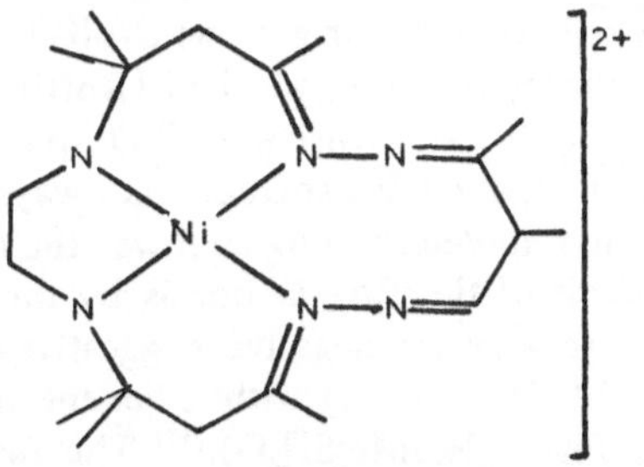

(4)

Scheme 5

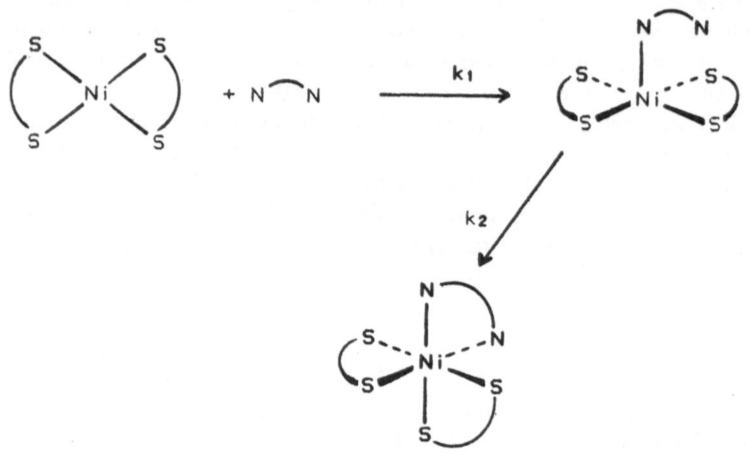

2(NCS$^-$), 2(N$_3^-$), acac$^-$, or en. In dmso or dmf, the planar form slowly converts to an equilibrium mixture dominated by octahedral bis-solvates. Activation enthalpies are about $100 \, kJ \, mol^{-1}$. The slow rates and high ΔH^+ indicate reactions involving Ni—N bond breaking.[52]

Lastly, more reactions of planar nickel(I) macrocyclic compounds have been reported. These tend to be powerful reducing agents and readily transfer an electron to organic halides, for example, producing halide anion and an organic radical. The nickel(I) complex of decamethylcyclam (cyclam is 1,4,8,11-tetraazacyclotetradecane) reacts some 10^4 times more slowly than does its tetramethyl analog.[53] This supports the assignment of an inner-sphere mechanism, requiring the approach of the organic halide, since complexes of, for example, cobalt(III) amines which react by outer-sphere routes differ little in their reduction rates.

5.6. Reactions of Planar Ir(I), Rh(I), Au(III), and Cu(II) Complexes

The kinetics of the replacements of various β-diketonates from [Rh(β-dik)(cod)] by trifluorobenzoylacetone (tfba) to yield [Rh(tfba)(cod)] have been followed in petroleum ether by UV spectroscopy.[54] Despite the nonpolar nature of the solvent, two-term rate laws involving solvento pathways and direct replacement pathways were found. The rate-determining step was the ring opening, caused by replacement of the weakest of the Rh—O bonds by the incoming nucleophile. Entropies of activation were large and negative, suggestive of associative processes.

The reactions of [RhCl(cod)(PR$_3$)] with 1-hexene lead to replacements of PR$_3$ (PR$_3$ is PPh$_3$ or $\frac{1}{2}$[(Ph$_2$PC$_2$H$_4$)Me$_2$Si]$_2$O).[55] The reactions proceed in two reversible steps via detectable, presumably five-coordinate, intermediate

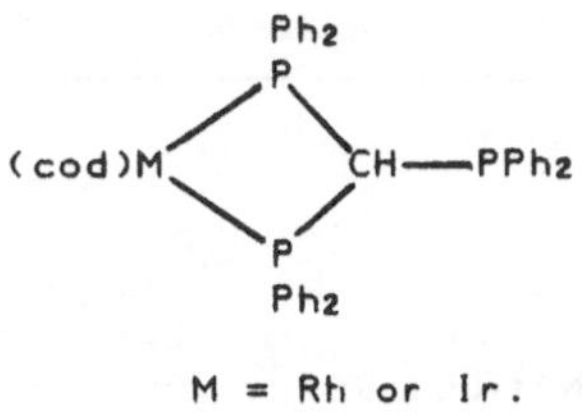

$M = Rh$ or Ir.

(5)

[RhCl(cod)(PR$_3$)(hexene)]. The fluxional nature of compounds (**5**) have been investigated by ^{1}H and ^{31}P NMR spectroscopy.[56] At room temperature the "dangling" and coordinated phosphines rapidly exchange positions. Retention of J_{Rh-P} above the high-temperature limit confirms this to be an intramolecular process. The iridium complex presumably behaves in the same way, but reaches coalescence at a lower temperature. The processes are probably analogous to ligand replacement by associative nucleophilic attack.

Equilibrium constants for the metathesis reactions (12) (L is PPh$_3$ or AsPh$_3$) have been used as a gauge of the relative affinities of [RhL$_2$(CO)]$^+$ for the various anions employed.[57] The trend found for the PPh$_3$ complexes was NCO$^-$ ≫ OAc$^-$,

$$[RhX(CO)L_2] + Y^- \rightleftharpoons [RhY(CO)L_2] + X^- \tag{12}$$

PhCOO$^-$ ≫ F$^-$, NCS$^-$ > Cl$^-$ > Br$^-$ > I$^-$ ≫ NO$_3{}^-$, CF$_3$COO$^-$ ≫ OTf$^-$, ClO$_4{}^-$. A similar series was established for the AsPh$_3$ complexes, but the positions of Cl$^-$ and NCS$^-$ were reversed.

The ready reduction of gold(III) to gold(I) compounds continues to complicate ligand exchange studies. The reactions of thiocyanate and [Au(NH$_3$)$_4$]$^{3+}$, *trans*-[AuCl$_2$(NH$_3$)$_2$]$^+$, and *trans*-[AuBr$_2$(NH$_3$)$_2$]$^+$ in aqueous acid have been examined by potentiometric pH measurements and sequential mixing stopped-flow spectrophotometry.[58] Equation (13) shows the overall stoichiometry for the reaction of the gold(III) tetra-ammine complex. The reaction proceeds by rate-determining substitution of ammine by SCN$^-$, followed by rapid reduction steps. The

$$3[Au(NH_3)_4]^{3+} + 6SCN^- + 4H^+ + 4H_2O \rightarrow 2[Au(SCN)_2]^-$$
$$+ [Au(SCN)(CN)]^- + SO_4^{2-} + 12NH_4^+ \tag{13}$$

kinetics are first order in each reagent, with no detectable solvento step. The *trans* diammine dihalide complexes undergo halide replacement by thiocyanate in two, rapid, reversible steps. *Trans* geometry was retained and again no solvento path was apparent. Negative enthalpies of activation indicate the operation of the expected associative pathways. Parameterization of the substitution rate constants to gauge the relative effects of *cis*, *trans*, entering and leaving groups indicates that here the nature of the entering ligand is more important than that of the *trans* ligand. This contrasts markedly with the situation in isoelectronic Pt(II) complexes, confirming a high discriminating power of Au(III) for entering ligands.

Reduction to gold(I) complexes takes place by three pathways operating in parallel from the substitution equilibria (Scheme 6) to a labile Au(I) intermediate. The reactions of the gold(III) complexes appear to proceed by outer-sphere,

Scheme 6

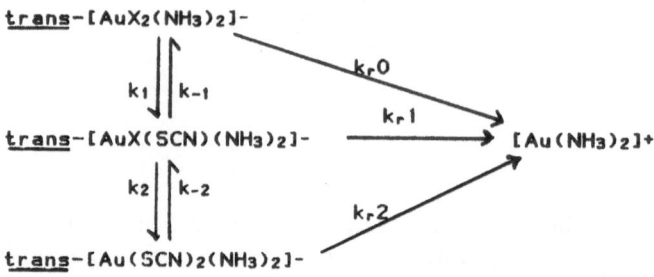

second-order reactions. For the reduction of the thiocyanate intermediates, the authors suggest a competitive electron transfer and ligand substitution, shown in Figure 5.3.

The reduction of $[AuCl_4]^-$ by sulfite (aqueous SO_2) also appears to involve formation of an initial complex species prior to the electron transfer step.[59] The reaction rate varies inversely with $[H^+]$, and free radical intermediates appear to be involved. Equations (14)–(16) summarize the processes.

$$SO_{2.aq} + Au(III) \rightleftharpoons [complex] \rightarrow HSO_3^- + Au(II) + H^+ \qquad (14)$$

$$HSO_3^- + Au(III) + H_2O \rightarrow SO_4^{2-} + Au(II) + 3H^+ \qquad (15)$$

$$SO_{2.aq} + Au(II) \rightarrow S_2O_6^{2-} + Au(I) + 2H^+ \qquad (16)$$

Structural distortions and associated high lability often prevent kinetic data on substitution reactions of four-coordinate copper(II) compounds being obtained. Slightly distorted square-planar coordination is found for the copper complexes [CuXL] (L is N-(2-(diethylamino)ethyl)salicylaldiminate, **6**; X are monodentate anionic ligands or water), confirmed by X-ray analysis for X = NCS^-.[60] The

(6)

kinetics of their ligand substitution reactions by H_2-salen (N,N'-disalicylidene-1,2-diaminoethane) and a methyl-substituted analog were followed spectrophotometrically in acetone. For most of the complexes, the product, [Cu(salen)], is formed according to a one-term second-order rate law, (17), and ΔS^+ values are negative,

$$rate = k[H_2\text{-salen}][CuXL] \qquad (17)$$

suggestive of an associative mechanism. When X is NO_3^-, NO_2^-, or H_2O, however,

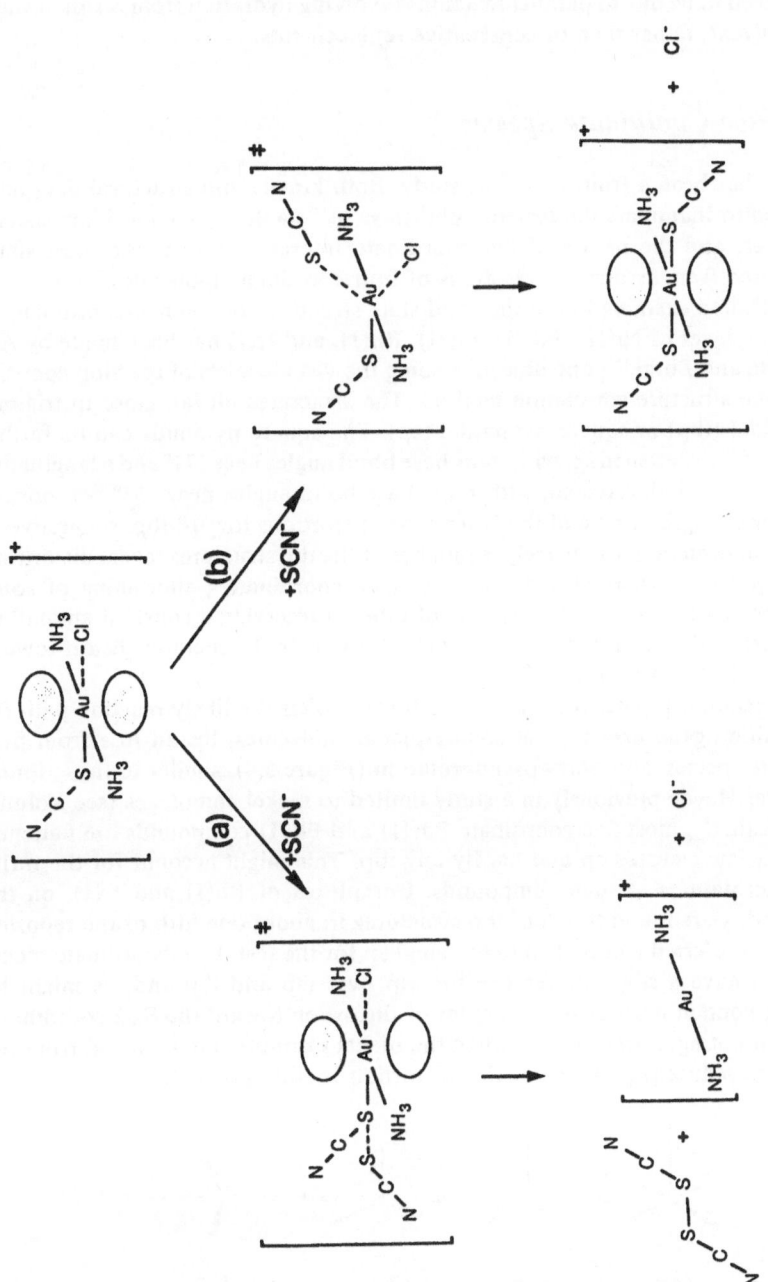

Figure 5.3. Competitive reductive elimination (a) and ligand substitution (b) in the reaction of *trans*-[AuCl(SCN)(NH₃)₂]⁺ with thiocyanate (Ref. 58). Reprinted with permission from *Inorganic Chemistry.* Copyright 1989 American Chemical Society.

two phase kinetics are observed, both processes being first order in H_2-salen. These are believed to be due to parallel reactions involving hydration from residual water in the solvent, rather than to consecutive replacements.

5.7. Five-Coordinate Species

This has been a fruitful area of study. Both kinetic and structural data have continued to illuminate the reaction pathways of "stable" five-coordinate species themselves, and the nature of five-coordinate intermediates or transition states encountered from associative reactions of four-coordinate molecules.

A detailed examination of the solid-state structures of 196 five-coordinate d^8 metal complexes of Ni(II), Pd(II), Pt(II), Rh(I), and Ir(I) has been made by Auf der Heyde and Bürgi,[61] one objective being the visualization of reaction coordinates by the structure correlation method. The structures all fall close to trigonal bipyramidal (tbp) or square pyramidal (sp). The square pyramids can be further subdivided into flattened sp, with *trans* base bond angles near 171° and a lengthened apical bond, and elevated sp, with *trans* base bond angles near 163° and normal apical bond lengths. Some of the latter show distortions toward tbp, suggestive of Berry pseudorotation. Conversely, a number of the tbp structures reveal distortions toward sp. The authors also describe a "glue coordinate"; shortening of some bonds appears to occur at the expense of others preserving a constant amount of bond order at the metal ion. A few trigonal bipyramidal structures distort toward a classical S_N2 transition state.

The relationship between these structures depicts the likely reaction path for nucleophilic ligand exchange at square-planar molecules, ligand loss from five-coordinate species, and Berry pseudorotation (Figure 5.4), similar to those found by Auf der Heyde previously in a study limited to nickel complexes (see Volume 4). Interestingly, most five-coordinate Pd(II) and Pt(II) compounds are flattened sp, with a few elevated sp and hardly any tbp. This might account for the rarity of pseudorotations at such compounds. Complexes of Rh(I) and Ir(I), on the other hand, were found to adopt tbp structures in about one-fifth of the reported cases, and preferred elevated sp to flattened sp for the rest. Five-coordinate nickel complexes have a slight preference for esp over fsp and tbp and, as might be expected, contain most of the examples of distortion toward the S_N2 coordinate. It is worth noting, however, that when the nickel examples are removed from the data set, a (reduced) presence of this distortion is still apparent.

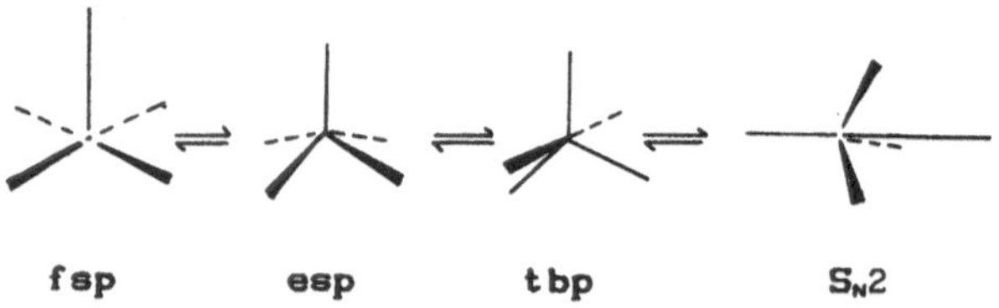

fsp   esp   tbp S_N2

Figure 5.4.   Geometrical relationships in five-coordinate species, from structure correlation studies (Ref. 61).

An *ab initio* MO study of the five-coordinate $[RhH(CO)_2(C_2H_4)(PH_3)]$ predicted tbp ground-state structures, but with only a small intrinsic energy barrier between these and sp structures, confirming that intramolecular transfers between the structures—the Berry pseudorotation—should be a low-energy process.[62] Five-coordinate $[Os(PMe_3)_5]$ has been synthesized. Its NMR spectra show it to be tbp and fluxional.[63] In solution it readily loses PMe_3 and converts to the previously isolated $[(Me_3P)_4HOsPMe_2CH_2]$. Osmium pentacarbonyl has also been obtained. This undergoes ligand replacement reactions by carbonyl loss according to a first-order rate law, and has a large positive value for ΔH^+ and a positive ΔS^+ (Scheme 7).[64] The reactivity of this compound is greater than $Fe(CO)_5$, but less than $Ru(CO)_5$, another example of second-row element compounds reacting fastest.

Scheme 7

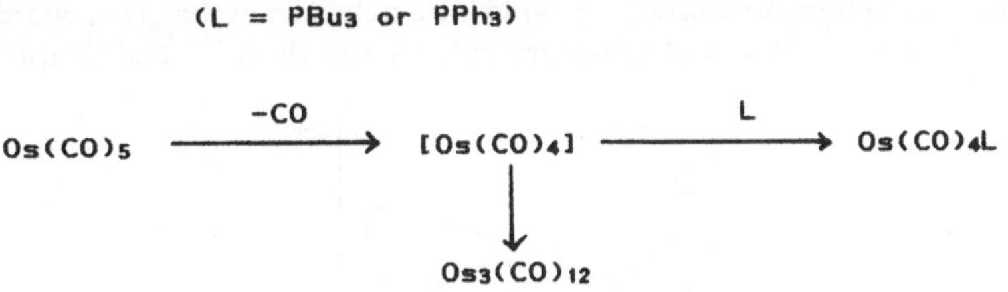

(L = PBu₃ or PPh₃)

$$Os(CO)_5 \xrightarrow{-CO} [Os(CO)_4] \xrightarrow{L} Os(CO)_4L$$

$$[Os(CO)_4] \downarrow$$

$$Os_3(CO)_{12}$$

Rare examples of five-coordinate Ag(III) complexes, $[Ag(py)_2(NCMe)_3][UF_6]_3$ and $[Ag(py)_4(NCMe)][MoF_6]_3$, have been isolated as orange solids.[65] The d^8 Ag(III) ions are diamagnetic and the complexes are substitution labile. Their structures have not been determined but the temperature-dependent NMR spectra are consistent with sp arrangements with a long Ag—NCMe apical link. The five-coordinate $[Pd(PMe_2Ph)_3(biL)](BF_4)_2$ (biL is 2,9-dimethyl-1,10-phenanthroline) has a flattened sp structure with a long Pd—N apical link. ¹H NMR spectroscopy in solution reveals fluxionality which averages the two halves of biL.[66] "Tripod" ligands such as $P(C_2H_4PPh_2)_3$ (tetraL) continue to be a source of isolable five-coordinate species, frequently with tbp geometry, and a number of complexes with structures (7) have been examined. The complex

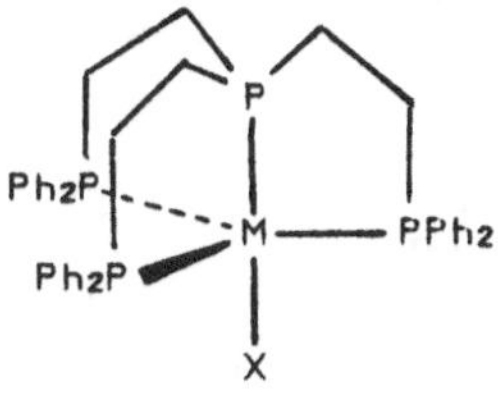

(7)

$[Pt\{P(OMe)_3\}(tetraL)]Cl_2$, with $P(OMe)_3$ in axial position X, reacts in water to eliminate MeCl producing $[Pt\{PO(OMe)_2\}(tetraL)]Cl$.[67] ³¹P NMR spectra of rhodium complexes of type (7) are of the type AMX_3 or $AKMX_3$, and comparisons of the coupling of Rh to the apical phosphorus atom, and of the chemical shifts of all the phosphorus atoms, have enabled *trans* influence series for several ligands X at such five-coordinate species to be drawn up.[68] The overall series for anionic and neutral X works out at Br^-, I^-, $Cl^- < MeCN < NCS^- < NO_2^- < py < CN^-$, $H^- < PR_3 < CO$.

More tbp platinum compounds with equatorial olefins and bidentate amino ligands have been described, along with their chiroptical properties in some cases,[69] and discussion of the factors which stabilize them.[70] The four-coordinate complexes [$PtCl_2(N,N'$-$R_2en)$] (R = *i*-Pr, *t*-Bu, *R*-CHMePh, and *R,S*-CHMePh) have stable chirality at coordinated nitrogen and do not exchange NH protons in $D_2O/CDCl_3$.[71] Their five-coordinate olefin adducts, on the other hand, undergo rapid inversion at N and exchange NH in a few seconds. Both processes take place via a facile dissociation of the Pt—N bond, rather than by N—H dissociation.

Complexes of the type (8), though four-coordinate in their ground state, are fluxional in solution, presumably via rapid intramolecular exchange of coordinated free nitrogen atoms by way of five-coordinate intermediates.[72] The complexes are

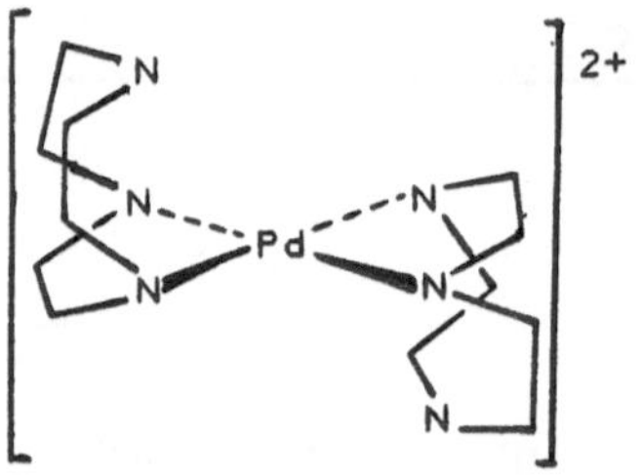

(8)

much more readily oxidized to their Pd(III) analogs than is usual for square-planar tetra-N donor compounds, and the ready attainment of five-coordinate geometry [part way to the octahedra most stable for Pd(IV) compounds] is thought to be responsible. The effect of ligand structure on coordination geometry can be quite subtle, however. While the palladium complex of an N_2S_4 macrocycle with tertiary nitrogens is square planar (9) and exhibits a ready, reversible Pd(II)/Pd(I) couple, the analog with secondary nitrogens is best described as N_2S_2 coordination with two additional, long Pd—S links approaching a distorted octahedral structure, (10).[73] This latter complex shows a ready Pd(II)/Pd(III) couple.

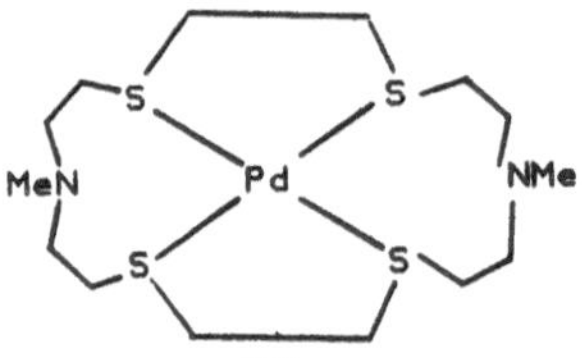

(9)

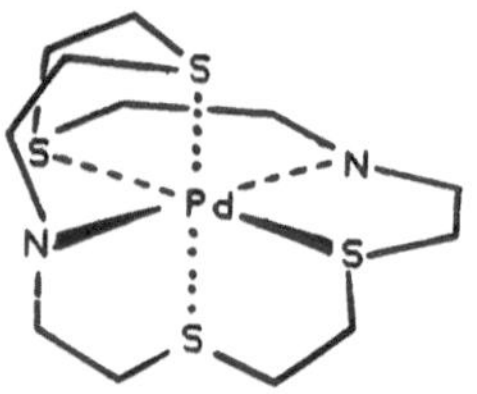

(10)

Another adduct of a four-coordinate d^8 species, which can be regarded as a "trapped intermediate" *en route* to oxidative addition, has been structurally characterized, (11).[74] Related to this, the MeI adduct $[Rh(\eta^5\text{-}C_5H_5)(MeI)(dppe)]^+$ reveals enhanced electrophilicity of the methyl group as a result of coordination to Ru(II).[75] Although the complex ion does not undergo oxidative addition of the MeI, several nucleophiles react with it by abstracting Me^+.

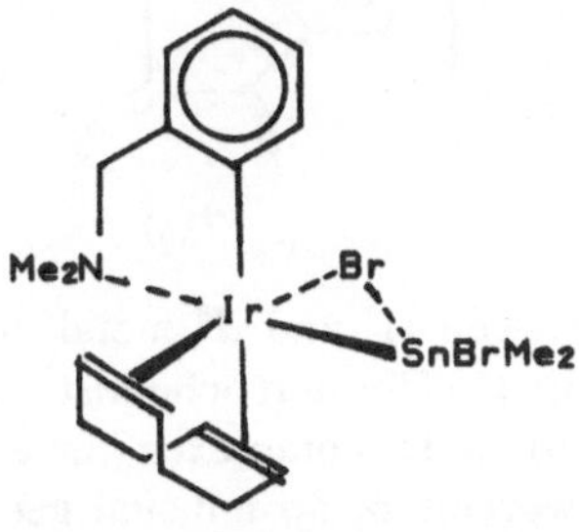

(11) 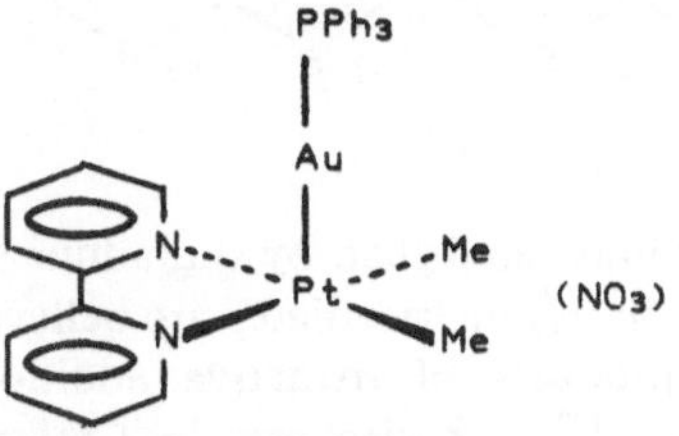   (12)

Adducts like the latter two complexes can be regarded as being formed by electrophilic attack at the metal ion rather than the more conventional nucleophilic attack. The bonding is more difficult to define when the bonds are between metals, however, and such an example is compound (12), formed from $[PtMe_2(bipy)]$ and $[Au(NO_3)(PPh_3)]$.[76] It resembles a number of platinum–silver complexes described in Volume 6, more of which have emerged very recently.[77] The symmetrical nature of compound (13) defeats any atttempt to assign donor–acceptor properties to its metal–metal bond. Nevertheless, the pairs of bromide ions are folded back toward a tbp geometry by the approach of the other metal, as would be expected for approach of a nucleophile.[78]

The structure of $Tl_2Pt(CN)_4$ also reveals expanded coordination, (14).[79] The octahedral units are not unlike the local geometry in the so-called one-dimensional

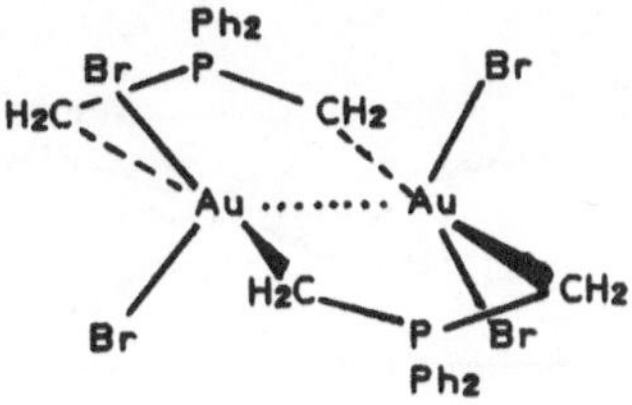

(13) 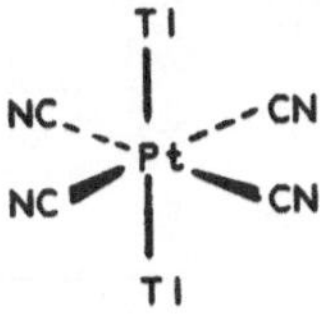   (14)

metals, in which the square-planar units of some complexes crystallize by forming chains of metal–metal bonds (sometimes enhanced by partial oxidation). Compound (15) and its monoanion are rare examples of complexes of second-row ligand atoms which exhibit this phenomenon.[80] The potential for mechanistic complexity in substrates of this type is illustrated by anion (16) which, as the potassium salt, remains yellow and monomeric over a wide range of temperatures and concentrations, but without these limits begins to stack to a red, metal–metal bonded form.[81]

The link between nucleophilic attack at square-planar complexes *en route* to ligand exchange, and electrophilic attack *en route* to oxidative additions, is nowhere

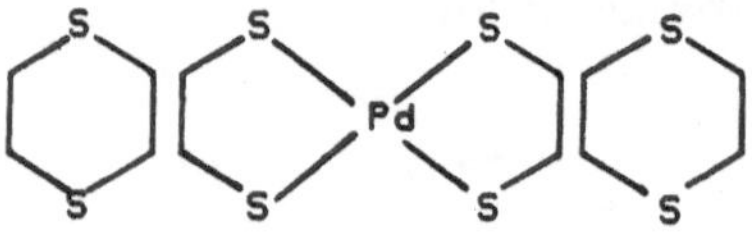

(15) 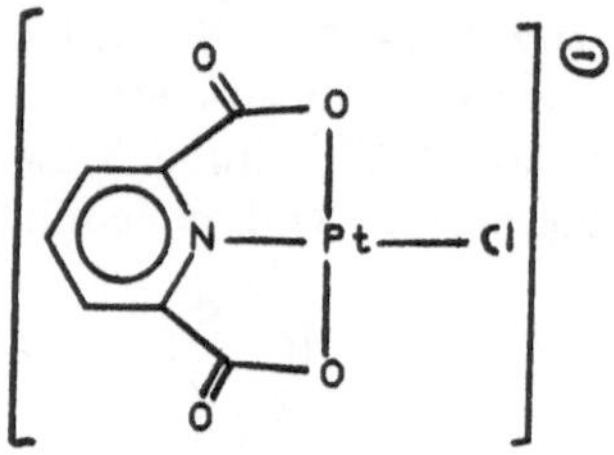   (16)

better illustrated than by the adducts formed between H_2 and d^8 metal ions, in which η^2-H_2 intermediates are believed to participate. The stereochemistry of the final products of oxidative additions of H_2 to Ir(I) complexes have been reviewed.[82] The discovery by Crabtree and co-workers of an unusual geometry of one such addition, compared with the geometry of ligand addition following nucleophilic attack at the same substrates (Scheme 8), has led to the postulation of three sets of conditions which appear to determine the product geometry in

Scheme 8

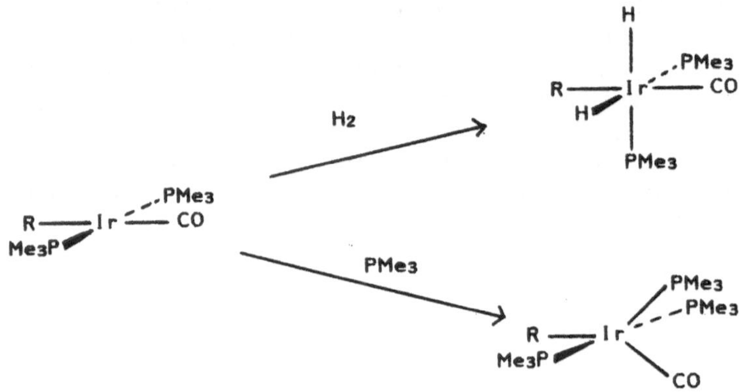

both cases.[83] These conditions are in agreement with, but usefully extend, previous explanations for the observed geometries,[82] and which explain the *trans* directing effect for ligand replacements. In each case, two *trans* ligands of the original square plane must adopt the equatorial position in an intermediate, which relates to a trigonal bipyramid. The ligands most likely to fall into such equatorial positions should have the following properties: (1) They should be good π-acceptors. This will reduce the energy of the filled d_{z^2} orbital, and so reduce the repulsion between it and the incoming L or H_2. (2) They should be good σ-donors, making the metal more π-basic by raising the energy of the d_{yz} orbital. This will more effectively overlap with π-acceptor orbitals on L or σ^* orbitals on H_2. (3) When the equatorial ligands are good π-donors, the metal p_z orbital is stabilized and oxidative additions are encouraged.

Finally, *ab initio* MO calculations on the products of H_2 and $[Rh(PH_3)_4]^+$ found two energy minima.[84] One corresponds to a structure of C_{2v} symmetry, resulting from addition of two H atoms (17). The other is based on a C_{3v} pyramidal structure, where an η^2-H_2 occupies an axial site (18).

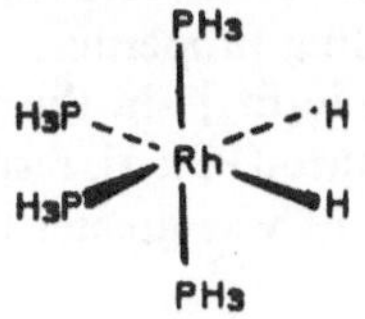

(17) 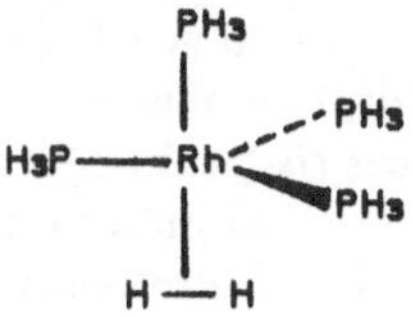   (18)

5.8. Trans Effect

The isolation of a large number of hydridonickel complexes *trans*-$[NiHX(PCy_3)_2]$ has enabled a *trans* influence series for the ligands X at nickel to be drawn up from values of $\nu(Ni—H)$ derived from IR spectra and from $\delta(Ni)H$ from their ^1H NMR spectra.[85] The series from high to low values runs Ph, Me > CN > SH, I, SR, SCN > Br, Cl > OCOR, OPh. It is almost 30 years since similar series were first compiled from more stable hydridoplatinum complexes, but such series are still being added to and refined. Measurements on a new set of hydridoplatinum compounds, *trans*-$[PtH(CF_3)L_2]$ (L are tertiary phosphines), have established a place for CF_3 in the series, but illustrate the need to be aware of the method of determination.[86] When compiled from the magnitude of $^1J_{Pt—H}$ from NMR measurements the order is CF_3 > Ph > C_6H_9 > cyanoalkyl > Me > Cl, but compilations from $\nu(Pt—H)$ from the IR spectra result in the order C_6H_9 > cyanoalkyl > Ph, Me > CF_3 > Cl. The very different positions of CF_3 in the two series reflect the fact that $^1J_{Pt—H}$ results mainly from the amount of s-orbital character in the Pt—H bond, while $\nu(Pt—H)$ is also sensitive to overall electrostatic effects induced by the electron-withdrawing fluorides. Compilations of $^1J_{Pt—P}$ data from the same compounds gave *cis* influences as Cl > CH_2CN > CF_3 > Me, Ph > cyanoalkyl.[86]

The *trans* directing ability of MeCN at Pt(II) has been compared with its *trans* influence, by examining replacement reactions of $[PtCl_3(NCMe)]^-$ by amines for the former, and the X-ray crystal structure for the latter.[87] The directing influence (dynamic *trans* effect) is high while the bond-weakening *trans* influence is low.

The kinetics of the reactions of $[PtCl_3L]^-$ (L = dmso, Et_2S, PMe_3, PEt_3, PPh_3, or $AsEt_3$) with many neutral or anionic nucleophiles, L', to produce *trans*-$[PtCl_2LL']$ have been followed in 95% methanol and the second-order rate constants k_2 (the reactions follow the usual two-term rate law) used to make comparisons between nucleophilic discrimination of the substrates and the *trans* effects of L.[88] Comparisons with n_{Pt}^o scales were poor, the anionic nucleophiles in particular being much less reactive than predicted. An anionic nucleophilicity scale based on $[PtCl_3(AsEt_3)]^-$ was therefore established instead. The presence of any of Et_2S, PMe_3, PEt_3, or PPh_3 in the *trans* position has much the same effect on the nucleophilic discrimination of $[PtCl_3L]^-$; $AsEt_3$ produces a greater discrimination, but dmso produces the greatest of all. This is interpreted as confirming that the *trans* effect of dmso acts largely through its π-accepting properties.

The hydrolysis reactions of *cis*- and *trans*-$[AuMe_2Et_2]^-$ in the presence of PPh_3 produces *cis*-$[AuEt_2Me(PPh_3)]$ and *trans*-$[AuEtMe_2(PPh_3)]$, respectively.[89] This would seem to indicate that the *trans* effect of Et was greater than that of Me in these Au—C protonolysis reactions.

5.9. Isomerizations

Many of the most interesting cases this time appear to involve dissociative activation, an unusual situation. *Cis* to *trans* isomerizations of $[PtRCl(PEt_3)_2]$ (R = Me, Et, Pr, Bu, or CH_2SiMe_3) take place spontaneously in 2-propanol. The reactions are all strongly inhibited by free Cl^-, indicating the operation of the dissociative route of Scheme 9, including a geometry change of the T-shaped intermediates.[90] The propyl and butyl reactions are accompanied by competing

Scheme 9

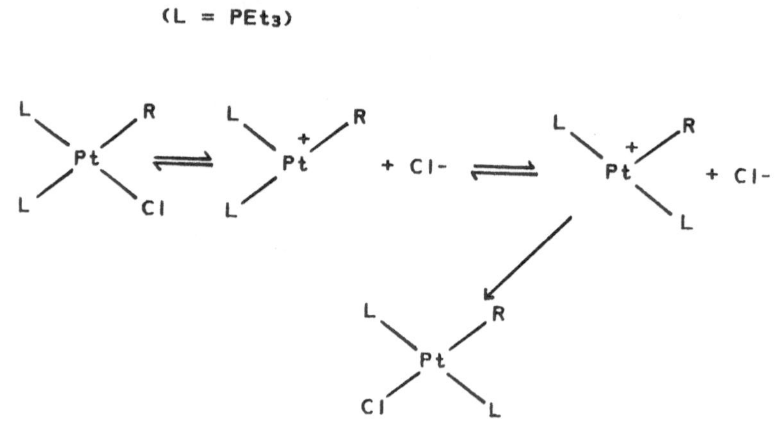

(L = PEt₃)

β-elimination processes. The isomerization reactions when R is Et, Pr, or Bu all proceed much faster than those when R is Me or CH_5SiMe_3, and an incipient β-hydrogen interaction with the metal, promoting the Cl^- loss, is proposed as an explanation. The low deterium kinetic isotope effect for the isomerization of the alkyl complexes ($k_H/k_D = 1.12$) suggests that H (or D) transfer does not develop very far, however.

Valuable, though indirect, support for the operation of such dissociative isomerization processes comes from protonation studies of the complex [RhH(PPh$_3$)$_4$] by H$_2$C(SO$_2$CF$_3$)$_2$.[91] This results in hydride abstraction to afford [Rh(PPh$_3$)$_3$][HC(SO$_2$)$_2$]. When dissolved in CH$_2$Cl$_2$, the red, diamagnetic 14-electron rhodium(I) cation is shown to be T-shaped by ^{31}P NMR spectroscopy, but a dynamic process at ambient temperature mixes the ligands by an intramolecular route. This is clearly related to the geometry change of the T-shaped intermediates, necessary to bring about the isomerizations discussed above.

Another dissociative process leads to a different sort of isomerization in the bis-*N:O* bridged "head to head" α-pyridonate complex, (**19**).[92] It isomerizes in

Scheme 10

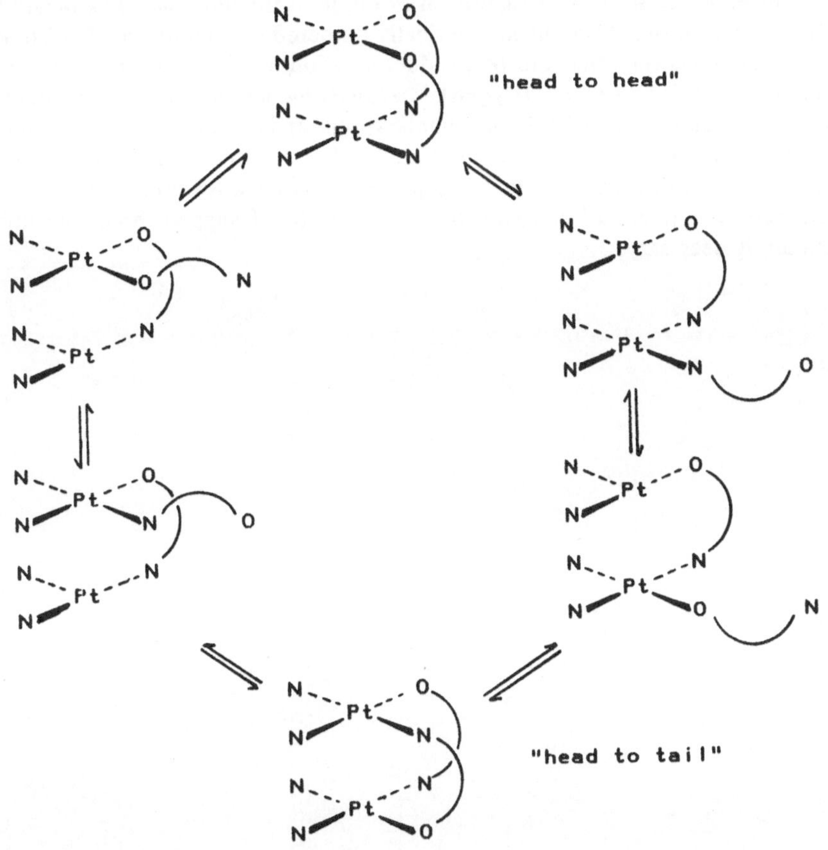

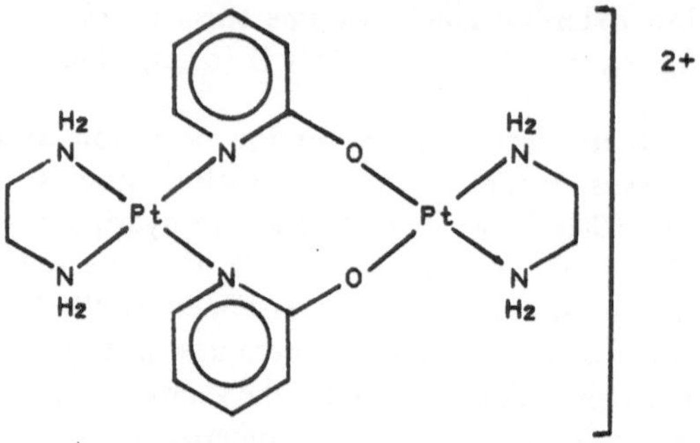

(19)

aqueous solution to an equilibrium with the "head to tail" isomer. The temperature-dependent kinetics of the approach to equilibrium yield values for ΔH^+ of 114(5) kJ mol^{-1} and for ΔS^+ of 40(10) J K^{-1} mol^{-1}, values typical of dissociative activation. Moreover, added dmso fails to accelerate the isomerization, which would be expected if the process was an example of the more usual associatively activated mechanism. Also, the less sterically hindered bis-ammine analog isomerizes 70 times slower. The rate or product distribution is unaffected by added 6-methyl-α-pyridone or [Pt(OH$_2$)$_2$en]$^{2+}$, suggesting an intramolecular process. Scheme 10 was proposed as the mechanism, one or both of the routes depicted fulfilling all the criteria described. The change in bonding mode of the intermediate monodentate ligand presumably takes place via associative attack at the four-coordinate platinum, and no geometry change at the T-shaped three-coordinate platinum is necessary.

Chapter 6

Substitution Reactions of Inert-Metal Complexes— Coordination Numbers 6 and Above: Chromium

6.1. Introduction

The literature available from the end of the last report[1] (June 1988) to December 1989 is covered in this chapter. The general chemistry of chromium is reviewed in two continuing series[2,3] and other reviews on the use of $[(TPP)Cr(V)(O)(X)]$[4,5] and amine chloro[6] or fluoro chromates[7] as oxidizing agents have appeared.

More closely related to kinetics and mechanism are three reviews from Danish chemists, one on the state of the art with respect to Cr(III) substitution reactions[8] and the other two[9,10] survey μ-diol dinuclear Cr(III) systems. Chromium chemistry is also discussed in a general review on metal-ion-assisted aquation reactions[11] and in a review on the effect of pressure on reaction rates.[12] In the following sections, units for E_a and ΔH^+ are in kJ mol^{-1}, ΔS^+ in J K^{-1} mol^{-1}, ΔV^+ in cm^3 mol^{-1}, visible absorption spectral maxima (λ_{max}) in nm, and molar extinction coefficients (ε) in M^{-1} cm^{-1}. Unless otherwise stated, all rate constants are reported at 25.0 °C (298.2 K).

Table 6.1. Kinetic Parameters and Aquation Rate Constants for $[CrL(NH_3)_5]^{n+}$ Ions

L	T (°C)	P (bar)	$10^4\,k_{obs}$ (s^{-1})	Solvent	$\Delta H^{\ddagger}$ (kJ mol^{-1})	$\Delta S^{\ddagger}$ (J K^{-1} mol^{-1})	$\Delta V^{\ddagger}$ (cm^3 mol^{-1})	Ref.
trif$^-$[a]	25.0	1	4.5	CH$_3$CN	89	−11	−8.9[b]	16
	25.2	5	5.15					
		50	6.00					
		100	7.46					
		150	8.55					
trif$^-$[a]	20.0	5	2.36	MeOH				16
		50	2.95					
		100	3.73					
		150	4.37					
	20.2	1	2.79					
	24.9	1	4.37					
	25.0	1	4.37		69.5	−76	−10.5[c]	
	31.5	1	8.26					
	36.5	1	13.2					
OS(CH$_3$)$_2$	25.0	1	0.195	0.01 M HClO$_4$	95.3	−15	−3.2	14
	36.5	1	0.82					
	42.5	1	1.73					
	45.0	50	1.87					
		500	1.98					
		1000	2.11					
	48.8	1	3.51					
	54.1	1	6.49					
OCH(NH$_2$)	25.0	1	0.51		94.0	−12	−4.8	14
	27.3	1	0.68					
	34.0	1	1.57					
	43.6	1	5.13					
	45.0	50	5.41					
		500	5.19					
		1000	6.33					
		1500	7.12					
	49.9	1	10.05					

Ligand	T/°C	P	k		$\Delta V^{\ddagger}$		Ref.
OC(NH$_2$)$_2$	25.0	1	0.20	93.5	−22	−8.2	14
	37.0	1	0.89				
	44.5	1	2.08				
	50.8	1	4.37				
	50.9	1	4.80				
		345	5.43				
		690	5.80				
		1035	5.80				
		1380	6.70				
			6.86				
OC(NHCH$_3$)$_2$	55.1	1	0.145	93.1	−25	−7.4	14
	25.0	1	0.30				
	30.8	1	0.61				
	36.3	1	1.40				
	43.3	1	1.44				
	45.0	50	1.64				
		500	1.88				
		1000	3.27				
OC(CH$_3$)N(CH$_3$)$_2$	51.4	1	0.19	91.9	−30	−6.2	14
	25.0	1	0.43				
	32.0	1	0.94				
	38.5	1	1.85				
	44.9	1	1.73				
	45.0	50	1.95				
		500	2.17				
		1000	4.56				
OP(OCH$_3$)$_3$	52.8	1	0.60	89.7	−23	−8.7	14
	25.0	1	2.34				
	34.4	1	5.58				
	42.0	1	4.39				
	45.0	50	5.12				
		500	6.03				
		1000	9.04				
Cl$^-$	46.3	1	13.9			ethyleneglycol/ water	17
	50.4	1					

^{a} trif$^-$ = CF$_3$SO$_3^-$. $\quad$ ^{b} $\Delta V^{\ddagger} = -3.2\ \mathrm{cm^3\ mol^{-1}}$ for the Co(III) analog. $\quad$ ^{c} $\Delta V^{\ddagger} = -3.1\ \mathrm{cm^3\ mol^{-1}}$ for the Co(III) analog.

6.2. Aquation and Solvolysis of Chromium(III) Complexes

6.2.1. $[Cr(III)(L_5)X]^{n+}$ Systems $(L = OH_2, NH_3)$

The accepted textbook wisdom, that the mechanism for aquation of pentaammine chromium(III) complexes is I_a, is again questioned.[13] It is proposed that for such complexes with neutral leaving groups the mechanism is indeed I_a, but, with charged leaving groups, solvation effects may discriminate in favor of the I_d process. However, while pressure-dependence studies using complexes with neutral leaving groups[14] certainly support a mechanistic differentiation between Co(III) $[I_d]$ and Cr(III) $[I_a]$, even with charged leaving groups (trifluoromethanesulfonato) this mechanistic differentiation remains.[14,15] Kinetic data are presented in Table 6.1. With neutral leaving groups, the average ΔV^+ is approximately -6 cm^3 mol^{-1} for Cr(III) and approximately $+2$ cm^3 mol^{-1} for Co(III); with triflate as the leaving group, the corresponding data are about -9 and about -3, respectively.

6.2.2. Cr(III)—C Bond Rupture

The decomposition of pentaaqua(organo)chromium(III) complexes in aqueous solution has been observed to occur by homolysis of the Cr—C bond [Eq. (1)] or by heterolysis (acidolysis) [Eq. (2)] with assistance from some proton source.

$$[(H_2O)_5CrR]^{2+} \underset{k_r}{\overset{k_f}{\rightleftharpoons}} [Cr(OH_2)_6]^{2+} + R^{\cdot} \tag{1}$$

$$[(H_2O)_5CrR]^{2+} \longrightarrow [Cr(OH_2)_6]^{3+} + HR \tag{2}$$

$$k_{obs} = k_1 + k_2[H_3O^+] \tag{3}$$

In the presence of excess Cr(II), the $[(H_2O)_5Cr—CR_1R_2R_3]^{2+}$ alkyl chromium(III) complexes decompose according to Eq. (2), where the observed rate constant (k_{obs}) depends on the $[H_3O^+]$ according to Eq. (3). Values for k_{obs} (s^{-1}) at pH = 1 are 1×10^{-4}, 4.9×10^{-4}, 3.1×10^{-3}, 2.1×10^{-5}, and 1.2×10^{-2} for $-CR_1R_2R_3 =$ CH$_2$OH, CH(CH$_3$)OH, C(CH$_3$)$_2$OH, CH$_2$CO$_2$H, and CHO(CH$_2$)$_2$OCH$_2$, respectively.[18] A more detailed acid-dependence study ($I = 1.0$ M) for $[(H_2O)_5Cr—CH_2Ph]^{2+}$ gives $k_1 = 3.8 \times 10^{-6}$ s^{-1}, $\Delta H^+ = 89.5$, $\Delta S^+ = -49.7$ and $k_2 = 2.0 \times 10^{-6}$ M^{-1} s^{-1}, $\Delta H^+ = 84.8$, $\Delta S^+ = -69.0$.[19] In the presence of O$_2$, the reaction shown by Eq. (2) is suppressed and, when R = $-$CH$_2$Ph, $k_f = 2.32 \times 10^{-3}$ s^{-1} ($I = 1.0$ M, independent of [H$^+$]). The spontaneous decomposition of $[(H_2O)_5CrCH_2Ph]^{2+}$ in the absence of reagents other than HClO$_4$ is a complicated process from a kinetic (and product analysis) point of view. The decrease of absorbance with time (at 41.0 °C) is faster than predicted by a first-order process and an analysis in terms of a half-order dependence on $[(H_2O)_5CrCH_2Ph^{2+}]$ and an inverse dependence on [H$^+$] has been made.[19] Further studies, especially on the time dependence of the product distribution, are required before a complete understanding of the homolytic reaction is possible.

6.2.3. Amine and Other Complexes

Table 6.2 present hydrolysis rate data reported for this class of complex.

6.2.4. Dechelation/Chelation Processes

The first formed Cr(III) product from the Cr(II) reduction of [Co(pyruvato)-$(NH_3)_5]^{2+}$ (Section 6.9.1) is now believed to be chelated $[(H_2O)_4Cr(pyruvato)]^+$ and the protonated form of this chelate has a ring-opening rate of $1.12 \times 10^{-4} \, s^{-1}$ ($I = 0.041-0.55 \, M$).[25]

The rate of decomposition of $[Cr(mal)_3]^{3-}$ and $[Cr(Memal)_3]^{3-}$ in the presence of EDTA has been investigated in basic solution ($[OH^-] = 0.04-0.12 \, M$). The reaction is catalyzed by HCO_3^-, and in both cases there is an EDTA dependent (k') and independent (k'') path. Kinetic parameters are summarized in Table 6.3. It should be noted that final product is not $[Cr(EDTA)(OH)]^{2-}$ and mal or Memal may still be coordinated. Under identical conditions, the Memal complex reacts more rapidly than the mal analog.[26] Rate constants for the thermal loss at 60 °C for the first and second ends of a tn ligand from $[Cr(tn)_3]^{3+}$ are 3.1×10^{-6} and $5.8 \times 10^{-5} \, s^{-1}$, respectively, with $E_a = 104$ for the second step.[27]

6.2.5. Metal-Ion-Assisted Aquation

This general area has recently been reviewed,[11] however, the mechanism of Hg^{2+} assisted chloride release reactions from $[CrCl(N_5)]^{2+}$ is now believed to be I_a rather than D or I_d.[28] The spontaneous acid hydrolysis of $[Co(ox)_2(hist)]^-$ [hist = (1)] proceeds via histamine loss,[29] but in the presence of Fe^{3+} the two

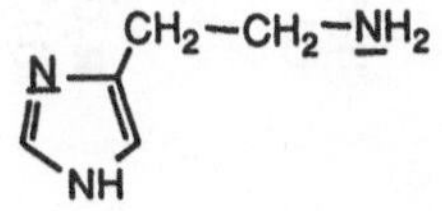

histamine (donor atoms underlined)

(1)

oxalato ligands are replaced successively [Eqs. (4) and (5)] and the new Cr(III) cations can be isolated in solution by ion-exchange chromatography.[30] The rates

$$[Cr(ox)_2(hist)]^- + Fe^{3+} \rightarrow [Cr(ox)(hist)(OH_2)_2]^+ + [Fe(ox)]^+ \quad (4)$$

$$[Cr(ox)(hist)(OH_2)_2]^+ + Fe^{3+} \rightarrow [Cr(hist)(OH_2)_4]^{3+} + [Fe(ox)]^+ \quad (5)$$

$$-d[Cr(III)]/dt = \{k_1 + k_2[Fe^{3+}]\}\{Cr(III)\} \quad (6)$$

$$Cr(III) = [Cr(ox)_x(H_2O)_{4-2x}(hist)]^n \quad (x = 1; n = +1: x = 2; n = -1)$$

of Fe^{3+}-assisted oxalate release for reactions (4) and (5) can be studied separately and plots of k_{obs} vs. $[Fe^{3+}]$ are linear with nonzero intercept (at constant $[H^+]$). The data are analyzed using the rate law (6), where the k_1 terms are composite.

Table 6.2. Kinetic Parameters for the Acid Hydrolysis of Some $[CrX(L_5)]^{n+}$ Complexes

Complex	X	Solvent (pH)	T (°C)	$10^5 k$ (s^{-1})	$\Delta H^{\ddagger}$ (kJ mol^{-1})	$\Delta S^{\ddagger}$ (J K^{-1} mol^{-1})	Ref.
[Cr(ox)(NCS)(H$_2$O)]	NCS$^-$	H$_2$O (2)	25.0	1.40[b]			20
			30.0	2.35[b]			
			35.0	4.75[b]			
			40.0	8.05[b]			
			45.0	7.50[c]			
			50.0	15.9[b]			
[Cr(NCS)(hist)(H$_2$O)$_3$]$^{2+}$	NCS$^-$	H$_2$O (1)	25.0	0.019[a]	110.5	−9.4	21
			30.0	3.42			
			70.0	10.9			
			80.0	30.2			
			90.0				
[CrCl(en)(Medpt)]$^{2+}$	Cl$^-$	H$_2$O (1)[d]	25.0	0.0014[a]	121	+2	22
			70.4	0.837			
			80.2	3.55			
			84.8	5.14			
			90.4	7.29			
[CrCl(tn)(Medpt)]$^{2+}$	Cl$^-$	H$_2$O (1)[d]	25.0	0.0031[a]	112	−20	
			70.4	1.18			
			80.2	3.31			
			84.8	6.09			
			90.4	10.9			
[CrCl(Me$_2$tn)(dpt)]$^{2+}$	Cl$^-$	H$_2$O (1)[d]	25.0	0.0208[a]	95.8	−60	
			70.5	3.53			
			74.8	5.01			
			80.2	8.19			
			84.8	14.1			

Complex	Leaving group	Solvent	T (°C)	k			Ref.
[CrCl(tn)(2,3-tri)]$^{2+}$	Cl$^-$				117	−5	
		H$_2$O (1)^d	25.0	0.0029^a			
			70.4	1.33			
			80.2	4.33			
			85.0	7.33			
trans-[CrCl(CN)(NH$_3$)$_4$]$^+$	Cl$^-$	H$_2$O (3)	90.4	15.0			23
			10.0	2.7			
trans-[CrF$_2$(tn)$_2$]$^+$	F$^{-\,e}$				+21.7		24
trans-[CrFCl(tn)$_2$]$^+$	Cl$^{-\,f}$				−69.5		24

^{a} Calculated from cited activation parameters.
^{b} Measured from the intercept of plots of k_{obs} vs. [NCS]$^-$ for the anation reaction (Section 6.3.2).
^{c} Measured directly.
^{d} 0.1 M HClO$_4$.
^{e} $\Delta V^{\ddagger} = -2$ to -3.
^{f} $\Delta V^{\ddagger} = -8$ to -7.

Table 6.3. Kinetic Parameters for the Reaction of $[Cr(AA)_3]^{3-}$ with EDTA at pH = 8.3 and I = 1.0 M $(NaClO_4)^a$

T (°C)	OH⁻		HCO₃⁻	
	$10^4 k'$ $(M^{-1} s^{-1})$	$10^5 k''$ (s^{-1})	$10^4 k'$ $(M^{-1} s^{-1})$	$10^5 k''$ (s^{-1})
55.0	1.9 (2.6)	1.6 (0.1)	3.2 (19.0)	4.0 (1.8)
60.0	2.7	3.0	3.6	6.3
65.0	3.7 (8.3)	4.9 (0.6)	3.9 (25.0)	9.1 (4.4)
75.0	(13.0)	(6.6)	(38.5)	(12.7)
ΔH^+	55.7 (79.5)	104 (193)	12.6 (31.0)	73.3 (80.4)
ΔS^+	−148 (−79.1)	−22.2 (−228)	−275 (−91.7)	−106 (−92.5)

^{a} Data for AA = Memal and mal, in parentheses.[26]

Kinetic parameters associated with k_2 are, for $x = 2$: $k_2 = 3 \times 10^{-3} M^{-1} s^{-1}$, $\Delta H^+ = 72.0$, $\Delta S^+ = -51.6$; and for $x = 1$: $k_2 = 6 \times 10^{-5} M^{-1} s^{-1}$, $\Delta H^+ = 83.8$, $\Delta S^+ = -44.8$.

The final product, $[Cr(OH_2)_4(hist)]^{3+}$, has been used in thiocyanate anation studies[21] (Section 6.3.2) and the relative resistance of this complex to Cr—N bond rupture illustrates the increased stability of six-membered N-donor chelate rings when compared to five-membered analogs.

6.2.6. Porphyrins

Just as in conventional substitution reactions in Cr(III) complexes, there is considerable controversy as to the mechanism of substitution in Cr(III) *meso*-substituted porphyrins; negative ΔS^+ values suggest associative activation of the axial sites while positive ΔV^+ values imply dissociative activation. A thorough study of the kinetic and equilibrium properties of the axial ligation reaction between imidazole[31] and pyridine[32] with $[Cr(TPPS)(H_2O)_2]^{3-}$ has now been made. Data were analyzed according to Scheme 1 and the rate and equilibrium constants are reported in Table 6.4. Data for the toluene soluble [CrCl(TPP)X] (X = py, quinoline, or PPh$_3$) with 1-methyl imidazole (Meim) are not quite comparable, as the reaction is not first order in Meim.[33] Kinetic parameters for the dissociation of X from [CrCl(TPP)X] in toluene are, in order: X, k (s^{-1}), ΔH^+, ΔS^+, ΔV^+; py, 8.0, 86.1, +61, +25.7; quinoline, 7.0, 85.3, +57, +23.8; and PPh$_3$, 4.2, 78.2, +29, +19.6.

6.3. Formation of Chromium(III) Complexes

6.3.1. The Nature of the Cr^{3+} Cation in Aqueous Solution

A detailed time-dependent product distribution of the hydrolytic oligomers formed when up to 0.8 equivalents of base is added to Cr^{3+}(aq) is now available.[34]

Scheme 1

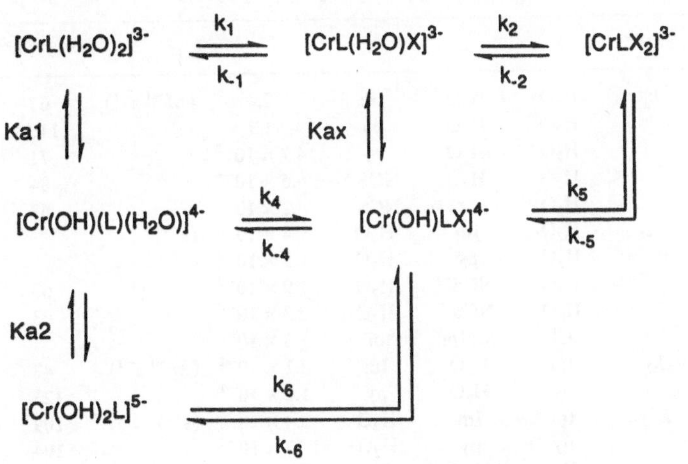
[CrL(H2O)2]3-
k1
k-1
[CrL(H2O)X]3-
k2
k-2
[CrLX2]3-
Ka1
Kax
[Cr(OH)(L)(H2O)]4-
k4
k-4
[Cr(OH)LX]4-
k5
k-5
Ka2
[Cr(OH)2L]5-
k6
k-6

After the addition of base, oligomerization causes a relatively rapid (hours) drop in pH. The concentration of the dimer (**2**) reaches a maximum after a few days ($I = 1.0\ M$, $NaClO_4$, $T = 25.0\,°C$) while the trimer (**3**) concentration increases continuously throughout the reaction and finally becomes the dominant product. The tetramer (**4**) concentration eventually (months) becomes constant and the

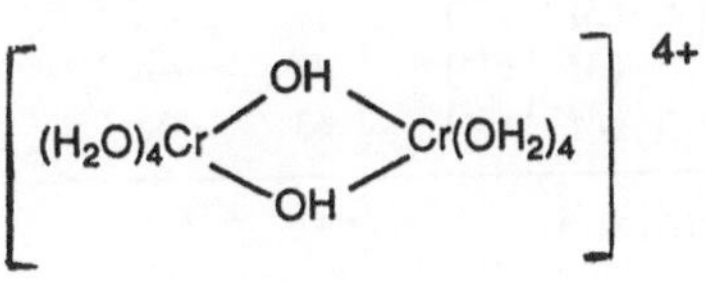
OH
(H2O)4Cr
Cr(OH2)4
OH
4+

dimer

(**2**)

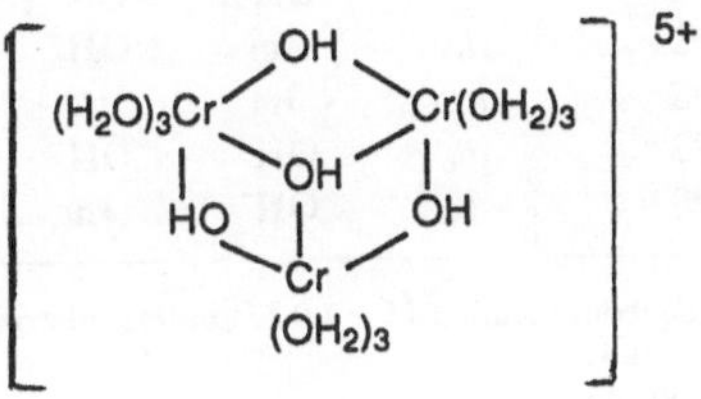
OH
(H2O)3Cr
Cr(OH2)3
OH
HO
OH
Cr
(OH2)3
5+

trimer

(**3**)

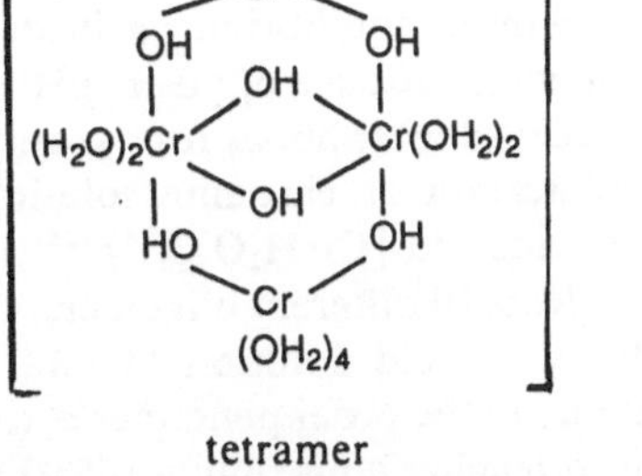
(OH2)4
Cr
OH
OH
OH
(H2O)2Cr
Cr(OH2)2
OH
HO
OH
Cr
(OH2)4
6+

tetramer

(**4**)

Table 6.4. Activation Parameters for the Forward (k_i) and Reverse (k_{-i}) Reaction trans-[Cr(porphyrin)XL] + Y $\rightleftharpoons$ [Cr(porphyrin)XY] + L[a]

Porphyrin	Path[b]	X	L	Y	$k_{\pm i}$		ΔH^+	ΔS^+	ΔV^+
TPPS	k_1	H_2O	H_2O	Im	2.4	$(M^{-1}s^{-1})$	67	−8	+7.4
		H_2O	H_2O	py	1.3		110	+125	
		H_2O	H_2O	py	4.7×10^{-3}		71	−54	+7.4
TMPP		H_2O	H_2O	NCS^-	6.6×10^{-4}		84	−23	
TAPP		H_2O	H_2O	NCS^-	4.2×10^{-3}		67	−64	+9.2
TPPS	k_{-1}	H_2O	Im	H_2O	$<4 \times 10^{-4}$	(s^{-1})			
		H_2O	py	H_2O	1.5×10^{-3}				
		H_2O	NCS^-	H_2O	1.9×10^{-3}		67	−79	
TMPP		H_2O	NCS^-	H_2O	2.7×10^{-4}		63	−100	
TPP[c]		Cl	MeIm	tol[c]	3.3×10^{-2}				
TPPS	k_2	Im	H_2O	Im	3.1×10^{-1}	$(M^{-1}s^{-1})$	42	−113	
		py	H_2O	py	1.8×10^{-1}		123	+149	
	k_{-2}	Im	Im	H_2O	2.4×10^{-4}	(s^{-1})	109	+54	
		py	py	H_2O	1.72×10^{-2}		104	+71	
TPPS	k_4	OH^-	H_2O	Im	9.8×10	$(M^{-1}s^{-1})$	67	+12	
		OH^-	H_2O	py	2.4×10^2	(pH = 10.31)	54	−17	
		OH^-	H_2O	py	1.87×10^2	(pH = 12.79)	49	−39	
		OH^-	H_2O	NCS^-	2.8×10		71	+17	
TMPP		OH^-	H_2O	NCS^-	4.0		63	−29	
		OH^-	H_2O	CN^-	7.9×10^{-1}		42	−113	
	k_{-4}	OH^-	Im	H_2O	6.0×10^{-2}	(s^{-1})	—	—	
		OH^-	py	H_2O	1.91×10	(pH = 10.31)	50	−51	
		OH^-	py	H_2O	8.73	(pH = 12.79)	54	−45	
TPPS		OH^-	CN^-	H_2O	2.8×10^{-3}		71	−59	
TPPS	k_5	Im	OH^-	Im	8.0×10^{-3}	$(M^{-1}s^{-1})$	75	−33	
TPPS	k_{-5}	Im	Im	OH^-	3.0×10^{-2}	$(M^{-1}s^{-1})$	79	−4	
TPPS	k_6	OH^-	OH^-	Im	3.8	$(M^{-1}s^{-1})$	71	−4	
TPPS	k_6	OH^-	Im	OH^-	9.3×10^{-1}	$(M^{-1}s^{-1})$	63	−38	

[a] In aqueous solution ($I = 1.0\ M$) unless otherwise stated. Data from Refs. 31–33.
[b] Scheme 1.
[c] In toluene.

initially formed (days) higher oligomers (pentamers and hexamers) decay in concentration, contributing to the stabilization of the pH after a few weeks. However, constant pH does not imply equilibrium as the decay of higher oligomers and the condensation of monomers produce opposite pH effects. Product analysis suggests that equilibrium is reached after about four years. The trimer (3) is more efficiently generated by acidification of chromite solutions (the green solution formed when excess base is added to $[Cr(H_2O)_6]^{3+}$).[35] Immediate acidification gives a solution containing at least 10 different oligomers, and some $[Cr(OH_2)_6]^{3+}$. These higher oligomers decay in acid solution (1.2 M, HClO$_4$) to produce a monomer, dimer, trimer mixture in the order pentamer ≤ tetramer < hexamer. The proportion of trimer produced reaches a maximum (65%) after about 27 hours. If chromatographically concentrated solutions of the trimer are now made alkaline (pH ~ 7), a light green precipitate of "active trimer hydroxide" $[Cr_3(\mu\text{-}OH)_4(OH)_5(OH_2)_4]\cdot 4H_2O$ is formed. This solid dissolves to give >90% of the trimer (3) when acidified with $HClO_4$.

6.3.2. Anation Reactions

Although $[Cr(OH_2)_6]^{3+}$ and $[Cr(OH)(OH_2)_5]^{2+}$ continue to be the main starting species for anation studies (Table 6.5), there is a growing interest in the use of other aqua ions, such as cis-$[Cr(big)_2(OH_2)_2]^{3+}$ or $[Cr(ox)_2(OH_2)_2]^-$. In view of the interest in the biochemical role of Cr(III) in the Glucose Tolerance Factor (GTF), several Cr(III) aqua ions have been reacted with insulin and the products characterized after purification by dialysis.[42] The chymotrypsin catalyzed hydrolysis of these Cr(III)–insulin complexes has also been investigated.[43]

Many anation studies can be described in terms of the mechanism shown in Eqs. (7) and (8), where there is an interchange of an H_2O ligand with the anating ligand after the formation of an outer-sphere ion pair. Under such circumstances, the observed rate constant (k_{obs}) can be described by the expression (9) where a and b are functions of K_{os}, $[H^+]$, and the various protonated forms of the anating ligand and the coordinated aqua ions.

$$[CrL(H_2O)]^{n+} + A^{y-} \overset{K_{os}}{\rightleftharpoons} \{CrL(H_2O)\cdot A\}^{(n-y)+} \tag{7}$$

$$\{CrL(H_2O)\cdot A\}^{(n-y)+} \overset{k}{\longrightarrow} products \tag{8}$$

$$k_{obs} = a[A]_T/(1 + b[A]_T) \tag{9}$$

Table 6.5. Anation Reactions

Substrate	Ligand	pH	$10^5 k_{an}$ (M^{-1} s^{-1})	$\Delta H^{\neq}$	$\Delta S^{\neq}$	Mech.	Ref.
$[Cr(OH_2)_6]^{3+}$	(S)-phenylalanine		18.5 (35 °C)	54	−142	I_a	36
	(2S,4R)-hydroxyproline		10 (40 °C)	73	−98	I_a	37
$[Cr(OH)(OH_2)_5]^{2+}$	(S)-phenylalanine	5.0	71 (45 °C)			I_d	36
	(2S,4R)-hydroxyproline	5.0	21 (40 °C)	63	−114	I_d	37
	bipy^a	5.0	12.1 (50 °C)	66		D	38
	phen^a	5.0	12 (40 °)				38
	Salicylate^d	5.0	9.6 (25 °C)				39
	Phthalate^d	5.0	16.1 (25 °C)				39
	Oxalate^d	5.0	16.4 (25 °C)				39
	Tartrate^d	5.0	18.0 (25 °C)				39
	Citrate^d	5.0	28.8 (25 °C)				39
c-$[Cr(ox)(hist)(OH_2)_2]^+$	ox	3.0^c	75 (65 °C)	81	−48		40
c-$[Cr(big)_2(OH_2)_2]^{3+}$	(S)-histidine	4.0^b	40 (45 °C)	26	−251	I_a	41

^{a} In 40% EtOH/H_2O.
^{b} [hist] = 0.1 M, I = 0.5 M.
^{c} [ox] = 0.1 M, I = 1.0 M.
^{d} [carboxylate] ~ 0.01 M, I ~ 0.13 M.

Table 6.6. Observed Pseudo-First-Order Rate Constants $(\times 10^6, s^{-1})$ for the Anation of $[Cr(H_2O)_4(AA)]^{n+}$ with NCS^- $[I = 1.0\,M\ (H^+, Na^+, ClO_4^-, NCS^-);\ pH = 1\text{--}2]^{(20,21)}$

T (°C)	[NCS$^+$]^a (M)						
	0.05	0.1	0.2	0.3	0.35	0.4	0.5
30.0	(0.653)	(1.37)	(2.67)		(4.38)		(6.10)
		0.322	0.502	0.741		0.897	1.07
35.0		0.523	0.901	1.23		1.56	1.83
37.0	(1.76)	(3.38)	(6.64)		(10.8)		(15.1)
40.0		0.861	1.39	1.90		2.28	2.64
44.0	(4.12)	(7.72)	(16.2)		(26.5)		(35.1)
45.0		1.44	2.31	2.88		3.76	4.27
50.0		2.54	3.57	4.83		6.09	6.49
51.0	(9.21)	(18.4)	(36.3)		(58.1)		(82.3)

^{a} Data in parentheses are for AA = hist, other entries are for AA = ox.

Thiocyanate is a popular anating ligand (Table 6.6) because: the products are relatively inert, the products are strongly absorbing in the UV–visible region, and the ligand is free from protolytic equilibria over a reasonable pH range. Possible disadvantages are: the potential for ambidentate coordination, and slow decomposition in strongly acidic media. In many cases the simple rate law (10) is obeyed with linear plots of k_{obs} vs. $[NCS^-]$ with a non-zero intercept. The interpretation of such plots, in terms of an ion-association preequilibrium (K_{os}) followed by a reversible rate-determining ligand interchange process [Eqs. (11) and (12)], has usually been oversimplified because the formation of ion pairs (K'_{os}) between the substrate and the supporting electrolyte (e.g., ClO_4^-) [Eq. (13)] has been neglected.[20] However, until K'_{os} data become readily available, this neglect will

$$k_{obs} = k_1 + k_2[NCS^-] \tag{10}$$

$$[Cr(L)_x(H_2O)_{6-x}]^{n+} + NCS^- \xrightleftharpoons{K_{os}} \{[Cr(L)_x(H_2O)_{5-x}]^{n+}\cdot NCS^-\} \tag{11}$$

$$\{[Cr(L)_x(H_2O)_{6-x}]^{n+}\cdot NCS^-\} \mathrel{\mathop{\rightleftharpoons}^{k_2}_{k_1}} [Cr(L)_x(NCS)(H_2O)_{6-x-1}]^{(n-1)+} + H_2O \tag{12}$$

$$[Cr(L)_x(H_2O)_{6-x}]^{n+} + ClO_4^- \xrightleftharpoons{K'_{os}} \{[Cr(L)_x(H_2O)_{6-x}]^{n+}\cdot ClO_4^-\} \tag{13}$$

probably continue. An alternative approach is to assume that $K_{os}(NCS^-) = K_{os}(ClO_4^-)$ and $[NCS^-] + [ClO_4^-] = I$, but this procedure seems to have no particular advantage over neglect, apart from recognition of the problem. More important is to verify that k_1 (obtained from the intercept of k_{obs} vs. $[NCS^-]$ plots) does indeed correspond to the measured rate of aquation of the product.[21] UV–visible absorption spectroscopy is the usual method adopted to follow reactions of this type, but low concentrations of Cr(III) $(\sim 10^9\,M)$ can be detected by chemiluminescence. Cr(III) bound to chelating ligands does not exhibit this property and the chemiluminescence method has been used to determine the rates of reaction of $[Cr(OH)(OH_2)_5]^{2+}$ $(10^{-6}\,M, pH \sim 5)$ with chelating carboxylic acids $(0.1\,M)^{(39)}$ (see Table 6.5).

The *trans*-labilization of a normally inert water molecule in penta-aquaalkylchromium(III) complexes[44] has now been explored with the prospect of forming Cr(III) complexes of phosphate derivatives such as AMP and ADP.[45] Using $[(H_2O)_5CrCH_2CN]^{2+}$ and phosphate or pyrophosphate as models, there is first a very rapid reaction (sec) followed by two sequential slower (hours) reactions. The final product has lost the Cr—C bond and appears to be $[(H_2O)_4Cr(OPO_x)_2]^{n-}$. The proposed reaction sequences are substitution, isomerization, and Cr—C bond rupture, with the isomerization step being followed by rapid incorporation of a second phosphate residue. Rate constants for the substitution $(M^{-1}\,s^{-1})$ followed by the isomerization (in parentheses) (s^{-1}) are $H_2P_2O_7^{2-}$, 0.63 (5.3×10^{-4}); $H_3P_2O_7^-$, 0.50 (1.4×10^{-3}); $H_2PO_4^{4-}$, 0.47 (1.08×10^{-3}); H_3PO_4, 0.068 (—); $H_2PO_2^-$, 0.19 (3×10^{-4}); H_2PO_2, 0.081 (—); $HC_2O_4^-$, 1.6 (2.8×10^{-2}).[45]

6.4. Base Hydrolysis

The rates of proton exchange have now been measured[46] for *trans*-$[Cr(NCS)_4(NH_3)_2]^{-}$ [47] to supplement the previously determined base hydrolysis data.[48] The value of $k_{exch} = 8 \times 10^2\,M^{-1}\,s^{-1}$ is about 2×10^8 times greater than that of base hydrolysis. From a comparison with other systems where both k_{OH} and k_{exch} data are available it is concluded that: the reactivity of a conjugate base derived from an anionic complex is greater than that derived from a cationic complex, and conjugate base formed by Co(III) complexes are about 10^4 times more reactive than those obtained for their Cr(III) analogs. It is also apparent that the proton exchange rate is dependent on the type of NH proton and the nature of the coordinated anionic ligands.

One other structural feature that has recently been explored in Cr(III) chemistry is the influence of the "flat" *sec*-NH proton on the rate of hydrolysis. The incorporation of this feature into the skeleton of a Co(III) chloro amine complex causes rate increases of up to 10^4 with respect to analogous systems where this feature is absent. However, *mer*-$[CrCl(en)(dpt)]^{2+}$ has a base hydrolysis rate of only 5 times that of *mer*-$[CrCl(en)(5Medpt)]^{2+}$ [49] (Table 6.7). On the basis of the activation volume data now becoming available for base hydrolysis it is possible that this process is more associative than previously thought and an interchange conjugate base mechanism for haloaminechromium(III) complexes has been proposed.[49,50]

The kinetics of hydrolysis of $[(L)Cr(OH)_2Cr(L)]^{4+}$ [L = tmpa (5)] have now been measured in basic solution.[51] The process has been described in terms of

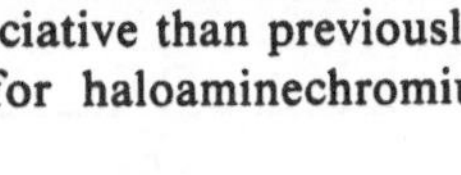

$$N-\left(CH_2-\left\langle\!\!\!\bigcirc\!\!\!\right\rangle\right)_3$$

tmpa

(5)

Table 6.7. *Activation Parameters for the Base Hydrolysis of Some $[CrX(N_5)]^{2+}$ Complexes*[a]

N_5	X	k_{OH}[b] $(M^{-1}s^{-1})$	$\Delta H^{\ddagger}$	$\Delta S^{\ddagger}$	$\Delta V^{\ddagger}$
$(NH_2CH_3)_5$	Cl	6.10×10^{-2}			+34.8
$(NH_3)_5$	Cl	5×10^{-4} (33.5 °C)			+17.0
$(NH_3)_5$	I	7.72×10^{-2} (17.2 °C)			+22.0
$(NH_3)_5$	$OCHNH_2$[c]	3.07×10^{-1}	54.3	−73	−7.6
$(NH_3)_5$	$OC(NH_2)_2$[c]	9.5×10^{-2} (23.5 °C)	62.2	−58	−9.6
$(NH_3)_5$	$OC(CH_3)N(CH_3)_2$	1.5×10^{-1} (29.0 °C)			+25.0

[a] Reference 50.
[b] At 5 kPa pressure.
[c] O- to N-bonded isomerization.

Scheme 2

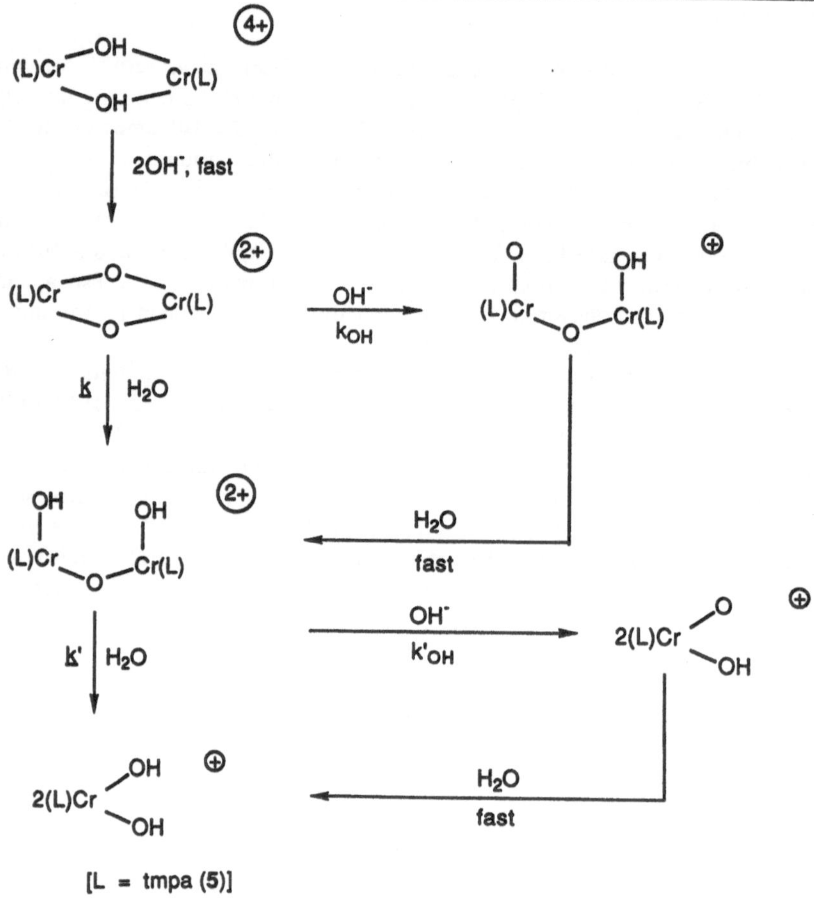

[L = tmpa (5)]

Table 6.8. Activation Parameters for the Rate Constants Associated with Scheme 2 (L = tmpa (5), I = 1.0 M)^a

Rate constant		ΔH^+	ΔS^+
k_{OH} (M^{-1} s^{-1})	5.8×10^{-4}	94	+8
k'_{OH} (M^{-1} s^{-1})	2.7×10^{-5}	117	+58
k (s^{-1})	3.0×10^{-4}	83	-33
k' (s^{-1})	4.7×10^{-5}	96	-4

^{a} Reference 51.

Scheme 2 as the observed first-order rate constants (at fixed pH) show fast (k_{obs}) and slow (k'_{obs}) components, both of which have a hydroxide-dependent (k_{OH}, k'_{OH}) and a hydroxide-independent (k, k') contribution (Table 6.8).

6.5. Oxdiation and Reduction of Cr(III) Complexes

The kinetics of the oxidation of $[Cr(OH_2)_6]^{3+}$ by Ce(IV)[52] and IO_4^- [53] as well as a series of aminocarboxylate [(edta,[54] 2-hydroxy-ethylene diaminetriacetate (toh),[55] nta[56,57]] chromium(III) complexes by IO_4^- have been reported. In all cases, the product is Cr(VI) [Eqs. (14) and (15)] and two of the studies describe the effect of mixed solvents[53,54] (Table 6.9). An inner-sphere electron transfer process is proposed for reaction (14).

$$Cr(III) + I(VII) \rightleftharpoons \{Cr(III)\cdot I(VII)\} \tag{14a}$$

$$\{Cr(III)\cdot I(VII)\} \rightarrow Cr(V) + I(V) \tag{14b}$$

$$Cr(V) + I(VII) \rightarrow Cr(VI) + I(VI) \tag{14c}$$

$$2I(VI) \rightarrow I(VII) + I(V) \tag{14d}$$

$$2Cr(III) + 3I(VII) \rightarrow 3I(V) + 2Cr(VI) \tag{14}$$

$$Cr(III) + Ce(IV) \rightarrow Ce(III) + Cr(IV) \tag{15a}$$

$$2Cr(IV) \rightarrow Cr(III) + Cr(V) \tag{15b}$$

$$Cr(V) + Ce(IV) \rightleftharpoons Ce(III) + Cr(VI) \tag{15c}$$

$$Cr(III) + 3Ce(IV) \rightarrow 3Ce(III) + Cr(VI) \tag{15}$$

With excess IO_4^-, the oxidations[53-56] proceed by a pseudo-second-order process (k_2), but the Ce(IV) oxidation, which is initially second order, is slowed by negative catalysis from the Ce(III) product, to give a first-order process (k_1).[52] Rate data are presented in Table 6.9.

$[(H_2O)_5CrO_2]^{2+}$ (λ_{max}247 nm; $\varepsilon = 7400$ M^{-1} cm^{-1}, λ_{max}290 nm; $\varepsilon = 3100$ M^{-1} cm^{-1}) formed from $[Cr(OH_2)_6]^{2+}$ with the stoichiometric amount of O_2 is a semistable Cr(III) superoxide complex ($t_{1/2} \sim 15$ min in 0.1 M $HClO_4$ at 25 °C, under anaerobic conditions). The reaction between this species and $[Cr(OH_2)_6]^{2+}$ was reported previously[1] and studies have now been extended to include a wider range of metal-ion-reducing agents[58] (Table 6.10). Data from other studies using

Table 6.9. Kinetic Parameters for the Oxidation of Cr(III) Species by IO_4^- or Ce(IV)

Cr(III) species	Medium	Oxidant	k_2 $(M^{-1}s^{-1})$	k_1 (s^{-1})	ΔH^+	ΔS^+	Ref.
$[Cr(OH_2)_6]^{3+}$	$HClO_4{}^a$	Ce(IV)	23.4	3.90×10^{-2}			52
	$H_2SO_4{}^b$	Ce(IV)	6.12	5.9×10^{-3}			52
	$H_2O^{c,d}$	IO_4^-	23.1		42.4	43.7	53
$[Cr(edta)(OH_2)]^-$	$H_2O^{e,d}$	IO_4^-	7.73		60.2	-63.6	54
$[Cr(nta)(OH_2)_2]$	H_2O^f	IO_4^-	6.32×10^{-2}		14.0	-220	56
$[Cr(toh)(OH_2)]$	H_2O^g	IO_4^-	2.95×10^{-3}		76.0	-39	55

^{a} $[Cr^{3+}] = 16.7 \, mM$, $[Ce^{4+}] = 50.0 \, mM$, $[H^+] = 2.27 \, M$, $I = 4.08$. Both k_2 and k_1 increase with increasing I and $[H^+]$.
^{b} $[Cr^{3+}] = 9.98 \, mM$, $[Ce^{4+}] = 30.2 \, mM$, $[SO_4^{2-}] = 30 \, mM$, $I = 4.08$. Both k_2 and k_1 decrease with increasing $[SO_4^{2-}]$.
^{c} $[Cr^{3+}] = 0.5 \, mM$, $[IO_4^-] = 30 \, mM$, $[H^+] = 2.85 \, mM$, $I = 0.5 \, M$.
^{d} Data in EtOH : H_2O also given.
^{e} $[Cr^{3+}] = 0.05 \, mM$, $[IO_4^-] = 5 \, mM$, $I = 0.2 \, M$.
^{f} $I = 0.5 \, M$.
^{g} $[Cr^{3+}] = 0.1 \, mM$, $I = 0.5 \, M$, pH = 7.30.

Ni^{3+}, Ni^+, Cu^+, or $MV^{+\cdot}$ as reactants are also given in Table 6.10. The redox potentials for a number of (diamino polycarboxylato) chromium II/III redox couples have recently been determined.[62]

6.6 Isomerization and Racemization

The *trans* → *cis* isomerization kinetics of bis(oxalato)diaquachromium(III) have now been measured in water : ethanol and water : *t*-butanol mixtures.[63] An oxalate dissociative mechanism has been proposed, as the rates in H_2O and D_2O are identical. When alcohol is added to water, there is a marked decrease in K_{isom}. Plots of $\ln(k_{isom})$ vs. $\ln[H_2O]$ have breaks at positions corresponding to boundary values for water : alcohol ratios determined for other systems (~0.05 mole fraction of *t*-butanol). In water, $10^4 k_{isom} (T \, °C) s^{-1} = 7.96$ (30.0), 9.33 (31.0), 11.2 (35.0), 22.2 (41.0), giving $E_a = 74$, $\Delta S^+ = +54$.

The configuration adopted by $Cr\{(R)(R)\text{-}(-)\text{-}bdtp\}_3$ [bdtp = (6)] is solvent dependent.[64] If the synthesis is performed in THF, the Δ enantiomer is first

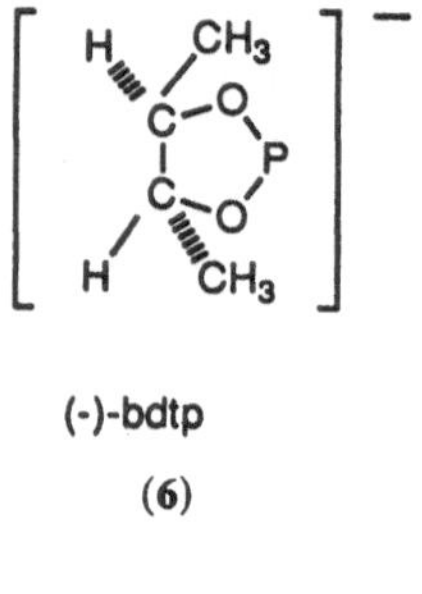

(-)-bdtp

(6)

Table 6.10. Second-Order Rate Constants (k, $M^{-1} s^{-1}$) Obtained for the Oxidation or Reduction of Cr(III) Complexes

Reductant	Oxidant	k ($M^{-1} s^{-1}$)	Mech[d]	Ref.
$[Ru(NH_3)_6]^{2+}$	$[(H_2O)_5Cr(O_2)]^{2+}$ [a]	9.5×10^5	os	58
$[Co(sep)]^{2+}$		8.5×10^5	os	58
$[V(OH_2)_6]^{2+}$		2.3×10^5	os	58
$[Co(cyclam)]^{2+}$		$\sim7 \times 10^6$	is	58
$[Co([15]aneN_4)]^{2+}$		6.2×10^5	is	58
$[Fe(OH_2)_6]^{2+}$		4.5×10^3	is	58
$[Cr(OH_2)_6]^{2+}$		$\sim8 \times 10^8$	os	58
$[(H_2O)_5Cr(CH_3)]^{2+}$	$[Ni(cyclam)]^{3+}$ [c]	<1	os	59
$[(H_2O)_5Cr(CH_2CH_3)]^{2+}$		1.3×10	os	59
$[(H_2O)_5Cr(CH(CH_3)_2)]$		2.2×10^2	os	59
$[(H_2O)_5Cr(CH_2COCH_3)]$		2.45×10^3	os	59
$[(H_2O)_5Cr(CH(CH_3)OCH_2CH_3)]^{2+}$		3.44×10^3	os	59
$[(H_2O)_5Cr(CH_2Ph)]^{2+}$		1.88×10^4	os	59
$[(H_2O)_5Cr(CH(CH_3)_2)]^{2+}$	$[Ni(tach)_2]^{3+}$ [c]	9.5×10	os	59
$[(H_2O)_5Cr(CH_2Ph)]^{2+}$		1.31×10^4	os	59
$MV^{\cdot+}$ [e]	$[Cr(bipy)_3]^{3+}$ [b]	1.95×10^9	os	60
$[Ni(Me_6[14]dieneN_4)]^+$		3.1×10^8	os	61
$[Ni(Me_6[14]tetreneN_4)]^+$		8.0×10^8	os	61
$[Ni(Me_2pyo[14]trieneN_4)]^+$		5.1×10^8	os	61
$[Cu(Me_6[14]dieneN_4)]^+$		3.7×10^6	os	61

[a] $I = 0.1\ M = [HClO_4]$.
[b] $I = 10^{-2}\ M$, pH = 4.
[c] $T = 23.4\ °C$, $I = 1.0\ M = [HClO_4]$.
[d] os = outer sphere; is = inner sphere.
[e] $MV^{\cdot+}$ is the methyl viologen radical cation.

formed (maximum concentration after 20 h at room temperature) but this reverts to the Λ isomer, and after 70–80 h an equilibrium system (with excess Λ) is reached. If EtOH is used as a solvent, the Λ isomer precipitates as it is formed. In THF, the rate of $\Delta \rightleftharpoons \Lambda$ equals $14 \times 10^{-4}\ s^{-1}$, a value similar to that obtained in $CHCl_3$. The CD intensity of the equilibrium $\Lambda \rightleftharpoons \Delta$ mixture in THF is temperature dependent and thermodynamic parameters for the equilibrium constant have been obtained. These parameters are also solvent dependent.[64]

6.7. Photochemistry and Photophysics of Chromium(III) Complexes

The $[Cr(bipy)_3]^{3+}$ cation is an important model compound in the photochemistry and photophysics of Cr(III) systems.[65-71] Consequently the solid-state structure is of considerable interest. An error in the space group previously reported[72] for $[Cr(bipy)_3](PF_6)_3$ has now been corrected,[73,74] but the single-crystal X-ray structural analysis of the perchlorate salt reveals that frequently used

Table 6.11. Analytical Data for [Cr(bipy)$_3$]$^{3+}$ Salts

Salt	Formula	M_R	Calculated %			
			C	H	N	Cr
[Cr(bipy)$_3$](Hbipy)(ClO$_4$)$_4$	C$_{40}$H$_{38}$N$_8$CrCl$_4$O$_{16}$	1075.55	44.7	3.1	10.4	4.8
[Cr(bipy)$_3$](ClO$_4$)$_3$	C$_{30}$H$_{24}$N$_6$CrCl$_3$O$_{12}$	818.91	44.4	3.0	10.3	6.3

preparative routes can yield a material which co-crystallizes with Hbipy·ClO$_4$, viz triclinic [Cr(bipy)$_3$](Hbipy)(ClO$_4$)$_4$.[73] Such a double salt is not readily detected on the basis of C, H, and N analytical data (Table 6.11) and only Cr data are definitive. The use of the double salt, under the impression that it was the pure perchlorate, would result in a reduction of the molar absorptivity coefficients by about 25%.

The chemistry of the excited state of [Cr(polypyridyl)$_3$]$^{3+}$ (polypyridyl = phen, bipy, etc.) is becoming quite sophisticated as [*Cr(polypyridyl)$_3$]$^{3+}$, generated by laser flash photolysis, have lifetimes of 250–420 μs in aqueous solution of various electrolytes. The chemistry can now proceed in two stages: (1) the addition of a quenching (reducing) agent, such as [Co(cyclam)]$^{2+}$,[70] oxalate,[65,71] ferrocenium ions,[69] or Ti(III),[67] to produce [Cr(polypyridyl)$_3$]$^{2+}$ and a one-electron transfer oxidized product, or (2) subsequent reaction of [Cr(polypyridyl)$_3$]$^{2+}$ with a suitable oxidizing agent, such as O$_2$,[65] SO$_2$,[70] or even [*Cr(polypyridyl)$_3$]$^{3+}$.[67] Data for the quenching kinetics are reported in Table 6.12 and data for the [Cr(poly-pyridyl)$_3$]$^{2+}$ reductions in Table 6.17 (see below).

The production of Cr(VI) as a byproduct in the 254 nm photolysis of Cr(III) ammine complexes has been studied in detail.[76] In HCl or HClO$_4$, O$_2$ is necessary

*Table 6.12. The Rate of Quenching [k_q (M^{-1} s^{-1})] of [*Cr(polypyridyl)$_3$]$^{3+}$ by Various Reducing Agents*

Polypyridyl complex	Quenching agent	$10^{-8} k_q$ ($M^{-1} s^{-1}$)	Ref.
[*Cr(bipy)$_3$]$^{3+}$	[Fe(Cp)$_2$]$^+$	4.5^a	69
	[Fe(Cp)(C$_5$H$_4$-nBu)]$^+$	3.2^a	69
	[Fe(Cp)(C$_5$H$_4$CH$_2$NMe$_2$)]$^+$	1.89^a	69
	[Fe(Cp)(C$_5$H$_4$CH$_2$OH)]$^+$	1.62^a	69
	[Fe(C$_5$H$_4$CH$_3$)$_2$]$^+$	1.40^a	69
	[Fe(C$_5$H$_4$-nBu)$_2$]$^+$	0.30^a	69
	[Fe(C$_5$Me$_5$)$_2$]$^+$	0.23^a	69
	[Fe(Cp)$_2$]	102^b	69
	[Fe(Cp)(C$_5$H$_4$-nBu)]	100^b	69
	[Fe(C$_5$H$_4$Me$_3$)$_2$]	96^b	69
	[Fe(Cp)(C$_5$H$_4$CHO)]^b	83.5	69
	[Fe(Cp)(C$_5$H$_4$CH$_2$OH)]^b	80	69
	[Fe(Cp)(C$_5$H$_4$CO$_2$H)]^b	71.5	69

(continued)

Table 6.12. (continued)

Polypyridyl complex	Quenching agent	$10^{-8} k_q (M^{-1} s^{-1})$	Ref.
	[Fe(C$_5$H$_4$CO$_2$H)$_2$][b]	61	69
	[Fe(Cp)(C$_5$H$_4$CH$_2$NMe$_2$)][b]	45	69
	[Ru(Cp)$_2$][b]	60	69
	[Os(Cp)$_2$][b]	96	69
	[Co(cyclam)(OH$_2$)$_2$]$^{2+}$ [c]	1.8	70
	Ti(III)[d]	0.199	67
	[Ti(OH$_2$)$_6$]$^{3+}$ [e]	0.19	67
	[Ti(OH)(OH$_2$)$_5$]$^{2+}$ [e]	6.0	67
	C$_2$O$_4^{2-}$	3.59×10^{-3} [f]	71
		1.57×10^{-3} [g]	71
[*Cr(phen)$_3$]$^{3+}$	C$_2$O$_4^{2-}$ [f]	4.14×10^{-3}	71
	Ti(III)[d]	0.212	67
	[Ti(H$_2$O)$_6$]$^{3+}$ [e]	0.16	67
	[Ti(OH)(H$_2$O)$_5$]$^{2+}$ [e]	4.6	67
	[Co(cyclam)(OH$_2$)$_2$]$^{2+}$ [c]	3.3	70
[*Cr(4,4'-Me$_2$bipy)$_3$]$^{3+}$	[Fe(C$_5$Me$_5$)$_2$]$^{+}$ [a]	0.137	69
	[Co(cyclam)(OH$_2$)$_2$]$^{2+}$ [c]	0.7	70
	Ti(III)[d]	0.0334	67
	[Ti(OH$_2$)$_6$]$^{3+}$ [e]	0.026	67
	[Ti(OH)(OH$_2$)$_5$]$^{2+}$ [e]	1.0	67
	C$_2$O$_4^{2-}$ [f]	2.1×10^{-4}	71
[*Cr(5-Clphen)$_3$]$^{3+}$	C$_2$O$_4^{2-}$ [f]	8.54×10^{-2}	71
	Ti(III)[d]	0.387	67
	[Ti(OH$_2$)$_6$]$^{3+}$ [e]	0.37	67
	[Ti(OH(OH$_2$)$_5$]$^{2+}$ [e]	9.7	67
	[Co(cyclam)(OH$_2$)$_2$]$^{2+}$ [c]	5.2	70
	[Fe(C$_5$Me$_5$)$_2$]$^{+}$ [a]	0.154	69
[*Cr(5-Mephen)$_3$]$^{3+}$	[Co(cyclam)(OH$_2$)$_2$]$^{2+}$ [c]	4.04	70
	Ti(III)[d]	0.169	67
	[Ti(OH$_2$)$_6$]$^{3+}$ [e]	0.14	67
	[Ti(OH(OH$_2$)$_5$]$^{2+}$ [e]	4.7	67
	C$_2$O$_4^{2-}$ [f]	3.18×10^{-3}	71
[*Cr(5,6-Me$_2$phen)$_3$]$^{3+}$	Ti(III)[d]	0.112	67
	[Ti(OH$_2$)$_6$]$^{3+}$ [e]	0.12	67
	[Ti(OH)(OH$_2$)$_5$]$^{2+}$ [e]	3.0	67
	[Co(cyclam)(OH$_2$)$_2$]$^{2+}$ [c]	2.55	69
[*Cr(4,7-Me$_2$phen)$_3$]$^{3+}$	Ti(III)[d]	0.0311	67
	[Ti(OH$_2$)$_6$]$^{3+}$ [e]	0.028	67
	[Ti(OH)(OH$_2$)$_5$]$^{2+}$ [e]	0.81	67
	[Co(cyclam)(OH$_2$)$_2$]$^{2+}$ [c]	1.28	69
[*Cr(tacn)$_2$]$^{3+}$	[Cr(CN)$_6$]$^{3+}$	1.6	75
	[Cr(tacn)$_2$]$^{3+}$	0.014	75

[a] In 70:30 CH$_3$CN:H$_2$O, $\mu = 0.05\ M$, HClO$_4$.
[b] In 70:30 CH$_3$CN:H$_2$O, $\mu = 0.05\ M$, LiClO$_4$.
[c] In 1 M H$_2$SO$_4$.
[d] In 1 M HCl.
[e] At $\mu = 1.0\ M$, HCl/LiCl.
[f] At $\mu = 0.75\ M$, Na$_2$SO$_4$.
[g] At $\mu = 2.0\ M$, Na$_2$SO$_4$.

for Cr(VI) production but in HNO_3 media Cr(VI) is produced in the absence of O_2. Cation:nitrate ion pairs are postulated when nitrate is present, and the interaction between O_2 and Cr(III) excited states when oxidizing anions are absent. Patterns for the ^2E excited-state lifetimes of several families of Cr(III) complexes have now been explored.[77,78] At least three relaxation channels are needed to account for the observed patterns. The shortest-lived complexes tend to have mainly N—H or O—H containing ligands; the next group contains ligands with strained or distorted environments followed by those containing NCS or oxygen donor ligands; complexes with the longest lifetimes contain CN^- or N—D groups. The data on which these generalizations are based are given in Table 6.13.

Table 6.13. Kinetic Parameters for (^2E) Excited State Lifetimes for Some Chromium(II) Complexes[a]

Complex	τ_{77K}^{N-H} (μs)	τ_{77K}^{N-D} (ms)	τ_{298K}^{N-H} (μs)	E_{app}[b]
$[Cr(CN)_6]^{3-}$		3.98	5×10^d	
$[Cr(NCS)_6]^{3-}$		4.17	1.7×10^{-3}	33.5
$[Cr(ox)_3]^{3-}$		0.90	7×10^{-4}	5.4
$[Cr(CN)_4(phen)]^-$		4.35		
trans-$[Cr(NCS)_4(NH_3)_2]^-$ [e]	349	2.2	5.5×10^{-3}	27, 42
$[Cr(acac)_3]$		0.476		
$[Cr(AcO)_3(tacn)]$	395			
$[Cr(CN)_3(tacn)]$	401			
$[Cr(NCS)_3(tacn)]$[f]	112			
cis-$[Cr(NCS)_2(phen)_2]^+$		0.82		
cis-$[Cr(NCS)_2(bipy)_2]^+$		0.75		
trans-$[Cr(NCS)_2(NH_3)_4]^+$	140	2.5		
cis-$[Cr(NCS)_2(NH_3)_4]^+$	123	2.5		
trans-$[Cr(NCS)(CN)(NH_3)_4]^+$			30 (293 K)	38
trans-$[Cr(NCS)_2(cyclam)]^+$	96		4.4	31
cis-$[Cr(NCS)_2(cyclam)]^+$	184			
trans-$[CrCl_2(cyclam)]^+$	88		8×10^{-4}	49
cis-$[CrCl_2(cyclam)]^+$	52		1×10^{-2}	27
cis-$[CrF_2(cyclam)]^+$	80			
trans-$[CrF_2(teta)]^+$	49		3×10	
trans$[Cr(CN)_2(cyclam)]^+$	356	3.03		
cis-$[Cr(CN)_2(bipy)_2]^+$		3.60		
cis-$[Cr(CN)_2(tetb)]^+$	207	1.86		
trans-$[Cr(CN)_2(teta)]^+$	379	5.78		
trans-$[CrCl_2([15]aneN_4)]^+$	68		1×10^{-7}	60
$[CrCl_2(cyclamN_3S)]^+$	66		2×10^{-7}	61
trans-$[CrBr_2(teta)]^+$	122		2×10^{-5}	38
$[CrCl(NH_3)_5]^{2+}$	41, 43	3.3	2.2×10^{-6}	38
$[Cr(CN)(NH_3)_5]^{2+}$	79, 100	7.46	1.4×10	52
$[Cr(NCS)(NH_3)_5]^{2+}$	90, 70	3.23	2.7×10^{-1}	34
$[Cr(ONO)(NH_3)_5]^{2+}$	62		6×10^{-3}	27

(continued)

Table 6.13. (continued)

Complex	τ_{77K}^{N-H} (μs)	τ_{77K}^{N-D} (ms)	τ_{298K}^{N-H} (μs)	E_{app} [b]
$[CrF(NH_3)_5]^{2+}$	50		2×10^{-3}	30
$[Cr(NH_3)_5(OH_2)]^{3+}$	56.8, 49	3.45	4×10^{-2}	32
$[Cr(NH_3)_6]^{3+}$	78	5.32	2.2, 1.6 [c]	47
$[Cr(en)_3]^{3+}$	120	3.03		
$[Cr(tn)_3]^{3+}$	138	4.76	2.6 [c]	
$[Cr(sen)]^{3+}$	96		2.4×10^{-2}	42
$[Cr(ditn)_2]^{3+}$	208	4.17		
$[Cr(tacn)_2]^{3+}$	400	3.03, 3.21	1.62×10^2	
$[Cr\{Et(tacn)_2\}]^{3+}$	50.3	0.114		
cis-$[Cr(NH_3)_2[cyclam)]^{3+}$	112, 116	1.59	1.0	43
trans-$[Cr(NH_3)_2(cyclam)]^{3+}$		3.74	1.36×10^2	67
$[Cr(en)(cyclam)]^{3+}$	135, 136	2.22	3.0	44
trans-$[Cr(NH_3)_2(teta)]^{3+}$	186		3.12×10	31
$[Cr(NH_3)_2(cyclamN_3S)]^{3+}$	116		2.7×10^{-1}	35
cis-$[Cr(NH_3)_2(Me_2pyo[14]eneN_4)]^{3+}$	101		1.42	
cis-$[Cr(cyclam)(OH_2)_2]^{3+}$	92			
trans-$[Cr(cyclam)(OH_2)_2]^{3+}$	106		5×10^{-2}	32
$[Cr(NH_3)_3(tacn)]^{3+}$	104			
$[Cr(tacn)(OH_2)_3]^{3+}$	649			
$[Cr(phen)_3]^{3+}$		5.32		
$[Cr(bipy)_3]^{3+}$		5.0		
$[Cr(4,7-Ph_2phen)_3]^{3+}$		3.51		
$[Cr(4,7-Me_2phen)_3]^{3+}$		4.29		
$[Cr(5-Cl-phen)_3]^{3+}$		5.41		
trans-$[Cr(NH_3)_4(OH_2)_2]^{3+}$		2.40		
$[Cr(terpy)_3]^{3+}$		5.44		

[a] Data from Refs. 77 and 78 in DMSO:H_2O, 1:1. These references also give the ligand abbreviations used in this table except [14]aneN$_4$, [9]aneN$_3$, rac-Me$_6$[14]aneN$_4$, and ms-Me$_6$[14]aneN$_4$ have been changed to cyclam, tacn, tetb, and teta, respectively.
[b] Based on an Arrhenius analysis of $\ln \tau^{-1}$ vs. $1/T$ (K). This relationship is often quite matrix dependent.[1]
[c] Reference 27.
[d] See Ref. 79.
[e] Reference 80.
[f] Reference 81.

As mentioned previously,[82] intramolecular energy transfer can become a very efficient process if the donor and acceptor can be incorporated into the same molecule. The polynuclear complexes $[(NC)Ru(bipy)_2(\mu\text{-}CN)Cr(CN)_5]^{2-}$ and $[(CN)_5Cr(\mu\text{-}CN)Ru(bipy)_2(\mu CN)Cr(CN)_5]^{4-}$ have been synthesized and visible light absorption by the Ru(II) leads to phosphoresence from the Cr(III) with fast (<10 ns) and efficient ($n = 1$) intramolecular energy exchange.[79] Intermolecular self-quenching seems to occur in relatively concentrated solutions with high-energy laser pulses.

6.8. The Solid State

6.8.1. Single-Crystal X-Ray Structures

Although this section may seem out of place in a review of kinetics and mechanism, a knowledge of the structure of the starting material is fundamental to a meaningful interpretation of kinetic data. For example, the problems of distinguishing between $[Cr(bipy)_3](ClO_4)_3$ and $[Cr(bipy)_3](Hbipy)(ClO_4)_4$ have been discussed in Section 6.7. These and other "mechanistically interesting" single-crystal X-ray structures for chromium compounds are reported with bond lengths (in angstroms) and bond angles (in degrees).

6.8.1.1. Chromium(III) Mononuclear Six-Coordinate

1. $(\pm)[C(NH_2)_3]_3[Cr(ox)_3] \cdot H_2O^{(83)}$: $Cr-O = 1.967(3)-1.980(3)$. Crystal data are also given for the Na^+, K^+, Rb^+, NH_4^+, and $N(Me)_4^+$ salts.
2. $Na_3[Cr(HCO_2)_6] \cdot 4H_2O^{(84)}$: unidentate hexaformato complex; $Cr-O = 1.973(1)-1.992(1)$.
3. $[K(18\text{-crown-6})(H_2O)][K(18\text{-crown-6})][Cr(NCS)_5(H_2O)] \cdot H_2O.^{(85)}$
4. cis-β-$[Hpip][Cr(salen)(ox)]^{(86)}$: salen = (7). $Cr-O$ (ox) $= 1.986(3), 2.019(3)$; $Cr-O$ (phenol) $= 1.923(4), 1.937(4)$; $Cr-N = 2.010(4), 2.033(5)$; $C=N = 1.287(7), 1.280(7)$.

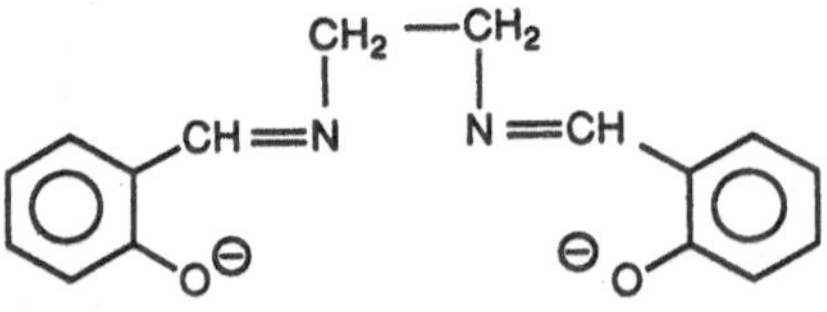

(7)

5. Δ-$(-)_{589}$-$[Co(en)_2(ox)][ucis$-$\Delta\Delta\Delta\Lambda$-$(S,S)$-$(-)_{456}$-$Cr(1,3\text{-pdda})(mal)] \cdot H_2O^{(87)}$: 1,3-pdda = (8). $Cr-O$ (mean) $= 1.959(2)$; $Cr-N$ (mean) $= 2.065(3)$.

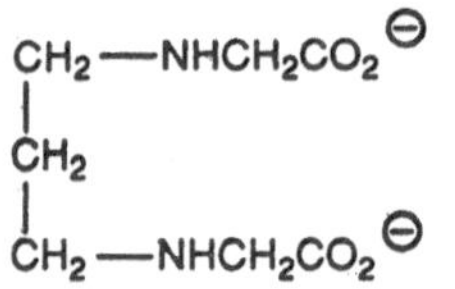

(8)

6. $Na[Cr(cyda)] \cdot 4.5H_2O^{(88)}$:   Six-coordinate, $Cr-O = 1.951(3)-1.998(3)$; $Cr-N = 2.058(3), 2.060(3)$.
7. $[Cr(cida)(H_2O)_2] \cdot 3H_2O^{(89)}$: cida = (9). The unsymmetrical quadridentate tripodal ligand adopts configuration (10) rather than (11). $Cr-N = 2.082(4)$;

Cr—OH$_2$ = 1.970(4), 1.968(4); Cr—O (carboxylate) = 1.930(3), 1.949(3), 1.963(3).

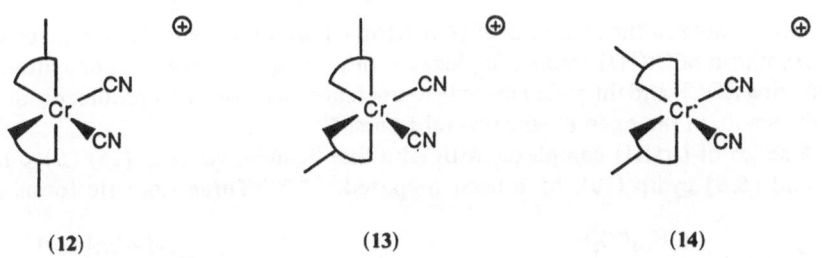

$\qquad$ (9) $\qquad$ (10) $\qquad$ (11)

8. *trans*-[Cr(CH$_2$Cl)(acac)$_2$(H$_2$O)]·EtOH[44]: Cr—C = 2.06(1); Cr—O (acac) = 1.943(5)–1.968(5); Cr—OH$_2$ = 2.134(6); C—Cr—O = 175.1(3).

9. *trans*-[Cr(CH$_2$Cl)(acac)$_2$(MeOH)][44]: two independent molecules per unit cell; Cr—C = 2.049(9), 2.052(9); Cr—O (acac) = 1.941(6)–1.974(5); Cr—O (MeOH) = 2.156(6), 2.138(6); C—Cr—O = 178.0(3), 177.4(4).

10. [Cr{S$_2$P(C$_2$H$_5$)$_2$}$_3$][90]: Cr—S = 2.442(8).

11. [Cr(NCS)$_3$(H$_2$O)$_3$] (18-crown-6).[91]

12. (−)-*cis*-Λ($\delta\delta$)-eq, eq-[Cr(CN)$_2${R-(+)-pn}$_2$]Cl[92]: One of two isolated isomers, with configuration (12). The others are (13) and (14).

$\qquad$ (12) $\qquad$ (13) $\qquad$ (14)

13. (±)-[Cr(en)$_3$]$_2$[HgCl$_4$]$_3$ [93]: Λ($\delta\delta\delta$) plus Δ($\lambda\lambda\lambda$); Cr—N (mean) = 2.074(9).

14. (±)-*mer*-[Cr(dien)$_2$][HgCl$_4$][Cl]·DMSO[93]: $\delta\lambda$, $\delta\lambda$; Cr—N (mean) = 2.077(8).

15. (±)-*ufac*-[Cr(dien)$_2$][HgCl$_4$][HgCl$_3$][93]: $\lambda\delta$, $\lambda\delta$; Cr—N (mean) = 2.075(22).

16. *sfac*-[Cr(dien)$_2$]$_4$[HgCl$_4$]$_2$[Hg$_2$Cl$_6$][Hg$_4$Cl$_{14}$][93]: $\lambda\delta$, $\lambda\delta$: Cr—N (mean) = 2.072(16).

17. (±)-*cis*-[CrCl(en)$_2$(Hen-*N*)][HgCl$_4$][Cl][93]: $\delta\lambda$; Cr—Cl = 2.302(3), Cr—N (mean) = 2.079(11).

18. (±)-*sfac*-[CrCl(dien)(Hdien-*N*,*N*)][Hg$_2$Cl$_7$][93]: $\lambda\lambda$,λ; Cr—Cl = 2.313(3), Cr—N (mean) = 2.092(24).

19. (±)-[Cr(bipy)$_3$](PF$_6$)$_3$ [74]: a redetermination in the centrosymmetric space group R3̄c; Cr—N (mean) = 2.042; bite angle = 79.1(3).

20. (±)-[Cr(bipy)$_3$](ClO$_4$)$_3$·[Hbipy](ClO$_4$)[73]: Cr—N (mean) = 2.032(6); bite angle = 79.9(4).

21. [Cr(urea)$_6$][Pt(OH)Cl$_5$]ClO$_4$·H$_2$O.[94]

6.8.1.2. Chromium(III) Dinuclear

22. trans-[(NH$_3$)(cyclam)(Cr)(μ-CO$_3$)Cr(cyclam)(NH$_3$)]I$_4$·2H$_2$O.[95]
23. [(tmpa)Cr(OH)$_2$Cr(tmpa)]Br$_4$·8H$_2$O[51,96]: tmpa = (5); *trans* centrosymmetric dimer; Cr—N (tert) = 2.041(1), Cr—N(py) = 2.048(11); Cr—O = 1.937(8), Cr—O—Cr = 101.0; O—Cr—O = 78.9.
24. cis-α-[{(OH)(pic)$_2$Cr(μ-OH)}$_2$Mn(H$_2$O)$_4$](S$_2$O$_6$)$_2$·2H$_2$O[97]: pic = 15; isomorphous with Co(II) analog; Mn···Cr = 3.728(1); Mn—O = 2.110(4); Cr—O = 1.935(4); Mn—O—Cr = 134.3(2): Co···Cr = 3.669(2); Co—O = 2.034(10); Cr—O = 1.934(10); Co—O—Cr = 135.2(5).

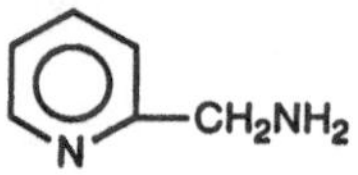

(15)

6.8.1.3. Chromium(V)

25. Na[Cr(EABA)$_2$O]·1$\frac{1}{2}$H$_2$O[98] (21): Two independent 5-coordinate anions in the unit cell; Cr=O = 1.547(5), 1.557(3); Cr—O (carboxylate) = 1.900(3) − 1.905(3); Cr—O (alkanolate) = 1.784(3)-1.798(3).

6.8.2. Synthesis and Solid-State Decomposition

A new route to the synthesis of [CrCl$_3$(thf)$_3$], an important starting material in the synthesis of Cr(III) amine complexes, is described. CrCl$_3$·6H$_2$O is dehydrated in refluxing SOCl$_2$ and thf added directly to the reaction mixture. The violet product is quite stable in the open air for several hours.[99]

A series of Cr(III) complexes with edta-like ligands, viz edtp (16) (S)-pdtp (17) and (S,S)-cydtp (18), have been prepared.[100–102] Three isomeric forms of

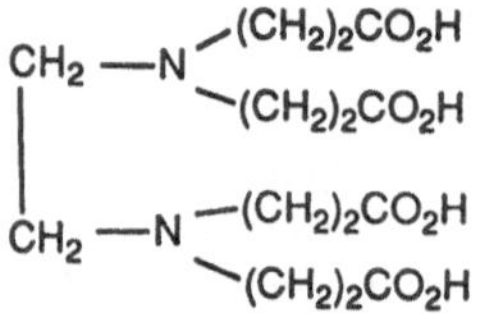

(16)

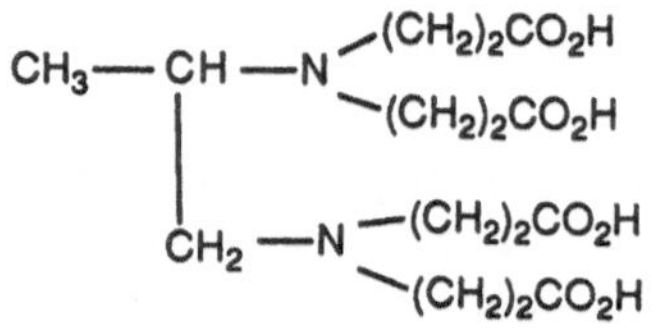

(17)

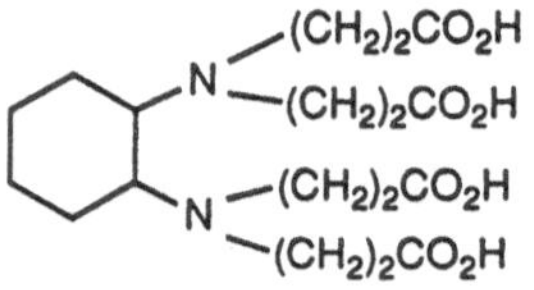

(18)

each hexadentate ligand can be obtained, and two of these undergo hydrolytic C—N bond rupture ($t_{1/2}$ ~ several hours at 60 °C) to form complexes with only three chelate "arms." ^{2}H-NMR spectroscopy continues to be a useful tool in

Table 6.14. Synthetic Descriptions of Some Cr(III) Complexes

Complex	Comments	Ref.
[CrCl$_3$(NH$_3$)$_3$]	*mer* and *fac* isomers	103
[Cr(en)$_3$]$_3$[FeCl$_6$][Cl$_6$]·H$_2$O	[FeCl$_6$]$^{3-}$ salts	104
[Cr(dien)(glygly)]ClO$_4$·H$_2$O	*mer* dipeptide complexes	105
M[F(AA)$_2$Cr(CN)Cr(CN)$_5$]	AA = en, tn; M = Li, Na, K, Rb	106
[(tmpa)XCrOCrX(tmpa)](ClO$_4$)$_2$	tmpa = (5), X = Cl, CN, N$_3$, NCS, NCO	96
[Cr(acac)$_3$]		107
[Cr$_2${(N-2aminoethyl)$_4$cyclam}Cl$_4$]	air stable Cr(II)	108
[Cr(ox)$_3$]$^{3-}$	resolution	109
[Cr$_2$(acetate)$_4$]·2H$_2$O	Cr(II)	110

probing the structure of this type of complex. Li[Cr(edtp)]·4H$_2$O has been resolved using (−)-[Co(en)$_2$(ox)]Br to give (+)-[Cr(edtp)]$^-$ as the least soluble diasteroisomer.[102] A brief description of other Cr(III) complexes that have been recently synthesized is given in Table 6.14.

6.9. Other Oxidation States

6.9.1. Chromium(II)

A bridging ligand reduction model vs. the outer-sphere mechanism for electron transfer has been tested using rate constants from Cr(II) systems. A correlation between the rate constant and the gas-phase electron affinity of the bridging group implies an inner-sphere mechanism. If such a correlation is absent an outer-sphere mechanism is assumed.[111]

The reaction between the outer-sphere electron acceptor, [Co(sep)]$^{3+}$ and Cr(II), in various electrolyte media, has been investigated by two independent groups.[112,113] Direct comparison of data is not readily made, but both studies indicate that ClO$_4^-$, CF$_3$SO$_2^-$, and NH$_2$SO$_3^-$ contribute only the "normally expected" positive salt effects, but halides (especially F$^-$) give a dramatic rate increase (Table 6.15). The cause of this effect is not obvious, as [CrX(OH$_2$)$_5$]$^{2+}$ is *not* observed as a product. Anion ion pair hydrogen bonding to the *sec*-NH protons of the sepulcate ligand is proposed.[112] Rate data for other systems using Cr(II) as the reductant are listed in Tables 6.16–6.19.

6.9.2. Chromium(IV)

The most stable Cr(IV) complexes are the diperoxy(amine) chelates and [Cr(O$_2$)$_2$(dien)] has now been investigated as an oxidant in glycine or acetate buffers.[121-125] With Ti(III) and [Fe(CN)$_6$]$^{4-}$, both the peroxo groups and the Cr(IV) center are reduced but with VO^{2+} only the Cr(IV) center. The Cr(III) products contain coordinated dien, but not all the N-donor atoms are coordinated.

Table 6.15. Second-Order Rate Constants $(10^3 k_{obs}, M^{-1} s^{-1})$ for the Reaction
Between $[Co(sep)]^{3+}$ and $Cr(II)$ in the Presence of Different Concentrations of
Added Salts in 0.01 M $HClO_4$ at 25 °C ^{a}

Concentration (M)	NaClO$_4$	NaCl	NaBr	NaI	NaF
0.01					52
0.02					152
0.03					272
0.04					415
0.05	0.348	6.40			705
0.06					910
0.07					1250
0.1	0.50	11.2	5.8	5.8	
0.2	0.61	18.2	9.1	8.1	
0.3	0.72				
0.4		27.4			
0.5	0.77				
0.5	0.87^b				

^{a} Reference 113, variable ionic strength.
^{b} Reference 112.

The reactions between Cr(IV) and hydroxylamine,[122] ascorbic acid,[123,124] or mercapto acids[125] are strong catalyzed by metal ions, such as Fe^{2+}, Fe^{3+}, Cu^{2+}, or V(IV). However, the reaction between Fe(II) and Cr(IV) is rapid and has a bimolecular rate constant of 3×10^3 $M^{-1} s^{-1}$ (22 °C, pH = 4.4, I = 0.4 M). With H_2O_2 and Cr(IV) [either the (dien) or (en)(H_2O) complex],[126] there is evidence for a blue diperoxo Cr(VI) intermediate.

6.9.3.   Chromium(V)

There is currently considerable interest in this oxidation state of chromium as Cr(V) intermediates play an important role in the mechanisms of Cr(VI) oxidations commonly used in organic chemistry.[127-130] Among the more stable Cr(V) complexes[131] is Na[Cr(EABA)$_2$O] (20) (Scheme 3; structure 24, Section 6.8.1.3) and, as reported previously,[1] the kinetics of the reaction between this and a variety of reducing agents has been investigated. It is now apparent that the reaction between (20) and oxalate is more complicated than previously thought, and ligand exchange occurs on the microsecond timescale (Scheme 3). The same sort of Cr(V) oxalato complexes are observed in the Cr(VI)–oxalate reaction.[127] It is also believed that Cr(V) is the active reagent in the production of Cr(VI)-induced cancers.[132,133]

Another Cr(V) complex that has kinetic potential is [Cr(phen)$_2$(O)$_2$]$^+$ (25), prepared by PbO_2 or PhIO oxidation of *cis*-[Cr(phen)$_2$(H_2O)$_2$]$^{3+}$. In aqueous acidic solution, (25) can oxidize Cu(II)L to Cu(III)L and the second-order rate constant (k_{obs}) varies with [H^+] according to Eq. (3).[134]

Table 6.16. Kinetic Data for the Reduction of Various Oxidants by Cr(II) Species[a,b] at 25.0 °C

Oxidant	Comments	k_2 ($M^{-1}\,s^{-1}$)	Mechanism[j]	Ref.
$[Ag(bipy)_3]^{2+}$	$[Cr(bipy)_3]^{2+}$	3.7×10^{8}	os	61
$[Ag(cyclam)]^{2+}$	$[Cr(bipy)_3]^{2+}$	1.3×10^{8}		114
$[Ni(CH_3)(cyclam)(OH_2)]^{2+}$	Product is $[(H_2O)_5CrCCH_3]^{2+}$	1.1×10^{5}	is	112, 113
$[Co(sep)_3]^{3+}$	Halide catalyzed[d]	8.7×10^{-4} [c]	os	115
$[Co(NH_3)_5(thiocarbamate)]^{+}$	$I = 1.0\ M$ (NaClO$_4$, HClO$_4$)[e]	$k_1 = 2.73$; $k_2 = 25.5$	is	25
$[Co(NH_3)_5(pyruvato)]^{2+}$	[H$^+$] independent	3.29×10^{-1} (15.8 °C); 3.71×10^{-1} (25.0 °C)[f]; 5.30×10^{-1} (35.3 °C)	is	116
(19)[g] $X,Y = 2,6$	$I = 1.0\ M$ (LiClO$_4$)	5×10^{-3}	os	117
(19)[g] $X,Y = 3,5$	$I = 1.0\ M$ (LiClO$_4$)	6.9×10^{-3} [h]	os	117
(19)[g] $X,Y = 2,4$	$I = 1.0\ M$ (LiClO$_4$)	1.34×10^{-2}; 3.9×10^{-2} [h]; 1.44	is	117
(19)[g] $X,Y = 2,5$	$I = 1.0\ M$ (LiClO$_4$)	7.7×10^{-2} [h]; 0.393	os; is	117
$Mo_4S_5^{5+}$	$I = 2.0\ M$ [H$^+$] independent	0.46×10^{3} (13.0 °C); 0.61×10^{3} (25.0 °C)[i]; 0.85×10^{3} (35.3 °C)	os	118

[Structure (19), charge 6+]

$(NH_3)_3Co$ — O — — O — $Co(NH_3)_3$ with pyridine ring (N), carboxylate bridges to $Co(NH_3)_3$, $Co(NH_3)_2(OH)_2$, and $(HO)_2Co(NH_3)_3$ units.

(19) 2,4-py(CO$_2^-$)$_2$

[a] Cr(aq)$^{2+}$ unless otherwise stated.
[b] See also Tables 6.15, 6.17-6.19.
[c] $\Delta H^{\ddagger} = 38$, $\Delta S^{\ddagger} = -180$ ($I = 0.5\ M$, LiClO$_4$).
[d] Same rate constant for X = Cl$^-$, Br$^-$, or I$^-$ ($I = 0.5\ M$, LiX, LiClO$_4$); $\Delta H^{\ddagger} = 46$, $\Delta S^{\ddagger} = -117$.
[e] $k_{obs} = k_1[H^+] + k_2[Cr^{2+}]$.
[f] $\Delta H^{\ddagger} = 15.7$, $\Delta S^{\ddagger} = -200$ ($I = 1.0\ M$ NaClO$_4$, HClO$_4$).
[g] (19) X,Y-py(CO$_2$)$_2$.
[h] Protonated at the py-nitrogen.
[i] $\Delta H^{\ddagger} = 17$, $\Delta S^{\ddagger} = -135$.
[j] os = outer sphere; is = inner sphere.

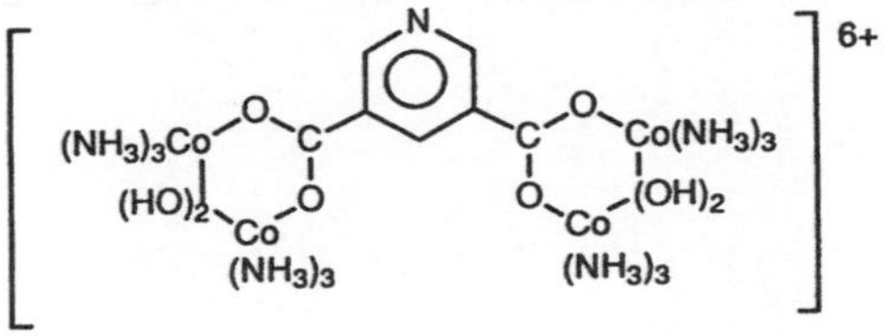

Table 6.17. Second-Order Rate Constants (k_{expt}, M^{-1} s^{-1}) for the Outer-Sphere Reduction of Some $[Co(N_6)]^{3+}$ Complexes with $[Cr(polypyridyl)_3]^{2+}$ at 23 ± 2 °C and $I = 0.15$ M [a,b]

Polypyridyl	$[Co(en)_3]^{3+}$ $10^{-3} k_{expt}$	$[Co(bipy)_3]^{3+}$ $10^{-8} k_{expt}$	$[Co(phen)_3]^{3+}$ $10^{-8} k_{expt}$	SO_2 $10^{-7} k_{expt}$	$[Co(cyclam)(OH_2)_2]^{3+}$ $10^{-8} k_{expt}$	O_2 $10^{-5} k_{expt}$
bipy	2.2	1.1	1.3	2.97	0.44	6.0
5-Cl-phen		1.2	1.6	0.683	0.170	2.5
phen		2.0	2.0	9.5	1.33	15.0
5-Me-phen	16			17	1.05	22.0
4,4'-Me$_2$-phen	66	2.8	3.6		~7	150
4,7-Me$_2$-phen	110	3.5	4.1		~7	260
5,6-Me$_2$-phen				28	1.58	

[a] The $[Cr(NN)_3]^{2+}$ species were generated by quenching the flash-photolysis-generated excited state $[*Cr(NN)_3]^{3+}$ with oxalate or $[Co(cyclam)]^{2+}$.
[b] References 65, 70, and 119.

Table 6.18. Second-Order Rate Constants for the Reaction of Some Carbon Centered Radicals with $[Cr(OH_2)_6]^{2+}$ at $24 \pm 1\,°C$ and $[H^+] = 2\text{-}20\,mM$ [a,b,c]

Radical	$10^{-8} k_{obs}$ ($M^{-1}s^{-1}$)
CH_3	2.2
C_2H_5	1.9
C_3H_7	2.2
CH_2OCH_3	2.3
CH_2Cl	2.4
CH_2Br	2.2

[a] Reference 120.
[b] The radicals (2-4 μM) were generated by flash photolysis of $[R-Co(cyclam)(OH_2)]^{2+}$.
[c] The product is $[Cr(OH_2)_5R]^{2+}$.

Table 6.19. Second-Order Rate Constants for the Electron Transfer Reactions between Ferrocenium Ions and $[Cr(bipy)_3]^{3+}$ at $25\,°C$, $I = 0.05\,M$ ($HClO_4$) in $70:30\,CH_3CN:H_2O$ [69]

Ferrocenium ion	$10^{-9} k_{obs}$ ($M^{-1}s^{-1}$)
$[Fe(C_5H_4CO_2H)_2]^+$	8.9
$[Fe(Cp)(C_5H_4CHO)]^+$	8.8
$[Fe(Cp)(C_5H_4CO_2H)]^+$	6.2
$[Fe(Cp)(C_5H_4CH_2OH)]^+$	6.0
$[Fe(Cp)(C_5H_4n\text{Bu})]^+$	5.3
$[Fe(C_5H_4CH_3)_2]^+$	4.5
$[Fe(Cp)(C_5H_4CH_2NMe_2)]^+$	3.6
$[Fe(Cp)_2]^+$	3.5

6.9.4. Chromium(VI)

The results from traditional redox studies involving Cr(VI) are summarized in Table 6.20. There is an increasing amount of data now being accumulated where $Cr_2O_7^{2-}$, pyridine chlorochromate (PCC), or pyridinefluorochromate (PFC) are being compared as oxidizing agents.[135,138] The pyridinehalochromates have the advantage of being soluble in a variety of nonaqueous solvents[7] and solvent effects can be quite marked. For example, the second-order rate constants for the oxidation of methylphenyl sulfide by PFC vary from 1.63 $M^{-1}s^{-1}$ in DMSO to

Scheme 3

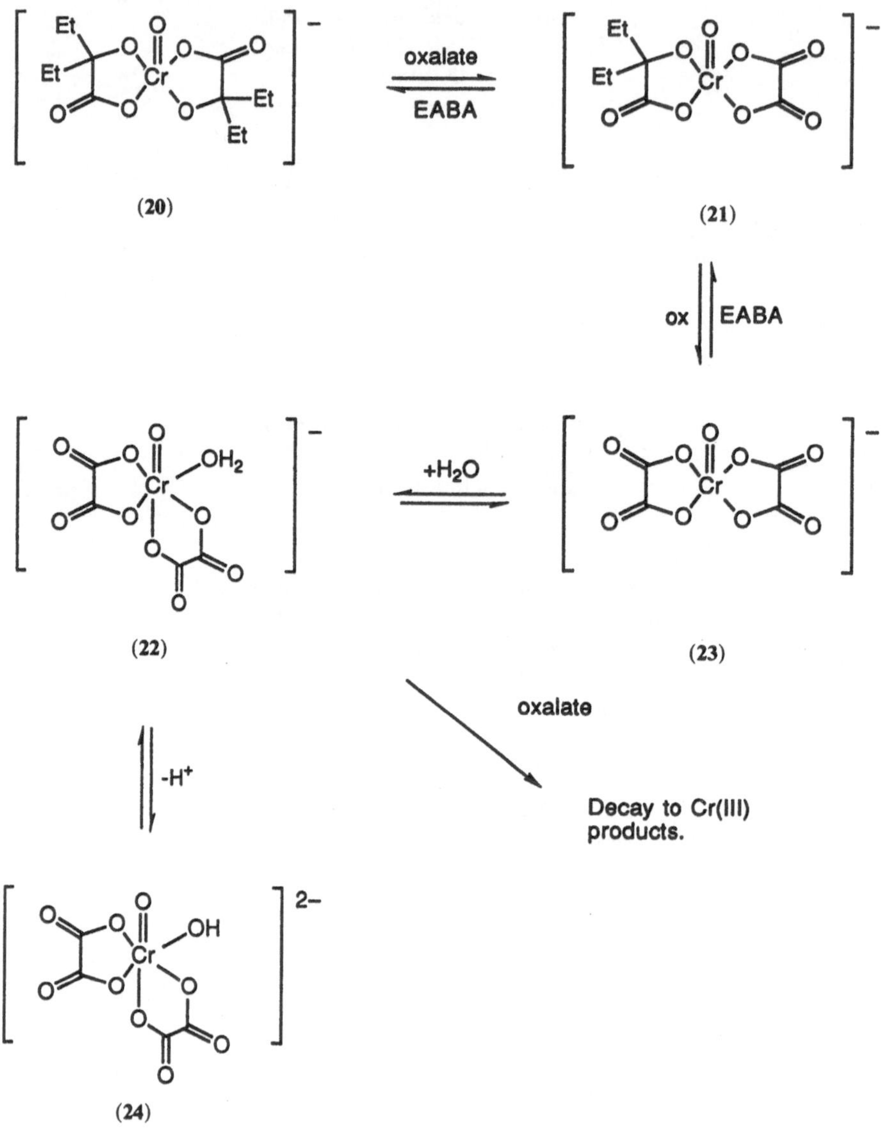

(20)

oxalate
⇌
EABA

(21)

ox ⇅ EABA

(22)

+H₂O
⇌

(23)

oxalate

Decay to Cr(III) products.

-H⁺ ⇅

(24)

$7.93 \times 10^{-2} \ M^{-1} \ s^{-1}$ in toluene.[135] The reaction between Fe(II)aq and Cr(VI) is nearly instantaneous, but when the only source of Fe(II) is in hematite or biotite, the initial rate of Cr(VI) reduction is dependent on the rate of mineral dissolution. These dissolution rates can be increased by low pH or by the addition of anions that complex Fe(II).[154]

Table 6.20. Oxidations Involving Chromium(IV)

Cr(VI) species	Reductant	Solvent	Comments	Ref.
PFC	Alcohols	DMSO	Aldehydes and Cr(III) produced, pseudo-first-order with excess alcohol	135
PCC	Alcohols	DMSO		136
PFC	Sulfides	DMSO	Sulfoxides and Cr(III) produced, pseudo-first-order with excess sulfide	137
$Cr_2O_7^{2-}$	L-ascorbic acid	aq acid	Rate order is $HClO_4 > H_2SO_4 > HCl$	138
CrO_4^{2-}	L-ascorbic acid	aq acid	$HCrO_4^-$ postulated	139
PCC	L-ascorbic acid	aq acid	Acid cat. but indep. of anion	138
$Cr_2O_7^{2-}$	Acetone	aq H_2SO_4		130
$Cr_2O_7^{2-}$	D-glucopyranose-1-phosphate	aq $HClO_4$		140
PCC	Sulfides	aq HAc		141
CrO_3Cl^-	TlCl	aq HAc + HCl	Second-order overall	142
$Cr_2O_7^{2-}$	1,2-Glycols		Cat. by V(V)	143
$Cr_2O_7^{2-}$	Acetaldehyde			144
$Cr_2O_7^{2-}$	$[(NH_3)_5CoL]^{2+}$ [a]	aq $HClO_4$	Addn. of Ce(III) changes the product ratio	145
$Cr_2O_7^{2-}$	Oxalate	aq H^+		146
$Cr_2O_7^{2-}$	Malonate	aq $HClO_4$	Accelerated by H^+	129
$Cr_2O_7^{2-}$	D-galacturonic acid	aq $HClO_4$		147
CrO_4^{2-}	Monosaccharides	aq CN^- plus OH^-		148
$Cr_2O_7^{2-}$	$Mn(CN)_6^{2-}$	aq H_2SO_4		149
$Cr_2O_7^{2-}$	Biguanides			150
$Cr_2O_7^{2-}$	HSO_3^-			151
$Cr_2O_7^{2-}$	Thiols	DMF	Cr(VI) "thioesters" initially formed	152
$Cr_2O_7^{2-}$	Piperidines	aq H^+		153
$Cr_2O_7^{2-}$	Oximes	aq H^+		

[a] $L = \alpha$-hydroxyacid.

6.10. Catalysis

Cr(III) is not normally regarded as a good catalytic species, probably because it is substitution inert and not easily oxidized or reduced. Nevertheless, addition of Cr(III) does accelerate the Ce(IV) oxidation of allyl alcohol.[155] There is also a report that $[Cr(NH_3)_5(OH_2)]^{3+}$ increases the rate of 2,4-dinitrophenylethyl-methylphosphate hydrolysis in aqueous solution.[156]

6.11. Miscellaneous

The rate of extraction of Cr(III) in aqueous $NaClO_4$ solution into CCl_4 as the $Cr(acac)_3$ complex has been determined. The rate is first order in Hacac and inverse order in $[H^+]$. Thus, $[Cr(OH)(H_2O)_5]^{2+}$ and Hacac are proposed as the active species with complex formation as the rate-controlling step, but the rate may also be dependent on the nature of the extracting solvent.[157]

Chapter 7

Substitution Reactions of Inert-Metal Complexes— Coordination Numbers 6 and Above: Cobalt

Reviews of interest include a general review of anation reactions of cobalt(III) complexes[1] and a discussion of the solvation of transition metal complexes.[2] The nature of the solvent can have a very large effect on such properties as solubilities, reactivities, redox potentials, formation constants, and various types of spectra. Such solvent effects reflect changes in the solvation of ions, complexes, initial states, transition states, and excited states.

7.1. Aquation

A number of cobalt(III) complexes of the tripodal quadridentate amine, 3,3′,3″-triaminotripropylamine [trpn = $N(CH_2CH_2CH_2NH_2)_3$], including [Co(trpn)Cl$_2$]ClO$_4$, [Co(trpn)Br$_2$]ClO$_4$, [Co(trpn)F$_2$]ClO$_4$·1.5H$_2$O, [Co(trpn)-(CH$_3$CN)$_2$]ClO$_4$, and [Co(trpn)(CH$_3$CN)Br]BrClO$_4$, have been prepared.[3] The acid hydrolysis of [CO(trpn)Cl$_2$]$^+$ and the rate of bromide substitution in [Co(trpn)CH$_3$CN]$^{2+}$ are fast. The aquation of [Co(trpn)F$_2$]$^+$ was studied in both neutral and acidic media. In neutral solution for loss of the first F$^-$ at 25 °C and $I = 0.5\ M$, $k_1 = 1.28 \times 10^{-3}\ s^{-1}$ ($\Delta H^{\ddagger} = 53.6\ kJ\ mol^{-1}$; $\Delta S^{\ddagger} = -125\ J\ K^{-1}\ mol^{-1}$), while k_2 for the loss of the second fluoride is $1.4 \times 10^{-4}\ s^{-1}$. For acid aquation $k_{aq} = 1.4 \times 10^{-4}\ s^{-1}$ ($\Delta H^{\ddagger} = 89.1\ kJ\ mol^{-1}$; $\Delta S^{\ddagger} = -19.7\ J\ K^{-1}\ mol^{-1}$). The acid hydrolysis of $[Co(trpn)(CH_3CN)_2]^{3+}$ over the pH range 1-4 reaction proceeds in two steps, as shown in Eqs. (1) and (2), with $k_1 = 1.2 \times 10^{-2}\,s^{-1}$ at 20 °C ($\Delta H^{\ddagger} = 66.9\,kJ\,mol^{-1}$; $\Delta S^{\ddagger} - 53\,J\,K^{-1}\,mol$) and $k_2 = 2.38 \times 10^{-3}\,s^{-1}$ at 20 °C ($\Delta H^{\ddagger} = 62.3\,kJ\,mol^{-1}$; $\Delta S^{\ddagger} = -81\,J\,K^{-1}\,mol^{-1}$).

$$[Co(trpn)(CH_3CN)_2]^{3+} \xrightarrow{k_1} [Co(trpn)(CH_3CN)(OH_2)]^{3+} + CH_3CN \quad (1)$$

$$[Co(trpn)(CH_3CN)(OH_2)]^{3+} \xrightarrow{k_2} [Co(trpn)(H_2O)_2]^{3+} + CH_3CN \quad (2)$$

The pressure dependence of the spontaneous aquation of a series of $[Cr(NH_3)L]^{3+}$ complexes (L = formamide, dimethyl formamide urea, N,N'-dimethylurea, dimethyl sulfoxide, trimethyl phosphate, and N,N'-dimethyl acetamide) have been studied and comparisons made with the analogous cobalt(III) complexes.[4] Values of $\Delta V^{\ddagger}$ lie in the range −3.2 to −8.7 cm^3 mol^{-1} and are not clearly related to leaving-group size. An approximate correlation between $\Delta V^{\ddagger}$ and $\Delta S^{\ddagger}$ exists. The average $\Delta V^{\ddagger}$ of *ca* −6 cm^3 mol^{-1} is indicative of a dissociative interchange (I_a) mechanism for Cr(III) contrasting with an average $\Delta V^{\ddagger}$ of *ca* +2 cm^3 mol^{-1} in the analogous Co(III) complexes where an I_d mechanism applies.

Volume profiles have been determined[5] for the aquation of $[Co(en)_2(NH_3)X]^{2+}$ ions (X = *trans*-Cl$^-$, *trans*-Br$^-$, *cis*-Br, *cis*-NO$_3^-$). Values of $\Delta V^{\ddagger}$ and the reaction volume ΔV are summarized in Table 7.1. It is concluded that the entering H$_2$O and the leaving group X participate almost equally in the transition state and that an I mechanism applies.

The synthesis of *cis*-$[Co(en)_2SO_3)Cl]$ by SO_3^{2-} substitution on aqueous *cis*- or *trans*-$[Co(en)_2Cl_2]^+$, *cis*- or *trans*-$[Co(en)_2(NCS)Cl]^+$, or *cis*-$[Co(en)_2ACl]^+$ (A = N_3^-, NO_2^-) has been described.[6] For aquation of *cis*-$[Co(en)_2(SO_3)Cl]$, $k_{aq} = 5.0 \times 10^{-4}\,s^{-1}$ at 25 °C ($\Delta H^{\ddagger} = 86 \pm 3\,kJ\,mol^{-1}$, $\Delta S^{\ddagger} = -20 \pm 9\,J\,K^{-1}\,mol^{-1}$), the product is $53 \pm 3\%$ *cis*- and $46 \pm 3\%$ *trans*-$[Co(en)_2(SO_3)OH_2]^+$, but $62 \pm 2\%$ of the *cis*-complex is observed in the Hg(II)-induced hydrolysis. *Trans*- activation by S-bonded sulfite is well documented; the effects of *cis*-SO_3^{2-} are shown to be quite unremarkable both kinetically and stereochemically.

Table 7.1. Volume Profiles (cm³ mol⁻¹) for Aquation Reactions[5]

Complex	$\Delta V^{\ddagger\,a}$	$\Delta V^{\circ\,b}$
$[Co(NH_3)_5Cl]^{2+}$	−6.0 (65)	−15.6
$[Co(NH_3)_5Br]^{2+}$	−6.4 (54)	−14.4
$[Co(NH_3)_5NO_3]^{2+}$	−5.7 (40)	−12.3
trans-$[Co(en)_2(NH_3)Cl]^{2+}$	−5.0 (70)	−17.3
trans-$[Co(en)_2(NH_3)Br]^{2+}$	−3.1 (60)	−14.7
cis-$[Co(en)_2(NH_3)Br]^{2+}$	−5.6 (60)	−16.1
cis-$[Co(en)_2(NH_3)NO_3]^{2+}$	−6.1 (45)	−13.2

a Temperatures (°C) in parentheses.
b At 25 °C.

Kinetic and structural *trans* effects are well documented for cobalt(III) complexes containing thio ligands such as sulfite, SO_3^{2-}, and thiosulfato, $S_2O_3^{2-}$. The crystal structures of seven bis(ethylenediamine)cobalt(III) complexes with monodentate oxalato, azido, nitro, sulfito, and thiosulfato ligands have now been determined.[7] The structural *trans* effect, that is, the difference between the Co—N(en) bond length *trans* to a ligand and the average of the two Co—N(en) bond lengths that are both *cis* to the ligand and *trans* to each other, decreases in the series SO_3^{2-} (0.059 Å) > SSO_3^{2-} (0.036 Å) > NO_2^- (0.019 Å) > N_3^- (0.005 Å) > $C_2O_4^{2-}$ (0.000 Å).

The aquation of $[Co(NH_3)_5Cl]^{2+}$ has been studied in ethylene glycol–water mixtures over the temperature range 35 to 55 °C.[8] The effects of solvent composition on k_{aq} and the activation parameters are discussed in detail. The effects of ion pairing on the aquation kinetics of $[Co(NH_3)_5Br]^{2+}$ in the presence of maleate and phthalate anions has been studied at a series of temperatures.[9] The rate-accelerating effects of the ions are attributed to the formation of more labile ion pairs. The rate and association constants of the ion pairs have been calculated.

7.2. Catalyzed Aquation

Some twenty years ago[10] it was shown by ^1H NMR spectroscopy that the NO^+ and Hg^{2+} induced aquation of *trans*-$[Co(NH_3)_4(ND_3)X]^{2+}$ took place with essentially complete retention of configuration. A reexamination[11] of the Hg^{2+}, Ag^+, and Cl_2 induced and spontaneous aquation of the pentaamminecobalt(III)-acido complexes has now been carried out. The findings can be summarized: (1) small, but real, stereochemical change occurs in the $CoOH_2^{3+}$ products derived from the Hg^{2+}- and Ag^+-induced reactions (2–3%); (2) some stereochemical change (1.4%) accompanies water exchange in $CoOH_2^{3+}$; and (3) no significant stereochemical change accompanies anation (by Cl^- or Br^-) of $CoOH_2^{3+}$ in acidic solution.

The Hg^{2+}-induced reaction of t-$[Co(tren)(NH_3)SCN]^{2+}$ (1) (abbreviated $CoSCN^{2+}$) in the presence and absence of added anions Y^- (NO_3^-, ClO_4^-, $CF_3SO_3^-$)[3] has been studied kinetically and the products determined by HPLC.[12]

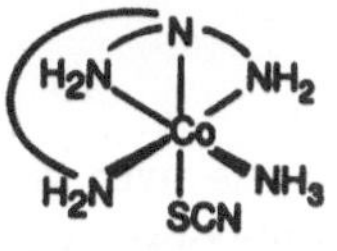

(1)

The Hg^{2+} reaction is interpreted in terms of equilibrium attachment of Hg^{2+} or HgY^+ to the sulfur atom of coordinated thiocyanate followed by rate-determining cleavage of the Co—SCN bond. For the Ag^+ reactions the rate and product data are discussed in terms of the binding of Ag^+ to both the S and N ends of coordinated thiocyanate followed by Co—SCN^{2+} bond cleavage and competition by NCS^-, OH_2, and Y^- (ClO_4^-, NO_3^-) for the intermediate of reduced coordination number.

Product distributions have been accurately determined or redetermined for the nitrosation reaction of $[Co(NH_3)_5N_3]^{2+}$ in acidic solution at 25 °C and $I = 1.0\ M$.[13] In chloride media, $[Co(NH_3)_5Cl]^{2+}$ is formed along with the aqua complex and the product ratio $[Co(NH_3)_5Cl]^{2+}/[Co(NH_3)_5OH_2]^{3+}$ is accurately linear in $[Cl^-]$ even when the coanion is the strongly ion-pairing SO_4^{2-}. The present work strongly suggests that the product distribution and the nitrosation rate law are independent, that is, the products arise by a process subsequent to the rate-determining step, and an intermediate is involved.

7.3. Base Hydrolysis

Complexes of the type $[Co(picdien)X](ClO_4)_2$ and $[Co(picdith)X](ClO_4)_2$ (picdien = 1,9-bis-12'-pyridyl)-2,5,8-triazanonane **(2)**, picdith = 1,11-bis-(2'-pyridyl)-2,6,10-triazaundecane **(3)**; X = Cl^-, Br^-, NO_2^-, NCS^-, N_3^-, $CH_3CO_2^-$, or H_2O) have been prepared.[14] All of the complexes have the $\alpha\beta$ configuration, those of picdien existing in either *syn* **(4)** or *anti* forms **(5)** while only the *anti* forms of the picdith complexes have been isolated. All of the complexes are

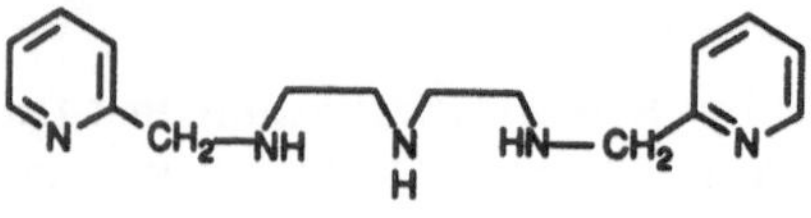

(2) picdien

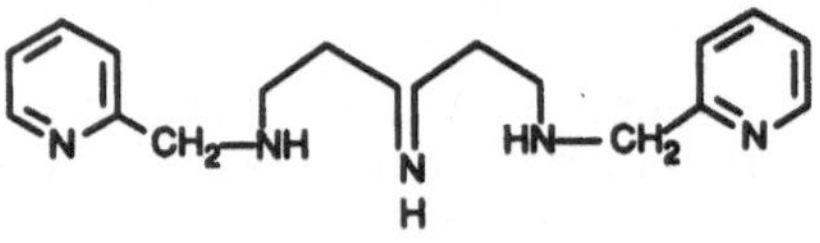

(3) picdith

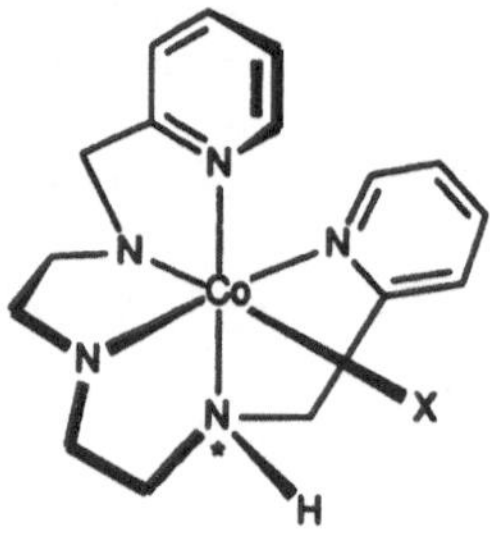

(4)

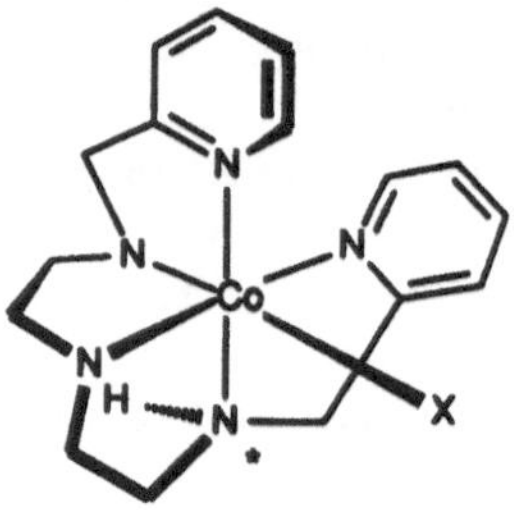

(5)

unusually sensitive to base hydrolysis over very wide ranges of pH. The pH-independent contribution (k_{aq}) is only observed at high temperatures and at high $[H^+]$ in the case of the chloro and bromo complexes of picdien. However, there is an important pH-independent contribution to the solvolysis of the corresponding picdith species, which are also somewhat more sensitive to base hydrolysis. Proton exchange studies show that proton transfer is faster than substitution in even the most labile systems.

Rate constants for the displacement of Cl^- from *trans*-$[Co(tn)_2(B)Cl]^{n+}$ (tn = 1,3-diaminopropane; B = OH^-; NCS^-, N_3^-, N_3^-, NO_2^-, NH_3, or CN^-) have been determined in aqueous solution over a range of temperature and pH, and comparisons made with the analogous 1,2-diaminoethane complexes.[15] For B = OH⁻, NCS⁻, and Cl⁻ where substitution is accompanied by stereochemical change and a large positive ΔS^+, values of k_{aq} are much greater than observed with the 1,2-diaminoethane complexes. For B = CN⁻, NO₂⁻, NH₃ (and SO₃⁻) where ΔS^+ is much smaller and even negative, substitution occurs with retention of configuration and the increase in reactivity is about tenfold. Base hydrolysis of all complexes studied is only increased by a factor of 10 to 20 in spite of the common occurrence of stereochemical change.

The kinetics of aquation and base hydrolysis of the *cis*-[Co(en)₂(NH₂Et)O₂CR]²⁺ ions (R = H or Me) have been studied in detail.[16] Aquation is strongly acid-catalyzed and rate and activation parameters for this process are reported. Similar rates are observed for both complexes in spite of the differences in basicity of coordinated formate and acetate. Aquation rates (k_{aq}) are also very similar, but base hydrolysis of the formato complex is some five times faster than that of the acetato complex, consistent with a dissociative SN_1CB mechanism and cleavage of the Co—O bond.

The kinetics of base hydrolysis of the *trans*-bis(malonato) complex (6) have also been investigated in the 15-35 °C temperature range with [OH⁻] = 0.015 mol dm⁻³ to 0.29 mol dm⁻³ at $I = 0.3\,M$.[17] For loss of the first malonato ligand $k_{OH} = 8.5 \times 10^{-3}\,dm^3\,mol^{-1}\,s^{-1}$ at 30 °C ($\Delta H^+ = 117\,kJ\,mol^{-1}$; $\Delta S^+ = 99\,J\,K^{-1}\,mol^{-1}$).

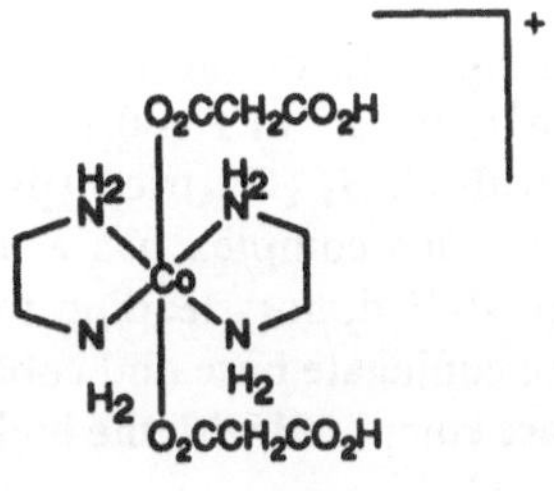

(6)

Volumes of activation (ΔV^+) and reaction volumes for the base hydrolysis of 16 different cobalt(III) amine complexes have recently been determined from high pressure kinetic and dilatometric measurements.[18] The results obtained are summarized in Table 7.2. The values of ΔV^+ range from 43 to 19 cm³ mol⁻¹ while reaction volumes range from 24 to 9 cm³ mol⁻¹ depending on the charge on the complex determined by the leaving group X. The results are consistent with earlier reported data for closely related systems, and indicate clearly both the minor role played by the nonparticipating ligands and the influence of complex charge. Noteworthy is the observation that reactions involving rate-limiting deprotonation rather than the usual rate-determining ligand dissociation display similar values of ΔV^+.

The complexes *cis*-α-[Co(trien)(ImH)C₁]²⁺ (ImH = imidazole, trien = 1,8-diamino-3,6-diazaoctane, *cis*-α-[Co(trien)BuⁿNH₂)Cl]²⁺, *cis*-α-[Co(trien-(NH₂CH₂CH(OMe)₂Cl]²⁺, and *cis*-β₂-[Co(trien)(py)Cl]²⁺ (py = pyridine) have been characterized and their kinetics of base hydrolysis studied.[19] The *cis*-α-

Table 7.2. *Volumes of Activation for the Base Hydrolysis of Cobalt(III) Complexes*[18]

Complex[a]	$\Delta V^{\ddagger}$ (cm³ mol⁻¹)
$[Co(NH_3)_5(DMF)]^{3+}$	+43.2 ± 1.7
$[Co(NH_2Me)_5Cl]^{2+}$	+32.8 ± 1.7
$[Co(NH_2Et)_5Cl]^{2+}$	+31.1 ± 0.5
cis-$[Co(en)_2(NH_3)Cl]^{2+}$	+31.8 ± 0.6
cis-$[Co(en)_2(NH_3)Br]^{2+}$	+30.8 ± 1.0
trans-$[Co(en)_2Cl_2]^+$	+24.8 ± 0.5
cis-$[Co(en)_2Cl_2]^+$	+27.9 ± 0.7
trans-$[Co(en)_2(N_3)Cl]^+$	+26.7 ± 0.4
cis-β-$[Co(trien)Cl_2]^+$	+35.7 ± 1.2
trans-$[Co(RSSR$-cyclam$)Cl_2]^+$	+20.1 ± 0.9
cis-$[Co(en)_2(NO_2)Cl]^+$	+20.8 ± 0.3
cis-$[Co(tet)Cl_2]^+$	+23.1 ± 0.8
cis-$[Co(tet)(Cl)OH]^+$	+25.9 ± 2.1
trans-RS-$[Co(tet)Cl_2]^+$	+22.1 ± 0.7
trans-RR(SS)-$[Co(tet)Cl_2]^+$	+24.4 ± 1.2
trans-RS-$[Co(tet)Cl(OH)]^+$	+23.8 ± 0.9

[a] tet = 2,3,2-tet = 1,9-diamino-3,7-diazanonane.

isomers with the *fac–fac* arrangement of the trien ligand (**7, 8**) have values of k_{OH} in the range 73 to 253 dm³ mol⁻¹ s⁻¹ at $I = 0.1$ M and 25 °C. Extremely rapid base hydrolysis is observed with *cis-β₂*-$[Co(trien)(py)Cl]^{2+}$ where k_{OH}^{25} is 6.65 × 10⁶ dm³ mol⁻¹ s⁻¹ at $I = 0.1$ M. This complex has a *mer–fac* arrangement of the trien ligand (**8**) with "flat" *sec*-NH donors leading to rapid base hydrolysis due to good π-overlap between the conjugate base and cobalt(III). The pyridine ligand causes a *ca* 30-fold rate increase compared with the hydrolysis of *cis-β₂*-$[Co(trien)-(NH_3)Cl]^{2+}$.

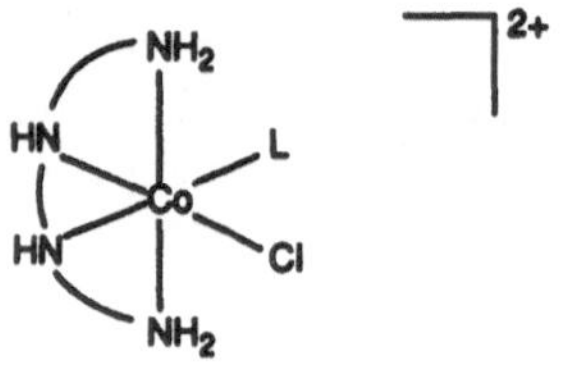

cis-α

(7)

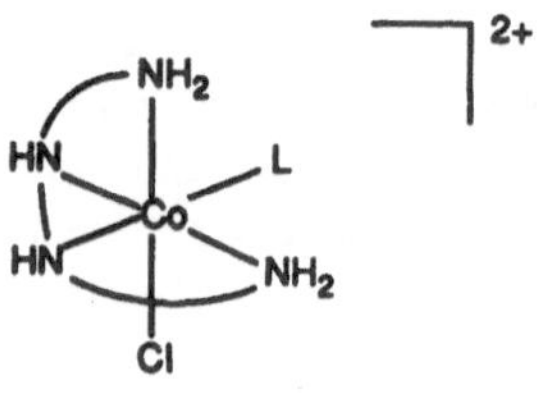

cis-β₂

(8)

7.4. Anation

The anation kinetics of *cis*-$[Co(en)_2(OH_2)_2]^{3+}$ by L-proline have been investigated using a series of solvent systems (aqueous methanol, ethanol, isopropanol, *t*-butanol, and dioxane), and rate law (3) established,[20] in which ProH = the

$$\text{rate} = \frac{k_a K_E[cis\text{-}[Co(en)_2(H_2O)_2^{3+}]_T[ProH]}{(1 + K_E[ProH])} \tag{3}$$

zwitterionic form of proline, and K_E is the ion-pair equilibrium constant. At high proline concentrations the reaction becomes independent of the proline concentration due to ion-pair formation. An increase in the organic component of the solvent increases K_E, and values of k_a, the anation rate constant, decrease as the dielectric constant of the solvent decreases. A good linear relationship is observed between k_a and the Grunwald–Winstein solvent parameter Y. A similar study of the anation of *cis*-diaqua-bis(biguanide)cobalt(III) and chromium(III) by aspartic acid in EtOH–H_2O media has also been published.[21] For both of these complexes rate law (4) holds, where K_E and k_a are the ion-pair and anation rate constants, respectively. The results are consistent with a mechanism involving outer-sphere

$$\text{rate} = \frac{k_a K_E[M(bigH)_2(H_2O)_2^{3+}]_T[AspH^-]}{(1 + K_E[AspH^-])} \tag{4}$$

association of the reactants followed by associative interchange of this outer-sphere complex into the products.

The rapid nitrosation reaction of $[Co(NH_3)_5{}^{17}OH_2]^{3+}$ with ice-cold $NaNO_2$–$HClO_4$ leads to the oxygen-bound nitrito complex $[Co(NH_3)_5ON^{17}O]^{2+}$ having one ^{17}O per cobalt but distributed equally between inner and outer sites.[22] This result is contrary to an earlier report where oxygen scrambling was not observed. A π-bonded HNO_2 intermediate (9) is proposed to account for the results.

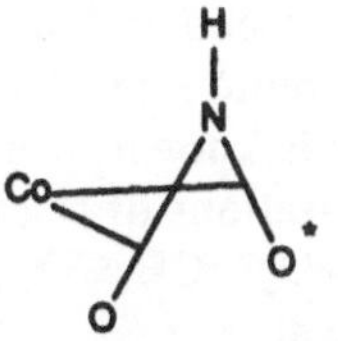

(9)

The kinetics and mechanism of the decomposition and anation of *trans*-$[Co(py)_4(H_2O)_2]^{3+}$ in aqueous acidic chloride solutions has been studied in some detail.[23] Two reactions occur: the faster reaction involves replacement of coordinated pyridine by water or chloride ion in parallel pathways, while the slower reaction involves spontaneous reduction of $[Co(py)_3(H_2O)_2Cl]^{2+}$ and $[Co(py)_3(H_2O)_3]^{3+}$ with the chloro complex decomposing at a faster rate than the aqua complex.

7.5. Solvolysis

The first step in the solvolysis of $[Co(trpn)X_2]^+$ (trpn = 3,3',3"-triaminotripropylamine = $N(CH_2CH_2CH_2NH_2)_3$; X = F, Cl, Br) has been investigated[24] using acetonitrile as solvent. At 25 °C, the rate constants for the release of the first halide are 7.8×10^{-4}, 1.66×10^{-3}, and 1.87×10^{-2} s^{-1} for X = F, Cl, and Br, respectively. Activation parameters for these reactions were also determined and the complex [Co(trpn)(CH$_3$CN)Br]BrClO$_4$, the solvolysis product from [Co(trpn)Br$_2$]ClO$_4$, isolated and characterized. No evidence was found for a second solvolysis step. The results are discussed in terms of a dissociative I_d mechanism. The extreme lability of [Co(trpn)X$_2$]$^+$ complexes is attributed to the flexibility of the six-membered chelate rings allowing ready formation of a dissociative transition state.

The solvolysis of *trans*-[Co(4-Etpy)$_4$Cl$_2$]$^+$ in water–isopropanal at various temperatures has been studied in detail.[25] The activation energy varied nonlinearly with the mole fraction of the cosolvent. The plot of log k vs. the reciprocal of the dielectric constant was also nonlinear. The influence of the solvent structure on the complex ion in the transition state dominates over that in the initial state. A similar study has also been carried out using water–ethanol mixtures.[26]

7.6. Isomerization

Activation volumes (ΔV^+) and reaction volumes (ΔV) have been obtained for the *trans* $\rightleftharpoons$ *cis* isomerizations of [Co(en)$_2$(H$_2$O)X]$^{n+}$ (X = H$_2$O, OH$^-$, NCS$^-$, NH$_3$, NO$_2^-$, Br$^-$, N$_3^-$) and [Co(en)$_2$(OH)Y]$^{n+}$ (Y = OH$^-$, NH$_3$).[27] The values of ΔV are close to zero (-6 to $+2$ cm^3 mol^{-1}), and a relationship exists between ΔV for [Co(en)$_2$(H$_2$O)X]$^{n+}$ and the position of X in the spectrochemical series. Values of ΔV^+ lie in the range 5 to 8 cm^3 mol^{-1} for isomerizations of [Co(en)$_2$(H$_2$O)X]$^{n+}$ (X = H$_2$O, OH$^-$, NCS$^-$, NH$_3$, N$_3^-$). This set includes three reactions (X = H$_2$O, OH$^-$, NCS$^-$) that are known to proceed with exchange of coordinated water with the solvent. The value of ΔV^+ is quite large (18 cm^3 mol^{-1}) for the isomerization of [Co(en)$_2$(OH)(NH$_3$)]$^{2+}$ which is known to occur without exchanging OH$^-$ or NH$_3$. It is concluded that isomerizations of [Co(en)$_2$(H$_2$O)X]$^{n+}$ (X = H$_2$O, OH$^-$, NCS$^-$, NH$_3$, Br$^-$, N$_3^-$, SeO$_3$H$^-$, SeO$_3^{2-}$, CH$_3$CO$_2^-$, Cl$^-$) occur via Co—OH$_2$ dissociation by an interchange mechanism.

The effect of placing thioether linkages *trans* to a site of nitrito substitution and its effect on the subsequent nitrito $\rightarrow$ nitro isomerization has been studied[28] using the [CoL(OH$_2$)]$^{3+}$ cation where L = 1,11-diamino-3,6,9-trithiaundecane (**10**).

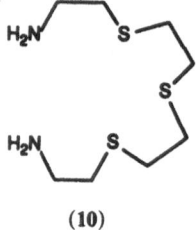

(**10**)

The kinetic data are consistent with the mechanism first proposed by Basolo and Pearson in which N$_2$O$_3$ is the nitrosation reagent. Nitrosation of [CoL(OH$_2$)]$^{3+}$ is 300 times faster than that of [Co(NH$_3$)$_5$OH$_2$]$^{3+}$ under identical conditions. Isomerization to the nitro complex is, however, very slow with $k_{isom} = 1.21 \times 10^{-4}$ s^{-1} at 25 °C with $\Delta H^+ = 111.3$ kJ mol^{-1} and $\Delta S^+ = +53$ J K^{-1} mol^{-1}.

7.7. Carbonato Complexes

The μ-amido-μ-carbonato complex (11) has been prepared and characterized, and the kinetics of acid hydrolysis leading to the μ-amido-μ-hydroxo-bis(bis(ethylenediamine)cobalt(III)) ion studied in detail.[29] Values of k_{obs} are independent of $[H^+]$ over the acidity range $[H^+] = 0.9\ M$ to $0.01\ M$. The kinetics of base hydrolysis were also studied over the range $[OH^-] = 0.7\ M$ to $0.025\ M$. The results are consistent with a kinetic scheme involving a rapid preequilibrium followed by a rate-determining step resulting in a μ-amido–dihydroxo complex.

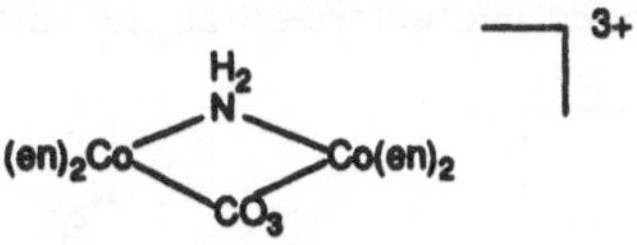

(11)

Extended Hückel molecular orbital calculations on modified structures of $[Co(NH_3)_4CO_3]^+$ have been carried out to investigate the effect of strain caused by the chelated carbonato ligand on the complex and on the decarboxylation of carbonato cobalt(III) complexes with tetraamine ligands.[30] It has been suggested that CO_3^{2-} chelation enlarges the N—Co—N angle (θ) *trans* to the O—Co—O angle. The most stable structure of $[Co(NH_3)_4CO_3]^+$ has $\theta = 94°$, somewhat larger than the normal octahedral angle of $90°$. The existence of d electrons in the d_{xy} orbital appears to be crucial in controlling the N—Co—N angle in this type of complex.

7.8. Vitamin B₁₂ and Cobaloximes

Several new organocobalt complexes of the type $[LCo(DO)-(DOH)Me_2pnCH_3]X$ have been prepared from $[Co(DO)(DOH)Me_2pn]Br_2$ where L = a neutral N- or P-donor ligand and $(DO)(DOH)Me_2pn$ has the structure (12).[31] The complexes were characterized by ^1H NMR, and in a few cases by ^{13}C NMR. The rates of dissociation of the ligand L were found to be greater than those of the analogous Costa complexes, where the equatorial ligand (DO)-

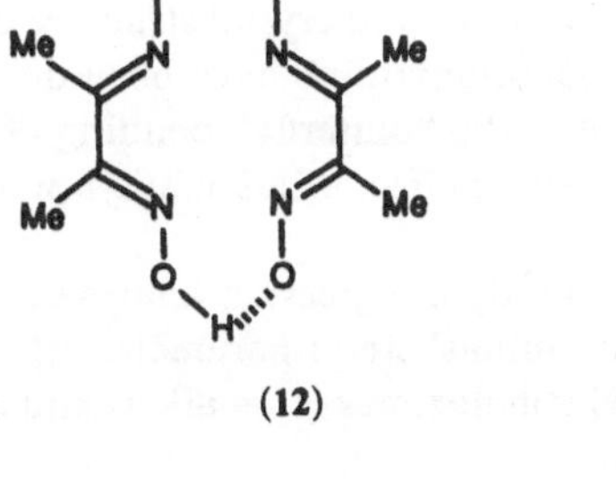

(12)

(DOH)pn has a propylene bridge in place of the 2,2-dimethylpropylene bridge of the present complexes, or the analogous cobaloxime complexes.

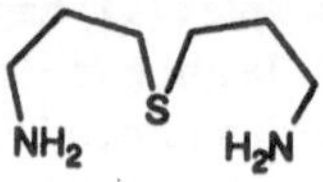

(13)

The μ-fac isomer of $[Co(aeaps)_2]^{3+}$ [aeaps = 2-aminoethyl 3-aminopropyl sulfide (13)] reacts with strong base, as shown in Eq. (5), to give the alkyl cobalt(III) complex (14) which has been characterized as the dithionate salt and its crystal

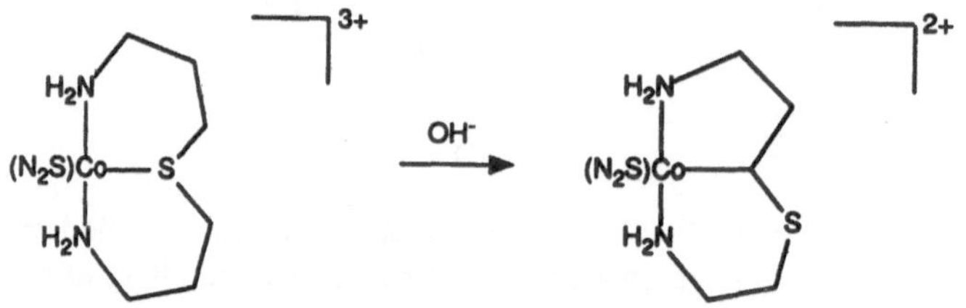

(5)

(14)

structure determined.[32] Spectrophotometry shows that the alkyl–cobalt(III) complex is only one of several species present at equilibrium in 1 M sodium hydroxide solution.

The course of the reaction of the "supernucleophile" $[pyCo(DH)_2]^-$ (py = pyridine, DH = dimethylglyoximato) with $CF_2{=}CFX$ depends on the nature of X giving $[pyCo(DH_2(CF_2CFClH)]$ (X = Cl) and $[pyCo(DH)_2(CF{=}CF_2)]$ (X = Br) as the major isolated products.[33] Aquocobalamin and $[Fe(CN)_6]^{4-}$ react to give the binuclear anion $[cobalamin-\mu\text{-NC-Fe(CN)}_5]^{3-}$, which can be isolated as the zinc(II) salt. The anation reactions of aquocobalamin by N_3^-, $[Fe(CN)_5NO]^{2-}$, $[Fe(CN)_5OH_2]^{2-}$, and $[Fe(CN)_6]^{4-}$ have been studied[34] as a function of ligand concentration, pH, temperature, and pressure. The observed second-order rate constants vary between 930 and 3600 $M^{-1}s^{-1}$ at pH ca 6 and 25 °C. Activation entropies and activation volumes are significantly positive, supporting a dissociative mechanism.

An important finding in organocobalt B_{12} model chemistry is that the Co—C bond lengthens and its bond dissociation energy decreases as the bulk of the alkyl group R increases. The longest bonds have been found with the adamantyl (ad) ligand in the B_{12} model system $[Co(Hdmg)_2(ad)L]$ (Hdmg = dimethylglyoximate, L = neutral ligand). The synthesis and crystal structure of two new complexes with L = NH_2Ph or 4-dimethylaminopyridine have been described.[35] The steric effect of the ad ligand is evident in the "butterfly" bending of the Hdmg ligands toward L. It is suggested that the steric effect of adamantyl on rates is mainly reflected in the transition state.

It is claimed[36] that multiple regression analyses, involving Taft's polar substituent constant σ^* and Dubois' steric parameter E'_s, can satisfactorily account for the properties of alkyl cobaloximes (R = alkyl) and alkyl cobalamines in terms of the effects of the R groups. The properties studied were the redox potential of the Co^{IV}-Co^{III} couple, Co $\rightarrow$ C charge transfer energy, the *trans* influence and *trans* effect of R in alkyl cobaloximes, and the *trans* influence of R in alkyl cobalamins. The kinetics of the reaction of aquacobalamin (vitamin B$_{12a}$) with imidazole, substituted imidazoles, histamine, and histidine have been studied as a function of pH at 25 °C.[37] The rate of substitution of water varies linearly with the pK_a of the incoming ligand L (L = anionic histidine, neutral histamine, 1-methyl imidazole, 1,5-dimethyl imidazole). The positive deviation of imidazole itself is attributed to the possibility of its reaction via two annular coordination sites. The negative deviation of 4(5)-methyl imidazole 4(5)-lactic acid, cationic histamine, and neutral histidine from the linear relationship is attributed to the presence of two tautomeric forms in solution and provides a means of estimating values of the tautomerization constant, K_T, for these ligands.

Chapter 8

Substitution Reactions of Inert-Metal Complexes— Coordination Numbers 6 and Above: Other Inert Centers

8.1. Introduction

The coverage and arrangement of this chapter are very much the same as in previous volumes of this series. Although activity in some areas has diminished, it has undergone a comparable increase in others. Thus classical kinetic studies of substitution at, e.g., rhodium(III) and platinum(IV) centers have been rather rare during the period covered by this report. On the other hand, there has been increasing activity in substitution at the lower oxidation states of osmium, thanks largely to Taube's novel studies, at ruthenium(II), largely in a photochemical context, and at technetium in various oxidation states, wholly in relation to the medical uses of a variety of its complexes. Perhaps the biggest increase in activity in the field defined by the title of this chapter has been in studies of systems in which ligand engineering has resulted in the dramatic slowing of substitution at normally labile centers. Such studies have often been impelled by the needs of radiotherapy, where a nucleus with specific radiochemical properties is required to be delivered to the site of action. An appropriate ligand is needed to transport the nucleus to its intended site, and of course it is necessary that this ligand shall not be lost en route if the metal ion happens to be substitution-labile. Uncharacteristically inert behavior of such centers as copper(II) and lanthanides(III) will be discussed in the following paragraphs, before the customary systematic treatment of the usual set of inert elements.

The tailoring of ligands to confer relative substitutional lability on a metal center is not new, and indeed has been mentioned before in this report, for instance in relation to cadmium(II).[1] Tetraazamacrocycles confer considerable inertness on nickel(II) and copper(II); the 14-membered cyclam (1) ring is particularly effective, increasing the half-life with respect to dissociation of [Ni(cyclam)]$^{2+}$ to a matter of years at low pH values.[2] Recently the encapsulating ligands (2) and (3) have been shown to cause a marked decrease in reactivity for their nickel(II)

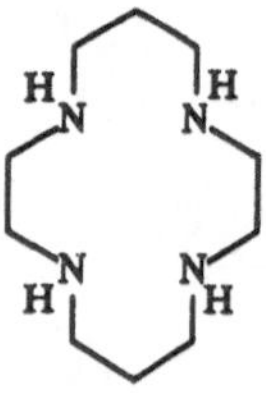

(1)

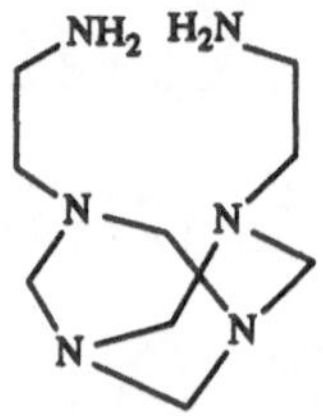

(2)

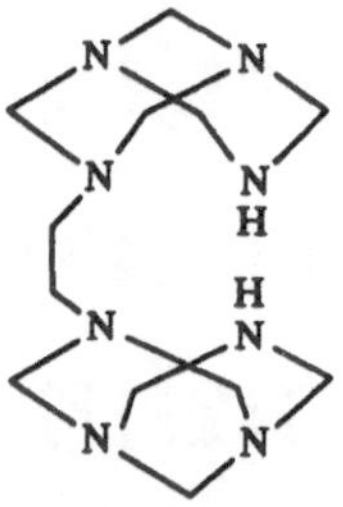

(3)

and copper(II) complexes,[3] as indeed have the simpler analogs (4) and (5). The trimethyl derivative of (5), viz. (6), gives a particularly inert nickel(II) complex, with a rate constant for dissociation of around 10^{-7} s^{-1}.[4] Methyl substitution in

(4) 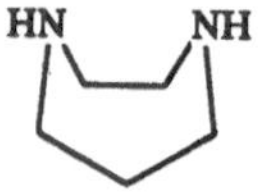   (5) 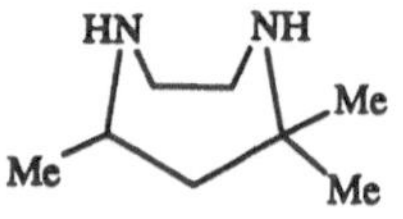   (6)

doubly deprotonated tripeptidocopper(II) complexes also decreases lability enormously. Rate constants for reaction of such complexes with trien span eight orders of magnitude. Large negative activation entropies for the polymethyl derivatives are attributed to the statistical difficulty of maneuvering the incoming trien around the protruding methyl groups so that it can bond to the copper.[5] Such great reductions in lability are particularly important for copper, since ^{67}Cu is a potential antigen for targetting cancer cells. Another such potential antigen is ^{90}Y. Again Y^{3+} is normally very labile, but can be made inert by complexing with appropriate multidentate garland or macrocyclic ligands. The aminocarboxylate ligand dtpa, (7), is fairly effective, but its butyl and nitrobenzyl derivatives are better. The rate constant for Y^{3+} loss from butyl-dtpa is 3.1×10^{-6} s^{-1}, corresponding to a half-life of four days, in serum at 37 °C, from the nitrobenzyl–dtpa complex nearly fifty times smaller,[6] and from tetraazamacrocycles with acetato-pendant arms, e.g., (8), even slower.[6,7] The Y^{3+} complex of (9) is also inert. It can only be prepared by a template reaction, and that takes many hours.[8] Related complexes whose ligands, (10), consist of two pyridine-2,6-dicarbaldehyde or 2,6-diacetylpyridine

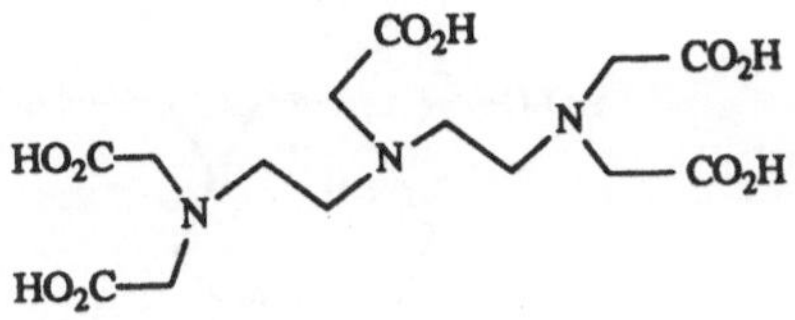

(7)

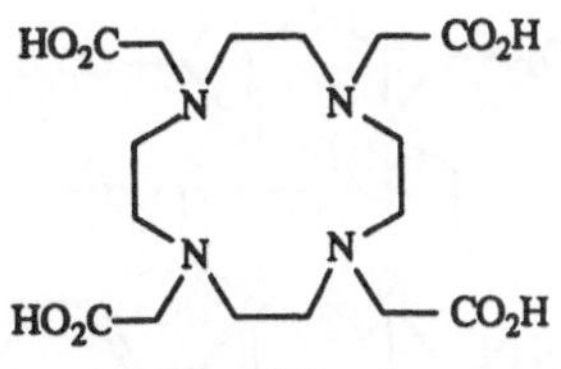

(8)

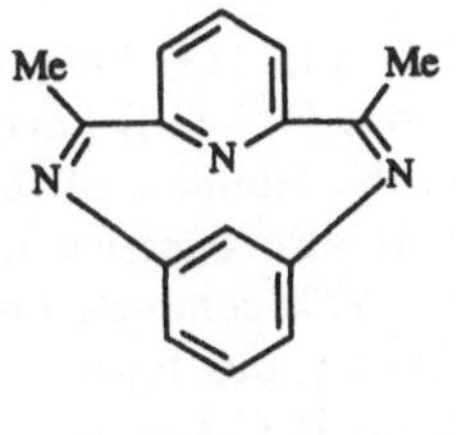

(9)

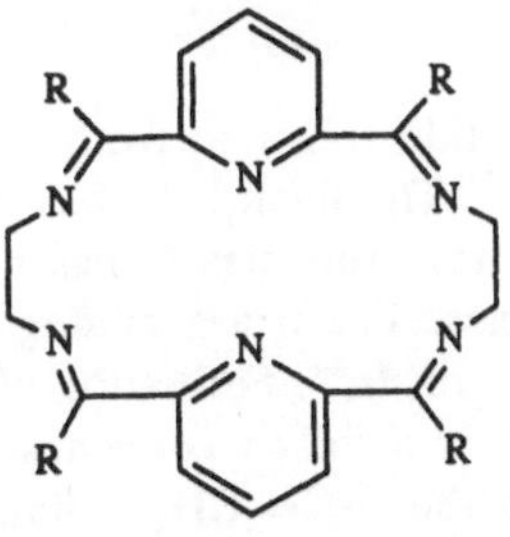

(10)

molecules condensed with two molecules of ethane-1,2-diamine are again very inert with respect to displacement of these macrocycles, but they also contain two water or acetate, or three thiocyanate, ligands which are very easily replaced.[9]

The situation is rather similar for gadolinium, whose dota (**8**) complex is a valuable contrast agent for magnetic resonance imaging (MRI) and whose dtpa complex is now in clinical use in the United States for magnetic resonance imaging and tumor detection. However, this dtpa complex is not quite inert enough; recent work has led to an expanded porphyrin ligand (the expansion involves the incorporation of a 2,2'-bipyridyl unit in place of one of the pyrrole rings) whose gadolinium complex has a half-life of approximately 37 days.[10] The ligand moieties dtpa and edta have been attached to monoclonal antibodies for the complexing and transport of ^{111}In, another radiotherapeutic nucleus. The half-life for the release of In^{3+} from lym-1-dtpa is about 9 days at 37 °C, from lym-1-benzyl-dtpa or lym-1-benzyl-edta an order of magnitude longer (lym-1 is an anti-B cell lymphoma monoclonal antibody).[11] In^{3+} is eight-coordinate in its dtpa complex, both in the solid state and in solution,[12] so is presumably eight-coordinate in these derivatives too. Macrocycles (**8**) and (**11**)–(**13**) confer somewhat greater inertness on In^{3+} than dtpa, especially at low pH values such as that of the stomach. The complexes with (**8**) and (**12**) transfer their In^{3+} to transferrin with a half-life of about a day at

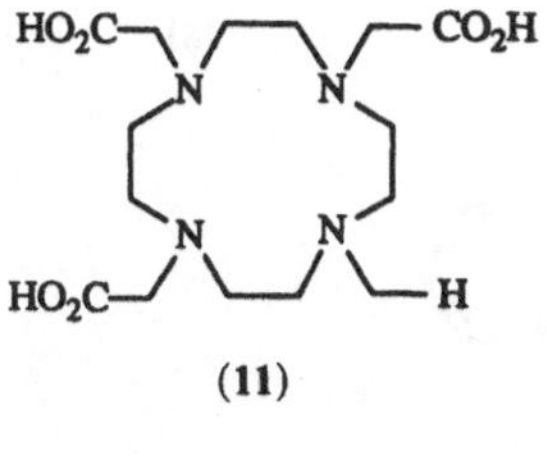

(11)

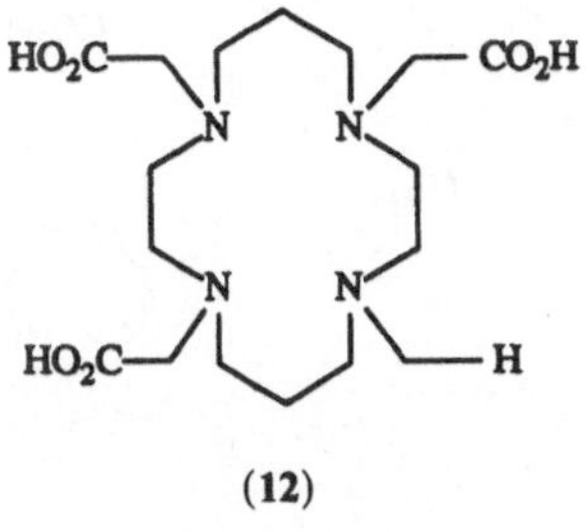

(12)

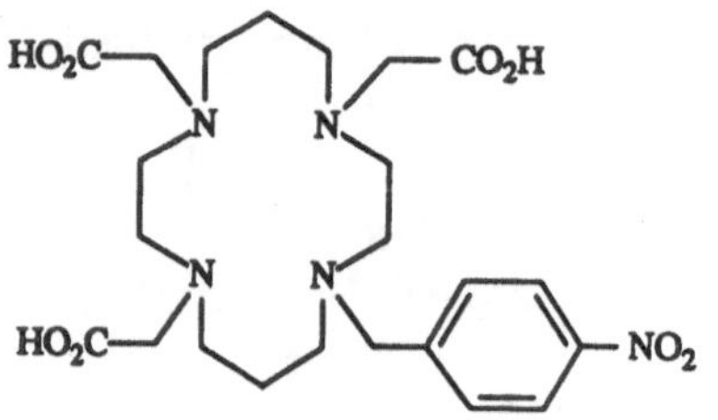

(13)

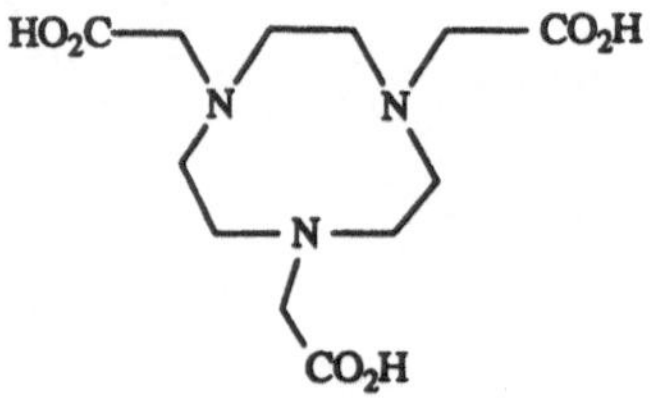

(14)

37 °C, while transfers from the complexes of (11) and (13) have half-lives of weeks.[13] The ligand (8) has eight donor atoms, but (11)–(13) have only seven donor sites—this clearly has no marked effect on the labilities of the respective complexes. The triaza analog of dota, viz. (14), is also effective in conferring inertness on In^{3+}. Formation of this In^{3+} complex is 97% complete only after half an hour at a ligand concentration of 20 μM.[14] Analogous ligands with one or more of the $-CH_2CH_2-$ links replaced by $-CH_2CH_2CH_2-$ react even more slowly.[15] The Ga^{3+} complex of (14) has a half-life of more than two weeks at pH = 0.[15] Rate constants for dissociation of this type of complex, including the yttrium(III) complexes mentioned above,[7] also decrease as $-CH_2CH_2-$ links are replaced by $-CH_2CH_2CH_2-$.[13,15]

There have been several nonpharmacologically oriented reports of slow substitution processes involving lanthanide(III) ions and complexes. The formation of the cerium complex of dota (8) has a rate constant of 1.7×10^{-8} s^{-1}; its dissociation is slow and follows complicated kinetics.[16] Displacement of Eu^{3+} by Cu^{2+} from the [Eu(egta)]$^-$ anion, egtaH$_4$ = (15), is also slow; this time there has been a detailed kinetic study.[17] Half-lives for dissociation of [Ln(tmdta)]$^-$ anions, tmdtaH$_4$ = (16), are around a minute, rather shorter than for dissociation of the closely analogous [Ln(edta)]$^-$ anions.[18] Dissociation of the La^{3+} complex of 2,6-dicarboxy-4-hydroxypyridine, monitored by ^{139}La NMR, is rather faster, with $k_{25} = 207$ s^{-1}, but still very slow in comparison with most lanthanide complexes.[19] Reaction of the chromium(II) complex *cis*-[CrCl$_2$(pap)$_2$], pap = 2-phenylazopyridine, (17), with 2,2′-bipyridyl in acetonitrile has a rate constant as low as 4×10^{-7} at 50 °C.[20] The skeptical reader should bear in mind that this astonishing inertness for a d^4 complex does have some precedent in an earlier study of the stability of [Cr(CNR)$_6$](PF$_6$)$_2$ solutions.[21]

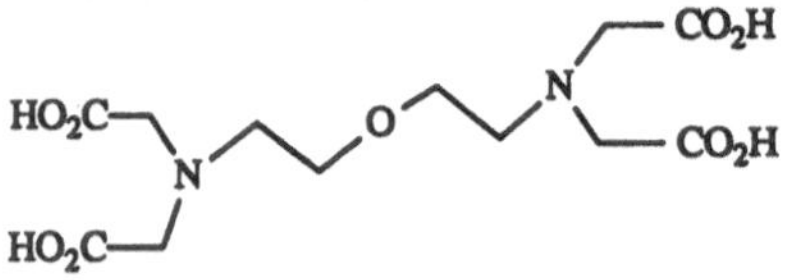

(15)

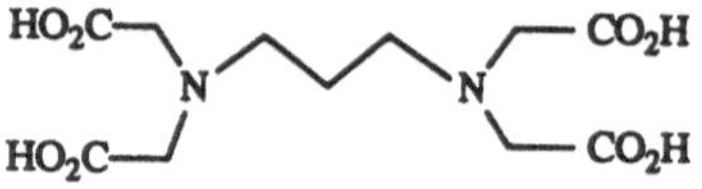

(16)

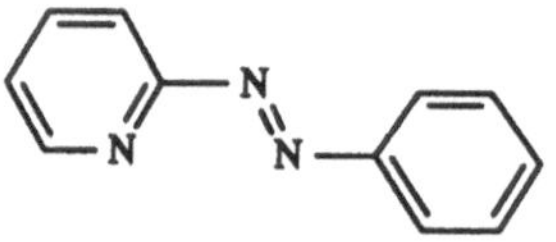

(17)

Another way to slow substitution is to covalently bond a ligand to silica—compare bonding ligands to monoclonal antibodies above. Oxine bound to silica reacts much more slowly with Al^{3+}, for example, than when it is in solution.[22] Another example of slow substitution at Al^{3+} is related to the indium chemistry mentioned above, involving its slow transfer from its transferrin complex by reaction with iron(III)–citrate to form the much more stable combination of iron–transferrin and aluminum–citrate complexes.[23] Further examples of slow substitution kinetics involving ferritin will be found in the iron(III) section (Section 8.3.4).

8.2. Groups 5 to 7

8.2.1. Vanadium

Vanadium(II) has a d^3 electron configuration, but rarely shows the high crystal field activation energy which might be expected. This has been demonstrated once again, in the course of electrochemical studies of various polypyridyl complexes. Stopped-flow techniques are indicated for substitution here.[24]

8.2.2. Molybdenum

The kinetics and mechanism of nitrogen–nitrogen bond breaking in the system $[Mo\{NN(CH_2)_4CH_2\}(dppe)_2]/NHEt_3BPh_4/RCN$ provide another contribution to the unraveling of mechanisms of dinitrogen fixation.[25] There have been two reports of kinetics of substitution at tetramolybdenum clusters. Thiocyanate substitution at $[Mo_4S_4(OH_2)_{12}]^{5+}$, a mixed-valence cluster with average $Mo^{3.25+}$, exhibits multiphasic kinetics,[26] while kinetics of thiocyanate substitution at the all-molybdenum(IV) cluster $[Mo_3S_4(OH_2)_9]^{4+}$, giving $[Mo_3S_4(NCS)_9]^{5-}$, show a significant contribution from $-SCN$ to $-NCS$ linkage isomerization. This tetra-sulfur cluster shows remarkably different behavior from its tetra-oxygen analog.[27] Thiocyanate substitution at $[Mo_3(\mu_3\text{-}S)(\mu\text{-}O)(\mu\text{-}S)_2(OH_2)_9]^{4+}$, (18), shows biphasic characteristics.[28] All three molybdenums are 4+ in this latter cluster, but of the three atoms two are sterically equivalent, the other inequivalent. Rate constants only differ by a factor of approximately four for substitution at the two nonequivalent types of molybdenum. For the series of trinuclear molybdenum(IV) species $[Mo_3O_xS_{4-x}(OH_2)_9]^{4+}$ reacting with thiocyanate there is a 570-fold spread in rate constants, with a similar spread for the reverse aquation rate constants, as x varies. The effects of replacing μ_2-O by μ_2-S are, surprisingly, the opposite of replacing μ_3-O by μ_3-S.[29] Still at molybdenum(IV), kinetics of fluoride substitution at *trans*-$[MoO_2(CN)_4]^{4-}$ have been described.[30] Finally, oxygen atom transfer from the hydrotris(3,5-dimethyl-1-pyrazolyl)borate-oxomolybdenum(IV) complex (19) to dimethyl sulfide, giving dimethyl sulfoxide, proceeds by an associative mechanism.[31] This molybdenum complex also catalyzes the oxygen transfer reaction shown in Eq. (1).

$$Ph_3P + Me_2SO \rightarrow Ph_3PO + Me_2S \tag{1}$$

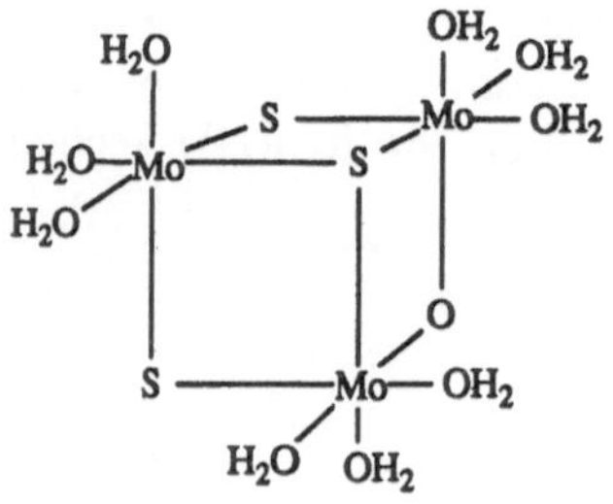

(18)

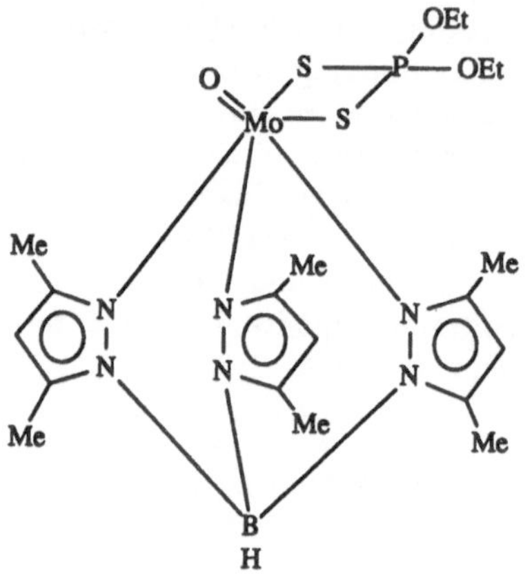

(19)

8.2.3. Tungsten

Complementing the *trans*-$[MoO_2(CN)_4]^{4-}$ plus fluoride study mentioned in the preceding section, kinetics of reaction of *trans*-$[WO_2(CN)_4]^{2-}$ with pyridine and with thiocyanate have been reported. These results, with published data for fluoride and azide attack, give a good linear free-energy relationship, and are discussed in terms of respective ligand *trans*-effects.[32] *Anti* → *syn* isomerization of the $W_2O_4^{2+}$ core in its 1,4,7-triaza-cyclononane complex has been monitored. This process has a timescale of ~10^4 s; it is acid-catalyzed.[33]

8.2.4. Manganese

The stepwise unwrapping of edta from its manganese(III) complex by cyanide is reflected in a variable-order dependence on the incoming ligand. The kinetically observable steps follow initial rapid reaction to $[Mn(CN)_3(edta)]^{4-}$ or $[Mn(CN)_4(edta)]^{5-}$. The cyanide dependence of the reverse reaction helped in the elucidation of the overall formation mechanism.[34]

8.2.5. Technetium

There is a continuing interest based on the widespread use and development of technetium complexes for radioimaging. Substitution kinetics are relevant in several ways. First, an administered agent should be inert with respect to ligand loss before reaching the required site. It should also not undergo any undesirable coordinated ligand modification before being imaged. The latest much-publicized imaging agent, Ceretec (hexamethylpropyleneamineoxime, cf. below), appears to be only just acceptable in this respect, since it undergoes an as yet uncharacterized reaction to a less suitable species within half an hour or so of generation. Substitution kinetics are also relevant to the successful preparation of the required agent in the standard reductive method from pertechnetate in the presence of the potential ligand.[35] Fortunately technetium has at least three oxidation states which are substitution-inert. Technetium(I) is d^6 and generally low-spin; technetium(IV) is d^3. Both therefore have high crystal field activation energies. Technetium(V) complexes are also fairly inert, presumably from electrostatically strong ligand binding. The growing family of technetium(I) complexes $[Tc(CNR)_6]^+$ are good myocardial imaging agents, while several TcO^{3+} complexes are in use or undergoing trials for imaging. Kinetic and mechanistic studies of technetium complexes will now be reviewed in increasing oxidation state order.

Despite its limited crystal field stabilization, the d^5 complex bis(4-chloroben-zenethiolato)bis(1,2-bis{dimethylphosphino}ethane)technetium(II), (20), under-goes relatively slow *trans* to *cis* isomerization, with a rate constant of $1.6 \times 10^{-4}\,s^{-1}$ in dichloromethane at ambient temperature.[36] The low-spin d^4 complex $[Tc(acac)_3]$ shows a similar reactivity with respect to exchange with ^{14}C-labeled acac to $[Cr(acac)_3]$. Activation parameters for the technetium complex are $\Delta H^+ = 119\ kJ\ mol^{-1}$ and $\Delta S^+ = -27\ J\ K^{-1}\ mol^{-1}$ in acetylacetone.[37]

Half-lives for hydrolysis of Tc—Cl to Tc—OH have been determined for four so-called "BATO complexes." These contain a semiencapsulating tris-dioxime ligand with a borate cap and the chloride ligand; the technetium is seven-coordinate, (21), with Y = OH or Me and either R = Me or pairs of R comprise cyclohexyl rings. Rate constants lie in the range 19 to $1.2 \times 10^{-4}\,s^{-1}$ for the four complexes examined, at 20 to 37 °C.[38] There is also some very qualitative kinetic information on base hydrolysis and ligand exchange (Cl for Br and *vice versa*) in this type of complex. These reactions have half-lives of minutes in media such as aqueous acetonitrile.[39]

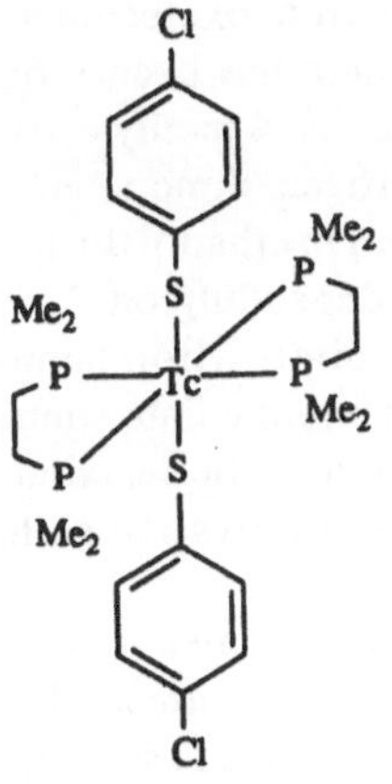

(20)

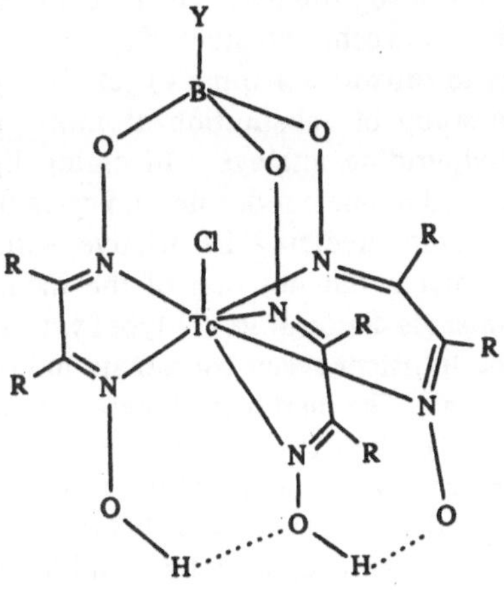

(21)

Technetium(IV) is an oxidation state where high crystal field activation ener-gies may be expected, and indeed substitution at $[TcBr_6]^{2-}$ [35] and at $[TcF_6]^{2-}$ [40] is very slow. Under forcing conditions the latter gives di-μ-oxo-di-technetium species, indicating a complex mechanism. The much-studied and much-used technetium–methylenediphosphonate complex is probably a tetranuclear technetium(IV) species with di-μ-oxo bridges linking each pair of technetium atoms. Both structural factors and crystal field effects confer great stability, and it comes as no surprise to read that this complex is stable for many months in aqueous media.[41]

Replacement of the glucoheptonate ligands in $[Tc^VO(glucohept)_2]$ by the appropriate tetradentate aminothiolate ligand, to give the potential imaging agent (22), follows a second-order rate law, but the kinetics are complicated by the existence of *syn* and *anti* forms of the ligand.[42] The currently much vaunted imaging agent Ceretec, the hexamethylpropyleneamineoxime complex (23), undergoes dissociation extremely slowly (though redox-catalyzed aquation is possible),[43] but undergoes some intramolecular rearrangement to give a less lipophilic and less active species with a half-life of about half an hour at ambient temperatures.

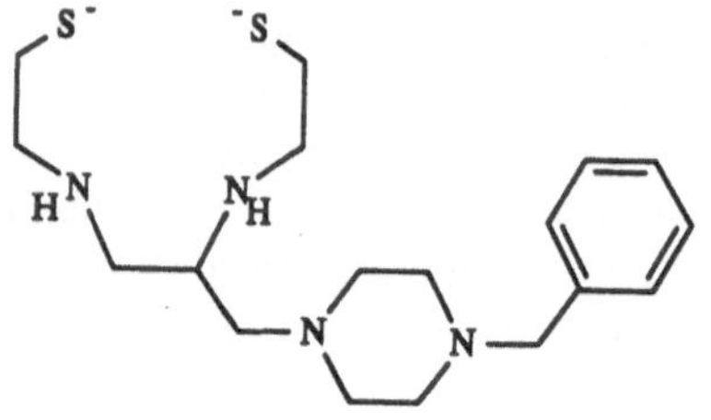

(22) 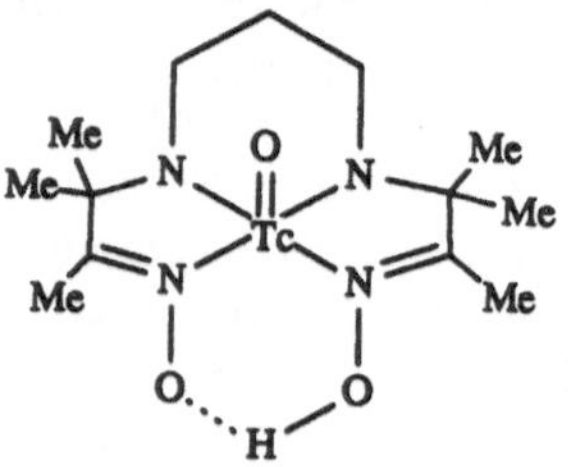   (23)

Reaction of the halide derivatives $[TcOX_4]^-$ with pyridine results not in simple substitution but in the formation of a mixed valence Tc—O—Tc species; pyridine N-oxide is also produced. The reaction is second-order in $[TcOX_4]^-$, with oxygen transfer as the central feature of the redox step.[44] Turning from oxotechnetium(V) species to dioxotechnetium(V) (cf. $[WO_2L_4]^{2-}$ above), there has been a detailed kinetic study of substitution at *trans*-$[TcO_2(py)_4]^+$ and its 4-methyl- and 3,5-dimethylpyridine analogs. Incoming ligands were nitrogen donors, including 1,2-ethanediamine, imidazole, and cyclam. In methanol and in ethanol the mechanism is solvent-mediated dissociative, with rate constants depending on the nature of the solvent but not that of the incoming group. In dimethylformamide the mechanism is dissociative (*D*-type) with a five-coordinate intermediate contrasting with the transient seven-coordinate intermediate postulated in the reactions with the alcohols. The inertness of these complexes is reflected in activation enthalpies in the range 75 to 100 kJ mol^{-1}.[45]

The nitrogen analogs of the well-known $[TcOX_4]^-$ anions mentioned above are technetium(VI) species $[TcNX_4]^-$. The chlorides in $[TcNCl_4]^-$ are labile, being easily and quickly replaced by bromide[46] or by fluoride.[47] There is some qualitative information on the conversion of $[TcNCl_4]^-$ into $[TcO_4]^-$, whose mechanism involves both substitution and electron transfer.[48]

Finally two reactions of technetium complexes which show clean first-order kinetics, despite ill-defined structures and unknown oxidation states. The first involves the dissociation or solvolysis of the technetium–spiperone dithiocarbamate complex (spiperone, (24), is involved in brain dopamine receptor behavior),[49]

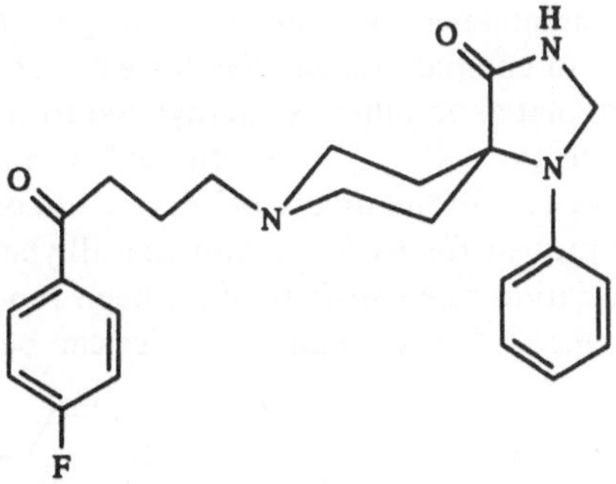

(24)

the second the reaction of an uncharacterized anionic technetium-formate species with dtpa (7). It is, perhaps naively, hoped that such kinetic behavior may be used in evidence for characterizing this species, which does seem to have potential as a synthetic starting material.[50]

8.2.6. Rhenium

End-to-end rotation, i.e., intramolecular linkage isomerization, of the coordinated dinitrogen in $[Re(\eta^5\text{-}C_5Me_5)(CO)_2(N_2)]$ occurs slowly [Re(I) is d^6], with a rate constant of $2.6 \times 10^{-4}\,s^{-1}$ in acetone at 287 K. The barrier of $90\,kJ\,mol^{-1}$ is similar to that estimated for $[Ru(NH_3)_5(N_2)]^{2+}$. Replacement of one of the carbonyl ligands by trimethyl phosphine raises the barrier sufficiently to slow rotation to the point of nonobservability.[51] Hydride derivatives of the $ReH_7(LL)$ type tend to undergo very rapid fluxional behavior, which makes it difficult to sort out the question of hydride vs. $\eta^2\text{-}H_2$ character. However the introduction of steric constraints, e.g., by using LL = 1,1'-bis(diphenylphosphino)ferrocene, slows down this intramolecular fluxionality enough to permit structural ^{1}H NMR studies at low temperatures.[52]

Kinetic (rate constant) and equilibrium data have been reported for the equilibrium (2) which is dissociatively activated in both directions.[53] Kinetics of

$$[ReO(OH_2)(CN)_4]^- + NCS^- \rightleftharpoons [ReO(NCS)(CN)_4]^{2-} + H_2O \qquad (2)$$

replacement of the first and second chlorides in $[Re_2Cl_8]^{2-}$ by phosphines or arsines give strong evidence in favor of the transmission of ligand labilization effects across the Re≡Re quadruple bond. As intermediates $[Re_2Cl_7L]^-$ can be isolated, it proved possible to get good kinetics on both stages of the forward, and indeed also of the reverse, reaction. All four reactions are believed to be associative.[54]

8.3. Iron

8.3.1. Penta- and Tetracyanoferrates

Kinetic parameters for formation of $[Fe(CN)_5L]^{n-}$ from $[Fe(CN)_5(OH_2)]^{3-}$ have been reported for a variety of incoming ligands L. One study has dealt with

L = 4,4'-bipyridyl and the ethane- and ethene-bipyridyls (**25**) and (**26**). The particular interest here was to compare reactivities for each of these three ligands as such, in their mono-protonated or mono-N-methylated forms, and in their derivatives with $[Fe(CN)_5]^{3-}$ or $[Co(NH_3)_5]^{3+}$ coordinated to one of the nitrogens.[55] The kinetic consequences of coordinating $[Co(NH_3)_5]^{3+}$ to one end of the entering ligand have also been probed for pyrazine and the alkyne-bipyridyl derivatives (**27**) and (**28**).[56] Dissociation rate constants have been reported for many of the products of these reactions.[55,56] These kinetic studies can be put into context with

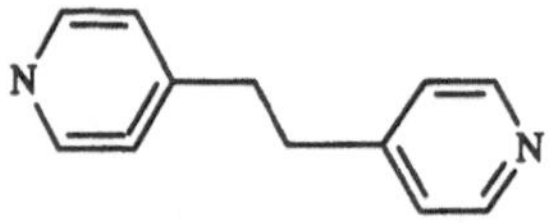

(**25**)

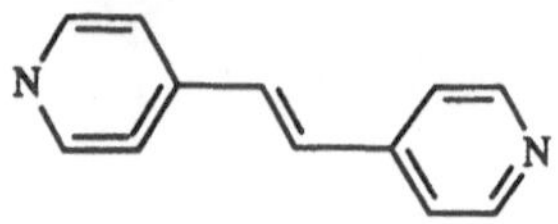

(**26**)

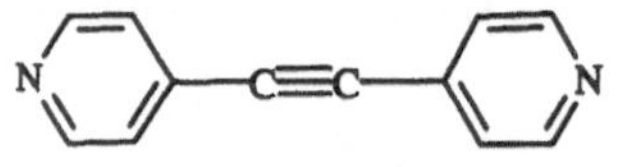

(**27**)

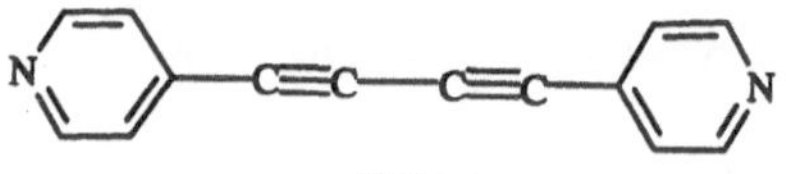

(**28**)

the help of a recent review of pentacyanoferrates(II).[57] The question of how closely these reactions approximate to limiting dissociative $[S_N1(\text{lim})$ or $D]$ processes has been discussed in these reports before.[58] An activation volume of $+7 \text{ cm}^3 \text{ mol}^{-1}$ for water exchange at the cobalt(III) analog $[Co(CN)_5(OH_2)]^{2-}$ suggests I_d rather than D activation here,[59] but the difference in charge on the central metal may still permit D activation in dissociation of the pentacyanoferrates(II). However, the question of D vs. I_d mechanisms in this area is further complicated by the remarkably small activation volumes for what appears, from kinetic evidence (mass law retardation), to be D substitution at ternary Group VI carbonyls. In fact photosubstitution at such species is characterized by rather less small activation volumes, as established recently for $[W(CO)_5(py)]$ and related compounds.[60] The pentacyanoferrate(II) moiety plays a key role in the $[Fe(CN)_5(OH_2)]^{3-}$ catalyzed diazotization of butylamine by nitrite.[61] Time dependences of spectra give qualitative kinetic information on the substitution-redox reactions of $[Fe(CN)_5(OH_2)]^{3-}$ and related complexes with nitric oxide in alkaline aqueous solution.[62]

Kinetic parameters have been established for solvolysis of the pentacyanoferrate(III) derivative $[Fe(CN)_5(NO_2)]^{3-}$. For aquation, which is acid-catalyzed at pH < 5, $\Delta H^+ = 43 \text{ kJ mol}^{-1}$, $\Delta S^+ = -80 \text{ J K}^{-1} \text{ mol}^{-1}$, and $\Delta V^+ = +2 \text{ cm}^3 \text{ mol}^{-1}$. Intrinsic and solvational contributions are presumably closely balanced in the case of ΔV^+.[63] Rate constants for solvolysis of $[Fe(CN)_5(NO_2)]^{3-}$ in water, methanol, dimethyl sulfoxide, and dimethylformamide correspond with the electron-donating abilities of the respective solvents. Activation volumes for the nonaqueous solvents, between $+20$ and $+27 \text{ cm}^3 \text{ mol}^{-1}$, reflect the dissociative nature of these solvolyses.[64] Rate constants for dissociation of the $[Fe(CN)_5(2,6\text{-Me}_2\text{pyrazine})]^{3-}$ anion in binary aqueous solvents containing methanol, acetone, or acetonitrile correlate well with acceptor numbers for the respective media, though with a very different slope for each cosolvent.[65] Photochemical studies in this area are represented by determination of quantum yields for photosubstitution at $[Fe(CN)_5L]^{3-}$ with $L = PPh_3$, $AsPh_3$, $SbPh_3$, $P(OMe)_3$, or CO. In all cases L leaves, rather than cyanide. For the first four complexes the quantum yields are around 0.15, but for the carbon monoxide complex the quantum yield was 0.38. This much higher value can be attributed to the fact that the higher ligand field of CO than of CN^- results in the lowest excited state being $^3A_2$ here, which has a longer lifetime than the ^3E state which is the lowest excited state for the other four complexes.[66]

Activation parameters for $\delta \rightleftharpoons \lambda$ interconversion of the 1,2-diamine rings in complexes $[Fe^{III}(CN)_4(diamine)]^-$, containing, e.g., cis-1,2-diaminocyclohexane or meso-butanediamine, have been determined by ^1H NMR spectroscopy. Activation enthalpies are in the range 25 to 40 kJ mol^{-1}, activation entropies between 0 and -8 J K^{-1} mol^{-1}.[67,68] These iron(III) complexes disproportionate according to third-order kinetics, with rate constants which reflect the steric strains in the respective diamine chelate rings.[68]

8.3.2. Iron(II)–Diimine Complexes

Although rate constants for high-spin $\rightleftharpoons$ low-spin interconversion for $[Fe(pyim)_3]^{2+}$, pyim = 2-(2'-pyridyl)imidazole (**29**), are extremely fast, they are relevant to discussions of racemization and dissociation of iron(II) diimine complexes, as in both cases significant participation by the high-spin form is often implicated in reactions involving the substitution-inert low-spin form.[69] Thus

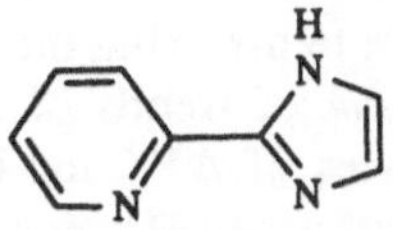

(**29**)

activation volumes for dissociation of such complexes are large and positive, e.g., $+23$ cm^3 mol^{-1} for tris(5-bromo-1,10-phenanthroline)iron(II) in aqueous solution,[70] suggesting large bond extensions, perhaps to the high-spin limit, on transition state formation.

Kinetic studies of aquation of $[Fe(phen)_3]^{2+}$ and derivatives in binary aqueous media remain popular. A group additivity approach has been applied to aquation of $[Fe(5NO_2phen)_3]^{2+}$ in aqueous alcohols (faster reaction) and formic and acetic acids (slower), to investigate its potential for mechanism diagnosis.[71] Rate constants for dissociation of the parent complex increase tenfold on going from water to 100% dimethylformamide.[72] Aquation rate constants and activation parameters have also been reported for the 5-nitro, 5-phenyl, and 4,7-diphenyl derivatives in water–dioxan mixtures.[73] Both papers contain obscure discussions of solvolysis mechanisms in DMF-rich and dioxan-rich media. In the latter media it seems that ion pairs play a key role, as evidenced by activation entropies. The discussion of reactivities in terms of hydrophobicities of the complexes and their respective transition states represents a qualitative initial state-transition state analysis.[73] An explicit analysis of this type has been published for the iron(II) complexes of the

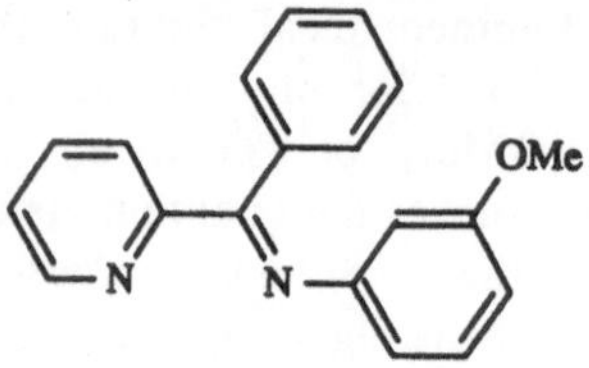

(30)

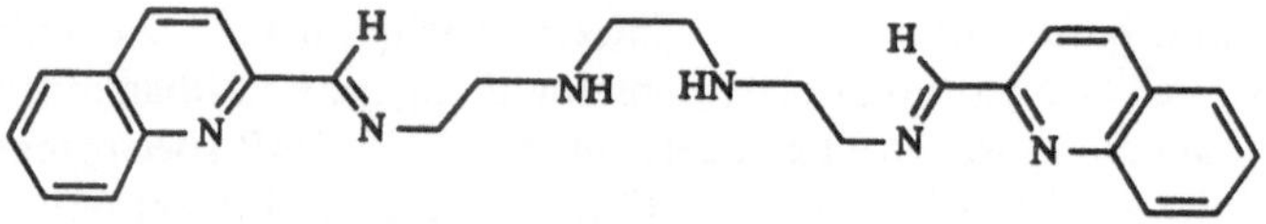

(31)

Schiff base ligands (30) and (31); ligand hydrophobicities again play a key role.[74] An initial state-transition state analysis for base hydrolysis of $[Fe(phen)_3]^{2+}$ in ethanol- and in 2-propanol–water mixtures is based on new sets of single ion transfer chemical potentials, on the TATB ($Ph_4As^+ = BPh_4^-$) assumption, for these two cosolvents.[75] The topics of initial state-transition state analyses and of single ion thermodynamic parameters, on which they are perforce generally based, have been reviewed.[76] Having carried out such an analysis for a reaction in mixed solvents, then Kirkwood–Buff theory can be pressed into service to obtain quantitative characterization of selective solvation of reactants and of transition states, as has recently been achieved for base hydrolysis of $[Fe(phen)_3]^{2+}$ in aqueous methanol.[77] An alternative route to assessing the role of solvation in reactions of this type is through establishment of trends of activation volumes with mixed solvent composition. Dependences of ΔV^+ for base hydrolysis of the iron(II) complex of the hexadentate Schiff base (32) on solvent composition for alcohol-water mixtures reflect solvent structural changes rather well.[78] The hexadentate encapsulating ligands (33) contain diamine rather than diimine groups, but they exert sufficiently large ligand field effects to render their iron(II) complexes low-spin when Y = H, CH_3, or NH_2. When Y = NH_3^+ the complex is on the low-spin/high-spin crossover borderline. Presumably all these complexes will prove to be substitution-inert.[79]

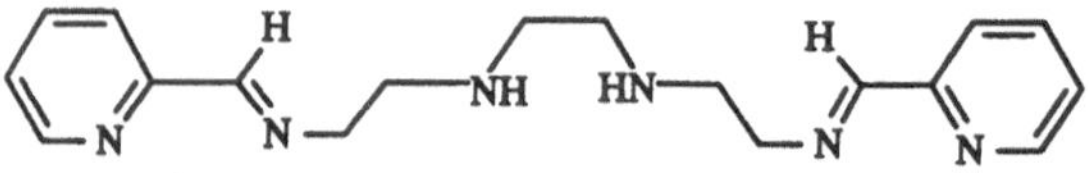

(32)

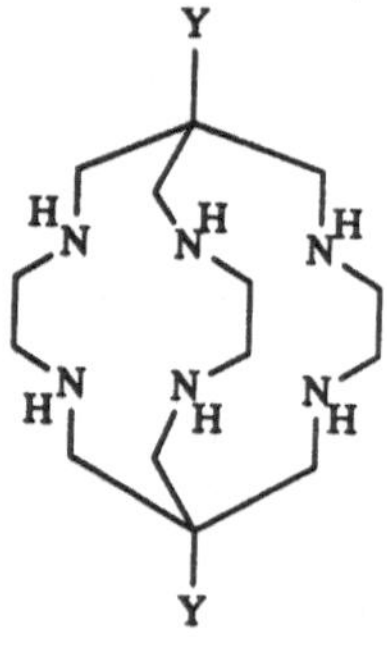

(33)

Micellar and microemulsion effects on reactivity in aquation and base hydrolysis reactions of iron(II)–diimine complexes have been much studied.[80] The latest contribution deals with the effects of added potassium chloride or bromide to micelles of the respective cetyltrimethylammonium halides. Effects on base hydrolysis of $[Fe(phen)_3]^{2+}$ and its 4,7-diphenyl and 3,4,7,8-tetramethyl derivatives can be interpreted in terms of competitive binding to the micelles in a pseudophase-ion exchange model.[81] In connection with these secondary effects of added halides it should be mentioned that further studies of kinetics of aquation of $[Fe(bipy)_3]^{2+}$ and of $[Fe(phen)_3]^{2+}$ in strong aqueous solutions of chlorides have been interpreted in terms of water and of chloride attack, with the postulation of transient diimine-chloride–iron(II) intermediates.[82]

The debate on hydroxide attack at coordinated diimine and pyridine ligands coordinated to transition metal ions such as Fe^{2+} continues. Catalysis of hydrolysis of the diimine ligand tptz, (**34**), by Cu^{2+} has been suggested to proceed by attack by hydroxide or by water at one of the pyridine rings of tptz when it is coordinated to the copper.[83]

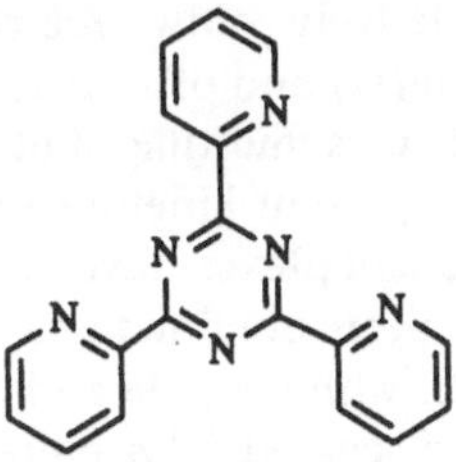

(34)

8.3.3. Other Iron(II) Complexes

The tris-ligand complex of iron(II) and 3-hydroxyiminopentane-2,4-dione (isonitrosoacetylacetone) (**35**) is low-spin, but must be near the spin-crossover region since it is formed from its components on the stopped-flow timescale.[84]

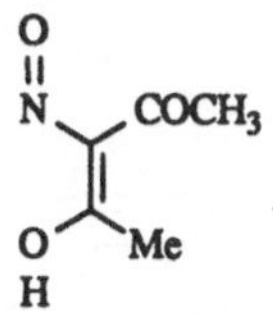

(35)

The rate constant for loss of nitric oxide from its complex with deoxyhemerythrin is just about in the "inert center" range; it is considerably smaller than that for dissociation of the analogous dioxygen complex. Kinetic parameters for formation and dissociation of a number of deoxyhemerythrin adducts are compared in this paper.[85] Steric effects on kinetics of addition of dioxygen and of carbon monoxide to four single-face hindered iron(II)–porphyrin complexes have been compared[86] and discussed.[87] Rather subtle kinetics are involved in the internal motion of carbon monoxide attached to haem in myoglobin. Such motion has been probed by ^{17}O NMR.[88] Formation of pyrazine-bridged iron(II) protoporphyrin IX species from the bis-pyrazine mononuclear parent is very fast,[89] but is mentioned here as the discussion is relevant to the question of transient five coordinated intermediates in pentacyanoferrate substitutions (cf. Section 8.3.1 above). The discussion of the hydrophilicity/hydrophobicity of the surface of this and two substituted derivatives, and the consequences in terms of solvation, and solvent and micellar effects on reactivities, is of interest in relation to some of the systems mentioned in Section 8.3.2. Kinetic patterns of thermal and photochemical substitution at tetraazamacrocycle complexes of iron(II) have been reviewed.[90]

8.3.4. Iron(III) Complexes

Most substitutions at iron(III) are fast, and are therefore discussed elsewhere in this report (see Chapter 9), but several are slow enough to monitor by conventional techniques and are therefore mentioned here (though pentacyanoferrate(III) complexes are in Section 8.3.1). The first system bridges this and the preceding sections, for it involves relatively slow *fac* $\rightleftharpoons$ *mer* isomerization for trishydroxamato complexes of iron(II) and of iron(III). These complexes containing ligand (36) have been known for some time, but isomer details have only been sorted out in the course of the present kinetic study.[91] Kinetics of formation of several iron(III)–hydroxamate complexes have also been reported.[92]

Kinetic parameters have been obtained for reaction of $[Fe(par)_2]^-$, par = 4-(2-pyridylazo)resorcinol (37), with cyanide and the mechanism of formation of the product $[Fe(CN)_3(par)]^{2-}$ discussed.[93] A more complicated sequence of steps is involved in the fairly slow displacement of edta from $[Fe(edta)]^-$ by hexadentate Schiff base ligands derived from trien and substituted salicaldehydes (38). Kinetic studies were carried out for two such ligands, and it was demonstrated the gallium(III) complexes of such ligands are also fairly inert (cf. Section 8.1).[94]

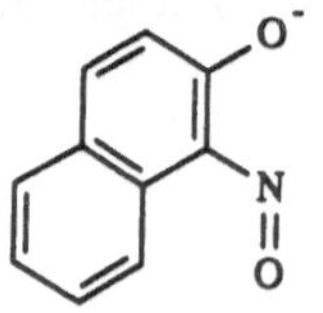

(36) 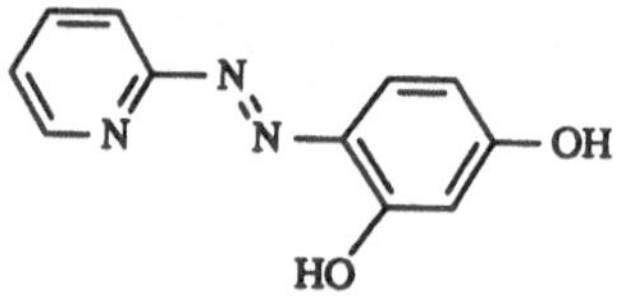   (37)

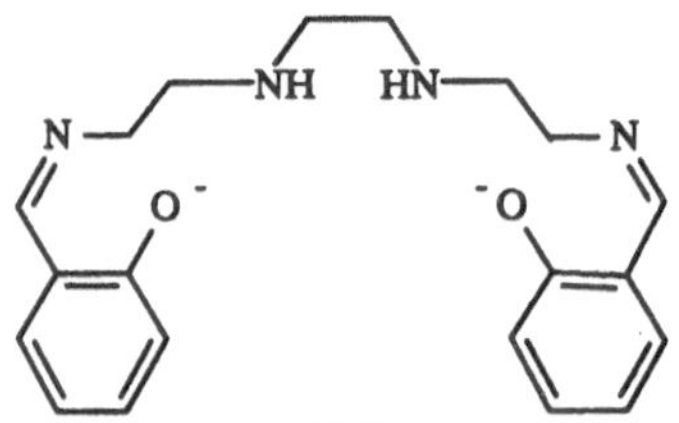

(38)

Rate constants and activation parameters have been obtained for dissociation of the axial ligands from tetraphenylporphyrin and tetramesitylporphyrin complexes $[Fe^{III}(porph)L_2]^+$. Steric factors were assessed by the comparison of results for L = 1- or 2-methylimidazole, 2-ethylimidazole, and 1,2-dimethylimidazole.[95] One presumes that the mechanisms of such reactions is limiting dissociative (*D*), like those of the chromium(III) analogs, where ΔV^+ lies between +20 and +26 cm^3 mol^{-1}.[96] Kinetics of formylation of the iron(III) complex of tetra-4-tolylporphyrin, a Vilsmeier reaction, may be complicated by anion generation.[97] Dimerization and hydrolysis kinetics form a small part of an extensive discussion of the chemistry of three water-soluble iron(III) porphyrin cations.[98]

The biphasic kinetic pattern described for the removal of iron from transferrin by pyrophosphate can be ascribed to the two different iron-containing sites in transferrin.[99] Various anions and acids can assist such removal.[100] Details of iron removal have been probed by studying the kinetics of metal removal from transferrin derivatives containing iron and cobalt variously distributed between the two inequivalent binding sites, and from transferrins containing iron in only one of the two sites. The kinetics of iron removal from the Fe$_C$ sites show a first-order dependence on pyrophosphate concentration, from the Fe$_N$ sites show saturation kinetics.[101] The current situation with respect to mechanisms of iron removal from, and incorporation into, transferrin have been reviewed.[102]

8.4. Ruthenium

8.4.1. Ruthenium(II)

A number of papers deal with photosubstitution, some dealing primarily with this topic, others only tangentially in connection with photoredox or photoactivation. A recent discussion of photodissociation of the intensively studied [Ru(bipy)$_3$]$^{2+}$ cation warns against neglecting the very slow but nonetheless nonzero rate of thermal dissociation, and recommends the use in photochemical studies of complexes which are even more inert with respect to thermal dissociation. Rigid phen is considerably more reluctant to undergo dissociation than flexible bipy. Indeed in the sequence [Ru(bipy)$_n$(phen)$_{3-n}$]$^{2+}$ only bipy is lost when the complexes with $n = 3, 2$, or 1 dissociate (thermally or photolytically); phen is lost only for $n = 0$, i.e., from [Ru(phen)$_3$]$^{2+}$.[103] It is interesting that rate constants for photosubstitution at this cation increase steeply with increasing acidity. Presumably this indicates the intermediacy of a protonated monodentate-phen species, which may be expected to be easier to generate in these circumstances than in thermal substitution at [Fe(phen)$_3$]$^{2+}$.[104] Photosubstitution at [Ru(bipy)$_3$]$^{2+}$ in acidic aqueous media has been suggested to involve associative activation, on the basis of observed added anion effects.[105] Photolysis of [Ru(bipy)$_3$]Cl$_2$ in DMF gives no fewer than six products, including a binuclear, presumably chloride-bridged, species.[106] Anions often affect the products of photosubstitution, especially when they are potential ligands such as chloride. Thus photolysis of [Ru(bipy)$_2$(*N*-vinylimidazole)$_2$]$^{2+}$ in acetonitrile in the presence of lithium chloride gives a mixture of chloride and acetonitrile substituted intermediates en route to the [Ru(bipy)$_2$(MeCN)Cl]$^+$ product.[107] Efficient photosubstitution by acetonitrile or by chloride has also been described for a series of complexes of the type $[Ru(terpy)L_3]^{2+}$ or *cis-* or *trans-*$[Ru(terpy)L_2Cl]^+$, where L = pyridine, 4-methyl-pyridine, or propionitrile.[108] Quantum yields have been determined for photosubstitution by chloride or acetonitrile at a dozen complexes of the $[Ru(diimine)_3]^{2+}$ type, in connection with efforts to monitor energy gap control of this process and to establish a redox potential correlation.[109]

The quantum yield for ammonia loss from $[Ru(NH_3)_4(dpp)]^{2+}$ and its binuclear analog $[(H_3N)_4Ru(dpp)Ru(NH_3)_4]^{4+}$, where dpp = 2,3-bis(2-pyridyl) pyrazine (**39**), is wavelength-dependent and low. This behavior is typical of ruthenium(II) ammine complexes in which the lowest excited state is MLCT.[110]

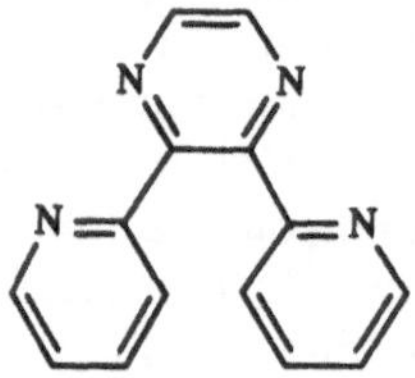

(**39**)

Hardly surprisingly, photodissociation of cage complexes, for example, those incorporating three 2,2'-bipyridyl units, is much slower than of $[Ru(bipy)_3]^{2+}$. The encapsulation of the ruthenium increases the photochemical inertness and the lifetime of the photoexcited state.[111]

Turning to thermal substitution, kinetic parameters (k, ΔH^+, and ΔS^+) are reported for the forward and reverse directions of the equilibria involving reaction of $[Ru(5NO_2phen)_2(bipy)]^{2+}$ or of $[Ru(5NO_2phen)(bipy)_2]^{2+}$ with hydroxide or with cyanide. Activation entropies suggest that solvation effects on the transition states may be less important for cyanide than for hydroxide.[112] Steric and electronic effects of varying the nature of the alkyl or aryl groups in the PR_3 ligand have been monitored for substitution at *cis-*$[Ru(bipy)_2(PR_3)(OH_2)]^{2+}$. The kinetic pattern is said to be satisfactorily consistent with the expected I_d mechanism,[113] though the attribution of the observed negative activation entropies to a dissociative process is less convincing. For substitution at *cis-*$[Ru(bipy)_2LX]^+$, where L = MeCN or py, by chloride or by MeCN, rate constants are much larger when $X = NO_3^-$ than when $X = NO_2^-$, $CF_3CO_2^-$, or Cl^-. This special behavior of nitrate is attributed to its ability to act in a bidentate manner.[114] There are a few sentences of mechanistic relevance to substitution at complexes of this type in a general review of the chemistry of (poly)pyridine complexes of ruthenium, rhodium, and palladium,[115] and some speculation about intramolecular or fluxional processes in a paper about complexes $[Ru(bipy)_2(4,4'-X_2bipy)]^{2+}$ containing substituents X of the type $-NH(CH_2)_2NMe_2$.[116]

A large number of rate constants and some activation parameters are reported for axial ligand substitution in the benzoquinonedioxime complexes $[Ru(bqdH)_2L_2]$, where bqdH = (**40**) and L is one of a range of nitrogen, phosphorus, and sulfur donor ligands. The mechanism is limiting dissociative (D), but the transient five-coordinate intermediates have very little discriminating ability.[117]

(40)

Kinetic and equilibrium data are reported for the reaction between [Ru^{II}(uroporphyrin I)(CO)] and cyanide. The rate law is first order in cyanide for addition of the first cyanide, but zero order in cyanide for the second step, which must therefore involve rate-limiting loss of carbon monoxide.[118] Reaction of [Ru(NH$_3$)$_5$(OH$_2$)]$^{2+}$ with diethyl phosphite also takes place in two kinetically distinct stages, to give *trans*-[Ru(NH$_3$)$_4$(OH$_2$){P(OEt)$_2$(OH)}]$^{2+}$. This study was carried out to investigate the effect of the coordinated ruthenium on the phosphate $\rightleftharpoons$ phosphonate equilibrium. The Ru(II) favors the phosphite form, suggesting that Ru(II)—P back bonding has a greater effect than O—P back bonding.[119] Kinetics of the forward and reverse reaction of *trans*-[Ru(NH$_3$)$_4$L(OH$_2$)]$^{2+}$ with pyrazine, where L = a phosphine or phosphite, were examined in order to assess the *trans*-effects and *trans*-influences of these ligands, in particular to probe the role of π-bonding in the *trans*-influence.[120] Reaction of [Ru(CN)$_5$(OH$_2$)]$^{3-}$ with 3- or 4-cyanopyridine gives a mixture of cyanide-N and pyridine-N isomers. Rate constants for formation, dissociation, and isomerization of these complexes have been determined. Rate constants for isomerization are close to 10^{-3} s^{-1} for both isomers at 25 °C. Dissociation rate constants are 6.0×10^{-5} and 2.5×10^{-5} for the 3- and 4-cyanopyridine complexes (N-bonded isomers), but activation enthalpies and entropies are markedly different for the two complexes. The kinetic patterns for these reactions correspond to limiting dissociative (D) mechanisms.[121] Substitution in [Ru(CN)$_5$(en)]$^{3-}$ also occurs by the D mechanism; formation of this complex from [Ru(CN)$_5$(OH$_2$)]$^{3-}$ occurs by the expected I_d mechanism. Kinetic parameters are given for both reactions, and these are compared with values for the analogous pentacyanoferrate(II) system.[122] Reactivity comparisons have also been made between Ru and Fe for reactions of the respective [M(CN)$_5$(NO)]$^{2-}$ anions with such reagents as hydroxide, piperidine, dithionite, and hydrosulfide.[123]

Change of coordination number is involved in a kinetic and mechanistic study of the ruthenium(II)–salen–triphenyl phosphine system. The rate constant for dissociation of [Ru(salen)(PPh$_3$)$_2$], by what must be a dissociative mechanism, is 1.2×10^{-3} s^{-1} in benzene at 25 °C.[124]

Qualitative comments on relative reactivities of complexes in the [RuX$_2$(R$_2$SO)$_m$(nitroimidazole)$_n$] series, where X = Cl or Br, m = 1, 2, or 3, and n = 1 or 2, have been made in relation to the role of such complexes as radio-sensitizers,[125] while some kinetic data for *cis* $\rightleftharpoons$ *trans* isomerization of chloride and bromide complexes [RuX$_2$(DMSO)$_4$] have been obtained as spin-off from antitumor activity studies on these species.[126] Isomerization kinetics have also been described for a slightly more complicated sequence involving *cis,cis,cis*, *cis,trans,trans*, and *trans,trans,trans* forms of [RuCl$_2$(CO)$_2$(PR$_3$)$_2$]. Activation entropies and volumes suggest dissociative mechanisms.[127] A mechanism has been suggested for the very slow conversion of *trans*-[RuCl(NO)(bipy)$_2$]$^{2+}$ in acetone into [RuCl(N≡CH)(bipy)$_2$]$^+$ through the intermediate (41).[128] Attempts to aquate

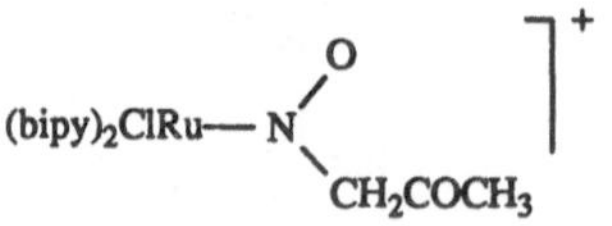

$$(bipy)_2ClRu—N\underset{CH_2COCH_3}{\overset{O}{\diagup}}\Bigg]^+$$

(41)

[RuCl$_2$(di-2-pyridyl ketone)$_2$] by the use of Ag$^+$ failed; instead the coordinated di-2-pyridyl ketone underwent hydrolysis.[129]

8.4.2. Ruthenium(III)

The kinetic scheme for intramolecular rearrangements within [Ru(acac)-(PhCOCHCOCF$_3$)$_2$] consists of a triangle of three reversible reactions. Methods for computing the six rate constants involved have been discussed, though in practice it was found that for this particular compound only four made significant contributions to the observed kinetics. The rearrangements involving the change from both phenyl groups *trans* to acac to both CF$_3$ groups *trans* to acac, and *vice versa*, do not contribute.[130] Complexes cis-[Ru(NH$_3$)$_4$(H$_2$NCH$_2$CO$_2$R)$_2$]$^{2+}$, with R = Me, Et, or Bz, which form rapidly in aqueous solution on mild oxidation of their ruthenium(II) forebears, undergo pH-dependent first-order rearrangement to give the corresponding chelates [Ru(NH$_3$)$_4$(H$_2$NCH$_2$CO$_2$)]$^{2+}$. Linkage isomerization of one of the coordinated ester molecules is believed to be the first step.[131] A further instalment of aquation and anation kinetic data on chloroaquaruthenium(III) species has appeared.[132] Some qualitative information has appeared on the ease of substitution by sulfur, phosphorus, arsenic, nitrogen, and oxygen donor ligands in *mer*- and *fac*-[RuCl$_3$(dmso)$_3$]. All three dmso ligands are S-bonded in the *fac*-isomer, but the *mer*-isomer contains both S- and O-bonded dmso ligands.[133] Similar information is available for [RuBr$_3$(dmso)$_3$], and indeed for its ruthenium(II) analog.[134] Kinetics of substitution have also been studied for replacement of N-, P-, and S-donor ligands in [Ru(bqd)(bqdH)L$_2$], bqdH = (40) above.[117]

Activation entropies and volumes for anation of [Ru(edta)(OH$_2$)]$^-$ by azide, thiocyanate, and (substituted) thioureas lie in the ranges −99 to −105 J K^{-1} mol^{-1} and −7 to −12 cm^3 mol^{-1}. The associative mechanism thus indicated is attributed to hydrogen-bonding interactions between carboxylate groups and coordinated water pulling the latter away from its expected position and thus leaving more space for the entering group, thereby facilitating formation of the transition state. Reactivity is very much higher than usual for ruthenium(III) complexes thanks to this effect.[135] Associative activation has also been established for analogous reactions of the [Ru(hedtra)(OH$_2$)] complex, where hedtra is N-(hydroxyethyl)ethylenediaminetriacetate. An I_a mechanism seems more likely than a limiting A mechanism.[136] The reaction of [Ru(edtaH)(OH$_2$)] with carbon monoxide brings us to the frontiers of organometallic chemistry. A kinetic and equilibrium study of this system was complicated by proton equilibria and by the formation of significant amounts of binuclear species.[137]

8.4.3. Ruthenium(IV), Ruthenium(V), and Ruthenium(VI)

Aquation of the binuclear $[Ru_2OCl_{10}]^{4-}$ anion has been studied spectrophotometrically. At least three chlorides can be lost from each ruthenium before the $Ru-O-Ru$ bridge ruptures. Two axial chlorides are lost first, followed by an equatorial chloride.[138]

Qualitative information is available on the timescale for loss of bipy from cis-$[Ru^{VI}O_2(bipy)_2]^{2+}$. Extremely slow ligand loss from cis-$[Ru^VO(OH)(bipy)_2]^{2+}$ probably takes place via initial disproportionation to Ru(IV) and Ru(VI).[139] These complexes are closely related to some of the complexes discussed in the preceding two sections, since they are generated simply by oxidation of cis-$[Ru(bipy)_2(OH_2)_2]^{2+}$.

8.5. Osmium

Several papers treat various isomerization processes of osmium(II) and osmium(III) species $[Os(NH_3)_5L]^{n+}$. For L = $PhNH_2$, rate constants (at 20 °C) for isomerization of the N-bonded to the η^2-bonded form are $2.3 \times 10^{-5}\,s^{-1}$ in acetone and $2.0 \times 10^{-4}\,s^{-1}$ in DMSO for the osmium(II) complex, 8×10^{-11} for the osmium(III) complex.[140] The isomerization of $[Os(NH_3)_5(\eta^2\text{-PhCN})]^{2+}$ to $[Os(NH_3)_5(NCPh)]^{2+}$ is a slow intramolecular process.[141] Metal tautomerization, $2,3\text{-}\eta^2 \rightleftharpoons 4,5\text{-}\eta^2$, takes place on a timescale of seconds for the complex $[Os(NH_3)_5(pyrrole)]^{2+}$ at -25 °C.[142] The osmium(II) complexes $[Os(NH_3)_5(LH)]^{3+}$, where LH^+ = pyridinium, N-methylpyridinium, lutidinium, or N-methyl-4-picolinium, all of which are η^2-bonded, are fluxional. The N-methylpyridinium complex loses the *trans*-ammonia ligand relatively easily, through an η^2-intermediate, with a half-life of a matter of hours at ambient temperatures.[143] The rate constant for arene → alkyne bonded isomerization for (42) on $[Os(NH_3)_5]^{3+}$ is $1.8 \times 10^4\,s^{-1}$ at 20 °C, which is two or three orders of magnitude faster than analogous isomerizations involving aniline or anisole.[144]

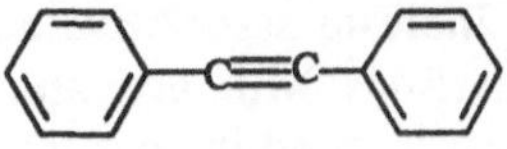

(42)

Alkyne addition reactions involving $[Os(NH_3)_5]^{2+}$ proceed via π-enol and π-vinyl ether complexes,[145] while treatment of $[Os(NH_3)_5(\eta^2\text{-}C_6H_6)]^{2+}$ with hydrogen in methanol in the presence of Pd/C catalyst gives, selectively, $[Os(NH_3)_5(\eta^2\text{-}hexene)]^{2+}$.[146]

To turn to classical substitution, the rate constant for aquation of $[Os(NH_3)_5(O_3SCF_3)]^{2+}$ is $8 \cdot 8 \times 10^{-4}\,s^{-1}$ at 25 °C in 1 M CF_3SO_3H. Rate constants for solvolysis in triethyl phosphate and in acetone are similar. All these solvolysis rates are relatively fast for osmium(III), since fluorosulfonate is a good leaving group.[147] Loss of triphenyl phosphine oxide from cis-$[Os(bipy)_2(OPPh_3)_2]^{2+}$ is very slow, with a half-life of many hours in acetonitrile. This observation is a byproduct of a redox investigation, which also provided information on reaction

(3) that may be considered a coordinated ligand reaction of osmium(VI).[148] Solvent and temperature kinetic control are both important in the preparation of new osmium(II) polypyridyl complexes from cis-[Os(bipy)$_2$Cl$_2$].[149] A detailed

$$\text{cis- or trans-}[Os(bipy)_2(O)_2]^{2+} + 2PPh_3 \rightarrow [Os(bipy)_2(OPPh_3)_2]^{2+} \qquad (3)$$

mechanism for the reaction of [Os(O$_2$CCF$_3$)$_2$(CO)$_4$] with pyridine thionate, pytH = (43), to give [Os(pyt)$_2$(CO)$_2$] has been elucidated through the isolation and characterization of a sequence of intermediates. The pyt ligand first forms a monodentate S-bonded intermediate, which then ring closes with the loss of one CF$_3$CO$_2^-$ ligand. The second CF$_3$CO$_2^-$ is replaced by the second pyt, again with subsequent ring closure.[150]

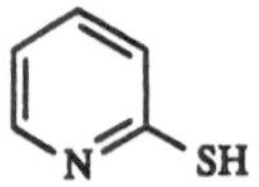

(43)

[Os(NH$_3$)$_6$]$^{2+}$ condenses with acetone in a redox-catalyzed coordinated ligand reaction, impelled by the Os(II)–imine intermediate being strongly stabilized by π-bonding.[151] Indeed one of the most distinctive features of the chemistry of osmium(II) is its unusually high tendency to π-bond. This has a large effect on reactivity, manifested in the various substitution and tautomerization reactions of [Os(NH$_3$)$_5$L]$^{2+}$ complexes discussed in this section. Kinetic parameters for reactions of the L = η^2-arene series have been conveniently assembled.[152]

8.6. Rhodium

Activation volumes for solvolysis of [Rh(NH$_3$)$_5$(OSO$_2$CF$_3$)]$^{2+}$ in methanol and in acetonitrile are −6.6 and −7.8 cm^3 mol^{-1}. These values, like that of −4.1 cm^3 mol^{-1} for aquation, indicate associative activation, even for this easily lost leaving group. In comparisons with activation volumes for solvolysis of [Rh(NH$_3$)$_5$Cl]$^{2+}$ one has to bear in mind the partial molar volumes of the leaving groups as well as electrostriction effects.[153] Base hydrolysis of [Rh(en)$_2$(ox)]$^+$ obeys a simple second-order rate law, with $k_2 = 4.6 \times 10^{-3}$ dm^3 mol^{-1} s^{-1} at 60 °C.[154] Base hydrolyses of [Rh(LLLL)Cl$_2$]$^+$, where LLLL = optically active methyl-substituted trien ligands, follow a two-term rate law ($k_1 + k_2$[OH$^-$] type). Rate constants and stereochemical courses of reaction are given.[155]

Inversion of the tetrahydrothiophen ligands in the complexes [Rh(tht)$_3$X$_3$], where X = Cl, Br, or I, takes place intramolecularly,[156] as does enantiomer interconversion of the semiencapsulated complex of the ligand (44). The barrier to the latter process is ~46 kJ mol^{-1} (at −55 °C).[157]

A detailed ^{1}H and ^{13}C NMR study of the relatively rapid, and specific, deuteration of [Rh(bipy)$_3$]$^{3+}$ in (CD$_3$)$_2$SO—CD$_3$OD reveals marked differences in behavior in comparison with the analogous osmium(II) and ruthenium(II) com-

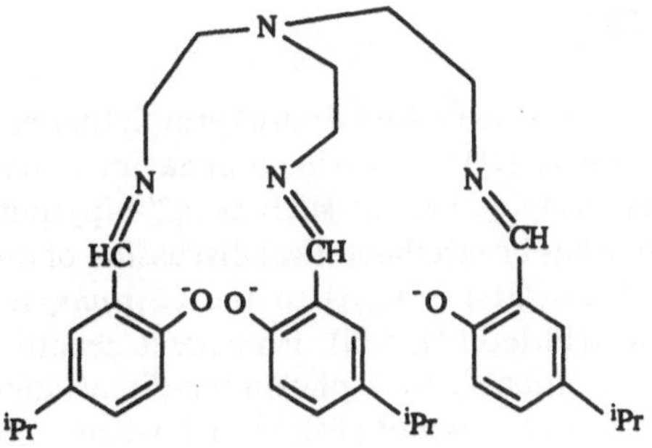

(44)

plexes. These differences are attributed to a different balance of δ- and π-bonding effects in the course of a detailed mechanistic discussion.[158]

Quantum yields and product ratios for photolysis of cis- and trans-$[Rh(en)_2(NH_3)Cl]^{2+}$ are compatible with the generally accepted dissociative mechanism for this type of reaction. Photolysis of $[Rh(en)_2(NH_3)(OH_2)]^{3+}$ leads to a photostationary state in which isomerization, racemization of the cis isomer, and water exchange at both cis and trans isomers occur concurrently. This investigation also established parameters for thermal water exchange at the two $[Rh(en)_2(NH_3)(OH_2)]^{3+}$ cations.[159] The product of photolysis of $[Rh(tren)Br_2]^+$ has a different structure from the product of thermal aquation of this cation.[160] A combination of thermal and photochemical activation has been explored for the trans-$[Rh(en)_2Br_2]^+$ cation.[161]

Kinetics of reactions of binuclear rhodium(II) complexes are perhaps rather fast for this chapter, but mention may be made of a kinetic determination of the rhodium–rhodium bond strength in an octaethylporphyrinato species[162] and of kinetics of axial substitution in the tetraacetato complex $[Rh_2(OAc)_4(OH_2)_2]$.[163]

8.7. Iridium

A study of aqua ions of iridium(III), iridium(IV), and iridium(V) established some qualitative kinetic data. Thus the iridium(III) dimer $[(H_2O)_4Ir(OH)_2Ir(OH_2)_4]^{4+}$ requires some two hours at 60 °C in 7 M CF$_3$SO$_3$H to become converted into $[(H_2O)_5Ir(OH)Ir(OH_2)_5]^{5+}$, which in turn requires about 10 hours for complete conversion into mononuclear Ir^{3+}aq. What is almost certainly the iridium(IV) dimer $[(H_2O)_4Ir(OH)_2Ir(OH_2)_4]^{6+}$ gives what is probably the oxide-bridged species $[(H_2O)_5IrOIr(OH_2)_5]^{6+}$ considerably more rapidly, with a rate constant of 10^{-2} s^{-1} at 25 °C. Finally, it is suggested that there is quite rapid interconversion of various dinuclear species in aqueous solutions containing iridium(V).[164] The question of linkage isomerism for dimethyl sulfoxide as ligand (cf. Section 8.4.2) at iridium(III) is of considerable importance to reactivity, especially in relation to the role of Ir/Cl$^-$/DMSO complexes in homogeneous catalysis.[165]

8.8. Platinum(IV)

A Vilsmeier reaction of tetra-4-tolylporphyrinplatinum(IV) has already been cited in the section on iron(III).[97] Fluxional behavior of binuclear platinum(IV) complexes containing bridging ligands such as 2,2'-bipyrimidine has been documented,[166] while there has been a theoretical discussion of the facile isomerization of *trans,trans,trans*-[PtCl$_2$(OH)$_2$(NH$_3$)$_2$] to the *cis,trans,cis* form.[167] The *mer*-isomer of [Pt(S$_4$N$_4$)Cl$_2$(PMe$_2$Ph)], (45), isomerizes readily to the *fac* form on heating in chloroform, probably by a phosphine-dissociation mechanism.[168] In contrast, *cis* ⇌ *trans* isomerization of [PtXMe$_3$L], where L = the thio-ligand (46) which can form four- or five-membered chelate rings, takes place intramolecularly through a very flexible seven-coordinate transition state. This provides the first example of a "metal-pivot" process for a mononuclear platinum(IV) complex.[169]

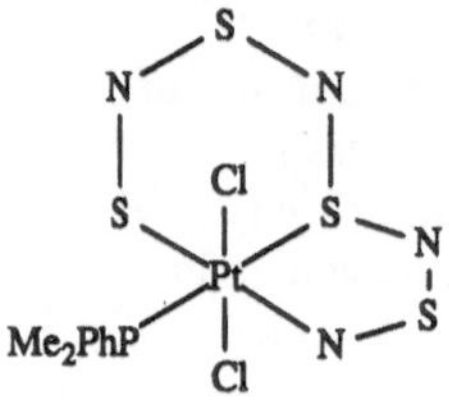

(45)

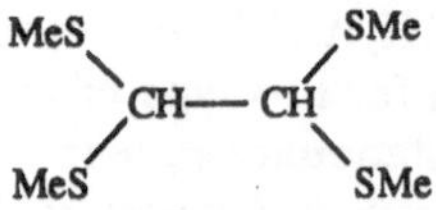

(46)

At lower temperatures, these *cis* and *trans* complexes exhibit inversion at the coordinated sulfur atoms,[170] as do the complexes [MX$_2${MeS(CH$_2$)$_n$CH=CH$_2$}] where M = Pt or Pd, and n = 2 or 3.[171] Inversion at sulfur in thioether complexes of platinum(II) has been documented,[172] and an example of such inversion at rhodium(III) was mentioned in Section 8.6[156]. The whole area of fluxional behavior of metal complexes of organic sulfides and selenides has been reviewed.[173]

Chapter 9

Substitution Reactions of Labile Metal Complexes

9.1. General

The range of fast ligand substitution processes considered in this chapter range from those of the lightest metal ion, lithium(I), to the trivalent lanthanide ions, and the majority of the systems explored fall into the kinetic range of the stopped-flow (SF) technique, which is usually considered to be the lower end of the "fast" reaction timescale. The use of activation volumes, ΔV^+, in mechanistic interpretation is becoming increasingly widespread as evidenced by many of the studies discussed in this chapter and two recent and comprehensive reviews of the determination and interpretation of this parameter in mechanistic studies.[1,2] Although reports of excellent basic mechanistic studies of fast ligand substitution processes appear with regularity (and hopefully will continue to do so), an increasing number of studies is concerned with various aspects of environmental or biological chemistry, to a greater or less extent. This is well illustrated by a spectrophotometric temperature-jump (TJ) and SF study of the formation and deoxygenation of oxymyoglobin (MbO$_2$).[3,4] Thus for the reaction (1) $k_c(298.2 \text{ K}) = 1.3 \times 10^7 \text{ dm}^3 \text{ mol}^{-1} \text{ s}^{-1}$, $\Delta H_c^+ = 23.0 \text{ kJ mol}^{-1}$, $\Delta S_c^+ = -30.0 \text{ J K}^{-1} \text{ mol}^{-1}$, and

$$Mb + O_2 \underset{k_d}{\overset{k_c}{\rightleftharpoons}} MbO_2 \tag{1}$$

$\Delta V_c^+ = 7.8 \text{ cm}^3 \text{ mol}^{-1}$; and $k_d(298.2 \text{ K}) = 21.8 \text{ dm}^3 \text{ mol}^{-1} \text{ s}^{-1}$, $\Delta H_d^+ = 76.1 \text{ kJ mol}^{-1}$, $\Delta S_d^+ = 36 \text{ J K}^{-1} \text{ mol}^{-1}$, and $\Delta V_d^+ = 23.3 \text{ cm}^3 \text{ mol}^{-1}$. These data compare with $k_c(298.2 \text{ K}) = 3.8 \times 10^5 \text{ dm}^3 \text{ mol}^{-1} \text{ s}^{-1}$, $\Delta H_c^+ = 17.1 \text{ kJ mol}^{-1}$, $\Delta S_c^+ = -81.1 \text{ J K}^{-1} \text{ mol}^{-1}$, and $\Delta V_c^+ = -8.9 \text{ cm}^3 \text{ mol}^{-1}$ for the reaction (2) involving carbon

$$Mb + CO \underset{k_d}{\overset{k_c}{\rightleftharpoons}} MbCO \tag{2}$$

monoxide. It is deduced from these data that for the MbO_2 equilibrium, movement of O_2 along the heme cavity is rate-determining, while for the MbCO equilibrium Fe—CO bond breaking and making are rate-determining. Although this chapter considers no other study as specifically biological as this, biological inferences are drawn from a significant number of the studies discussed below.

9.2. Ligand Substitution on Complexes of Uni- and Bivalent Metal Ions

9.2.1. Ligand Substitution on Complexes of Alkali and Alkaline Earth Metal Ions

The alkali and alkaline earth metal ions are labile and preferentially bond with oxygen donor atoms. As a consequence the majority of ligand substitution studies tend to be of reactions with multidentate ligands bearing oxygen donor groups, which produce complexes that are of moderate stability and are characterized by labilities falling within the time range of available kinetic techniques. Such ligands are crown ethers, cryptands, catenands, antibiotics, and similar species.

The exchange of alkali metal ions in crown ether complexes can proceed through either a unimolecular or a bimolecular mechanism as shown in Eqs. (3) and (4), respectively, and the factors affecting the operation of these mechanisms have been investigated in a ^{23}Na NMR study.[5] When the crown ether (C) is 18C6

$$M^+ + C \underset{k_d}{\overset{k_c}{\rightleftharpoons}} [M \cdot C]^+ \tag{3}$$

$$[M \cdot C]^+ + {}^*M^+ \underset{k_2}{\overset{k_2}{\rightleftharpoons}} [{}^*M \cdot C]^+ + M^+ \tag{4}$$

(1,4,7,10,13,16-hexaoxacyclooctadecane) it is found that the two mechanisms operate in parallel in acetonitrile, propylene carbonate, and pyridine, while in acetone the exchange mechanism is almost exclusively unimolecular. The ΔH^+ characterizing the unimolecular mechanism follow the trend acetonitrile < acetone < pyridine ≈ propylene carbonate, but ΔG^+ at 301.5 K does not show a large variation due to the compensation effect of ΔS^+ (Table 9.1). There is no apparent relationship between the Gutmann donor number of the solvent, D_N, and the activation parameters for exchange, indicating that several factors, including crown ether conformational rearrangement and reorganization of the solvent cage, contribute to the exchange parameters.

The role of counterions in the bimolecular exchange of Na^+ in $[Na \cdot DB24C8]^+$ in nitromethane solvent has also been investigated by ^{23}Na NMR methods.[6] As Na^+ has been shown to be completely isolated from direct solvent and counterion interaction when it is complexed by DB24C8 (2,3,5,6,8,9,14,15,17,18,20,21-dodecahydro-1,4,7,10,13,16,19,22-octaoxadibenzocyclotetracosadiene),[7,8] it is considered that the influence of the counterion is a consequence of it forming an ion pair with solvated Na^+ ($K_{ion\,pair} = 10^3$ and 80 for NaNCS and NaI) and the consequent decrease in electrostatic repulsion between Na^+ in the ion pair and

Table 9.1. Kinetic Parameters for Na$^+$ Exchange in 18C6 Crown Ether Complexes[5]

Solvent	D_N	Rate constant^a	ΔH^+ (kJ mol^{-1})	ΔS^+ (J mol^{-1} K^{-1})	ΔG^+(301.5 K) (kJ mol^{-1})
Acetonitrile	14.1	$k_d = 3.8 \times 10^3$	32	−65	53.2
		$k_2 = 2.6 \times 10^5$			42.6
Propylene	15.1	$k_d = 1.6 \times 10^3$	54	−5	55.4
carbonate		$k_2 = 6.9 \times 10^5$	35	−16	40.2
Acetone	17.0	$k_d = 1.48 \times 10^4$	44	−21	49.8
Pyridine	33.1	$k_d = 5.3 \times 10^2$	49	−30	58.1
		$k_2 = 2.0 \times 10^4$			49.0

^{a} Units of k_d are s^{-1} and those of k_2 are dm^3 mol^{-1} s^{-1}.

[Na·DB24C8]$^+$. At 294 K Na$^+$ exchange on [Na·DB24C8]$^+$ is characterized by $10^{-5} k_2 = 2.8$, 3.5, 4.5, and 13 dm^3 mol^{-1} s^{-1} in the presence of the counterions BPh$_4^+$, PF$_6^-$, I$^-$, and SCN$^-$, respectively. As the variation of k_2 with the nature of the counterion is less than expected, it is concluded that any stabilization of the transition state by ion pairing is offset by similar ion pairing in the ground state. During the course of this study a small amount of the [Na$_2$·DB24C8]$^{2+}$ species was detected in solution, which makes an interesting comparison with the analogous species observed in the solid state.[9] The influence of the ligand on mechanism has been demonstrated by a ^{209}Tl NMR study of the complexation of Tl$^+$ by 18C6 and its linear analog, pentaglyme, in acetonitrile which shows that in the former case only the bimolecular mechanism operates, while both the unimolecular and bimolecular mechanisms operate in the latter case.[10]

The ultrasonic relaxation kinetic technique measures faster reactions than the NMR method and thus is able to detect more steps in the crown ether complexation sequence. This is exemplified by a study of the complexation of LiClO$_4$ by 18C6 in propylene carbonate in which two relaxations are detected which are attributed to the operation of an Eigen–Winkler mechanism for Li$^+$ exchange as in Eq. (5).

$$\text{Li}^+ + 18\text{C6} \underset{k_{-1}}{\overset{k_1}{\rightleftharpoons}} \text{Li}^+\cdots18\text{C6} \underset{k_{-2}}{\overset{k_2}{\rightleftharpoons}} [\text{Li}\cdot18\text{C6}']^+ \underset{k_{-3}}{\overset{k_3}{\rightleftharpoons}} [\text{Li}\cdot18\text{C6}]^+ \qquad (5)$$

Li$^+\cdots$18C6 represents an outer-sphere complex and [Li·18C6$'$]$^+$ represents a complex in which the solvation and the crown ether conformation are in an intermediate state on the way to the final ground state, [Li·18C6]$^+$. The exchange process is characterized by $k_2 = 1.45 \times 10^8$ s^{-1} and $k_3 = 6.3 \times 10^6$ s^{-1} at 298.2 K.[11]

There is considerable interest in incorporating crown ethers, cryptands, and similar species into artificial membranes in the development of ion-selective electrodes. As part of a study in this area the kinetics of the complexation of Na$^+$ by several propeller crowns (dinaphtholpolycycloalkins), (1)–(6), in pyridine/dichloromethane solvent have been determined by ^{23}Na NMR.[12] It is found that a Na$^+$ unimolecular exchange process dominates at high temperatures with a bimolecular mechanism becoming significant at lower temperatures, but it was not possible to separately determine their kinetic parameters.

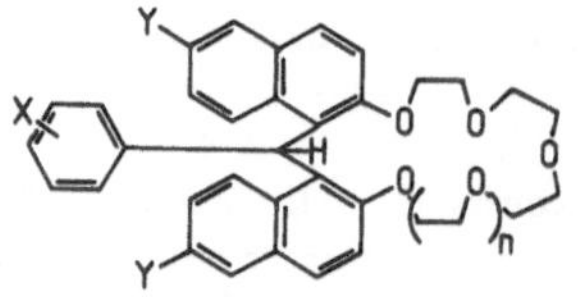

(1)–(6)

crown	X	Y	n
1	2-F	H	1
2	2,6-Cl$_2$	Bu^t	1
3	3,5-Me$_2$	Bu^t	2
4	3,4,5-(MeO)$_3$	Bu^t	1
5	2-MeO	H	1
6	3,4,5-(MeO)$_3$	Bu^t	2

Thallium(I) is toxic to humans and is known to activate several K$^+$ activated enzymes[13] and accordingly the determination of its complexation characteristics with ligands, such as cryptands, which also coordinate the alkali metal ions, are of substantial interest. The rates of decomplexation of thallium(I) cryptates have been determined in a range of solvents using a conductimetric SF method.[14,15] It is found that the thallium(I) cryptates are more stable and less sensitive to cryptand cavity size than the corresponding potassium(I) and rubidium(I) cryptates. The stability constants (K) are strongly sensitive to the nature of the solvent and the cryptand, and this is reflected in the variation of the cryptate decomplexation rate constant (k_d) and the lesser variation of the complexation rate constant (k_c). Thus k_d and k_c vary over approximately six and three orders of magnitude, respectively (Table 9.2), which results in a variation of about nine orders of magnitude for $K = k_c/k_d$. For all of the cryptates k_d is greatest in water, and it is seen that [Tl·C222]$^+$ decomplexes most slowly, while in the other solvents [Tl·C221]$^+$ decomplexes most slowly (C222 = 4,7,13,16,21,24-hexaoxa-1,10-diazabicyclo-[8.8.8]hexacosane and C221 = 4,7,13,16,21-pentaaoxa-1,10-diazabicyclo[8.8.5] tricosane). It is thought that this may be a consequence of the ability of water to form hydrogen bonds with the cryptand, and also its ability to interact directly with Tl$^+$ in the cryptate. In acetonitrile/water mixtures the stability of [Tl·C222]$^+$ increases with increasing mole fraction of acetonitrile as a consequence of a decrease in k_d and an increase in k_c.[14]

The diaza crown ethers may be viewed either as crown ethers in which two ether oxygens have been replaced by amine nitrogens, or as monocyclic precursors of the cryptands, which renders them of particular interest in comparisons of the complexing abilities of crown ethers and cryptands. Thus a potentiometric titration and SF conductimetric study of the complexation of Ca^{2+}, Sr^{2+}, and Ba^{2+} by a range of diaza crown ethers in methanol shows that, as for the corresponding crown ethers, there is no strong selectivity between these metal ions, or large variations in complex stability with diaza crown ether ring size and structure, except when large differences between the size of the cation and the ligand cavity

Table 9.2. Decomplexation (k_d s^{-1}) and Complexation (k_c dm^3 mol^{-1} s^{-1}) Rate Constants for Thallium(I) Cryptates in a Range of Solvents at 298.2 $K^{(14,15)}$

Solvent	Rate constant	Cryptand				
		C211	C221	C222	C2$_B$22	C2$_B$2$_B$2
Water	k_d	—	1.10×10	5.99	8.06×10	1.6×10^2
	k_c	—	6.9×10^7	2.4×10^7	5.58×10^7	6.5×10^6
Methanol	k_d	—	2.58×10^{-2}	5.21×10^{-2}	1.24	4.32
	k_c	—	1.49×10^9	9.93×10^8	2.11×10^8	8.62×10^8
Ethanol	k_d	$\leq 4 \times 10^{-3}$	5.99×10^{-3}	1.34×10^{-2}	0.28	0.99
	k_c	$<5.3 \times 10^2$	6.13×10^8	1.37×10^9	2.55×10^7	3.85×10^8
Acetonitrile	k_d	—	4.28×10^{-1}	1.16×10^{-3}	2.63×10^{-3}	5.32×10^{-2}
	k_c		3.56×10^8	2.12×10^9	4.57×10^7	9.04×10^8
Propylene	k_d	15.9	6.22×10^{-3}	1.34×10^{-2}	1.88×10^{-2}	2.51×10^{-2}
carbonate	k_c	6.05×10^7	8.39×10^9	8.07×10^9	1.01×10^9	1.62×10^8
Dimethyl-	k_d		1.41×10^{-1}	1.82×10^{-1}	4.07	2.10×10
formamide	k_c		5.74×10^7	2.1×10^7	2.51×10^7	2.90×10^7
Dimethyl	k_d	—	2.03	1.87	8.6	ca 0.3
sulfoxide	k_c		1.28×10^7	3.73×10^6	4.02×10^5	ca 1×10^4

exist.[16] The complexes of C22 (1,4,10,13-tetraoxa-7,16-diazacyclooctadecane) and C2$_B$2 (2,3-benzo-1,4,10,13-tetraoxa-7,16-diazacyclooctadecane) are up to seven orders of magnitude less stable than the corresponding cryptates formed by C222 and C2$_B$22 (5,6-benzo-4,7,13,16,21,24-hexaoxa-1,10-diazabicyclo[8.8.8]hexa-cosane), as seen for the first pair in Table 9.3. This arises almost entirely from the differences in decomplexation rates of the two classes of complexes while the formation rates are almost identical for the mono- and bicyclic ligands for a given alkaline earth metal ion.

The binding of Mg^{2+} and Ca^{2+} by the inophores X-14885A (7) and A-23187 (8) has been studied in methanol using a spectrophotometric SF method.[17] The rate-determining steps for the formation and decomplexation of MA$_2$ (where A$^-$ is an ionophore ligand) are the formation and decomplexation of MA$^+$, the addition or loss of the second ionophoric ligand being faster. Complexation of M^{2+} by the

Table 9.3. Parameters for the Complexation of Alkaline Earth Metal Ions by the Diaza Crown Ether C22 and the Cryptand C222 in Methanol at 298.2 K [16]

M^{2+}	$\log(K/dm^3 \, mol^{-1})$		$\log(k_c/dm^3 \, mol^{-1} \, s^{-1})$		$\log(k_d/s^{-1})$	
	C22	C222	C22	C222	C22	C222
Ca^{2+}	3.5	8.1	5.0	4.5	1.48	-3.66
Sr^{2+}	5.7	11.8	5.7	5.5	0.04	-6.26
Ba^{2+}	5.9	12.9	6.8	6.7	0.90	-6.20

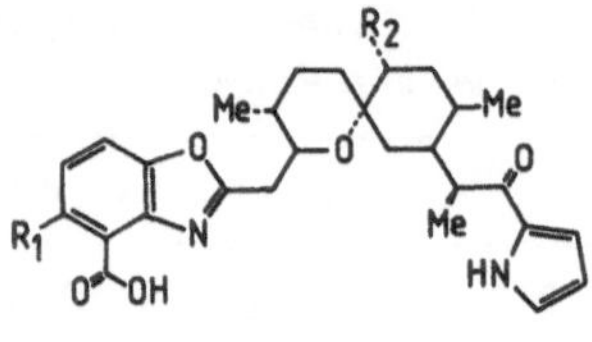

(7)–(8)

(7) X-14885A $R_1 = OH$ $R_2 = H$
(8) A-23187 $R_1 = NHMe$ $R_2 = Me$

anionic and the protonated ionophore occur as in Eqs. (6) and (7). The acid catalysed decomplexation of X-14885A and A-23187 occur through different sequences, as shown in Eqs. (8) and (9), respectively. The derived rate constants are given in Table 9.4.

$$2A^- + M^{2+} \underset{k_{-1}}{\overset{k_1}{\rightleftharpoons}} MA^+ + A^- \overset{fast}{\rightleftharpoons} MA_2 \tag{6}$$

$$2AH + M^{2+} \underset{k_{-2}}{\overset{k_2}{\rightleftharpoons}} MA^+ + AH + H^+ \overset{fast}{\rightleftharpoons} MA_2 + 2H^+ \tag{7}$$

$$MA_2 + 2H^+ \overset{K_H}{\underset{fast}{\longrightarrow}} MA^+ + AH + H^+ \overset{k_H}{\longrightarrow} 2AH + M^{2+} \tag{8}$$

$$MA_2 + 2H^+ \overset{K_H}{\underset{fast}{\longrightarrow}} MA^+ + AH + H^+ \overset{K_H'}{\underset{fast}{\longrightarrow}} MAH^{2+} + AH \overset{k_H'}{\longrightarrow} 2AH + M^{2+} \tag{9}$$

9.2.2.   Ligand Substitution on Complexes of Gold(I) and Copper(I)

When the alkali metal ions are excluded, there are very few reported studies of ligand substitution on monovalent metal ions remaining. However, this may well change in the case of Au(I) as a consequence of the increasing interest in Au(I) compounds in the treatment of rheumatoid arthritis.[18] Thus a spectrophotometric SF study of Au(I) ligand substitution shown in Eq. (10) (mpt = 1-methylpyridine-2-thione) yields k_f and $k_r = 1.4$ and $1.9 \times 10^8 \, dm^3 \, mol^{-1} \, s^{-1}$,

$$[Au(CN)_2]^- + mpt \underset{k_r}{\overset{k_f}{\rightleftharpoons}} [Au(CN)(mpt)] + CN^- \tag{10}$$

respectively, and the rate constant for the reaction of HCN on [Au(CN)(mpt)] is estimated to have an upper limit of $25 \, dm^3 \, mol^{-1} \, s^{-1}$ at 298.2 K in aqueous

Table 9.4. *Rate Parameters for the Complexation of* Mg^{2+} *and* Ca^{2+} *by the Ionophores X-14885A and A-23187 in Methanol at* $I = 0.1$ *(Bu$_4$NClO$_4$) at 298.2 K* [17]

Ionophore	M^{2+}	k_1 ($dm^3 \, mol^{-1} \, s^{-1}$)	k_2 ($dm^3 \, mol^{-1} \, s^{-1}$)	k_{-1} (s^{-1})	k_H ($dm^3 \, mol^{-1} \, s^{-1}$)	$k_H' K_H$ ($dm^3 \, mol^{-1} \, s^{-1}$)
X-14885A	Mg^{2+}	1.1×10^7	1.8×10^5	3.0	1.15×10^4	
	Ca^{2+}	$>10^8$	1.2×10^6	7	2.6×10^5	
A-23187	Mg^{2+}	4.6×10^6	1.2×10^5	0.3		3.3×10^6
	Ca^{2+}	$>5 \times 10^7$	6×10^5	2		1.1×10^8

0.1 mol dm^3 HClO$_4$/NaClO$_4$.[19] There is no evidence of an intervening solvolysis step in reaction (10) and accordingly the 8×10^6-fold difference in the relative reactivities of CN$^-$ and HCN is taken as evidence for the operation of an A mechanism.

Ligand substitution studies of Cu(I) have also been little reported, and the reactions shown in Eqs. (11) and (12) appear to represent the first quantitative

$$[CuL]^+ \underset{k_{-d}}{\overset{k_d}{\rightleftharpoons}} [CuL']^+ \tag{11}$$

$$[CuL']^+ + Zn^{2+} \xrightarrow{k_x} Cu^+ + [ZnL]^{2+} \tag{12}$$

ligand substitution study (spectrophotometric SF) of five-coordinate Cu(I).[20] In CH$_3$CN solution this metal ion exchange process shows saturation kinetics consistent with the rate-determining step $[k_d(298.2\,K) = 310\,s^{-1}]$ being the partial dissociation of the pentadentate ligand, L, to give a reactive intermediate, $[CuL']^+$, which may either reassociate to regenerate $[CuL]^+$, or may react with solvated Zn^{2+} to produce [ZnL]$^{2+}$ and solvated Cu$^+$. Structure (9) is that of [CuL]$^+$ where L = 2,6 bis[1-((2-pyridin-2-yl-ethyl)imino)ethyl]pyridine.

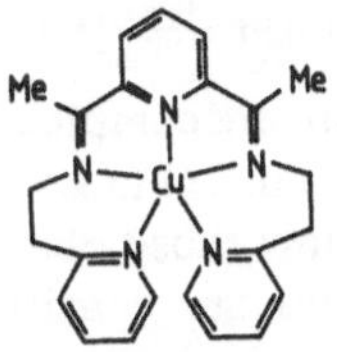

(9)

9.2.3. Ligand Substitution on Complexes of Uni- and Bivalent First-Row Transition Metal Ions

The ligand substitution reactions of the bivalent first-row transition metal ions are the most studied of those of the labile metal ions, probably because the visible $d-d$ spectra of the transition metal ions make them particularly amenable to spectrophotometric study, and also because their reaction timescale is usually well within those of the SF and NMR techniques. Thus it has been shown that the mechanism of dimethylformamide (dmf) exchange on [M(dmf)$_6$]$^{2+}$ (M = Mn—Ni) varies systematically from I_d to D, in contrast to the analogous [M(solvent)$_6$]$^{2+}$ in water, methanol, and acetonitrile where the mechanism varies from I_a to I_d as the number of d electrons increases.[21,22] This has occasioned a spectrophotometric SF study of the closely related substitution of the bidentate ligands *trans*-pyridine-2-azo(p-dimethylaniline) (Pada) and diethyldithiocarbamate (Et$_2$DTC) on [M(dmf)$_6$]$^{2+}$ shown in Eq. (13) (where L-L represents a bidentate ligand) which

$$[M(dmf)_6]^{2+} + L\text{-}L \underset{k_d}{\overset{k_c}{\rightleftharpoons}} [M(dmf)_4(L\text{-}L)]^{2+} + 2dmf \tag{13}$$

Table 9.5. *Kinetic Parameters for Ligand Substitution on* $[M(dmf)_6]^{2+}$ *(Ref. 23 and references cited therein)*

	M = Mn L = Et$_2$DTC$^-$	L = Fe L = Et$_2$DTC$^-$	L = Fe L = Pada	L = Co L = Et$_2$DTC$^-$
k_c(298 K) (dm^3 mol^{-1} s^{-1})	6.4×10^7	1.2×10^7	4.3×10^5	4.3×10^6
k_d(298 K) (s^{-1})	651	39.1	2.7×10^3	102
$\Delta H_c^{\ddagger}$ (kJ mol^{-1})	47.5	38.8	52.1	39.1
$\Delta H_d^{\ddagger}$ (kJ mol^{-1})	30.1	17.9	48.9	27.3
$\Delta S_c^{\ddagger}$ (J K^{-1} mol^{-1})	64	20	38	26
$\Delta S_d^{\ddagger}$ (J K^{-1} mol^{-1})	−90	−154	−15	−115
$\Delta V_c^{\ddagger}$ (cm^3 mol^{-1})	9.5 (228 K)	12.3 (238 K)	7.5 (238 K)	12.1 (238 K)
$\Delta V_d^{\ddagger}$ (cm^3 mol^{-1})			9.5 (238 K)	

is characterized by the kinetic data in Table 9.5.[23] It is found that the chelation step is fast and that the reaction sequence in Eqs. (14) and (15) applies, where

$$[M(dmf)_6]^{2+} + \text{L-L} \underset{\text{fast}}{\overset{K_o}{\rightleftharpoons}} [M(dmf)_6\cdots(\text{L-L})]^{2+} \underset{k_{-1}}{\overset{k_1}{\rightleftharpoons}} [M(dmf)_5(\text{L-L})]^{2+} + dmf \quad (14)$$

$$[M(dmf)_5(\text{L-L})]^{2+} + dmf \overset{\text{fast}}{\rightleftharpoons} [M(dmf)_4(\text{L-L})]^{2+} + dmf \quad (15)$$

$[M(dmf)_6\cdots(\text{L-L})]^{2+}$ is an outer-sphere complex in which L-L resides in the second coordination sphere. Thus the data in Table 9.5 pertain to the first bond formation. The trends in lability closely follow those observed earlier for dmf exchange on $[M(dmf)_6]^{2+}$, and the $\Delta V_c^{\ddagger}$ values are consistent with a dissociative mode of activation for the ligand substitution. In contrast the formation of ternary complexes of $[Ni(nta)(H_2O)_2]^-$ with amino acids is thought to involve a fast first bond formation with the chelating step being slow.[24]

A change in mechanism with solvent is also postulated for solvent exchange on square-pyramidal mono(solvento)(3,7,11-tribenzyl-3,7,11,17-tetraazabicyclo-[11.3.1]heptadeca-1(17),13,15-triene)nickel(II) in which three nitrogens of the tetraaza ring and a solvent molecule occupy the basal plane.[25] When the solvent is dmf and acetonitrile, respectively, a ^{1}H NMR study yields $10^{-2} k$(298.2 K) = 32.7 and 3.82 s^{-1}, $\Delta H^{\ddagger}$ = 36.3 and 68.9 kM mol^{-1}, $\Delta S^{\ddagger}$ = −55.8 and 35.6 J K^{-1} mol^{-1}, and $\Delta V^{\ddagger}$(323 K) = 10.6 and −3.5 cm^3 mol^{-1}, from which it is concluded that dmf exchanges through an I_d or D mechanism and acetonitrile exchanges through an I_a mechanism. This mechanistic change is attributed to the greater size of dmf inducing too much steric crowding for the operation of an associative activation mode. The decreased rate of the exchange process relative to other nickel(II) amine complexes probably arises from the steric crowding caused by the three benzyl groups of the tetraaza ligand.

The two acetic acid exchange processes in Eqs. (16) and (17) have been studied in deuterated dichloromethane diluent by ^{17}O NMR spectroscopy, and

$$[Mn(AcOH)_6]^{2+} + {}^*AcOH \overset{k_a}{\rightleftharpoons} [Mn(AcOH)_5({}^*AcOH)]^{2+} + AcOH \quad (16)$$

$$[Mn(AcO)_2(AcOH)_4] + {}^*AcOH \overset{k_b}{\rightleftharpoons} [Mn(AcO)_2(AcOH)_3({}^*AcOH)] + AcOH \quad (17)$$

are characterized by: $k_a(298.2 \text{ K}) = 1.6 \times 10^7 \text{ s}^{-1}$, $\Delta H_a^+ = 29 \text{ kJ mol}^{-1}$, $\Delta S_a^+ = -10 \text{ J mol}^{-1} \text{ K}^{-1}$, and $\Delta V_a^+ = 0.4 \text{ cm}^3 \text{ mol}^{-1}$; and $k_b(298.2 \text{ K}) = 4.8 \times 10^7 \text{ s}^{-1}$, $\Delta H_b^+ = 32 \text{ kJ mol}^{-1}$, $\Delta S_b^+ = 9 \text{ J mol}^{-1} \text{ K}^{-1}$, and $\Delta V_b^+ = 6.7 \text{ cm}^3 \text{ mol}^{-1}$.[26] On the basis of their volumes of activation the two exchange processes are assigned I and I_d mechanisms, respectively. The $[\text{Mn}(\text{AcO})_2(\text{AcOH})_4]$ species may be likened to a hydrolyzed hexaaquametal ion, and it is proposed that the enhanced electron donation to AcO^- (by comparison to that of AcOH) labilizes the AcOH ligands toward dissociatively activated exchange. It appears that for both $[\text{Mn}(\text{AcOH})_6]^{2+}$ and $[\text{Mn}(\text{AcO})_2(\text{AcOH})_4]$ the large size of AcOH causes substantial steric crowding at the Mn(II) center and thereby decreases the tendency toward associative activation in a similar manner to dmf in $[\text{Mn}(\text{dmf})_6]^{2+}$.[21]

Copper(II) and zinc(II) are two of the more labile divalent metal ions and as a consequence the former is too labile for its water exchange rate to be determined by the NMR methods which utilize the paramagnetism of other divalent first-row transition metal ions, while the latter is diamagnetic and such NMR methods cannot be applied. However, it has been shown that water exchange rates and mechanisms can be deduced with reasonable reliability from simple ligand substitution studies, and this is one of the reasons for a recent variable-pressure spectrophotometric SF study of the substitution of 2-chloro-1,10-phenanthroline on Cu(II) and Zn(II).[27] The observed rate constants for the complexation reaction (k_c) and the decomplexation reaction (k_d) and their associated activation parameters for Cu(II) and Zn(II) are: $k_c(298 \text{ K}) = 1.1 \times 10^7$ and $1.1 \times 10^6 \text{ dm}^3 \text{ mol}^{-1} \text{ s}^{-1}$, $\Delta H^+ = 33.6$ and 37.9 kJ mol^{-1}, $\Delta S^+ = 3$ and $-2 \text{ J K}^{-1} \text{ mol}^{-1}$, $\Delta V^+ = 7.1$ and $5.0 \text{ cm}^3 \text{ mol}^{-1}$, $k_d(298 \text{ K}) = 102$ and 887 s^{-1}, $\Delta H^+ = 60.6$ and 57.3 kJ mol^{-1}, $\Delta S^+ = -3$ and $4 \text{ J K}^{-1} \text{ mol}^{-1}$; and $\Delta V^+ = 5.2$ and $4.1 \text{ cm}^3 \text{ mol}^{-1}$. These data are consistent with the operation of an I_d mechanism for the rate-determining first bond formation by 2-chloro-1,10-phenanthroline with the subsequent chelation step being faster [Eq. (18)]. For this mechanistic sequence (in which $[\text{M}(\text{H}_2\text{O})_6 \cdots \text{L-L}]^{2+}$ is an outer-sphere complex) it may be shown that the relationships in Eq. (19) apply,

$$[\text{M}(\text{H}_2\text{O})_6]^{2+} + \text{L-L} \underset{}{\overset{K_o}{\rightleftharpoons}} [\text{M}(\text{H}_2\text{O})_6 \cdots \text{L-L}]^{2+} \overset{k_i}{\longrightarrow}$$

$$[\text{M}(\text{H}_2\text{O})_5\text{L-L}]^{2+} + \text{H}_2\text{O} \overset{\text{fast}}{\longrightarrow} [\text{M}(\text{H}_2\text{O})_4\text{L-L}]^{2+} + 2\text{H}_2\text{O} \quad (18)$$

$$k_f = K_o k_i = K_o F k_{ex} \quad (19)$$

where k_i is the interchange rate constant, k_{ex} is the water exchange rate constant, K_0 is the outer-sphere association constant, and F is a statistical factor. On this basis $k_{ex}(298 \text{ K})$ is calculated to be $7 \times 10^8 \text{ s}^{-1} < k_{ex} < 5 \times 10^9 \text{ s}^{-1}$ for $[\text{Cu}(\text{H}_2\text{O})_6]^{2+}$, and $3 \times 10^7 \text{ s}^{-1} < k_{ex} < 6 \times 10^8 \text{ s}^{-1}$ for $[\text{Zn}(\text{H}_2\text{O})_6]^{2+}$. The latter data set affords an interesting comparison with an ultrasonic study[28] of ligand substitution on $[\text{Zn}(\text{H}_2\text{O})_6]^{2+}$ by carboxylate ligands, which is interpreted through the reaction sequence in Eq. (20). When $\text{L}^- = \text{HCO}_3^-$, CH_3CO_2^-, respectively, $K_0 = 1.1$, 1.2, and 1.8; $10^{-7} k_1 = 11.8$, 9.1, and 6.7 s^{-1}; and $10^{-7} k_{-1} = 3.3$, 1.9, and 1.5 s^{-1} at 298.2 K, with $I = 2.0 \text{ mol dm}^{-3}$ (NaNO$_3$). The k_1 are identical within the reported experimental errors.

$$[\text{Zn}(\text{H}_2\text{O})_6]^{2+} + \text{L} \underset{}{\overset{K_o}{\rightleftharpoons}} [\text{Zn}(\text{H}_2\text{O})_6 \cdots \text{L}]^+ \underset{k_{-1}}{\overset{k_1}{\rightleftharpoons}} [\text{Zn}(\text{H}_2\text{O})_5\text{L}]^+ + \text{H}_2\text{O} \quad (20)$$

There is evidence that the coordination of metal ions by macrocyclic ligands can be accelerated by the attachment to the macrocycle of a coordinating pendant arm capable of capturing the metal ion with subsequent transfer of the metal ion to the macrocyclic cavity.[29-31] Entry of metal ions into the cavities of macrocycles lacking such pendant arms can be slow partly as a consequence of conformational changes which often accompany the coordination of the metal ion. A spectrophotometric SF study of the coordination of metal ions by two polyaza

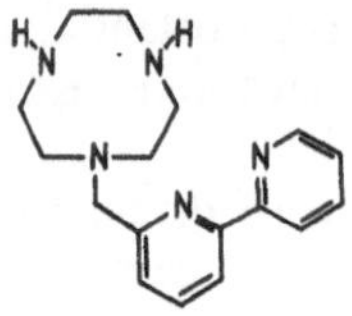

(10) 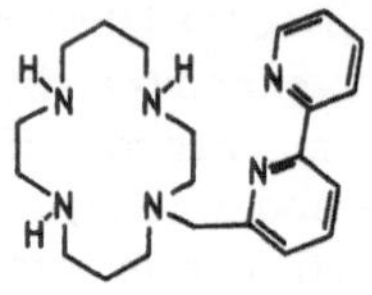   (11)

macrocycles functionalized with a pendant 2,2'-bipyridyl-6-yl-methyl arm (10) and (11) in dmso shows the reaction to proceed through at least two steps:

$$[M(Me_2SO)_6]^{2+} + L \underset{k_{-1}}{\overset{k_1}{\rightleftharpoons}} \text{intermediate} \xrightarrow{k_2} \text{product} \tag{21}$$

characterized by the rate constants in Table 9.6.[32] The first step is identified as the initial coordination by the pendant 2,2'-bipyridyl-6-yl-methyl arm, and it is seen that the k_1 values for (10) and (11) are quite similar for a given metal as expected for a pendant group which is almost independent of the macrocycle. This identification is further strengthened by the observation that the ratios k_1/k_{-1} for (10) and (11) are similar to that for bipy.

Table 9.6. Rate Constants (298.0 K) for the Reaction (Ref. 32 and references cited therein)

$$[M(Me_2SO)_6]^{2+} + L \underset{k_{-1}}{\overset{k_1}{\rightleftharpoons}} \text{intermediate} \xrightarrow{k_2} \text{product}$$

M	L	$10^{-2} k_1$ $(dm^3\,mol^{-1}\,s^{-1})$	k_{-1} (s^{-1})	$10^{-2} k_2$ (s^{-1})
Zn	10	125.1	5.01	19.6
	11	123.1	5.05	76.0
	bipy	187 (292.2 K)	8.25 (292.2 K)	
Ni	10	0.415	ca 0	5.3
	11	1.627	ca 0	4.7
	bipy	0.69 (298.2 K)		
Co	10	27.73	0.54	11.0
	11	9.186	0.13	7.2
	bipy	36 (298.2 K)		
Cu	10	2456	20.0	133.6

The influence of steric, inductive, and ring-strain effects on the complexation of Cu(II) by four new tetradentate diamino diamide ligands (4-methyl-4,7-diazadecandediamide, 4-Me-L-2,2,2; 4,7-dimethyl-4,7-diazadecanediamide, 4,7-Me$_2$-L-2,2,2; 4-ethyl-4,7-diazadecanediamide, 4-Et-L-2,2,2; and 4-methyl-4,8-diazaundecanediamide, 4-Me-L-2,3,2) has been investigated in a spectrophotometric SF study.[33] For the complexation and decomplexation reactions, respectively, Eqs. (22) and (23) hold, where k_c and k_{cH} refer to complexation by

$$k_{c\,obs} = k_c + k_{cH}K_H[H^+] \tag{22}$$

$$k_{d\,obs} = k_d + k_{dH}[H^+] \tag{23}$$

the neutral and protonated ligand, respectively, K_H is the ligand protonation constant, and k_d and k_{dH} refer to the acid independent and dependent decomplexation reactions. Thus k_c (dm^3 mol^{-1} s^{-1}), k_{cH} (dm^3 mol^{-1} s^{-1}), k_d (s^{-1}), and k_{dH} (dm^3 mol^{-1} s^{-1}) for 4-Me-L-2,2,2 are 9.46×10^7, 1.72×10^4, 2.01×10^3, and 2.51×10^2; for 4,7-Me$_2$-L-2,2,2 are 8.45×10^6, 3.51×10^3, 4.75×10^{-3}, and 3.21×10^2; for 4-Et-L-2,2,2 are 9.82×10^7, 1.96×10^4, 8.34×10^{-4}, and 8.76×10^1; and for 4-Me-L-2,3,2 are 4.86×10^7, 3.24×10^4, 3.28×10^{-1}, and 2.30×10^5 at 298.2 K and $I = 0.10$ mol dm^{-3} (NaClO$_4$). The complexation process involves several steps and it is deduced that k_c characterizes the first metal–nitrogen bond formation, and that k_{cH} characterizes a rate-determining step involving proton loss from the ligand. A related study of the effect of ring size on the decomplexation of Cu(II) tetraamine complexes has also been reported.[34]

The displacement of tripeptides from doubly deprotonated (tripeptido)cuprate(II) complexes [Cu(H$_{-2}$L)$^-$] by triethylenetetramine (trien) has been studied as a function of the number and position of methyl groups in the amino acid residues in twelve tripeptides that consist of glycyl (G), L-alanyl (A), and α-aminoisobutyryl (Aib) residues.[35] The rate law for displacement, derived from spectrophotometric SF data, is given by Eq. (24), where k_d characterizes a path in which the tripeptide is partially unwrapped prior to rapid displacement by trien,

$$k_{obs} = k_d + k_{trien}[trien] \tag{24}$$

and k_{trien} characterizes the direct displacement shown in Scheme 1. The [Cu(H$_{-2}$Aib$_3$)]$^-$ complex ($k_d = 2.2 \times 10^{-4}$ s^{-1} and $k_{trien} = 0.127$ dm^3 mol^{-1} s^{-1}) reacts 8 orders of magnitude more slowly with trien than does the [Cu(H$_{-2}$G$_3$)]$^-$ complex ($k_d = 0.12$ s^{-1} and $k_{trien} = 1.1 \times 10^7$ dm^3 mol^{-1} s^{-1}) at 298.2 K. Methyl groups on the α-carbon of the second and third amino acid residues (from the amino terminus) decrease the rate of displacement by trien to a much greater extent than methyl groups on the first residue. An empirical correlation between the k_{trien} and the position and number of methyl groups is observed (at 298.2 K and pH ca 11) as is shown in Eq. (25), where C_i denotes the number of methyl groups in the ith amino acid residue. This variation is attributed to steric effects.

$$\log k_{trien} = 6.7 - 0.2C_1 - 2.2C_2 - 1.6C_3 \tag{25}$$

Scheme 1

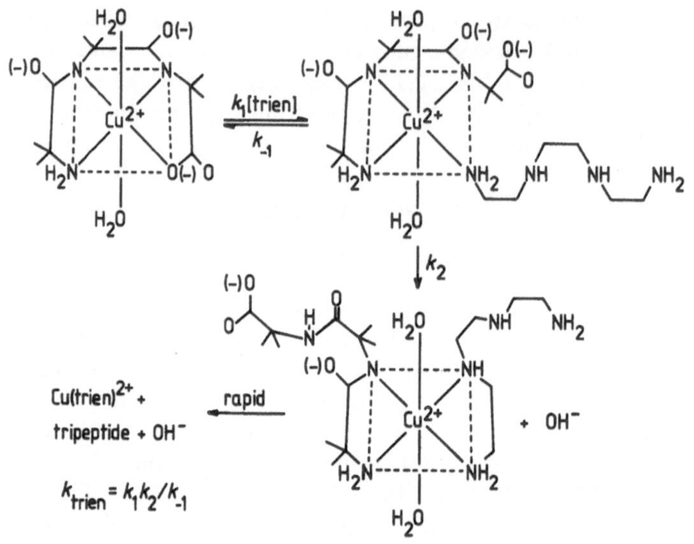

The reaction exhibits a substantial pH dependence because trien and Htrien$^+$ are more reactive than H$_2$trien^{2+}. Above pH $\approx$ 11 the rate of displacement decreases due to the formation of $[Cu(H_{-2}L)(OH)]^{2-}$ which is less reactive than $[Cu(H_{-2}L)]^-$, except in the case of $[Cu(H_{-2}Aib_3)]^-$ which does not form a hydroxo complex, and whose reactivity is so low that hydroxide ion acts as a second nucleophile to displace the tripeptide and increase the overall rate of displacement by trien.[36] This path is not observed for the other complexes as direct tripeptide displacement by trien is much more favorable.

The $[Cu(H_{-2}Aib_3)]^-$ complex also differs from the other tripeptide complexes in exhibiting a displacement rate independent of [trien] $\geq 5 \times 10^{-3}$ mol dm^{-3} at pH ≤ 9 as a consequence of proton transfer to the Cu—N peptide bond adjacent to the terminal carboxylate becoming rate determining in catalyzing the breaking of that bond prior to trien substitution. The variation of the observed rate constant, k_{obs}, for displacement by trien under these conditions is given by Eq. (26), where $k_d = 2 \times 10^4$ s^{-1}, $k_H = 2.5 \times 10^6$ dm^3 mol^{-1} s^{-1}, and $k_{HA} \approx 5 \times 10^{-2}$ dm^3 mol^{-1} s^{-1} in

$$k_{obs} = k_d + k_H[H^+] + k_{HA}[HA] \tag{26}$$

B(OH)$_3$ buffer.[36] A k_{HA} general acid catalysis term is often found for reactions carried out in the presence of buffers, and a study of the acid-catalyzed decomplexation of oligopeptide complexes of Cu(II) and Ni(II) has shown that the 2,6-lutidine type buffers have a very small general acid catalysis effect which is 2000 times smaller than that of acetate buffers.[37]

The square-planar d^9 complexes of the type $[Cu(sal-enNEt_2)X]$ (**12**) resemble square-planar d^8 Pt(II) and Pd(II) where the first coordination sphere is occupied by a tridentate and a monodentate ligand, in that the site occupied by the monodentate ligand is most labile toward substitution. However, a spectrophotometric SF study of the H_2salen (13) reaction (27) shows the Cu(II) complex to be *ca* 3

$$[Cu(sal\text{-}enNEt_2)X] + H_2salen \xrightarrow{k} [Cu(salen)] + H_2sal\text{-}enNet_2 + X^- \qquad (27)$$

orders of magnitude more labile than similar Pd(II) complexes.[38] Thus for reaction (27) in acetone k ($dm^3\,mol^{-1}\,s^{-1}$) (298.2 K) has the values shown in parentheses when $X^- = Cl^-$ (1.88), Br^- (5.46), NCS^- (2.15), N_3^- (1.83), NCO^- (0.660), and thiourea (256). For substitution by H_2Mesalen $k = 0.616$ and $0.570\,dm^3\,mol^{-1}\,s^{-1}$ when $X = Cl^-$ and NCS^-, respectively. For $[Cu(sal\text{-}enNHEt)X]$ (14) $k = 8.79$ and $1.41\,dm^3\,mol^{-1}\,s^{-1}$, respectively, characterize substitution by H_2salen and H_2Mesalen (15) when $X^- = Cl^-$. These reactions are assigned an A mechanism on the basis that the less sterically hindered $[Cu(sal\text{-}enNHEt)X]$ reacts substantially

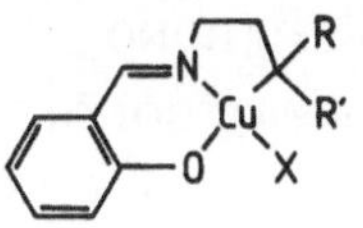

(12)–(13) 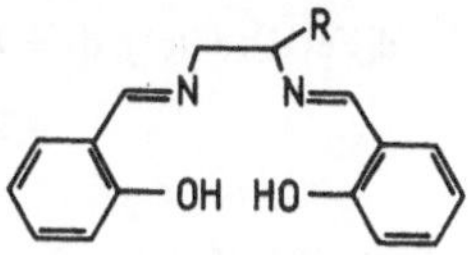   **(14)–(15)**

(12) R = R' = Et; $[Cu(sal\text{-}enNEt_2)X]$ (13) R = H; H_2salen
(14) R = H; R' = Et; $[Cu(sal\text{-}enNHEt)X]$ (15) R = Me; H_2 Mesalen

more rapidly than the more hindered $[Cu(salenNEt_2)X]$, and that the more hindered H_2Mesalen reacts more slowly with both complexes than does the less hindered H_2salen.

In a similar spectrophotometric SF study of the reaction of the bis(N-alkylsalicylaldiminato)nickel(II) complexes [Eq. (28)] in acetone k ($dm^3\,mol^{-1}\,s^{-1}$)

$$[Ni(Rsal)_2] + H_2salen \xrightarrow{k} [Ni(salen)] + 2RsalH \qquad (28)$$

(293 K), ΔH^+ ($kJ\,mol^{-1}$) and ΔS^+ ($J\,K^{-1}\,mol^{-1}$) are, respectively, 0.00476, 112, and 92 when R = $CH(CH_2OH)CH(OH)Ph$; 1.39, 40.7, and -103 when R = $CH(CH_3)CH(OH)Ph$; and 56.4, 35.7, and -89 when R = CH_2CH_2Ph.[39] The large differences in lability are attributed to the formation of a substitution inert octahedral complex [as a consequence of coordination through the hydroxyl groups of $CH(CH_2OH)CH(OH)Ph$ and $CH(CH_3)CH(OH)Ph$], which is in equilibrium with the more labile square-planar complex, and which undergoes substitution through an A mechanism for all three $[Ni(Rsal)_2]$ complexes.

There is evidence that ligands containing aromatic rings can assume coplanar orientations, or "stack," and thereby cause an aggregation of metal complexes containing such ligands.[40] Evidence for such stacking has been sought in a spectrophotometric SF study of the displacement of Rsal from $[Ni(Rsal)_2]$ [where R = CH_2Ph, $(CH_2)_2Ph$, $(CH_2)_3Ph$, $(CH_2)_4Ph$, and $(CH_2)_2(4\text{-}hydroxyphenyl)$] by H_2salen in acetone.[41] The rate constants characterizing the slow displacement of the first Rsal (the second displacement being faster) are 48.8, 43.4, 64.0, 87.0, and

49.5 dm^3 mol^{-1} s^{-1} at 298 K, respectively, as R varies in the sequence above. As no stacking is considered possible when R = CH$_2$Ph, and maximum stacking should occur when R = (CH$_2$)$_4$Ph, it is deduced that the variation of the rate constants is too small to be indicative of the occurrence of significant stacking which is expected to lead to a significant rate reduction.

An investigation into potential catalytic cycles for the removal of oxides of nitrogen from power station flue gases has resulted in the observation that in aqueous solution NO, HONO, and NO$_2^-$ react with Fe(II)(edta) to produce Fe(II)(edta)NO.[42] Although the reaction with NO is too fast for SF and TJ techniques, the reactions with HONO/NO$_2^-$ are within the SF timescale. This reaction follows two parallel paths which are pseudo-first and pseudo-zero order in Fe(II)(edta) concentration, respectively. The contribution of each path depends on the pH and the total nitrite concentration, and the overall rate law is given by Eq. (29), where k_a and k_b = 62 and 90 dm^3 mol^{-1} s^{-1}, respectively, at 298.2 K and

$$-d[\text{Fe(II)(edta)}]/dt = k_a[\text{HONO}]^2 + k_b[\text{Fe(II)(edta)}][\text{HONO}]$$

$$= 2k_1[\text{HONO}]^2 + 2k_2[\text{Fe(II)(edta)}][\text{HONO}] \qquad (29)$$

$I = 0.5$ mol dm^{-3} (Na$_2$SO$_4$), while k_1 and k_2 refer to the reaction sequences in Eqs. (30)–(34). Reactions (31) and (33) are redox reactions, and (32) and (34) are substitution reactions.

$$\text{HONO} + \text{HONO} \xrightarrow{k_1} \text{NO}^+ + \text{OH}^- + \text{HONO} \qquad (30)$$

$$\text{Fe(II)(edta)} + \text{NO}^+ \xrightarrow{\text{fast}} \text{Fe(III)(edta)} + \text{NO} \qquad (31)$$

$$\text{Fe(II)(edta)} + \text{NO} \xrightarrow{\text{fast}} \text{Fe(II)(edta)NO} \qquad (32)$$

$$\text{HONO} + \text{Fe(II)(edta)} \xrightarrow{k_2} \text{Fe(III)(edta)} + \text{NO} + \text{OH}^- \qquad (33)$$

$$\text{Fe(II)(edta)} + \text{NO} \xrightarrow{\text{fast}} \text{Fe(II)(edta)NO} \qquad (34)$$

9.2.4. Ligand Substitution on Complexes of Platinum(II) and Palladium(II)

The determination of solvent exchange rates by NMR spectroscopic methods is well established and widely applied. Nevertheless, the exchange of Me$_2$SO on [Pt(Me$_2$SO)$_4$]$^{2+}$, in which two Me$_2$SO are bonded through oxygen (and bear a *cis* relationship to each other) and the other two Me$_2$SO are bonded through sulfur, appears to be the first reported example of the complete kinetic characterization of the fast exchange of an ambidentate solvent on a metal ion.[43] This ^{1}H NMR study shows that the resonance arising from oxygen-bonded Me$_2$SO exhibits exchanged induced modification in a lower temperature range than does that of sulfur-bonded Me$_2$SO, so that their exchange processes can be separately characterized. It is seen from Table 9.7 (which contains all of the solvent exchange data for square-planar complexes so far reported) that oxygen-bonded Me$_2$SO is much

Table 9.7. Kinetic Parameters for Solvent Exchange on Square-Planar Pt(II) and Pd(II) Species (Ref. 43 and references cited therein)

Complex	k (298 K) (kg mol^{-1} s^{-1})	$\Delta H^{\neq}$ (kJ mol^{-1})	$\Delta S^{\neq}$ (J K^{-1} mol^{-1})	$\Delta V^{\neq}$ (cm^3 mol^{-1})
$[Pt(Me_2SO)_4]^{2+}$ oxygen bonded	3.2×10^3	32.8	−62.0	−2.5
$[Pt(Me_2SO)_4]^{2+}$ sulfur bonded	2	47	−74	−5
$[Pt(H_2O)_4]^{2+}$	7.1×10^{-6}	89.7	−43	−2.2
$[Pd(H_2O)_4]^{2+}$	41	49.5	−48	−2.2
trans-$[Pd(Me_2S)_2Cl_2]$	197	38.1	−73.2	−5.9

more labile than sulfur-bonded Me_2SO as expected for a ligand *trans* to sulfur by comparison to a ligand *trans* to oxygen. Even so, the difference in lability is not as great as would be expected from the differences in *trans* effects exerted by sulfur and oxygen donor ligands and may reflect a lower efficiency of sulfur-bonded ligands as leaving groups. The much lower lability of $[Pt(H_2O)_4]^{2+}$ is more consistent with the anticipated relative labilities expected for ligands *trans* to oxygen and sulfur. The negative $\Delta V^{\neq}$ values indicate the operation of an associatively activated mechanism, but it is not possible to assign an I_a or an A mechanism with reasonable certainty.

For some time the tendency for square-planar Pt(II) complexes to add a fifth ligand, and thereby undergo ligand substitution through an associatively activated process, obscured the possibility of a dissociatively activated ligand substitution mechanism operating in such complexes. In recent studies, however, it has been observed that strong σ-donor *trans* ligands can facilitate dissociative activation as exemplified by *cis*-$[PtR_2L_2]$ (R = Me or Ph, L = Me_2SO or Me_2S) which generates the coordinatively unsaturated $[PtR_2L]$ species.[44] This type of substitution has been particularly closely examined in the substitution of L by bipy in $[Pt(Ph)_2L_2]$, where L is a thioether.[45] It is found that the reaction proceeds through parallel dissociative (k_1) and associative (k_2) paths shown in Scheme 2, which are characterized by the rate constants (determined by conventional and SF spectrophotometry)

Scheme 2

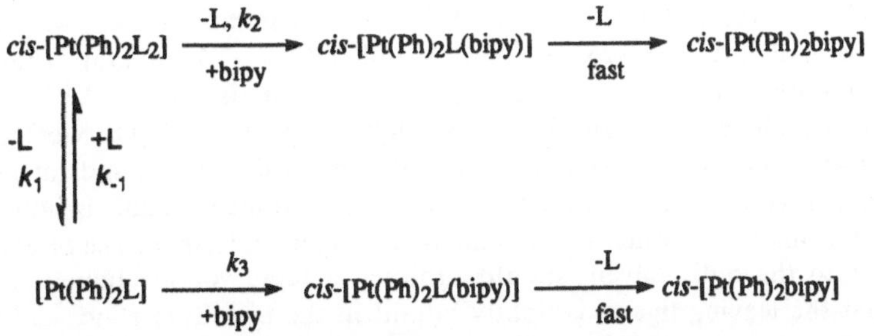

Table 9.8. Rate Constants for the Substitution of L in $[Pt(Ph)_2L_2]$ by Bipy in Dichloromethane at 298.16 K [45]

L	$10^2 k_1$ (s^{-1})	k_3/k_{-1}	k_2 $(dm^3 \, mol^{-1} \, s^{-1})$
Me_2S	0.528	0.246	0.0023
Et_2S	0.378	0.467	0.00
MeEtS	0.432	0.332	0.0064
$n\text{-}Pr_2S$	0.356	0.590	0.0065
$i\text{-}Pr_2S$	19.0	3.74	0.43
$n\text{-}Bu_2S$	0.272	0.602	0.0017
Bz_2S	1.25	0.318	0.0042
MeBzS	0.542	0.333	0.006
MePhS	17.6	0.484	0.94
Et_2Se	0.601	0.270	0.012

in Table 9.8. The k_1 path dominates the substitution and the variation of lability with L is $Bz_2S > MeBzS > Me_2S > MeEt > Et_2S > n\text{-}Pr_2S > n\text{-}Bu_2S$. In the absence of steric effects, this is consistent with the lability of the sulfide being dependent on the amount of electron density transmitted to the sulfur atom and finally to the Pt—S bond strength. The greater reactivity when L = $i\text{-}Pr_2S$ or MePhS is attributed to these ligands causing greater steric crowding.

In contrast, a mechanism close to the extreme *A* case appears to operate for substitution of water by acetonitrile in $[Pd(H_2O)_4]^{2+}$ in reactions (35) and (36),

$$[Pd(H_2O)_4]^{2+} + MeCN \underset{k_{-1}}{\overset{k_1}{\rightleftharpoons}} [Pd(H_2O)_3(MeCN)]^{2+} + H_2O \qquad (35)$$

$$[Pd(H_2O)_3(MeCN)]^{2+} + MeCN \underset{k_{-2}}{\overset{k_2}{\rightleftharpoons}} [Pd(H_2O)_2(MeCN)_2]^{2+} + H_2O \qquad (36)$$

which are characterized at 298 K by $k_1 = 309 \, dm^3 \, mol^{-1} \, s^{-1}$ ($\Delta H^+ = 46 \, kJ \, mol^{-1}$, $\Delta S^+ = -43 \, J \, K^{-1} \, mol^{-1}$, and $\Delta V^+ = -4.0 \, cm^3 \, mol^{-1}$), $k_{-1} = 16 \, s^{-1}$ ($\Delta H^+ = 59 \, kJ \, mol^{-1}$, $\Delta S^+ = -24 \, J \, K^{-1} \, mol^{-1}$, and $\Delta V^+ = -1.5 \, cm^3 \, mol^{-1}$), $k_2 = 38.7 \, dm^3 \, mol^{-1} \, s^{-1}$, and $k_{-2} = 33.2 \, s^{-1}$, determined by spectrophotometric SF.[46] (It should be noted that k_2 and k_{-2} are composite parameters referring to both *cis* and *trans* isomers.) These data compare with $k_1 = 2.45 \, dm^3 \, mol^{-1} \, s^{-1}$ ($\Delta H^+ = 58 \, kJ^{-1}$, $\Delta S^+ = -44 \, J \, K^{-1} \, mol^{-1}$, and $\Delta V^+ = -9.2 \, cm^3 \, mol^{-1}$), $k_{-1} = 0.24 \, s^{-1}$ ($\Delta H^+ = 69 \, kJ \, mol^{-1}$, $\Delta S^+ = -25 \, J \, K^{-1} \, mol^{-1}$, and $\Delta V^+ = -1.7 \, cm^3 \, mol^{-1}$) for the mono-substitution of Me_2SO on $[Pd(H_2O)_4]^{2+}$.[47] A comparison of the ΔV^+ values for the substitution of H_2O on $[Pd(L)(H_2O)_3]^{2+}$ (where L = H_2O, MeCN, or Me_2SO) shows them to be very similar ($-1.9 \pm 0.4 \, cm^3 \, mol^{-1}$), which indicates that the difference in volume between the transition state and the reactants is independent of the leaving ligand when water is the incoming ligand. As release of MeCN or Me_2SO into the bulk solvent should produce a large volume increase, this suggests that the leaving ligand is tightly bound in the transition state, and that the substitution may closely approach the extreme *A* mechanism.

An A mechanism is assigned to water exchange on both $[Pd(Me_5dien)H_2O]^{2+}$ and $[Pd(Et_5dien)H_2O]^{2+}$ on the basis of negative ΔV^+ values (determined by ^{17}O NMR), and their decreased labilities (by comparison to $[Pd(H_2O)_4]^{2+}$ and $[Pd(Me_5dien)H_2O]^{2+}$) are attributed to their greater steric crowding causing a decrease in the ease of associative attack by the incoming ligand (Table 9.9). A spectrophotometric SF study of the substitution of $[Pd(Me_5dien)H_2O]^{2+}$ and $[Pd(Et_5dien)H_2O]^{2+}$ by thiourea, dimethylthiourea, and tetramethylthiourea is characterized by $\Delta V^+ = -9.3$, -9.1, and -13.4; and -8.3, -10.2, and -12.7 cm^3 mol^{-1}, respectively, indicative of an A mechanism operating for these substitution processes also.[48] The negative ΔV^+ values obtained for substitution of both complexes by Cl$^-$, Br$^-$, I$^-$, N$_3^-$, and C$_2$O$_4^{2-}$ are also interpreted in terms of the operation of an A mechanism.[49] A linear relationship exists between the partial molar volumes of the ground state and transition state of $[Pd(R_5dien)X]^{(2-n)+}$ (R = Me or Et) for the X^{n-} substitution process in Eq. (37).

$$[Pd(R_5dien)H_2O]^{2+} + X^{n-} \underset{k_d}{\overset{k_c}{\rightleftharpoons}} [Pd(R_5dien)X]^{(2-n)+} \tag{37}$$

Thus, intrinsic and solvational volume contributions affect the partial molar volumes of the product and transition states in a similar manner and are governed by the nature of R$_5$dien and X^{n-}. It is also deduced that the transition state is more compact than either the reactant or product ground state.

The effect of solvent variation on ligand substitution of Pd(II) complexes has been examined in a spectrophotometric SF study of the solvolysis of $[Pd(Me_5dien)pyridine]^{2+}$ in which pyridine is displaced by the solvent.[50] The $10^5 k$ (dm^3 mol^{-1} s^{-1}) (298.2 K), and ΔV^+ (cm^3 mol^{-1}) values, respectively, characterizing the solvolysis for each solvent are given in parentheses: H$_2$O (1.87, -3.1), MeOH (4.1, -6), EtOH (7.4, -4.1), Me$_2$SO (52.8, ≈ 0), dmf (90.1, -3.4), and MeCN (1310, -2.4). There is no obvious correlation between the ΔV^+ magnitudes and the nature of the solvent, but their negative or zero values, in combination with the negative ΔS^+ values characterizing each system, are consistent with the operation of an A mechanism. In aqueous solution in the range $7 \leq pH \leq 12$, reaction (38) proceeds as shown with no involvement of conjugate base or hydroxy bridged species.[51] The effects of varying R (= H, Me, and Et) are substantial as

$$[Pd(R_5dien)H_2O]^{2+} + I^- \underset{k_d}{\overset{k_c}{\rightleftharpoons}} [Pd(R_5dien)I]^+ + H_2O \tag{38}$$

Table 9.9. *Kinetic Parameters for Water Exchange on Aqua Complexes of Pd(II) (Ref. 48 and references cited therein)*

Complex	$k(298.1\ K)$ (s^{-1})	ΔH^+ (kJ mol^{-1})	ΔS^+ (J K^{-1} mol^{-1})	ΔV^+ (cm^3 mol^{-1})
$[Pd(H_2O)_4]^{2+}$	560	49	-26	-2.2
$[Pd(dien)H_2O]^{2+}$	5100	38	-47	-2.8
$[Pd(Me_5dien)H_2O]^{2+}$	187	62	8	-7.2
$[Pd(Et_5dien)H_2O]^{2+}$	2.9	63	-25	-7.7

is seen from k_c ($dm^3 mol^{-1} s^{-1}$) and k_d (s^{-1}), respectively, obtained at 298.2 K by spectrophotometric SF methods, and given in parentheses after each variant of R_5dien: 1,4,7-Me$_3$dien (1.48×10^6, 20.8), 1,4,7-Et$_3$dien (5.44×10^5, 5.48), 1,1,4-Et$_3$dien (2.38×10^4, 0.231) 1,1,7,7-Me$_4$dien (7.25×10^3, 0.653), 1,1,4,7,7-Me$_5$dien (3.38×10^3, 0.344), 1,1,7,7-Et$_4$dien (2.96, 2.37×10^{-4}), and 1,1,4,7,7-Et$_5$dien (1.21, 1.95×10^{-4}). The systematic decrease in k_c and k_d coincides with increasing steric hindrance at the substitution site.

Two other spectrophotometric SF studies have increased the amount of data available on Pd(II) ligand substitution processes. In one the formation of tetra-chloropalladate(II) from bis(oxalato)- and bis(malanato)palladium(II) in aqueous acid chloride solution has been studied.[52] It is found that $[Pd(mal)_2]^{2-}$ is much more labile than $[Pd(ox)_2]^{2-}$, and that in both cases the formation of $[PdCl_4]^{2-}$ proceeds in two consecutive steps both of which are characterized by the two-term rate law (39). The rate-determining steps characterized by k_0 and k_H are thought to be the formation of reactive intermediate species in which the leaving ligand

$$k_{obs} = k_0 + k_H[H^+][Cl^-] \tag{39}$$

and the protonated ligand, respectively, are monodentate and a water molecule has entered the first coordination sphere so that Pd(II) remains four-coordinate. In the second study the reactions of $[Pd(par)OH]^-$ and $[Cd(par)_2]^{2-}$ [par = 4-(2-pyridylazo)resorcinol] with CN^- to form $[Pd(CN)_4]^{2-}$ and $[Cd(CN)_4]^{2-}$ have been characterized.[53]

9.2.5. Ligand Substitution on Complexes of Mercury(II)

Sulfhydryl groups have a very high affinity for Hg(II), and as a consequence it is thought that all inorganic mercury in mammalian blood and tissue is complexed by sulfhydryl groups of cysteine-containing peptides and proteins. This has prompted a ^{13}C NMR study of the complexation of Hg(II) by the sulfhydryl-containing ligands glutathione, cysteine, and penicillamine.[54] Thus for the equilibria of the glutathione (GS^-) complexes:

$$Hg(SG)_2 + GS^- \underset{k_{-1}}{\overset{k_1}{\rightleftharpoons}} Hg(SG)_3^- \tag{40}$$

$$Hg(SG)_2 + GSH \underset{k_{-2}}{\overset{k_2}{\rightleftharpoons}} Hg(SG)_3^- + H^+ \tag{41}$$

k_1, k_2, and $k_{-2} = 1.3 \times 10^7$, 1.7×10^4, and 6.3×10^9 $dm^3 mol^{-1} s^{-1}$, respectively, and $k_{-1} = 6.3 \times 10^3 s^{-1}$ at 298.2 K in aqueous solution. It is considered that, as a consequence of the ubiquity of glutathione in cellular systems, reactions similar to (40) and (41) play a major role in the mobility of Hg(II) in biological systems. The mechanism of ligand substitution on CH_3Hg(II) complexes is of both chemical and toxicological interest, and as a consequence has been the subject of substantial investigation. The kinetic data from a spectrophotometric TJ and SF study of reaction (42) (where Y = 4-nitro-2-sulfonato-thiophenolate or 4-nitro-3-carboxy-

$$CH_3HgY^+ + X \underset{k_{-1}}{\overset{k_1}{\rightleftharpoons}} CH_3HgX^+ + Y \qquad (k_1/k_{-1} = K_1) \tag{42}$$

Table 9.10. *Rate Constants at $I = 0.1\ mol^{-1}\ dm^3$ (NaClO$_4$) and 293.2 K for the Reaction*[55]

$$CH_3HgY+ + X \underset{k_{-1}}{\overset{k_1}{\rightleftharpoons}} CH_3HgX = +Y$$

X	k_1 $(dm^3\ mol^{-1}\ s^{-1})$	k_{-1} $(dm^3\ mol^{-1}\ s^{-1})$	log K
	Y = 4-nitro-2-sulfonato-thiophenolate		
Br$^-$	2.8×10^3	3.6×10^9	-6.10
Imidazole	1.8×10^4	7.0×10^9	-5.58
SO$_3^{2-}$	2.5×10^4	1.0×10^9	-4.61
I$^-$	$\approx 2 \times 10^5$	$\approx 3 \times 10^9$	-4.12
OH$^-$	3.3×10^5	7.3×10^8	-3.35
S$_2$O$_3^{2-}$	2.1×10^7	1.3×10^9	-1.82
CN$^-$	3.2×10^8	1.7×10^7	1.28
HOCH$_2$CH$_2$S$^-$	1.4×10^9 $(1.2 \times 10^3)^a$	5.6×10^5	3.40
	Y = 4-nitro-3-carboxylato-thiophenolate		
Cl$^-$	7.9×10^2 $(4.4 \times 10^6)^a$	4.9×10^9	-6.82
NCS$^-$	1.8×10^3 $(1.3 \times 10^7)^a$	1.8×10^9	-6.02
S$_2$O$_3^{2-}$	6.7×10^7	1.0×10^9	-1.17
CN$^-$	1.2×10^9	1.4×10^7	1.93
HOCH$_2$CH$_2$S$^-$	1.7×10^9	1.5×10^5	4.05

^{a} Reaction of the protonated free ligand.

lato-thiophenolate, and X is one of the ligands in Table 9.10) is interpreted in terms of an I_a mechanism, and shows that the exchange process is close to diffusion control.[55] The variation of k_1 and k_{-1} with K_1 is a good example of the correlation between lability and stability expected for a diffusion-controlled exchange process. In cases where the incoming ligand is protonated, it is seen that there is a substnatial decrease in reactivity by comparison with that of the ligand conjugate base.

9.3. Ligand Substitution on Complexes of Trivalent Metal Ions and Metal Ions of Higher Valency

9.3.1. Ligand Substitution on Complexes of the Trivalent Main Group Metal Ions

The readily protolyzed light trivalent metal ions have provided a continuous challenge to mechanistic investigators through the ambiguity which can arise in distinguishing between the substitution of a ligand on the hexaaqua metal ion and the substitution of a protonated ligand on the hydroxopentaaqua metal ion. In a study of F^- substitution on Al(III), the use of a fluoride ion-selective electrode to directly measure the variation of $[F^-]$ has been claimed to eliminate this "proton ambiguity."[56] It has been deduced that the substitution of $[Al(H_2O)_6]^{3+}$ by F^- is characterized by $k_f(298.2 \text{ K}) = 1.55 \text{ s}^{-1}$, $\Delta H^+ = 96.3 \text{ kJ mol}^{-1}$, and $\Delta S^+ = 82.9 \text{ J K}^{-1} \text{ mol}^{-1}$, and the substitution of $[Al(H_2O)_5OH]^{2+}$ by F^- by $k_f(298.2 \text{ K}) = 2196 \text{ s}^{-1}$, $\Delta H^+ = 83.7 \text{ kJ mol}^{-1}$, and $\Delta S^+ = 99.8 \text{ J K}^{-1} \text{ mol}^{-1}$. In the derivation of these rate constants it has been assumed that the Eigen–Wilkins mechanism applies and outer-sphere constants, K_o, of 1.57 and 0.919 $\text{dm}^3 \text{ mol}^{-1}$ were employed [calculated using the Fuoss equation and assuming that the Al(III) center and F^- are separated by two water molecules to give an interaction distance of 750 pm]. An I_d mechanism is assigned to the substitution of $[Al(H_2O)_5OH]^{2+}$. In contrast, an I_a mechanism has been assigned to the substitution of $[Al(H_2O)_6]^{3+}$ on the basis that the derived, statistically corrected interchange rate constant for F^- substitution, $k_i = 2.06 \pm 0.47 \text{ s}^{-1}$, is larger than that characterizing water exchange ($k = 1.29 \pm 0.04 \text{ s}^{-1}$) which is considered to proceed through an I_d mechanism.[57] In view of the necessary assumptions made in the derivation of k_i this difference in the two rate constants may not represent unambiguous evidence for a change from the I_d mechanism which usually operates for ligand substitution on $[Al(H_2O)_6]^{3+}$. Environmental studies of the complexation of Al(III) by F^- have also been reported.[58,59]

Although the chlorothallium(III) complexes are among the most stable chloro metal complexes, there has been very little information available on the kinetics of their ligand exchange processes as a consequence of their high lability, until recently. A ^{205}Tl NMR study in aqueous 3 mol dm^3 perchloric acid at 298.2 K has now yielded a substantial number of rate constants.[60] Thus for reactions (43) $k_{01} = 4.5 \times 10^4$ ($\Delta H^+ = 49 \text{ kJ mol}^{-1}$, $\Delta S^+ = 12 \text{ J K}^{-1} \text{ mol}^{-1}$), $k_{12} = 5.2 \times 10^4$, $k_{23} = 2.7 \times 10^7$, and $k_{34} < 3 \times 10^7 \text{ dm}^3 \text{ mol}^{-1} \text{ s}^{-1}$. For reaction (44) $k_{02} = 6.4 \times 10^5 \text{ dm}^3 \text{ mol}^{-1} \text{ s}^{-1}$, and for reactions (45) $k'_{01} < 1 \times 10^8$, $k'_{12} = 3.3 \times 10^8$, $k'_{23} = 1.3 \times 10^9$, and $k'_{34} = 4.7 \times 10^8 \text{ dm}^3 \text{ mol}^{-1} \text{ s}^{-1}$ ($\Delta H^+ = 6.6 \text{ kJ mol}^{-1}$, $\Delta S^+ = -107 \text{ J K}^{-1} \text{ mol}^{-1}$). It is deduced that the reactions characterized by k_{01} and k'_{34} proceed through I_d and I_a mechanisms, respectively.

$$TlCl_m^{3-m} + {}^*TlCl_n^{3-n} \underset{}{\overset{k_{nm}}{\rightleftharpoons}} {}^*TlCl_m^{3-m} + TlCl_n^{3-n} \tag{43}$$

$$Tl^{3+} + TlCl^{2+} \underset{}{\overset{k_{02}}{\rightleftharpoons}} 2TlCl \tag{44}$$

$$TlCl_n^{3-n} + Cl^- \underset{}{\overset{k'_{m(n+1)}}{\rightleftharpoons}} TlCl_{n+1}^{2-n} \tag{45}$$

9.3.2. Ligand Substitution on Complexes of the Trivalent Transition Metal Ions

Hexasolventochromium(III) species are generally inert, but the presence of an alkyl or porphyrin ligand in the first coordination sphere usually results in a considerable labilization to ligand substitution of the Cr(III) center, as is illustrated in Eq. (46). In aqueous ($I = 1.00 \text{ mol dm}^3$ $HClO_4/NaClO_4$) solution a spectrophotometric SF study yields k_1 ($dm^3 \text{ mol}^{-1} \text{ s}^{-1}$) (298.2 K) = 0.63, 0.50, 0.47, and

$$[Cr(H_2O)_5CH_2CN]^{2+} + X^{n-} \xrightarrow{k_1} trans\text{-}[Cr(H_2O)_4(X)CH_2CN]^{(2-n)+} + H_2O \quad (46)$$

0.07 when $X^{n-} = H_2P_2O_7^{2-}$, $H_3P_2O_7^-$, $H_2PO_4^-$, and H_3PO_4, respectively.[61] This substitution is followed by an isomerization to the *cis* isomer, a rapid second substitution, and the subsequent loss of the alkyl ligand.

The axial substitution reactions of metalloporphyrins have been the subject of considerable interest over the last two decades. A very detailed spectrophotometric SF study of the reaction of imidazole with [*meso*-tetrakis(*p*-sulfonatophenyl)porphinato]diaquochromate(III) in the range pH 6.00 to 0.100 mol dm^3 NaOH at $I = 1.00 \text{ mol dm}^3$ (NaClO$_4$) shows several paths for the imidazole substitution:

$$[Cr(TPPS)(H_2O)_2]^{3-} \underset{-Im,k_{-1},+H_2O}{\overset{+Im,k_1,-H_2O}{\rightleftharpoons}} [Cr(TPPS)(H_2O)Im]^{3-} \underset{-Im,k_{-2},+H_2O}{\overset{+Im,k_2,-H_2O}{\rightleftharpoons}} [Cr(TPPS)(Im)_2]^{3-}$$

$$(47)$$

$$[Cr(TPPS)(H_2O)(OH)]^{4-} \underset{-Im,k_{-3},+OH^-}{\overset{+Im,k_3,-OH^-}{\rightleftharpoons}} [Cr(TPPS)(H_2O)(Im)]^{4-}$$

$$\underset{-Im,k_{-4},+OH^-}{\overset{+Im,k_4,-OH^-}{\rightleftharpoons}} [Cr(TPPS)(Im)_2]^{3-} \quad (48)$$

$$[Cr(TPPS)(OH)_2]^{5-} \underset{-Im,k_{-5},+OH^-}{\overset{+Im,k_5,-OH^-}{\rightleftharpoons}} [Cr(TPPS)(H_2O)(Im)]^{4-} \quad (49)$$

where k_i (298.2 K) ($dm^3 \text{ mol}^{-1} \text{ s}^{-1}$), ΔH_i^+ ($kJ \text{ mol}^{-1}$), and ΔS_i^+ ($J \text{ K}^{-1} \text{ mol}^{-1}$) are, respectively, $k_1 = 2.40$, 68.6, -7.87; $k_2 = 0.305$, 42.3, -113; $k_3 = 98.2$, 65.3, 12.0; $k_4 = 7.96 \times 10^{-3}$, 75.7, -31.6; and $k_5 = 3.78$, 69.0, -4.3; while k_{-i} (298.2 K), ΔH_i^+ ($kJ \text{ mol}^{-1}$), and ΔS_i^+ ($J \text{ K}^{-1} \text{ mol}^{-1}$) are, respectively, $k_{-2} = 2.44 \times 10^{-4} \text{ s}^{-1}$, 110, 56.0; $k_{-4} = 3.01 \times 10^{-2} \text{ dm}^3 \text{ mol}^{-1} \text{ s}^{-1}$, 81.2, -2.7; and $k_{-5} = 9.29 \times 10^{-1} \text{ dm}^3 \text{ mol}^{-1} \text{ s}^{-1}$, 61.9, 39.3.[62] It is apparent that the porphyrin and OH^- ligands labilize Cr(III) to axial substitution, but there are no obvious trends in the activation parameters.

In toluene, the axial substitution of chloro(5,10,15,20-tetraphenylporphinato)chromium(III) by 1-methylimidazole shown in Eq. (50) proceeds through the *D* mechanistic sequence in Eqs. (51) and (52) that is characterized by k_1

$$[Cr(TPP)(Cl)(L)] + 1\text{-MeIm} \rightleftharpoons [Cr(TPP)(Cl)(1\text{-MeIm})] + L \quad (50)$$

$$[Cr(TPP)(Cl)(L)] \underset{k_{-1}}{\overset{k_1}{\rightleftharpoons}} [Cr(TPP)(Cl)] + L \quad (51)$$

$$[Cr(TPP)(Cl)] + 1\text{-MeIm} \underset{k_{-2}}{\overset{k_2}{\rightleftharpoons}} [Cr(TPP)(Cl)(1\text{-MeIM})] \quad (52)$$

(298.2 K) (s^{-1}), ΔH^+ ($kJ \text{ mol}^{-1}$), ΔS^+ ($J \text{ K}^{-1} \text{ mol}^{-1}$) and ΔV^+ ($cm^3 \text{ mol}^{-1}$) values of 8.0, 86.1, 61, and 25.7 for L = pyridine; 7.0, 85.3, 57, and 23.8 when L = quinoline; and 4.2, 78.2, 29, and 19.6 when L = triphenylphosphine (determined by variable-pressure spectrophotometric SF methods).[63] For the variation of L in the same sequence $10^2 k_{-2}$ (s^{-1}) = 2.8, 3.6, and 3.6.

The solvolysis reaction (53) studied by SF and conventional spectrophotometry is characterized by k (298.2 K) (s^{-1}), $\Delta H^{\ddagger}$ (kJ mol^{-1}), $\Delta S^{\ddagger}$ (J K^{-1} mol^{-1}) and $\Delta V^{\ddagger}$

$$[Fe(CN)_5NO_2]^{3-} + \text{solvent} \xrightarrow{k} [Fe(CN)_5(\text{solvent})]^{2-} + NO_2^- \tag{53}$$

(cm^3 mol^{-1}) values of, respectively, 1.25×10^{-3}, 121.8, 180.3, and 26.9 in dmf; 1.18×10^{-4}, 137.4, 140.5, and 25.9 in Me$_2$SO; and 5.25×10^{-5} (308.2 K), 116.5, 50.7, and 19.6 in MeOH. The lability increases with the electron-donating ability of the solvent and the solvolysis is considered to proceed through a D mechanism.[64] The solvolysis of $[Fe(CN)_5NO_2]^{3-}$ is substantially slower in water at neutral pH, but is subject to acid catalysis such that in the pH range 0.3–2.0, k(298.2 K) = 12.9 dm^3 mol^{-1} s^{-1}, $\Delta H^{\ddagger}$ = 43 kJ mol^{-1}, $\Delta S^{\ddagger}$ = -80 J K^{-1} mol^{-1}, and $\Delta V^{\ddagger}$ = 2.2 cm^3 mol^{-1}.[65]

It has been observed that the edta complexes of Ru(III) are at least seven orders of magnitude more labile toward ligand substitution than nonchelated Ru(III) complexes.[66,67] For the reactions (54)–(56), k_1, k_2, and k_3 = 130, 1420,

$$[Ru(Hedta)(H_2O)] + L \xrightarrow{k_1} [Ru(Hedta)L] + H_2O \tag{54}$$

$$[Ru(edta)(H_2O)]^- + L \xrightarrow{k_2} [Ru(edta)L]^- + H_2O \tag{55}$$

$$[Ru(edta)(OH)]^{2-} + L \xrightarrow{k_3} [Ru(edta)L]^{2-} + OH^- \tag{56}$$

and 75 dm^3 mol^{-1} s^{-1}, respectively, for substitution by dimethylthiourea, indicating the effect of the three different protonation states on the lability of the complex. More detailed kinetic data (spectrophotometric SF) for the $[Ru(edta)(H_2O)]^-$ system is presented in Table 9.11, from which it is concluded that the substitution mechanism is I_a, largely on the basis of the decrease in the rate of substitution as the steric hindrance increases with increase in ligand size in the sequence thiourea < dimethylthiourea < tetramethylthiourea, and of the substantially negative $\Delta V^{\ddagger}$ values. It is thought that the increased lability of the edta complexes is a consequence of hydrogen bonding between a free acetate arm of edta and coordinated water.[66] This has prompted a study of ligand substitution reactions of related complexes of Ru(III) in which edta is replaced by N-(hydroxyethyl)ethylenediaminetriacetate (hedtra), and which lack the pendant acetate group of the edta complexes.[68] These reactions (57) and (58) also appear to proceed through

$$[Ru(hedtra)(H_2O)] + L \xrightarrow{k_4} [Ru(hedtra)L] + H_2O \tag{57}$$

$$[Ru(hedtra)(OH)]^- + L \xrightarrow{k_5} [Ru(hedtra)L] + OH^- \tag{58}$$

an associatively activated ligand substitution mechanism. A comparison with $[Ru(Hedta)(H_2O)]$, $[Ru(edta)(H_2O)]^-$, and $[Ru(edta)(OH)]^{2-}$ indicates that these complexes are 1–2 orders of magnitude more labile than Ru(hedtra)(H$_2$O)] and $[Ru(hedtra)(OH)]^-$, which is attributed to the greater labilizing effect of the extra acetate arm in edta (by comparison with the effect of the N-hydroxyethyl group in hedtra).

Table 9.11. Kinetic Parameters for the Reaction[66]

$$[Ru(edta)(H_2O)]^- + L \xrightarrow{k_2} [Ru(edta)L]^- + H_2O$$

L	k_2 $(dm^3\ mol^{-1}\ s^{-1})$	$\Delta H^{\ddagger}$ $(kJ\ mol^{-1})$	$\Delta S^{\ddagger}$ $(J\ K^{-1}\ mol^{-1})$	$\Delta V^{\ddagger}$ $(cm^3\ mol^{-1})$
Thiourea	2970	22.3	−105	−6.8
Dimethylthiourea	1450	25.3	−107	−8.8
Tetramethylthiourea	154	28.9	−107	−12.2
Azide	2070	26.4	−94	−9.9
Thiocyanate	270	37.2	−75	−9.6

The opportunities to study ligand substitution processes on square-planar trivalent metal complexes formed with monodentate ligands are limited. Hence a study of such reactions on Au(III) by pH measurement and sequential mixing SF spectrophotometry is of considerable interest.[69] Reaction (59) is characterized by

$$[Au(NH_3)_4]^{3+} + SCN^- \xrightarrow{k} [Au(NH_3)_3(SCN)]^{2+} + NH_3 \qquad (59)$$

$k(298.2\ K) = 7.6\ dm^3\ mol^{-1}\ s^{-1}$, $\Delta H^{\ddagger} = 61\ kJ\ mol^{-1}$, and $\Delta S^{\ddagger} = -26\ J\ K^{-1}\ mol^{-1}$, and this is followed by rapid reduction to Au(I). For $trans$-$[Au(NH_3)_2X_2]^+$ ($X^- =$ Cl^- or Br^-) the following ligand substitutions occur:

$$trans\text{-}[Au(NH_3)_2X_2]^+ + SCN^- \underset{k_{-1}}{\overset{k_1}{\rightleftharpoons}} trans\text{-}[Au(NH_3)_2(SCN)X]^+ + X^- \qquad (60)$$

$$trans\text{-}[Au(NH_3)_2(SCN)X]^+ + SCN^- \underset{k_{-2}}{\overset{k_2}{\rightleftharpoons}} trans\text{-}[Au(NH_3)_2(SCN)_2]^+ + X^- \qquad (61)$$

and are characterized by $k_1 = 9.0 \times 10^3$ and 8.9×10^4; $k_{-1} = 0.6$ and 1.32×10^3; $k_2 = 1.56 \times 10^5$ and 1.4×10^5; $k_{-2} = 3.4 \times 10^2$ and $1.0 \times 10^4\ dm^3\ mol^{-1}\ s^{-1}$, respectively, when $X^- = Cl^-$ and Br^- at 275.2 K. In contrast to the isoelectronic Pt(II) complexes, the nature of the entering ligand is more important in determining substitution rate than is the nature of the $trans$ ligand, which confirms earlier indications[70] of a very high discriminating power for entering ligands for the soft Au(III) center and explains the large range of substitution rates observed.

9.3.3. Ligand Substitution on Complexes of the Trivalent Lanthanide Ions

The tripositive lanthanide ions constitute the longest range of chemically similar metal ions and in consequence provide an opportunity to study mechanistic variations as their ionic radii systematically decrease with the lanthanide contraction. However, in water in particular, the absence of direct determinations of the number of solvent molecules in the first coordination sphere of these ions has

Table 9.12. Parameters for Water Exchange on $[Ln(H_2O)_8]^{3+}$ [72]

Ln	$10^{-7} k$ (298 K) (s^{-1})	ΔH^+ (kJ mol^{-1})	ΔS^+ (J K^{-1} mol^{-1})	ΔV^+ (cm^3 mol^{-1})
Tb	55.8	12.1	−36.9	−5.7 (269.1 K)
Dy	43.4	16.6	−24.0	−6.0 (268.5 K)
Ho	21.4	16.4	−30.5	−6.6 (268.8 K)
Er	13.3	18.4	−27.8	−6.9 (268.8 K)
Tm	9.1	22.7	−16.4	−6.0 (269.1 K)
Yb	4.7	23.3	−21.0	

hampered mechanistic interpretation of ligand substitution data. Thus the determination by neutron scattering of the stoichiometries $[Dy(H_2O)_8]^{3+}$ and $[Yb(H_2O)_8]^{3+}$ in aqueous solution, and the observation that all eight water molecules are equidistant from the metal center consistent with cubic or dodecahedral, square antiprismatic, or cubic (less favored) stereochemistry, is of considerable importance.[71] The water exchange parameters, obtained from a variable-temperature and pressure ^{17}O NMR study, are presented in Table 9.12, and present an opportunity to reinterpret earlier ligand substitution studies.[72,73] The negative ΔV^+ values are interpreted in terms of an I_a water exchange mechanism proceeding through a nine-coordinate transition state, probably of tricapped trigonal prismatic stereochemistry. The earlier-determined sulfate substitution rates[74] and the new water exchange rates for $[Ln(H_2O)_8]^{3+}$ are very similar, which at first sight is unexpected for an associatively activated mechanism. However, it appears that the selectivity of the hard Lewis acid Ln^{3+} ions for entering ligands is small.

A ^{139}La NMR study[75,76] of reaction (62) is probably the first reported use of this nuclear spin in quantitative kinetic studies (2,6-dicarboxy-4-hydroxypyridine = $Hdcp^{2-}$). The derived rate parameters are k_c and k_d(298 K) = 8×10^8 cm^3 mol^{-1} s^{-1}

$$[La(H_2O)_n]^{3+} + Hdcp^{2-} \underset{k_d}{\overset{k_c}{\rightleftharpoons}} [La(Hdcp)(H_2O)_{(n-3)}]^+ + 3H_2O \tag{62}$$

and 207 s^{-1}, respectively, with corresponding ΔH^+ = 29 and 38.3 kJ mol^{-1}, ΔS^+ = 23 and −72 J K^{-1} mol^{-1}, and ΔV^+ = 7 and −4.9 cm^3 mol^{-1}. The complexation reaction (k_c) is considered to proceed through an Eigen–Wilkins mechanism, and $K_o = 1700$ dm^3 mol^{-1} is calculated from the Fuoss equation, on which basis the interchange rate constant for the substitution by tridentate $Hdcp^{2-}$ (k_i) is calculated to be 5×10^5 s^{-1}, which is small compared to the water exchange rate constant for $[La(H_2O)_n]^{3+}$. This suggests that chelation is the rate-determining step in the formation of $[La(Hdcp)(H_2O)_{(n-3)}]^+$.

As some trivalent lanthanides (Ln^{3+}) specifically displace Ca^{2+} in biological systems and in consequence are used as paramagnetic probes for Ca^{2+} sites in NMR studies, it is of interest to examine the relative labilities of Ca^{2+} and Ln^{3+} in similar coordination sites. The potentially octadentate ligand 2-[(2-bis[carboxymethyl]amino-5-methylphenoxy)methyl]-6-methoxy-8-bis[carboxymethyl]aminoquinoline, or quin2, a fluorescent indicator for Ca^{2+} which also coordinates Ln^{3+}, provides an opportunity to do this. Thus a fluorimetric SF study of the displacement of quin2 from its Ca^{2+} and Ln^{3+} complexes by edta is characterized by the rate law (63) where for Ca^{2+}, Pr^{3+}, Tb^{3+}, Dy^{3+}, and Yb^{3+}, $k_1 = 44.8$, 2.87×10^{-3}, 1.99×10^{-4}, 1.27×10^{-4}, and $9.02 \times 10^{-6} \text{ s}^{-1}$, respectively, and $k_2 = 1.41 \times 10^3$, 9.70×10^{-1}, 8.48×10^{-2}, 7.40×10^{-2}, and $2.40 \times 10^{-3} \text{ dm}^3 \text{ mol}^{-1} \text{ s}^{-1}$, respectively.[77]

$$k_{obs} = k_1 + k_2[\text{edta}] \tag{63}$$

Thus the Ca^{2+} quin2 complex is several orders of magnitude more labile than its Ln^{3+} analogs, which show a systematic decrease in lability with decrease in ionic radius consistent with the concomitant increase in surface charge density increasing the Ln^{3+}-quin2 charge–dipole interaction and decreasing their lability in the multistep displacement process. The ionic radii of Ca^{2+} (118 pm) and Pr^{3+} (117.9 pm) are similar, and accordingly their relative labilities may also be rationalization on the basis of surface charge density.

The decomplexation of trivalent lanthanide complexes of 1,10-diaza-4,7,13,16-tetraoxacyclooctadecane-N-acetic acid (K22MA) and 1,10-diaza-4,7,13,16-tetraoxacyclooctadecane-N,N'-di-β-propionic acid (K22DP) (which are diaza crown ethers with one and two carboxylic acid pendant arms attached to the diaza nitrogens, respectively) has been studied by a spectrophotometric SF method, and is characterized by the rate law (64). When the diaza crown ether is K22MA, $k_d = 0.88$, 0.25, and 3.99 s^{-1}, respectively, for La^{3+}, Eu^{3+}, and Lu^{3+}; and $k_H = 1.42 \times 10^4$ and $4.93 \times 10^3 \text{ dm}^2 \text{ mol}^{-1} \text{ s}^{-1}$ for La^{3+} and Eu^{3+}; and when the diaza

$$k_{obs} = k_d + k_H[\text{H}^+] \tag{64}$$

crown ether is K22DP, $k_d = 0.206$, 0.011, and 0.612 s^{-1}, respectively, for La^{3+}, Eu^{3+}, and Lu^{3+}; and $k_H = 2.45 \times 10^3$ and $43 \text{ dm}^3 \text{ mol}^{-1} \text{ s}^{-1}$ for La^{3+} and Eu^{3+} [at 298.2 K and $I = 0.10 \text{ mol dm}^{-3}$ (LiClO$_4$)].[78]

9.3.4. Ligand Substitution on Complexes of Oxo Metal Ions

The formation of peroxovanadium(V) in strongly acidic aqueous solution is characterized by the rate law (65) where, at 298.2 K, $k_1 = 60.2 \text{ s}^{-1}$, $\Delta H^+ = 21.1 \text{ kJ mol}^{-1}$, $\Delta S^+ = -140 \text{ J K}^{-1} \text{ mol}^{-1}$ and $\Delta V^+ = 9.9 \text{ cm}^3 \text{ mol}^{-1}$; $k_2 = 3.47 \times 10^3 \text{ dm}^3 \text{ mol}^{-1} \text{ s}^{-1}$, $\Delta H^+ = 46.5$, $\Delta S^+ = -21$, and $\Delta V^+ = 2.8$; $k_3 = 1.63 \times 10^3 \text{ dm}^6 \text{ mol}^{-2} \text{ s}^{-1}$, $\Delta H^+ = 20.1$, $\Delta S^+ = -116$, and $\Delta V^+ = 14.2$, as shown from a high-pressure spectrophotometric SF study.[79] The k_1, k_2, and k_3 paths are attributed to substitution on a protonated species (presumably protonated at the

$$d[\text{VO}(\text{O}_2)^+]/dt = (k_1[\text{H}^+]^{-1} + k_2 + k_3[\text{H}^+])[\text{VO}_2^+][\text{H}_2\text{O}_2] \tag{65}$$

are attributed to substitution on a protonated species (presumably protonated at the oxo group of VO_2^+), direct substitution on VO_2^+, and substitution on a hydrolyzed VO_2^+ species, respectively. The positive ΔV^+ values are thought to indicate the formation of an expanded transition state with a distorted bipyramidal structure. The formation of the diperoxo species is characterized by the rate law (66) where

$k_4 = 3.49 \times 10^3$ dm^3 mol^{-1} s^{-1}, $\Delta H^{\ddagger} = 40.2$ kJ mol^{-1}, $\Delta S^{\ddagger} = -42$ J K^{-1} mol^{-1}, and $\Delta V^{\ddagger} = 0.0$ cm^3 mol^{-1}. It is deduced from $\Delta V^{\ddagger}$ that the penetration of the oxygen

$$d[VO(O_2)_2^-]/dt = k_4[VO(O_2)^+][H_2O_2] \tag{66}$$

atom of H_2O_2 into the first coordination sphere is compensated for by V—OH$_2$ bond elongation in an interchange substitution mechanism.

The formation of the tungsten(VI) oxo diperoxo species in aqueous HClO$_4$ solution is governed by the rate law (67), where $k = 2.4 \times 10^3$ dm^3 mol^{-1} s^{-1} at 298.2 K.[80] It is thought that the reaction sequence occurs in steps (68)–(70).

$$d[WO(O_2)_2(OH)(H_2O)^-]/dt = k[W(VI)][H_2O_2]] \tag{67}$$

$$[WO_2(OH)_2(H_2O)_2] + H_2O_2 \xrightarrow{\text{slow}} [WO_2(O_2)(H_2O)_3 + H_2O] \tag{68}$$

$$[WO_2(O_2)(H_2O)_3] \underset{}{\overset{\text{fast}}{\rightleftharpoons}} [WO_2(O_2)(OH)(H_2O)_2] + H^+ \tag{69}$$

$$[WO_2(O_2)(OH)(H_2O)_2]^- + H_2O_2 \xrightarrow{\text{fast}} [WO(O_2)_2(OH)(H_2O)]^- + 2H_2O \tag{70}$$

9.4. Ligand Substitution Processes in Dimeric and Higher Oligomeric Metal Complexes

The dirhodium(II) tetrakis(μ-carboxylate) complexes have been the subject of numerous studies as a consequence of their catalytic and antitumor activities, quite apart from their own intrinsic interest. The four bridging carboxylate ligands are inert to substitution, but the two axial ligands at either end of the dirhodium axis are labile. Thus it has been shown that the axial ligand substitution reactions of the diaqua dirhodium(II) tetracetate species proceeds in the two steps shown in Eqs. (71) and (72), where the rate-determining formation of

$$[Rh_2(O_2CCH_3)_4(H_2O)_2] + PR_3^{n+} \underset{k_{-1}}{\overset{k_1}{\rightleftharpoons}} [Rh_2(O_2CCH_3)_4(H_2O)(PR_3)]^{n+} + H_2O \tag{71}$$

$$[Rh_2(O_2CCH_3)_4(H_2O)(PR_3)]^{n+} + PR_3^{n+} \underset{k_{-2}}{\overset{k_2}{\rightleftharpoons}} [Rh_2(O_2CCH_3)_4(PR_3)_2]^{2n+} + H_2O \tag{72}$$

$[Rh_2(O_2CCH_3)_4(H_2O)(PR_3)]^{n+}$ is characterized by the kinetic parameters (determined by spectrophotometric SF) given in Table 9.13.[81] The small dependence of the kinetic parameters on the nature of the substituting ligand is considered to be evidence for the operation of an I_d mechanism.

The use of soluble rhodium(II) carboxylate catalysts in the cyclopropanation of olefins is the origin of a very detailed crystallographic and spectrophotometric SF study of dirhodium(II) tris- and tetrakis(tritolylbenzoate), and tetrakis(pivalate).[82] The Rh(II)–Rh(II) moiety at the Rh$_2$(O$_s$C)$_4$ core resides inside a parallelepipedic box formed by the carboxylate ligands such that smaller ligands, such as pyridine, can coordinate at either end of the dirhodium axis. The formation

Table 9.13. Kinetic Parameters for the Axial Ligand Substitution Reactions of Dirhodium(II) Tetraacetate in Aqueous Solution at $I = 0.10 \, mol \, dm^{-3}$ (LiClO$_4$) and 298.2 K [81]

Ligand	$10^{-5} k_1$ (dm^3 mol^{-1} s^{-1})	ΔH^+ (kJ mol^{-1})	ΔS^+ (J K^{-1} mol^{-1})
P(CH$_2$CH$_2$CN)$_3$	8.85	36	−10
Ph$_2$P(CH$_2$CH$_2$COOH)	7.4		
Ph$_2$P(m-SO$_3$Ph)$^-$	6.48	35	−16
Ph$_2$P(CH$_2$CH$_2$COO)$^-$	4.41	40	−3.3
P(CH$_2$CH$_2$COO)$_3^{3-}$	1.32	43	−2.9
Ph$_2$P(CH$_2$CH$_2$NH(CH$_3$)$_2$)$^+$	5.83	33	−22
PhP(CH$_2$CH$_2$CH$_2$NH$_3$)$_2^{2+}$	3.01	41	−0.8
Imidazole	71		
L-Histidine	3.6		
5'-AMP	29		
Isonicotinic acid	10		
Nicotinic acid	8		

of mono and pyridine adducts according to Eqs. (73) and (74) in dichloroethane

$$A(H_2O)_2 + py \underset{k_{-1}}{\overset{k_1}{\rightleftharpoons}} A(H_2O)(py) + H_2O \tag{73}$$

$$A(H_2O)(py) + py \underset{k_{-2}}{\overset{k_2}{\rightleftharpoons}} A(py)_2 + H_2O \tag{74}$$

at 298.2 K are characterized for A = Rh$_2$(tolylbenzoate)$_3$(O$_2$CCH$_3$), Rh$_2$(tolylbenzoate)$_4$ and Rh$_2$(pivalate)$_4$, respectively, by $k_1 = 5 \times 10^4$, 4×10^3, and 1.5×10^7 dm^3 mol^{-1} s^{-1}; $k_{-1} = 5 \times 10^{-5}$, 3×10^{-4}, and 1×10^1 s^{-1}; $k_2 = 1.7 \times 10^5$, 1.5×10^5, and $>5 \times 10^7$ dm^3 mol^{-1} s^{-1}; and $k_{-2} = 2.0$ and 7 s^{-1} (no determination in the third case). The variation of k_1 is attributed to the large steric hindrance the tritolylbenzoate bridges cause to substitution of water by pyridine, which could be attributable to either steric hindrance of the leaving ligand, or of the entering ligand, or both. Similar trends in relative lability are observed for the other three substitution processes.

Regioselective ligand exchange of in-plane acetate ligands on octakis(μ-acetato-O,O')tetraplatinum(II) complex shown in Eq. (75) is characterized by the two-term rate law (76) where, in CDCl$_3$ at 298.2 K, $k_0 = 0.36$ s^{-1} and $k_1 = 4.4$ dm^3 mol^{-1} s^{-1}, as determined by a ^{1}H NMR spin saturation transfer method, but no exchange is observed for the out-of-plane acetates.[83] Molecular orbital

$$[Pt_4(CH_3COO)_8] + {}^*CH_3COOH \rightleftharpoons [Pt_4(CH_3COO)_4({}^*CH_3COOH)] + CH_3COOH \tag{75}$$

$$k_{obs} = 4(k_0 + k_1[CH_3COOH]) \tag{76}$$

calculations indicate that the in-plane acetates are less strongly bound and their greater lability is attributed to this cause.

Part III

Reactions of Organometallic Compounds

Chapter 10

Substitution and Insertion Reactions

10.1. Substitution Reactions

10.1.1. Introduction

The systematic behavior of the rates of associative reactions of metal carbonyls with P-donor nucleophiles has continued to receive attention.[1] Quantification of the standard or intrinsic reactivity of the carbonyl and its sensitivity to electronic and steric properties of the nucleophiles can provide a full dynamic characterization of each compound. A thorough analysis of a large number of P-donor ligands according to their purely σ-donating capacity or their combined σ-basicity and π-acidity has been effected[2] by correlation of reduction potentials and ν_{co} values for $[(\eta^5\text{-Cp})\text{FeL(CO)COMe}]$, $[(\eta^5\text{-MeCp})\text{FeL(CO)Me}]$, and $[(\eta^5\text{-Cp})\text{FeL(CO)Me}]$. Π-acids show higher values of ν_{co} than are predicted by the good correlation between ν_{co} and $E°$ values for those ligands concluded to be pure σ-donors. Surprisingly $P(OMe)Ph_2$ and $P(OEt)Ph_2$ are not π-acids but $P(CH_2CH_2CN)_3$, $P(CH_2CH_2CN)_2Ph$, and PPh_2H are. A good correlation of χ (relative ν_{co} values obtained by Bartik et al. for $[\text{Ni(CO)}_3\text{L}]$ complexes) and pK_a values is obtained for most σ-donor ligands if a steric correction term is applied to the pK_a values in terms of Tolman cone angles [Eq. (1)] and χ values or sterically corrected pK_a values ($pK_a + 0.047\theta$) should be used when correlating rate constants for associative reactions with the σ-donicity of P-donor nucleophiles.

$$pK_a = 18.9 - 0.675\chi - 0.047\theta \tag{1}$$

Estimates of the geometrical shielding of metal clusters by the CO ligands have been made.[3] $[\text{Os}_3(\text{CO})_{12}]$ is shielded by 99% for nucleophiles approaching along an axis perpendicular to the Os_3 plane as compared to only 85% shielding for approach in the Os_3 plane.

Systematics of dissociative reactions of metal carbonyls as a function of the size of ancillary substituents have also been considered.[4] Variously shaped steric profiles (plots of log k, corrected for electronic effects, against θ) can be obtained depending on the onset (steric threshold) and magnitude of steric effects on the ground state compared with the same effects in the transition state. The most likely behavior would seem to be an earlier onset of steric effects in the ground state together with a larger magnitude, when they come into play. Other possibilities are considered, however, and an example is claimed for the situation where the steric threshold is lower in the ground state but the steric effect on the transition state is greater when the steric thresholds are exceeded. The choice of electronic parameters for ancillary ligands in such reactions is limited largely to values of χ_{CO} in the complexes studied or to values of χ_{CO} or $\delta(^{13}CO)$[5] for $[Ni(CO)_3L]$. The first is more suitable in principle because it would more accurately reflect the net electron donor capacity (σ-donicity minus π-acidity) of the substituent. However, they are not always known with suitable precision so the latter two parameters are used instead in spite of differences in the balance of σ-donicity and π-acidity that might be expected to operate in the compound being examined and the $[Ni(CO)_3L]$ standard. That this can certainly be the case is shown by the π-acidity of PMe_3 in $[(\eta^5\text{-}Cp)MoX_2(PMe_3)_2]$,[6] and of PPh_3 in $[PtCl_3(PPh_3)]^-$,[7] where there are no CO ligands to compete with the P-donor.

Doubt has been raised regarding the D nature of apparently dissociative reactions. A growing number of studies[8] confirm there is appreciable bonding of solvent molecules to "vacant" coordination sites. Lifetimes of coordinatively unsaturated intermediates are arguably so short that they would not exist long enough to wait for the approach of a ligand and dissociative reactions would have to be formulated as in reactions (2) and (3) where *both* reactions would be I in nature.

$$M-L+S \rightleftharpoons M-S+L \qquad (2)$$

$$M-S+L' \rightarrow M-L'+S \qquad (3)$$

A particularly striking illustration of this is provided by enthalpies for CO dissociation from $[M(CO)_6]$ (M = Cr, Mo, and W), obtained by photoacoustic calorimetry.[9] For M = Cr, $\Delta H° = 28$, 27, 26, and 24 kcal mol^{-1} in pentane, heptane, iso-octane, and cyclohexane; and for M = Mo and W, values in heptane were 32 and 33 kcal mol^{-1}. Comparison with activation enthalpies for CO dissociation in the gas phase provides estimates of 10, 9, and 13 kcal mol^{-1} for the strength of agostic bonding of heptane in $[M(CO)_5(heptane)]$ (M = Cr, Mo, and W, respectively). Activation parameters for CO dissociation in alkane solvents are very similar (M = Cr) or smaller by about 8 and 6 kcal mol^{-1} (M = Mo and W) than those in the gas phase, and it is suggested CO dissociation is accompanied by entry of a solvent molecule in an I_d process for $[Cr(CO)_6]$ and I_a processes for $[Mo(CO)_6]$ and $[W(CO)_6]$. These results mean assignments of D mechanisms may have to be reformulated quite generally as I_d or even I_a solvations. Our ideas about mechanisms of substitution reactions of organometallics would then be reverting to those held for a long time for aquation and anation reactions etc. of Werner complexes in aqueous solution.

The calculated preference for *cis* vs. *trans* substitution of CO in $[Cr(CO)_5L]$ (L = PH_3 or NH_3) and $[Mn(CO)_5X]$ (X = Cl or H) is between 3 and 9 kcal mol^{-1}.[10] This is made up of ground and transition state effects, and relaxation of the products occurs though it falls short of forming a trigonal bipyramidal geometry. Relaxation becomes more important with increasing π-donicity of the substituent and this increases the *cis* effect by strengthening the bond to the *trans* CO. Conversion of the *trans*-loss intermediate to the *cis*-loss intermediate requires ~10 kcal mol^{-1}, but fluxionality of the latter can occur in such a way that ^{13}CO ligands introduced into a *cis* position by *cis* labilization can readily end up in a *trans* position. This is fully in accord with microreversibility restrictions.[11]

An important set of papers directly relevant to our ideas about dissociative reactions have been published as a "Polyhedron Symposium in Print" on metal—ligand bonding energetics in organotransition metal compounds.[12] Time-resolved photoacoustic calorimetry has provided Cr—L bond dissociation energies (BDE) in $[Cr(CO)_5L]$ where L = a variety of solvents or more conventional ligands.[13] The Cr-heptane bond strength in $[Cr(CO)_5(heptane)]$ is ~10 kcal mol^{-1} while the value of ΔH^+ for displacement of the heptane by pyridine is only 5.2 ± 0.4 kcal mol^{-1} and this indicates an associative mechanism. The negligible value of ΔS^+ implies bond making is offset by a high degree of bond breaking. A number of BDEs for metal carbonyls in the gas phase have been derived from laser powered homogeneous pyrolysis.[14] The value for $[Cr(CO)_6]$ of 36.8 kcal mol^{-1} supports the I_d mechanism for the solution reactions, and the 38 ± 2 kcal mol^{-1} for $[Mn_2(CO)_{10}]$ agrees with solution phase kinetics[15] and implies the solvent does not contribute to the energetics. Many metal-C and other BDEs are reported for a wide variety of metals and ligands.[16] A detailed analysis of the importance of cage and other solvent effects that cloud the significance of BDEs derived from kinetics is also reported.[17] The enthalpies of addition (in kcal mol^{-1}) of $H_2(-9.9)$, $N_2(-13.5)$, MeCN(-15.1), py(-18.9), $P(OMe)_3(-26.5)$, and CO(-30.4) to $[(PCy_3)_2W(CO)_3]$ in toluene have been measured.[18] These are much lower than expected from gas phase studies and imply agostic interaction (~10 kcal mol^{-1}) of a cyclohexyl group with the apparently vacant coordination site.

10.1.2. Mononuclear Complexes

10.1.2.1. Displacement of Carbon Monoxide

Kinetic parameters for [nucleophile]-independent reactions of $[Mn(CO)_5X]$ (X = Cl, Br) in toluene have been measured.[19] $\Delta V^+ = +20.6$ cm^3 mol^{-1} for X = Cl but the reactions are somewhat faster than those in 1,2-dichloroethane, and the values of ΔS^+ in toluene are significantly less positive. A D mechanism is assigned, although the solvent effects may indicate an I mechanism with possibly some perceptible I_a character in toluene.

Substitution kinetics of $[Os(CO)_5]$ in decalin have been reported and compared with those of $[Fe(CO)_5]$ and $[Ru(CO)_5]$.[20] These show the same "triad effect" ($[Fe(CO)_5] \ll [Ru(CO)_5] > [Os(CO)_5]$) seen in the group-six hexacarbonyls. Calculations of M—CO bond strengths are quoted that agree with relative labilities. However in view of the possible assignment of I_a mechanisms to $[Mo(CO)_6]$ and $[W(CO)_6]$,[9] and the negligible value of ΔS^+ for $[Os(CO)_5]$, some associative solvation may be causing the triad effect. The extreme inertness of $[Fe(CO)_5]$ is ascribed to the paramagnetic nature of the $[Fe(CO)_4]$ intermediate. However, if the diamagnetic form of $[Fe(CO)_4]$ has a higher enthalpy, then the value of ΔH^+ for formation of this intermediate would be even higher than the observed value. The value of ΔH^+ for CO loss from $[Fe(CO)_5]$ is not much higher than that for $[Mn_2(CO)_{10}]$ or $[Cr(CO)_6]$. It may be we should ask why ΔH^+ values for $[Ru(CO)_5]$ and $[Os(CO)_5]$ are so low, and the answer may lie again in the associative nature of a rate-determining solvation. The effects of substituents in $[Ru(CO)_4L]$ on CO dissociation have been described.[21] Electronic effects are small and rates increase systematically with the cone angles of L. Values of ΔH^+ are closely similar for most of the substituents, so steric accelerations manifest themselves through more positive entropies. The quantified steric effects of L, compared with those for mer-$[Ru(CO)_3L(SiCl_3)_2]$, cis-$[Mn(CO)_4LBr]$, and $[Ni(CO)_3L]$, show that they increase with increasing coordination number and, possibly, with increasing bulkiness of other ligands present.

Reactions of trans-$[W(CO)_4(NO)X]$ (X = Cl, Br, I) with PPh_3 are dissociative in nature,[22] and about 100 times faster than corresponding reactions of the isoelectronic $[Re(CO)_5X]$ because of slightly more favorable values of ΔH^+ and ΔS^+. Rates decrease, and values of ΔH^+ increase, along the series X = Cl > Br > I as is observed also for $[Re(CO)_5X]$. Reactions with the more basic nucleophile $P(n\text{-}Bu)_3$ include a term first order in $[P(n\text{-}Bu)_3]$ not seen in reactions of $[Re(CO)_5X]$, and this is presumably due to the presence of the NO ligand. Values of ΔH_2^+ increase quite substantially, and values of ΔS_2^+ become substantially less negative, along the series X = Cl < Br ≤ I. This indicates a decrease of associative character with increasing polarizability of X, possibly because Br and I are effective in diluting the helpful electron-withdrawing ability of the NO. Associative rates are generally much slower than for $[V(CO)_5(NO)]$, which has a weakly bonded CO trans to the NO unlike trans-$[W(CO)_4(NO)X]$. $[W(CO)_3(NO)P(n\text{-}Bu)_3Br]$ reacts with $P(n\text{-}Bu)_3$ only by a dissociative path and ΔH^+ is much higher, and ΔS^+ more positive, than for the dissociative reaction of the parent complex.

Reactions of $[M(CO)_5SR]^-$ (M = Cr, Mo, W; R = H, Ph) in THF with P-donors lead to facile replacement of cis CO ligands at ambient temperatures, thus illustrating the very strong cis-labilization by the thiolate ligands.[23] In the absence of P-donors the complexes dimerize rapidly and reversibly to form $[M(CO)_4SR]_2^{2-}$ which also undergo reactions with P-donors to form substituted mononuclear products. Rates of substitution into $[M(CO)_5SR]^-$ increase along the series M = Mo > Cr > W in accordance with the triad effect. No kinetic studies were carried out to investigate the effects, if any, of [P-donor] on the rates.

The complex $[(\eta^5\text{-}Cp^*)_2TiClCO]$ disproportionates according to Eq. (4) via reactions (5)–(7)[24] with $k_1 = 5 \times 10^{-3}\ s^{-1}$ at 5 °C, $\Delta H_1^+ \sim 20\ kcal\ mol^{-1}$, $\Delta S_1^+ \sim$

$$2[Cp^*_2TiCl(CO)] \longrightarrow [Cp^*_2TiCl_2] + [Cp^*_2Ti(CO)_2] \tag{4}$$

$$[Cp^*_2TiCl(CO)] \underset{k_{-1}}{\overset{k_1}{\rightleftharpoons}} [Cp^*_2TiCl] + CO \tag{5}$$

$$[Cp^*_2TiCl] + [Cp^*_2TiCl(CO)] \overset{k_2}{\longrightarrow} [Cp^*_2TiCl_2] + [Cp^*_2Ti(CO)] \tag{6}$$

$$[Cp^*_2Ti(CO)] + CO \overset{fast}{\longrightarrow} [Cp^*_2Ti(CO)_2] \tag{7}$$

25 cal K^{-1} mol^{-1}, and $k_{-1}/k_2 \sim 2$. These values can be compared with $\Delta H^+ = 29$ kcal mol^{-1} and $\Delta S^+ = 14$ cal K^{-1} mol^{-1} reported for dissociative loss of CO from $[Cp^*_2Ti(CO)_2]$ and the results are compared with earlier work on the CO-induced disproportionation of $[Cp_2TiCl]_2$.

Isomerizations of *ttt*-$[RuCl_2(CO)_2L_2]$ complexes [L = $P(CH_2Ph)_3$, PPh_2Me, $PPhMe_2$, and PMe_3] are strongly inhibited by the presence of free CO.[25] They are proposed to occur by CO dissociation as shown in Scheme 1 where $k_3 > k_4$,

Scheme 1

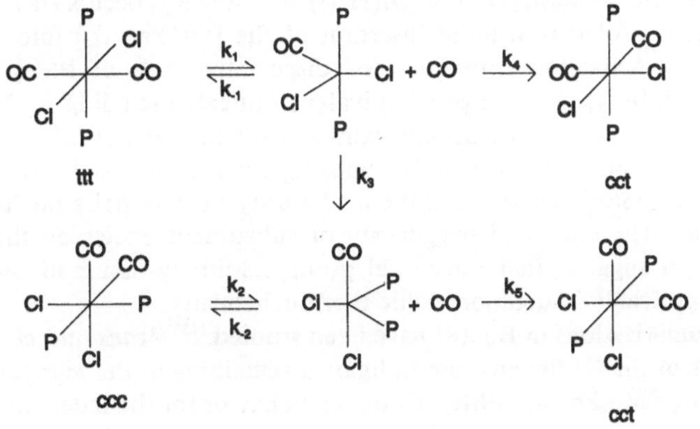

so that the *cct* isomer is formed indirectly after Berry rotation of the first trigonal bipyramidal intermediate. For L = $P(CH_2Ph)_3$, $k_5 > k_{-2}$ so that little *ccc* is formed, but for other P-donors $k_5 < k_{-2}$ so that *ccc* is formed in appreciable amounts before the final *cct* product appears. Activation parameters for isomerization of the complexes are shown in Table 10.1, and the rates are clearly entropy controlled and likely to be determined largely by the relative populations of the two TBP intermediates and the relative probabilities of attack by CO at the various accessible sites in the trigonal plane. The dependence of the rates on the nature of the substituents is, therefore, quite difficult to rationalize. The large positive values for the two measurements of ΔV^+ imply that the transition states do correspond to the loss of CO, with quite late transition states, and that the pressure dependencies of the subsequent competing steps are quite small.

Table 10.1. Activation Parameters for Isomerization of [RuCl$_2$(CO)$_2$L$_2$] Complexes

L	Isomerization	$10^5 k_{obs}$ (s^{-1}) at 100 °C	ΔH^+ (kcal mol^{-1})	ΔS^+ (cal K^{-1} mol^{-1})	ΔV^+ (cm^3 mol^{-1})
P(CH$_2$Ph)$_3$	ttt → cct^a	1200	30.1 ± 0.5	+13 ± 2	
PPh$_2$Me	ttt → ccc^a	1500	27.0 ± 1.0	+5 ± 3	+19 ± 2
	ccc → cct^b	50	25.7 ± 1.0	−5 ± 4	
PPhMe$_2$	ttt → ccc^c	150	27.2 ± 1.0	+1 ± 4	+19 ± 2
	ccc → cct^b	2.4	22.8 ± 0.5	−19 ± 3	
PMe$_3$	ttt → ccc^c	30	25.7 ± 1.0	−6 ± 10	

^{a} In chloroform.
^{b} In tetrachloroethane.
^{c} In chlorobenzene.

10.1.2.2. Displacement of Other Monodentate Ligands

[Os(PMe$_3$)$_5$] undergoes dissociative exchange[26] with P(CD$_3$)$_3$; $\Delta H^+ \sim$ 27 kcal mol^{-1} and it is much more labile than [Os(CO)$_5$].[20] In the absence of free PMe$_3$, metallation to form [Os(PMe$_3$)$_3$(H)(η^2-CH$_2$PMe$_2$)] occurs via a dissociative path in various solvents without insertion of the [Os(PMe$_3$)$_4$] into any solvent C—H bonds. Activation enthalpies for dissociative loss of PMe$_3$ from [(η^5-Cp*)Ru(X)(PMe$_3$)$_2$] (X = range of univalent radicals including C, N, O, and S donors, halides, and H) provide approximate estimates of BDEs.[27] They cover a range of 19 kcal mol^{-1} with generally opposing changes in $T\Delta S^+$ covering a range of 6 kcal mol^{-1}. Steric effects, and the availability of lone pairs on X, accelerate the reactions. The scale and irregularity of substituent effects on the Ru–PMe$_3$ bond strength suggests that functional group additivity assumptions cannot be generally applicable in organometallic thermochemistry.

The isomerizations in Eq. (8) have been studied.[25] *Trans* and *cis* refer to the dispositions of the C1 ligands, the L$_3$ ligands remaining in the *mer* configuration; L = PMePh$_2$, PMe$_2$Ph, and PMe$_3$. Complex behavior for the latter two complexes

$$trans\text{-}[RuCl_2(CO)L_3] \rightarrow cis\text{-}[RuCl_2(CO)L_3] \qquad (8)$$

indicated the generation of varying but undetectably small quantities of free L in solution. Rates are decreased by the presence of added free phosphine. A similar mechanism to that described above for the bis-phosphine complexes was proposed and the values of ΔS^+ (8–14 cal K^{-1} mol^{-1}) and ΔV^+ (+15 to +16 cm^3 mol^{-1}) confirmed the importance of L-dissociation in the rate-determining step. The intermediates generated by dissociation of L from *trans*-[RuCl$_2$(CO)L$_3$], or of CO from *ttt*-[RuCl$_2$(CO)$_2$L$_2$], are good catalysts for hydrogenation of alkenes, their effectiveness decreasing with increasing basicity and decreasing size of L.[28] Solid-phase isomerizations of [RuCl$_2$(CO)$_2$L$_2$] and [RuCl$_2$(CO)L$_3$] complexes have also been reported.[29]

Reactions of the methylidene complex [(η^5-Cp)$_2$Ti(CH$_2$)PMe$_3$] proceed via reversible dissociation of PMe$_3$ to form [(η^5-Cp)$_2$Ti(CH$_2$)], which then reacts with other P-donors, alkenes, alkynes, and CO to form a variety of organometallic products.[30] The rate of loss of PMe_3 is $\sim 9 \times 10^{-3}$ s^{-1} at 25 °C in C_6D_6 and addition of PMe_3 to the intermediate is ~ 10 times faster than addition of 3-hexyne. The relative stabilities of different P-donor complexes is $PMe_3 > PMe_2Ph \gg PEt_3$, which shows the importance of steric destabilization. Reaction with $Me_2C=CH_2$ forms the metallacyclobutane derivative $[(\eta^5-Cp)_2Ti(\eta^2-CH_2CMe_2CH_2)]$ in a mixture governed by an equilibrium constant of $\sim 2 \times 10^{-2}$.

The dissociative loss of py from *fac*-$[Mo(CO)_3(diphos)(py)]$ in toluene is governed by $k_1(25\,°C) = 4 \times 10^{-3}$ s^{-1}, $\Delta H_1^{\ddagger} = 25.47$ kcal mol^{-1}, and $\Delta S_1^{\ddagger} = 16.1$ cal K^{-1} mol^{-1}.[31] Dissociative loss of py from $[W(CO)_3(PCy_3)_2py]$ in toluene to form the apparently five-coordinate complex $[W(CO)_3(PCy_3)_2]$ has been shown[32] to be governed by $k_1(25\,°C) = 0.13$ s^{-1}, $\Delta H_1^{\ddagger} = 23.4 \pm 0.4$ kcal mol^{-1}, and $\Delta S_1^{\ddagger} = 15.9 \pm 1.3$ cal K^{-1} mol^{-1}. When PCy_3 is replaced by the corresponding per-deuterated ligand, a temperature-independent kinetic isotope effect, $k_H/k_D = 1.21 \pm 0.01$, is observed, this indicating substantial agostic interaction between a cyclohexyl group and the developing "vacant" coordination site. Rates of *addition* of a number of ligands to $[W(CO)_3(PCy_3)_2]$ were measured independently (Table 10.2) and the importance of steric effects is evident. Inverse kinetic isotope effects ($k_H/k_D = 0.86$ and 0.77, respectively) were found for reactions with $P(OMe)_3$ and PPh_2Me and are consistent with a strengthening of the C—H and C—D bonds as the agostic bonds between the cyclohexyl groups and the metal are broken. The approximate equality of k_H/k_D for loss of py with k_D/k_H for addition of $P(OMe)_3$ or PPh_2Me might suggest that the agostic bonding is about 50% complete in the transition states. The values of ΔH^+ (~ 4 kcal mol^{-1}) and ΔS^+ (~ -25 cal K^{-1} mol^{-1}) for addition of $P(OMe)_3$ reflect the relative weakness of the agostic bonding compared with the incipient W—P bond.

Flash photolysis of $[Cr(CO)_5(Z\text{-cyclooctene})]$ in toluene and in the presence of 4,4'-dimethyl-2,2'-bipyridine (dmbpy) forms $[Cr(CO)_5(\mu\text{-dmbpy})Cr(CO)_5]$ initially and, subsequently, $[Cr(CO)_6]$ and $[Cr(CO)_4(\eta^2\text{-dmbpy})]$ via reactions (9)-(12).[33] Relative yields of $[Cr(CO)_4(\eta^2\text{-dmbpy})]$ and $[Cr(CO)_6]$ depend on [CO] and [dmbpy] as would be expected from this scheme. Rates for loss of the

$$[(OC)_5Cr(\mu\text{-dmbpy})Cr(CO)_5] \rightleftharpoons [Cr(CO)_5(\eta^1\text{-dmbpy})] + [Cr(CO)_5] \qquad (9)$$

$$[Cr(CO)_5(\eta^1\text{-dmbpy})] \rightleftharpoons dmbpy + [Cr(CO)_5] \qquad (10)$$

$$[Cr(CO)_5(\eta^1\text{-dmbpy})] \rightarrow [Cr(CO)_4(\eta^2\text{-dmbpy})] + CO \qquad (11)$$

$$[Cr(CO)_5] + CO \rightarrow [Cr(CO)_6] \qquad (12)$$

dinuclear complex in the absence of any free added CO or excess dmbpy lead to $k_{obs} = 3.3 \times 10^{-4}$ s^{-1} at 23 °C, $\Delta H^+ = 26.3$ kcal mol^{-1}, and $\Delta S^+ = 12$ cal K^{-1} mol^{-1}, although the significance of these values in view of the complexity of the reaction

Table 10.2. Rates of Addition of Ligands to $[W(CO)_3(PCy_3)_2]$ *in Toluene at 25 °C*

L	py	$P(OMe)_3$	PPh_2Me	2,5-Me_2py
$10^{-4}k_2\,(M^{-1}\,s^{-1})$	100	5.5	1.7	0.86

scheme is not clear. It is unfortunate that the temperature dependence of the rates at high (~ 0.1 M) [dmbpy] were not measured, since the rates then appear to be independent of the values of [dmbpy] and consistent with rate-determining reaction (9) and efficient scavenging of the $[Cr(CO)_5]$ by the dmbpy.

Rates of dissociative loss of L from the complexes $trans$-$[Co(Hdmg)_2RL]$ have been measured where R = Me, Et, i-Pr, or adamantyl, and L = aniline, 1,5,6-trimethylbenzimidazole, 4-cyanopyridine, and $P(OMe)_3$.[34] Structural data were also obtained. The electron-donor capacity, EP, of R was estimated from ^{13}C NMR data and rates generally increase with EP as do the Co—L bond lengths, suggesting that the $trans$ effect operates largely on the ground state. When R = adamantyl, however, the rate is slower than expected from its EP. The steric effect on the ground state, caused by interaction of the adamantyl group with the Hdmg$^-$ ligands, is so great that further out-of-plane bending of the Hdmg$^-$ ligands is difficult and small in the transition state, i.e., the steric effect has effectively reached near-saturation in the adamantyl complex. The existence of steric effects for complexes containing the other R groups is evident from the observed extent of Hdmg$^-$ bending in the structures.

Reductive elimination of ethane from $[PtMe_4L_2]$ (L = MeNC or 2,6-di-Me-C_6H_3NC) proceeds in benzene mainly after L dissociation.[35] For loss of MeNC at 60 °C the rate constant is 1.7×10^{-3} s^{-1}, and for $Me_2C_6H_3NC$ it is 0.9×10^{-3} s^{-1}. The values of $\Delta H^{\pm}$ and $\Delta S^{\pm}$ for loss of MeNC are 16.3 ± 0.8 kcal mol^{-1} and -21 ± 3 cal K^{-1} mol^{-1}, the low $\Delta H^{\pm}$ and negative $\Delta S^{\pm}$ being ascribed to agostic interactions, between a CH$_3$ ligand and the Pt atom, that develop as the MeNC leaves. The addition of the isonitrile ligands to the intermediate is faster than reductive elimination by a factor of $\sim 10^6$.

Some interesting complexes $[(H_3N)_5Os(arene)]^{2+}$ have been synthesized.[36a] The arenes are η^2 bonded and contain a wide variety of substituents. Rates of tautomerization at 20 ± 2 °C vary from 9×10^{-1} s^{-1} for 2,3-η^2-PhOMe $\rightleftharpoons$ 5,6-η^2-PhOMe to 1×10^4 s^{-1} for 1,2-η^2-PhH $\rightleftharpoons$ 2,3-η^2-PhH. These rather slow rates are ascribed to losses of π-bonding when the presumed arenium intermediate is formed. The rate of replacement of η^2-C_6H_6 is virtually independent of solvent or incoming ligand with $t_{1/2} \sim 20$ h at 25 ± 3 °C so the reactions are largely dissociative. Both electron-withdrawing and electron-donating substituents stabilize the η^2-bond and σ and π bonding of the arene is implied. 1,2,3,4-Tetramethylbenzene is more labile than benzene but none of these substituent effects exceeds a factor of ~ 20. Cyclic voltammetry allows measurement of the dynamics of the corresponding Os(III) complexes. For these complexes electron-withdrawing substituents increase the rates and electron-donating ones decrease them. This reflects the enhanced role of σ-bonding and diminished role of π-bonding when the oxidation state is increased. Arene groups with substituents containing σ donor atoms such as N or O can undergo isomerization to N or O donor isomers at rates competitive with, or greatly in excess of, rates of arene dissociation. Catalytic hydrogenation[36b] of benzene in $[(H_3N)_5Os(\eta^2-C_6H_6)]^{2+}$ in the presence of a Pd(0)/C catalyst leads quantitatively to the very inert $[(H_3N)_5Os(\eta^2$-cyclohexene$)]^{2+}$; oxidation to Os(III) increases lability. $[(H_3N)_5Ru(\eta^2-C_6H_{10})]^{2+}$ is much more labile than its Os analog, the cyclohexene dissociating with $t_{1/2} \sim 15$ min at 20 °C. Although alkenes are usually more labile than P-donor ligands, reactions of $[(Ph_3P)_2Pt(\eta^2\text{-}C_2H_4)]$ with bulky P-donors ($\theta \geq 142°$) at $-30\,°C$ lead to displacement of a PPh_3 ligand.[37] At room temperature, loss of C_2H_4 competes with PPh_3 replacement showing that the former process involves a higher activation enthalpy.

The coordinated alkyne in $[(\eta^5\text{-}Cp)Ru(PMe_3)_2(\eta^2\text{-}MeC_2H)]^+$ has been shown[38] to undergo first-order isomerization to the vinylidene ($=C=CHMe$) complex in acetonitrile with $\Delta H^+ = 23.4 \pm 0.3\,kcal\,mol^{-1}$ and $\Delta S^+ = +3.9 \pm 0.9\,cal\,K^{-1}\,mol^{-1}$ ($t_{1/2}$ at $40\,°C \sim 5\,min$). At higher temperatures this undergoes replacement of the vinylidene by acetontrile to release MeC_2H with $\Delta H^+ = 26.8 \pm 0.7\,kcal\,mol^{-1}$ and $\Delta S^+ = -49\,cal\,K^{-1}\,mol^{-1}$. The intermediacy of a $Ru(H)$-$(\eta^1\text{-}C{\equiv}CMe)$ complex is considered to be possible but unlikely. Reactions of the corresponding ethyne complex in acetonitrile are much slower ($t_{1/2}$ at $60\,°C > 5\,h$) and lead to competitive substitution and isomerization to the vinylidene product. The latter is very stable to ethyne loss with $t_{1/2} > 1$ day at $90\,°C$. A related study[39] has shown that the presumed η^2-alkyne compounds, formed by displacement of N_2 from $trans\text{-}[ReCl(dppe)_2(N_2)]$, are destabilized by repulsive interactions between $d\pi$ metal electrons and the alkyne π-electrons not involved in coordination. This destabilization is sufficiently strong that the η^2-alkyne compounds are not seen, even as intermediates, and the only products are the vinylidene isomers. Theoretical studies are quoted to support an isomerization mechanism in which slippage of the alkyne occurs to form an η^1-alkyne followed by a 1,2-H^+ migration.

Although dissociation of the N_2 ligand has not received the kinetic studies it deserves, dissociative displacement of N_2 from $[(Cy_3P)_2W(CO)_3(N_2)]$ by py in toluene has $k(25\,°C) = 75\,s^{-1}$, $\Delta H^* = 17.8 \pm 0.7\,kcal\,mol^{-1}$, and $\Delta S^+ = 9.8 \pm 2.3\,cal\,K^{-1}\,mol^{-1}$.[40] It is assisted by intramolecular generation of agostic interactions with a cyclohexyl group. These results contrast with undetectable loss of N_2 from $[(diphos)_2W(N_2)_2]$ after $36\,h$ at $70\,°C$. Addition of N_2 to $[(Cy_3P)_2W(CO)_3]$ has $k(25\,°C) = 5.0 \pm 1.0 \times 10^5\,M^{-1}\,s^{-1}$ and $\Delta H^+ = 4.3 \pm 1.7\,kcal\,mol^{-1}$. This value can be compared with $\Delta H^+ = 5.3 \pm 1.2\,kcal\,mol^{-1}$ for displacement of cyclohexane from $[(OC)_5Cr(C_6H_{12})]$ and the latter reaction occurs only seven times faster than addition of N_2 to $[(Cy_3P)_2W(CO)_3]$. Rates of loss of H_2 from $[(Cy_3P)_2W(CO)_3(H)_2]$ have been resolved[40] into rates of isomerization to a dihydrogen intermediate and subsequent dissociation of H_2. Isomerization has $\Delta H^+ = 16.9 \pm 2.2\,kcal\,mol^{-1}$ and $\Delta S^+ = 5 \pm 7\,cal\,K^{-1}\,mol^{-1}$, the equilibrium constant being ~ 0.5 at $25\,°C$. Subsequent dissociation of the H_2 molecule is somewhat more facile with $\Delta H^+ = 14.4 \pm 0.5\,kcal\,mol^{-1}$ and $\Delta S^+ = 1 \pm 2\,cal\,K^{-1}\,mol^{-1}$, the equilibrium constant for this loss being $\sim 2 \times 10^{-4}\,M$. The H/D isotope effect for isomerization is negligible but for H_2 dissociation k_H/k_D is 1.7 ± 0.1. The negligible effect for the isomerization indicates a close balancing of a normal ($k_H/k_D > 1$) isotope effect for weakening the $W{-}H$ and $W{-}D$ bonds, and an inverse effect for formation of $H{-}H$ and $D{-}D$ bonds. The normal isotope effect for loss of H_2 is close to those found generally for reductive elimination of $(H)_2$. An interesting feature is in the process of oxidatively adding H_2 to the W atom; the intermediate $W(H_2)$ complex is formed and loses H_2 many times before oxidative addition occurs. It is also concluded that the reason why oxidative addition of H_2 is slower than CO coordination is due to the energy needed to break the H—H bond rather than due to slower formation of an (H_2) intermediate. Related theoretical studies[41] have estimated $[Co(CO)_5(H_2)]$ to be 37 kcal mol^{-1} more stable than $[Cr(CO)_5(H)_2]$ and ΔH^+ for dissociation of the (H_2) ligand to be ~10 kcal mol^{-1}. $[Cr(CO)_4(H_2)_2]$ is only 29 kcal mol^{-1} more stable than $[Cr(CO)_4(H_2)(H)_2]$, because of the greater π-basicity of the $[Cr(CO)_4(H_2)]$ moiety.

The rates of displacement of chloride from $[(\eta^5\text{-Cp})_2TiCl_2]$ by $X^- = Br^-$, I^-, SCN^-, and CN^- in acetonitrile have been measured.[42] There is no evidence for mixed intermediates and $k_{obs} = k_1 + k_2[X^-]$. The rates increase along the series $X = Br < I < SCN < CN$ showing that the hard Ti(IV) has been made kinetically soft by the Cp ligands. The Cp* analog has much lower k_2 values but only slightly lower k_1 values. The presence of H_2O in the MeCN increases k_1 for reaction with SCN^- because of incipient solvation of the leaving Cl^- but decreases k_2 because of solvation of the entering SCN^-. Addition of toluene to the acetonitrile decreases k_1 and increases k_2 for the opposite reasons. The reactions are all much faster than those of the d^1 $[(\eta^2\text{-Cp})_2VCl_2]$ complexes.

10.1.2.3. Displacement of Bidentate Ligands

A substantial and systematic study of the replacement of dithiaalkanes (DTA) from $[(DTA)W(CO)_4]$ by some P-donor ligands in chlorobenzene has been reported.[43] DTA (S⌢S) is $Me_3CS(CH_2)_xSCMe_3$ ($x = 1-5$), and the reactions generally proceed in two steps as in Eq. (13). The initially formed *cis* product subsequently isomerizes to a more stable *trans* form. When $x = 1$ it was shown that chelate ring

$$[(\eta^2\text{-S}⌢\text{S})W(CO)_4] \xrightarrow{\text{L}} [(\eta^1 - \text{S}⌢\text{S})W(CO)_4L] \xrightarrow{\text{L}} cis\text{-}[W(CO)_4L_2] \qquad (13)$$

opening and addition of L are concerted processes not involving the formation of $[(\eta^1\text{-S}⌢\text{S})W(CO)_4(\text{solvent})]$. The displacement of $(\eta^1\text{-S}⌢\text{S})$ in the second stage of reaction (13) is a straightforward dissociative process in which the rates vary systematically with L according to Eq. (14) where a, b, and c are constants

$$\log k = a\theta_L + b\nu_L + c \qquad (14)$$

characteristic of the system, and θ_L and ν_L are Tolman cone angles and electronic parameters, respectively; $a = 0.0472$ deg^{-1} and $b = -0.0102$ cm, the negative value of the latter showing that rates increase with increasing L → W electron donation and the consequently decreasing W—S bond strengths. The steric effect is comparable to that observed in *mer*-$[Ru(CO)_3L(SiCl_3)_2]$.[21] Values of $|a|/(|a| + |b|)$ and $|b|/(|a| + |b|)$ are discussed as measures of the relative importance of steric and electronic effects but the differences in the units make this of little quantitative value. What can be said is rate constants change by a factor of ~30 over the cone angle range of 31°, and by a factor of less than 2 over the 25 cm^{-1} range of ν available. The steric effects of L show up mainly in $T\Delta S^+$, which changes by 5.4 kcal mol^{-1} at 50 °C for a 29° change in cone angle, rather than in ΔH^+ which changes by only 3.1 kcal mol^{-1}. The *cis* ⇌ *trans* isomerization for L = P(n-Bu)$_3$ has $\Delta H^+ \sim 31$ kcal mol^{-1} and $\Delta S^+ \sim 22$ cal K^{-1} mol^{-1} for the forward reaction. Considerable loosening in the transition state occurs, but there was no ligand dissociation. When $x = 2$, ring opening is followed by competitive ring closure and L addition, and comparatively rapid loss of the monodentate DTA occurs from $[(\eta^1\text{-}S\frown S)W(CO)_4L]$. When $x = 3$, initial reversible ring opening again occurs to form $[(\eta^1\text{-}S\frown S)W(CO)_4L]$ in detectable amounts when L = $P(OCH_2)_3CMe$, and subsequent loss of the DTA ligand is of comparable rate to the first step. However, for L = other P-donors both ring opening and concerted attack by L occur in the first step. For $x = 4$ and 5, reversible ring opening and addition of L occurs in the first step and this is followed by relatively slow loss of the monodentate DTA by a dissociative mechanism. For $x = 2\text{-}5$, the relative rates of ring opening in $[(\eta^2\text{-}S\frown S)W(CO)_4]$ increase with x along the series 1 $(x = 2):40$ $(x = 3):500$ $(x = 4):7600$ $(x = 5)$ while the rates of ring closure decrease, and only by a factor of 9, along this series. Thus the transition states for ring opening can be concluded to be "late" while those for ring closure are consequently "early."

Generation of $[(\eta^1\text{-}S\frown S)W(CO)_4(CB)]$ by flash photolysis in chlorobenzene (CB) enabled rates of ring closure to be obtained and the activation parameters are shown in Table 10.3. For $x = 1$ (in dichloroethane) and 2, ring closure is probably associative but for $x = 3\text{-}5$ the values of ΔH^+ correspond to those expected for dissociative loss of CB and the values of ΔS^+ are probably related to entropy changes imposed by freezing of internal rotations when the transition state is formed.

Rates of ring closure for $[(\eta^1\text{-}N\frown N)W(CO)_5]$ have been reported where $N\frown N$ is 2-CHNR, 6-X-pyridine (R = n-Bu, i-Pr, Ph, and t-Bu, and X = H, Me).[44] The complexes were formed photochemically from $[W(CO)_6]$, and rate constants are given in Table 10.4. The steric effects of the R groups when X = H show that it is the py nitrogen atom that is coordinated in the $[(\eta^1\text{-}N\frown N)W(CO)_5]$. The absence of steric effects in the reactions of complexes with X = Me suggests that it is the aliphatic nitrogen that is the donor in these cases. Scavenging of the solvated $[W(CO)_5S]$ product of photolysis by the pyridine nitrogen when X = H is much faster than the scavenging by the aliphatic nitrogen atom when X = Me. The rates of ring closure are all much greater than those expected for simple CO dissociative processes.

Table 10.3. Activation Parameters for Ring Closure
in $[(\eta^1\text{-}DTA)W(CO)_4(CB)]$ in
Chlorobenzene $[DTA = Me_3CS(CH_2)_xSCMe_3]$

x	ΔH^+ (kcal mol^{-1})	ΔS^+ (cal K^{-1} mol^{-1})
1[a]	5.6 ± 0.15	−12.7 ± 0.7
2	8.8 ± 0.5	−3.5 ± 1.6
3	11.5 ± 0.3	2.6 ± 1.5
4	11.11 ± 0.06	0.2 ± 0.2
5	10.5 ± 0.05	−2.3 ± 0.16

[a] Solvent is dichloroethane, not chlorobenzene, because rates in CB were too fast for precise measurement.

Table 10.4. Rate Constants for Ring Closure of Some [(η^1-2-CHNR, 6-X-pyridine)W(CO)$_5$] Complexes in Benzene[44]

R	n-Bu	i-Pr	Ph	t-Bu
k_{obs} (s^{-1}) (X = H)	1.7×10^{-2}	2.8×10^{-3}	2×10^{-3}	1.5×10^{-5}
k_{obs} (s^{-1}) (X = Me)	8.2×10^{-3}	8.6×10^{-3}	3.2×10^{-3}	2.6×10^{-3}

The displacement of a series of β-diketones from [Rh(β-diketone)(COD)] (COD = 1,5-cyclooctadiene) by TFBA (trifluorobenzoyl acetone) in petroleum ether shows the rate equation: $k_{obs} = k_1 + k_2$[TFBA].[45] The values of k_1 are relatively small, occasionally show a negative temperature dependence, and are not discussed. The values of k_2 are discussed in terms of the effects of the substituents in the β-diketone ligands on the strengths of the Rh—O bonds. However, the values of $\Delta H_2^{\ddagger}$ and $\Delta S_2^{\ddagger}$ fit on an excellent linear isokinetic plot with an isokinetic temperature of about -7 °C, so the relative rates are determined more by entropic factors than enthalpic ones and it is likely, therefore, that the discussion should be formulated more in terms of $\Delta H_2^{\ddagger}$ and $\Delta S_2^{\ddagger}$ values and differential solvation of the departing O-donor atoms than in terms of the Rh—O bond strengths. If the relative rates had been measured at only $\sim$20 °C lower than those chosen for discussion there would have been no change in the rates to discuss! Displacement of COD in mer-[Cr(CO)$_3$P(OMe)$_3$COD] by isoprene in deuterobenzene is clearly first order in [isoprene][46] and a highly reversible breaking of one (η^2-alkene) bond of the COD must compete with addition of the isoprene to the "vacant" coordination site so created. In the presence of THF this vacant coordination site is completely scavenged, no reformation of the η^4-COD complex occurs, and the rate becomes essentially independent of [isoprene]. While the Cr(CO)$_4$ moiety prefers to bond to nonconjugated dienes, Cr(CO)$_3$P(OMe)$_3$ prefers to bond to conjugated dienes because of its greater π-basicity.

10.1.2.4. Photochemically Initiated Reactions

A number of papers[8] have appeared on rates of reaction of solvent molecules with vacant coordination sites generated by fast flash photolysis. Earlier work claiming that solvated Cr(CO)$_5$ is formed within 1 ps is queried,[8a] and it is claimed that solvation occurs on a 100 ps timescale. Others still believe that solvation is complete after a few ps[8b] and the slower processes observed[8a] are ascribed to dissipation of vibrational energy in the electronically excited states produced by the flash,[8c] or to vibrational deexcitation of electronically excited solvated products.[8d] Formation of [W(CO)$_5$S] (S = perfluoromethylcyclohexane) over a 20 ps timescale is ascribed to "rotational correlation times for reorganization of primary and secondary coordination spheres."[8e] Rates of linkage isomerization of alkane-bonded THF or ROH to the O-donor forms have also been investigated.[8a,b] While the details of these temporally esoteric processes need further investigation, it does at least seem necessary for the whole question of the existence of coordinatively unsaturated species, in thermal dissociative loss of ligands in solution, to be reevaluated. If they do not exist long enough to be scavenged directly by ligands entering from bulk solvent, then the reactions of these ligands have to be considered as *I* processes, the vacant coordination sites being accessible to entering ligands only if these ligands are already present in the outer coordination sphere of the solvated intermediates. There will, however, be cases where the apparently vacant coordination sites are actually occupied by agostically bonded groups from other ligands or by bridging CO groups in di- or poly-nuclear carbonyls. The question of the actual donor groups involved in solvation (e.g., η^2-arene interactions or M $\leftarrow$ Cl interactions in solvation by chlorobenzene) may also be important. These subtle but fundamental points will not be explicitly considered in the analysis of the following kinetic studies.

[W(CO)$_5$S] (S = a variety of solvents) is formed more rapidly from [W(CO)$_5$L] (L = pyridine or piperidine) than from [W(CO)$_6$],[47] and this is associated with a rather small wavelength dependence of quantum yields across the LF singlet excitation region. Direct reaction occurs from these states. No direct reaction occurs from the CT states but internal conversion to a LF state is probably efficient. There is direct reaction from the LF triplet but yields are smaller. The rates are discussed in terms of excited states that are not vibrationally equilibrated, but that have to decay along paths determined by vibrational selection rules governing the excitation.

The use of a mixed heptane–perfluoromethylcyclohexane solvent leads to the conclusion that displacement of heptane from [Cr(CO)$_5$(heptane)] by hexene proceeds about equally by associative and dissociative paths.[48] The value of ΔH^+ for reaction in pure heptane is only a little lower than previous estimates of the Cr–heptane bond strength, but the negligible value of ΔS^+ does allow for competitive associative and dissociative paths. Replacement of chlorobenzene from cis-[W(CO)$_4$(PPh$_3$)CB] is concluded[49] to be dissociative because the relative rates of attack by py, P(O—i-Pr)$_3$, and a group of six alkenes differ only by a factor of ~3. Activation enthalpies for attack by four of the alkenes are essentialy constant at ~11.5 kcal mol^{-1} but the values of ΔS^+ are all close to zero. The latter might be taken to indicate mixed dissociative and associative paths as was concluded for the [Cr(CO)$_5$(heptane)] reactions[48] and reaction of the sterically inhibited 2,3-dimethyl-2-butene is a factor of 10 slower than the other, less sterically inhibited, alkenes. However, the rates of displacement of CB are clearly not related to the strengths of the W—N, W—P, and W-alkene bonds that are being formed. In connection with this, it is worth remarking that earlier work[50] on the displacement of CB from [W(CO)$_4$LCB] showed that rates assigned to the *addition* of nucleophiles to [W(CO)$_4$L] actually increase with the cone angle of L as do the rates of displacement of py from [W(CO)$_4$Lpy]. Since addition of nucleophiles to the vacant coordination site in [W(CO)$_4$L] might be expected to become *slower* as the size of L increases, it has to be wondered whether reactions assigned as additions to [W(CO)$_4$L] should not be reassigned to I_d reactions of [W(CO)$_4$LCB].

Another extensive study[51] involves the displacement of THF from [Co(CO)L(NO)(THF)] [L = CO, PPh$_3$, and P(OPh)$_3$] generated by flash photolysis of [Co(CO)$_2$L(NO)] in THF. Reactions with various nucleophiles, L', were strictly first order in [L']. Reactions in mixed THF-cyclohexane solvents with

L = PPh$_3$ and P(OPh)$_3$ were found to behave as expected for simple reversible dissociation of THF but, when L = CO, an additional term corresponding to direct associative attack of L' on [Co(CO)L(NO)(THF)] had to be included. The values of $10^{-3} k_{ass}$ (M^{-1} s^{-1}) for this path varied with L' as follows: PPh$_3$ (4.06) > AsPh$_3$ (2.41) ~ P(OBu)$_3$ (2.36) > P(O—i-Pr)$_3$ (1.59) ≫ P(OPh)$_3$ (0.46) ≫ 1-hexene (0.007), while the second-order rate constants assigned to the highly reversible loss of THF and subsequent addition of L' varied with L' according to $10^{-3} k_{diss}$ (M^{-1} s^{-1}) = P(O—i-Bu)$_3$ (0.31) ~ PPh$_3$ (0.34) > P(O—i-Pr)$_3$ (0.28) > AsPh$_3$ (0.24) > P(OPh)$_3$ (0.13) ≫ 1-hexene (0.008). Values of $\Delta H^{\ddagger}_{ass}$ and $\Delta S^{\ddagger}_{ass}$ for direct attack by L' are 5–7 kcal mol^{-1} and −(23–25) cal K^{-1} mol^{-1}, respectively, and are consistent with associative processes. The dependence of k_{ass} on the P-donor nucleophiles was given by Eq. (15) where ν and θ are Tollman's electronic parameters and cone angles, respectively, for L'. The dissociative processes follow Eq. (16) where the relative values of k_{diss} correspond to the relative values for rates of addition of L' to the coordinatively unsaturated [Co(CO)$_2$(NO)]. The negative

$$\log k_{ass} = 135 - 0.0632\nu - 0.00549\theta \tag{15}$$

$$\log k_{diss} = 73.5 - 0.0329\nu - 0.006623\theta \tag{16}$$

values for the coefficients of the electronic parameter show that the rates increase with increasing net electron-donor capacity of the nucleophiles, as expected, and the negative coefficients for the steric parameter show that rates decrease with increasing size of the nucleophile, also as expected. Activation parameters for the overall dissociative path are $\Delta H^{\ddagger}_{diss} \sim 7$–8 kcal mol^{-1} for L' = P(O—i-Pr)$_3$, PPh$_3$, and P(OPh)$_3$ and $\Delta S^{\ddagger}_{diss} = -(13$–16) cal K^{-1} mol^{-1}. Although the lower electronic discrimination of [Co(CO)$_2$(NO)] is not unexpected, the similar dependence on the size of the nucleophiles and the negative values of $\Delta S^{\ddagger}_{diss}$ present more of a problem. This is rationalized in terms of geometrical rearrangements involved in dissociation of THF from [Co(CO)$_2$(NO)(THF)] and/or in addition of THF or L' to [Co(CO)$_2$(NO)]. Further investigations of these systems is clearly called for.

Flash photolysis of [Mo(CO)$_4$(diphos)] in toluene, T, produces fac-[Mo(CO)$_3$(diphos)T] as a predominant intermediate although cis-[Mo(CO)$_4$(η^1-diphos)] is also formed.[31] The latter undergoes first-order ring closure with $\Delta H^{\ddagger} = 7.7$ kcal mol^{-1} and $\Delta S^{\ddagger} = -9.3$ cal K^{-1} mol^{-1} by an intramolecular associative process. The former undergoes displacement of T to form fac-[Mo(CO)$_3$(diphos)L] [L = py, P(OCH$_2$)$_3$CMe, P(OMe)$_3$, P(OEt)$_3$, P(O—i-Pr)$_3$, PPh$_3$, and SbPh$_3$] by an overall second-order process ascribed to slow reversible dissociation of T followed by fast readdition of T which competes very successfully with addition of L. Rates of the overall reaction with L are in the range 4–11 × 10^6 M^{-1} s^{-1} and there is therefore very little discrimination. The competition between py and the other ligands for the [Mo(CO)$_3$(diphos)T] is the same whether it is produced by flash photolysis of [Mo(CO)$_4$(diphos)] or by thermal loss of py from fac-[Mo(CO)$_3$(diphos)py]. The nature of the bonding of toluene in this and similar intermediates is discussed in terms of Mo-(η^2-arene) interactions vs. Mo—H$_3$CC$_6$H$_5$ agostic interactions.

Flash photolysis of $[(\eta^5\text{-Cp})\text{Co}(\text{CO})_2]$ in the gas phase produces $[(\eta^5\text{-Cp})\text{Co}(\text{CO})]$.[52] This reacts with CO and C_2H_4 at rates near the diffusion limit but reaction with N_2 is 200 times slower. Reaction with CO may produce $[(\eta^3\text{-Cp})\text{Co}(\text{CO})_3]$. In the absence of a reactant the CO bridged dimer $[(\eta^5\text{-Cp})_2\text{Co}_2(\text{CO})_3]$ is formed. In cyclohexane, CH, flash photolysis produces $[(\eta^5\text{-Cp})\text{Co}(\text{CO})(\text{CH})]$ from which the weakly bound CH can be displaced by N_2, MeCN, or $\text{P}(n\text{-Bu})_3$ with rate constants varying from $4\text{--}30 \times 10^9 \ M^{-1}\,s^{-1}$. Reaction with $[(\eta^5\text{-Cp})\text{Co}(\text{CO})_2]$ to form the dimer proceeds at similar rates. In mixed benzene–CH solvents the main product is $[(\eta^5\text{-Cp})\text{Co}(\text{CO})(\text{C}_6\text{H}_6)]$ which exists in equilibrium with $[(\eta^5\text{-Cp})\text{Co}(\text{CO})(\text{CH})]$, the equilibrium constant being $\sim 2 \times 10^{-4}$. The rate constant for displacement of benzene in the mixed solvents by $\text{P}(n\text{-Bu})_3$ is $7 \times 10^5 \ M^{-1}\,s^{-1}$ and the major process is replacement of benzene by CH, direct attack on the benzene complex being relatively unimportant. Other kinetics in this complex system are discussed in detail.

UV photolysis of $[(\eta^5\text{-Cp})\text{V}(\text{CO})_4]$ in frozen gas matrices at 12 K leads to loss of CO to form $[(\eta^5\text{-Cp})\text{V}(\text{CO})_n]$ $(n = 1\text{-}3)$.[53] $[(\eta^5\text{-Cp})\text{V}(\text{CO})_2]$ exists as two different isomers and the coordinatively unsaturated species can react with CO or other small molecules doped into the matrix. Long-wavelength photolysis leads to $[(\eta^3\text{-Cp})\text{V}(\text{CO})_4]$ which, in a CO matrix, can form $[(\eta^3\text{-Cp})\text{V}(\text{CO})_5]$. The quantum yields for formation of $[\text{Fe}(\text{CO})_4(E\text{-cyclooctene})]$ from $[\text{Fe}(\text{CO})_5]$ in n-hexane are 0.8 (302 nm) and 0.77 (254 nm), but those for formation of the bis-alkene from the mono-alkene complex are 0.59 (302 nm) and 0.82 (254 nm), the wavelength effect being ascribed to there being two LF excited states, loss of alkene being more favored, compared to loss of CO, from the lower-energy state.[54] Formation of the bis-alkene complex is an indication of the strong bonding of this particular alkene. Substitution of PF_3 into $[\text{Fe}(\text{CO})_5]$ in the gas phase can be accomplished by infrared sensitization by SF_6.[55] Many metal carbonyls quench the triplet states of biacetyl and benzil in benzene at rates that vary from quite small ($[\text{Mo}(\text{CO})_6]$, $k_q \sim 4 \times 10^5 \ M^{-1}\,s^{-1}$) to $1\text{-}10 \times 10^9 \ M^{-1}\,s^{-1}$ for several di-, tri-, and tetra-nuclear carbonyls.[56] It seems quite possible that substitution processes can be sensitized in this way.

Ligand field excitation of $[\text{W}(\text{CO})_5\text{L}]$ (L = py, 4-cyanopyridine, and 4-acetyl-pyridine) in toluene leads to displacement of L by $\text{P}(\text{OEt})_3$, and values of $\Delta V^+ = 5.7, 6.3,$ and $9.9 \ cm^3\,mol^{-1}$ for dissociation of L from the LF states were obtained.[57] The quantum yields for L = 4-Ac-py and 4-NC-py were much lower than those for L = py itself because of depopulation of the labile LF states through efficient crossing into lower-energy MLCT states which are not labile to ligand loss. The positive values of ΔV^+ are ascribed to the volume increase that accompanies the dissociative loss of ligand, and the rather larger values for the 4-NC- and 4-Ac-py are ascribed to the fact that the dissociative process has to originate by back reaction of the CT state to the LF state, the CT state being smaller than the LF state. Rather small values of ΔV^+ accompany the luminescence of the 4-NC- and 4-Ac-py complexes. The value of ΔV^+ for high-energy photosubstitution reactions of $[\text{W}(\text{CO})_4(\text{phen})]$ with PEt_3 in toluene[58] is $+8 \ cm^3\,mol^{-1}$ and corresponds to dissociative loss of CO from the LF state. This is less than the value of $\sim +20 \ cm^3\,mol^{-1}$ usually seen for thermal CO dissociative reactions, and this is explained by the fact that LF excitation has already increased the size of the complex from which the CO is dissociating. Lower-energy MLCT excitation has $\Delta V^+ = -12 \text{ cm}^3 \text{ mol}^{-1}$ and this corresponds to associative substitution into the excited state which is essentially a 17-electron oxidized form of W that is expected to undergo rapid associative reactions (see below).

Photolysis of $[Ru(CNMe)_6]^{2+}$ in MeCN leads to photosolvation with high quantum yield (0.75) while photodimerization occurs in H_2O or CH_2Cl_2 with somewhat lower quantum yields (~ 0.3).[59] Photodissociation of MeCN from $[Ru(CNMe)_5(NCMe)]^{2+}$ leads to the dimer in H_2O or CH_2Cl_2 ($\phi \sim 0.2$). The dimer is not photosensitive but reacts thermally with MeCN to form $[Ru(CNMe)_5(NCMe)]^{2+}$. $[(\eta^3\text{-allyl})Ru(CO)_3Br]$ exists as *exo* and *endo* isomers.[60] Photolysis enriches the *exo* form while the *endo* form is thermally more stable, the rate of *exo* → *endo* having a $t_{1/2}$ of ~ 10 min at room temperature. The photo-isomerization proceeds via the dimer $[(\eta^3\text{-allyl})Ru(CO)_2(\mu\text{-Br})]_2$ (formed by CO dissociation) and re-uptake of CO to form the *exo* complex. Photolysis in a low-temperature matrix leads to CO dissociation and room-temperature photolysis in the presence of PPh$_3$, C_2H_4, or ^{13}CO leads to $[(\eta^3\text{-allyl})Ru(CO)_2LBr]$. The dimer does not react with C_2H_4 while it does react with CO or PPh$_3$ which must therefore attack it directly and not simply react with $[(\eta^3\text{-allyl})Ru(CO)_2Br]$ formed by its dissociation into the mononuclear intermediate. No evidence for $(\eta^1\text{-allyl})$ intermediates is seen in contrast to thermal reactions of $[(\eta^3\text{-allyl})Ru(CO)_3Cl]$.

Photolysis of $[(\eta^5\text{-Cp})Ir(CO)(H)_2]$ in neopentane leads to oxidative addition of the neopentane to the $[(\eta^5\text{-Cp})Ir(CO)]$ moiety.[61] However, extensive H/D scrambling occurs when $[(\eta^5\text{-Cp})Ir(CO)(D)_2]$ is used so the simple photorelease of D_2 (or H_2) followed by oxidative addition cannot be the only reaction path. Photolysis in a 12 K CD_4 matrix leads cleanly and very efficiently to $[(\eta^5\text{-Cp})Ir(CO)(D)(CD_3)]$ while reaction of $[(\eta^5\text{-Cp})Ir(CO)(D)_2]$ in a CH_4 matrix leads only to the $[Ir(H)CH_3]$ complex. The room-temperature photolysis is considered not to go via photodissociation of CO, photoinduced slippage to the $(\eta^3\text{-Cp})$ isomer, or photoinduced H transfer to form the $[(\eta^4\text{-}C_5H_6)Ir(CO)(H)]$ isomer being considered as more likely possibilities. UV photolysis of $[(\eta^5\text{-Cp})Mn(CO)_3]$ in supercritical Xe, pressurized with H_2, leads cleanly to the dihydrogen complex $[(\eta^5\text{-Cp})Mn(CO)_2(H_2)]$.[62] This is quite stable under H_2 but the H_2 is replaced readily by N_2 ($t_{1/2} \sim 25$ min at 25 °C under 100 atm N_2). Photolysis of $[(\eta^5\text{-Cp})Re(CO)_3]$ leads to the dihydride $[(\eta^5\text{-Cp})Re(CO)_2(H)_2]$ via what is thought to be a *cis*-$(H)_2$ intermediate which then photoisomerizes to the *trans*-$(H)_2$ product.

The 366 nm photolysis of $[(\eta^5\text{-MeCp})Mn(CO)_{3-n}(PPh_3)_n]$ ($n = 0$ to 2) has been studied.[63] The reaction $n = 0 \rightarrow n = 1$ in the presence of free PPh$_3$ has $\phi = 0.77$ while $n = 1 \rightarrow n = 2$ has $\phi = 0.59$; $\phi = 0$ for $n = 2 \rightarrow n = 3$ and for $n = 1 \rightarrow n = 0$ under CO. Photolysis of $n = 2 \rightarrow n = 1$ does, however, occur efficiently ($\phi = 0.88$); $k(+CO)/k(+PPh_3) = 0.5$ for $[(\eta^5\text{-MeCp})Mn(CO)_2]$ and 3.5 for $[(\eta^5\text{-MeCp})Mn(CO)(PPh_3)]$. *Cis*-$[(\eta^5\text{-Cp*})Re(CO)_2Me_2]$ undergoes photodissociation of CO in 13 K matrices and uptake of released CO at 100 K leads to the *trans*-isomer.[64] Photolysis of the *trans*-isomer leads to the same photodissociated intermediate as produced from the *cis*-isomer. Similar processes occur for the dihalide $[(Cl)_2, (Br)_2,$ and $(I)_2]$ complexes. The $[(\eta^5\text{-Cp*})Re(CO)Me_2]$ intermediate does not appear to be coordinated by the glass, and the reactions occur independently of irradiating wavelength.

Extensive photochemical studies have been reported on a series of R—N=CH—CH=N—R (R-DAB) and other α-diimine complexes.[65-69] For $[Fe(CO)_3(\eta^2\text{-}R\text{-}DAB)]$ (1) [R = (a) Cy, (b) p-tol, (c) 2,6-diisopropylphenyl (i-Pr$_2$)Ph, (d) t-Bu] the R-DAB coordinates as a chelate ligand and (1a)-(1c) undergo CO photodissociation as a primary process while (1d) undergoes Fe—N bond breaking with formation of $[Fe(CO)_3(\eta^1\text{-}R\text{-}DAB)L]$ as an intermediate detectable at low temperatures.[65] For (1a) ϕ increases from 0.05-0.20 with increasing energy of irradiation and the temperature dependence of ϕ from low-energy excitation leads to $\Delta E_a = 525$ cm^{-1}. Photoreaction of (1a) or (1b) in n-pentane at 150 K, in the absence of a substituting ligand, leads to a mixture of a CO-bridged dimer $[Fe_2(R\text{-}DAB)_2(CO)_4(\mu\text{-}CO)]$ and an isomer of $[Fe(CO)_3(R\text{-}DAB)]$ in which the R-DAB is η^4 coordinated as in a butadiene complex. These results indicate that all photosubstitutions occur from a slightly higher ^3LF state while photoisomerization occurs from a lower-energy ^3ML state, the latter referring to excitations within the Fe(R-DAB) metallacycle rather than to MLCT states. $[M(CO)_3(R\text{-}DAB)]$ complexes [M = Fe, Ru; R = C(i-Pr)$_2$H, (i-Pr)$_2$Ph] show photodissociation of CO from a ^3LF state.[66] $[Fe(CO)_2(R\text{-}DAB)P(OPh)_3]$ undergoes Fe—N bond breaking from the ^3LF state and isomerization to the η^4 butadiene-like isomer from the ^3ML state. The latter process occurs also with $[Ru(CO)_2(PR_3')(R\text{-}DAB)]$ while ^3LF excitation of the latter leads to complete dissociation of R-DAB and formation of $[Ru(CO)_3(PR_3')_2]$ and $[Ru(CO)_2(PR_3')_3]$. The $[Ni(CO)_2(R\text{-}DAB)]$ is pseudoplanar when R = (i-Pr)$_2$Ph and tetrahedral when $R = t$-Bu.[67] Low-temperature (170 K) photochemistry for R = t-Bu into its low-energy ^3MLCT state leads to Ni—N bond breaking and a series of thermal reactions leading to various products depending on the entering group. MLCT excitation does not cause any differential weakening of the Ni—N or Ni—CO bonds in this tetrahedral complex so it is the weaker, Ni—N, bond that breaks. Higher-energy excitation leads to production of metal $\rightarrow \pi^*(CO)$ transitions and consequent decomposition. The pseudoplanar structure for R = (i-Pr)$_2$Ph changes the nature of the lowest-energy transition, so that it leads to differential weakening of Ni—CO bonds and CO dissociation.

Photochemistry of a series of $[Ph_3Sn\text{-}M(CO)_3(\alpha\text{-}diimine)]$ complexes (M = Mn, Re) into low-energy ^3MLCT states shows that CO loss is the primary process for M = Mn while Sn—M bond breaking occurs when M = Re.[68] Efficient intersystem crossing occurs from the ^3MLCT state into $\sigma\sigma^*$ Sn—M states. This state involves weakening of both the Sn—M and M—CO bonds, but because the Sn—Mn bond is stronger than the Sn—Re bond, and the Mn—CO is weaker than the Re—CO bond, the results are as observed. The Mn(t-Bu-DAB) complex is unique in showing such a high quantum yield ($\sim$0.8) for CO loss following 600 nm irradiation. Irradiation of this complex is also unique in producing, in the absence of added ligands, a product in which the coordination site vacated by the CO is filled by an agostically bound Me group from the t-Bu substituent in the diimine. Irradiation into the MLCT band of $[(OC)_4CoM(CO)_3(bipy)]$ (M = Mn, Re) in toluene leads to efficient crossing into a Co—M $\sigma\sigma^*$ state and consequent Co—M homolysis.[69] In the absence of radical scavengers (t-BuNO or CBr$_4$) no product is seen for M = Re because of very fast back reaction with [Co(CO)$_4$], but for M = Mn the only product is [Mn$_2$(CO)$_6$(bipy)$_2$]. Photolysis in the presence of P-donors or donor solvents forms [M(CO)$_3$(bipy)L]$^+$[Co(CO)$_4$]$^-$. These are also formed by initial homolysis to form the "16-electron" [M(CO)$_3$(bipy)] in which the 17th electron is on the bipy ligand. This product is a strong reducing agent that reduces [(OC)$_4$CoM(CO)$_3$(bipy)] and leads to a chain reaction that can be quite long when M = Mn (ϕ = 7–60) depending on the steric and electronic nature of the P-donors. A very unusual feature of the M = Re complex is the formation of the contact ion pair [Re(CO)$_3$(bipy)$^+$Co(CO)$_4^-$] in toluene at 230 K. Toluene is only a very weak base and is assumed not to be able to form a 19-electron adduct [Re(CO)$_3$(bipy)S], so the [Re(CO)$_3$(bipy)] radical itself is believed to reduce either the parent compound or [Co(CO)$_4$].

10.1.2.5. Reactions Involving "Slippage"

Photoelectron spectroscopy of [(η^5-Ind)Mn(CO)$_3$], [(η^5-Ind)$_2$M] (M = Fe, Ru), and [(η^5-Ind)(η^5-Cp*)Ru] suggests that the indenyl ligand is a better electron-releasing ligand than Cp, comparable to MeCp, but less than Cp* [70] in contrast to values of M-ring force constants which had suggested that the C$_6$ ring in the indenyl causes electron withdrawal from the C$_5$ ring. The better electron-releasing nature of indenyl is part of the reason why [(η^5-Ind)$_2$M] is more easily oxidized than [(η^5-Cp)$_2$M] (M = Fe, Ru) although another contribution is the fact that the electrons being removed come from a d_{xy} orbital. This is of higher energy in the indenyl complexes because the lower symmetry results in a repulsive interaction between the d_{xy} electrons and the HOMO of the ring. The importance of the latter repulsion in the so-called indenyl effect was not discussed, but it might be weaker in the (η^3-Ind) isomer and this could be an important contribution to the effect. In addition, since the main bonding between (η^5-ring)$^-$ compounds and metals is donation of *four* electrons into d_{xz} and d_{yz} orbitals, the significance of the 18-electron rule in discussion of the indenyl effect might possibly not be as fundamental as it appears.

The energetics of slippage have been measured by the values of ΔH° = -5.7 ± 0.7 kcal mol^{-1} and ΔS° = -20.6 ± 2.9 cal K^{-1} mol^{-1} for the equilibrium in Eq. (17) for reaction in toluene under 130 bar CO.[71] Rates of addition of CO are governed by ΔH^+ = 13.9 ± 1.5 kcal mol^{-1} and ΔS^+ = -21.3 ± 3.6 cal K^{-1} mol^{-1}.

$$[(\eta^5\text{-Cp})_2\text{Cr(CO)}] + \text{CO} \rightleftharpoons [(\eta^3\text{-Cp})(\eta^5\text{-Cp})\text{Cr(CO)}_2] \tag{17}$$

These values are comparable to the values of ΔH^+ and ΔS^+ for CO exchange in [(η^5-Cp)Co(CO)$_2$]. Very similar values are obtained for the H$_4$C$_5$—CMe$_2$CMe$_2$—C$_5$H$_4$ complex, but in this case the kinetics are complicated by oxidative catalysis involving a 17-electron species [possibly due to the adventitious presence of Cr(III)]. Addition of CO to this species is believed to proceed via slippage rather than formation of a three-electron two-center Cr—CO bond. In the case of [(η^5-Cp)$_2$Cr(CO)] the 17-electron oxidized species loses CO rapidly and this path does not contribute to the addition kinetics.

Reduction of $[(\eta^6\text{-}C_6H_6)Cr(CO)_3]$ with naphthalide at room temperature produces $[(\eta^4\text{-}C_{10}H_8)Cr(CO)_3]^{2-}$, presumably via rapid exchange of naphthalene with $[(\eta^4\text{-}C_6H_6)Cr(CO)_3]^{2-}$.[72] The product undergoes protonation in H_2O to form $[(\eta^5\text{-}C_{10}H_9)Cr(CO)_3]^-$ and oxidation by O_2 leads to $[(\eta^6\text{-}C_{10}H_8)Cr(CO)_3]$. When 1-Me- or 1,4-Me$_2$-naphthalene is used, the Cr is bonded to the *unsubstituted* ring. The displacement of C_6H_6 by $C_{10}H_8$ can be ascribed to a thermodynamic "naphthalide effect." The pentadienyl compounds $[(\eta^5\text{-}C_5H_7)Mo(CO)_2L]^+$ (L = dppe or dmpe) add MeCN to form $[(\eta^3\text{-}C_5H_7)Mo(CO)_2L(NCMe)]^+$ in equilibrium with the η^5-complex.[73] Values of K are $0.8 \pm 0.06\ M^{-1}$ (L = dmpe) and (14.3 ± 0.6) M^{-1} (L = dppe) in $CDCl_3$. The effect of dmpe vs. dppe may be due to the better electron-donor capacity of the former or to the larger size of the latter. Slippage appears to be easier for $\eta^5\text{-}C_5H_7$ than for $\eta^5\text{-}C_5H_5$ ligands.

Flash photolysis of $[(\eta^5\text{-}Cp)Fe(CO)_2(\eta^1\text{-}CH_2Ph)]$ in hexane is proposed to lead to $[(\eta^5\text{-}Cp)Fe(CO)(\eta^3\text{-}CH_2Ph)]$ which adds CO or PPh$_3$, to revert to the $(\eta^1\text{-}CH_2Ph)$ complex, with rate constants of $151 \pm 10\ M^{-1}\,s^{-1}$ and $1.31 \pm 0.05 \times 10^3\ M^{-1}\,s^{-1}$, respectively.[74] These values seem too low to be compatible with simple displacement of weakly solvated hexane in an $[(\eta^5\text{-}Cp)Fe(CO)(\eta^1\text{-}CH_2Ph)\text{-}$ (hexane)] intermediate. The ratio $k(+PPh_3)/k(+CO) = 8.9$ is somewhat larger than is common for attack at so-called "vacant" coordination sites and may therefore be compatible with concerted ligand addition and $\eta^3 \rightarrow \eta^1$ slippage of the benzyl ligand. $[(\eta^5\text{-}Ind)Fe(CO)_3]^+$ reacts with BH_4^- in acetone at $-60\,°C$ to give $[(\eta^5\text{-}Ind)Fe(CO)_2CHO]$, which undergoes disinsertion of CO at $-10\,°C$ and formation of $[(\eta^5\text{-}Ind)Fe(CO)_2H]$.[75] This very facile reaction may also involve an $\eta^5 \rightarrow \eta^3$ indenyl change concerted with H$^-$ migration, followed by an $\eta^3 \rightarrow \eta^5$ change concerted with CO dissociation.

Associative substitution reactions of $[(\eta^5\text{-}C_5H_4X)Rh(CO)_2]$ with PPh$_3$ increase in rate with the electron-withdrawing power of X, as measured by Hammett σ_p values (X = Me, H, CF$_3$, and NO$_2$).[76] This is a consequence of decreasing electron density at the metal, a trend confirmed by good linear plots of ν_{CO} against σ_p, including X = NMe$_2$ or Cl. However, the rates of reaction of these two complexes exceed, by about 3 and 1.5 orders of magnitude, respectively, those predicted from the trend shown by the other ligands. These two-electron donors are believed to stabilize the η^3-allyl-ene transition state or intermediate by direct interaction with the ene group to form a four-electron three-center π system. The generally rapid increase in rates, for the other four ligands, with increasing σ_p is quantified by a rather large value of $\rho = 4$, and this is ascribed to transition state stabilization. The evident and pronounced weakening of the Rh—CO and, by inference, ($\eta^5\text{-}$ C$_5$H$_4$X) $\rightarrow$ Rh bonds in the ground state as σ_p increases suggests ground-state destabilization must be considered as an additional or alternative explanation. Another indication of the preferential slippage of indenyl compared to cyclopentadienyl is shown[77] by the complete displacement of indenyl$^-$, and C$_2$H$_4$, in reaction (18) which is much faster than displacement of $(\eta^5\text{-}Cp)^-$ and C$_2$H$_4$ from $[(\eta^5\text{-}Cp)Rh(C_2H_4)_2]$.

$$[(\eta^5\text{-}Ind)Rh(C_2H_4)_2] + 2\ dmpe \rightarrow [Rh(dmpe)_2]^+ + Ind^- + 2C_2H_4 \qquad (18)$$

Two-electron reduction of $[(\eta^5\text{-}Cp^*)Rh(\eta^6\text{-}C_6Me_6)]^{2+}$ to $[(\eta^5\text{-}Cp^*)Rh(\eta^4\text{-}C_6Me_6)]$ shows[78] the major slippage of C$_6$Me$_6$ occurs during the second stage of reduction, from Rh(II) → Rh(I), and it is probably concerted with the reduction process. Some smaller distortion of the C_6Me_6 coordination during the Rh(III) → Rh(II) step may also occur. The stabilization of the η^4-C_6Me_6 compared to the η^6-C_6Me_6 in the Rh(I) complex is estimated to be at least 6 kcal mol^{-1}. Replacement of PPh$_3$ in $[(\eta^5\text{-Ind})Ir(H)(PPh_3)_2]^+$ by PMe$_3$, t-BuNC, or CO has been shown to occur rapidly but only PMe$_3$ displaces PPh$_3$ from $[(\eta^5$-PhC$_5$H$_4$)Ir(H)(PPh$_3$)$_2$]$^+$ and no reaction occurs with $[(\eta^5$-Cp)Ir(H)(PPh$_3$)$_2]^+$.[79] Examples of fully coordinated η^3 or even η^1 intermediates are detected by NMR at low temperatures. $[(\eta^5\text{-Ind})Ir(PPh_3)_2Et]^+$ undergoes β-elimination of C$_2$H$_4$ at 60 °C but only in donor solvents which may induce slippage by an initial addition reaction. It is not clear how the fully coordinated $[(\eta^3\text{-Ind})Ir(PPh_3)_2(S)Et]^+$ could undergo β-elimination which requires a vacant coordination site, but perhaps the $\eta^3 \rightarrow \eta^1$ change in this complex is more facile than the $\eta^5 \rightarrow \eta^3$ slippage in the parent compound. In nondonor solvents β-elimination occurs photochemically from the parent compound, so photoinduced $\eta^5 \rightarrow \eta^3$ slippage may occur readily under these more energetic conditions.

An interesting and unusual invocation of slippage has been made[80] for the scrambling of the protons in $[(\eta^5\text{-Cp})(\eta^5\text{-Cp*})Ta(\mu\text{-H})_2BH_2]$. The terminal hydrides exchange individually with the bridging hydrides and formation of an intermediate in which either Cp or Cp* has "slipped" to form $[(Cp)(Cp*)Ta(\mu\text{-}H)_3BH]$ is proposed. $\Delta H^+ = 17.0 \pm 0.8$ kcal mol^{-1}, and the data lead to a value of ΔS^+ ($+2 \pm 2$ cal K^{-1} mol^{-1}) which is not unreasonable for a concerted process in view of the positive entropy expected for the "dissociative" $\eta^5 \rightarrow \eta^3$ change and the negative one expected for the $(\mu\text{-H})_2 \rightarrow (\mu\text{-H})_3$ process. The 12-electron $[Os(\eta^1\text{-}o\text{-tolyl})_4]$ undergoes an associative reaction with PMe$_3$ at 20 °C in hexane to give the 18-electron $[(\eta^6\text{-arene})Os(\eta^1\text{-}o\text{-tolyl})_2(PMe_3)]$. The arene is the substituted ditolyl ligand 2-(2-MeC$_6$H$_4$)MeC$_6$H$_4$ which is coordinated to the Os through one of the rings.[81] In effect this is an $\eta^1 \rightarrow \eta^6$ "reverse slippage" caused by ligand addition, although the sequence of steps involved is not at all obvious. An unusual process is the relatively facile (100 °C) replacement of the η^6-ditolyl ligand in $[(\eta^6\text{-ditolyl})Os(\eta^1\text{-tolyl})_2(PMe_3)]$ by other arenes, and its much more facile displacement from the analogous $[(\eta^6\text{-ditolyl})Os(\eta^1\text{-tolyl})_2CO]$. The latter displacement is caused by t-BuNC and probably proceeds by a succession of slippages begining with $\eta^6 \rightarrow \eta^4$ to form eventually $[(t\text{-BuNC})_3Os(\eta^1\text{-tolyl})_2(CO)]$ in which the CO ligand has been retained.

$[(\eta^3\text{-CH}_2\text{CMeCH}_2)Pt(acac)]$ very rapidly adds PPh$_3$ to form an equilibrium mixture of $[(\eta^1\text{-CH}_2\text{CMeCH}_2)Pt(acac)(PPh_3)]$ and $[(\eta^3\text{-CH}_2\text{CMeCH}_2)$-Pt(PPh$_3$)$_2]^+$[acac]$^-$.[82] In benzene the equilibrium constant to form the ion pair is small, but in CH$_2$Cl$_2$ it is large. These differences affect the kinetics of the subsequent reductive coupling reaction between the acac$^-$ and the allylic group. Reactions of the triple sandwich $[(ring)Rh(ring)Rh(ring)]$ (ring = η^5-C$_4$H$_4$BPh) in CH$_2$Cl$_2$ with various nucleophiles are essentially associative and produce a mixture of $[(ring)RhL_3]^+$ and $[(ring)_2Rh]^-$.[83] Activation parameters (ΔH^+ in kcal mol^{-1}; ΔS^+ in cal K^{-1} mol^{-1}) are 9.9 ± 0.3; −23.7 ± 1.2 for L = PMe$_3$, 12.8 ± 0.2; −17.7 ± 0.7 for L = PEt$_3$, and 10.6 ± 0.4; −26.0 ± 1.2 for L = P(OMe)$_3$. Both steric and electronic effects are important because the large ligands P(i-Pr)$_3$ and

PCy$_3$ do not react at all and the smaller but less basic P(OMe)$_3$ reacts at about the same rate as the larger and much more basic PMe$_3$. Lateral attack by the nucleophile at one metal atom, coupled with $\eta^5 \to \eta^3$ slippage, is proposed.

10.1.2.6. Substitution Involving 17- and 19-Electron Species

Associative substitution of 17-electron complexes was proposed[84] to proceed via three-electron two-center bonding some time ago and this proposal has received theoretical justification.[85] Sites of nucleophilic attack were considered and the importance of steric accessibility of the half-filled orbital emphasized. 19-Electron radicals can be different depending on whether they are produced by associative attack on 17-electron radicals or by one-electron reduction of 18-electron species, and reaction paths for 19-electron radicals can differ depending on the position of the odd electron.

Rates of very fast associative reactions of radical anions such as [Fe(CO)-(NO)$_2$]$^{\pm}$, [Co(CO)$_2$(NO)]$^{\pm}$, [(η^3-C$_3$H$_5$)Co(CO)$_2$]$^{\pm}$, and [(η^3-Cp)Co(CO)$_2$]$^{\pm}$ with PF$_3$, PMe$_3$, NO, SO$_2$, alkenes, ketones, and O$_2$ in the gas phase have been reported.[86] Intermediate adducts were observed. [(η^1-C$_3$H$_5$)Co(CO)$_3$]$^{\pm}$ reacts with PF$_3$ to form [Co(CO)$_3$PF$_3$]$^{\pm}$ via the proposed insertion product [allylC(O)Co(CO)$_2$(PF$_3$)]$^{\pm}$. [Fe(CO)(NO)$_2$]$^{\pm}$, [Co(CO)$_2$(NO)]$^{\pm}$, and [(η^3-C$_3$H$_5$)Co(CO)$_2$]$^{\pm}$ undergo associative CO displacement by NO to form 18-electron products via one-electron donor NO adducts. $\eta^3 \to \eta^1$ slippage is much harder for Cp than for allyl groups. Substitution reactions of [Fe(CO)$_4$]$^{\pm}$ with PF$_3$ are relatively slow and no reaction occurs with PMe$_3$. [Fe(CO)$_4$]$^{\pm}$, [Cr(CO)$_5$]$^{\pm}$, and [Mn(CO)$_4$H]$^{\pm}$ all lose CO associatively with NO to give 18-electron products. The importance of electron transfer in reactions with some alkenes is emphasized. The complex reactions of [Mn(CO)$_4$H]$^{\pm}$ with O$_2$ are believed to proceed via an intermediate Mn—CHO complex but this insertion does not occur with PF$_3$.

A detailed study[87] of the intricate processes following generation of [(η^5-Cp)Fe(CO)$_2$Me]$^+$ (2$^+$) by oxidation of the neutral parent in acetone (Ac) has been reported. Addition of solvent produces [(η^5-Cp)Fe(CO)Ac(COMe)]$^+$ (3$^+$) which decomposes, probably back to (2$^+$), with $\Delta H^+ = 13.1 \pm 1.8$ kcal mol^{-1} and $\Delta S^+ = -2.8 \pm 7.8$ cal K^{-1} mol^{-1}. At $-75\,°$C (3$^+$) reacts with PhSMe to form [(η^5-Cp)Fe(CO)(PhSMe)(COMe)]$^+$ with $k = 0.004\ M^{-1}$ s^{-1} but [(η^5-Cp)Fe(CO)-(ClO$_4$)(COMe)] does not react in this way. Other kinetic details of this complex system are described. Reduction of [(η^4-C$_4$H$_6$)Fe(CO)$_3$] in the gas phase produces [(η^2-C$_4$H$_6$)Fe(CO)$_3$]$^{\pm}$, which loses CO to form [(η^4-C$_4$H$_6$)Fe(CO)$_2$]$^{\pm}$ with $E_a \sim$ 20 kcal mol^{-1} and $k \sim 600$ s^{-1}.[88] This rate constant can be compared with the 3.5 s^{-1} obtained for the formation of the 19-electron [Mn(CO)$_2$(dppe)$_2$] in acetonitrile from [Mn(CO)$_2$(η^2-dppe)(η^1-dppe)],[89] the reverse of which has $k \sim 10^6$ s^{-1}. Reduction of [Mn(CO)$_6$]$^+$ provides no evidence for [Mn(CO)$_6$]. Even a variety of substituted [Mn(CO)$_5$L] complexes are undetectable and their lifetimes must be less than 0.1 μs, although the lifetime of [Mn(CO)$_2$(dppe)$_2$] is of the order of 1 μs. The 19-electron [(η^5-Cp)Fe(η^6-arene)] loses arene by a second-order reaction in THF with some P-donors to form [(η^5-Cp)FeL$_2$] at rates $\sim 10^7$ times faster than the corresponding 18-electron cationic complex.[90] When L = P(OMe)$_3$,

$\Delta H^+ = 13.3 \pm 1.0 \, \text{kcal mol}^{-1}$ and $\Delta S^+ = -22 \pm 3 \, \text{cal K}^{-1} \, \text{mol}^{-1}$, and the reactions are believed to follow a sequence of η^6-arene $\rightleftharpoons$ η^4-arene $\rightleftharpoons$ $(\eta^4$-arene)L $\rightarrow$ $(\eta^2$-arene)L $\rightarrow$ $(\eta^2$-arene)L$_2$, with alternating 19- and 17-electron configurations, before final loss of arene. The first equilibrium lies well toward the 19-electron species, so leading to second-order kinetics. $[(\eta^5$-Cp)Cr(CO)$_3]_2$ undergoes rapid reversible homolysis, and reactions with P-donors suggest[91] that $[(\eta^5$-Cp)Cr(CO)$_3]$ undergoes associative substitution as do most other 17-electron organometallics. $[(\eta^5$-Cp)Mo(CO)$_3]$ is also substitutionally labile.[92] $[Co(CO)_3L_2]^+$ reacts with $[Co(CO)_3L]^-$ via ion pairing and electron transfer (L = P donors).[93] The $[Co(CO)_3L]$ radical undergoes L exchange on the ms timescale. The "19-electron" $[Co(CO)_3L_2]$ [L = 2,3-bis(diphenylphosphino)maleic anhydride] has a small part of the unpaired electron density in a Co—CO antibonding orbital.[94] This is sufficient to enhance its CO-dissociative lability to $\Delta H^+ = 23.8 \pm 0.6 \, \text{kcal mol}^{-1}$ and $\Delta S^+ = +11.1 \pm 2.2 \, \text{cal K}^{-1} \, \text{mol}^{-1}$, the corresponding cation being totally inert under these conditions (10–30 °C). Gas-phase reactions of $[Cr(CO)_5]^-$ with a variety of organic molecules have shown the importance of electron transfer, which is made easier by destabilization of the odd electron in the 19-electron adduct.[95] The "19-electron" $[ArN(O)Re(CO)_5]$ undergoes loss of the ArNO with $\Delta H^+ = 4.9 \pm 0.5 \, \text{kcal mol}^{-1}$ and $\Delta S^+ = -47 \, [\pm 34 \, \text{(sic)?}] \, \text{cal K}^{-1} \, \text{mol}^{-1}$, while the O-bonded isomer $[ArN—O—Re(CO)_5]$ undergoes bimolecular coupling through the N atoms with $\Delta H^+ \sim 1 \, \text{kcal mol}^{-1}$ and $\Delta S^+ = -44 \, \text{cal K}^{-1} \, \text{mol}^{-1}$.[96] The redox behavior of a large number of $[RuCl_2(CO)_2L_2]$ and $[RuCl_2(CO)L_3]$ complexes has been studied.[97] The kinetics of some of the 17-electron Ru(III) complexes are related to catalytic activity of these species.

The 17-electron $[(\eta^5$-Cp)(η^5-C$_5$H$_7$)V(CO)] has been found to undergo CO *dissociative* reactions at similar rates ($k = 2.73 \times 10^{-4} \, \text{s}^{-1}$ at 60 °C) to those of the 18-electron $[(\eta^5$-Cp)(η^5-C$_5$H$_7$)Cr(CO)].[98] The energy difference between 17-electron ground states and 15-electron intermediates can therefore be quite similar to the difference between 18-electron ground states and 16-electron intermediates in these cases where the metals have sufficient d electrons to provide adequate π-bonding to the CO ligands. On the other hand, the 17-electron dmCH complex $[(\eta^5$-dmCH)$_2$V(CO)] (dmCH = dimethylcyclohexadienyl) undergoes slow associative CO exchange, which is possible because the dmCH–V–dmCH angle is 155°. The bis-pentadienyl complex $[(\eta^5$-C$_5$H$_7$)$_2$V(CO)] has a pd–V–pd angle of $\sim$180° and undergoes mainly dissociative exchange. $[(\eta^5$-Cp*)$_2$V(CO)] also has a (ring)–V–(ring) bond angle of 154° but it undergoes much more rapid associative CO exchange than the bis-dmCH complex. The dmCH ligands are much weaker net electron donors to the V atom as compared with Cp* ligands, and that would be expected to enhance associative reactions of the dmCH complex. However, EPR measurements suggest that the odd electron in the Cp* complex is much more delocalized onto the organic ligand than is the case for the dmCH complex, so that an empty orbital, rather than a half-filled one, is available for nucleophilic attack. The $[(\eta^5$-C$_5$H$_7$)$_2$V(CO)] does not delocalize the odd electron in this way, and is also more sterically unfavorable to nucleophilic attack, and therefore reacts dissociatively. Ring slippage is not believed to be a factor in these reactions.

Finally, several important papers on the role of 17- and 19-electron organometaalic radicals in labile substitution and other reactions have been published as a Proceedings of a NATO Advanced Research Workshop.[99]

10.1.2.7. Substitution Induced by Other Methods

Substitutions and isomerizations can be enhanced by simple one-electron oxidations of 18-electron complexes. Substitution of arene groups in $[(arene)Cr(CO)_3]$ by MeCN can be enhanced in this way,[100] and formation of $[(\eta^5\text{-Cp})(\eta^3\text{-Cp})W(CO)_2]$ by addition of CO at atmospheric pressure can be similarly effected,[101] as can substitution into $[W(CO)_5py]$.[102] *Cis* → *trans* isomerization of $[Cr(CO)_4L_2]$ [$L = P(OMe)_3$] is effectively catalyzed by one-electron oxidation, the 17-electron *trans*-$[Cr(CO)_4L_2]^+$ oxidizing *cis*-$[Cr(CO)_4L_2]$ very effectively.[103] Substitution of PR_3 into the very inert $[Fe(CO)_5]$ to form pure $[Fe(CO)_3L_2]$ can be readily effected by BH_4^- in n-BuOH[104] by successive formation of $[Fe(CO)_4CHO]^-$, $[HFe(CO)_4]^-$, and $[HFe(CO)_3L]^-$ which reacts with solvent to form H_2 and $[Fe(CO)_3L(OBu)]^-$, the latter undergoing facile substitution of BuO^- by L. Substitution of various ligands L into $[(\eta^5\text{-CpMe})$ or $(\eta^5\text{-Ind})Ru(CO)_2I]$ is catalyzed by $[(\eta^5\text{-Cp})_2Fe_2(CO)_4]$ and this provides a convenient way of making a wide range of $[(\eta^5\text{-ring})Ru(CO)L(I)]$ complexes.[105]

Quantitative kinetic studies of labilization of inert complexes by Me_3NO have been extended to the $[M(CO)_5]$ complexes (M = Fe, Ru, and Os).[106] Reactions are first order in $[M(CO)_5]$ and $[Me_3NO]$ but independent of [L], the entering ligand, and values of ΔH^+ are low (10–15 kcal mol^{-1}) and ΔS^+ values are negative (−17 to −28 cal K^{-1} mol^{-1}), the variation of ΔH^+ being almost balanced by variation in ΔS^+ so that rates only increase four-fold along the series Fe < Ru < Os. This contrasts with a fortyfold decrease observed earlier for $M_3(CO)_{12}$ (Fe > Ru > Os). The order for $[M(CO)_5]$ does not seem to correlate particularly well with ν_{CO} values, which increase from $[Fe(CO)_5]$ to $[Ru(CO)_5]$ but decrease slightly to $[Os(CO)_5]$. The large decrease in ΔH^+ and ΔS^+ along the series shows that nucleophilic attack at the C atoms must become much more pronounced. The contrast in the lability order between the mono- and tri-nuclear carbonyls is ascribed to CO-bridge formation in the latter.

10.1.3. Dinuclear Complexes

The highly nucleophilic[107] carbonylate $[Re(CO)_5]^-$ displaces $[Mn(CO)_5]^-$ rapidly from $[Mn_2(CO)_{10}]$ in THF (and $[Co(CO)_4]^-$ and $[CpMo(CO)_3]^-$ from $[Co(CO)_4]_2$ and $[CpMo(CO)_3]_2$, respectively) in second-order reactions that probably involve nucleophilic attack at a C atom of a CO ligand.[108] The ultimate dinuclear product is $[Re_2(CO)_{10}]$ and no intermediate mixed metal carbonyls are detected. Nucleophilic attack on $[M(CO)_6]^+$, $[M(CO)_5L]^+$, and $[Mn(CO)_4(PPh_3)_2]^+$ (M = Mn, Re; L = various P-donors) by a range of Co-, Mn-, Fe-, and Re-containing carbonylates has been studied.[109] The ultimate thermodynamic products are a mixture of dinuclear carbonyls formed by one-electron transfer reactions, but in over half the reactions the first step appears to involve rapid nucleophilic attack at the C of a CO on $[M(CO)_6]^+$, for example, with effective transfer of a CO^{2+} to the nucleophile. The reactions (19) occur slowly when M = Re and very rapidly when M = Mn so the rates are not related to the thermodynamic driving force, which is the same in each reaction, or simply to the relative nucleophilicity of the carbonylate itself.

$$[M(CO)_6]^+ + [M(^{13}CO)_5]^- \rightleftharpoons [M(CO)_5]^- + [M(^{13}CO)_5(CO)]^+ \qquad (19)$$

Low-temperature (77 K) matrix photolysis of $[MnRe(CO)_{10}]^{(110)}$ leads to $[MnRe(CO)_9]$ and warmup to 298 K in the presence of PPh_3 leads to $[(CO)_5ReMn(CO)_4(PPh_3)]$ consistent with earlier work showing that room-temperature photolysis leads to CO loss from the Mn atom. Thermal reaction of the latter complex under conditions where $[MnRe(CO)_{10}]$ reacts with PPh_3 to form $[(OC)_5MnRe(CO)_4(PPh_3)]$ suggests $[(OC)_5ReMn(CO)_4(PPh_3)]$ could be the initial product of substitution into $[MnRe(CO)_{10}]$ but that it cannot accumulate in substantial amounts. It has also been shown[111] that CO is lost preferentially from the Mn during mass spectrometry but that reaction with t-BuNC, catalyzed by PdO in benzene at ambient temperature, leads to substitution at Re.[112]

Reaction of dppm with $[Cp_2Fe_2(CO)_4]$ leads to disproportionation to form $[CpFe(CO)(dppm)]^+$ and $[CpFe(CO)_2]^-$ by a radical chain reaction involving substitution into $[CpFe(CO)_2]$ radicals formed photochemically or by adventitious thermal processes.[113] A chain length of ~460 was deduced. A novel electron transfer catalysis induced by two-electron reduction of $[Fe_2(CO)_6(\mu\text{-}SR)_2]$ leads to $[Fe_2(CO)_5L(\mu\text{-}SR)_2]$ and $[Fe_2(CO)_4L_2(\mu\text{-}SR)_2]$ [L = $P(OMe)_3$ etc.] by a chain reaction via the dianion $[Fe_2(CO)_6(\mu\text{-}SR)_2]^{2-}$ which contains two labile Fe^- centers and no Fe—Fe bond.[114] In the presence of CO, substitution occurs via associative attack, at an Fe^- center, that is allowed by partial decoordination of a bridging sulfur ligand. For L = $P(OMe)_3$ a rate constant of $3 \times 10^2\ M^{-1}\,s^{-1}$ is inferred. Substitution into $[Fe_2(CO)_6(\mu\text{-}SR)(\mu\text{-}R'CS)]$ occurs at the Fe atom that has an Fe=C bond to the bridging R'CS ligand.[115] Slow breaking of one of the Fe—$(\mu\text{-}SR)$ bonds is possibly the first step followed by addition of L in competition with reformation of the SR bridge. Activation parameters are $\Delta H^+ = 10.2 \pm 0.4$ kcal mol^{-1} and $\Delta S^+ = -30 \pm 1.4$ cal K^{-1} mol^{-1}. The competition ratio for attack at the intermediate by $P(OMe)_3$ is 6 M^{-1} which is surprisingly high if addition of $P(OMe)_3$ is competing with simple closure of the SR bridge. While thermal substitution in these complexes leads to a kinetic product containing an equatorial substituent, ETC substitution can occur at low temperatures and leads to the thermodynamically stable isomer with an axial substituent.[116] Thermal and ETC substitution by t-BuNC leads eventually to a bis substituted product with one substituent on each Fe atom. The second stage of thermal substitution is not much slower than the first, but this is because a higher ΔH^+ is almost balanced by a less negative ΔS^+. The $[Fe_2(CO)_6(\mu\text{-}CRCR'COCR''CR''')]$ complex containing a dienone "flyover" bridge undergoes photodissociation of a CO ligand ($\lambda \geq 400$ nm, $\phi \sim 10^{-2}$) that can lead to substitution or very efficient catalytic isomerization of 1-enes.[117] The flyover is believed to act as a photosensitizer as well as ensuring the stability of the dimer.

Associative substitution kinetics of the 33-electron complex $[Fe_2(CO)_7(\mu\text{-}PPh_2)]$ by various P-donors have been studied electrochemically.[118] The first substituent enters *trans* to the bridging P atom on the six-coordinate Fe atom, but a second substitution leads to a mixture of isomers. Evidence for 35-electron adducts is obtained, and the values of the second-order rate constants are sensitive to the basicity and size of the phosphine nucleophiles according to Eq. (20). The

$$\log k_2 = 6.93 + 0.060 \, pK_a - 0.036\theta \qquad (20)$$

phosphites $P(OMe)_3$ and $PPh(OMe)_2$ react more slowly than predicted by Eq. (20), and an attempt to account for this by adding a π-acidity parameter to the equation is successful in a formal sense. However, it inevitably implies that π-acidity actually decreases the rate which is not what would be expected. The data can, in fact, be perfectly well analyzed by the methods described elsewhere,[1] without recourse to any π-bonding considerations, the steric profile curving down from linearity at low θ values in accordance with the disappearance of the steric effect for the smaller phosphite ligands in the way seen for many other such reactions. Rates are 10^{5-6} faster than for the 34-electron parent anion, which is not isostructural with the radical, and for the 34-electron analog $[FeCo(CO)_7(\mu\text{-}PPh_2)]$ which is. This rate difference is considerably smaller than that observed for 17- and 18-electron mononuclear systems, and this may be related to the fact that the odd electron is distributed mainly between the five-coordinate Fe atom and two of its CO ligands,[118,119] while substitution occurs at the six-coordinate Fe atom.

Facile dissociation (in THF at 60 °C) of two CO ligands from $[Cp_2Mo_2(CO)_4(\mu\text{-dppm})]$ occurs with fission of a P—C bond in the dppm to form $[Cp_2Mo_2(CO)_2(\mu\text{-}PPh_2)(\mu\text{-}Ph_2PCH_2)]$ that contains a Mo=Mo bond.[120] It is not clear at what stage P—C bond fission occurs. Various alkynes react with $[Cp(OC)NiW(CO)_3(MeCp)]$ to form bridged alkyne products at 20 °C with rate constants of $2-5 \times 10^{-3} \, M^{-1} \, s^{-1}$ at low [alkyne].[121] Rate saturation occurs at higher [alkyne] and attack occurs at the Ni atom after reversible formation of a reactive isomer which may be $[(\eta^3\text{-Cp})(OC)NiW(CO)_3(MeCp)]$. The corresponding Mo complex reacts via a similar intermediate, but this can also lead to a nickelacyclobutanone product as well as the bridged alkyne. Rates decrease with increasing size of the alkyne. Homolysis of the complexes occurs in the presence of small amounts of impurities but not in rigorously purified complexes. The coordinatively unsaturated complex $[RuCl(dppb)(\mu\text{-Cl})]_2$ [dppb = 1,4-bis(diphenylphosphino)butane] undergoes equilibrium addition of H_2, under 1 atm H_2, to convert ~60% of the complex to $[(H_2)(dppb)Ru(\mu\text{-Cl})_3Ru(dppb)Cl]$, and acts as a very efficient catalyst for hydrogenation of alkenes as well as for H_2/D_2 exchange.[122] Since the H_2 adduct is coordinatively saturated, breaking a Cl^- bridge or one Ru—P bond would seem to be required in order to allow coordination of the substrate. The H_2 ligand can be replaced by N_2 very readily. $[Cp_2Rh_2(CO)_2(\mu\text{-}CO)]$ undergoes photodissociation of CO to form $[CpRh(\mu\text{-CO})_2RhCp]$ which contains a planar $Rh(\mu\text{-CO})_2Rh$ structure and, presumably, a Rh=Rh bond.[123] Thermal reversal of this reaction occurs by second-order addition of CO with

$k(19\,°C) = 1.2 \pm 0.1 \times 10^3\ M^{-1}\,s^{-1}$, $\Delta H^{+} = 5.3 \pm 0.6$ kcal mol^{-1}, and $\Delta S^{+} = -26 \pm 3$ cal K^{-1} mol^{-1}.

10.1.4. Trinuclear and Larger Clusters

[Ru$_3$(CO)$_{12}$] undergoes reactions with L [L = P(n-Bu)$_3$] in toluene to form successively [Ru$_3$(CO)$_{11}$L], [Ru$_3$(CO)$_{10}$L$_2$], and [Ru$_3$(CO)$_9$L$_3$] by a mixture of dissociative (S$_N$1) and associative (S$_N$2) processes.[1a,124] Each cluster can also undergo fragmentation to form [Ru(CO)$_{5-n}$L$_n$] ($n = 1$ or 2) by associative, so-called F$_N$2, processes. The relative importance of the S$_N$2 and F$_N$2 processes can be established quantitatively by measurements of the final yields of [Ru(CO)$_4$L], [Ru(CO)$_3$L$_2$], and [Ru$_3$(CO)$_9$L$_3$]. About 40% of the associative reactions of [Ru$_3$(CO)$_{12}$] result in fragmentation at 25 °C but this decreases with increasing temperature to a negligible amount at 60 °C. About 30–40% of the associative reactions of [Ru$_3$(CO)$_{10}$L$_2$] lead to fragmentation between 15 and 40 °C, but [Ru$_3$(CO)$_{11}$L] is the most susceptible to fragmentation, virtually all of its associative reactions leading to fragmentation from 25–60 °C. This can be related to the unusually long Ru—Ru bond that is *cis* to the substituent in the Ru$_3$ plane of the cluster.[125] [Ru$_3$(CO)$_{12}$] is known to undergo similar sequences of reactions with other nucleophiles and it should therefore be possible to apply this approach quite generally so as to classify a wide range of clusters according to their susceptibility to F$_N$2 attack by various nucleophiles.

[Ru$_3$(CO)$_{12}$] undergoes thermal decomposition on oxygen-free carbon supports to form Ru metal in high dispersion.[126] The values of ΔH^{+} are significantly lower than for CO dissociation in solution, and this may be due to attack by nucleophilic centers on the surface of the carbon. Incipient formation of metallic Ru—Ru bonds was also postulated as a possible explanation. Similar studies on [Os$_3$(CO)$_{12}$] have been reported.[126] The Cl$^-$ ion substitutes into [Ru$_3$(CO)$_{12}$] to form the very reactive [Ru$_3$(CO)$_{11}$Cl]$^-$ by an associative process that is very much faster than reactions with P-donors.[127] The sequence of reactions following this first step has also been examined,[128] and these results all help to explain why halide ions act as promoters of homogeneous catalysis. The H$^-$ ligand can be substituted into [Ru$_3$(CO)$_{12}$] by means of Et$_3$BH$^-$ which reacts rapidly to form [Ru$_3$(CO)$_{11}$CHO]$^-$ at -78 °C in THF, and subsequently to form [Ru$_3$(CO)$_{11}$H]$^-$ at -50 °C.[129a] Reaction of Et$_3$BH$^-$ with [Co$_3$(CO)$_9$(μ_3-CPh)] at -78 °C also leads to a formyl product; warmup does not lead to formation of [Co$_3$(CO)$_8$(H)(μ_3-CPh)]$^-$ but to [Co$_3$(CO)$_9$(μ_3-CPh)]$^{\doteq}$ and [Co(CO)$_4$]$^-$.[129b]

Detailed photokinetic studies of [Ru$_3$(CO)$_{12}$] with $\lambda = 366$ nm have provided additional evidence in favor of a previously proposed mechanism for this and the related [Os$_3$(CO)$_{12}$] cluster.[130] An important feature of the reaction scheme is the formation of an adduct [Ru$_3$(CO)$_{12}$L] with a variety of ligands of very wide donor strength. These adducts show the unusual feature of being able to revert directly to [Ru$_3$(CO)$_{12}$] in competition with reactions leading to fragmentation of the cluster. The balance between these processes varies greatly with the nature of L and results in a wide range of limiting quantum yields at high [L]. Steric and electronic effects on the rates of formation and further reaction of [Ru$_3$(CO)$_{12}$L] are discussed. Flash photolysis of $[Ru_3(CO)_{12}]$ in iso-octane with higher energy light ($\lambda = 308$ nm) leads to immediate formation of $[Ru_3(CO)_{11}]$, which reacts with CO and THF at ambient temperatures with rate constants of $2.4 \times 10^9 \ M^{-1} s^{-1}$ and $6.1 \times 10^9 \ M^{-1} s^{-1}$, respectively.[131] The THF dissociates with a rate constant of $2 \times 10^6 \ s^{-1}$, and these data provide an equilibrium constant of $3 \times 10^3 \ M^{-1}$ for addition of THF to $[Ru_3(CO)_{11}]$. The near diffusion controlled rates of addition to $[Ru_3(CO)_{11}]$ suggests that it is only very weakly solvated, if at all.

Photolysis of $[Ru_3(CO)_{10}(\mu\text{-H})(\mu\text{-COMe})]$ [in which the H and COMe ligands bridge two $Ru(CO)_3$ moieties] leads, in the presence of donor ligands, L, to formation[132] of an adduct $[Ru_3(CO)_{10}(\mu\text{-H})(\mu\text{-COMe})L]$ similar to the $[Ru_3(CO)_{12}L]$ adduct discussed above. When L = CO this can revert to the starting complex or can form an isomer containing a Ru—C(Me)—ORu bridge. Similar isomers can be formed by CO loss from the adduct when L = $P(n\text{-Bu})_3$, $P(OMe)_3$, or py but the main products are simple substituted forms of the reactant cluster. This system differs from $[Ru_3(CO)_{12}]$ in that higher-energy light leads to adduct formation and lower-energy light leads to no photoproducts while low-energy light leads to adducts of $[Ru_3(CO)_{12}]$ and high-energy light leads to photodissociation of CO. The photoreaction in the presence of H_2 was also discussed.

$[Ru_3(CO)_{10}(\mu\text{-dppm})]$ undergoes associative reactions with more nucleophilic P-donor ligands by $S_N 2$ and $F_N 2$ processes.[133] The cluster is sensitive to the basicity of the nucleophiles to about the same extent as $[Ru_3(CO)_{12}]$ in spite of the presence of the two good P-donor atoms. The sensitivity to the steric effects of the nucleophiles is significantly greater than that of $[Ru_3(CO)_{12}]$, but the standard or intrinsic reactivities in the absence of steric effects are about the same. The activation enthalpies for attack by $P(n\text{-Bu})_3$ on $[Ru_3(CO)_{10}L_2]$ decrease substantially along the series L = $P(n\text{-Bu})_3 > CO > \frac{1}{2}$ dppm and the importance of steric strain in determining the reactivity of the $[Ru_3(CO)_{10}(dppm)]$ cluster is emphasized. The kinetics of associative and dissociative reactions of $[Ru_3(CO)_{11}P(OEt)_3]$ with some P- and As-donor nucleophiles have been reported.[134] Electronic and steric profiles for the associative reactions show the substituent increases the sensitivity of the Ru_3 cluster toward the basicity and size of the nucleophiles, but decreases the standard and intrinsic reactivities of the cluster. Combined with data for other monosubstituted clusters, the activation parameters for the CO dissociative process reveals a wide and systematic variation in ΔH^+ and ΔS^+ with variation of the substituent. This suggests systematic changes in the structures of the coordinatively unsaturated clusters formed by CO dissociation.

The Me_3NO-induced substitutions of a number of $[M_3(CO)_{11}L]$ clusters in $CHCl_3$ have been studied [M = Fe; L = $P(OEt)_3$, $P(OMe)_3$; M = Ru, Os; L = $P(OMe)_3$, $P(OEt)_3$, $P(n\text{-Bu})_3$, PPh_3, $AsPh_3$, and $SbPh_3$].[135] Rates are first order in $[Me_3NO]$ as is usual for such reactions. Only disubstituted products are formed which shows that these are relatively insensitive to Me_3NO. The substituents all decrease the rates compared with those for $[M_3(CO)_{12}]$, and the rates for $[M_3(CO)_{11}L]$ decrease along the series M = Fe > Ru > Os, although the activation enthalpies alone would lead to the sequence Fe < Ru $\gg$ Os. Attack by the Me_3NO occurs at an unsubstituted metal atom and, for M = Ru and Os in $CHCl_3$, values of $\log k_2$ increase systematically with the decreasing net electron-donor capacity of L (measured by the increase in the highest value of ν_{CO} for the various complexes) along the series L = P(n-Bu)$_3$ < PPh$_3$ < AsPh$_3$ < SbPh$_3$. The P(OMe)$_3$ and P(OEt)$_3$ substituents react more slowly than expected from this trend. The data can all be encompassed quite well by Eq. (21), where β_L and γ_L represent the dependence

$$\log k_2 = \alpha_L + \beta_L \nu_{CO} + \gamma_L \theta \tag{21}$$

of $\log k_2$ on the electronic and steric properties of the substituent. Values of β_L and γ_L are given in Table 10.5, where they are compared with corresponding values obtained from rate constants for the spontaneous loss of CO from [Ru$_3$(CO)$_{11}$L] clusters.[136]

The increasing rates with decreasing net electron-donor capacity of L for Me$_3$NO-induced reactions is clearly quantified by the β_L parameters and this contrasts with the decreasing rates, and negative β_L parameter, found for the spontaneous reactions. The latter clearly involve transition state stabilization by the better donor substituents. The positive values of γ_L for all the reactions show CO loss, whether induced or spontaneous, is facilitated by the steric strain caused by larger substituents. As expected, the benefit of larger substituents is not so pronounced when nucleophilic attack at a CO ligand is involved. The fact that substituent effects are negligible when M = Os and L = P(n-Bu)$_3$, PPh$_3$, and AsPh$_3$, and when the reactions are carried out in CH$_2$Cl$_2$, is in accord[135] with the lower H-bonding power of CH$_2$Cl$_2$ compared with CHCl$_3$ and the consequent higher activity of the Me$_3$NO. This results in generally faster rates and, we suggest, greater extents of Me$_3$NO$\cdots$CO bond making and Os$\cdots$CO bond breaking. These will tend to change the electronic effect of better donors from ground-state stabilization (positive β_L) toward transition-state stabilization (negative β_L), but should increase the steric effect toward that for the spontaneous reactions. The balance between these two effects could lead to the observed absence of any substituent effect on the rates.

The cluster [Fe$_2$Os(CO)$_{12}$] undergoes dissociative reactions with PPh$_3$ etc. and the substituent appears at the Os atom.[137] Nevertheless, it is argued that CO dissociates from an Fe atom since the rate parameters are closer to those for [Fe$_3$(CO)$_{12}$] than for [Os$_3$(CO)$_{12}$]. Reactions of [Os$_3$(CO)$_{11}$(NCMe)] and [Os$_3$(CO)$_{10}$(NCMe)$_2$] with H$_2$ in 1,2-dichloroethane proceed by rate-determining dissociation of the labile MeCN ligands to form [(μ-H)(H)Os$_3$(CO)$_{11}$] and [(μ-H)$_2$Os$_3$(CO)$_{10}$], respectively.[138] Relative rates of attack on the [Os$_3$(CO)$_{11}$] inter-

Table 10.5. Electronic and Steric Parameters for Loss of CO from [M$_3$(CO)$_{11}$L] Clusters [see Eq. (21)]

M	β_L (cm)	γ_L (deg^{-1})	RMSD^a
Ru^b	0.062 ± 0.009	0.016 ± 0.003	0.063
Os^b	0.040 ± 0.009	0.014 ± 0.003	0.062
Ru^c	−0.025 ± 0.012	0.028 ± 0.005	0.172

^{a} Root-mean-square deviation of $\log k$ (calc) from $\log k$ (expt).
^{b} Loss of CO induced by Me$_3$NO in CHCl$_3$.
^{c} Parameters for spontaneous thermal dissociation of CO in hexane.

mediate can be obtained and vary along the series $MeCN > PPh_3 \sim CO > H_2$. The value of ΔH^+ for attack by H_2 is ~ 4 kcal mol^{-1} larger than for attack by CO, but addition of CO is entropically more favored. The value of k_H/k_D for attack by H_2 and D_2 is 1.36 ± 0.03, which is compatible with a normal three-center transition state. Thermodynamic parameters ($\Delta H° = -3.7 \pm 0.3$ kcal mol^{-1} and $\Delta S° = -11 \pm 1$ cal K^{-1} mol^{-1}) were obtained for addition of MeCN to $[(\mu\text{-}H)_2Os_3(CO)_{10}]$ to form $[(\mu\text{-}H)(H)Os_3(CO)_{10}(NCMe)]$, but these values represent the combination of the effects of forming a new Os—NCMe bond, lengthening the Os—Os bond by a considerable amount, and changing one bridging H into a terminal one.

An extensive study of the electrochemical behavior of a large group of $[Co_3(CO)_{9-n}L_n(\mu_3\text{-}CMe)]$ clusters ($n = 0\text{-}3$; L = a variety of phosphine ligands) includes kinetic data for the rapid loss of L from the 49-electron clusters $[Co_3(CO)_8L(\mu_3\text{-}CMe)]^{-}$.[139] Reduction does not result in fission of any of the Co—Co bonds and the 49-electron anions relieve their excess of ligands by ligand dissociation. The rates of loss of L at -43 °C increase in the sequence $PMe_3 < PMe_2Ph < PMePh_2 < PPh_3$ and it is concluded that steric effects are dominant. However, the activation enthalpies follow the sequence $PMe_3 \leq PMePh_2 < PMe_2Ph$, covering a range of 3.6 kcal mol^{-1}, while the values of $T\Delta S^+$ vary in the same order and cover a range of 3.3 kcal mol^{-1}. The rates at -13 °C would have all been approximately the same, so inferences based on the rates at one temperature slightly below this do not have much significance. The small positive, or quite substantially negative, values of ΔS^+ also suggest some complexity in the dissociative processes. Elimination of CO from the various 49-electron intermediates competes with loss of L less effectively the higher the value of n, and the relative importance of these processes, and whether CO loss is followed by L addition or L loss is followed by CO addition (both sequences leading to substitution), or whether ligand loss is followed by reduction to the 48-electron dianions $[Co_3(CO)_{9-n}L_{n-1}(\mu_3\text{-}CMe)]^{2-}$, all combine to lead to a rationalization of the complex electrochemical behavior in a detailed general scheme.

Substitution reactions of $[Fe_2Co(CO)_9(\mu_3\text{-}CCO)]^{-}$ by P-donors occur at the Co atom by associative processes involving metal–metal or metal–C bond breaking.[140] Electronic and steric profiles were obtained, the latter curving down at lower θ as the steric effect disappears. The transition state is more polar than the ground state but there are only small cation effects. After substitution at the Co atom the P-donors slowly undergo pairwise interchange with the CO in the bridging CCO^{-} ligand. Rates decrease with increasing size of the P-donor but $P(OMe)_3$ and $P(OPh)_3$ react very much more slowly than expected from a simple steric trend, possibly because the P-donors are simultaneously leaving and entering groups. Those complexes that undergo slower P-donor migration to the bridging C atom react by an alternative path involving introduction of a second substituent onto the Co atom followed by migration of one of the substituents. The substitution process shows increases in k to a limiting value at high [L], and are believed to involve reversible cluster opening followed by competing attack by L, followed by CO loss and cluster closing.

The replacement of CO by H_2 in $[Ru_{3-n}Os_n(CO)_{10}(\mu_2\text{-}H)(\mu_2\text{-}COMe)]$ to form $[Ru_{3-n}Os_n(CO)_9(\mu_2\text{-}H)_3(\mu_3\text{-}COMe)]$ ($n = 1$ and 2) has been studied, together with the reverse reaction, and the results compared with those for $n = 0$ and 3.[141] The reactions are dissociative and, for loss of CO or H_2, the rates for the Os_3 clusters are much the slowest and suggest that only one Ru atom is required for relatively rapid reaction to occur, and that reductive elimination of H_2 occurs through a $Ru(H)_2$ species. Some not unreasonable assumptions lead to the bond strengths Ru—H—Ru (98 kcal mol^{-1}), and Os—H—Os (102 kcal mol^{-1}), and to some bond strength differences including (Ru—H—Os) − (Ru—Os) = 71 kcal mol^{-1}. For $n = 0$, $\Delta V^+ = +9.6 \pm 0.6$ and $+20 \pm 2$ cm^3 mol^{-1} for dissociative loss of CO and H_2, respectively.[142] The low value for CO loss is taken to indicate contraction of the cluster, a (μ_3-COMe) form being generated, in which both the C and O atoms are bonded to the Ru atom that has lost the CO. Addition of H_2 to this intermediate leads to decoordination of the O atom without much change in volume, and a similarly small volume change occurs when H_2 is lost from the product and the O recoordinates. The value of ΔV^+ for loss of H_2 is therefore normal. Some interesting reactions of the analogous [Ru$_3$(CO)$_9$(μ-H)$_3$(μ_3-CSR)] cluster with alkynes, and isomerization and reactions with H_2 of the products, have been described.[143]

[Pd$_3$(μ_3-CO)(μ_2-dppm)$_3$]$^{2+}$ undergoes CO dissociative reactions with some SCE species (E = N$^-$ and S) after preliminary reversible addition of the SCE to a Pd atom.[144] (The three dppm ligands bridge the three sides of the Pd$_3$ triangle.) Adduct formation is effectively complete with SCN$^-$; $\Delta H^+ = 24.9 \pm 1.0$ kcal mol^{-1}, $\Delta S^+ = 4.1 \pm 1.2$ cal K^{-1} mol^{-1}, and $k_1(25\,°C) = 3.6 \times 10^{-5}$ s^{-1} for loss of CO from this adduct in MeOH. Reaction with CS$_2$ leads to small amounts of adduct formation, and displacement of CO becomes first order in [SCE] in this case with $k_2 = 6.8 \times 10^{-3}$ M^{-1} s^{-1} for the Pd complex and 3.6×10^{-2} M^{-1} s^{-1} for the Pt analog. Loss of CO is accompanied by development of S and C bonding to different metal atoms and, for E = N$^-$, eventual S—C bond fission and the formation of [M$_2$(μ-SMCE)]$^{2+}$ products which still have the three metal atoms bridged by dppm ligands.

A review of the substitutional reactivities of the clusters [H$_x$M$_4$(CO)$_{12}$] (M$_4$ = Fe, Re, Co, or Rh, or combinations thereof) and their structures has emphasized the importance of site selectivity,[145] and the reaction of [Co$_2$Rh$_2$(CO)$_{12}$] with CO (1.7 atm) at 0 °C in n-hexane to form [CoRh(CO)$_7$] proceeds with $t_{1/2} \sim 10$ min.[146] The dinuclear complex reacts with H_2 (0.7 atm H_2, 1.7 atm CO; $t_{1/2} \sim 20$ min) to form [HCo(CO)$_4$] and [Rh$_4$(CO)$_{12}$] showing a highly facile activation of H_2. A very thorough study of the substitution reactions of [M$_4$(CO)$_9$(tripod)] has been reported where M = Co, Rh, and Ir, and tripod = HC(PPh$_2$)$_3$.[147] Substitution occurs at the apical metal atom with $k_{obs} = k_1 + k_2[L]$, although k_1 is negligible for M = Ir. Reactions are generally slower than in the unsubstituted analogs because substitution strengthens M—CO bonding for electronic reasons and deters associative attack, perhaps for steric and electronic reasons. The Ir complex reacts much more slowly than [Ir$_4$(CO)$_9$(PPh$_3$)$_3$] because of more facile CO dissociation, attributed to the effects of CO bridging, in the latter cluster. Metal-dependence of the k_1 path shows the sequence Ir(10^{-5}) ≪ Co(1) < Rh(700) while the k_2 path shows the sequence Ir(10^{-3}) < Co(1) ≪ Rh(10^6). The product [Rh$_4$(CO)$_8$L(tripod)] reacts further with L when $\theta \leq 132°$ to introduce a second substituent at the apical metal, but this migrates subsequently to a basal position. The results are compared in detail with those for related complexes and the high susceptibility of these clusters, as with clusters generally, toward associative attack is discussed in terms of a variety of different factors including a favorably disposed LUMO, metal–metal bond strengths, and edge-opening of the clusters, etc.

The clusters $[Fe_nRu_{4-n}(CO)_{11}(\mu_4\text{-}PPh)_2]$ add two CO ligands reversibly over 30–60 min at room temperature under 1 atm CO when $n = 1$-3, but not when $n = 4$.[148] Addition is irreversible when $n = 0$. The product contains two μ_3-PPh and only two M—M bonds. Substitution with (p-tolyl)NC goes by addition of the nucleophile followed by CO dissociation, but P-donors do not show formation of the intermediate adduct.

The slow ($t_{1/2} \sim 5$ h at 90 °C) isomerization of the cluster $[CpMoCo_2(CO)_5\{\mu_3\text{-}Mo(CO)_2Cp\}\{\mu_2\text{-}o\text{-}(RP)_2C_6H_4\}]$ has been studied where the o-$(RP)_2C_6H_4$ ligand changes from bridging the two Mo—Co bonds in the basal MoCo$_2$ triangle to bridging one Mo—Co bond and one Co—Co bond in the triangle.[149] Values of $\Delta H^+ = 29.8 \pm 1.0$ kcal mol^{-1} and $\Delta S^+ = 3 \pm 4$ cal K^{-1} mol^{-1} were obtained, but whether the reaction is truly intramolecular or involves metal ligand dissociation is not yet clear. The high nuclearity carbonyl cluster (HNCC) *trans*-[$\{Rh(CO)_2\}_2$-$\{Fe(CO)_3\}_4B]^-$, containing an octahedral Rh$_2$Fe$_4$ array around an encapsulated B atom, undergoes very facile isomerization to the *cis* form at ambient temperatures but only in the presence of nucleophiles such as CO.[150] The reaction is first order in [CO] with $k_2 = 4.2 \times 10^{-2} M^{-1}s^{-1}$, $\Delta H_2^+ = 17.0$ kcal mol^{-1}, and $\Delta S_2^+ = -7.2$ cal K^{-1} mol^{-1}. The P-donor PMe$_2$Ph also promotes isomerization, but less efficiently. The mechanism is proposed to involve initial Fe—Fe bond breaking concerted with CO entry so that no change of electron count is involved. This weakens the Rh$_2$Fe$_4$ cluster so that rearrangement can occur followed by final loss of the added ligand. This type of facile cluster rearrangement induced by nucleophiles is of obvious general importance and interest.

$[Ir_6(CO)_{16}]$ reacts rapidly with Cl$^-$ in dry THF at room temperature to form $[Ir_6(CO)_{15}Cl]^-$ and then $[Ir_6(CO)_{14}Cl]^-$.[151] The Cl$^-$ can be replaced by SCN$^-$, Br$^-$, and I$^-$, and the Br$^-$ and I$^-$ products are stable to CO. The HNCC $[Rh_6C(CO)_{13}(AuL)_2]$ contains a distorted Ru$_6$ octahedron around an encapsulated C atom.[152] Two adjacent faces are capped by the AuL moieties which are joined by an Au—Au bond. This cluster adds two CO ligands at room temperature over 1–2 hours to form $[Rh_6C(CO)_{15}(AuL)_2]$, which contains a trigonal biprismatic Rh$_6$ cluster with the two triangular faces capped with the two AuL groups, a reaction of obvious mechanistic and kinetic interest. A number of $[Os_{10}C(CO)_{24}(ML)_2]$ clusters also undergo isomerization reactions that are proposed to go via cap $\rightleftharpoons$ edge bridge $\rightleftharpoons$ cap rearrangements (M = Au, Ag, or Cu, or combinations thereof, and L = PPh$_3$ or PMe$_2$Ph).[153]

Although substitution of HNCCs can sometimes be effected by Me$_3$NO activation, yields are not always satisfactory. An efficient alternative involves chemical oxidation of quite readily obtained anionic HNCCs such as $[Ru_6C(CO)_{16}]^{2-}$, $[Os_7C(CO)_{20}]^{2-}$, and $[Os_{10}C(CO)_{24}]^{2-}$ with Ag$^+$, Ph$_3$C$^+$, or C$_7$H$_7^+$ in CH$_2$Cl$_2$ or CH$_3$CN, and in the presence of P-donor ligands.[154] Excellent yields of the neutral monosubstituted HNCC products are obtained, the two electrons in the $[HNCC]^{2-}$ having been replaced by two electrons from the entering ligand.

10.2. Insertion Reactions

10.2.1. Carbon Monoxide Insertion

Enthalpies for reactions shown in Eq. (22) have been measured, where R = Me, Et, and L = several P-donors.[155] For phosphines they become more favorable as

$$[CpMo(CO)_3R] + L \rightarrow [CpMo(CO)_2L(COR)] \qquad (22)$$

the basicity of the phosphine increases and as the size decreases (R = Et: L = PPh_3, −15.4; $PMePh_2$, −18.2; PMe_2Ph, −19.7; PMe_3, −22.5 kcal mol^{-1}). Steric effects must be important since the values for L = $P(OMe)_3$ and PMe_2Ph are the same, the former being much less basic but considerably smaller. The values become slightly more favorable, by 3–4 kcal mol^{-1}, along the series R = Me < CH_2Ph < Et. Estimates of the enthalpies for CO insertion, Eq. (23), were also made. These

$$[CpMo(CO)_2LR] + CO \rightarrow [CpMo(CO)_2L(COR)] \qquad (23)$$

become increasingly favorable as the net electron-donor capacity of L increases [R = Et: L = $P(OMe)_3$, −15.2; PPh_3, −16.3; $PMePh_2$, −16.5; PMe_2Ph, −15.9; $P(n\text{-}Bu)_3$, −16.8; PMe_3, −17.5 kcal mol^{-1}]. The position of $P(OMe)_3$ in this series suggests that steric effects are not as important as for the reactions in Eq. (22), as might be expected. A thermodynamic reaction profile for the hydroformylation reaction is deduced which shows that the only endothermic stage (by 6 kcal mol^{-1}) is the reaction of $[Co_2(CO)_8]$ with H_2 to form $[HCo(CO)_4]$. Thermodynamics of hydride transfer in the gas phase enable an estimate of $\Delta H^\circ = +10 \pm 7$ kcal mol^{-1} to be made for reaction (24) and this leads to a value of ΔG° of about +21 kcal mol^{-1} for the reaction at room temperature under 1 atm CO.[156] The corresponding

$$[Fe(CO)_4H]^- + CO \rightarrow [Fe(CO)_4CHO]^- \qquad (24)$$

carbonylation reaction of $[HCo(CO)_4]$ has been calculated to have $\Delta H^\circ = +16$ kcal mol^{-1} compared to −5 kcal mol^{-1} for $[MeCo(CO)_4]$.[157] The migration of the axial Me group in the latter complex has a low activation barrier and $\Delta H^\circ \leq 17$ kcal mol^{-1} while direct insertion was calculated to have a barrier of *ca* 50 kcal mol^{-1}.

The kinetics of the reactions (25) have been reported to be first order in $[PPh_3]$ in toluene, THF, and CH_2Cl_2 [158] (R = Me and $p\text{-}XC_6H_4$; X = H, Me, Cl). Values

$$[(Cp^*)Rh(CO)IR] + PPh_3 \rightarrow [Cp^*Rh(PPh_3)I(COR)] \qquad (25)$$

of ΔH^+ and ΔS^+ for reactions in toluene are 12–15 kcal mol^{-1} and −10 to −19 cal K^{-1} mol^{-1}, respectively, and the reaction is faster, for entropic reasons, for

R = Ph than for R = Me. Reactions in MeCN and $MeNO_2$ are considerably faster and show limiting rates at high [PPh_3]. Reactions are faster with more electron-donating substituents, X.

The stereochemistry of the low-temperature insertion of CO into [$M(CO)_2(PMe_3)_2(X)Me$] in toluene has been examined.[159] (The PMe_3 groups are *trans* to each other and the CO ligands are *cis*; M = Fe or Ru, and X = I or CN.) There is clear evidence for the migration mechanism. The photochemical reaction of [$Cp^*Os(CO)_2(CH_2Ph)$] with PMe_2Ph leads to CO insertion but also to migration of the benzyl radical, formed by Os—C homolysis, to form [$(\eta^4-C_5Me_5CH_2Ph)Os(CO)_2L$].[160]

The ^{12}CO and ^{13}CO groups in [$(Ind)Fe(^{12}CO)_2(^{13}COMe)$] do not interchange except upon addition of [$CpFe(CO)_2$]$^-$ (Fp^-).[161] Fp^- is believed to attack a ^{12}CO ligand and this is followed by methyl migration to the central Fe atom, leaving all CO ligands terminal. The methyl migration requires slippage of the indenyl ring. Reactions of P-donors with [$Cp(OC)_2Fe(CH_2)_nFe(CO)_2Cp$] ($n = 3-7$) lead first to [$Cp(OC)_2Fe(CH_2)_nC(O)Fe(CO)LCp$] and then to [$CpL(OC)$-$FeC(O)(CH_2)_nC(O)Fe(CO)LCp$].[162] The corresponding reactions of [$Cp(OC)_3W(CH_2)_nW(CO)_3Cp$] are much slower, and substitution of one CO ligand by L on each W atom occurs as well. Migratory insertion of CO into the Fe—C bond of [$Cp(OC)_2Fe(CH_2)_3Si(OMe)_3$] is induced by PPh_3 but one-electron reduction in MeCN catalyzes the process which probably is accompanied by coordination of a solvent molecule.[163] Irreversible rapid insertion of CO into [$(PP_3)RhMe$] [$PP_3 = P(CH_2CH_2PPh_2)_3$] is probably facilitated by decoordination of one P atom.[164] Insertion of CO into [$(\eta^3-allyl)Pd(PMe_3)_2$]$^+X^-$ to form [$allylC(O)Pd(PMe_3)_2X$] is also facile.[165]

A detailed kinetic study of the migration of the benzyl group in [$RhCl_2(PPh_3)_2(COCH_2Ph)$] has been reported.[166] The benzyl group is axially bound to a planar *trans*-[$RhCl_2(PPh_3)_2$] moiety and the migration proceeds reversibly to form [$RhCl_2(CO)(PPh_3)_2(CH_2Ph)$], which is the *trans*-(PPh_3)$_2$, *cis*-(Cl)$_2$ isomer. This loses $PhCH_2Cl$ to form [$RhCl(CO)(PPh_3)_2$] with $k(5\,°C) = 6 \times 10^{-4}\,s^{-1}$, $\Delta H^+ = 17.0 \pm 0.4\,kcal\,mol^{-1}$, and $\Delta S^+ = -12.2 \pm 1.6\,cal\,K^{-1}\,mol^{-1}$. The reaction is complicated by a reversible isomerization of the initial [$RhCl_2(PPh_3)_2(COCH_2Ph)$] complex to form appreciable amounts of the isomer in which the benzoyl group is axially bound to a planar *cis*-[$RhCl_2(PPh_3)_2$] moiety, but this does not undergo migration of the CH_2Ph group and is a "dead end."

Insertion of CO into the Ta—Si bond in [$Cp^*Cl_3TaSiMe_3$] occurs to form a thermally unstable [$Cp^*Cl_3Ta(\eta^2-COSiMe_3)$] complex which undergoes some interesting further insertion processes.[167] Reaction of CO with [$(OEP)Rh$]$_2$ occurs very readily to form an equilibrium mixture of the adduct [$\{(OEP)Rh\}_2CO$] (in which the CO is terminally bound to a Rh atom), and the insertion products [$(OEP)RhC(O)Rh(OEP)$] and [$(OEP)RhC(O)C(O)Rh(OEP)$], where OEP = octaethylporphyrin.[168] Values of equilibrium constants, and $\Delta H°$ and $\Delta S°$ values for all the equilibria are reported and the formation of the adduct, the monoinsertion product, and the double insertion product are all exothermic with $\Delta H° = -10 \pm 1$, -12 ± 2, and $-21 \pm 2\,kcal\,mol^{-1}$, respectively. The major factor involved in these reactions is the weak Rh—Rh bond, due to repulsion between the OEP ligands.

10.2.2. Other Insertions

Insertion of the dienes $CH_2=CRCR'=CH_2$ into the Fe—H bond in [Cp(OC)$_2$FeH] occurs with rates first order in [complex] and [diene] but independent of CO.[169] An inverse primary H/D isotope effect is observed and the rate-determining step is shown to be H-atom abstraction by the diene to form the allyl radical, $CH_2-CR=CR'CH_3$, which can either react with the [Cp(OC)$_2$Fe] radical in the solvent cage, to form the insertion product [Cp(OC)$_2$Fe—CH$_2$CR=CR'Me], or it can diffuse away from the [Cp(OC)$_2$Fe] to form other products.

Photolysis of [(HBPz$_3^*$)Rh(CO)(C$_2$H$_4$)] in benzene [HBPz$_3^*$ = tris-(3,5-di-Me-pyrazol-1-yl)borato] leads to approximately equimolar amounts of [(HBPz$_3^*$)Rh(CO)(H)Ph] and [(HBPz$_3^*$)Rh(CO)(Et)Ph], probably via a common intermediate.[170] The thermodynamic instability of M—Et bonds toward β-elimination in some early transition metal compounds has been examined and discussed in terms of strong metallacyclopropane bonding of ethylene to the metals.[171]

Second-order insertion reactions of alkynes into the Ru—H bond in [Ru(CO)-(H)(NCMe)$_2$(PPh$_3$)$_2$]$^+$ have been reported[172] as has the unusual insertion of HC$_2$CN into the C—H bond in [Cp$_2$Fe$_2$(CO)$_2$(μ-CO)(μ-C=CHR)].[173] Reaction of alkynes, HC$_2$R', with [RMn(CO)$_5$] occur via initial CO insertion followed by alkyne attack at [(RCO)Mn(CO)$_4$] and insertion into the RC(O)—Mn bond to form [(OC)$_4$Mn-O=C(R)-CH=CR'] which contain aromatic five-membered rings.[174] The mer-[Rh(PMe$_3$)$_3$Cl(H)(η^1-OOCR)] (R = CH$_2$CH$_2$C≡CH) undergoes insertion of the pendant alkyne group into the Rh—H bond to form a product containing a six-membered rhodalactone, Rh-C(=CH$_2$)(CH$_2$)$_2$C(O)O, ring.[175] Insertions of C$_2$(COOMe)$_2$ into Ru—H and Ru—C$_6$H$_4$X bonds probably occur via initial replacement of a CO ligand by the alkyne.[176] In toluene the Ind$^-$ ligand is displaced from [(η^5-Ind)Ir(COD)] by PMe$_3$ and attacks the COD ring, but in THF the Ind$^-$ ligand is simply replaced by three PMe$_3$ ligands.[177]

Insertion of CO$_2$ into Cu—H and Cu—Me bonds has been examined theoretically and the former is very different because of the ready formation of a Cu—H—CO$_2$ adduct prior to formation of an η^1- or η^2-formato complex.[178] Formation of a [(OC)$_5$CrH—CO$_2$]$^-$ adduct is calculated to be exothermic by ~8 kcal mol^{-1} and this can be followed by concerted CO dissociation and formation of [(OC)$_4$Cr(η^2-O$_2$CH)].[179]

CE$_2$ inserts into the M—OPh bond in [(OC)$_5$MOPh]$^-$ with relative rates Cr > W and E$_2$ = S$_2$ > OS > O$_2$. CO$_2$ inserts into the M—OPh bond in [(OC)$_4$LMOPh]$^-$ with relative rates W > Cr and L = P(OMe)$_3$ > PMe$_3$ > PPh$_3$.[180] CS$_2$ inserts into the W—SR bond in [Cp(OC)$_2$(PPh$_3$)WSR] via formation of [Cp(OC)$_2$(CS$_2$)WSR] and then [Cp(OC)$_2$W(η^2-S$_2$CSR)] (R = CHMe$_2$, CH$_2$Ph, C$_6$H$_5$, and p-tolyl).[181] Rates increase with increasing [CS$_2$] and decrease with increasing [PPh$_3$], and are slower for [Cp(OC)$_3$WSR]. Reactions of the 16-electron [M(CO)Cl(PPh$_3$)$_2$(R)] (M = Ru, Os; R = o-tolyl) in liquid SO$_2$ lead to formation of $[M(CO)Cl(PPh_3)_2(SO_2R)]$ and then (M = Ru) $[M(CO)Cl(PPh_3)_2(OSOR)]$, but the detected adduct $[M(CO)Cl(PPh_3)_2(SO_2)R]$ (M = Os) does not seem to be an intermediate.[182] $[Cp(PPh_3)_2Ru(C_2Ph)]$ reacts with CS_2 to form $[Cp(PPh_3)Ru(\eta^2\text{-}S_2CC_2Ph)]$.[183] $[(OC)_5WER_3]^-$ readily undergoes insertion of SO_2, but not CO_2, to form $[(OC)_5WS(O)_2ER_3]^-$ (where E = Si and R = Me or Et; E = Sn and R = Me or Ph; E = C and R = H).[184] For $[(OC)_5WSnMe_3]^-$ reaction proceeds via instantaneous formation of $[(OC)_5WOS(O)SnMe_3]^-$, which subsequently forms the more stable S-bonded isomer.

The $[Cp^*IrN(t\text{-}Bu)]$ complex undergoes addition of CO_2 [and CO or $CN(t\text{-}Bu)$] across the Ir=N(t-Bu) bond but complete insertion does not occur.[185] The electrophiles NO^+, NS^+, and $p\text{-}NO_2C_6H_4N_2^+$ all insert into the Cr—R bond in $[Cp(ON)_2CrR]$ (R = Me, CH_2SiMe_3, Ph) by classical S_E2 processes.[186]

Acknowledgments

The invaluable help of Mr. Lezhan Chen and Ms. Ying Zheng in preparing this chapter is gratefully acknowledged.

Chapter 11

Metal-Alkyl and Metal-Hydride Bond Formation and Fission; Oxidative Addition and Reductive Elimination

11.1. Introduction

A premiere area of organotransition metal chemistry is the scission and formation of C—C, C—H, and H—H bonds of organic substrates. These transformations have been demonstrated to be accessible through the use of coordinatively unsaturated organometallic fragments as shown in Eq. (1), where R, R' = alkyl, aryl, H, etc. Mechanistic aspects of this reaction have still not been fully appreciated.

$$L_nM \ + \ R\text{-}R' \ \rightleftharpoons \ [L_nM\,(R)\,(R')] \tag{1}$$

However, the discovery of σ-coordination by H_2 is an important recent advance in this area. The three-center, two-electron "bond" is also recognized in intramolecular agostic C—H metal interactions. Recently evidence has been mounting in support of a C—H σ-interaction for "free" hydrocarbons (*vide infra*).

Notable during latter 1988 and 1989 was the publication of *Activation and Functionalization of Alkanes*,[1] which devotes much attention to homogeneous and heterogeneous organometallic C—H bond activation in its 372 pages. Also during the reporting period, two entire journal issues were dedicated to relevant topics: Alkane Activation and Functionalization [*New J. Chem.*, **13**, number 11 (1989)] and Metal–Ligand Bonding Energetics in Organotransition Metal Compounds [*Polyhedron*, **7**, number 16/17 (1988)]. An excellent review comparing alkyl and aryl C—H bond activation appeared,[2] as did reviews discussing reactions of metal–hydride complexes with hydrocarbons,[3] C—C bond formation via reductive elimination,[4] agostic complexes,[5] and H_2 addition to iridium(I) species.[6] Also published were reviews on gas-phase activation by metal ions,[7,8] cyclopalladation,[9] polynuclear iridium hydride complexes,[10] and the important photocarbonylation reaction of hydrocarbons catalyzed by [RhCl(CO)(PMe_3)_2].[11,12]

11.2. Dihydrogen Complexes

The literature concerning η^2-dihydrogen complexes grew substantially during the reporting period. The designation of "trihydrogen complexes" applied to $[CpIrLH_3]^+$,[13] $[Cp^*RuLH_3]$,[14] and $[(RCp)_2NbH_3]$, $R = Me$, Me_3Si,[15] was recanted. New work, published concurrently by two groups,[16,17] instead attributed the tremendous J_{H-H} scalar couplings (measuring up to 1400 Hz) seen in the above compounds to quantum-mechanical proton exchange occurring in otherwise "classical" trihydride species. The novel tunneling mechanism was confirmed by the observation of J_{H-T} in agreement with calculated values. Still open to speculation is the structure of (**2**), which exists [see Eq. (2)] in thermal equilibrium with the *cis*-$(H_2)(H)$ complex (**1**).[18] ^{31}P NMR data confirms the change in geometry

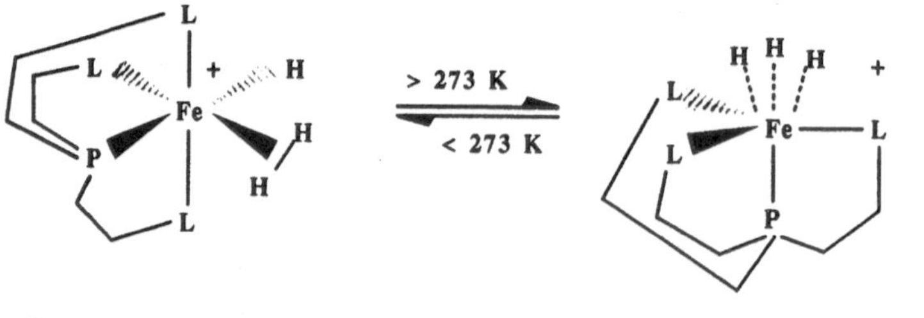

(**1**)   L = PPh$_2$ (**2**)

shown. The actual configuration of the hydrogen ligands in (**2**) is not yet known, but the compound is quite resistant to H_2 substitution. The corresponding $[(PP_3)MH_2]^+$, $PP_3 = (Ph_2PCH_2CH_2)_3P$, $M = Rh$, complex exists as a dihydride with octahedral geometry (like that of **1**) below 173 K and a trigonal bipyramid with an axial η^2-H_2 ligand above 173 K. This is borne out by experimental[19] and *ab initio*[20] results. For $M = Co$, only the dihydrogen species is accessible and only the classical dihydride for $M = Ir$.[20]

There has been mounting discussion on the criteria for evaluation of H—H distances and the formulation of fluxional polyhydrides as classical vs. nonclassical. It was pointed out[21] that the predicted H—H bond distances from NMR T_1 relaxation values[22] must be corrected for relaxation effects due to H_2 ligand rotation. This was supported by neutron diffraction data.[23] Crabtree[22] has suggested that temperature-dependent minimum T_1 times be used to identify molecular

Table 11.1. Rhenium Polyhydride Complex Structures

Species[a]	Solid-state structure[b]	Solution conformation	Comments	References
ReH_9^{2-}	Classical	Classical	High T_1	22
ReH_8^-	—	Classical	High T_1	22
ReH_7L_2 [c]	Classical	Probably Classical	Low T_1, but low IPR shift[d]	25–27
ReH_5L_3	Classical	Possibly Nonclassical	Low T_1, IPR not yet studied	28, 29
$ReH_4L_4^+$	Classical	Probably Classical	Moderate T_1, no D_2 exchange	28, 30, 31
$Re_2H_8L_4$	Classical	Probably Classical	Moderate T_1	32
$TpReH_6$ [e]	—	Probably Classical	Low T_1, but low IPR shift	28, 33
$TpReH_4L$	—	Probably Classical	Low T_1, but low IPR shift	27
$ReClH_2L_4$	Nonclassical	Nonclassical	Very low T_1	30, 34

[a] L = Phosphine unless otherwise noted.
[b] Neutron or X-ray diffraction.
[c] L = Phosphine or Bp, bis(pyrazolyl)methane.
[d] Isotopic perturbation of resonance (see text).
[e] Tp = tris(pyrazolyl)borate.

hydrogen complexes. He has offered T_1 (minimum) values of <80 ms (at 250 MHz) as indicating the presence of η^2-H_2 ligand(s), while those >150 ms would imply a probable lack thereof. These threshold values were called into question in one study on group VIII metal–dihydrides[24]; however, care was not taken to convert the T_1 data (which is field-dependent) to 250 MHz. Much effort was devoted to structural characterization of the diverse rhenium polyhydrides. A summary of this work is presented in Table 11.1. Solid-state structural determinations of these species have generally indicated classical metal–hydride configurations. The solution structures, however, are quite difficult to sort out. Very low T_1 values are often encountered. These may be artifacts of low moments of inertia in these complexes or they may reflect the existence of η^2-H_2 intermediates in the course of nonrigid behavior. New techniques are needed to assess these compounds. Electrochemical oxidation has been used to promote rapid H_2 loss from [$ReH_7(PPh_3)_2$], leading the authors to speculate a dihydrogen configuration.[29] More reliable evidence is provided by the isotope perturbation of resonance (IPR) technique. This predicts differing proton chemical shifts and temperature-dependent isotope fractionation in randomly deuterated polyhydrides. The IPR method indicated classical structures for [$TpReH_6$], Tp = tris(pyrazolyl)borate, [$TpReH_4(PPh_3)$], [$BpReH_7$], Bp = bis(pyrazolyl)borate and [$ReH_7diphos$], diphos = dppb and *bis*-1,1′-(diphenylphosphino)ferrocene.[26,27] These species all display very short T_1 values. One rhenium hydride that is nonclassical,

$[ReCl(H_2)(PMePh_2)_4]$,[30,34] would be predicted to have a classical configuration based on the corresponding dinitrogen complex N_2 stretching frequency.[35] This may be due to steric confinement of the hydrogens.

Kinetic studies of H_2 replacement by N_2 or other ligands have revealed a low binding enthalpy for $M-H_2$.[36-38] However, dihydrogen coordination is entropically favored over that of dinitrogen due to the lower mass of H_2. The disproportionation shown in Eq. (3) of 17-electron (3) to diamagnets (4) and (5) was observed by NMR.[39,40] Other new molecular hydrogen complexes include the first non-

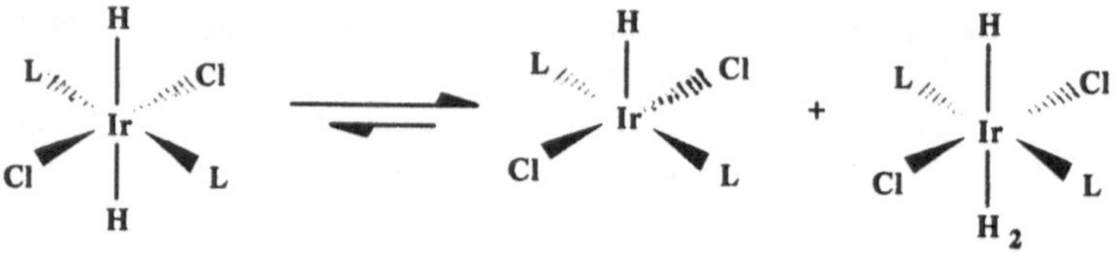

(3) $L = P(^iPr)_3$, PCy_3 (4)   (5)

phosphine-bearing species characterizable by NMR: $[Cp^*Ru(CO)_2(H_2)]^+$ and $[Cp^*Re(CO)(NO)(H_2)]^+$.[41] Both are stable only at low temperatures and, due to the presence of π-acceptor ligands, they are extremely acidic. These complexes will protonate diethyl ether. IR evidence has been obtained in matrix studies for the previously postulated species $[Ni(CO)_3(H_2)]$.[42] Visible light photoejection of NH_3 from $[Cr(CO)_5(NH_3)]$ has allowed the more efficient matrix generation of $[Cr(CO)_5(H_2)]$.[43] Photolysis of the parent tricarbonyl complex in supercritical xenon produced the novel $[CpMn(CO)_2(H_2)]$.[44] An array of other new molecular hydrogen complexes has appeared recently.[45-50]

As the structural chemistry of dihydrogen complexes becomes further elucidated, several investigators have directed their efforts toward studying the reactivity of these species. Hydrogen complexes are natural candidates for hydrogenation catalysis. Such behavior is observed for $[IrH_2(H_2)(PMe_2Ph)_3]^+$,[51,52] $[OsH_3(H_2)(PMe_2Ph)_3]^+$,[53] and $[RuH_2(H_2)(PPh_3)_3]$.[54] Another catalytic cycle involving the latter complex leads to the dehydrogenation of secondary alcohols to ketones.[55] The key to this reactivity is the well-known lability of H_2.[56] The 16-electron product produced by H_2 release from $[RuH_2(H_2)L_3]$, $L = hhhP(p-C_6H_5Me)_3$, has been found to be stable,[57] despite the fact that similar intermediates, including $L = PPh_3$, are too short-lived to be identified. An agostic interaction could be behind this increased stability, but a crystal structure could not be obtained. Thus far advantage has not been taken of the extreme acidity (*vide supra*) of many η^2-H_2 complexes.

11.3. Metal-Hydride Complexes

Reviews during late 1988 and 1989 relevant to metal-hydrides include a treatment of H_2 oxidative addition and dihydride transfer reactions involving $[Ir(X)(CO)L_2]$.[6] Stereochemical aspects are emphasized. A discussion of polynuclear iridium hydrides has also appeared.[10] The relative bond strengths of $M-H$

and M-alkyl have continued to attract attention. Valence-bond calculations indicate that increasing covalent bonding between the metal and ancillary ligands tends to destabilize M—H and M—R.[58] LCAO MO calculations have generated values for metal-hydride bond dissociation energies which are very consistent across the transition series.[59] This result is generally supported by gas-phase metal ion studies.[60,61] Ligand substitution kinetics,[62-65] e.g., iodinolytic isoperibol titration calorimetry, have been employed to determine relative M—R and M—H bond enthalpies for the early transition metals and lanthanides. In this technique, the metal-iodine bond enthalpy is used as a known calibration value, against which to compare other metal-ligand bond strengths. Another set of $[L_nM—H]$ bond dissociation enthalpies was cleverly calculated by measuring the oxidation potential of the conjugate base $[L_nM]^-$ and using the literature pK_a data.[66] The data are in good agreement with other literature values.

Several studies on "A-frame" complexes have demonstrated that, initially, hydrogen oxidatively adds across a single metal.[67-69] Iridium is always the favored site of attack in heteronuclear A-frames. Rearrangement may subsequently lead to monohydride ligation at both metal centers. Hydrogen addition to the iridium d^8 species in Scheme 1 occurs with perpendicular orientation to the carbonyl.[70] The initial product is observed only at low temperatures for R = Ph, however. Upon warming it rearranges to the "parallel" species, (7), by means of dissociation and reattack of H_2, rather than an η^2-benzene. Parallel attack may be induced by the use of L = π-acceptor, e.g., tetracyanoethylene. These facts have been rational-

Scheme 1

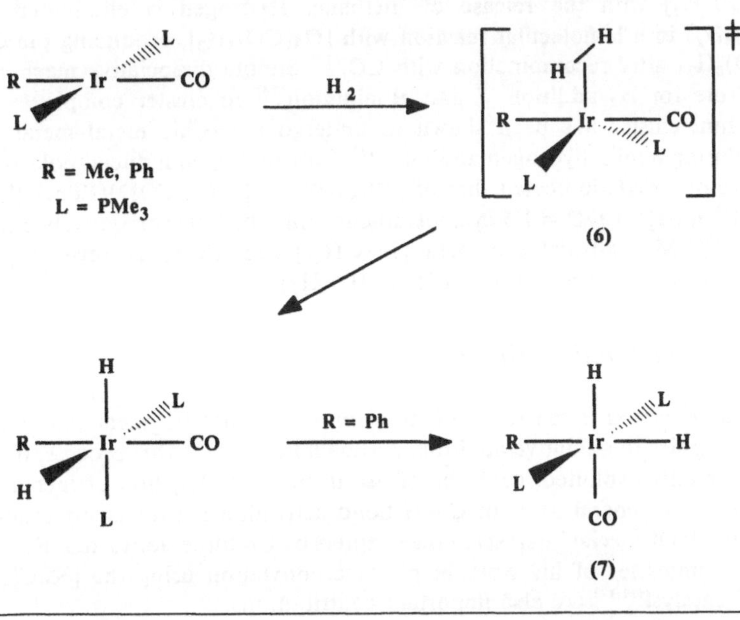

Scheme 2

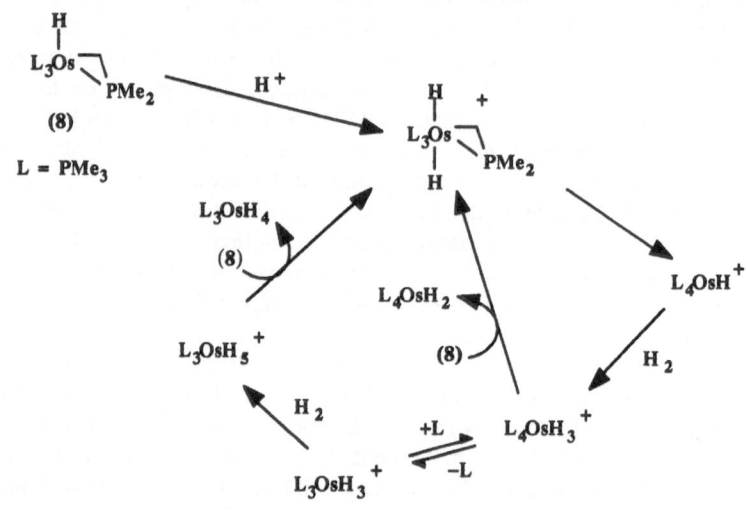

ized based on molecular orbital predictions for stabilization of the trigonal bipyramid transition state, (6).

Scheme 2 outlines an interesting catalytic cycle leading to activation of dihydrogen.[71] It is initiated by protonation and proceeds at less than 1 atm H_2. Under similar conditions, $[(PMe_3)_4Os(Me)(H)]$ catalytically produces $[(PMe_3)_4OsH_2]$ with the release of methane. Hydrogen is eliminated from $[Os(CO)_4H_2]$ in a bimolecular reaction with $[Os(CO)_3H_2]$, producing the dimer $[Os_2(CO)_8H_2]$ after recombination with CO.[72] Simple dissociative mechanisms are the rule for H_2 addition[73] and elimination[74] in cluster complexes. One triruthenium cluster has been shown to undergo reversible metal–metal bond rupture during a mild hydrogen addition.[75] *Para*-hydrogen induced polarization in the proton NMR indicated that of $[RhCl(PPh_3)_3]$, $[Rh(COD)(PPh_3)_2]^+$, and $[Rh(COD)dppe]^+$, COD = 1,5-cyclooctadiene, only the latter oxidatively adds H_2 reversibly.[76] Matrix-isolated carbene $[Fe(CH_2)]$ adds hydrogen reversibly with the loss of methane or formation of $[Fe(CH_3)(H)]$.[77]

11.4. C—H Bond Activation

Bond activation is the result of insertion of a metal fragment into a C—H bond, giving rise to oxidative addition as shown in Eq. (1). This process, if it can be commercially exploited, could be of use in the homologation of inert alkane feedstocks. The general topic of C—H bond activation is extensively treated in Hill's book.[1] Of special interest are the chapters by Crabtree, Jones, and Rothwell. Tanaka's summaries of his work in photocarbonylation using the $[RhCl(CO)(PMe_3)_2]$ catalyst[11,12] are also important contributions.

11.4.1. Unactivated C—H Bonds

While the activation of sp^2, sp, and heteroatom-substituted C—H bonds have become widely documented, the oxidative addition of alkane bonds remains restricted to only a few systems. Therefore the goal of understanding the chemistry of these complexes has attracted much study. The most studied systems involve the d^8 fragment [Cp*M(PMe$_3$)], (9), M = Rh, Ir. LCAO calculations based on nonlocal density functions reflected the more favorable reaction pathway for C—H bond activation by (9) than by [M(CO)$_4$], (10), M = Ru, Os.[78] Complexes (9) have empty σ-type d-based orbitals. These orbitals are partially full for (10). Thus (9) is a better σ acceptor. Moreover, back-donation into C—H σ^* is facilitated by the symmetry of the HOMO in (9). Jones and Feher[2] provide a discussion comparing alkyl and aryl bond activation using (9), M = Rh. Activation of hydrocarbon bonds is thermodynamically preferred according to: H—Ph > H-vinyl > H—Me > H—CH$_2$R > H—CHR$_2$ > H—CR$_3$ > H-benzyl, that is, favoring stronger C—H bonds. This is because M—C bond strengths follow the same ordering as C—H strengths, but span a wider energy range. Similar conclusions were reached in a thermodynamic analysis of (9), M = Ir.[79] Ligand substitution kinetics have been used to evaluate metal-alkyl bond dissociation enthalpies with the fragments [Pt(Me)dppe],[80] [Cp*Ru(PMe$_3$)$_2$],[80] [Cp$_2^*$M], M = Sm, Hf, Sc, and Zr,[62-65] and [Co(py)(dmg)], dmg = H(dimethylglyoximate)$^-$.[81] Molecular orbital calculations indicate larger bond energies for early transition and lanthanide metal-alkyls.[59] This is a result of the lesser number of valence electrons available for repulsive interaction with the carbon 2s electrons. Photoelectron spectroscopy of group V metallocene complexes in fact shows only a slight difference in M—Me and M—H strengths.[82]

There has also been interest in identifying intermediate species for C—H activation. Specifically an alkyl σ-complex, akin to the η^2-H$_2$ and agostic interactions, has been sought. There is ample precedent for σ-complexes in the organometallic complexes of the more polarizable silanes. Oxidative addition of Si—H is often incomplete.[83-85] Recently a manganese-coordinated Sn—H bond has shown the same effect.[86] Reductive elimination studies on metallocene-hydrido-methyl complexes have provided kinetic and isotope effect evidence for a σ-methane intermediate. Three groups have independently identified an inverse kinetic isotope effect during methane elimination from tungstenocene[87,88] and rhenocene.[89] Deuterium exchange between hydride and methyl sites was also noted with the latter species. The intermediate may actually resemble (11), since σ-coordination favors hydrogen due to its exclusive s character (e.g., in agostic interactions). Photolabilization of CO from [CpM(CO)$_2$] at low temperature in cyclohexane leads only to a very weakly bound solvent complex for M = Co.[90]

$$Cp_2M \overset{\text{\tiny H}}{\underset{H}{\longleftarrow}} \overset{+n}{CH_2}$$

M = W, n = 0
M = Re, n = 1

(11)

When M = Rh[91] oxidative addition occurs, but the alkyl and hydride ligands are readily replaced by [CpRh(CO)], leading to dimerization. Similar investigations have been conducted at reduced temperatures in liquified rare gases.[92-94] Liquid xenon especially is a good, inert solvent for alkane activation reactions. A weak solvent complex is formed with the rare gas (M—Kr < M—Xe, despite lower temperature for liquid Kr), according to Scheme 3.[94] This is replaced by alkane

Scheme 3

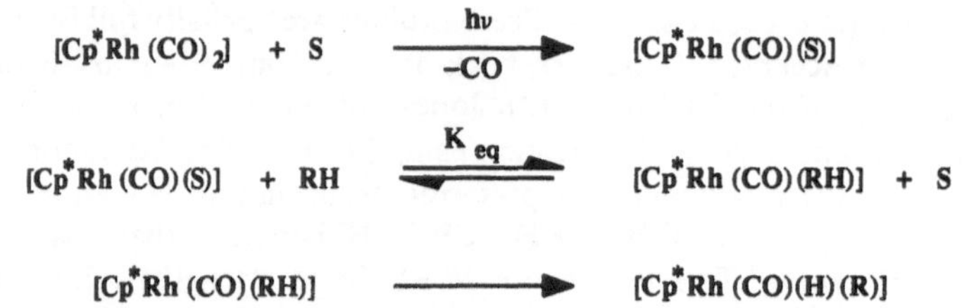

$$[Cp^*Rh(CO)_2] + S \xrightarrow[-CO]{h\nu} [Cp^*Rh(CO)(S)]$$

$$[Cp^*Rh(CO)(S)] + RH \underset{}{\overset{K_{eq}}{\rightleftharpoons}} [Cp^*Rh(CO)(RH)] + S$$

$$[Cp^*Rh(CO)(RH)] \longrightarrow [Cp^*Rh(CO)(H)(R)]$$

followed by oxidative addition. It should be noted that the alkane "solvent complexes" kinetically detected in the above work bear close scrutiny with regard to their bonding mode. A flash photolytic study involving [RhCl(CO)L$_2$], L = PMe$_3$ indicated simple first-order carbonyl dissociation, followed directly by metal attack on solvent alkane.[95] Consistent with synthetic work, when L = P(aryl)$_3$, aromatic, but not aliphatic, solvents are activated. In the latter case, recombination of [RhClL$_2$] with CO or dimerization occurs. When [CpIr(CO)D$_2$] is irradiated at room temperature in *neo*-pentane, [CpIr(CO)(CH$_2$CMe$_3$)X] is formed, X = H or D, with the liberation of H$_2$, HD, and D$_2$.[96] Intermediate species (12) or (13) are more likely than (14), since H$_2$ is known to be more photolabile than CO in this case.

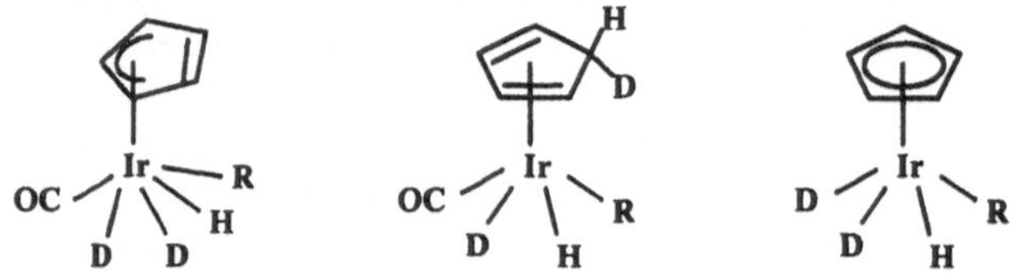

(12)   (13)   (14)

R = CH$_2$CMe$_3$

Reports of new alkane-activating systems have been scarce. [Cp$_2^*$Sm(Me)(THF)]cyclometallates a Cp* methyl, losing methane. The cyclometallate then attacks C—H bonds in alkanes, yielding a mixture of organosamarium products.[97] [Os(PMe$_3$)$_4$(H)(CH$_2$CMe$_3$)] releases CMe$_4$ and activates methane in low yields under high CH$_4$ pressure.[98] In agreement with the bond-strength conclusions above, weaker R may be replaced by strong R' in [('Bu$_3$SiNH)$_3$ZrR] according to R, R' = cyclohexyl < Me < Ph < H.[99] A polar mechanism involves the abstraction of an amido proton. Slow hexane H/D exchange is catalyzed by a cluster-supported iridium hydride complex.[100-102] Very small amounts of cyclohexane carbonylation to benzoic and cyclohexanoic acids occur in the presence of $[Pd(O_2CCH_3)_2]$ and CF_3CO_2H.[103]

Early transition[104] and lanthanide metal ions[105] tend to show the greater aptitude toward gas-phase alkane activation. However, middle and late transition metal ions[60,61,106,107] may be compelled to undergo these transformations. Bond enthalpies may be approximated by observing threshold energy levels for the endothermic reactions.

11.4.2. Activated C—H Bonds

Oxidative disruption of aryl hydrocarbon bonds is fast becoming a commonly encountered reaction. The relative importance of thermodynamic M–aryl bond strength (*vide supra*) and kinetic η^2-arene precoordination effects in this transformation has received some consideration. X-ray crystal structures of (15)[108] and (16)[109] have been put forward as examples of arrested precoordination. For

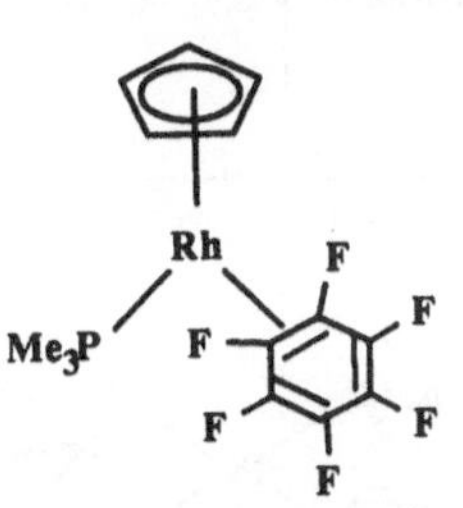

(15) 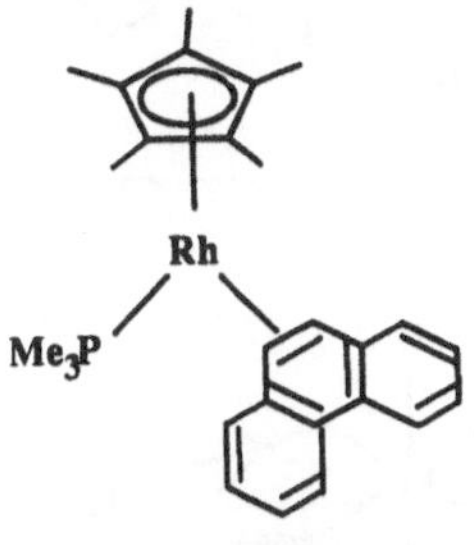   (16)

benzene itself, kinetic evidence supports such an intermediate[108] and in the case of naphthalene, equilibrium (4) was shown to exist ($K_{eq} = 2.0$).[109] Pyridinium

$$[Cp^*Ru(PMe_3)(H)(C_{10}H_7)] \rightleftharpoons [Cp^*Ru(PMe_3)(\eta^2-C_{10}H_8)] \qquad (4)$$

salts also form η^2-arene complexes with pentammine osmium(II).[110] On the other hand, known agostic complex (17) was compared to the benzoquinoline analog (18) prepared in Scheme 4.[111] In this instance, approach is forced to favor "end-on" attack. Despite the lack of allowable π-coordination, cyclometallation, rather than agostic coordination, occurs. Also interesting here is the proposed generation of a dihydrogen ligand via C—H oxidative addition to a metal–hydride complex.

Electrophilic arene bond activation has been demonstrated for $[(NP)_3Rh]^+$, $NP_3 = (PPh_2CH_2CH_2)_3N$,[112] $[Ir(PMe_3)_3]^+$,[113] $[('Bu_3SiNH)_3ZrR]$,[99] and $[PtTf(PMe_3)_2(CH_2CMe_3)]$, $Tf = O_3SCF_3$.[114] The mechanism for the platinum complex has been investigated. Oxidative addition intermediate species (19) is considered more likely than electrophilic aromatic substitution intermediate (20). $[Cp_2Ta(PMe_3)(Me)]$ will not activate benzene except in the presence of catalytic

Scheme 4

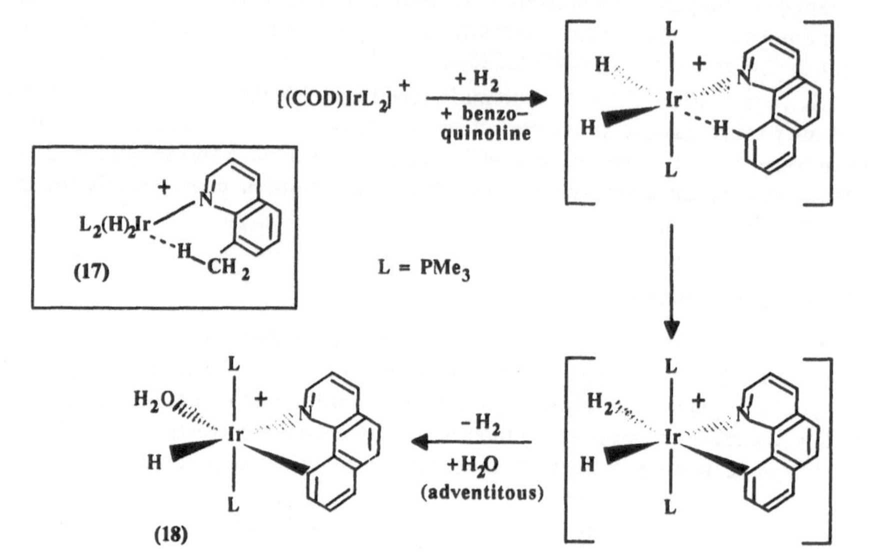

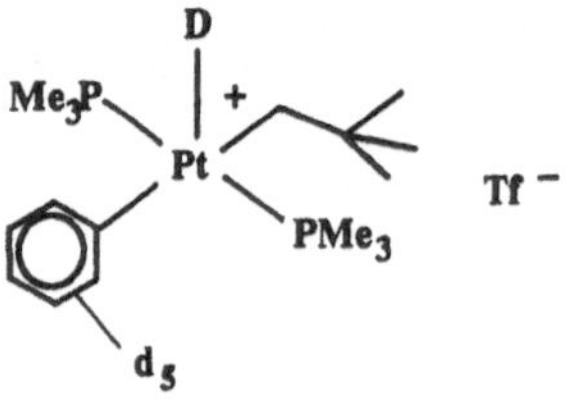

amounts of $H_2Si^iBu_2$.[115] Oxidative addition of a H—Si bond follows loss of phosphine. Methane may then be eliminated and benzene added. Activation of benzene hydrocarbon bonds has been demonstrated using other systems as well.[93,101,116-121]

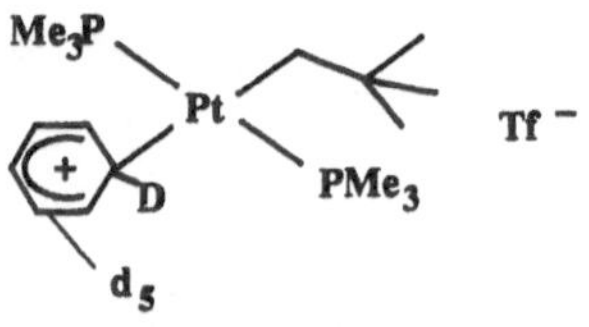

Unlike metallation of arenes, the oxidative addition of vinyl C—H bonds is believed to usually be mechanistically distinct from π-coordination. The factors favoring these competitive routes are not well understood. It has generally been found that olefin π-complexes are of lower energy. The reaction of $[Cp^*Ir(PMe_3)]$ with ethylene produces a temperature-independent 2:1 ratio of the σ-vinyl to π-olefin complexes.[122] These are thought to be produced according to separate (and yet energetically similar) pathways as shown in Scheme 5. Rearrangement to the thermodynamically preferred π-ethylene complex at elevated temperatures occurs by yet another pathway. Three investigators[123-125] have concurrently discovered the ability of $[TpIr(\pi\text{-olefin})L]$, L = CO, π-olefin, to form σ-vinyl-hydride

Scheme 5

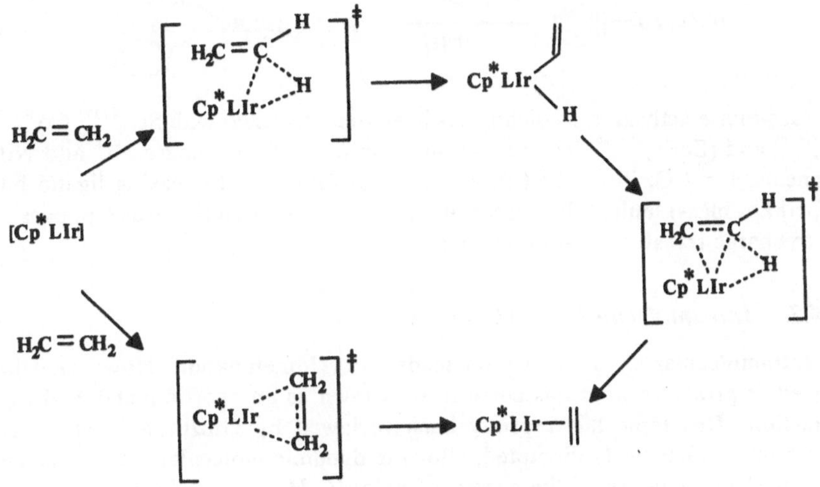

species photochemically or thermally. In other systems, geometric considerations such as cyclometallation[126] or metal-metal σ,π-interaction[127] aid σ-vinyl complex formation. Finally in other cases, reversible vinyl C—H activation must be invoked due to vinyl H/D scrambling.[128-130]

The initial oxidative attack on an ethylene oxide C—H by [Cp*Rh(PMe$_3$)] has been seen by low-temperature NMR.[131] This is remarkable in light of the well-known reactivity of the ethylene oxide ether linkage. Rhodium organometallics have been shown to add acetylinic and aldehydic hydrocarbon bonds.[132,133] An unelucidated sequence of reductive elimination, cyclometallation, and acetylinic C—H addition took place in reaction (5).[134] Carbon-carbon bond activation is very rare and mitigating factors, such as ring strain, usually play a part in these reactions. The strained C—C bond in fluorinated cyclopropene has been oxidatively

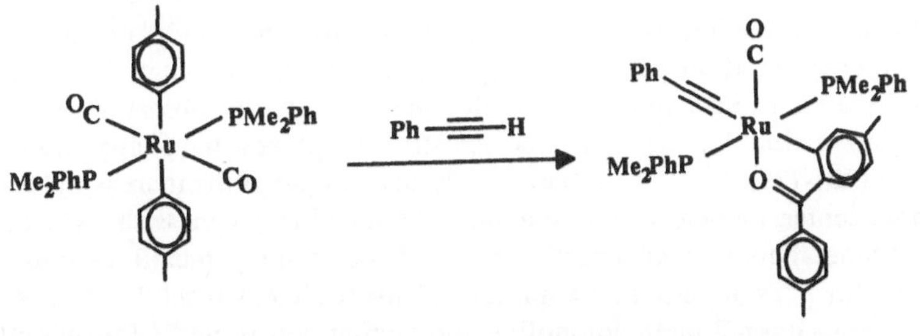

(5)

added across coordinatively unsaturated platinum[135] and iridium[136] complexes. Reaction (6) shows a rare example of nonstrained C—C activation.[121]

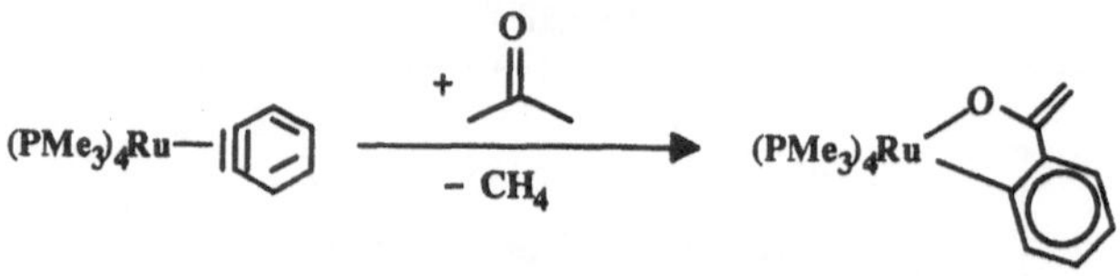

$$(PMe_3)_4Ru—\text{⬡} \quad \xrightarrow[-CH_4]{+ \text{⬠}} \quad (PMe_3)_4Ru\text{⬡} \qquad (6)$$

Gas-phase activation of olefins has been demonstrated with Sc^+,[137] Co^+,[138] Ni^+,[138] and $[ScFe]^+$.[139] The activation of arenes with gas-phase Ni^+ and NiL^+ fragments, $L = CO$, PF_3, and Cp, was demonstrated.[140] Increasing ligand bulk (or perhaps mass) limited the degree of reactivity. The exception was Cp, possibly due to charge transfer, creating $Ni^{2+}Cp^-$.

11.4.3. Intramolecular C—H Bonds

Intramolecular C—H activation leads to cyclometallation. However, when arrested, it produces what has come to be known as an agostic metal-hydrogen interaction. This topic has recently been reviewed by Ginzburg.[5] This weak interaction is often easily disrupted, allowing dynamic molecular rearrangements to occur. The dynamics of the agostic $[CpCoL(C_2H_5)]^+$ have been found to be extensive.[141] These include rotation of the C—C bond, inversion of stereochemistry at cobalt, and rotation of the entire C_2 fragment via an intermediate ethylene complex. All involve release of the agostic "bond." The propyl complex shows coupling constants J_{CH} and J_{PH} intermediate for those of agostic and terminal hydrides. Intermediacy of agostic species has been invoked to explain the rapid, intramolecular proton site exchange in (21)[142] and (22).[143] Agostic protons display acidities similar to those found for terminal metal-hydrides.[144] An attempt

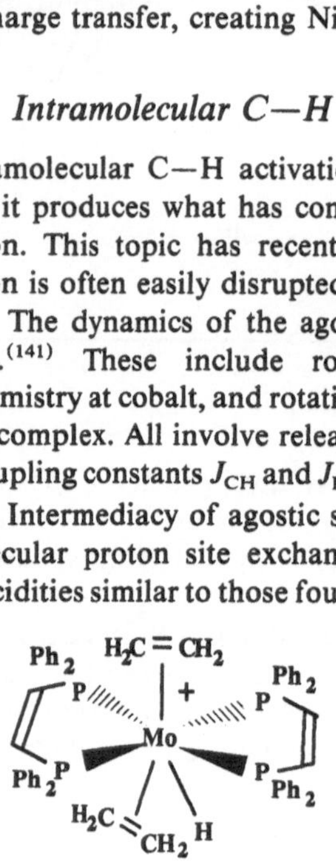

(21)

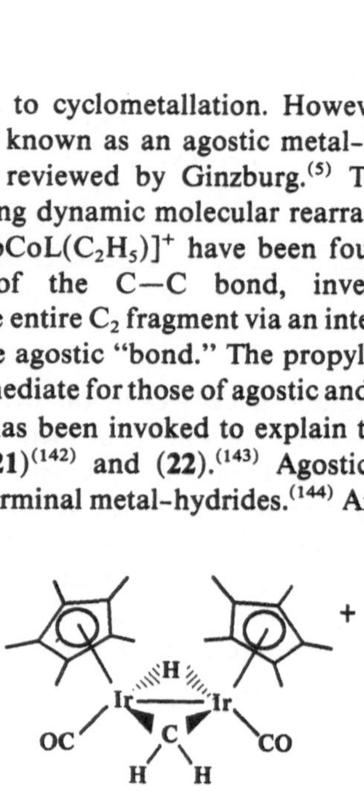

(22)

to insert an isonitrile into an agostic zirconium-hydrogen "bond," not surprisingly, led to replacement of the agostic hydrogen and insertion elsewhere.[145] Molecular orbital calculations have not yet succeeded in predicting the observed $[TiCl_3(Me)]$ agostic interaction.[146] Several new agostic complexes have appeared,[147-149] including (23)-(25).[150-152] Complex (25) is an example of the increasingly recognized three-center, two-electron interaction of a metal fragment with a C—Si bond.

Cyclometallation is an effective method for forming metallocarbon bonds, especially for d^8, square-planar complexes. This topic has recently been reviewed with the focus upon 8-methylquinoline and similar substrates.[9] Further study of the reversible Pd—N based systems has revealed that a combination of factors determines the preferred attack site of "bifunctional" substrates.[153,154] Five-

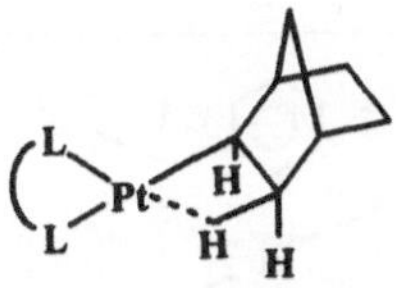

(23)   L⌒L = chelating diphosphine 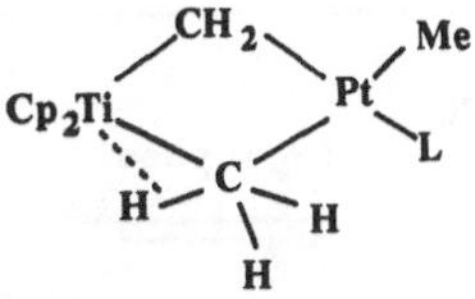   (24) 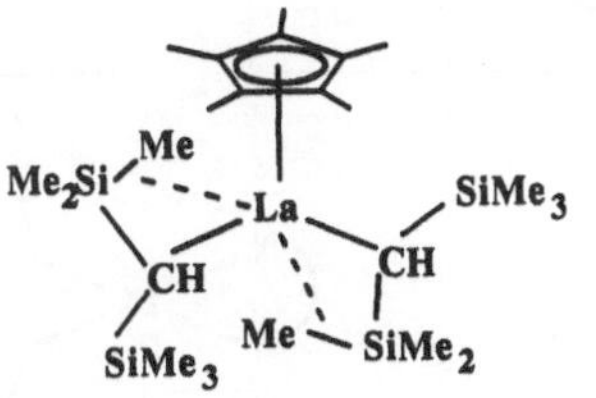   (25)

membered, rigid palladocycles are favored, presumably for geometric reasons. However, six-membered palladocycles are preferred to five-membered in flexible, sp³ carbon systems. Strongly basic nitrogen is disfavored. Several studies of cyclometallated [(COD)Ir(benzylamine)] have revealed interesting phenomena. Oxidative addition of hydrogen promotes either release of dimethylbenzylamine (for R = H) or the isomerization (for R = alkyl) shown in Scheme 6.[155] In the absence of hydrogen atmosphere, cyclometallation of an aminomethyl must precede isomerization, demanding more thermal energy.[156] Protonation by adventitious water has been credited for *ortho*-position H/D exchange in a similar system.[157]

Scheme 6

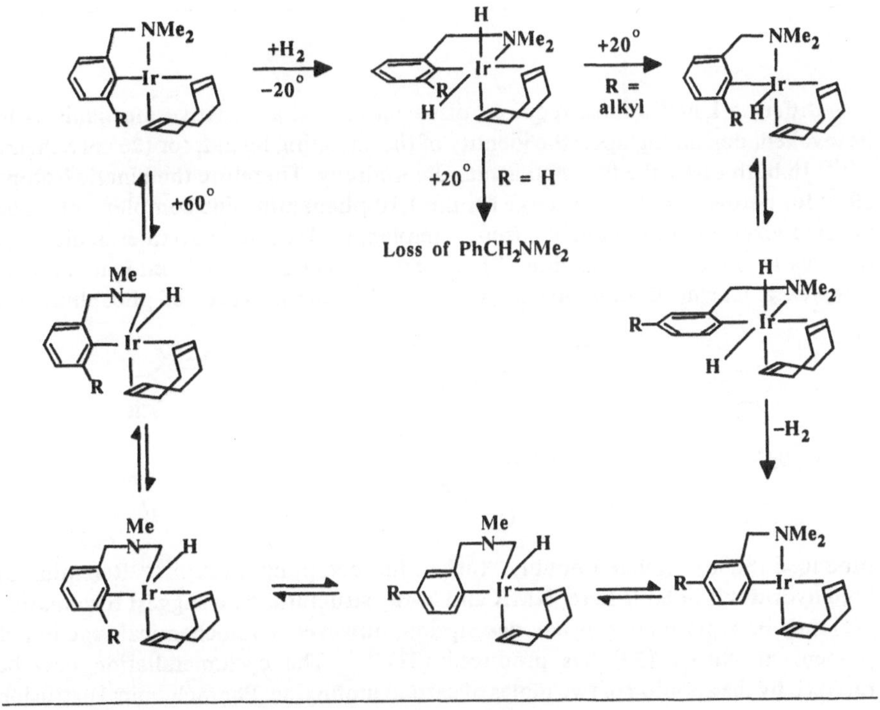

Scheme 7

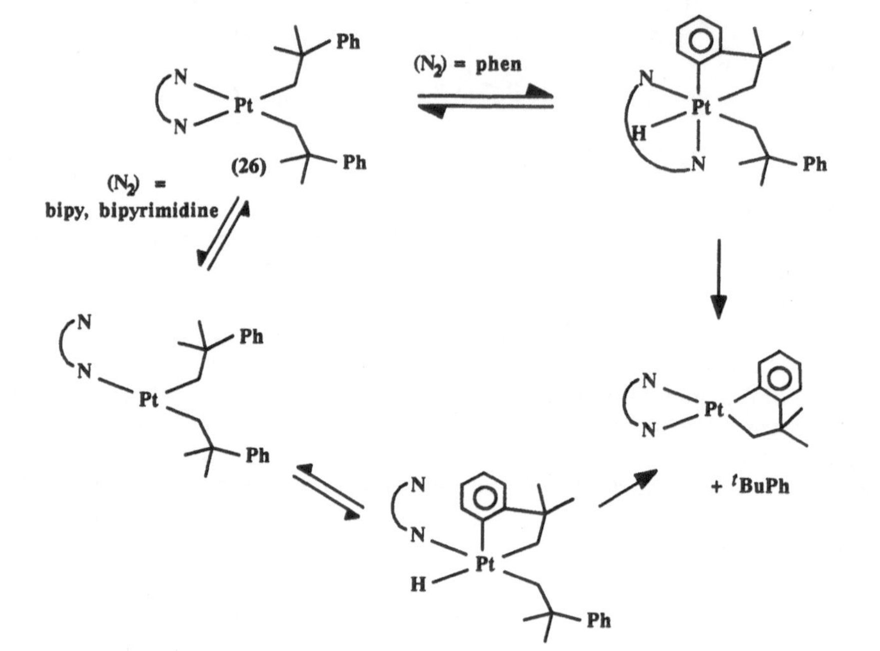

Differing kinetics have required dissociative and associative mechanisms to be invoked, depending upon the identity of the chelating ligand, for (26) in Scheme 7.[158] In both cases, the first step is rate-determining. Therefore the kinetic isotope effect for perdeuterophenyl is large for the 1,10-phenanthroline complex only. The novel "tucked-in" hexamethylbenzene complex, (27), must be an intermediate for the alkyne insertion in reaction (7).[159] Such species are well known for Cp* complexes. Chemical reduction of [Cp*TaCl₄] in the presence of phosphine has

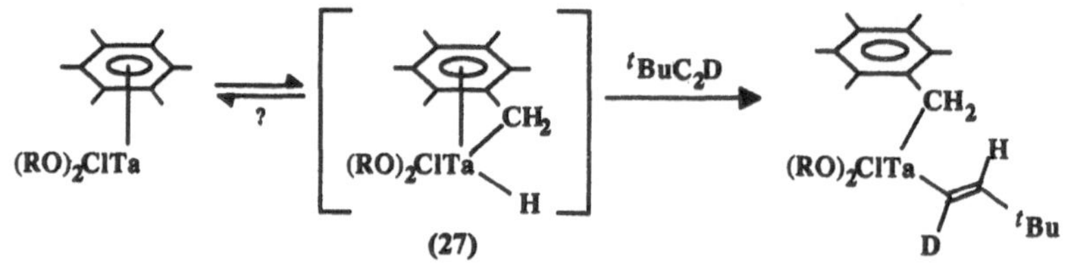

produced the first isolated doubly "tucked-in" compound, (28).[160] Retention of both hydrides is notable here. NMR and X-ray structural data suggest formulation (29) may be a more appropriate description, however. Photochemical reaction of μ_3-benzene cluster (30) has produced (31).[161] The cyclometallation may be reversed by the addition of two moles of carbon monoxide. Photochemical extrusion

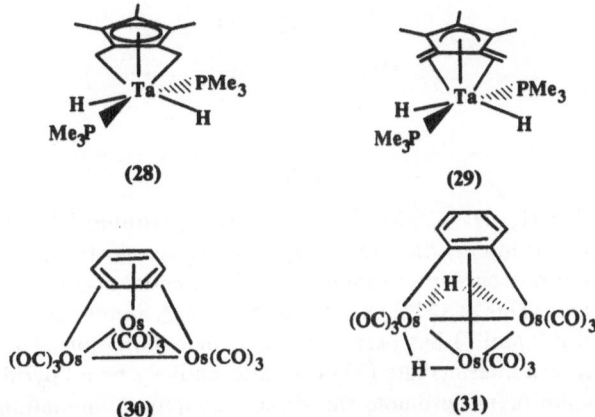

of CO_2 has been used to prepare cyclometallated compounds.[162] 2,6-Diphenylphenoxide ligands have been cyclometallated to group VI metals[163–165] and even tin.[166] There have been many other reports of cyclometallation, involving palladium(II),[167–174] triosmium clusters,[175–180] and other metals.[126,181–190]

The activation of intramolecular carbon–hydrogen bonds by metal ions in the gas phase first involves π-coordination, followed by "remote functionalization" of C—H or C—C several carbons distant from the initial coordination site. Schwarz[7] has reviewed his prolithic work in this area. High-resolution translational energy loss measurements on the remote functionalization of internal alkynes by Cr^+ have been made.[191] These indicate that the ions involved are in the ground state and that no added collisional energy transfer is needed for cyclometallation to proceed. The reactions of Fe^+, Co^+, and/or Ni^+ have been studied with alkyl-substituted amines,[192–194] alcohols,[194,195] ketones,[196] nitriles,[197–201] phosphaalkynes,[202] and allenes.[203] The kinetic energy spectra of the products released have also been used to evaluate transient metal ion–small molecule complexes and the transition states leading to their fragmentation.[8,204]

11.5. Reductive Elimination Forming Carbon–Carbon Bonds

Reductive elimination from late transition metal complexes may be used to form new carbon–carbon bonds. It is also important as it is the microscopic reverse of the very rare C—C bond activation reaction. A comprehensive review of this topic has been published.[4] Mechanisms requiring and those precluding preliminary ligand dissociation are encountered. The thermolytic process oftentimes does not yield a unique product, even in simple systems. For example, $[(\text{tmeda})\text{Pd}(\text{Me})_2]$, tmeda = tetramethylethylenediamine, yields mostly methane, along with ethane and a trace of ethylene, upon heating in benzene.[205] The key intermediate is postulated to be the agostic complex (32), which bears much hydridocarbene character. Elimination of methane from (32) competes with solvent C—H activation, as demonstrated by isotope labeling. Intramolecular attack on methyl must occur to a lesser extent following initial nitrile dissociation from

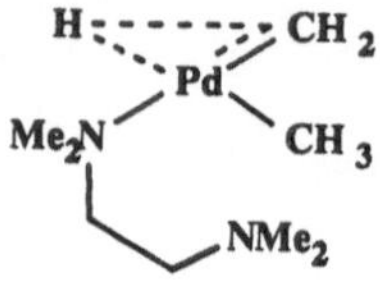

(32)

$[L_2PtMe_4]$, L = NCCH$_3$, NC(2,6-Me$_2$C$_6$H$_3$).[206] This platinum(IV) complex produces only ethane. The intermediate probably involves an agostic methyl. Thermal release of ethane also occurs in the solid state. Differential scanning calorimetry provided an enthalpy value of 32 kcal mol^{-1} for Pt—Me. A nondissociative mechanism ($\Delta S^{\ddagger} = 26$ cal K^{-1} mol^{-1}) has been proposed for elimination of biphenyls in Scheme 8.[207] Putative intermediate (33) may alternatively be a "pyridyne" complex. *Para*-CF$_3$ substituents promote the direct reductive elimination pathway.

Scheme 8

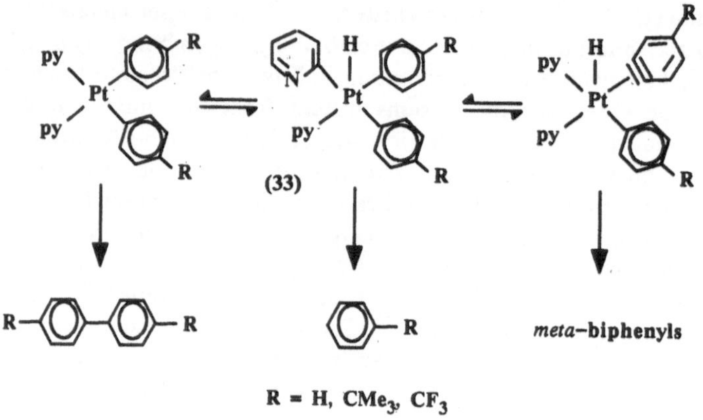

R = H, CMe$_3$, CF$_3$

A nondissociative mechanism was also claimed for *cis*-[(PPh$_3$)$_2$Pt(vinyl)-(alkynyl)].[208] The unusual rearrangement reaction (8) proceeds by initial dissociation of phosphine.[209] A formal β-methyl elimination must occur here, but the

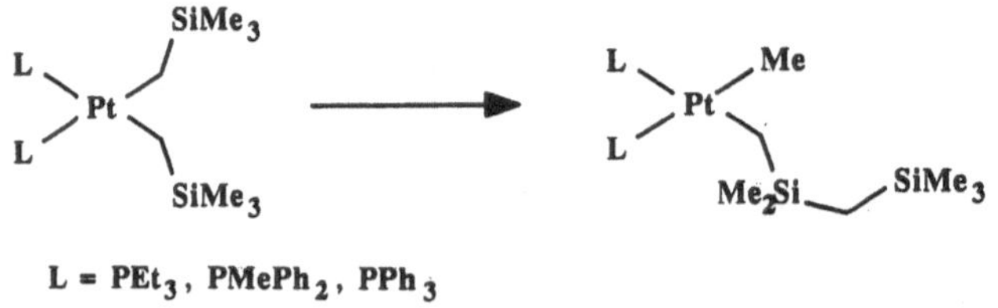

(8)

L = PEt$_3$, PMePh$_2$, PPh$_3$

existence of a deuterium isotope effect implies intermediate cyclometallation. The cross-coupling reaction of PhI and MeMgI to produce toluene is catalyzed by *trans*-[L$_2$Pd(Ph)(I)]. The cycle in Scheme 9 requires reversible generation of

Scheme 9

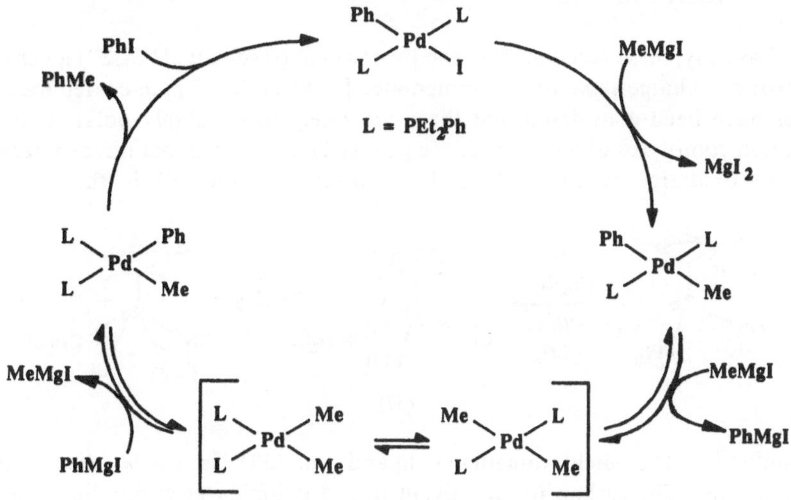

the dimethylpalladium intermediate, which can isomerize to a *cis* arrangement, permitting toluene release.[210] The elimination itself does not require release of phosphine.

The rate of elimination in (34) favors M = Ni > Pd.[211] The rate is independent of added phosphine and cannot therefore be dissociative. When PPh₃ is replaced with dppe in (34), M = Pd, (35) is formed. Reductive elimination occurs more rapidly for (34) than for (35). Upon addition of dppe to (34), M = Ni, (36) is

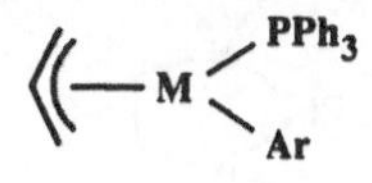

(34) M = Pd, Ni 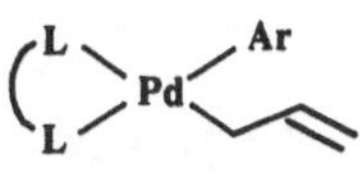   (35) 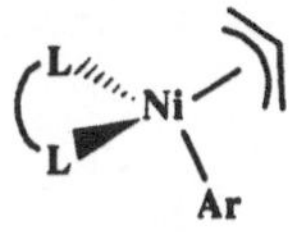   (36)

L⌢L = dppe

observed by low-temperature NMR. Elimination from this 18-electron d^8 system greatly exceeds that from 16-electron (34) or (35). Reductive elimination from [(η^3-allyl)PdL(Me)] favors the bimolecular production of ethane.[212] Photochemical reductive eliminations are known. *Para*-biphenyls are stereospecifically released upon photolysis of *cis*-[(diars)Pt(p-C₆H₄X)₂], X = alkyl, halide.[213,214] Ethane is photochemically eliminated from [CpCo(PPh₃)Me₂].[215] The resulting [CpCo(PPh₃)] attacks the starting material to form the [CpCo(PPh₃)Me] radical. Photogenerated [CpRh(C₂H₄)H₂] slowly eliminates ethane in liquid xenon at 170 K.[92]

11.6. Oxidative Addition and Reductive Elimination of Alkyl Halides

Typically, these reactions proceed by a two-step (so-called "S_N2") mechanism involving a charge-separated intermediate: $[L_nMR]^+X^-$. Three-center transition states have been considered, but these are likely to be highly polar. Lone-pair donation complexes of alkyl halides (e.g., CH_3I) are known, but not as intermediates to oxidative addition. Recently reported reaction (9) is the first such

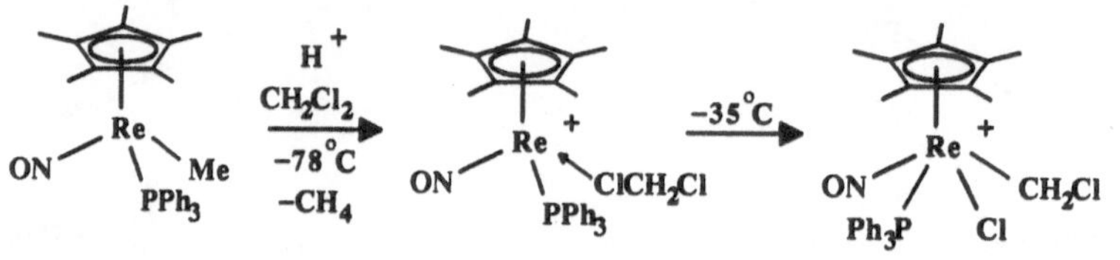

(9)

example.[216] The dichloromethane ligand in (37) is easily displaced by iodomethane. The carbon in the solvent ligand is highly electrophilic. Attack by $X^- = I^-$, Br^- produces $[Cp^*ReCl(NO)(PPh_3)]$ and $ClCH_2X$.

The intramolecular addition of C—X bonds has been considered by two groups.[217,218] The cyclometallation of complex (38) in Scheme 10[218] had little entropic component, indicating a concerted, rather than S_N2, pathway. Cyclometallation was preferred according to: C—Br > C—Cl > C—H > C—F. This is the reverse of the aryl–X bond energies. A perfluorinated aryl ring could be used to force formation of (39) for X = F. The reaction of vinyl triflates with the $[(PPh_3)_2Pt]$

Scheme 10

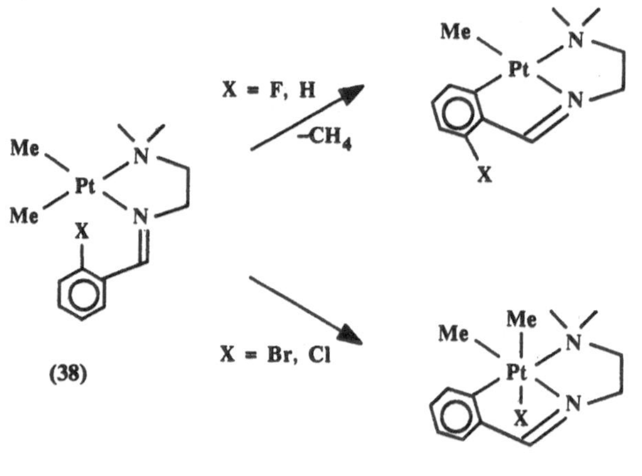

fragment proceeded by initial, rate-limiting π-coordination as determined by NMR spectroscopy.[219] Intramolecular addition then rapidly produced [(PPh$_3$)$_2$PtTf(vinyl)]. For vinyl halides, however, the latter step is rate-determining.

Addition of MeX to [(tmeda)PdMe$_2$] generates (40), X = I, Br, Tf or (41), X = Tf, S = NCCD$_3$, OC(CD$_3$)$_2$ in coordinating solvents.[205] Both rapidly and cleanly eliminate ethane. The favored elimination of ethane or p-BrC$_6$H$_4$Et from (42) may be controlled by the choice of (L$_2$) = bipy or phen.[220] S$_N$2-type intermediates were observed at low temperature in this reaction. A complex, but

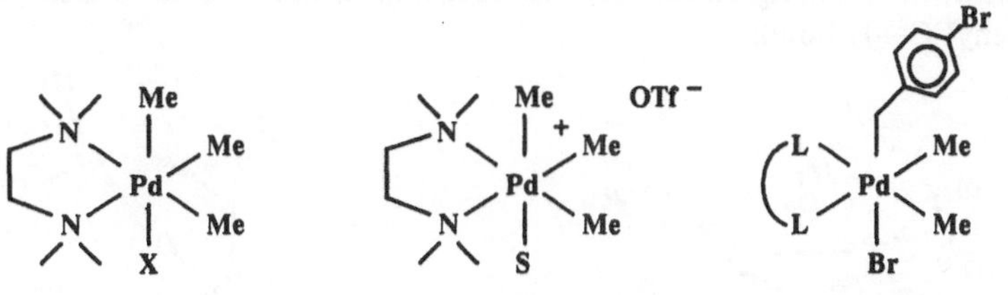

(40) (41) (42)

reasonable, set of products were realized in the treatment of bridged Pt$_2$ complexes with MeI.[221] Charge separation was sometimes dependent upon solvent polarity.

11.7. Oxidative Addition and Reductive Elimination Involving Two Metal Centers

A complex series of transformations was uncovered when the complex [Cp*Ir(η^3-allyl)(H)] was heated at high concentrations with various substrates.[222] The chemistry is based upon the diiridium intermediate (43) in Scheme 11. Complex

Scheme 11

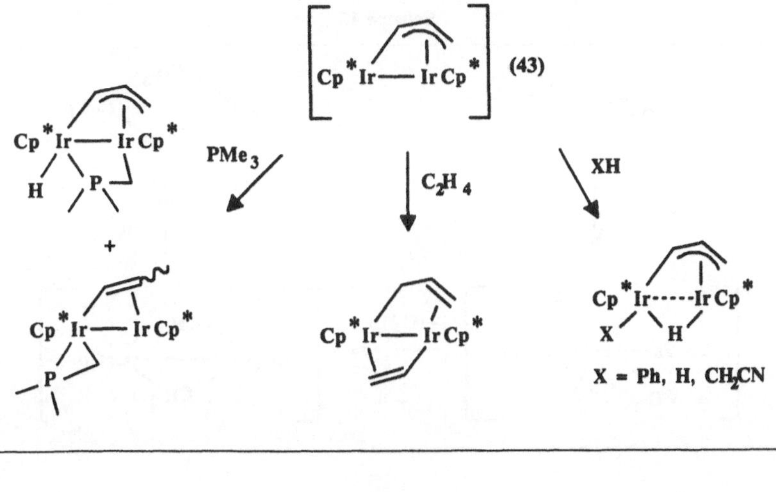

(43) adds various activated C—H bonds. Intramolecular reductive coupling of coordinated alkynes in (44) creates a coordinative unsaturation.[223] This leads to intramolecular oxidative addition of an alkyl C—H bond. Mechanistic aspects in the thermal methyl exchange between platinum and zirconium in (45) were elucidated.[224] Crossover in isotope labeling experiments has indicated a combination of inter- and intramolecular routes. Phosphine dissociation from platinum is probable. This leads to bridging of zirconium methyl. Reduction of [Cp$_2$ZrCl(CH$_2$PPh$_2$)] in coordinating solvent produces dimer (46).[225] The mechanism includes sequential alkyl transfer and intramolecular attack on cyclopentadienyl C—H bonds.

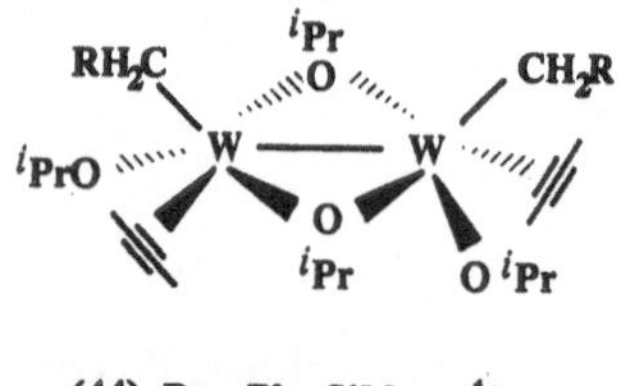

(44) R = Ph, SiMe$_3$, ^{t}Bu

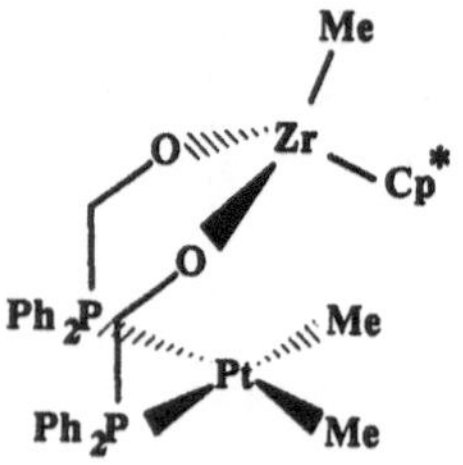

(45)

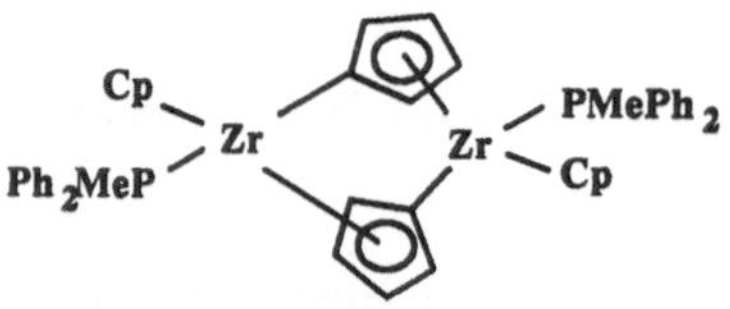

(46)

Scheme 12 presents a very unusual example of intramolecular oxidative addition, followed by the loss of a kinetically labile hydrogen radical.[226] ESR spectroscopy of radical (47) shows nuclear hyperfine interaction with cobalt but

Scheme 12

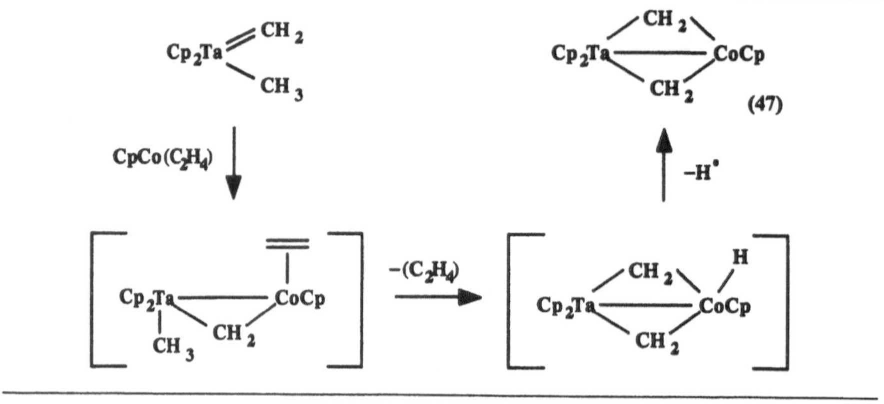

not with tantalum. $[(Cp^*Ru)_2(\mu\text{-}H)_4]$ produces the bond activation products (48) and (49) with ethylene[127] and PPh_3,[227] respectively. Palladium A-frame compounds (50) show reductive elimination products from external/external positions

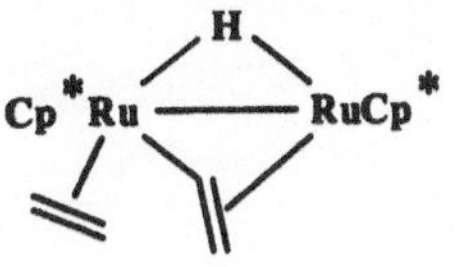

(48)

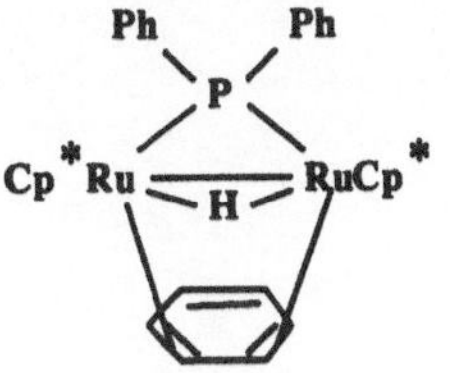

(49)

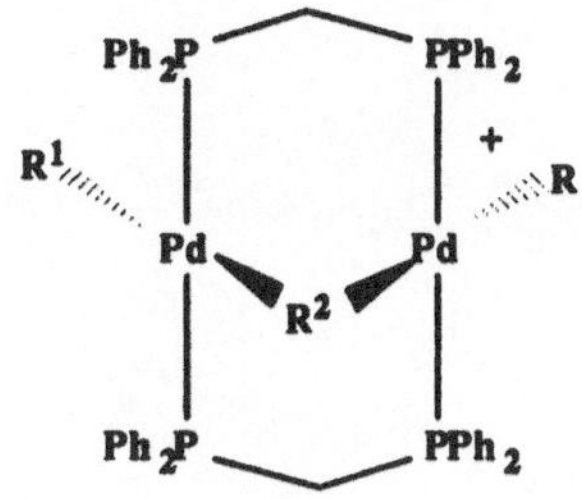

(50)

R^1	R^2	R^3	Products
H	Cl	Me	CH$_4$
Me	H	Me	CH$_4$, C$_2$H$_6$
COCH$_3$	Cl	H	CH$_3$CHO
COCH$_3$	H	COCH$_3$	CH$_3$CHO

as well as internal/external.[228] Reaction (10) demonstrates the formal elimination of W—Me from dinuclear species (51).[229] A typical S_N2 mechanism accompanies

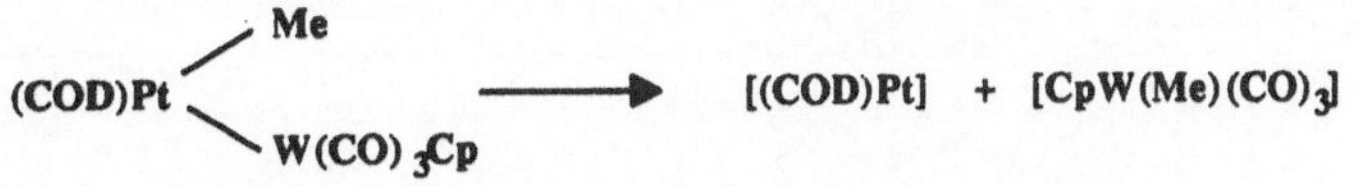

$$(COD)Pt\big\langle \begin{array}{l} Me \\ W(CO)_3Cp \end{array} \longrightarrow [(COD)Pt] + [CpW(Me)(CO)_3] \qquad (10)$$

(51)

the addition of MeI to produce (52).[230] However, the reversible reaction generating (53) shows no salt or solvent polarity effects. A rapid, radical cage mechanism may be occurring.

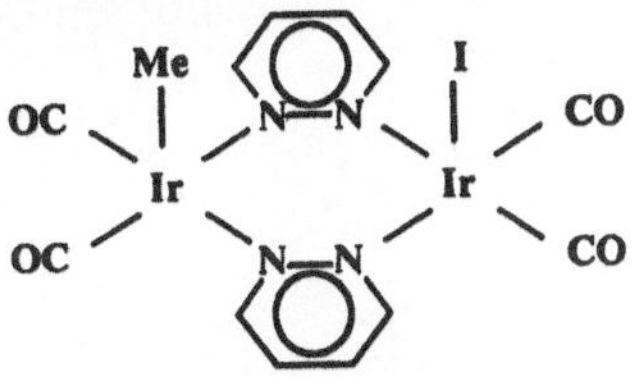

(52)

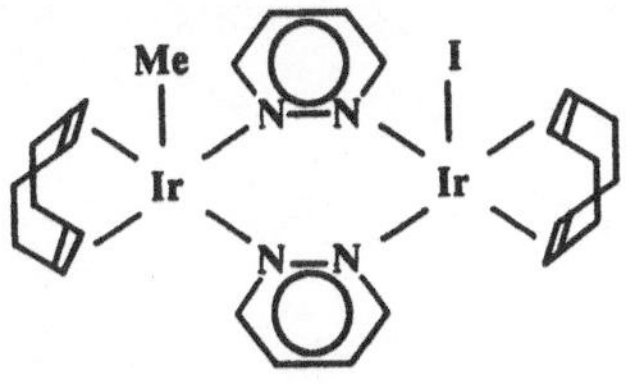

(53)

Chapter 12

Reactivity of Coordinated Ligands

12.1. Introduction

This chapter is concerned with coordinated ligand reactions, and notably enhanced reactivity resulting from coordination. It is divided into three sections. The first two deal with coordination complexes of cobalt(III) and other metal centers, while the last section discusses ligand reactivity of organometallic compounds. The emphasis is on stoichiometric reactions; catalytic processes are covered in Chapter 14.

12.2. Cobalt Complexes

12.2.1. Amino Acids and Peptides

Solid-phase peptide synthesis using a cobalt(III) spacer between the resin and peptide has been described.[1] This work extends the solution-phase peptide methodology developed with cobalt(III) protecting groups.[2] Reaction of *trans*-[Co(en)$_2$Cl$_2$]Cl with 4-(aminomethyl)benzoic acid gives the *cis*-complex (1) (Scheme 1) which is then converted to the aqua complex by base hydrolysis and acidification with CF$_3$CO$_2$H. Reaction of the aqua complex with any activated Boc-amino acid active ester gives (2). Compound (2) is then activated at the carboxyl group to give the active ester (3) which is then coupled to (aminomethyl) polystyrene to give (4). The paper describes the synthesis of the pentapeptide Leu-enkephalin starting from (4). The method can be used to prepare very large peptides by the technique of fragment condensation.

Scheme 1

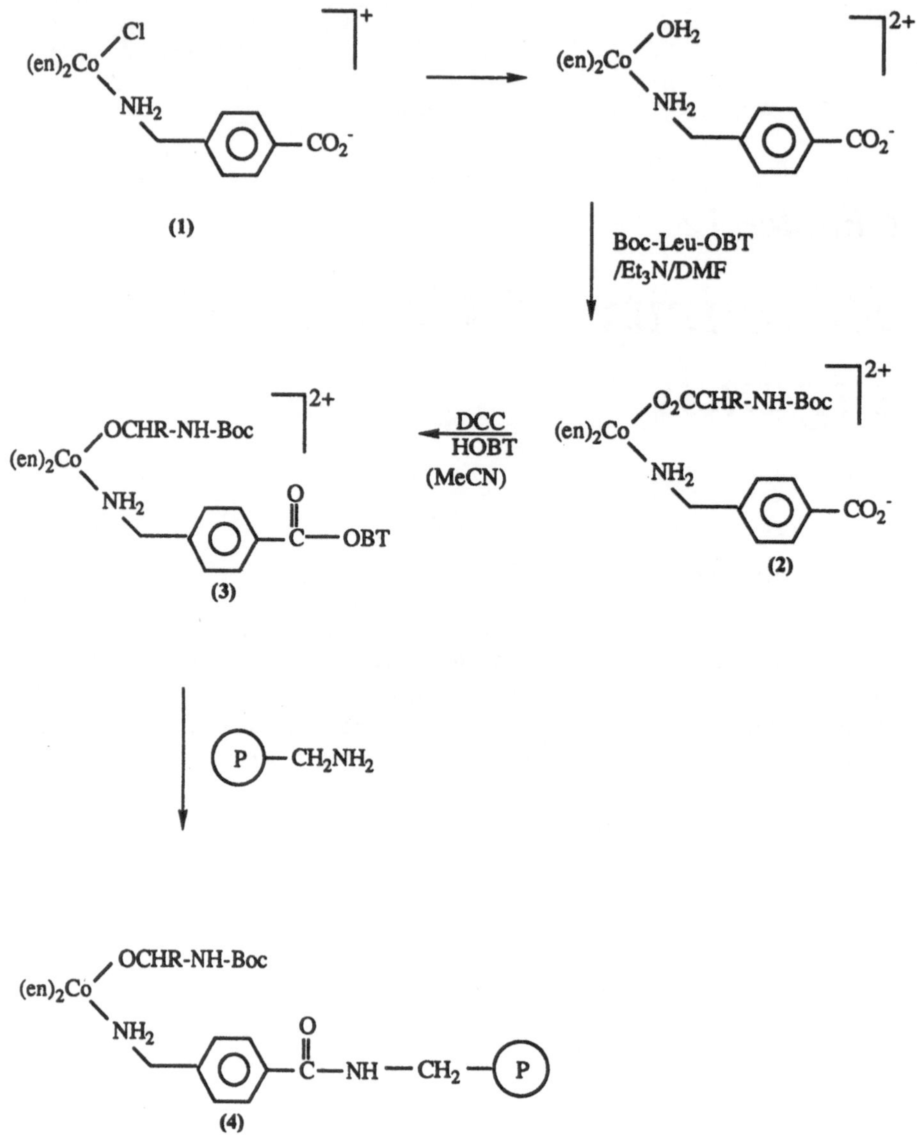

The synthesis, X-ray structure, and electronic spectra of *trans*-[Co(en)$_2$Cl(Boc-L-val)]BF$_4$ (Boc = *t*-butyloxycarbonyl) has been described.[3] The complex was prepared by reaction of *trans*-[Co(en)$_2$(OH)Cl]$^+$ and an active ester form of Boc-L-valine. Cobalt(III) is coordinated in a distorted octahedral fashion by four N atoms, a Cl atom, and a carboxylate oxygen from the Boc-L-valine ligand. The visible absorption spectrum has two bands at 447 and 578 nm characteristic of a *trans*-$[Co(en)_2L_2]^{n+}$ complex. Conversion of coordinated threonine in a cobalt(III) complex into 3-carboxyisothiazoline shown in Eq. (1) has been described.[4] Treatment of Δ-$[Co(en)_2(threoninato)]^{2+}$ (5) with thionyl chloride in DMF followed

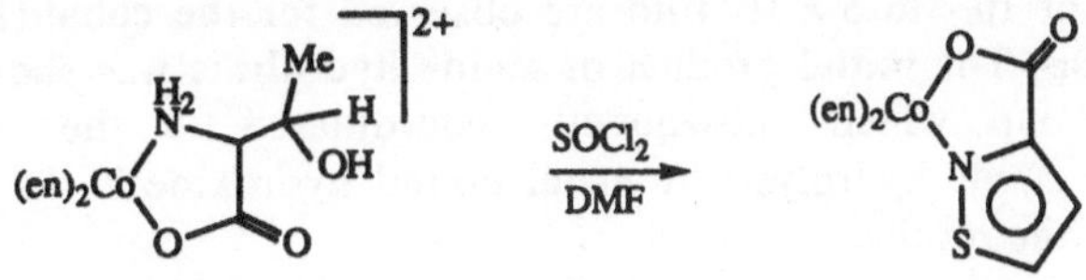

$$\text{(1)}$$

(5)   (6)

by cation-exchange chromatography gives the isothiazoline complex (6) which has the Δ(—)$_{589}$-configuration about cobalt. Rapid oxidation of chelated amino acids to imines and amides by thionyl chloride in DMF has also been described.[5] Oxidation of chelated hydroxy-L-proline by thionyl chloride gives substituted dehydropyrroles, which undergo subsequent base-induced elimination to give[6] chelated pyrrole-2-carboxylate complexes (7) as shown in Eq. (2).

$$\text{(2)}$$

(7)

Protection of the amino acid function of *C*-formylglycine by synthesis from the chiral $[Co(en)_2(glyO)]^{2+}$ complex (glyO = $NH_2CH_2CO_2^-$) has been described.[7] The Δ(—)$_{589}$-$[Co(en)_2(C$-formylglycinato$)]^{2+}$ complex condenses in aqueous pyridine buffer with *S*-penicillamine to give largely the (2*S*, 4*S*, α*S*) and (2*S*, 4*S*, α*R*) isomers of the corresponding cobalt(III) complexes of the coordinated penicilloates.

The rate of intramolecular amide hydrolysis in cobalt(III) complexes has been studied as models for zinc(II)-containing peptidases such as carboxypeptidase A, thermolysin, and angiotensin-converting enzyme.[8] The cobalt(III) complexes (8) and (9) were prepared by hydrogen peroxide oxidation of the corresponding

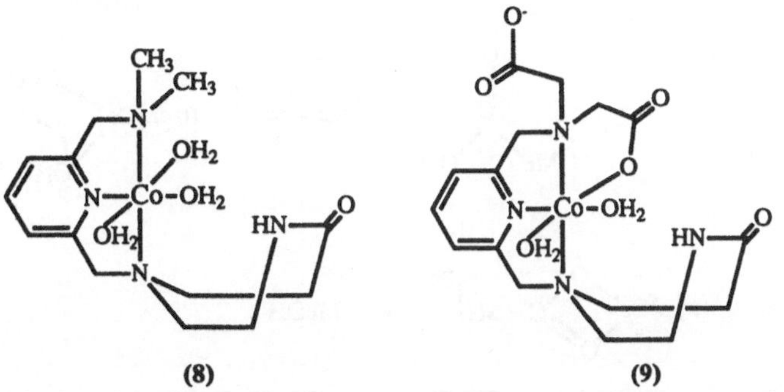

(8)   (9)

cobalt(II) complexes. The cobalt(III) is oriented perpendicularly to the amide plane in such a way that neither the lone pairs of the carboxyl oxygen nor the amide nitrogen can coordinate to the metal. A metal bound water or hydroxide has, however, ready access to the acyl carbon in these complexes. Large rate enhancements of 10^7- to 5×10^8-fold are observed for the cobalt(III)-mediated amide hydrolysis. The initial product of amide hydrolysis was shown to have a free amino group which subsequently coordinated to the metal center. Intramolecular amide hydrolysis by metal bound hydroxide is the only process consistent with the results.

12.2.2. Carboxylic and Phosphate Esters

Currently there is considerable interest in developing artificial esterases and peptidases that hydrolyse esters and amides. The first nonenzymatic, catalytic hydrolysis of methyl acetate and acetylcholine in neutral water at 25 °C has now been described.[9] The complex $[Co(trpn)(OH)(OH_2)]^{2+}$ (**10**) catalyzes the hydrolysis of methyl acetate at pH 7.6. The reaction is believed to involve the steps shown

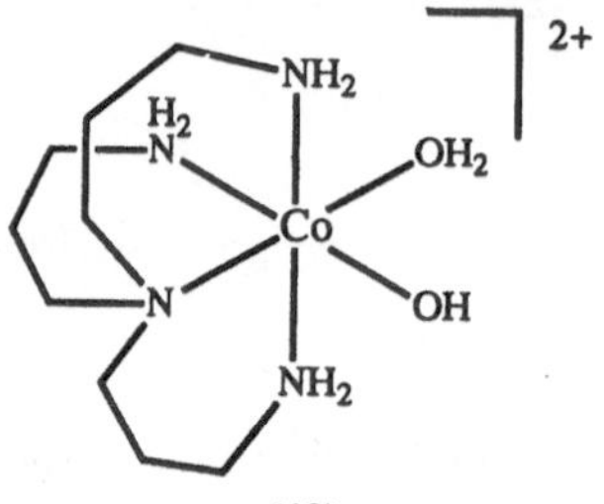

(**10**)

in Scheme 2. Complexation of the ester to cobalt is the rate-determining step. Water hydrolysis of methyl acetate is very slow ($k = 3.2 \times 10^{-10}\,\mathrm{s}^{-1}$ at 25 °C ($t_{1/2} = 70$ years). In a typical catalytic experiment methyl acetate (1 M) and $[Co(trpn)-(OH)(OH_2)]^{2+}$ (1 mM) was maintained at pH 7.6 using a pH-stat to monitor the

Scheme 2

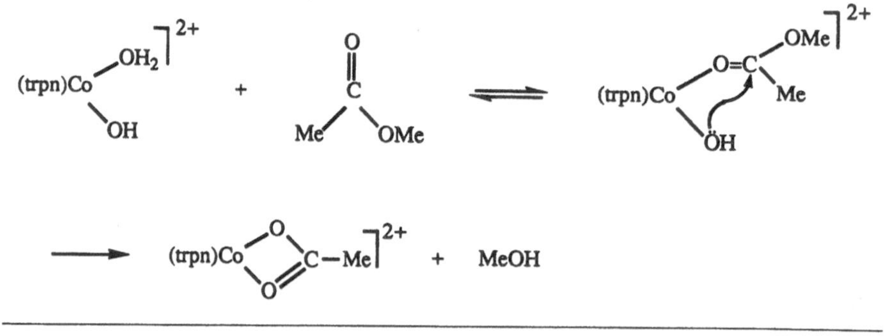

release of acetic acid, three turnovers of the catalyst took place within 150 minutes, and production of methanol was confirmed by ^{1}H NMR.

The phosphate diesters, ethyl p-nitrophenyl phosphate, and bis(4-nitrophenyl) phosphate in cis-$[Ir(en)_2(OH)O_2P(OR)_2]^+$ react with the cis hydroxo group at pH 8 to liberate nitrophenolate ion 10^6-fold faster than for the free ligand under the same conditions.[10] The chelate phosphate esters (11) were not observed, only the ring-opened monodentate monoesters (12) were obtained as a result of P—O bond cleavage. Although the cis-hydroxo ligand is a good nucleophile towa d

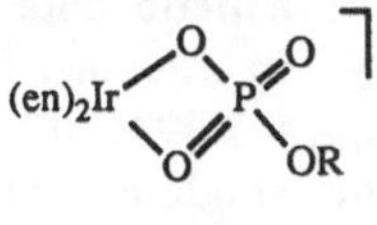

(11) 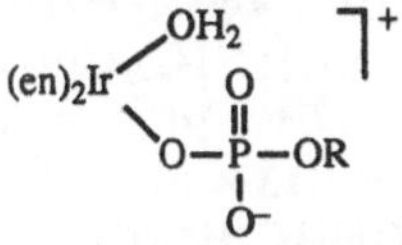   (12)

bound phosphate esters, the reactions of the Ir(III) complexes proceed about 10^3 times slower than the reactions of the analogous cobalt(III) complexes despite the more basic coordinated OH⁻ in the iridium(III) ion. This effect is attributed to the larger size of the Ir(III) ion compared with Co(III), which makes ring closure more difficult. As a result, it appears unlikely that reactions forming four-membered chelate phosphate esters will be relevant in biological systems as the metal ions involving Mg^{2+} and Zn^{2+} are larger than Ir(III). In addition, the observation that the chelated ester prefers to undergo ring opening rather than lose the alcohol group suggests that such chelate esters are unlikely to be effective in the enzymatic systems.

The hydrolysis of 2,4-dinitrophenylphosphate (DNPP) to orthophosphate and 2,4-dinitrophenolate is accelerated in the presence of excess $[Co(pn)_2(OH_2)_2]^{3+}$ or $[Co(trpn)(OH_2)_2]^{3+}$ at rates which vary with pH in a manner suggesting that the hydroxoaquatetraamine cobalt(III) complex is the active catalyst (pn = trimethylenediamine; trpn = 3,3′,3″-triaminotripropylamine).[11] Detailed mechanistic schemes are proposed. For the trpn complex at pH 6.0 and a 25:1 cobalt-to-DNPP ratio (5×10^{-5} M DNPP) the observed rate enhancement is ~3 × 10^3-fold. The calculated specific rate constant for hydrolysis in the reactive 1:1 complex ($k \sim 0.2$ s^{-1}) represents a rate acceleration of some 3×10^4-fold.

The nickel(II)-containing enzyme jack bean urease catalyzes the hydrolysis of urea to ammonia and carbon dioxide [Eq. (3)] by a factor of 10^{14} at pH 7.0 and 38 °C. As it is believed that coordination of urea to Ni(II) is fundamental to

$$H_2NCONH_2 + H_2O \rightarrow H_2NCO_2H + NH_3 \tag{3}$$

the mechanism of catalysis, there has been continuing interest in preparing metal complexes of urea that might model urease by undergoing ready decomposition to NH_3 and CO_2. The N- and O-bonded complexes of phenylurea, (13) and (14),

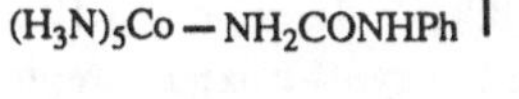

(13) 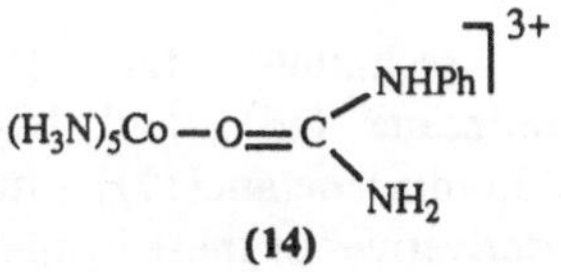   (14)

have been prepared and their reactivities studied.[12] The N-bonded isomer reacts in acidic solution to give mainly $[Co(NH_3)_6]^{3+}$, CO_2, and $PhNH_3^+$ (~70%), but parallel pathways leading to $[(NH_3)_5CoOC(NH_2)NHPh]^{3+}$ (linkage isomerization) and $[(NH_3)_5CoOH_2]^{3+}$ (aquation) are also observed. The major pathway involves an elimination reaction of $[(NH_3)_5CoNH_2CONHC_6H_5]^{3+}$ to give $Ph\tilde{N}H_3$ and $[(NH_3)_5CoNCO]^{2+}$. This latter complex, which has been isolated and characterized, is known to undergo rapid hydration (pH < 2) to give $[(NH_3)_5CoNH_2CO_2H]^{3+}$, which subsequently decomposes by three parallel paths to $[Co(NH_3)_6]^{3+}$ and traces of $[(NH_3)_5CoOH_2]^{3+}$ and $[(NH_3)_5CoOCONH_2]^{3+}$. Kinetic data for the three parallel reactions of $[(NH_3)_5CoNH_2CONHC_6H_5]^{3+}$ in aqueous $HClO_4$ (1.0 M, 25 °C) are $k_{elim} = 1.04 \times 10^{-2}\,s^{-1}$ (elimination), $k_{NO} = 2.39 \times 10^{-3}\,s^{-1}$ (isomerization), and $k_{aq} = 1.13 \times 10^{-3}\,s^{-1}$ (aquation). The single-crystal structure of $[(NH_3)_5CoNHCONHC_6H_5](ClO_4)_2 \cdot H_2O$ shows that the cobalt is in an approximately octahedral environment with phenylurea bonded as its anion through nitrogen. The urea *exo* C—N bond is 0.11 Å longer than the *endo* C—N bond and, as in uncoordinated phenylurea, the electron-withdrawing phenyl ring is neither planar with, nor conjugated to, the urea moiety. These features are probably retained in the protonated complex and account for the facility of the elimination pathway. In contrast, neither elimination nor hydrolysis of the O-bonded phenylurea were observed. The complex undergoes parallel aquation and O-to-N linkage isomerization with both processes being base-catalyzed.

12.2.3. Ligand Synthesis

The intramolecular condensation of monodentate aminomethylcarbonyl ligands, NH_2CH_2COR (R = H, Me, Ph), with quadridentate amines such as tren[tris(2-aminoethyl)amine] and tren(1,8-diamino-3,6-diazaoctane) bound to cobalt(III) has been studied.[13] Chloride bound to the pentamine cobalt(III) directs the condensation to the amine site *trans* to it by base-catalyzed removal of a proton on the amine. The resultant carbinol amine complexes give imines (15), according to Eq. (4), which can be isolated and then reduced by BH_4^- to the fully saturated amine complexes without significant reduction of the Co(III) center.

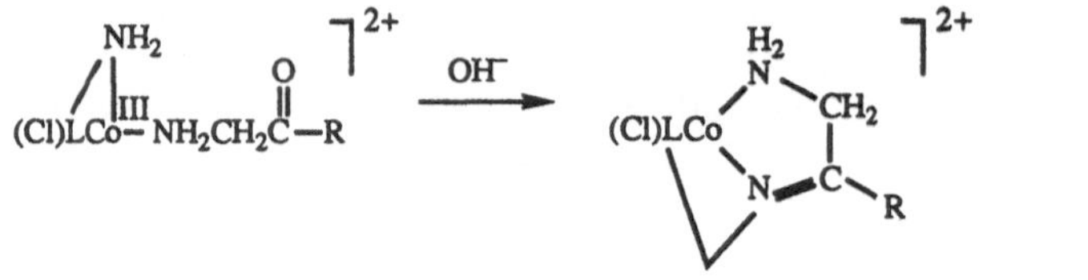

$$\qquad\qquad (4)$$

(15)

L⌐⌐NH₂ = trien or tren

Using this technique the $[Co(tetraen)Cl]^{2+}$ [tetraen = 1,11-diamino-3,6,9-triazaundecane (16)] and $[Co(trenen)Cl]^{2+}$ [trenen = 3-(2-aminoethyl)-1,8-diamino-3,6-diazaoctane(17)] have been synthesized regiospecifically. Substituted derivatives of these ligands and their cobalt(III) complexes have been

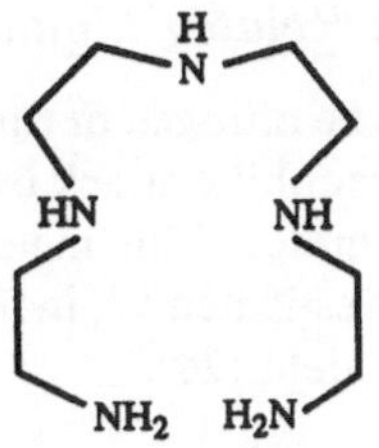

tetraen **(16)**

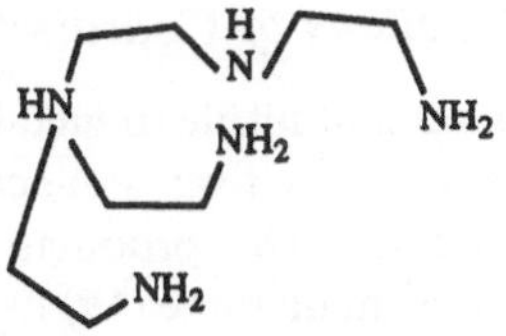

trenen **(17)**

prepared and characterized by X-ray crystallography. The kinetics of cyclization of the carbonyl residues were followed in some cases in conjunction with proton exchange at the amine sites. Limited aquation and base hydrolysis studies were also conducted for the $[Co(trenen)Cl]^{2+}$ ions where the ligand is substituted at C(7) by methyl, phenyl, or imine.

Meridional forms of the N_3S quadridentate complexes of the type **(18)** with R′ = Me and NH_2 are readily prepared from $[Co(en)_2(S(R)CH_2CH_2NH_2)]^{3+}$ (with R = CH_2COCH_3 and CH_2CN, respectively) in basic solution.[14] X-ray crystallography of the complex with R′ = NH_2 (monohydrate of the triflate salt) establishes the β-configuration of the quadridentate.

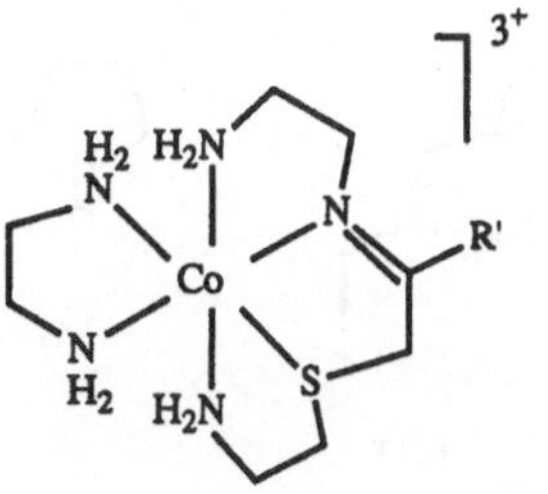

(18)

12.3. Other Metal Complexes

12.3.1. Introduction

This section reviews studies on the reactions of coordinated ligands bound to metals other than cobalt. The emphasis is upon reactivity at carbon in ligands which are bonded to metals through donor atoms of Groups V or VI. In general, the coverage is limited to stoichiometric reactions, although catalytic reactions have been included where they are of particular relevance or provide a particular insight into the ways in which coordination to a metal ion may modify reactivity. The reactions of ligands coordinated to cobalt centres are discussed earlier in this chapter. The second and final volume of a book dealing with the reactions of coordinated ligands has appeared.[15]

12.3.2. Reactions of Coordinated Nitriles and Related Ligands

Coordination of nitrile to metal directly through the nitrogen or through some other proximal atom activates the carbon toward nucleophilic attack by water and other nucleophiles. The products are commonly amides. The metal-mediated reactions of dicyanodiamide (19) have been further investigated[16]; in the presence of nickel or copper salts (19) reacts with alcohols to yield (20).

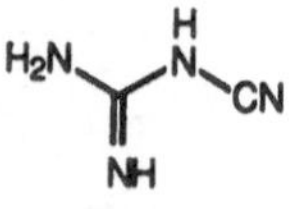

(19) 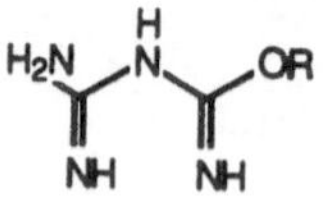   (20)

The hydrolysis of coordinated nitriles has recently attracted some attention. A very facile hydrolysis of nitriles to the corresponding amides at platinum(IV) centers has been described.[17] The complex [Cu(H$_2$NCOCH$_2$CONHNH$_2$)Cl] is formed from the reaction NCCH$_2$CONHNH$_2$; the copper(II) both promotes the hydrolysis and is reduced to copper(I).[18] The hydrolysis of 2-cyanopyridine to 2-pyridinecarboxamide is accelerated several hundred times by the copper(II) complexes of the ligands (21) and (22). In the case of the (22) some picolinic acid was formed, resulting from the intramolecular attack of alkoxide to yield an intermediate iminoester.[19]

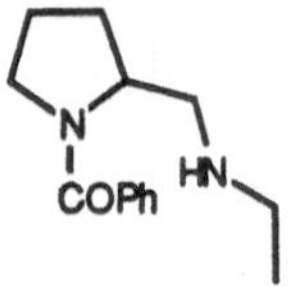

(21) 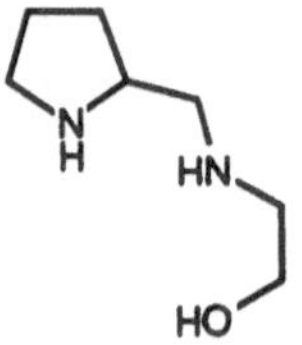   (22)

Related reactions at platinum(II) centers involve the methanolysis of *cis-* or *trans-*[PtCl$_2$(PhCN)$_2$] to yield various isomers of the iminoester complexes [PtCl$_2$LL'] (L or L' = PhCN or MeO(Ph)C=NH).[20] The reaction of the same nitrile complex with lithium phenylamidate yields[21] the triazadiketonate analog (23).

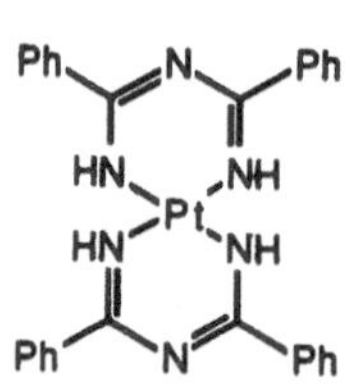

(23)

A number of catalytic reactions of isocyanates in the presence of ruthenium clusters has been reported. In the presence of [HRu$_3$(CO)$_{11}$]$^-$, Et$_3$SiH reacts with RNCO to yield the novel spiro compound (24), while hydrogenation in the presence of [H$_3$Ru$_4$(CO)$_{12}$]$^-$ yields RHNCONRCHO; mechanisms were proposed[22] for

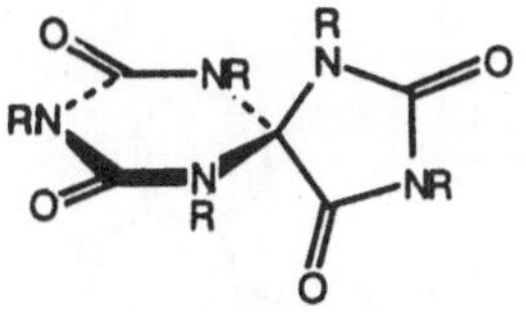

(24) 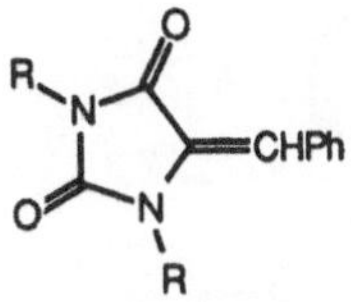   (25)

both of these reactions. No mechanisms were offered to explain the formation of (25) from the reaction of RNCO with PhC≡CH in the presence of $[Ru_3(CO)_{11}]^{2-}$.

12.3.3.   Nucleophilic Addition to Coordinated Carbonyls

A wide range of nucleophiles has been shown to attack coordinated carbonyl compounds. The products depend upon the fate of the tetrahedral hydroxy intermediate formed. In the reaction of carbonyl compounds and amines the loss of water results in the formation of imines, and this methodology has widespread synthetic application. A novel sexidentate macrocyclic ligand (26) has been prepared by template methods. The formation of the hydrazone from hydrazine and

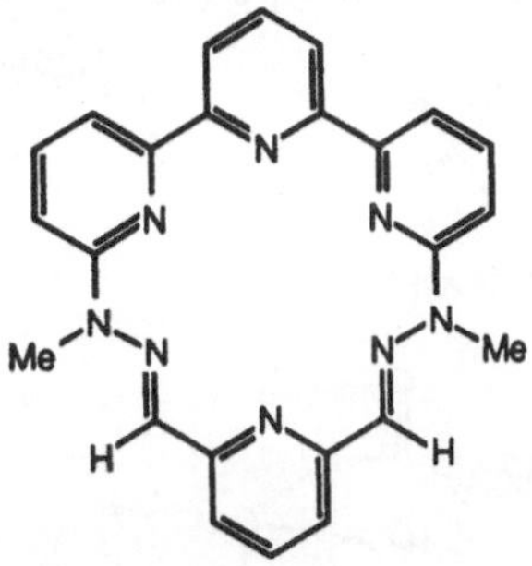

(26)

aldehyde is unremarkable, but the isolation of the free ligand from a template condensation deserves comment. This was rationalized on the basis of a mismatch in the hole-size of the ligand and the template ion (Ni^{2+}) labilizing the macrocyclic complex toward metal loss; this was presented as a further example of the transient template effect.[23]

Hole-size effects are also of importance in the metal-directed self-condensation of 2-aminobenzaldehyde; condensation in the presence of $[Mo_2(O_2CCF_3)_4(PR_3)_2]$ in ethanol results in the formation of complexes of (27). The ring size is controlled by the addition of ethanol, and complexes with varying numbers of ethoxy groups (0–2) are obtained with various metals.[24]

A particularly spectacular example of a metal-controlled reaction is seen in the reactions of 1,2-diaminobenzene with manganese(II) chloride in acetone solution. In the absence of air the expected macrocycle (28) is formed, but under an atmosphere of oxygen the novel compound (29) is obtained. Various intermediates in the formation of (29) were isolated, and a detailed mechanism proposed.[25]

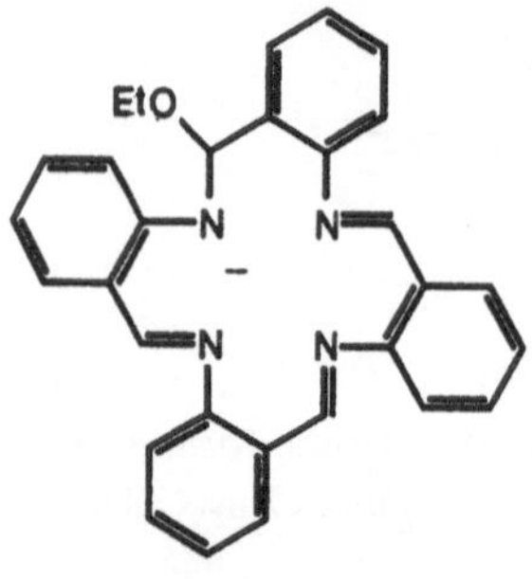

(27) 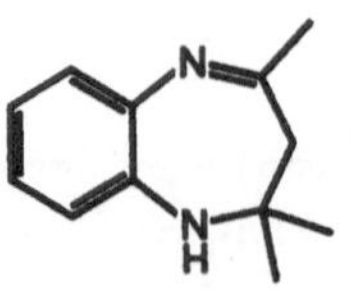   (28) 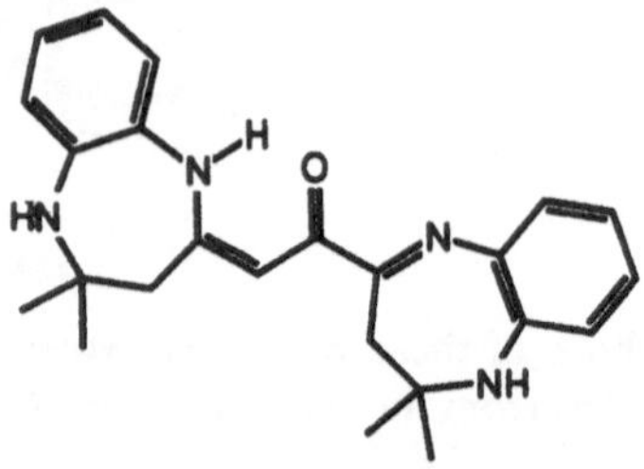   (29)

The hydrolysis of chelated amino acid esters, $H_2NCHRCO_2R$, is known to be accelerated by metal ions, most notably cobalt(III). Dramatic enhancements are also observed with copper(II). Mechanistic studies of the hydrolysis of amino acid esters with copper(II) complexes of glycyl-DL-valine[26] and dien $(H_2HCH_2CH_2NHCH_2CH_2NH_2)$ have been reported.[27] The hydrolysis of benzyl-penicillin (30) by copper(II) salts to give (31) has been further investigated, and it is proposed[28] that the key step involves intramolecular attack by metal-coordinated hydroxide in an intermediate of type (32).

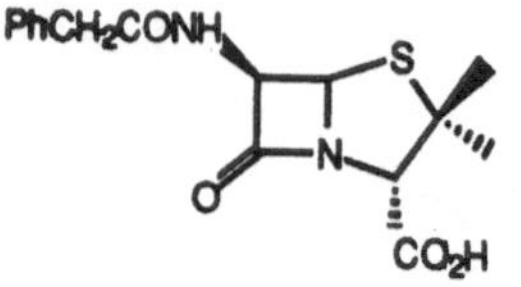

(30) 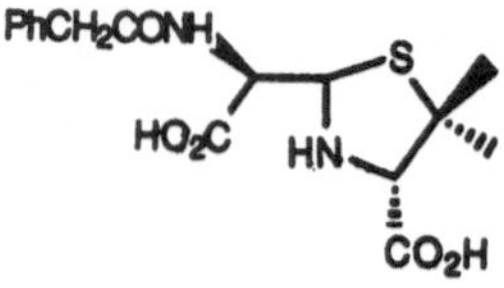   (31)

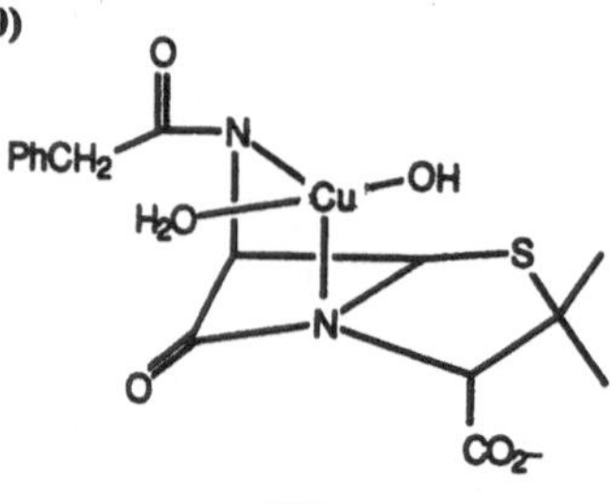

(32)

The hydrolysis of the monocarboxylate and phosphonate esters (33) derived from cyclam is enhanced by copper(II), which initially coordinates to the macrocyclic ligand. In the case of the phosphonate, at pH values above 11.5 the attack is

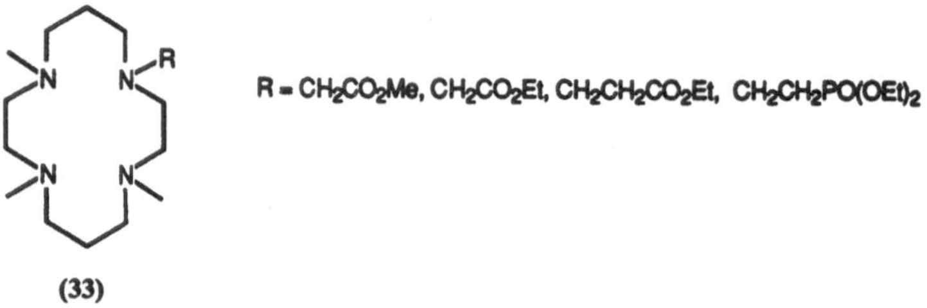

$R = CH_2CO_2Me, CH_2CO_2Et, CH_2CH_2CO_2Et, CH_2CH_2PO(OEt)_2$

(33)

by metal-coordinated hydroxide. The hydrolysis of the carboxylates obeys second-order kinetics. The carboxylate oxygen is coordinated to the metal, but it is not possible to distinguish between attack by inter- or intramolecular nucleophile.[29]

The reduction of carbonyl compounds by hydride transfer is also subject to metal ion catalysis. The reduction of *bis*(2-pyridyl)ketone (34) by $NaBH_4$ is accelerated by zinc(II) or cobalt(II). In the case of Zn^{2+} pseudo-first-order kinetics are observed, and the metal ion acts as a simple Lewis acid and activates the carbonyl by coordination to the oxygen. In the case of cobalt(II) a mechanism involving cobalt(I) intermediates is proposed.[30]

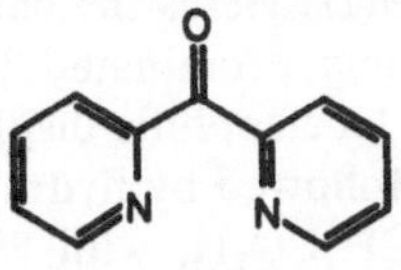

(34)

The hydrolysis of amides is also accelerated by transition metal ions, and in the case of cerium(IV)- and cobalt(III)-mediated processes, the hydrolysis of MeCONHMe is associated with the reduction of the metal. The mechanism proposed,[31] involving an amido radical, is shown in Scheme 3.

Scheme 3

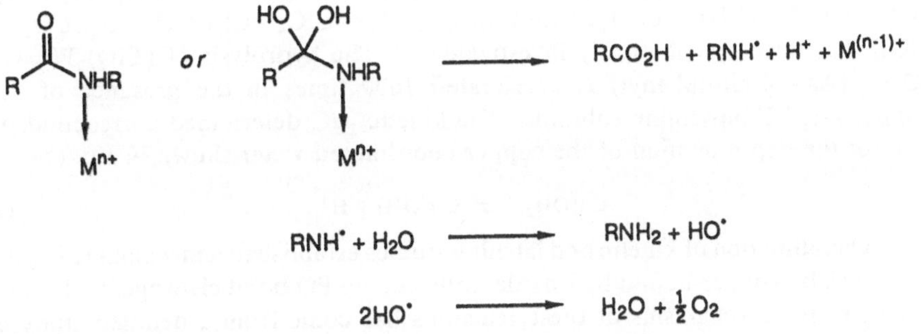

The racemization of amino acids in the presence of metal ions has also been studied; zinc, copper(II), cobalt(II), and palladium(II) ions all enhance the racemization of serine, while zinc and nickel(II) retard that of threonine.[32]

12.3.4. Electrophilic Attack at Coordinated Nitrogen and Oxygen

There has been considerable interest in metal-mediated reactions involving amino acids and their precursors. Most of these studies have involved cobalt(III) centers (see Section 12.2), but there is increasing interest in the use of other centers. The template condensation of (S)-o-[(benzylpropyl)amino]benzophenone with

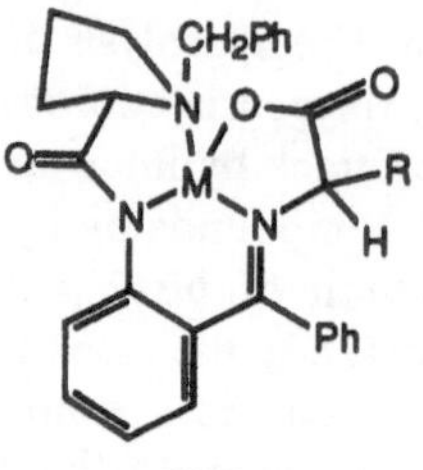

(35)

glycine in the presence of nickel(II) yields the chiral square-planar complex (35) (R = H). This complex is cleanly brominated to yield a mixture of the two diastereomers of (35) (R = Br). Nucleophilic displacement of halide is facile, and reaction with diethyl malonate followed by hydrolysis results in the formation of aspartic acid, $HO_2CCH(NH_2)CH_2CO_2H$, with 80% enantiomeric excess of the L-enantiomer.[33]

12.3.5. *Reactions of Coordinated Phosphorus Compounds*

In the same way that the hydrolyses of carboxylate esters are enhanced upon coordination to a metal, so are those of alkoxy phosphorus compounds. We have already seen an example of this for a cyclam derivative.[29] Copper(II) compounds are particularly effective, and the hydrolysis of $(^1PrO)_2P(=O)F$, by the copper(II) complexes [CuLX$_2$] or [CuL$_2$X$_2$] (L = $Me_2NCH_2CH_2NMe_2$, $H_2NCH_2CH_2CH_2NH_2$ or 1,2-$(Me_2N)_2C_6H_4$; X = ClO_4, Cl_3CCO_2, F_3CCO_2 or NO_3), has been systematically investigated.[34] The hydrolysis of $(EtO)_2P(=O)$-(OAr) (Ar = 4-nitrophenyl) is accelerated 1600 times in the presence of 1:1 copper(II):2,2'-bipyridine solutions. The kinetic pK_a determined corresponds to that for the deprotonation of the copper-coordinated water shown in Eq. (5).

$$Cu(OH_2)^+ \rightleftharpoons Cu(OH) + H^+ \tag{5}$$

A combination of kinetic and labeling studies established a mechanism involving attack by copper-bound hydroxide, followed by PO bond cleavage.[35] Further details of the mechanisms of these reactions has come from a detailed study of the hydrolysis of *cis*-[Ir(en)$_2$(OH){OP(=O)(OR)$_2$}]$^+$ (36) complexes (R = ethyl or 4-nitrophenyl). The reaction involves intramolecular attack by coordinated hydroxide, and rate enhancements of 10^6 are found. The products of the reactions are not the chelated phosphate esters (37) expected from a knowledge of cobalt(III) chemistry, but monodentate phosphate monoesters (38). This is assigned to relative differences in the sizes of the metal ions and the basicity of the coordinated

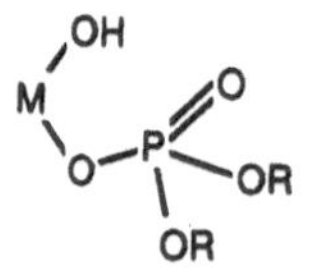

(36) 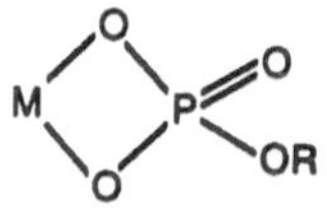   (37) 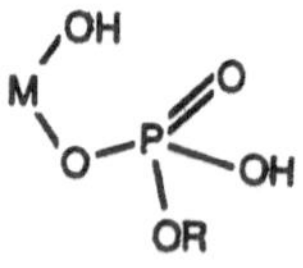   (38)

hydroxide. The intermediate chelated phosphate ester is attacked by *external* hydroxide; no loss of the second alkoxy group was observed.[36] Polymer bound copper(II) complexes are also effective in the hydrolysis of 4-nitrophenyl phosphates, and have a number of synthetic and work-up associated advantages.[37]

A dramatic metal-directed reaction of $Ph_2PO_2CCH=CH_2$ has been reported. Reaction with Wilkinson's catalyst yields $[RhCl(Ph_2POPPh_2)(PPh_3)]$, apparently via (39), which is the first-formed compound in the 1:1 reaction of $Ph_2PO_2CCH=CH_2$ with $[RhCl(PPh_3)_3]$. Most notably, the reaction of $[RhCl(PPh_3)_3]$ with Ph_2POPPh_2 does not give $[RhCl(Ph_2POPPh_2)(PPh_3)]$.[38]

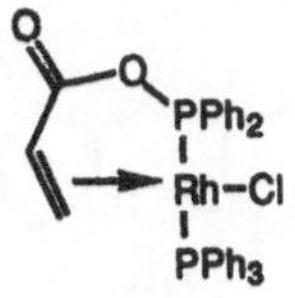

(39)

A similar transformation of P—O bonded compounds at a metal center is seen in the reaction of $[PtCl_2(dppm)]$ with aqueous sodium hydroxide to yield[39] $[(Ph_2MeP)\{Ph_2P(=O)\}Pt(\mu\text{-}OH)_2Pt(Ph_2MeP)\{Ph_2P(=O)\}]$. The key step in the reaction is thought to involve nucleophilic attack by hydroxide with cleavage of a P—C bond as shown in Eq. (6).

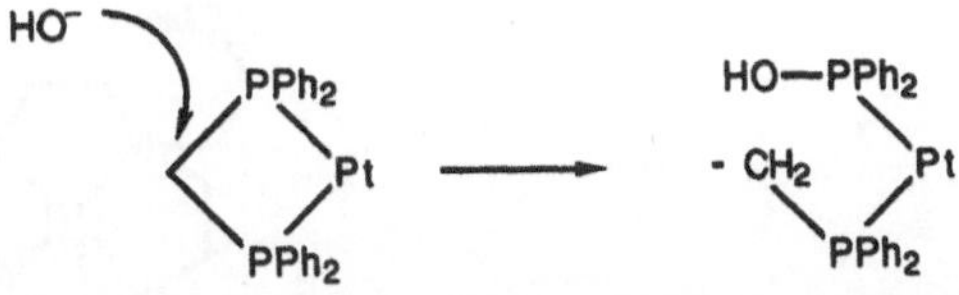

$$\hspace{10cm} (6)$$

Metal-directed cycloaddition reactions of vinylphosphines have recently been reported. Upon heating $[NiBr_2L_2]$ (L = diphenylvinylphosphine) in butanol, a retro-Michael reaction occurs; see Eq. (7), loss of acetylene is followed by an intramolecular addition to yield $[NiBr_2(dppe)]$.[40]

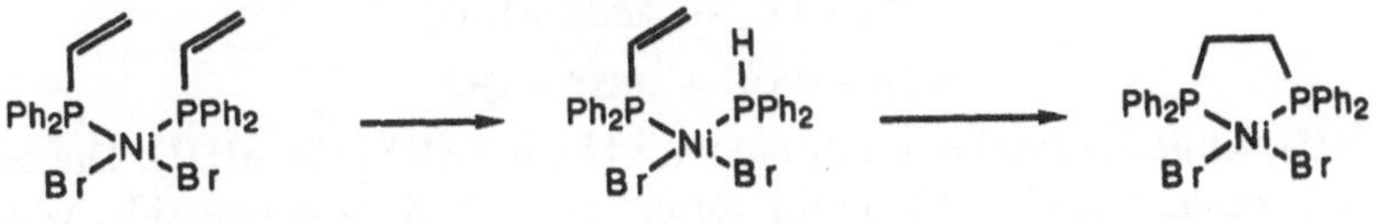

$$\hspace{10cm} (7)$$

Very closely related processes are observed[41,42] in the cycloaddition of diphenylvinylphosphine to nickel, platinum or palladium coordinated phospholes to yield complexes of the type (40) as in Eq. (8).

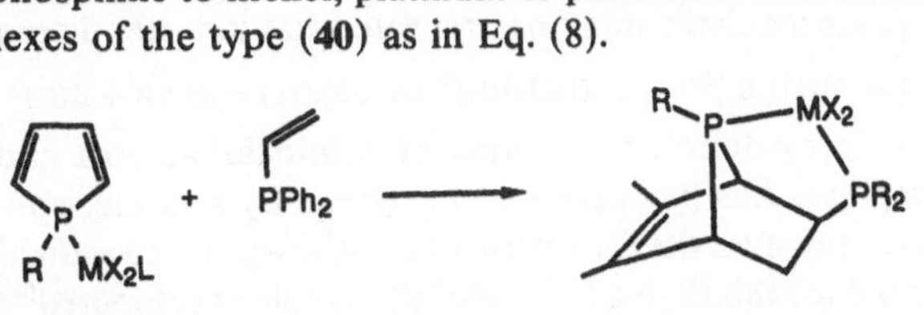

$$\hspace{10cm} (8)$$

(40)

12.3.6. Reactions of Coordinated Sulfur Ligands

Metal compounds have found some application as desulfurization catalysts, and a novel reaction of *N*-phenylthiourea has been reported. The reaction of $PhNCSNH_2$ with ethanolic $Cu(OAc)_2$ in the presence of 2,2'-bipyridine results in the formation of the phenylcarbodiimido complex $[(bipy)Cu(\mu\text{-}PhN{=}C{=}N)_2Cu(bipy)]$.[43] The same group have also described a related partial desulfurization of 3-methylimidazoline-2(1*H*)-thione to yield di(1-methyl-imidazolyl)sulfide **(41)** upon reaction with copper(II) salts.[44] These reactions are mechanistically obscure.

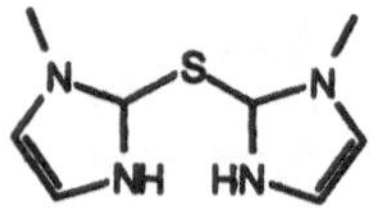

(41)

A more common reaction involves oxidation of a coordinated thiolate to disulfide or an S—O compound. The former reaction may occur with or without the involvement of an oxygen transfer reagent (or dioxygen) while the latter requires such a species. The oxidation of the thiolate complex **(42)** by dioxygen shown in Eq. (9) is first order in **(42)** and dioxygen, and it is proposed that the sulfenate **(43)** arises by attack of a noncoordinated thiolate on coordinated dioxygen.[45]

$$\tag{9}$$

(42)   **(43)**

The aerobic oxidation of cysteine, $HSCH_2CH(NH_2)CO_2H$, to $HO_2C(NH_2)CHCH_2SSCH_2CH(NH_2)CO_2H$ is catalyzed by metal ions. The reaction proceeds in two stages, Eqs. (10) and (11). The first reaction is catalyzed by

$$2RSH + O_2 \rightarrow RSSR + H_2O_2 \tag{10}$$

$$2RSH + H_2O_2 \rightarrow RSSR + 2H_2O \tag{11}$$

copper(II), while the second is only enhanced by iron(II) or iron(III). The copper-catalyzed reaction thus results in the formation of hydrogen peroxide, while the iron-catalyzed reaction does not. Detailed mechanistic studies of this system have been reported.[46] In contrast, the anaerobic oxidation by iron(III) involves hydroxy complexes.[47] A mechanistic study has revealed the key step is reaction (12).

$$[Fe(OH)RS_2]^{2-} + [Fe(OH)RS]^- \rightarrow 2Fe(II) + RSSR + RSH \tag{12}$$

The anaerobic oxidation of a range of dithiocarbamates and xanthates by inert metal complexes has been shown to proceed by an outer-sphere process, in which the key step involves the formation of a sulfur-centered radical. The standard electrode potential for the $Et_2NCS_2^{\cdot}/Et_2NCS_2^-$ couple is estimated[48] to be 0.425 ± 0.33 V vs. SCE.

An interesting displacement of thiolate in nickel(II) and palladium(II) complexes by secondary amines, Eq. (13), has been investigated.[49] The reaction is

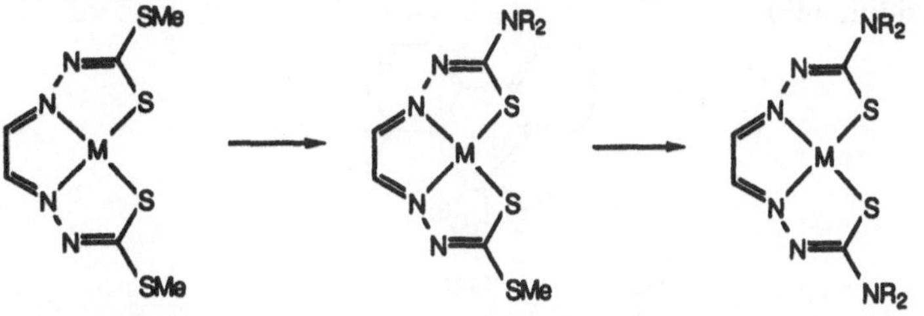

$$\tag{13}$$

third order overall (first order in complex, second order in amine). Surprisingly, complexes of the related ligand (44) do not undergo a similar reaction with amines.[50]

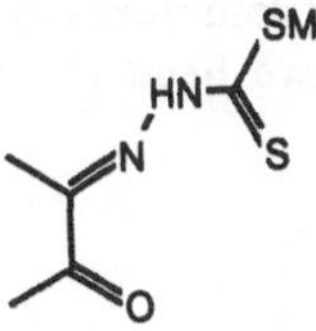

(44)

12.3.7. Reactions of Coordinated Heterocyclic Ligands

This area continues to be of interest, and the most basic of proposals continues to be controversial. The reactions of coordinated pyridines with nucleophiles has been the basis of a continuing series of papers by Gillard and co-workers, and has generated a heated literature over the years. The luminescence (but not the circular dichroism) of $[Cr(bipy)_3]^{3+}$ shows a pH dependence similar to that previously observed for the solvation, racemization, and photoracemization of this ion, and a mechanism involving the formation of a covalent hydrate by the nucleophilic attack of hydroxide upon the 6-position of the coordinated 2,2'-bipyridine is proposed.[51] The observation that zinc compounds are active catalysts for the thermal aerial oxidation of substituted pyridines to $2(1H)$-pyridones is undoubtedly related to these mechanistic suggestions.[52] A detailed analysis of the ^1H and ^{13}C NMR spectra of the cations $[M(bipy)_3]^{n+}$ (M = Ru, $n = 2$; M = Rh, $n = 3$) has been reported; these results have been correlated with the base-catalyzed deuteration of these ligands at the 3- and 6-positions of the 2,2'-bipyridine ring and interpreted in terms of a separability of the σ and π components to the metal–ligand interaction.[53] The analogy between coordination of a pyridine to a metal and quaternization is at the basis of the Gillard mechanism, and has been questioned by a number of workers. Fenske–Hall quantum-chemical studies have confirmed previous suggestions that relative σ and π components to the metal–ligand interaction may be separable and counteractive.[54]

Related to the above reports is the observation that the reaction of $[Zr(OAr)_2Bz_2]$ (Ar = 2,6-di-*tert*-butylphenyl) with 2,2'-bipyridine, 4,4'-dimethyl-

2,2'-bipyridine or 1,10-phenanthroline leads to the formation[55] of complexes derived from the nucleophilic attack of the benzyl group on the 6-position of the 2,2'-bipyridine (45).

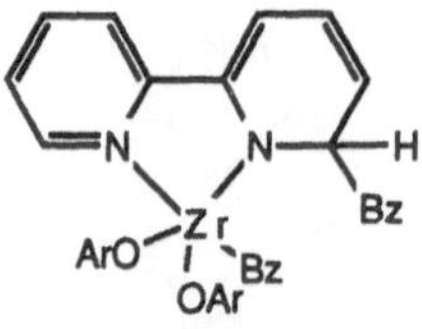

(45)

The pH-dependent ^{1}H NMR spectrum of $[Ru(bipy)_2(py)_2]^{2+}$ salts has been shown not to involve the formation of covalent hydrates.[56] The reaction of 2,4,6-tris(2-pyridyl)-1,3,5-triazine (46) with aqueous copper(I) salts results in a metal-promoted hydrolysis to yield complexes of 2-pyridinecarboxamide and (47). A number of metal complexes have been characterized structurally from these reactions.[57-59]

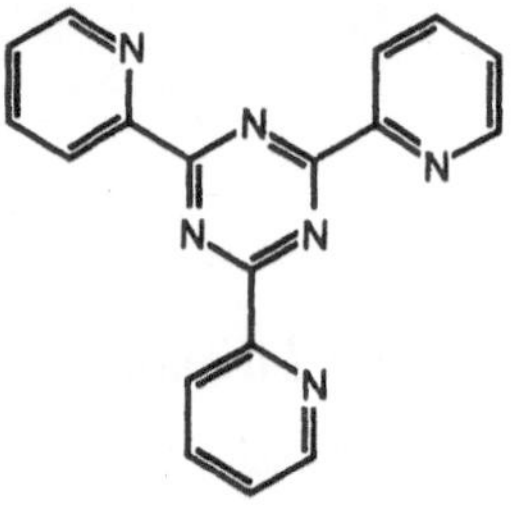

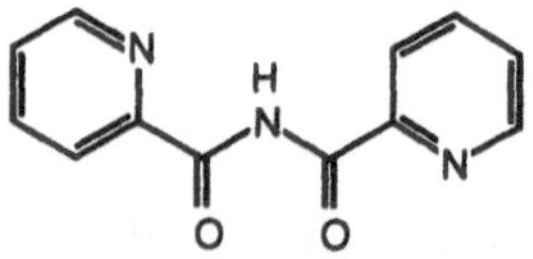

(46)

(47)

A detailed study of the effect of coordination to a $[Rh(NH_3)_5]^{3+}$ center of a range of imidazoles and pyrazoles has been reported.[60] The relative σ and π components controlling the pK_a of the coordinated heterocycle are discussed and compared with related cobalt(III) and ruthenium(III) complexes. The addition of hydroxide to the ligand in a range of 5-nitro-1,10-phenanthroline complexes has been described.[61] It is proposed that attack upon $[Ru(bipy)_n(5\text{-nitrophen})_{3-n}]^{2+}$ can occur at the 4- or 7-positions in addition to the expected 6-position of the Meisenheimer adduct. The spectroscopic data supporting this proposal do not appear to be unambiguous. Finally, the displacement of halide from coordinated 4-halo substituted pyridines has been developed[62] into a new synthetic method.

12.3.8.   Reactions Involving Cyclometalated Ligands

A useful application[63] of cyclometalated complexes is seen in the smooth and highly selective chlorination reaction (14).

A cyclometalated intermediate is implicated in the reaction of (48) with phosphines to yield (49). The reaction only occurs in the presence of nickel or copper salts, and is thought to involve M—C bonded species formed by insertion into the phenyl C—Br bond *ortho* to the 2,2'-bipyridine. Support for this proposal

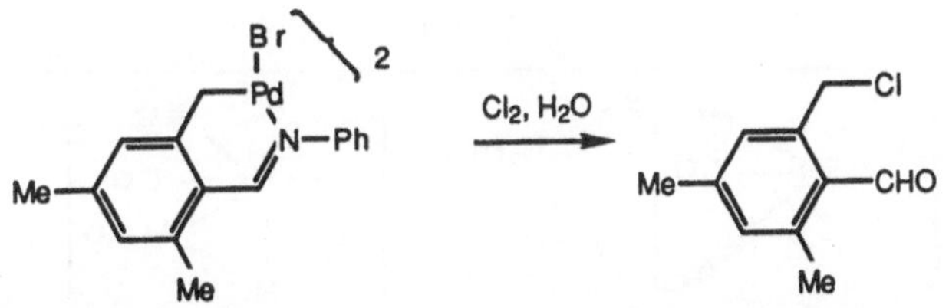

(14)

comes from the observation that the reaction of [{Cu(PPh₃)Br}₄] with (49) yields a red material which is thought to be a cyclometalated intermediate.[64]

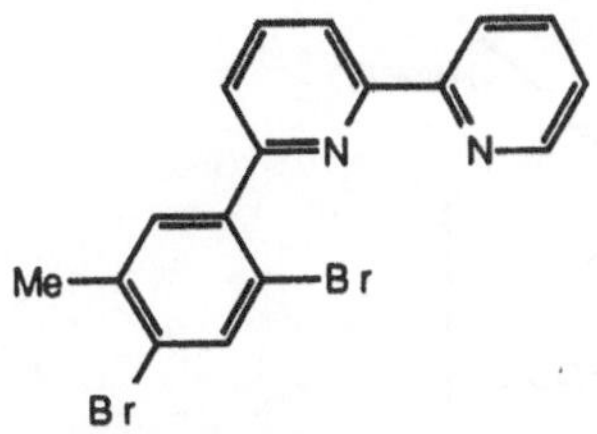

(48) 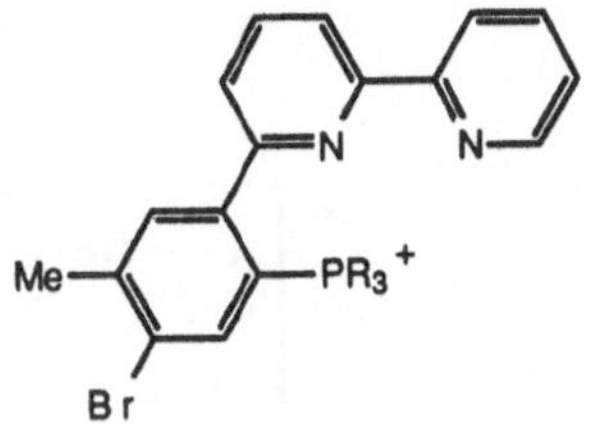   (49)

12.4. Organometallic Compounds

12.4.1. Reactions with Nucleophiles

12.4.1.1. Addition at Carbon Monoxide Ligands

Detailed kinetic studies have been reported[65] for the addition of the nucleophiles $N_2H_4 \cdot H_2O$, NH_2NHMe, NH_2NMe_2, NH_2OH, and N_3^- to a carbonyl ligand of the complex [ReL(NO)(CO)₂] [50], L = 1,4,7-triazacyclononane in methanol solvent. The reactions follow the rate law (15) which was rationalized

$$\text{Rate} = K_1 k_2[\text{Nuc}][\text{Re}]/(1 + K_1[\text{Nuc}]) \qquad (15)$$

in terms of a general mechanism involving rapid preequilibrium nucleophilic addition to a CO ligand, followed by rate-determining rearrangement to the product. This is illustrated in Scheme 4; for the case of reaction with hydrazine hydrate the carbazoyl intermediate (51) was identified by its $\nu(\text{CO})$ band at 1650 cm⁻¹. With azide as nucleophile, the preequilibrium constant, K_1, is sufficiently large that rate law (15) simplifies to Rate = $k_2[\text{Re}]$.

The first NMR (¹³C, ³¹P) characterization of a carbamoyl intermediate (52) has been described[66] in the related reaction of pyrrolidine with *trans*-[Pd(COPh)-(CO)(PMe₃)₂]BF₄. The O-protonated carbamoyl structure (53) was supported by an X-ray structural analysis of its O-alkylated analog.

A study[67] of nucleophilic addition by PhLi to the bimetallic complexes [(CO)₄M(μ-PPh₂)₂Pt(PCy₃)] (54), M = Mo or W, reveals differences in regiospecificity compared to mononuclear octahedral metal carbonyl substrates. Addition of PhLi occurs on an equatorial CO of (54) (*trans* to the phosphido bridge),

Scheme 4

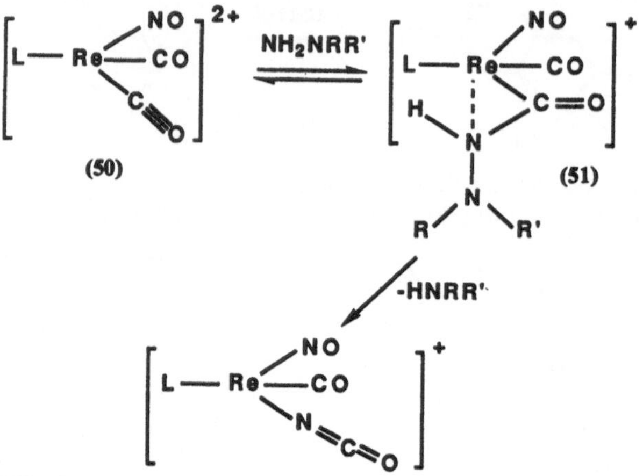

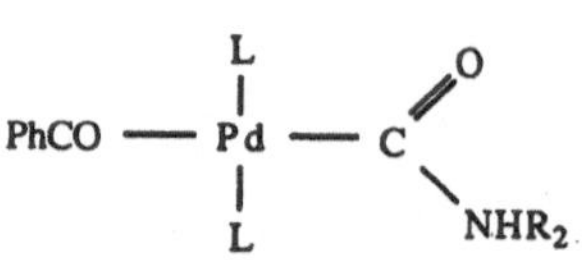

rather than on an axial CO (*trans* to CO) as previously found for related mononuclear complexes. The observed regiospecificity was shown from extended Hückel molecular orbital calculations to be a consequence of the M—Pt bonding, rather than steric effects.

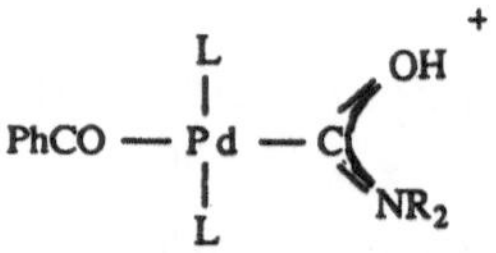

Several reviews have appeared on the role of carbon monoxide and metal carbonyl complexes in the commercial production of fuels and chemicals, with emphasis on Fischer–Tropsch-type hydrogenation processes.[68-70]

12.4.1.2. Reactions at Alkyl and Acyl Ligands

Deprotonation of the species MCH_3^+ (M = Fe, Co) with a series of reference bases has been monitored, yielding the proton affinities of the corresponding carbene MCH_2 species.[71] Detailed kinetic studies have also been reported for the hydrogenation of the osmium alkyl bond in the complex (**55**) (L = PMe_3) [see Eq. (16)] which was shown to proceed by an acid-catalyzed path.[72]

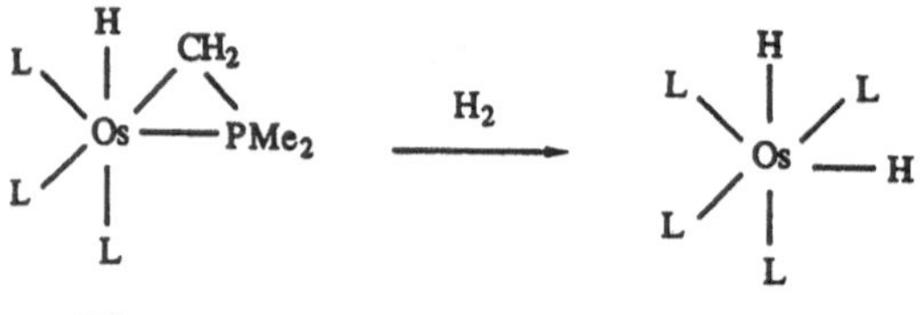

$$ \tag{16} $$

The zirconium acyl complexes [Cp$_2$ZrX(COR)] (X = Cl, Me) react rapidly with alkylaluminum reagents R$_2$AlY (Y = R', Cl) to give alkylaluminum adducts of ketone complexes [Cp$_2$Zr(μ-η^2-OCRR')(μ-Cl)AlR$_2$].[73] Deuterium labeling experiments using [Cp$_2$Zr(CD$_3$)(COCD$_3$)] as substrate demonstrate a mechanism involving the intramolecular migration of an alkyl group to a *cis*-acyl ligand. These results have implications for CO reduction over metal oxide catalysts, providing a mechanistic rationale for the formation of previously proposed[74] ketone intermediates.

12.4.1.3. Addition at Carbene and Related Ligands

Gladysz[75] has shown previously that the anticlinal isomers of carbene complexes [Cp(NO)(PPh$_3$)Re=CHR]$^+$ are much more thermodynamically stable than their synclinal isomers (anticlinal:synclinal > 99:1), and that nucleophilic addition at the carbene ligand occurs exclusively over the smaller NO ligand on the face opposite PPh$_3$. However, unlike these rhenium complexes which have substantial barriers to isomer interconversion, in the related iron benzylidene complexes [Cp(CO)(PR$_3$)Fe=CHPh] (56) (R = Ph, Et), the synclinal (56a) and anticlinal (56b) isomers are in very rapid equilibrium (ΔG^+ < 7 kcal mol^{-1}).[76] For these latter cases, where nucleophilic addition is slower than carbene isomer interconversion, nucleophilic addition by MeO$^-$ and H$^-$ has now been shown[76] to give mixtures of both diastereomeric products (Scheme 5), including significant amounts derived from the minor, but intrinsically more reactive, synclinal isomers. The higher reactivity of the synclinal isomers toward nucleophiles was rationalized in terms of the earlier conformational analysis of Davies and Seeman.[77]

Related diastereomeric benzylidene complexes ($S_{Fe}S_c$)- and ($R_{Fe}S_c$)-[Cp(CO)-(PPh$_2$R*)Fe=CHPh] [R* = (S)-2-methylbutyl] have been prepared.[78] Transfer

Scheme 5

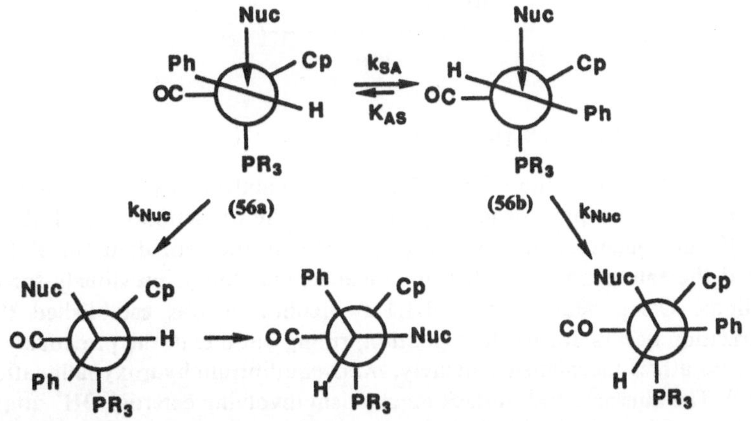

of the carbene ligand from these complexes to styrene and vinyl acetate gives phenylcyclopropane derivatives with moderate to high enantioselectivity. These stereochemical results are consistent with a mechanism involving alkene attack on the synclinal isomer of the carbene substrate, followed by backside attack of the developing electrophilic center at $C\gamma$ on the Fe—$C\alpha$ bond.

Nucleophilic attack by PMe_3 on the anticlinal (ac) and synclinal (sc) isomers of the vinylidene complex $[Cp(NO)(PPh_3)Re=C=CHMe]$ (**57a** and **57b**) has been shown[79] to give the *Z*- and *E*-vinyl phosphonium adducts $[Cp(NO)PPh_3]Re\{C(PMe_3)=CHMe\}]^+$ (**58a** and **58b**) respectively; see Eq. (17).

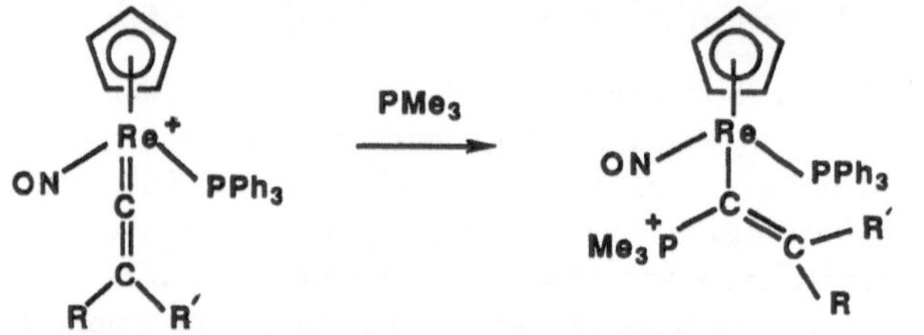

$$ac\text{-}(\mathbf{57a})\ (\ R = H,\ R' = Me)$$
$$sc\text{-}(\mathbf{57b})\ (\ R = Me,\ R' = H\ )$$

(Z) -(**58a**)
(E) -(**58b**)

(17)

This establishes that nucleophilic addition upon the $C\alpha$ center of (**57a**) and (**57b**) occurs stereospecifically from a direction *anti* to the bulky PPh_3 ligand, as illustrated by the transition state (**59**).

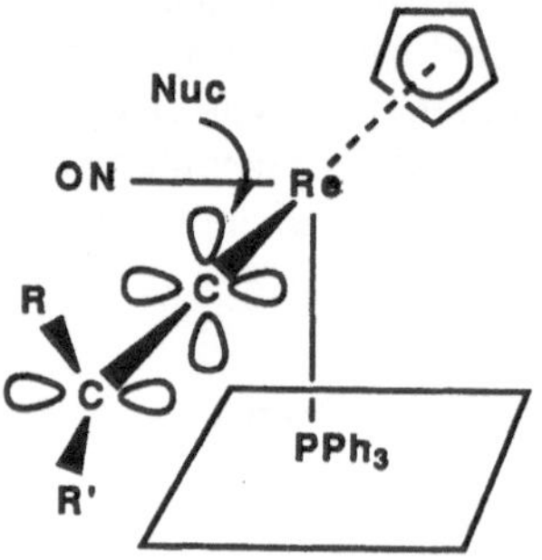

(**59**)

12.4.1.4.. Addition at η^2-Alkenes

The mechanism of the oxidation of olefins catalyzed by Pd(II) continues to be a controversial topic. An investigation[80] of the oxidation of allyl alcohol by $[PdCl_4]^{2-}$ in aqueous solution, using quinone as the reoxidant for Pd(0), has revealed the rate expression (18), identical to that found previously for ethene oxidations. Using deuterated allyl-1,1-d_2-alcohol, it was established that no isomerization occurs during the oxidation, ruling out one of the previously postulated *trans* attack mechanisms, namely, *trans* equilibrium hydroxypalladation; see Eq. (19). The alternate *trans* attack mechanism involving external OH⁻ attack was also eliminated on the basis of rate and equilibrium data indicating that it would have to be faster than a diffusion-controlled reaction. The results were therefore interpreted in terms of *cis* attack by a coordinated OH⁻ group [see Eq. (20)] as postulated earlier by Henry.[81]

$$\text{Rate} = k[\text{PdCl}_4^{2-}][\text{C}_6\text{H}_6\text{O}]/[\text{H}^+][\text{Cl}^-]^2 \tag{18}$$

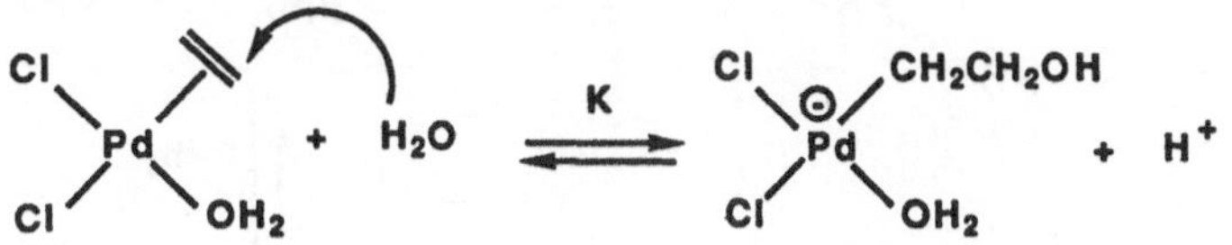

(19)

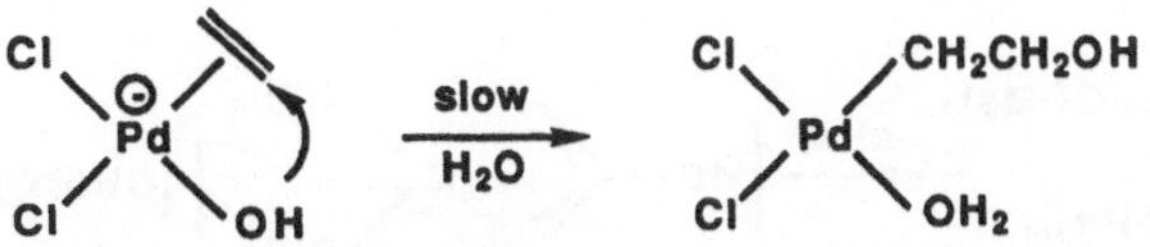

(20)

The mechanism of the related palladium-catalyzed asymmetric hydrocyanation of alkenes has also been examined[82] in detail using [Pd(RR-diop)(C₂H₄)] and related chiral species as precursor complexes. These additions gave the *exo* nitrile product with up to 40% enantiomeric excess, indicating stereoselective complexation of norbornene to the Pd(0) via the *exo* face. The plausible reaction intermediates [Pd(diop)(norbornene)] and the hydrido cyanide (60) were characterized by ¹H and ³¹P NMR spectroscopy. Oxidative addition of HCN to Pd(0) is believed to precede rate-determining alkene binding.

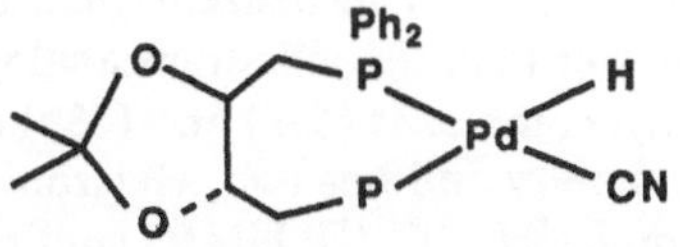

(60)

The first quantitative measure of the relative migratory aptitudes of hydride and alkyl groups has been obtained[83] by comparing the β-migratory insertion reactions of the complexes [Cp*{P(OMe)₃}RhH(C₂H₄)]⁺ and [Cp*{P(OMe)₃}Rh(C₂H₅)(C₂H₄)]⁺, e.g., Eq. (21). The difference in free energies of activation $(\Delta G_{\text{Etmig}}^{\ddagger} - \Delta G_{\text{Hmig}}^{\ddagger})$ is 10.3 kcal mol⁻¹, corresponding to a rate ratio $k_{\text{Hmig}}/k_{\text{Etmig}}$ of 10^7–10^8.

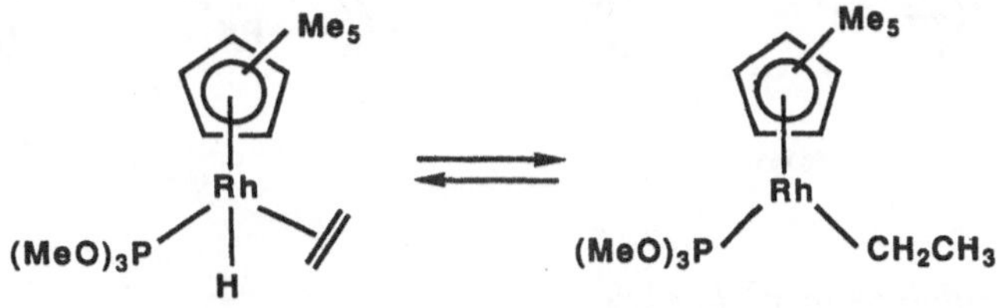

(21)

12.4.1.5. Addition at η^3-Enyls

The kinetics of the reactions of the allyl complexes [M(η^3-CH₂CMeCH₂)(acac)] (61) (M = Pd, Pt) with PPh₃, leading to C—C coupling products, have

Scheme 6

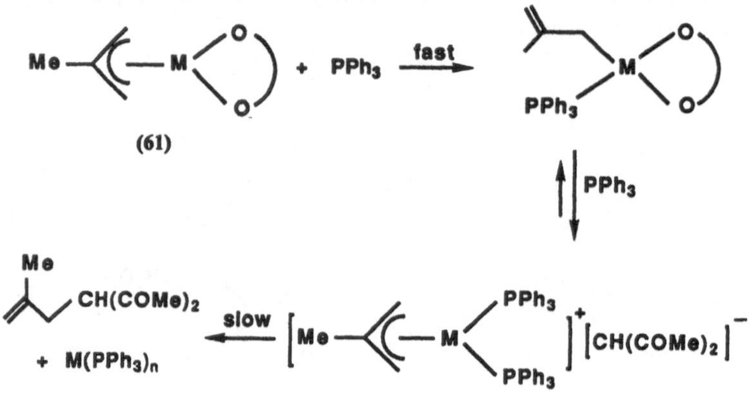

(61)

(62)

been measured in both dichloromethane and benzene as solvents.[84] The data are consistent with the general mechanism shown in Scheme 6. Low-temperature ^{1}H NMR measurements in CH_2Cl_2 confirm the formation of the ion-pair intermediate (62). The rate-determining collapse of these ion pairs to the product via nucleophilic addition to the enyl ligand was faster in benzene than in CH_2Cl_2. The platinum complex was somewhat slower than its palladium analog.

The diastereomeric allyl complexes (63a) and (63b) contain a chiral bidentate ligand which lacks C_{2v} symmetry and has two, electronically very different phosphorus donors. Spectroscopic (^{1}H, ^{13}C, ^{31}P NMR) studies of these complexes and their reactions with nucleophiles have been carried out[85] in order to throw further light on the factors controlling asymmetric catalytic allylation. A correlation was proposed between the NMR parameters of the η^3-allyl diastereoisomers, the donor properties of the ligand, and the regiochemistry of nucleophilic addition to the allyl termini.

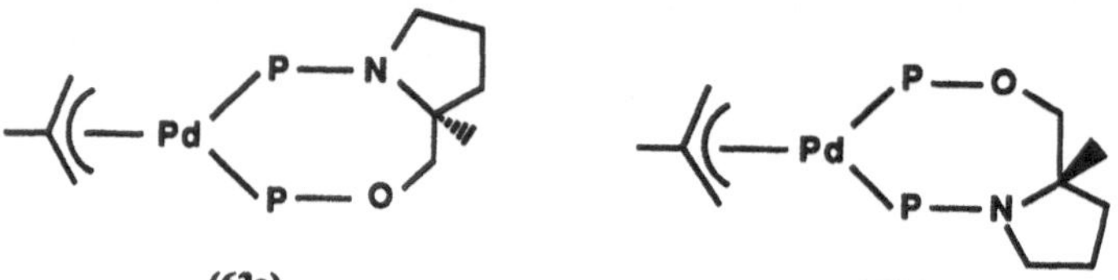

(63a)   (63b)

12.4.1.6. Addition at η^5-Dienyls

Contrary to the rules of Davies and co-workers,[86] attack by CN^- on the complexes (64a) (R = CH_2Ph) and (64b) [R = $CHS(CH_2)_3S$] occurs on the odd cyclohexadienyl ring rather than the even benzene ligand, despite the presence[87] of the bulky R group on the dienyl moiety; see Eq. (22). The heterobifunctional

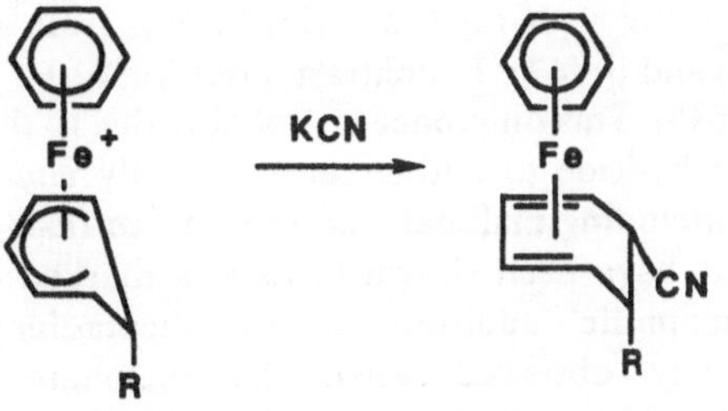

$$(22)$$

(64a, b)   **(65a, b)**

cis-1,2-cyclohexadiene structure of the product was confirmed in the case of (65a) by a single-crystal X-ray analysis.

Cis-difunctionalization of cyclohexadienes has also been previously achieved[88] by the sequence shown in Scheme 7 (M = Mn). A study of the related rhenium complexes has now shown[89] that addition of H^- to the dienyl ring of the complexes (66) (M = Re; X = 1,2,3,4,5-Me; R = Ph or Me) occurs in a stereo-specific *endo* fashion. This unusual behavior, also noted for the manganese analogs, is believed to arise via the intermediacy of metal formyl species. Unfortunately, other nucleophiles such as CN^-, $CH(CO_2Me)_2^-$, and PBu_3 do not add to the dienyl ring of (66), presumably due to steric hindrance by the ring methyl substituents, but instead cause substitution of a CO ligand.

Scheme 7

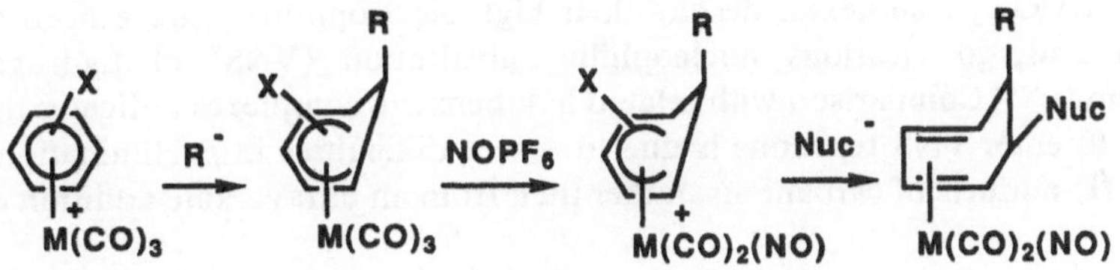

(66)

Addition of the carbanion $^-CH(CO_2Me)_2$ to the chiral dienyl complexes (67) (R = Ph, Me) occurs with appreciable asymmetric induction, giving a 2:1 ratio of the diastereomeric products (68a) and (68b); see Eq. (23). The major product (68a)

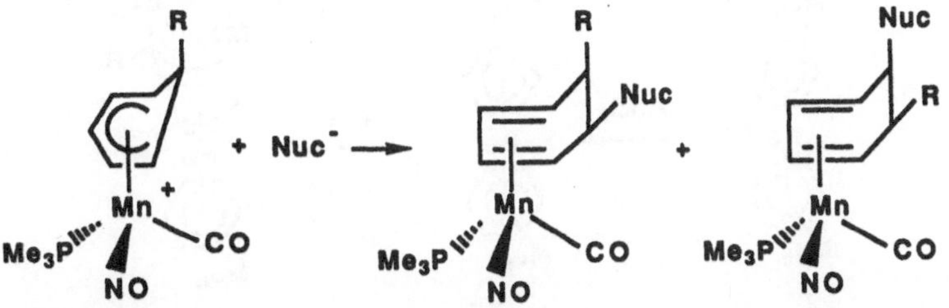

$$(23)$$

(67)   **(68a)**   **(68b)**

(characterized structurally by X-ray diffraction) arises from attack of malonate at C-5, *cis* to the NO ligand in (**67**). In contrast, addition of H⁻ gives predominantly (2:1) diastereomer (**68b**). This difference is probably due to the unusual preference mentioned above for hydride to add stereospecifically *endo* to complexes (**67**). Other nucleophiles, including malonate, add in an *exo* fashion.[90]

Ring substituents have been shown to have a marked effects on the regio-selectivity of nucleophilic addition to cycloheptadienyl ligands.[91] The unusual regioselectivity observed when the malonate ion adds to $[(1\text{-}MeO_2C\text{—}C_7H_8)Fe(CO)_3]^+$ and $[(3\text{-}MeO_2C\text{—}C_7H_8)Fe(CO)_3]^+$ was rationalized in terms of the electron-withdrawing methoxycarbonyl substituent causing only a weak frontier orbital interaction, resulting in charge control of the regioselectivity.

12.4.1.7. Reaction at η^6-Arene Ligands

The highly nucleophilic metal carbonyl anions $[CpFe(CO)_2]^-$ and $[(C_5Me_5)Fe(CO)_2]^-$ displace the halide group from the coordinated arene ligands in $[(\eta^6\text{-}XRC_6H_4)Cr(CO)_3]$ (X = F, Cl, I) substrates to yield novel $[(\eta^6\{CpFe(CO)_2\}(RC_6H_4)Cr(CO)_3]$ products.[92] However, certain haloarene complexes react largely through an apparent electron-transfer pathway to produce $[CpFe(CO)_2]_2$ and $[(\eta^6\text{-}RC_6H_5)Cr(CO)_3]$. The fraction of reduced products was shown to be dependent on (1) the reducing power of the metal carbonyl anion, (2) the electron-donating ability of the R group (electron donor ≫ electron acceptor), (3) the substitution pattern of the arene (*ortho* ≫ *meta* ~ *para*), and (4) the identity of the halogen leaving group (I ≫ Cl > F). In contrast, $[(\eta^6\text{-}$ arene)Cr(CO)_3]$ complexes, despite their high electrophilicity, have been shown not to undergo vicarious nucleophilic substitution (VNS) of hydrogen by carbanions.[93] Comparison with related halobenzene complexes indicates that the failure to enter VNS reactions is due to severe difficulties in β-elimination from the σ (H)-adducts of carbanions, rather than from an unfavorable addition equilibrium.

Reaction of AlEt₃ with the $[Fe(arene)_2]^{2+}$ complexes (arene = C_6H_6, mesityl-ene, pentamethylbenzene, and C_6Me_6) has provided the first examples of direct, high-yield alkylation at an alkylated arene ring location (Scheme 8).[94] *Exo*

Scheme 8

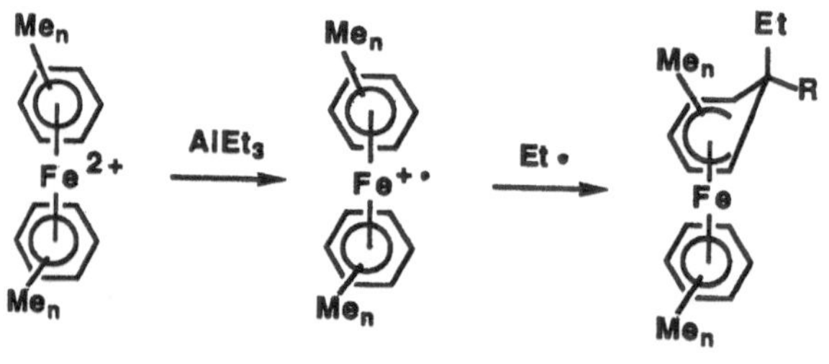

(69)

addition to the arene ring was confirmed by X-ray crystal structures for the products (69) (Me$_n$ = 1,3,5-Me$_3$; R = H$^+$; and Me$_n$ = Me$_5$, R = Me). An ET/radical coupling process was proposed (Scheme 8) rather than direct nucleophilic addition. In this mechanism, AlEt$_3$ acts initially as a one-electron reducing agent and alkylation subsequently occurs via radical coupling of the reactive 19-electron [Fe(arene)$_2$]$^+$ cation with an ethyl radical.

Studies of nucleophilic attack on the π-bound benzo[b]thiophene (BT) complexes [CpRu(BT)]$^+$ and [Cp*Ir(BT)]$^{2+}$ have been undertaken as potential models for the reactions of BT on hydrodesulfurization (HDS) catalysts.[95] In these complexes, the BT coordinates to the metal via the benzene rather than the thiophene ring. A range of nucleophiles, including hydrides and sulfides which are proposed to be present on HDS catalyst surfaces, add to the benzene ring to give primarily C-7 adducts; see Eq. (24). However, products that might be precursors to, or are similar to, proposed intermediates in the HDS of BT, such as

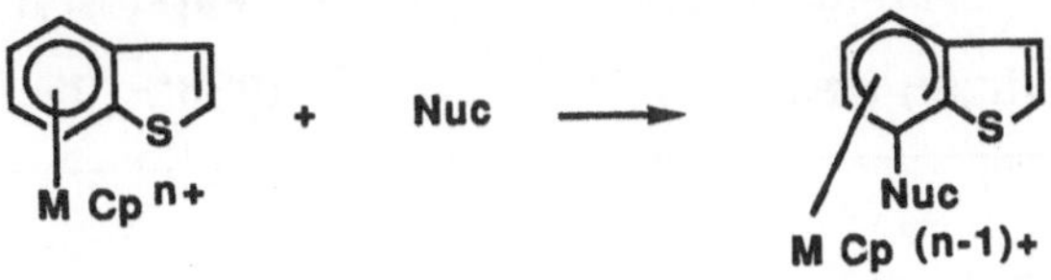

$$\text{(structure)} \quad + \quad \text{Nuc} \quad \longrightarrow \quad \text{(structure)} \tag{24}$$

M Cp $^{n+}$ → M Cp $^{(n-1)+}$, Nuc

2,3-dihydrobenzothiophene and thiophenols, were not observed; this indicated that the π-arene-bonded complexes are not good models for the HDS reactivity of benzothiophene.

12.4.2. Reactions with Electrophiles

12.4.2.1. Attack at Acyl Ligands

The new chiral iron acyl complex [CpFe(CO)PPh$_2$(C$_6$F$_5$)}(COMe)] (70) has been synthesized, and the stereospecificity of the aldol and imine condensations of its enolates (71) (lithium, tin, aluminum, and copper counterions) with benzaldehyde and benzylideneaniline examined (Scheme 9).[96] The reactions proceed with good to excellent stereoselectivities (89-99% de). Strikingly, the direction of asymmetric induction is independent of the metal enolate species employed, giving the (R^*, S^*) diastereomeric product (72a) selectively in all cases. This contrasts with the behavior of the extensively studied acyl complex [CpFe(CO)(PPh$_3$)-(COMe)] (73), where the stereoselectivity is very dependent on the metal species.[97,98] This marked difference between (70) and (73) was rationalized in terms of an electron donor–acceptor-type interaction between the pentafluorophenyl group (electron acceptor) and the enolate oxygen (electron donor), causing the highly selective generation of *endo* (or *syn*) enolate (71); see Eq. (25). In contrast (73) favors the selective generation of *exo* (or *anti*) enolate. From the *endo* enolate (71a), (R^*, S^*) isomers should be formed through a quasi-boat-like transition state previously proposed by Davies.[97]

Scheme 9

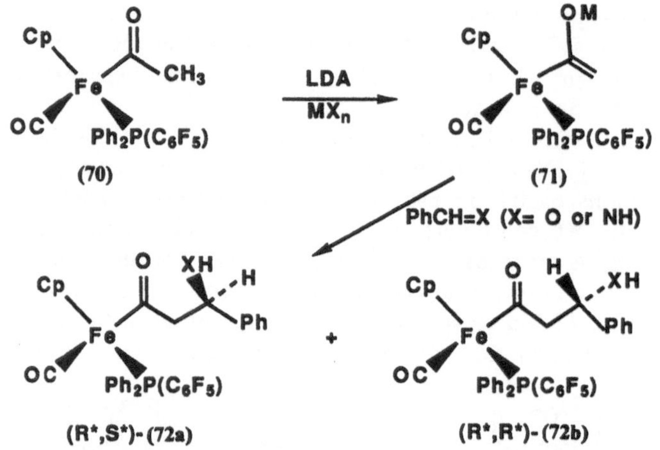

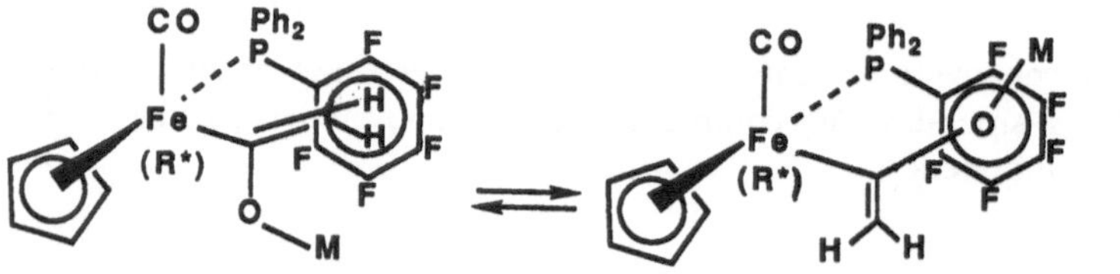

(70)   (71)

PhCH=X (X= O or NH)

(R*,S*)-(72a)   +   (R*,R*)-(72b)

                                                                    (25)

exo-enolate   endo-enolate

(71a)   (71b)

12.4.2.2. Attack at Carbene Ligands

The protonation of vinylidene ligands bound to electron-rich metal centers is well-known to give the corresponding alkylidyne complexes. The first kinetic study of such a process has been reported[99] for the reaction shown in Eq. (26). The kinetic data were rationalized in terms of the overall mechanism shown in Scheme 10. For example, in the presence of excess $NHEt_3BPh_4$, reaction (26) obeys the rate law (27), which is a limiting form of the general rate law for Scheme 10. Initial

$$trans\text{-}[ReCl(CCHPh)(dppe)_2] + NHEt_3^+ \rightarrow trans\text{-}[ReCl(CCH_2Ph)(dppe)_2]^+ + NEt_3 \quad (26)$$

$$k_{obs} = k_4 + k_3[NHEt_3^+] \quad (27)$$

protonation is believed to occur on the rhenium atom to give the species (74), rather than at the vinylidene ligand, i.e., equilibrium K_1 is established much faster than k_2. The conversion of (74) to the alkylidyne product (75) then proceeds via a combination of paths k_3 and k_4. The proposed initial protonation to form (74) is consistent with the isotope effects observed with $NDEt_3^+$ ($k_4^H/k_4^D = 1.16$ and $k_3^H/k_3^D = 1.34$).

Scheme 10

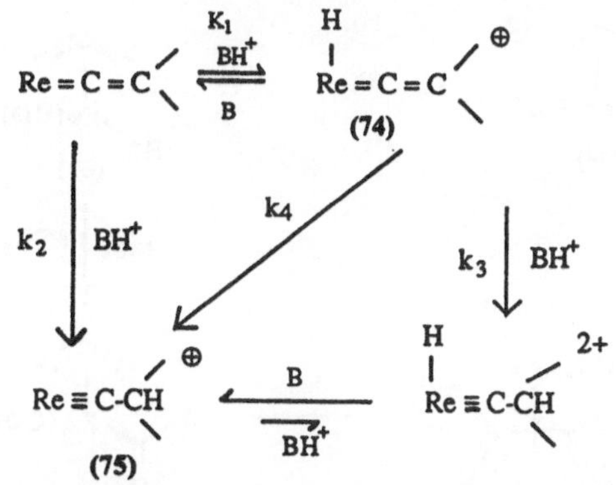

$$Re=C=C \underset{B}{\overset{K_1 \atop BH^+}{\rightleftharpoons}} \overset{H}{Re=C=C} \oplus \quad (74)$$

$$k_2 \Big| BH^+ \qquad k_4 \qquad k_3 \Big| BH^+$$

$$Re \equiv C\text{-}CH \oplus \quad \underset{BH^+}{\overset{B}{\rightleftharpoons}} \quad \overset{H}{Re \equiv C\text{-}CH} \ 2+$$

$$(75)$$

Protonation of the acetylide complexes (76) (R = Me, Ph, 1-C$_{10}$H$_7$) with CF$_3$SO$_3$H at low temperature ($-78\,°$C) has been shown[100] from ^{1}H NMR analysis to proceed with high stereoselectivity, giving preferentially (>98% de) the anticlinal diastereomeric vinylidene complexes (77a) (Scheme 11). Analogous C$_\beta$ methylations of (76) are similarly stereospecific. This high 1,3 asymmetric induction was

Scheme 11

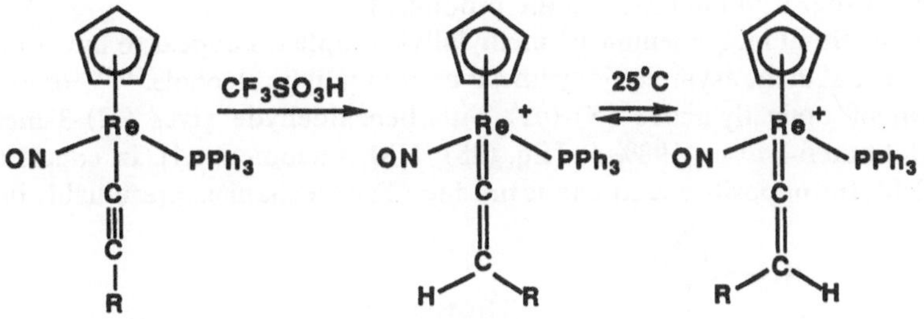

(76)　　　　　ac-(77a)　　　　　sc-(77b)

attributed to electrophilic attack upon the C$_\beta$ atom of (76) from the side opposite to the PPh$_3$ ligand, giving the less stable Re=C=C isomer with the C$_\beta$ substituent *syn* to the PPh$_3$ ligand. Equilibration of the two diastereomers occurs on warming to room temperature, with (77b) predominating.

12.4.2.3. Attack at η^3-Allyl Ligands

Synthetic and mechanistic studies have been reported[101] on the reaction of alkyl halides with the anionic allyl complex [Fe(CO)$_3$(η^3-C$_3$H$_5$)]$^-$ (78). Treatment of (78) with alkyl halides in THF followed by addition of PPh$_3$ gave α,β-unsaturated

Scheme 12

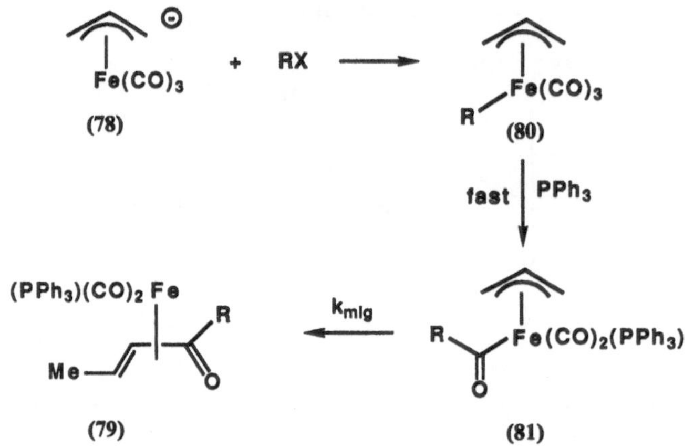

ketone complexes $[Fe(CO)_2(PPh_3)$ $(\eta^4\text{-enone})]$ (**79**) exclusively ($\geq 95\%$) as *trans* isomers. Evidence for the pathway shown in Scheme 12 included spectroscopic characterization of the thermally unstable intermediates (**80**). Treatment of (**80**) with PPh_3 at $-78\,°C$ causes immediate formation of the acyl complexes (**81**), which were also characterized by ^1H NMR spectroscopy. These latter complexes convert by a first-order acyl migration to the products (**79**).

Homochiral molybdenum η^3-methylallyl complexes appear to have considerable potential in the asymmetric synthesis of homoallylic alcohols.[102] For example, reaction of optically-active (S)-(**82**) with benzaldehyde gives (R)-3-methyl-1-phenyl-3-buten-1-ol in >98% ee [Eq. (28), NM = neomenthyl]. In contrast, (R)-(**82**) yields the opposite enantiomeric product. The mechanism presumably involves

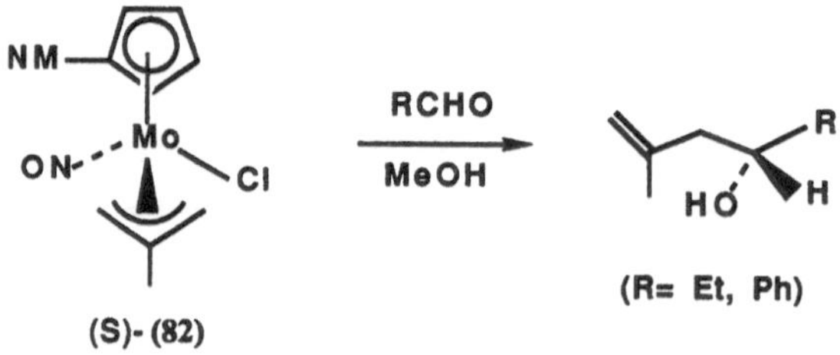

$$(28)$$

attack of an η^1-allyl on an η^1-aldehyde through a chair-like transition state, in which the allyl group adds to a specific enantioface of the aldehyde for a given absolute stereochemistry at the metal.

12.4.2.4. Attack at η^4-Dienes

Spectroscopic and synthetic studies have been described for the protonations of the series of anionic $[(\eta^4\text{-diene})Mn(CO)_3]^-$ complexes shown in Scheme 13.[103]

Protonations of (83) and (84) give agostic complexes. *Exo* addition to the η^4-cycloheptatriene ring of (85) was established from experiments in D_2O using ^1H and ^2D NMR for product identification. Similar *exo* addition of H^+ (D^+) occurs

Scheme 13

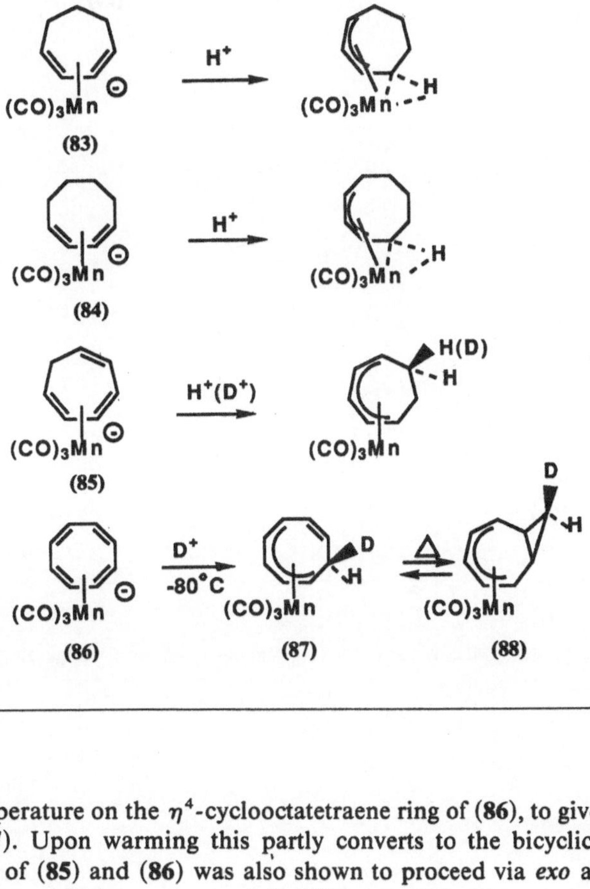

at room temperature on the η^4-cyclooctatetraene ring of (86), to give the fluxional complex (87). Upon warming this partly converts to the bicyclic species (88). Methylation of (85) and (86) was also shown to proceed via *exo* addition to the η^4-diene rings, with the structure of the [(7-exomethylcyclooctatrienyl)Mn(CO)$_3$] product being established by X-ray analysis.

12.4.2.5. Miscellaneous Reactions with Electrophiles

The regioselectivity of electrophilic addition to the 3-oxaplatina(II)cyclobutane complex (89) is very dependent on the nature of the electrophile.[104] Treatment of (89) with the strong electrophile acetyl chloride leads to exclusive attack at the "hard" metallacycle oxygen [Eq. (29)], to give the ring-opened product

(90). In contrast, the weakly electrophilic MeI reacts preferentially at the "soft" Pt(II) center, leading to oxidative addition.

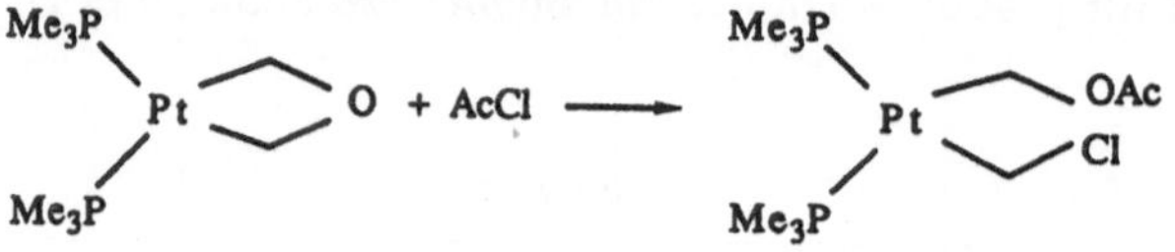

(29)

$$\text{(89)} \qquad\qquad\qquad\qquad \text{(90)}$$

Chapter 13

Rearrangements, Intramolecular Exchanges, and Isomerizations of Organometallic Compounds

13.1. Introduction

The organization of this chapter follows closely that used in the previous Volume.[1] Most of the internal molecular rearrangements reported here are based on the results of high-resolution dynamic NMR (DNMR) experiments performed in one and/or two dimensions. Recent developments of these techniques both in the general context of organometallic chemistry[2] and in the specific area of S and Se complexes of platinum[3] have been reviewed.

13.2. Mononuclear Compounds

13.2.1. Isomerizations and Ligand Site Exchange

As in the previous review[1] compounds in this section are ordered according to the position of the metal atom in the Periodic Table. Organolithium compounds exhibit a wide variety of fluxional rearrangements. Two recent reports have concerned 2-methylallyllithium[4] and the complex of neopentyllithium with penta-methyldiethylenetriamine (PMDTA).[5] The former compound, in diethylether, exists as two species of differing aggregation which undergo exchange and internal rotation, both processes being slow on the ^{1}H NMR timescale at 150 K. The complex of neopentyllithium with PMDTA is a 1 : 1 tridentate, monomeric complex which exhibits both rotation about the C, Li bond axis and Li, N bond exchanges. NMR bandshape analyses above 150 K gave ΔH^+ values of 32.2 and 36.8 kJ mol^{-1}, respectively for the two processes.

Among Group IV metal complexes, fluxional complexes of zirconium have been reported.[6,7] Halozirconocene dithiocarbamate complexes [Cp$_2$ZrX(RR'dtc)] (RR'dtc = S$_2$CNRR') (1) exhibit N-methyl group exchange.[6] When RR'dtc is replaced by SCZNMe$_2$ (Z = O,S) a mechanism involving rotation about the C—N bond is proposed. For the complexes (X = 4-pyridyloxy, OPh; Z = S) much higher

(1)

exchange rates occur, probably involving Zr—S bond rupture promoted by the π-donor properties of the aryloxy ligands. The complex [(RCp)$_2$Zr(CH$_2$OCH$_3$)$_2$] contains both η^1- and η^2-CH$_2$OCH$_3$ ligands which rapidly interconvert as evidenced by variable-temperature ^{1}H NMR spectra, $\Delta G^+(173 \text{ K}) = 32.6 \pm 1.7$ kJ mol^{-1}. This interconversion provides a kinetically feasible pathway for CH$_2$ insertion into the Zr—CH$_2$OCH$_3$ σ-bond, followed by β-OCH$_3$ elimination to give the observed formation of ethene and [Cp$_2$Zr(OCH$_3$)$_2$].[7]

The tris(norbornane-*exo*-2,3-dithiolate) complexes of niobium(V) and tantalum(V) [A][M(ndt)$_3$] (A = Ph$_4$P$^+$, Et$_4$N$^+$, Bu$_4$N$^+$) exist in solution (CD$_2$Cl$_2$) as two interconverting isomers, described as synclastic and anticlastic forms.[8] Activation parameters (ΔH^+) for the exchange are 38.5 ± 2.0 and 46.4 kJ mol^{-1} for the Nb and Ta complexes, respectively, and suggest non-bond-rupture pathways. The usual mechanism of *cis–trans* isomerization of octahedral M(A-B)$_3$ complexes is the rhombic twist, but the authors suggest a modified mechanism involving bicapped tetrahedral (BCT) intermediates (Scheme 1). This mechanism, like the rhombic twist, occurs with inversion of optical configuration.

Organochromium(0) complexes of general type [Cr(CO)$_2$(CX){(MeO)$_3$P}$_3$] (X = O, S, Se) are stereochemically nonrigid octahedral complexes which, in the case of the X = S and Se complexes, undergo rearrangement via trigonal-prismatic (Bailar) intermediates rather than bicapped–tetrahedral structures. The activation enthalpies were in the range 64.9–75.3 kJ mol^{-1}. The precise characterization of the mechanism was made by ^{31}P 2D-NOESY NMR experiments.[9] The cycloheptatrienyl complexes [MX(CO)$_2$(L-L)(η^3-C$_7$H$_7$)] (M = Mo, W; X = I, Cl; L-L = dppe) undergo a fluxional process which interconverts inequivalent phosphorus environments. A trigonal twist mechanism is proposed.[10] The complex [Cr(CO)$_2${P(OMe)$_3$}(C$_{10}$H$_{11}$)] where C$_{10}$H$_{11}$ = 5*H*-benzocycloheptene, exhibits three fluxional processes in solution, isomeric exchange, hindered rotation of the Cr(CO)$_2${P(OMe)$_3$} fragment, and a 1,5-H shift of H(5-*endo*) within the C$_{10}$H$_{11}$ ligand. Activation barriers were $\Delta G^+(220 \text{ K}) = 37.6$, $\Delta G^+(280 \text{ K}) = 57.8$, and $\Delta G^+(320 \text{ K}) = 61.1$ kJ mol^{-1}, respectively.[11] The W(CO)$_5$ moiety undergoes 1,3-

Scheme 1

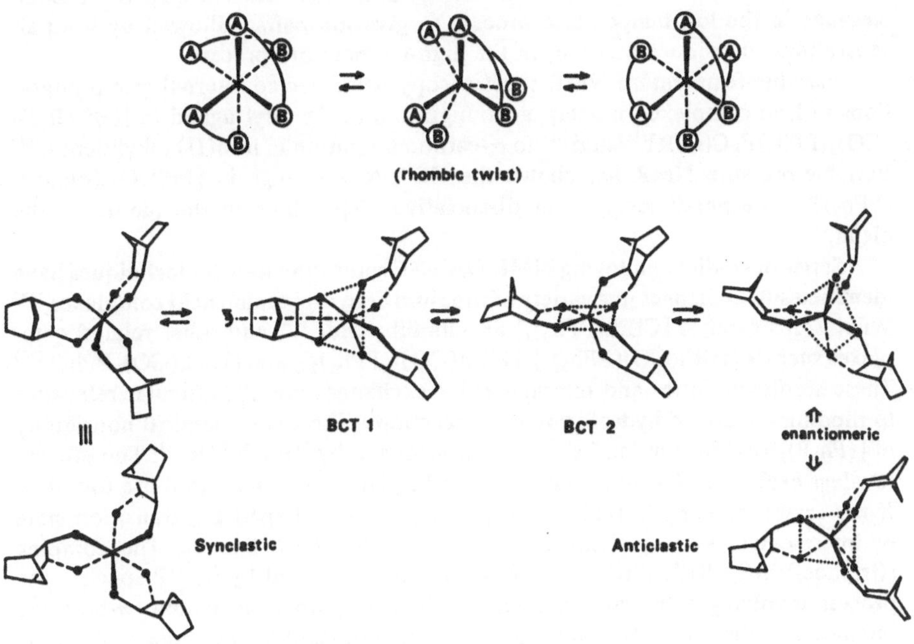

metallotropic shifts in its complexes with 1,4,5-triazanaphthalene and 1,3,5,8-tetraazanaphthalene.[12] The equilibrium mixtures of species depended on the competition between the more basic pyridine or pyrimidine nitrogens and the stronger π-back donation property of the pyrazine nitrogen. In a separate study, $[Os(CO)_4(PMe_3)]$ acts as a ligand to $M(CO)_5$ (M = Cr, W), forming an unbridged donor–acceptor metal–metal bond (Os—Cr or Os—W) with the phosphine *trans* to this bond in the solid state.[13] In solution, a small amount of the *cis* phosphine isomer was also detected. In the case of the W complex, NMR spin saturation transfer showed both isomers to be in rapid exchange, presumably by pairwise carbonyl exchange between metal atoms. Dynamic ^{31}P NMR methods have been used to measure ionization and ion-pair separation barriers in fluorine-bridged adducts of $[R_3P(CO)_3(NO)W]^+$ (2).[14] A low-temperature intramolecular process

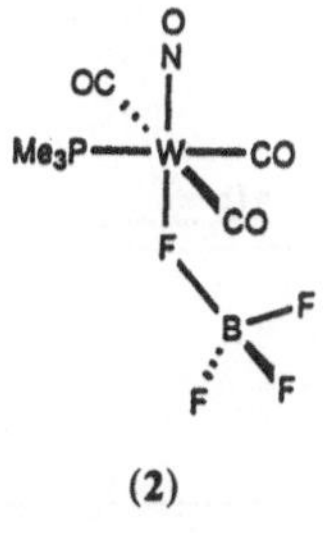

(2)

involves anion "spinning" while an intermolecular process results in the collapse of each R_3P multiplet to a singlet. The energy data are consistent with $W-F$ bond cleavage in the low-temperature process to give ion-pairs, followed by ion-pair separations to produce free ions in the high-temperature process.

Variable-temperature NMR spectroscopy has detected internal group migrations in iron complexes, namely, acyl migration to the allyl ligand in $[(\eta^3\text{-allyl})(CO)_2(PPh_3)FeC(O)R]^{[15]}$ and β- to α-carbon migration in iron(II) alkylidenes.[16] Variable-pressure NMR has shown the CO site exchange in $[Fe(CO)_2(olefin)(PPh_3)]$ to be nondissociative or dissociative, depending on the nature of the olefin.[17]

Series of studies employing NMR DANTE saturation transfer techniques have identified site exchanges in a variety of rhodium dehydro–amino acid complexes,[18] Wilkinson's catalyst $[ClRh(PPh_3)_3]$ and its dihydride,[19] and other related complexes such as $[HRh(CO)(PPh_3)_3]$, $[HRh(CO)_2(PPh_3)_2]$, and $[HRh(CO)_2PPh_3]$.[20] These studies of intra- and intermolecular exchanges are of particular relevance to rhodium-catalyzed hydroformylation reactions. The stereochemical nonrigidity of $[(Ph_3P)_3Rh]^+$ in solution has been characterized by ^{31}P DNMR.[21] The process involves exchange of ground-state T-shaped P_3Rh units, similar to those found by X-ray crystallography in $[(Ph_3P)_3Rh][ClO_4]$, via a Y-shaped C_{3v} transition state or intermediate and involving permutation of the Ph_3P ligands. The complex $[(triphos)Rh(S_2CS)][BPh_4]$ is also fluxional, as evidenced by its ^{31}P spectra, the process involving a turnstile rotation mechanism with considerable π-bonding interaction between S and Rh.[22] Heterometallic hydrides of type $[L_3MH_2ZnN(SiMe_3)_2]$ (M = Rh, Ir; $L_3 = (PMe_3Ph)_3$ etc.) have been synthesized. They all possess pseudooctahedral geometry with terminal hydrides and linear two-coordinate zinc. The observed nonrigidity is attributed to easy access to an $M(\mu\text{-H})_2ZnN(SiMe_3)_2$ transition state which possesses *fac* geometry of the three P centers. Site exchange barriers were lower for Rh than for Ir.[23]

Effective migration of the SiMe$_3$ group occurs in the palladium π-allyl complexes (Scheme 2). The SiMe$_3$ group occupies the *anti* position initially, but slow isomerization in solution produces the *syn*-SiMe$_3$ species.[24] The η^3-oxodimethylenemethane complexes of Pd and Pt bear a close resemblance to π-allyl complexes. The complex $[Pd\{CH_2C(O)CH_2\}(PH_3)_2]$ undergoes isomerization, thought to

Scheme 2

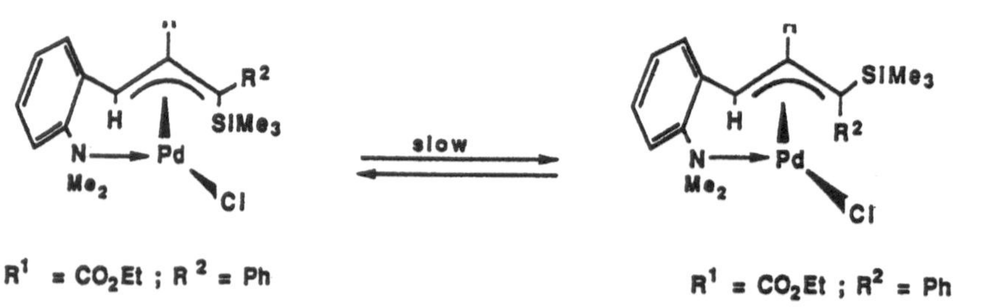

$R^1 = CO_2Et$; $R^2 = Ph$ $R^1 = CO_2Et$; $R^2 = Ph$

proceed via a planar C_{2v} transition state. The ΔG^+ value for the process was calculated from ^{1}H NMR spectra to be 38.5 kJ mol^{-1} at $-78\,°C$, the coalescence temperature of the *syn* and *anti* protons.[25] Restricted rotation about Pt—C σ bonds was examined in a series of complexes of general type *cis*-bis(phenyl)bis(ligand)platinum(II). *Ortho* substituents on the phenyl rings were the major cause of the isomerization involving E,Z conformers of the complexes.[26] Platinum–oxygen bonds were the cause of fluxionality in the Pi(II) chelate [PtCl(Ph$_2$PCH$_2$CHMeOH)(Ph$_2$PCH$_2$CHMeOH)] Cl and related alkoxo-bridged species.[27]

The fluxionality of organosilicon compounds has been the subject of two studies. At low temperature, the ^{1}H and ^{13}C NMR spectra of [(Me$_3$Si)$_2$C(SiMe$_2$OMe)(SiPh$_2$Cl)] indicate nine nonequivalent methyl groups. Two-dimensional NOESY and COSY NMR experiments showed the enantio topomerization involving correlated rotation of the Me$_3$Si groups to be of type ESSS (using Mislow's terminology).[28] The crystal structure of 2-(MeSiF$_2$)C$_6$H$_4$CH$_2$NMe$_2$ indicates a strongly distorted trigonal bipyramidal coordination with N and F atoms at apical positions.[29] DNMR studies on this and related complexes indicated positional exchange of substituents at Si. The exchange processes have energy barriers in the range 29–79 kJ mol^{-1} and can be considered in terms of regular processes (viz. pseudorotation, turnstile rotation) or irregular processes with bond ruptures/reformations. However, an MNDO simulation of the exchange indicates that no clear distinction between the mechanisms is possible, it being a question of which limiting Si–N distance constitutes a chemical bond. Fluxional actinide complexes include [(η-Cp)U(BH$_4$)$_3$L$_2$][30] and [Cp$_2^*$M(Cl){(CH$_2$)(CH$_2$)PRR'}] (M = U, Th; R = Me, R' = Ph, R = R' = Ph, Me).[31] The former complexes in solution (L = THF, DME, etc.) undergo ligand exchange according to Scheme 3,[30] while the latter complexes (M = Th) exhibit metal–carbon bond breaking, rotation, and recombination processes. These are fast on the ^{1}H NMR timescale at ambient temperatures producing equivalence of the two methylene environments.[31]

13.2.2. Ligand Rotations about the Metal–Ligand Bond

Commencing with alkyl group rotations, there have been two studies of the stereodynamics of chromium carbonyl complexes with η^6-hexaalkylbenzene. For the complex [{Cr(CO)$_2$L}$_2$(μ-N$_2$)] (L = η^6-hexaethylbenzene), ^{13}C NMR bandshape changes were attributed to slowed ethyl group rotation with a barrier, $\Delta G^+(300\text{ K})$, of 46.0 ± 3.0 kJ mol^{-1}.[32] The ligand hexa-*n*-propyl-benzene adopts a D_{3d} symmetry with the alkyl substituent alternately up and down with respect to the observer.[33] This geometry is retained in the chromium tricarbonyl complex and a decoalescence phenomenon observed in its ^{13}C-{^{1}H} NMR spectrum was attributed to slow propyl group rotation. NMR bandshape analysis gave a $\Delta G^+(300\text{ K})$ value of 49.6 kJ mol^{-1}.

Olefin rotation occurs in the complexes *mer*-tricarbonyl(η^2-olefin)(η^4-norbornadiene) tungsten (3) with Arrhenius activation energies of 58 and 54.8 kJ mol^{-1} for complexes, L = η^2-(Z)-cyclo-octene and η^2-ethene, respectively.[34] Restricted

Scheme 3

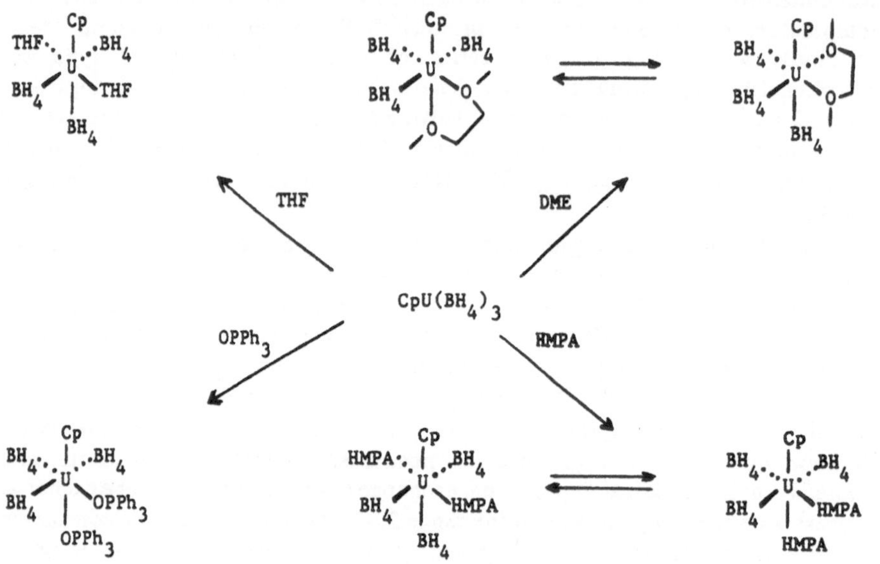

rotation of the butadiene ligand explains the fluxional properties of the niobium complex (**4**).[35] Proton NMR spin magnetization experiments showed no evidence

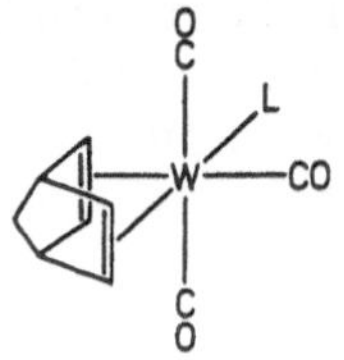

(3) 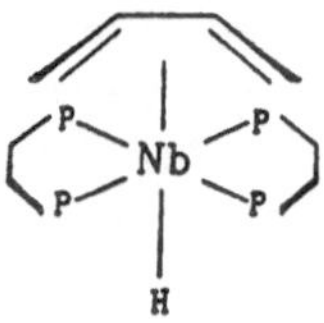   (4)

of intramolecular hydrogen exchange processes. Magnetization transfer experiments were also used to investigate the mechanism of the conversion of intermediate 16-electron tungstenocene alkyls into alkene hydrides, and of fluxionality within the complex $[W(\eta\text{-Cp})_2(CH_2=CHCH_3)H]PF_6$. The latter exists as 1^+-*endo* and 1^+-*exo* forms and interconversion between them involves alkene rotation.[36] Some new $[M(\eta^5\text{-}C_5H_5X)(\eta^2\text{-}C_2H_4)_2]$ (M = Rh, Ir) complexes have been synthesized.[37] They display ethene rotations with activation energies in the range 56–76 kJ mol^{-1}. The ^{1}H NMR spectra also indicate that the M-cyclopentadienyl ring bonding is not fully delocalized.

Monoalkyne tungsten(II) complexes of type $[W(LL')_2(CO)(RC\equiv CR')]$ (LL' = S_2CNMe_2, S_2PMe_2, etc.) (**5**) display fluxional behavior due to propeller rotation of the coordinated alkyne when R = R' = Me, LL' = S_2CNMe_2 or 2-SC_5H_4N, but when R = Ph, R' = Me, Ph, and LL' = S_2PMe_2 an alternative process

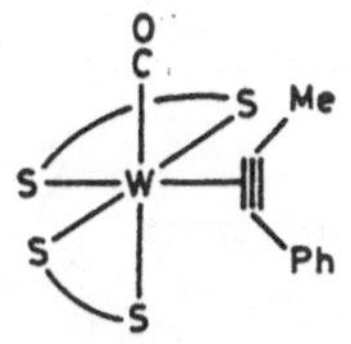

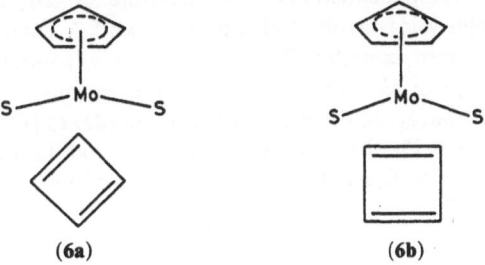

(5)

attributed to dechelation of the ligand is observed.[38] A similar dechelation process occurs in the but-2-yne complexes $[WBr(L_2)(CO)(MeC{\equiv}CMe)_2]$ ($L_2 = S_2CNMe_2$, S_2PMe_2).[39] In contrast, the ($L_2 =$ acac) complex is nonfluxional. The related complexes $[W(S_2PMe_2)_2(MeC{\equiv}CMe)_2]$ and $[W(L_2)_2(CO)(MeC{\equiv}CMe)]$ are fluxional, with the former involving dechelation of S_2PMe_2 and the latter involving alkyne propeller rotation plus, in the case of $L_2 = S_2P(OMe)_2$, two other processes one of which exchanges the methyl groups of one $S_2P(OMe)_2$ group. The barrier to but-2-yne rotation has been measured by variable-temperature ^1H NMR for a series of $[WI_2(CO)L_2(\eta^2\text{-}RC{\equiv}CR)]$ complexes.[40] Magnitudes of $\Delta G^+(T_c)$ were in the range 41–65 kJ mol^{-1} and were discussed in terms of electronic and steric effects of the phosphorus donor ligands. Analogous rotations in the complexes $[Cp^*V(CO)_2(RC{\equiv}CR')]$, $[(C_5H_4CoMe)V(CO)_2(RC{\equiv}CR')]$, and $[C_9H_7V(CO)_2(RC{\equiv}CR')]$[41] were in the range 40–50 kJ mol^{-1}, somewhat higher than in the corresponding Cp complexes.

There have been two reports of hindered rotation about metal–aryl bonds.[42,43] Palladium–C(aryl) rotation was restricted in the complexes $[Pd(C_6H_4\text{-}2\text{-}N{=}NPh)\text{-}(\eta^5\text{-}Cp)(PR_3)]$ when $PR_3 = PEt_3$, PMe_2Ph, and PEt_2Ph. The more crowded complex involving $\eta^5\text{-}C_5H_4Me$ and PCy_3 remained static up to the onset of decomposition.[42] In the complex $[(C_5Me_4H)_2Ti(p\text{-}C_6H_4Me)_2]$ the aryl rotation barrier was 61.9 ± 4 kJ mol^{-1}.[43] A recent example of cyclobutadiene rotation is the molybdenum complex $[Mo(S_2CNR_2)(\eta^4\text{-}C_4(CF_3)_4)Cp]$, where the CF$_3$ groups on the ring contribute to a high barrier to rotation between the two structures (**6a, 6b**).[44]

(6a)   (6b)

Pentadienyl (L) ligand rotation has been studied by $^{31}P\text{-}\{^1H\}$ NMR spectroscopy for the ruthenium–phosphine complexes $[(L)RuCl(PR_3)_2]$ (**7**).[45] The pentadienyl ligand rotates with respect to the RuClP$_2$ framework exchanging the two phosphines as well as the two sides of the ligand (Scheme 4).

A number of NMR studies have been directed toward restricted rotations of substituted cyclopentadienyl ligands in metallocenes. For instance, in octaisopropyl ferrocene[46] the activation barrier for tetraisopropyl cyclopentadienyl

Scheme 4

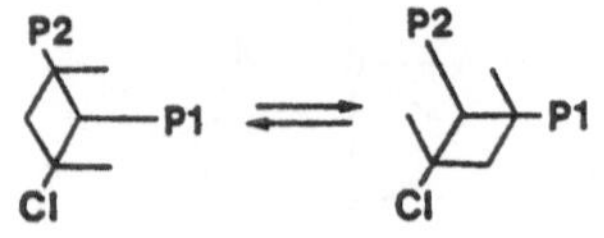

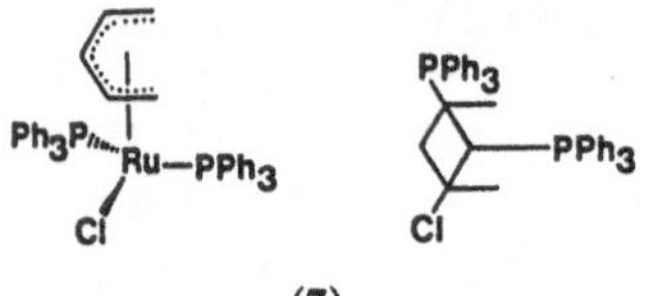

(7)

ring rotation was found to be 56.8 kJ mol^{-1}. Similar hindrance to ring rotation occurs in tetrakis (trimethylsilyl) metallocenes of Fe and Ti.[47] In the ferrocene compound the eclipsed structure (8) is favored while in the titanocene complex the chloride ligands lead to a preferred staggered arrangement of the Cp rings (9).

(8) 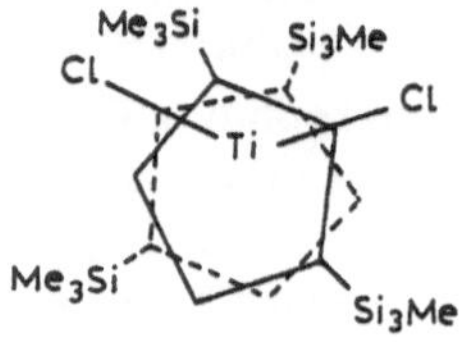   (9)

The latter are virtually eclipsed in 1,1',2,2',4,4'-hexakis (trimethylsilyl) ferrocene, which also shows restricted rotation of the two five-membered rings ($\Delta G^{\neq}$ = 46 kJ mol^{-1}).[48]

The dynamic stereochemistry of the complexes *cis*-[Mo(CO)$_2$L] (L = η^4-1,3,4-tetramethylsilole) involves change of Δ and Λ enantiomers according to Scheme 5. The more complex fluxionality of *cis*-[Mo(CO)(PPh$_3$)L$_2$] [L = η^4-1,1-dimethylsilole) was also discussed in terms of various possible mechanisms of isomerization.[49] Finally, in the mixed ligand complexes [(COD)M(C$_4$H$_4$BR)] (R = Me, M = Ni; R = Ph, M = Ni, Pd, Pt) rotational barriers for 1,5-cyclo-octadiene (COD)–borole (C$_4$H$_4$BR) ligand–ligand motions were estimated from ^{1}H NMR spectra.[50]

13.2.3. Migration of Metal Atoms between Different Ligand Sites

Fluxionality in allyl complexes is a widespread phenomenon. The η^3-allyl complexes [Ni(η^3-allyl)(aryl)(diphos)] (10) are nonrigid 18-electron species which exhibit *syn–anti* proton exchange and, in the case of the complex (R^1 = Me, R^2 = H), rapid site exchange of the two allylic termini. The *syn–anti* exchange is thought to occur via a short-lived η^1-allyl intermediate (11).[51]

Scheme 5

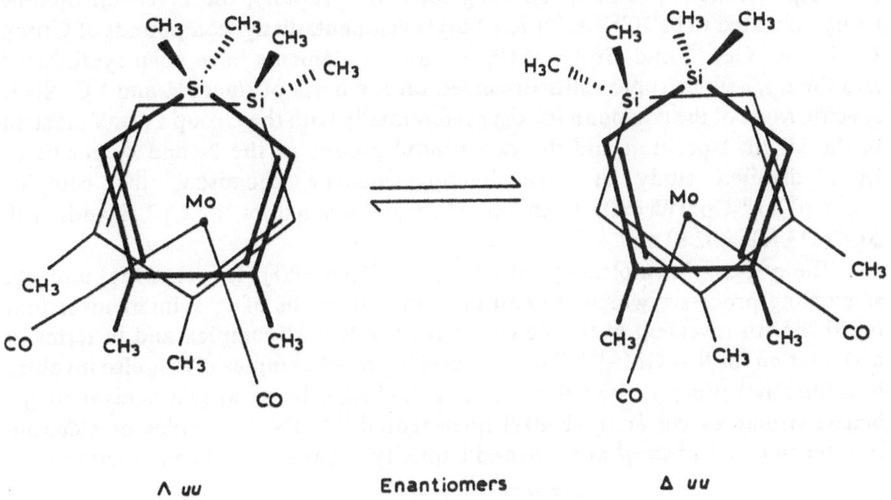

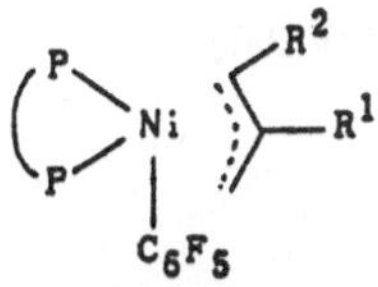

Λ *uu*   Enantiomers   Δ *uu*

(10) 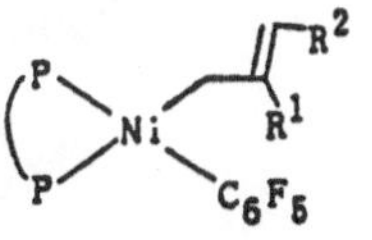   (11)

An $\eta^5 \rightleftharpoons \eta^3$ reversible interconversion has been observed by ^{31}P NMR for a pentadienyl ligand in the complexes $[(\eta^5\text{-}C_5H_7)Mo(PP)(CO)_2]BF_4$ where PP = dppe or dmpe.[52] Three mechanisms were proposed to explain the fluxionality, the metal–dienyl rotation mechanism (Scheme 6), and a pseudorotation-like mechanism being thought most likely on the basis of the measured ΔG^+ values.

Cyclopentadiene ligands display many types of fluxionality. A particularly novel type has recently been discovered in bis(cyclopentadienyl)beryllium.[53] X-ray data show the molecule to possess a "slip sandwich" structure and this structure persists in the solution and vapor phases. A recent NMR study based on partially

Scheme 6

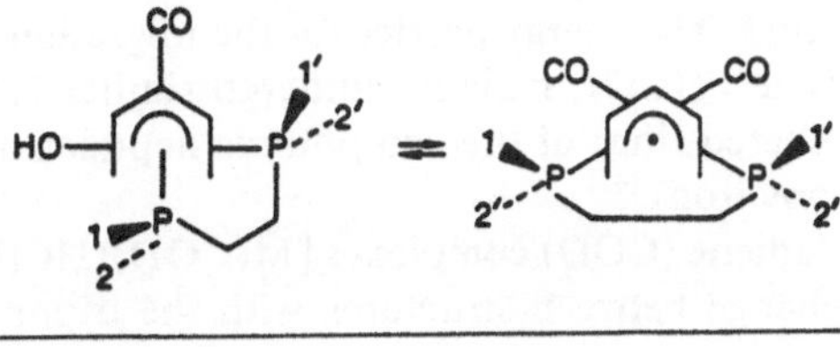

relaxed spin–spin coupling between ^{13}C and ^9Be has shown that the two Cp rings exchange between η^5 and η^1 bonding roles very rapidly, the inversion lifetime being estimated to be $10^{10\pm1}$ s. Di-*tert*-butyl cyclopentadienyl compounds of Group IV (Si, Sn, Ge)[54] and Group V (P, As, Sb)[55] elements have been synthesized and their 1,2-sigmatropic shifts discussed on the basis of their ^1H and ^{13}C NMR spectra. Most of the compounds exist preferentially with the Group IV or V element in the allylic 1-position and the two *t*-butyl groups in the 2- and 4- positions. In a detailed study of permethyltungstenocene chemistry, the complex $[(\eta^5\text{-Cp*})(\eta^1\text{-Cp*})W(=O)_2]$ exhibits $\eta^1 \rightleftharpoons \eta^5$ exchange of the Cp* ligands with $\Delta G^*(213\text{ K}) \approx 42$ kJ mol^{-1}.[56]

The η^3-benzyl complexes $[CpFe(CO)\{\eta^3\text{-CH(R)Ph}\}]$ (R = H, OMe) undergo scrambling processes which are best explained in terms of σ, π interconversions together with inversion at the Fe center for the R = H complex and in terms of aryl rotation for R = OMe.[57] The η^3-benzyl ligand of complex (12) is also involved in a fluxional process, thought to involve exchange between two equivalent η^3-benzyl structures via an η^1-benzyl intermediate.[58] Two examples of electron-transfer-induced $\eta^6 \rightleftharpoons \eta^4$ arene ligand hapticity changes have been reported.[59,60]

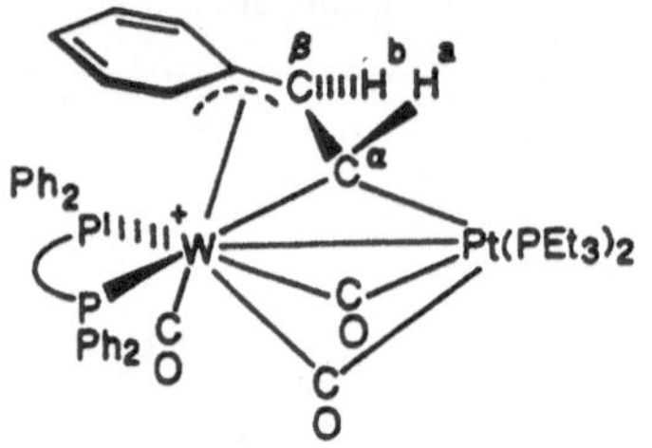

(12)

One of the complexes was $[(\eta^6\text{-C}_6Me_6)Rh(\eta^5\text{-Cp*})]^+$, a doublet state species, with a planar arene structure. Such a structure indicates that when a metal undergoes a two-electron-transfer-induced change from η^6 to η^4 hapticity, the one-electron intermediate can retain a planar arene structure even if it involves a 19-electron metal.[59] The analogous complexes $[(\eta^6\text{-C}_6Me_6)M(\eta^5\text{-Cp*})]^{2+}$ (M = Rh, Co) undergo similar arene hapticity changes, the kinetics of which were followed by NMR line-broadening and electrochemical methods.[60] The naphthalene ligand is η^2-bonded to nickel in $[(C_{10}H_8)(i\text{-Pr}_2P(CH_2)_n i\text{-Pr}_2P)Ni]$ (n = 2, 3).[61] In both the liquid and solid states the P$_2$Ni moiety moves between the 1,2- and 3,4-positions within one naphthalene ring, without interchanging the P atoms. Energy barriers were >96 kJ mol^{-1} (solid state) and <25 kJ mol^{-1} (solution). Solution 2D-EXSY NMR experiments also reveal two further fluxional processes involving rotation and migration of the P$_2$Ni moiety around the 1,2-, 3,4-, 5,6-, and 7,8-positions of both six-membered rings. The energy barrier for the migration process is approximately 60 kJ mol^{-1}. The Fe(CO)$_3$ moieties undergo similar 1,2-migration around the fused seven-membered rings of the compounds heptaleneirontricarbonyl and heptalenebis(tricarbonyliron).[62]

The 1,5-cyclooctadiene (COD) complexes $[M(COD)(HC(PPh_2)_3)]^+BF_4^-$ (R = Rh, Ir) undergo exchange between structures with the tripod ligand acting as a bidentate chelate. The exchange process causes equilibrium of all P nuclei as well as the vinyl and methylene protons of the diene. One proposed mechanism is shown (Scheme 7).[63]

Scheme 7

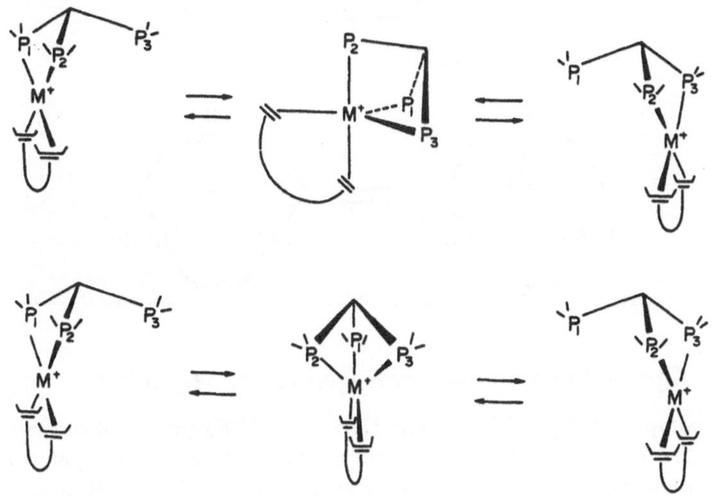

NMR magnetization transfer experiments have helped to clarify the mechanisms of fluxionality in the cyclo-octatetraene (COT) complexes $[Ru(\eta^6\text{-}COT)(\eta^4\text{-}C_7H_8)]$ (C_7H_8 = 2,5-norbornadiene) and $[Os(\eta^6\text{-}COT)(\eta^4\text{-}C_8H_{12})]$ (C_8H_{12} = 1,5-cyclo-octadiene). A 1,5-shift mechanism was shown to be dominant in both complexes with a 1,3-shift occurring at a slower rate.[64] The fluxionality of $[M(CO)_3(\eta^6\text{-}COT)]$ (M = Cr, W) has recently been reexamined using 2D-EXSY NMR.[65] Experiments showed that 1,2- and 1,3-shifts are almost equally favored for the W complex while for the Cr complex the 1,3-shift mechanism is slightly preferred. In both cases, rates of 1,4- and 1,5-shifts were negligible. Surprisingly, the perfluoro-COT complexes $[Mn(\eta^5\text{-}Cp^*)(\eta^6\text{-}C_8F_8)]$ and $[Mn(\eta^5\text{-}Cp^*)(\eta^6\text{-}C_8F_8)]$ show no fluxionality of the C_8F_8 ring on the ^{19}F NMR timescale, while the complex $[Mn(\eta^5\text{-}Cp^*)(\eta^6\text{-}COT)]$ exhibits dynamic behavior attributed to either a 1,3-shift or a random shift-mechanism for metal migration around the COT ring.[66]

Reports of fluxionality in N-bonded ligand complexes include octahedral manganese(I) complexes with 1,8-naphthyridine, $[Mn(\eta^1\text{-}naph)(\eta^2\text{-}phen)(CO)_3]ClO_4$.[67] At above-ambient temperatures there is a 1,3-shift of the Mn coordination between the N atoms on each ring ($\Delta G^+ = 63$ kJ mol^{-1}). The nitrogen macrocycle, 1,4,7-triazacyclononane, forms a palladium(II) complex which undergoes rapid ring hopping of the Pd(II) around the macrocycle (Scheme 8).[68]

Scheme 8

13.2.4. Agostic Bonding and Hydrogen Atom Migrations

The complexes $[C_5R_5(L)Co(CH_2CHR'-\mu-H)]^+BF_4^-$ possess three-center, two-electron $Co\cdots H_a\cdots C$ bonding at low temperatures. At higher temperatures three distinct dynamic processes were monitored by ^1H and ^{13}C NMR spectra. These were rotation of the agostic group, rotation of the C_2 fragment via the classical ethylene hydride complex, and inversion at the Co center via a six-electron alkyl species (Scheme 9).[69] The complex $[Ru(Cp^*)(C_8H_{12})]BF_4$ has been shown by ^1H and ^{13}C NMR to involve an agostic H interaction between Ru and the cyclooctadienyl ring.[70] Protonation of $[M(C_5R_5)(diene)]$ (M = Co, Rh, Ir; diene = 2,3-dimethylbutadiene, 1,3-cyclohexadiene) with HBF_4 gave fluxional cationic species. When M = Ir, classical M—H bonding was observed, while for M = Co, Rh, agostic $M\cdots H\cdots C$ bonding is preferred in the ground-state structures. NMR bandshape analyses provided energy data for methyl rotation and 1,4-H shifts in the agostic structures.[71]

Intramolecular hydrogen exchange was detected between the hydride and methyl ligands in isotopically labeled derivatives of $[Cp_2W(H)CH_3]$. At high solution concentrations intermolecular hydride exchange also occurs.[72] The rhenium complex $[(\eta^5-C_5H_4Li)Re(NO)(PPh_3)(H)]$ exhibits hydrogen migration, which is intramolecular and may proceed via three mechanisms, namely, (a) concerted migration, (b) initial $\eta^5 \rightarrow \eta^1$ isomerization of C_5H_4Li followed by hydride migration, and (c) initial hydride migration to give a η^4-cyclopentadiene complex followed by a 1,5-sigmatropic shift of the *exo* hydrogen (Scheme 10).[73] The rearrangement $Rh(\pi\text{-}HC{\equiv}CCO_2Et) \rightarrow Rh(H)(CCO_2Et)$ has been examined in detail.[74] It is preceded at lower temperature by a fluxional process which involves the π-alkyne species. The structural identities of the low-temperature solution species were not established but at ambient temperatures they appear to be trigonal bipyramidal.

Scheme 9

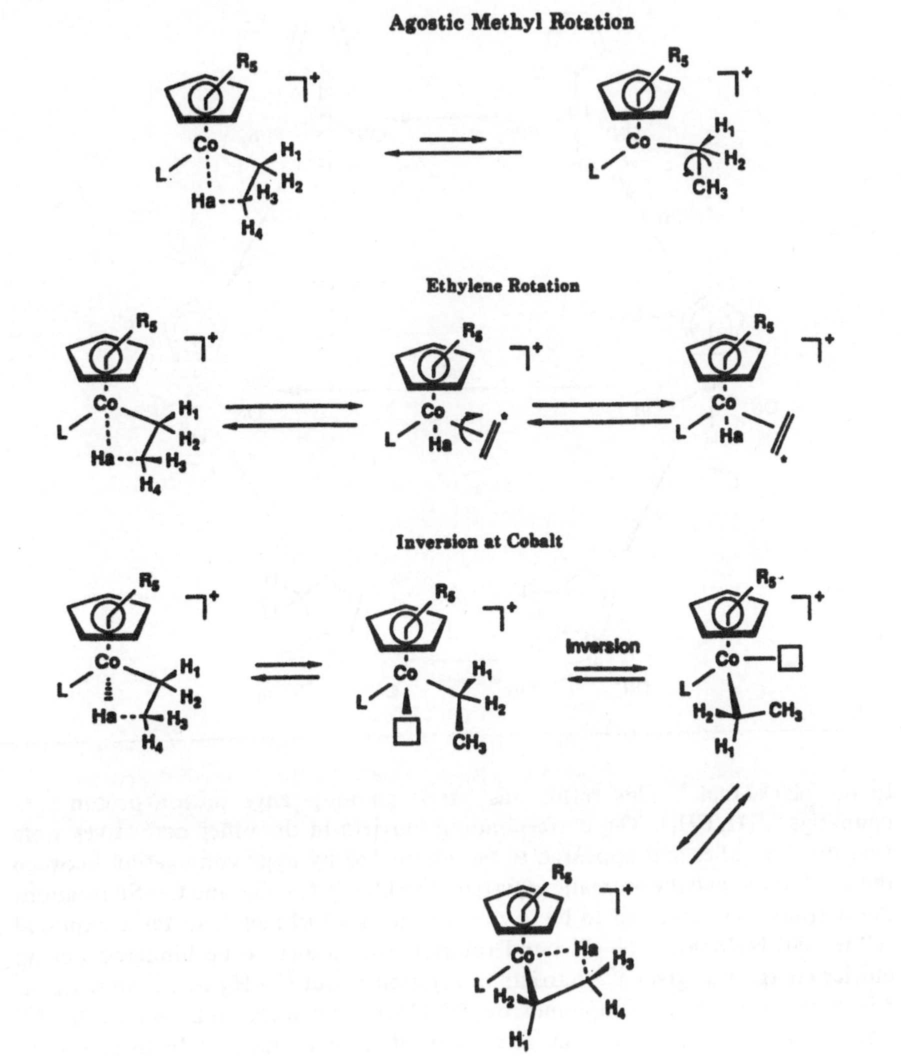
Agostic Methyl Rotation
Ethylene Rotation
Inversion at Cobalt
Inversion

13.2.5. *Internal Ligand Rearrangements*

13.2.5.1. *Restricted Rotations*

An NMR study of steric and hyperconjugative barriers in benzyltrimethyl X derivatives (X = C, Si, Ge, Sn, Pb) has been reported.[75] The steric barrier to rotation about the $C(sp^2)—C(sp^3)$ bond for the compound X = C was shown

Scheme 10

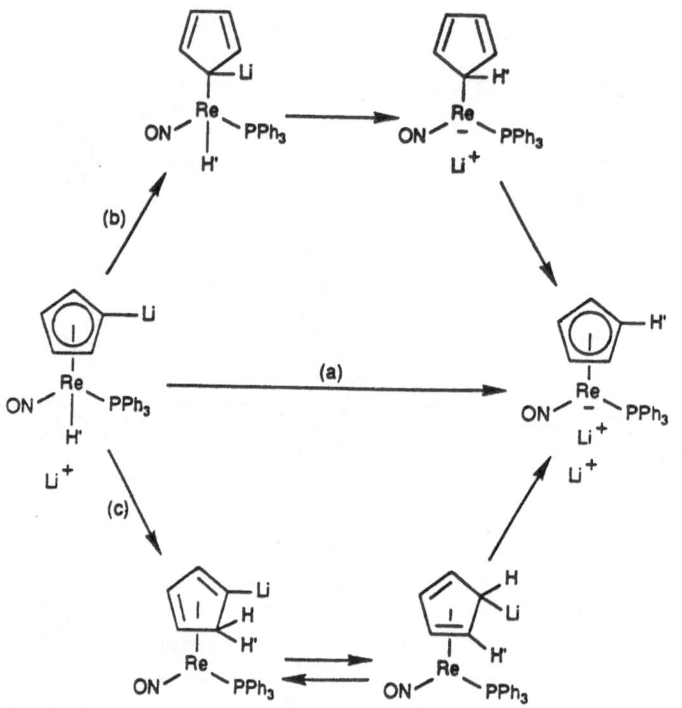

to be >20 kJ mol⁻¹. This result was based on long-range proton–proton spin couplings $^6J(\mathrm{H}, \mathrm{CH}_2)$. The corresponding barriers in the other derivatives were very much smaller and appeared to be dominated by hyperconjugation between the C—X bond and the aromatic π system. For C—Si, C—Ge, and C—Sn rotations the barriers were assessed to be in the region of 9.5 kJ mol⁻¹. In the compound $(\mathrm{CMe}_3)_2\mathrm{SiFNHSnMe}_3$, N—Si bond rotation was shown to be hindered but no barrier energy was given.[76] Barriers to rotation about C—Hg bonds were determined for arylmercury compounds by ^{13}C NMR spin–lattice relaxation times.[77] The average barrier height was 15 ± 5 kJ mol⁻¹. In a study of bis(organosilyl) amides, the compound (13) containing the five-membered silazane ring was formed.

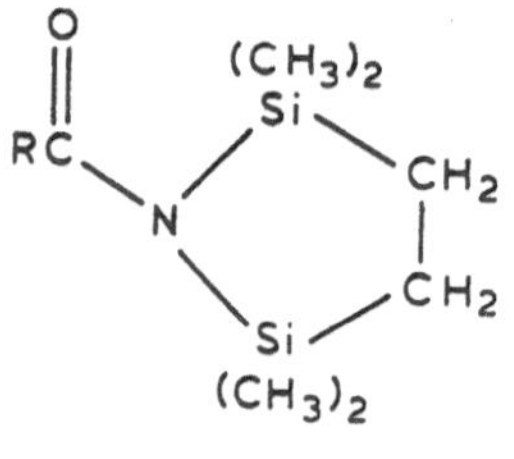

(13)

Variable-temperature ^{13}C or ^{29}Si NMR spectra enabled approximate magnitudes of C—N bond rotation barriers to be determined.[78] Fluxional behavior of cyclo-pentadienylcobalt(III) complexes with dithiocarbamate and xanthate ligands such as $[Co(\eta\text{-Cp})(S_2CNR_2)_2]$ also includes restricted C—N rotation. Complexes such as $[Co(\eta\text{-Cp})(\eta^1\text{-}S_2CNR_2)(\eta^2\text{-}S_2CNR_2)]$ also undergo $\eta^1\text{-}\eta^2$ ligand exchange.[79]

13.2.5.2. Tetrahedral and Pyramidal Inversions

Inversion of tetrahedral mercury(II) centers has been studied in mercury(II) pyrazolaldiminate chelates[80] and bis(chelates).[81] Activation energies were *ca* 54 kJ mol^{-1}. Ligand exchange reactions also occurred for which the activation barriers were in the range 65-69 kJ mol^{-1}. Pyramidal inversions of phosphorus atoms are often too slow to be monitored by normal NMR techniques. However, the secondary phosphido-iron complex (**14a,b**) exhibits inversion of the phosphorus stereocenter with an energy barrier ΔG^{+}(253 K) of 60 ± 4 kJ mol^{-1}. This relatively low barrier is consistent with the presence of the strongly electronegative iron(II) substituent on the P atom.[82]

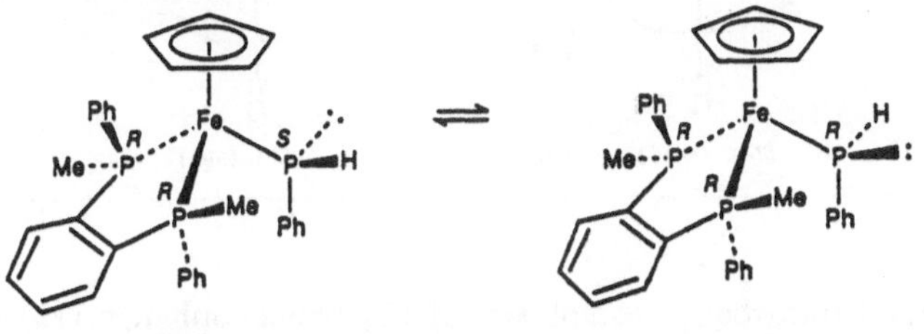

(14a) ⇌ (14b)

Pyramidal sulfur atom inversions fall suitably within the NMR timescale if the S atom is coordinated to transition metals. This is illustrated in recent studies of platinum(II) complexes with thioether ligands.[83,84] Activation parameters were as follows: $\Delta H^{+} = 73.5 \pm 17$ kJ mol^{-1} and $\Delta S^{+} = 16.7 \pm 5.9$ J K^{-1} mol^{-1}. Kinetic parameters were fairly solvent-independent and indicated that the inversion occurred without Pt—S bond breaking. Pyramidal inversion rates have been measured for S atoms attached to Rh and Ir in complexes $[M(\eta^5\text{-Cp*})Cl_2L]$ (L = SMe$_2$, S(CH$_2$Ph)$_2$, S(CH$_2$SiMe$_3$)$_2$, etc.)[85] and attached to Re atoms in the bridging complex $[(OC)_5Re\text{-}L\text{-}Re(CO)_5]^{2+}$ (L = 1,3-dithiane, 1,3,5,-trithiane).[86] Complexes of Mo(CO)$_4$ with chelating dithioethers are of type *cis*-$[Mo(CO)_4(RSCH_2CH_2SR')]$ (R = R' = Me, Et, ^{i}Pr, ^{t}Bu; R = Me, R' = ^{t}Bu).[87] The symmetrical dithioether complexes can exist as *DL* and *meso* forms in the absence of sulfur inversion, while the unsymmetrical ligand complex exists as two pairs of *DL* species which can be interconverted by inversion of the *t*-butyl- and methyl-sulfur atoms (Scheme 11). The relative populations of the *DL*-1 (*syn*) and *DL*-2(*anti*) species was 52/48. Variable-temperature ^{1}H NMR studies were performed on all the complexes. In the MeSCH$_2$CH$_2$S^{t}Bu ligand complex the inversion of the ^{t}Bu—sulfur dominated the spectral changes. The activation energies for S-alkyl inversion decreased in the order Me < Et < ^{i}Pr < ^{t}Bu reflecting the varying extent of distortion toward a planar sulfur geometry due to the size of the thioalkyl

Scheme 11

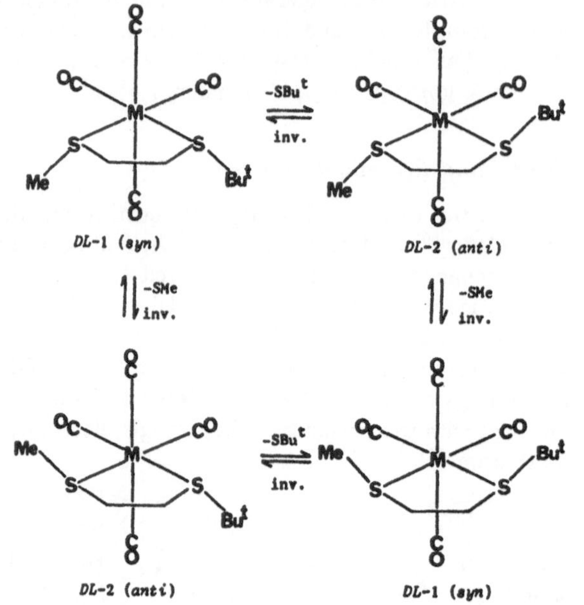

group. Metal tetracarbonyl complexes of [3]ferrocenophanes (15) (E = S, Se; M = Cr, Mo, W) undergo pyramidal inversion of the coordinated S or Se atoms at rates which are NMR detectable.[88] A bridge reversal fluxion also occurs, but

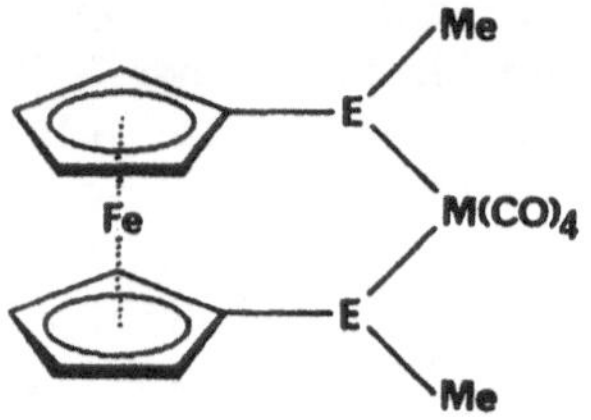

(15)

this is always fast on the NMR timescale. The static invertomers of $[M(CO)_4\{(C_5H_4SCH_3)_2Fe\}]$ are shown in Scheme 12. The *DL* species are dominant in solution (≥78% abundance) and activation parameters for bridge-reversal averaged *meso* → *DL* interconversions were calculated to be in the range 31–50 kJ mol^{-1}. Analogous [3]ferrocenophane complexes with trimethylplatinum(IV) halides, namely $[PtXMe_3\{(C_5H_4ECH_3)_2Fe\}]$ (E = S, Se), have been synthesized and their stereodynamics followed by DNMR.[89] When S inversion is slow on the NMR timescale the *DL* forms are again dominant in solution (≥90% abundance). Activation energy data for inversion were in the ranges 41.7–43.4 kJ mol^{-1} (sulfur) and 58.0–58.8 kJ mol^{-1} (selenium). At above-ambient temperatures the ligand

Scheme 12

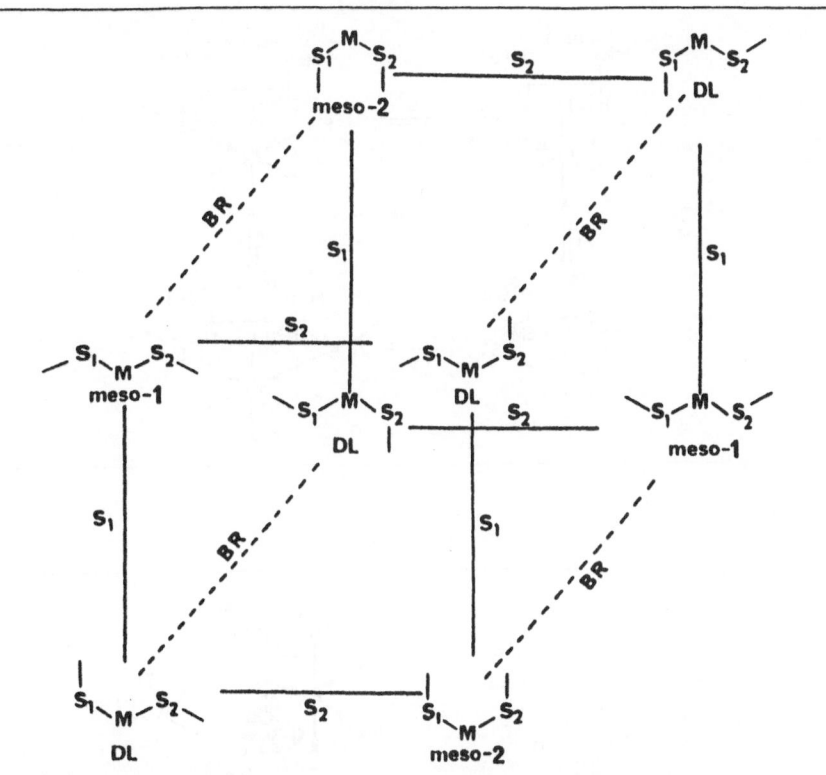

executes 180° "pancake" rotations with respect to the PtXMe$_3$ moiety, accompanied by exchange of the Pt–methyl environments. A detailed DNMR study of the low- and high-temperature fluxionality of trimethylplatinum(IV) halide complexes with 1,1,2,2-tetrakis (methylthio) ethane has been reported.[90,91] The complexes [PtXMe$_3${(MeS)$_2$CHCH(SMe)$_2$}] exist primarily as a *trans-DL* pair in CDCl$_3$ solution. Pyramidal sulfur inversion was found to be measurably different at the two nonequivalent S centers (Scheme 13), and full rate and activation energy data were reported.[90] At above-ambient temperatures the complex exhibits a novel 1,3-metal pivot, which involves the breaking of individual Pt—S bonds, followed by 109.5° (tetrahedral angle) pivots of the pendant —CH(SMe)$_2$ group about its attached C—C bond. Such a process brings a previously uncoordinated *gem* S-methyl into coordination with Pt and simultaneously interconverts the more populous *trans* isomer with the less abundant *cis* species (Scheme 14).[91]

Trimethylpalladium(IV) complexes are highly unstable but structures have been assigned to organo-palladium(IV) complexes with flexible bidentate ligands such as (py)$_2$CHMe. Conformational and fluxional effects have been reported[92] and compared with related Me$_3$Pt(IV) systems.

Scheme 13

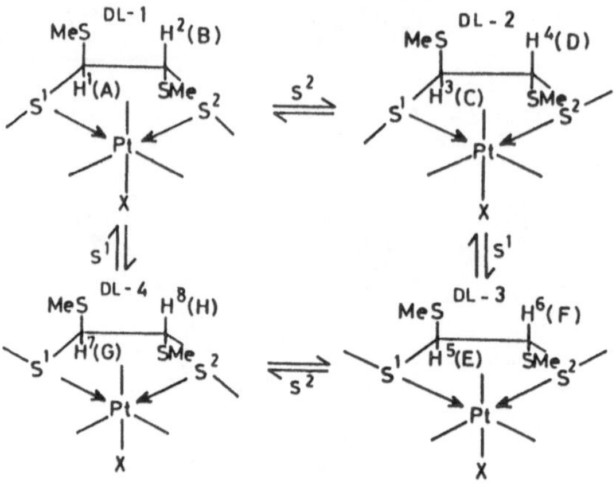

Scheme 14

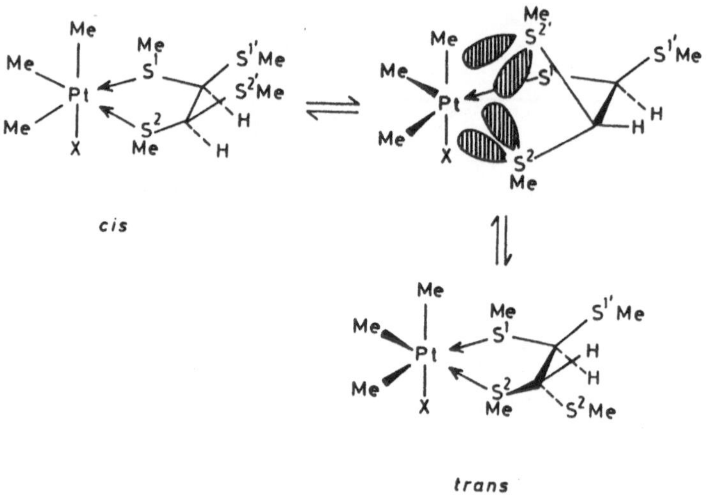

13.2.5.3. Ring Reversals

Five-membered ring pseudorotations are normally very facile and unable to be monitored by NMR. The hafnocene complex $[(\eta^5\text{-Cp})_2\text{Hf}(1,2\text{-Te}_2\text{C}_6\text{H}_4)]$ contains a five-membered HfTeC_2 ring which adopts an envelope conformation, but this undergoes rapid reversal at room temperature.[93] However, in the spiro compounds (16)[94] steric and electronic factors significantly raise the pseudorotation barriers to within the range 74-1–84.1 kJ mol^{-1}.

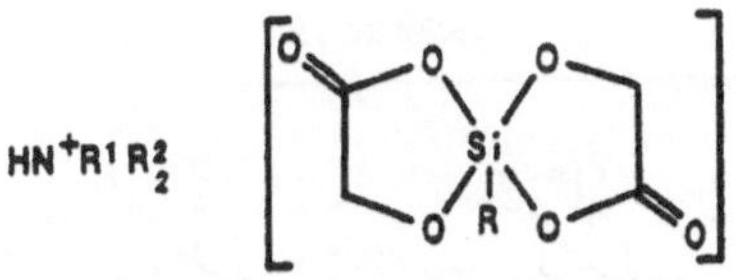

$$HN^+R^1R^2_2 \qquad \left[\begin{array}{c} \text{(structure 16)} \end{array} \right]$$

(16)

Six-membered ring reversals normally involve chair–chair interconversions. However, the central ring of 9,10-dihydrosilaazaanthracenes[95] and dibenzometal-lacyclohexa-2,5-dienes[96] adopt boat conformations and boat–boat interconversions were examined by 2D-EXSY[95] and variable-temperature 1D NMR.[96] The novel MSe_5 rings (M = Ti, Zr, Hf) do adopt preferred chair conformations and their ring-reversal activation energies were deduced from ^1H studies of $[Cp_2MSe_5]$.[97]

13.3. Dinuclear Compounds

13.3.1. Restricted Rotations

Rotational barriers of diiron μ-vinylcarbyne complexes have been measured by NMR bandshape analyses. The planar vinylcarbyne ligand (**17**) possesses three NMR measurable rotational barriers. Rotation of the entire ligand has the lowest barrier, $\Delta G^+(218\,K) = 44.4\,kJ\,mol^{-1}$, rotation of the NMe_2 group has a $\Delta G^+(210\,K)$ value of $42\,kJ\,mol^{-1}$, and rotation of the aryl group occurs with $\Delta G^+(275\,K)$ of $54\,kJ\,mol^{-1}$. Molecular orbital calculations indicate that diiron vinylcarbyne complexes have low ligand rotation barriers because the bridging carbyne carbon has orthogonal p orbitals which are able to accept electron density from the vinyl unit throughout the rotation.[98]

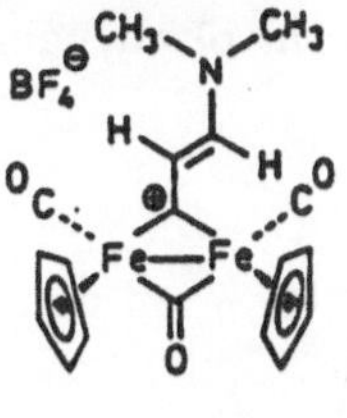

(17)

The chalcogen-bridged binuclear bent metallocenes, $[\{(Me_3CCp)_2Zr\}_2(\mu\text{-}O)\text{-}(\mu\text{-}Y)]$, where Y = Se, Te, exhibit hindered Me_3CCp—Zr rotation in solution caused by lateral interaction of Cp-bonded substituents.[99] Activation energy parameters are in the range 40–46 kJ mol^{-1}. In the dinuclear compound $[Mo_2(\eta\text{-}Cp)_2(CO)_4(\mu\text{-}dppm)]$ (dppm = $Ph_2PCH_2PPh_2$) the restricted rotation process is about the Mo—Mo bond, producing rotamers with *gauche* and *anti* arrangements of the Cp ligands which interconvert rapidly at room temperature (Scheme 15).[100] The halide-bridged dimer $[(tetramethylthiophene)RuCl_2]_2$ reacts with phosphine and amine donors to give complexes of type $[(tetramethylthiophene)RuCl_2L]$.

Scheme 15

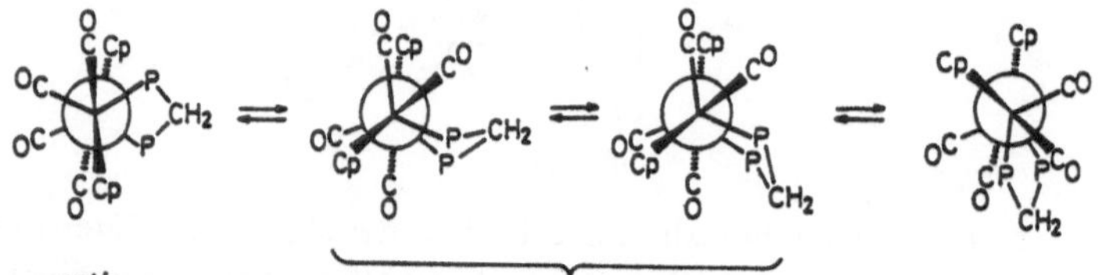

anti   P-atoms eclipsed   gauche

When L is bulky PPh_3 or $P(C_6H_4Me)_3$, hindered rotation about the Ru—P and Ru---TMT axes occurs.[101] Finally, in this section, a novel rotation about the C—Ar axis was detected in the μ-carbene complexes (18).[102]

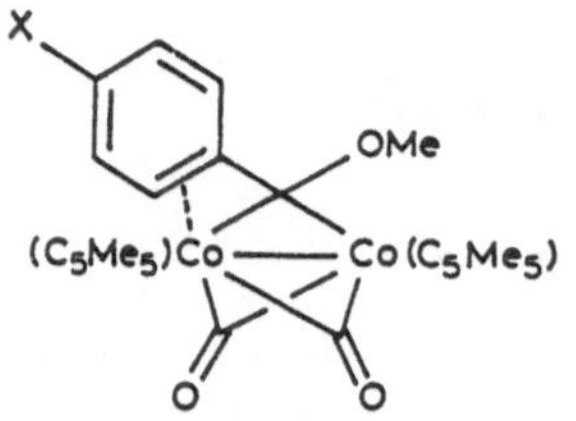

(18)

13.3.2. Carbonyl Ligand Migrations

Intramolecular exchange of carbonyl ligands has been detected in the novel ditungsten complexes (19).[103] Activation energy barriers are in the range 60–62 kJ mol^{-1}. A series of ditungsten, dimolybdenum, and mixed molybdenun/tungsten complexes (20)[104] (M,M' = Mo, W; R = H, Me) also exhibit carbonyl migration. In this case two processes are involved, namely, local scrambling of carbonyls

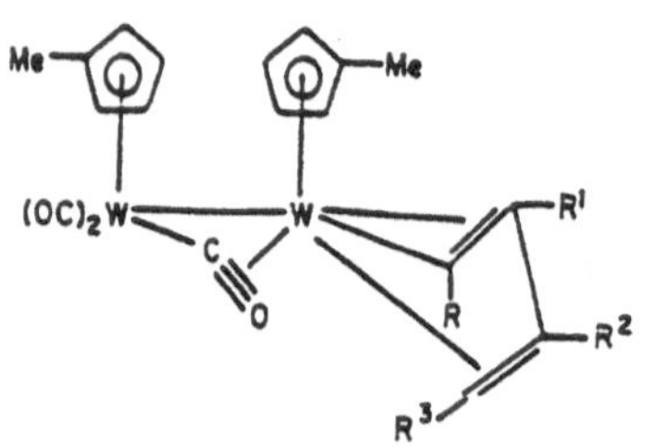

(19) 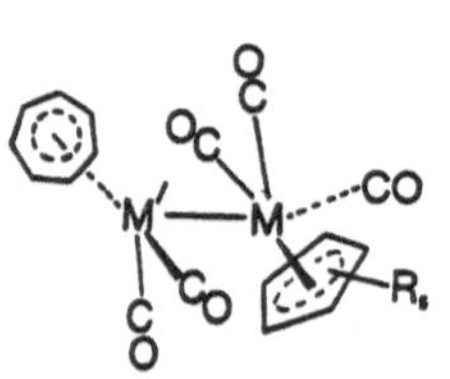   (20)

cis and *trans* to the M—M bond in the $(OC)_3(C_5Me_5)Mo$ unit, and internuclear exchange of the four carbonyls *cis* to the M—M bond.

13.3.3. Migration of Other Ligands between Metal Atoms

Isomeric bis(trimethylstannyl) dihydropentalenes have been shown by ^{13}C-{^1H} NMR spectra to exhibit suprafacial metallotropic rearrangements which

Scheme 16

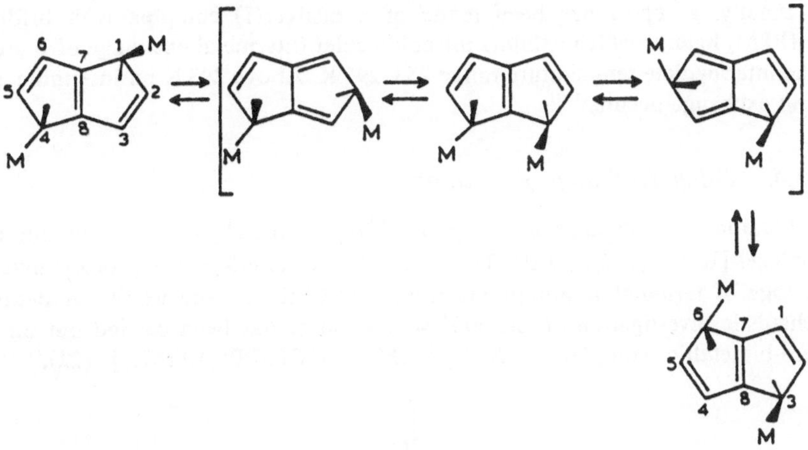

proceed as [1,5]-sigmatropic shifts of the SnMe group according to Scheme 16. The three intermediates were not detected in the spectra and their steady-state concentrations were insignificant.[105]

Intramolecular rearrangements in the slipped triple-decker complexes $[Cp_2Rh_2(COT)]^{2+}$ and $[Cp_2^*Co_2(COT)]^{2+}$ were studied in solution by ^{1}H NMR bandshape analysis.[106] The predominant rearrangement pathway for the Rh complex involves a 1,3-shift of one metal at a time along the COT ring periphery ($\Delta G^+ = 60.7$ kJ mol^{-1}) while the Co complex rearrangement involves 1,2-shifts of both metals ($\Delta G^+ = 58.6$ kJ mol^{-1}). The NMR changes rule out a true triple-decker transition state with a C_8 rotation axis. The dirhodium complexes $[(COD)_2Rh_2(\mu\text{-}R)_2]$ are also fluxional, this time involving the bridging alkyls (R = Me,Me$_3$SiCH$_2$).[107] Low-temperature NMR studies suggest that the dimers have idealized C_{2v} symmetry. The fluxional process appears to interconvert such structures via a planar species with D_{2h} symmetry, but the detailed mechanism is not yet fully established. The dirhodium tetrahydride complexes (21) undergo three separate exchange processes: intermolecular exchange with free H$_2$ and two intramolecular processes which exchange the four hydride ligands. The chelate

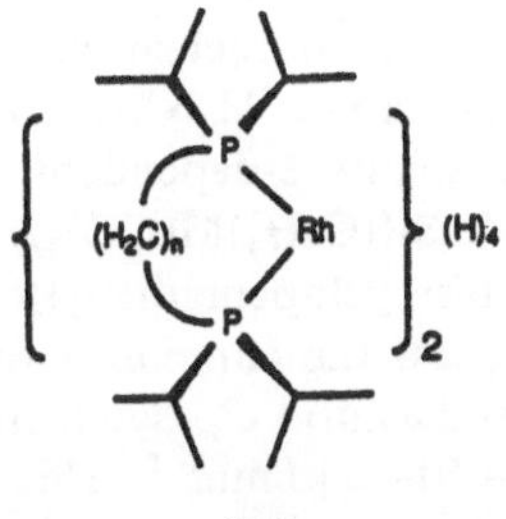

(21)

ring size affects the geometry about the Rh centers and a "rocking" fluxion is present at low temperatures.[108]

Finally, a report has been made of a disilver(I) complex with bridging $CH_2(PPh)_2$ ligands which exhibits intramolecular intermetal exchange of P atoms in the intermediate temperature range 253–293 K. Above 293 K an intermolecular ligand exchange occurs.[109]

13.3.4. Other Exchange Reactions

Fluxionality in main group complexes is represented by new organoaluminum complexes $[R_2AlY(CH_2)_mOAlR_3]$ ($m = 2,3$; $Y = OR'$, NR'_2). The process involves exchange of terminal R groups via transient R-bridged species.[110] A detailed mechanistic investigation of ZrMe/PtMe exchange has been carried out on the hetero-bimetallic complexes cis-$[Cp^*ZrMe(\mu\text{-}OCH_2PPh_2)_2PtMe_2]$ (22).[111] A

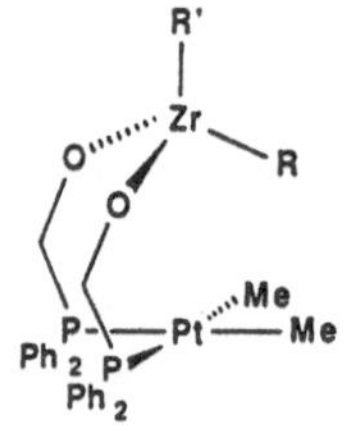

R = Me, R' = Cp*
R = Cp*, R' = Me

(22)

combination of deuterium labeling, kinetics, and ^{195}Pt NMR cross-over experiments showed that the most probable pathway involved intramolecular ZrMe/PtMe exchange and bimolecular ZrMe/ZrMe scrambling. The intramolecular process was the rate-determining step with $\Delta H^+ = 124 \pm 4\,kJ\,mol^{-1}$. The diruthenium zirconoxy carbene complex $[Cp_2Ru_2(CO)_2(\mu\text{-}CO)\{(\mu\text{-}CHOZrHCp_2^*\}]$ displays a fluxion which exchanges the inequivalent Cp and Cp* ligands (Scheme 17).[112] Metallocyclic zirconoxycarbene complexes were prepared by treating $M(CO)_6$ with [(butadiene)ZrCp$_2$]. Treatment with a ketone gave nine-membered metallocyclic complexes which exhibited ring conformational isomerism.[113] Magnitudes of ΔG^+ for the process were in the range 69.5–71.1 kJ mol^{-1}.

Numerous studies of intramolecular exchange phenomena in Group VI metal complexes have been reported. The dinuclear chromium compound $[(CpCr)_2(\mu\text{-}COT)]$ produces variable-temperature ^1H NMR spectra which are most convincingly interpreted as due to temperature-dependent singlet–triplet equilibria.[114] In the binuclear complexes $[Me_2Si\{(C_5H_4)M(CO)_3\}_2]$ (M = Cr, Mo, W) (23) the metals are bridged by the bis(cyclopentadienyl)dimethylsilane ligand. At low temperatures NMR spectra show the complexes to possess chiral C_2 symmetry, but this changes to an averaged achiral C_{2v} symmetry at higher temperatures with an activation barrier $\Delta G^+ = 50$–70 kJ mol^{-1}. This is attributed to exchange of equivalent *gauche* conformations via a *syn* intermediate.[115] The complex

Scheme 17

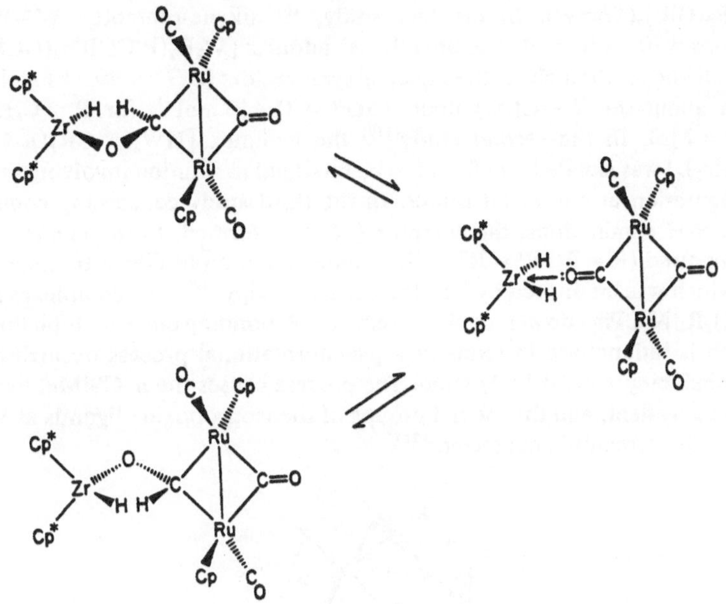

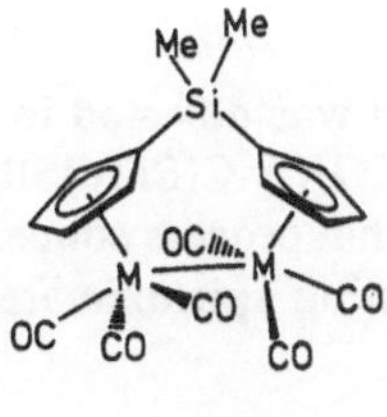

(23)

$[(\eta\text{-Cp})(CO)_3MoPtMe(PPh_3)_2]$ involving Mo—Pt bonding shows a *cis–trans* isomerization process in solution[116] while the complex $[(\eta^6\text{-azulene})Mo(\mu\text{-}\eta^6: \eta^4\text{-azulene})Rh(CO)_2]^+$ involving Mo—Rh bonding exhibits a more extensive reorganization of its bonding properties (Scheme 18).[117]

Scheme 18

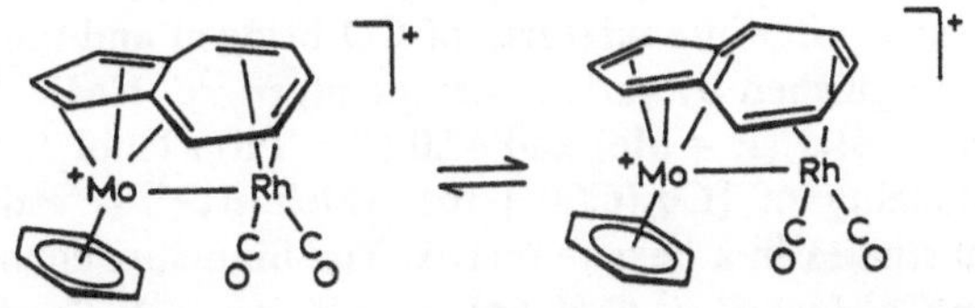

A series of papers[118-120] has recently examined the factors affecting the activation of W—C, C—H, and C—C bonds in the chemistry of the complexes $1,2\text{-}W_2R_2(OR')_4(W\equiv W)$. In the first study,[118] alkyne-promoted W—W alkyl migrations were examined. The bis(alkyne) adducts $[W_2R_2(R'CCR'')_2(OCHMe_2)]$ exhibit alkyne rotation about the metal–alkyne vectors ($\Delta G^+ \sim 40.2$ kJ mol^{-1}) and rotation about the W—C(Ar) bond ($\Delta G^+ = 31.4$ kJ mol^{-1}, for $R = C_6H_4Me\text{-}p$, $R' = R'' = Me$). In the second study[119] the complex $[HW_2(\mu\text{-}CR)(\mu\text{-}C_4Me_4)$-$(OCHMe_2)_4]$ was isolated and found to be fluxional in solution involving inversion of configuration at one metal center. In the third study concerning competitive α-H vs. β-H eliminations, the complex (24) was formed. In toluene-d_8 solution this compound ($R = Et$, $R' = R'' = Me$) underwent a rapid dimer–tetramer equilibrium which was monitored by ^{1}H NMR spectroscopy.[120] The complexes $[W_2(\mu\text{-}CSiMe_3)_2R_4.N_2CPh_2]$ do *not* involve direct W—W bonding but do exhibit fluxionality which is rationalized in terms of a pseudorotational process occurring about the trigonal-bipyramidal W(1) atom. The process causes the μ-CSiMe$_3$ ligands to become equivalent, and the methyl groups of the isopropoxide ligands at W(2) to lose their diastereotopic character.[121]

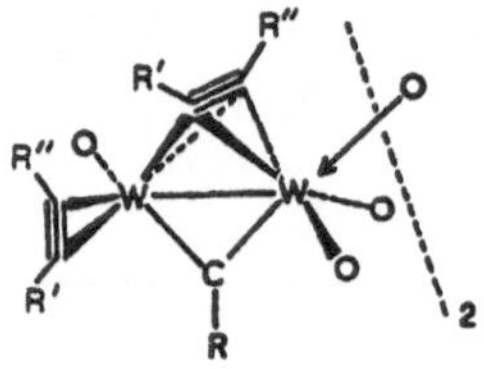

(24)

Rapid *cis–trans* isomerization was detected in bridging diiron carbene complexes of type $[Cp_2Fe_2(CO)_2(\mu\text{-}CO)(\mu\text{-}C(SMe)SR)]$[122] while in the bis(tricarbonyliron) complex (25) the fluxional process concerned $Fe(CO)_3$ moiety rotation, with one of the *exo* moieties rotating approximately three times faster than the other moiety.[123]

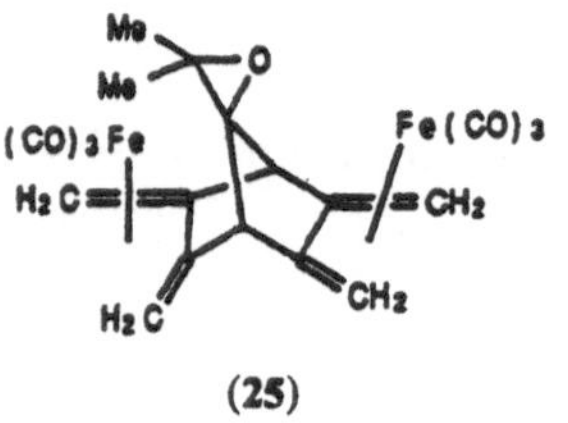

(25)

There have been two reports of fluxional dinuclear cobalt complexes. The first concerns the complexes $[Co_2(CO)_4(\mu\text{-}R_2PCH_2PR_2)_2]$ ($R = Me$, Ph).[124] These exist in solution as equilibrium mixtures of CO bridged and nonbridged isomers which can be distinguished by variable-temperature ^{13}C NMR. The ΔG^+ values for the process were 41.5 ($R = Me$) and 47.0 ($R = Ph$) kJ mol^{-1}, in contrast to the very rapid fluxionality of $[Co_2(CO)_8]$ for which ΔG^+ is estimated to be *ca* 27 kJ mol^{-1} from studies in a hexane matrix. The binuclear cobalt complexes are of type $[\{(\eta^5\text{-}Cp)Co\}_2(\mu_2\text{-}\eta^3,\eta^2\text{-}C_8H_xSe)]$ ($x = 6, 10, 12$). The fluxional process equilibrates the Cp groups without affecting the remainder of the molecule. This process is most likely to involve a rocking of the Co_2 unit such that Co_1 and Co_2 are alternately bonded to C_1.[125]

Cis–trans isomerization also occurs in the dinuclear chlorine-bridged palladium(II) complex $[\{Pd(CH_2SiMe_2C_5H_4N)Cl\}_2]$. NMR bandshape analysis gave the activation parameters $\Delta G^+ = 68.6 \text{ kJ mol}^{-1}$, $\Delta H^+ = 63.6 \text{ kJ mol}^{-1}$, and $\Delta S^+ = 7.0 \text{ J K}^{-1} \text{ mol}^{-1}$.[126]

13.4. Cluster Compounds

A number of review articles have described the applications of multinuclear NMR methods for studying structures and internal rearrangements of metal clusters. The more specific cluster types covered were transition metal carbonyl clusters,[127] chiral clusters,[128] and mixed metal clusters containing carbyne or ketenylidene bridges.[129]

13.4.1. Rearrangements Involving the Relative Motion of Metal Atoms in a Cluster

A novel fluxion described as the "Bloomington Shuffle" was detected in the C_{2h} rhomboidal 12-electron cluster $[W_4(O-i\text{-}Pr)_{12}]$.[130] The fluxion, depicted in Scheme 19, involves the W=W and W—W bonds migrating around the W_4 ring producing a time-averaged symmetrical rhombus, $W_4(\mu\text{-}O)_4(O)_8$, as the transition state. This motion of the metal atoms is coupled with a correlated rotation about the W—O atoms such that proximal/distal W—O—i-Pr group exchange occurs. 2D-EXSY NMR studies clearly identified the pairwise exchanges in the fluxion.

The reaction of $[Fe_3(CO)_{12}]$ with $P(OR)_3$ (R = Me, Et, Ph) in the presence of the bimetallic catalyst $[\{Fe(CO)_2(PPh_3)(SEt)\}_2]$ leads to two isomers of $[Fe_3(CO)_9\{P(OR)_3\}_3]$ which interconvert.[131] In polar solvents the major isomer exhibits a structure analogous to $[Fe_3(CO)_{12}]$ while the minor isomer contains only

Scheme 19

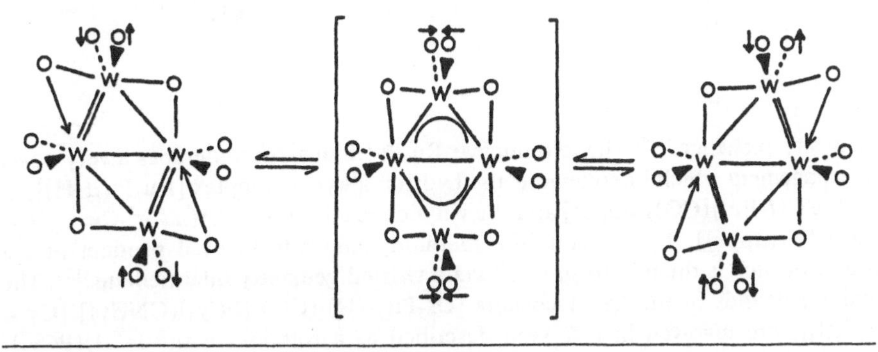

terminally bonded CO groups. The isomerization may be ascribed to the motion of the Fe$_3$ triangle inside the polyhedral cloud of ligands or as an interconversion between left-hand and right-hand screw structures.

The first direct evidence of stereochemical nonrigidity of the metal skeletons of Group IB metal heteronuclear clusters in solution has been obtained by ^{109}Ag-{^{1}H} INEPT NMR spectra on the clusters [Ag$_2$Ru$_4$(μ_3-H)$_2${μ-Ph$_2$P(CH$_2$)$_n$PPh$_2$}-(CO)$_{12}$] (n = 1–6) (**26**).$^{(132)}$ All the clusters adopt a capped trigonal–bipyramidal metal core geometry. In the ground-state structures there are two distinct Ag sites while ambient-temperature NMR spectra indicate a single averaged structure in the case of n = 1, 2, or 4. The mixed metal clusters [M$_2$Ru$_4$(μ_3-H)$_2${μ-Ph$_2$As(CH$_2$)$_n$EPh$_2$}(CO)$_{12}$] (M = Cu or Ag; E = As or P; n = 1 or 2) also display nonrigidity in solution.$^{(133)}$ They exist as interconverting pairs of isomers involving an intramolecular metal core rearrangement. At higher temperatures the Ag/Ru clusters undergo two additional dynamic processes. The first involves a novel intramolecular exchange of the As and P atoms in the Ph$_2$As(CH$_2$)$_n$PPh$_2$ ligands between the two Ag atoms in each cluster. The second process in a higher temperature range involves the bidentate ligands undergoing intermolecular exchange between clusters. The gold/ruthenium clusters [Au$_2$Ru$_4$(μ_3-H)(μ-H){μ-Ph$_2$P(CH$_2$)$_n$PPh$_2$}(CO)$_{12}$] (n = 3–6) all adopt the structure (**27**). Variable-temperature ^{1}H and ^{31}P-{^{1}H} NMR studies at ambient temperatures reveal Au

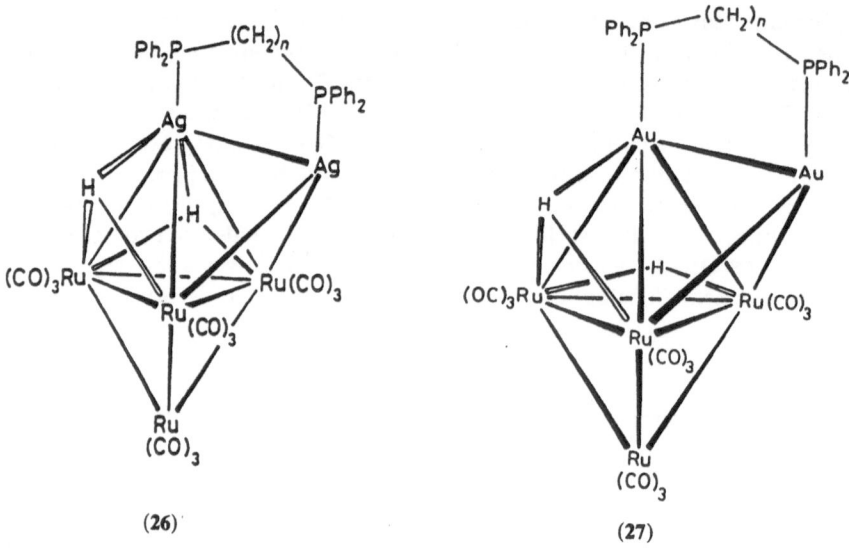

(26)   (27)

atom site exchange.$^{(134)}$ The tetranuclear Ru$_3$Pt framework undergoes a very novel rearrangement which interconverts the hydrido alkynyl complex [Ru$_3$Pt(μ-H){μ_4-η^2-C≡C(t-Bu)}(CO)$_9$(dppe)] and the vinylidene cluster [Ru$_3$Pt(μ_4-η^2-C=C(H)t-Bu)(CO)$_9$(dppe)]. The latter is the thermodynamically favored product in the process which is thought to proceed via a twisted geometry intermediate.$^{(135)}$ The hydride ligands in the Os$_3$Pt clusters [Os$_3$Pt(μ-H)$_2$(CO)$_9$(PCy$_3$)(CNCy)] (Cy = c-C$_6$H$_{11}$) are involved in a fluxion described as a rotation of the Pt(H)(PCy$_3$)-

(CNCy) group about the Os_3 triangle, coupled with a conrotatory migration of the $Os(\mu\text{-}H)Os$ hydride to an adjacent $Os-Os$ edge. The evidence for this mechanism came from ^{187}Os–hydride scalar coupling constants. One- and two-dimensional ^{13}C EXSY NMR spectra revealed two CO exchange processes, a lower-energy tripodal rotation in the two equivalent $Os(CO)_3$ groups and a higher-energy process involving total CO scrambling.[136] In the triosmium cluster $[Os_3(CO)_{10}(t\text{-}BuNC)_2(\mu\text{-}CH_2CO)]$, the three Os atoms and the μ-ketene ligand form a five-membered triosmacyclopentanone ring which undergoes inversion causing exchange of the methylene protons of the μ-ketene ligand.[137]

The skeletally chiral metal cluster $[\{(\mu_3\text{-}\eta^2\text{-}C_2-t\text{-}Bu)(CO)_9Ru_3\}_2(\mu_4\text{-}Hg)]$ racemizes at higher temperatures to a structure containing a face-bridging Hg atom (Scheme 20). Evidence for this came from ^{13}C NMR spectra at high temperatures. At somewhat lower temperatures localized and delocalized CO exchanges were also detected.[138]

Scheme 20

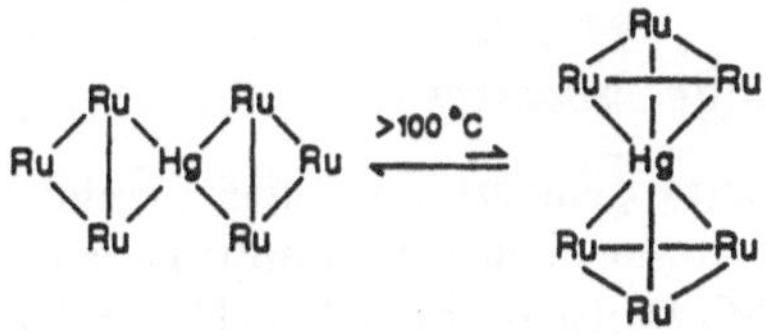

13.4.2. Carbonyl Ligand Exchanges

Two examples of carbonyl exchange in trinuclear clusters have appeared. In the triiron and triruthenium clusters $[M_3(CO)_9(dtn)]$ where dtn = 3,4-diazatricyclo$[4.2.1.0.^{2,5}]$non-3-ene, carbonyl scrambling is in evidence at room temperature but is arrested at low temperatures on the ^{13}C NMR timescale.[139] In triosmium clusters $[Os_3X(CO)_{11}]^-$ (X = Cl, Br, NCO) the lowest temperature fluxion involves complete exchange of all equatorial carbonyls. At higher temperatures this is followed by a series of trigonal twists and/or pairwise bridge-terminal CO exchange steps that lead to total carbonyl exchange.[140]

The new binary carbonyl $[Os_4(CO)_{14}]$ has recently been synthesized and shown to possess an irregular tetrahedral Os_4 skeleton with all the carbonyls essentially terminal ligands.[141] There is evidence from the broadness of the infrared carbonyl stretching bands that carbonyl fluxionality, while fast on the NMR timescale, might be observable on the IR timescale. Such an interpretation would make the fluxional frequency of the order of $9.4 \times 10^{11} \text{ s}^{-1}$. The mechanism of carbonyl scrambling in $[Ir_4(CO)_4(\text{phosphine})]$ complexes has been a source of contention for many years. A recent reappraisal of the mechanism for $[Ir_4(CO)_{11}(PEt_3)]$ has appeared.[142] Detailed NMR magnetization transfer experiments indicate a mechanism involving an intermediate retaining bridging carbonyls rather than a species containing only terminal carbonyls. The related complexes $[Ir_4(CO)_{11}Br]^-$,[143] $[Ir(CO)_{10}(diarsine)]$ and $[Ir(CO)_{10}(1,5\text{-cyclo-octadiene})]^{[144]}$ differ in their carbonyl fluxionality. In the bromo complex, the bridging and Br-site-sharing carbonyls do not exchange with other carbonyls. In the diarsine and cyclo-octadiene complexes the different fluxional mechanisms are thought to be a consequence of the different ground-state geometries; in the former case two carbonyls are semibridging while in the latter the basal carbonyls are symmetrical bridging ligands.

The mixed metal tetranuclear cluster $[RhRu_3(\mu\text{-}H)_2(\mu\text{-}CO)(CO)_9(\eta^5\text{-}Cp)]$ exhibits at least three distinct CO exchange processes.[145] The (η^5-Cp*) derivative exists in solution as three isomers in dynamic equilibrium, only two of which are significantly stereochemically nonrigid. In the trigonal bipyramidal pentanuclear clusters $[PtIr_4(CO)_9(\mu\text{-}CO)_n]^{2-}$ ($n = 3,5$) the extent of CO fluxionality increases appreciably with the temperature[146] while in the hexanuclear species $[Rh_6H(CO)_{13}C]^-$ complete CO and hydride migration occurs at $-20\,°C$.[147] Above that temperature decomposition occurs to give initially $[Rh_6H(CO)_{15}C]^-$ and then $[Rh_{12}(CO)_{24}(C)_2]^{2-}$.

13.4.3. Other Exchange Processes

A variety of other rearrangements in trinuclear metal clusters has been reported. The highly reactive benzyne molecule is trapped as a ligand in the triruthenium cluster $[Ru_3(CO)_7(\mu_3\text{-}\eta^2\text{-}C_6H_4)(\mu_2\text{-}PPhFc)_2]$ (FcH = $Fe(\eta^5\text{-}Cp)_2$) where it undergoes slow rotation which has been monitored by ^1H NMR spectra.[148] Two reports on nonrigid Os_3 clusters include arene ring whizzing and alkene rotation in $[Os_3(CO)_8(\eta^2\text{-}CH_2CHR)(\mu_3\text{-}\eta^2:\ \eta^2\text{-}C_6H_6)]^{[149]}$ and rapid enantiomerization, affecting hydride diastereotopic methylene ^1H NMR signals, in $[Os_3H_2L(CO)_9]$ (L = C=CHOEt, CH≡COEt).[150] The mixed metal cluster $[(OC)_5\text{-}Cr\{Os(CO)_3(PMe_3)\}_2]$ exists in solution as roughly equal concentrations of two isomers. In each isomer the radial carbonyls of the $Cr(CO)_5$ unit are equivalent but different to the fifth carbonyl, implying free rotation of the $Cr(CO)_5$ moiety in both isomers.[151]

Turning to tetranuclear clusters, complex, ill-defined fluxionality was reported for the novel arylcopper(I) compound $[Cu_4Br_2\{C_6H_3(CH_2NMe_2)_2\text{-}o,o'\}_2]$, involving exchange of the CH_2NMe_2 ligands and rotation of the aryl groups.[152] In the tetraplatinum cluster $[Pt_4(\mu\text{-}H)(\mu\text{-}CO)_2(\mu\text{-}Ph_2PCH_2PCH_2)_3(Ph_2PCH_2PCH_2)]^+$ the fluxionality involves a rapid rotation of a bridging hydride below, and a bridging $Pt(CO)_2$(dppm-P) unit above, a $Pt_3(\mu\text{-}dppm)_3^+$ triangle.[153] Trimethylplatinum(IV) halides are tetrameric in both solid and solution phases. Recently synthesized mixed tetramers of types $[(PtMe_3SMe)_2(PtMe_3X)_2]$ (X = Cl, Br, I) and $[(PtMe_3SMe)_3(PtMe_3Cl)]$ have been shown to undergo facile Pt–methyl scrambling.[154] In the case of $[(PtMe_3SMe)_3(PtMe_3X)]$ (28) the fluxion exchanges the two distinct methyls (Me_B and Me_D) of Pt^2Me_3 while the three equivalent environments of Pt^3Me_3 are unaffected. This observation suggests that fluxionality may exist in the symmetrical tetramers $[(PtMe_3Z)_4]$ (Z = halogen, OH, SMe, etc.) but this has yet to be proved conclusively.

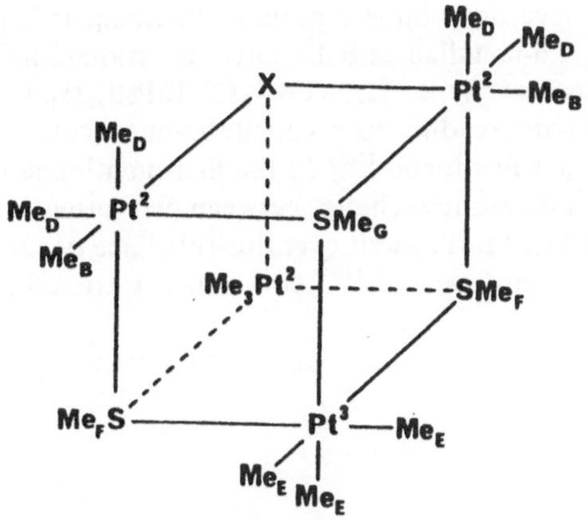

(28)

There have been a number of reports of hydride ligand fluxionality in metal clusters. In the triruthenium–rhodium cluster $[Ru_3Rh(\mu-H)_2(CO)_{10}(\eta^5-Cp)]$ hydride exchange occurs[155] while in the Ru_3 cluster $[HRu_3(C{\equiv}CBu^t)$-$(CO)_7(dppm)]$ hydride migration occurs concurrently with acetylide rotation.[156] Acetylide exchanges, $\pi \rightarrow \sigma$ and $\sigma \rightarrow \pi$ interchanges, between bridged Os atoms occur in $[Os_3H(CO)_9(L)(\mu-\eta^2-C{\equiv}CPh)]$ $(L = CO, PMe_2Ph)$ (Scheme 21)[157] while in the Os_3 cluster $[H_3Os_3(CO)_9(\mu_3-CC_6D_5)]$ reversible intramolecular H/D exchange between ring and metal sites is the dynamic process.[158]

Reports of hydride fluxionality in higher nuclearity metal clusters include $[Re_7(\mu-H)C(CO)_{21}]^{2-}$,[159] and $[Os_{10}H_n(CO)_{24}X_y]^{m-}$ $(n = 4, m = 2, y = 0; n = 4, m = 1, y = 1; n = 5, m = 1, y = 0)$.[160] These clusters show no $^1J(^{187}Os-^1H)$ satellites in the ^1H NMR spectra, suggesting rapid hydride fluxionality with the hydrides jumping between octahedral and tetrahedral sites. Some novel adamantane-like mercury–chalcogen clusters have been synthesized and subjected to detailed ^{31}P, ^{77}Se, ^{125}Te, and ^{199}Hg NMR studies. The compounds are of general type $[(\mu-ER)_6(HgL)_4]^{2+}$ $(E = S, Se, Te, L = tertiary phosphine)$.[161,162] In the Hg/Te clusters[162] NMR studies indicated that the $(\mu-TeR)_6Hg_4$ core is long-lived, though rapid Te inversion leads to a time-averaged tetrahedral symmetry.

Scheme 21

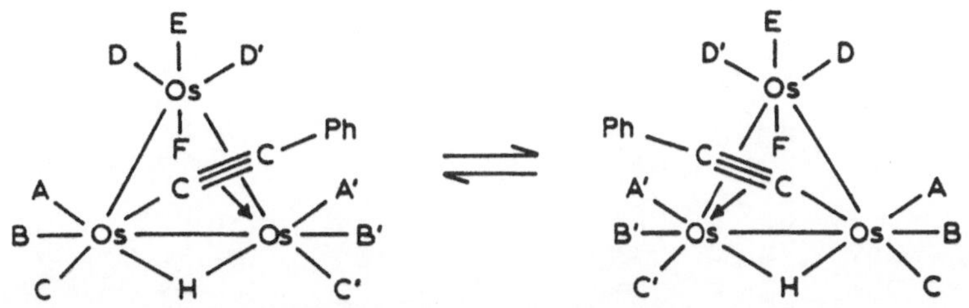

Finally, there have been three reports of fluxional metallaboranes, namely, ten vertex *closo*-2,1,6-metalladicarbaboranes of ruthenium and rhodium,[163] platinatelluraboranes of type $[2,2\text{-}(PR_3)_2 1,2\text{-}TePtB_{10}H_{10}]$ $(R_3 = Et_3,\ n\text{-}Bu_3,\ Me_2Ph)$,[164] and 12-vertex dinuclear cobaltaboranes containing the $Co_2(CO)_5$ moiety and SEt_2 cage substituents.[165] In the first-mentioned compounds,[163] the fluxionality is an enantiomeric exchange between two forms, in the second case[164] slow rotation of the $Pt(PMe_2Ph)$ unit over the TeB_4 face occurred $[\Delta G^+(328\ K) = 62\ kJ\ mol^{-1}]$, and the third example[165] concerned restricted B—S bond rotation.

Chapter 14
Homogeneous Catalysis of Organic Reactions by Transition Metal Complexes

14.1. General

The proceedings of a symposium on oxygen activation held at Texas A & M University, March 1987, have been published in a book, *Oxygen Complexes and Oxygen Activation by Transition Metals.*[1] The abstracts of the 4th International Symposium on Organometallic Chemistry Directed Towards Organic Synthesis, held at Vancouver, July 1987, have appeared in a special issue of *Pure and Applied Chemistry*[2]; they include contributions on chiral catalysis, carbonylations, oxidation–reduction reactions, and stereospecific cross-coupling reactions. A review on metal clusters in catalysis deals with homogeneous and heterogenized systems,[3] and another describes the hydroformylation, hydroesterification, hydrocarboxylation, and amidocarboxylation of fluoroolefins and fluoroalkyl halides.[4] Reviews have appeared to mark the appearance of Volume 50 of *Journal of Molecular Catalysis,* dealing with supported metals and supported organometallics,[5] and with the advantages and disadvantages of organometallic complexes as catalyst precursors, using mainly the nickel-catalyzed butene dimerization and ethylene oligomerization as examples.[6] Recent results on the mechanism of the Fischer–Tropsch reaction based on $(C_5Me_5)Rh$ complexes have been summarized.[7] The role of metal carbalkoxy complexes in the carbonylations of dienes and halides as well as the hydrogenation of CO has been discussed.[8] Evidence for the mechanism of sulfur removal from thiophene-containing substrates has been summarized, based on the chemistry of ruthenium complexes as models for heterogeneous hydrodesulfurization processes.[9] Cationic metal complexes may be important in a number of catalyzed reactions, and the role of some, such as $[Pd(NCMe)_4]^{2+}$ and $[Ni(NCMe)_6]^{2+}$, in reactions with C=C, C—C, and C—H bonds have been described.[10] Palladium-catalyzed metallo-en reactions (alkene dimerizations) have been reviewed, with emphasis on intramolecular cycliz-ations.[11] Another review describes the catalytic amination of olefins.[12] The palladium-catalyzed coupling reactions of enol triflates with activated olefins and with tin reagents have been the subject of a survey.[13] Another review also deals with the application of homogeneous catalysis in organic synthesis, directed to the synthesis and functionalization of indols, and includes palladium-catalyzed amination of olefins, reactions of azirines, alkynylations, cyclizations of haloani-lines, and the synthesis of alkaloids.[14] Hydroformylation, reduction, and oxidation are the subject of an annual survey covering the year 1987[15]; further surveys cover the years 1987[16] and 1988.[17] A special volume on organometallic compounds and optical activity includes papers on asymmetric catalysis.[18] The "chemical multiplication of chirality" has been reviewed, centered mainly on asymmetric hydrogenations with ruthenium "BINAP" catalysts.[19] The enantioselective homogeneous catalysis involving transition metal π-allyl intermediates has been summarized, with emphasis on the reactions of allyl complexes with nucleophiles, diene oligomerization, telomerization, codimerization, and hydrosilylation.[20] A further account has appeared on transition-metal-catalyzed epoxidations.[21]

14.2. Hydrogenations

Although not a proper catalytic cycle, a stepwise process involving η^2-coordi-nation of benzene to Os^{2+} achieves the selective reduction to cyclohexene. The reduction step involves H_2 and Pd on charcoal. The osmium complex can be recycled[22] as shown in Scheme 1. While the hydride complex $[Os(H)Cl(CO)L_2]$ does not react with alkenes, it reversibly takes up H_2 to give $[Os(H)(H_2)Cl(CO)L_2]$

Scheme 1

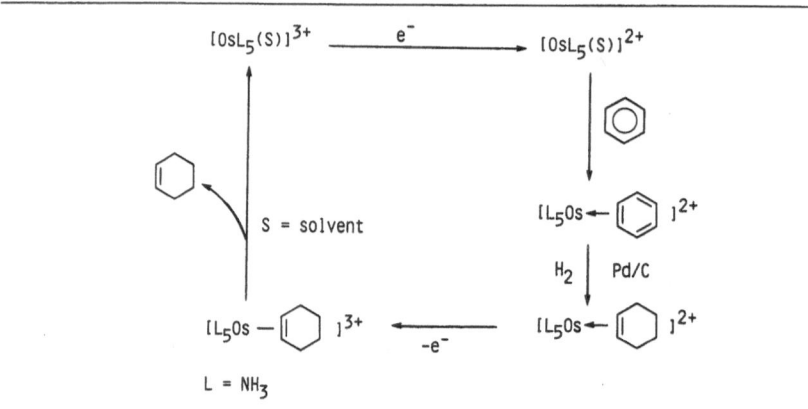

(L = PPh$_3$) which was characterized by NMR; it catalyzes the hydrogenation of styrene to ethylbenzene. The reaction is solvent-dependent, $C_6H_6 < C_2H_4Cl_2 <$ *i*-PrOH.[23] In a series of complexes of the tetradentate (p$_4$) ligand [p$_4$ = P(CH$_2$CH$_2$PPh$_2$)$_3$], [(p$_4$)M(η^2-H$_2$)]$^+$ (M = Co, Rh, Ir), the reactivity depends on the H—H bond strength. For M = Co, the H—H bond remains strong, and addition of diethylmaleate displaces the H$_2$ ligand, leading to isomerization to fumarate instead of hydrogenation. The Rh complex has a weaker H—H bond and gives mainly hydrogenation product, while the Ir complex is present as a classical hydride and does not react with maleate.[24] A number of dimeric Rh hydrides have been prepared, such as [(p$_3$)Rh(μ-H)$_3$Rh(p$_3$)]$^{n+}$ ($n = 2,3$) and [(p$_3$)Rh(H)(μ-H)$_2$Rh(H)(p$_3$)]$^{2+}$ (p$_3$ = MeC(CH$_2$-PPh$_2$)$_3$); these hydrogenate alkenes slowly, and it is suggested on the basis of NMR studies that the dimeric structure probably remains intact throughout the catalytic cycle.[25] The mechanism of ethylene hydrogenation with an iridium dihydrogen complex has been studied. A number of equilibria could be established in independent reactions (Scheme 2). Although the

Scheme 2

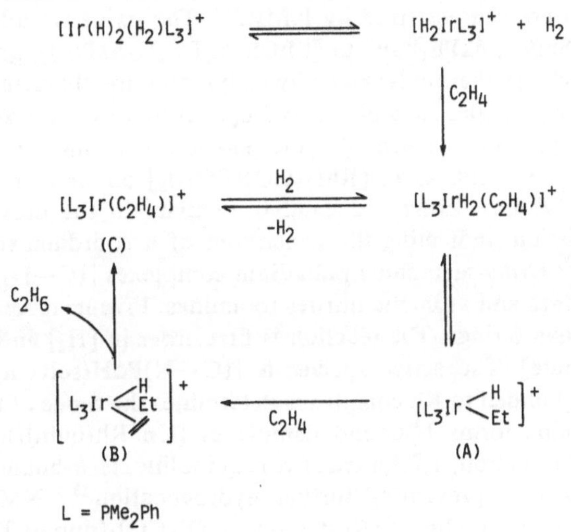

L = PMe$_2$Ph

intermediates (**A**) and (**B**) could not be observed spectroscopically, support for (**B**) was found in the reaction with methylacrylate which gives (**1**) and is seen as a model for (**B**). Intermediate (**C**) can reversibly react with ethylene to give [Ir(C$_2$H$_4$)$_2$L$_3$]$^+$ which was characterized by X-ray diffraction.[26]

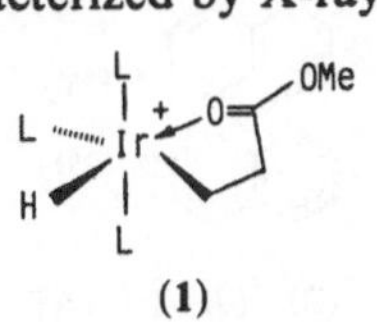

(**1**)

Para-hydrogen-induced polarization (PHIP) gives enhanced ^{1}H NMR absorptions and emissions and is indicative of a *pairwise* transfer of H_2 to the substrate. The effect was observed for the hydrogenation of perdeutero-styrene with *para*-enriched H_2 catalyzed by $[RuH_4(PPh_3)_3]$. The complex exchanges H_2 rapidly and transfers it in one step to the substrate. The hydrogenation mechanism with $[RuHCl(PPh_3)_3]$ is more complex; here, H exchange is suppressed by the olefinic substrate, and the PHIP effect is much weaker. Electron-withdrawing olefins which readily insert into Ru—H bonds to give alkyls prior to hydrogenation do not show any PHIP effect, since the hydrogen comes from two sources in this case.[27] In the hydrogenation of phenylacetylene catalyzed by $[RhCl(PPh_3)_3]$ the PHIP effect shows that H_2 addition to the metal is reversible, followed by substrate insertion and hydrogenolysis, while with $[Rh(COD)(dppe)]^+$ the substrate coordinates first, followed by irreversible H_2 additions.[28] The thermodynamics of the hydrogenation of cyclohexene by water-soluble catalysts, such as $Na[Ru(EDTA-H)N_3]$, in the presence of different π-acids X have been determined. The catalytic activity increases in line with the π-acidity of X; in the same sequence, the formation of ruthenium hydrides becomes more endothermic.[29] The data indicate that hydride complexes are involved in the rate-determining step (X = Cl, CO, $SnCl_3$, PPh_3); the complexes were characterized by NMR.[30] The oxidative addition of H_2 to $[RhClL_3]$ (L = $SbPh_3$, $AsPh_3$) and to $[RhCl(Ph_2PC_2H_4AsPh_2)]_2$ gives *cis*-Rh^{III} H_2 complexes, which are thermodynamically stable and are characterized by NMR. The thermodynamic parameters of the hydrogenation of cyclohexene were determined.[31] Photolytic enhancement of hydrogenation reactions is observed if photolabile ligands are present, as in $[RhH(CO)(PPh_3)_3]$ on irradiation by an XeCl excimer laser (308 nm). Enhanced catalytic activity in the dark was observed following irradiation, indicating the generation of a coordinatively unsaturated active species.[32] *Ortho*-metallated palladium complexes $[(C—N)PdX]_2$ in DMF reduce nitroalkenes and aromatic nitriles to amines. Five-membered metallacyles are less active than 6-rings. The reaction is first order in $[H_2]$ and $[Pd]$ and zero order in [substrate]. The active species is $[(C—N)PdH(solvent)]$.[33] The way heterocycles are bonded to Rh complexes determines the mode of their reduction. Although quinoline forms N-bound complexes $[Cp*Rh(quin)(MeCN)_2]^{2+}$, the product of hydrogenation, 1,2,3,4-tetrahydroquinoline, is π-bound to Rh via the benzene ring, thereby preventing further hydrogenation.[34] NMR studies and kinetics suggest that the cluster $[Ru_3(\mu\text{-}H)_2(\mu\text{-}O)(CO)_5(dppm)_2]$ remains intact during the hydrogenation of olefins at 80 °C/40 atm H_2; a mechanism for the reduction of the olefin by the trinuclear cluster is suggested.[35]

Ruthenium complexes of "BINAP" (2) are highly selective catalysts for asymmetric hydrogenations. It has now been shown that exposure of $[RuH(BINAP)_2]^+$ to H_2 gives an η^2-H_2 complex.[36] The catalyst $[Ru(OAc)_2\{(S)\text{-}$

(2), (*R*)-BINAP

Scheme 3

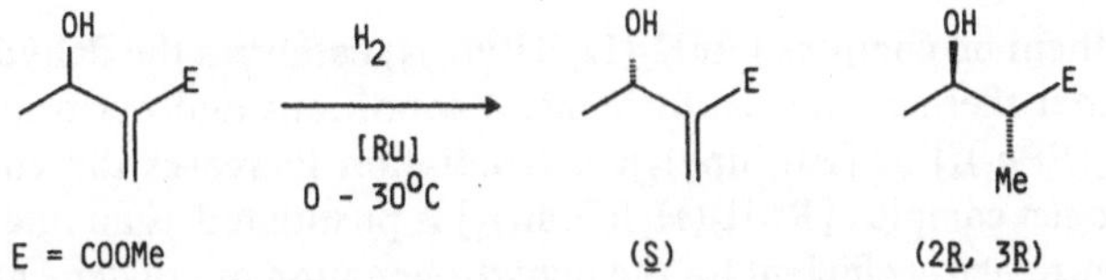

E = COOMe   (S)   (2R, 3R)

BINAP}] allows the kinetic resolution of racemic allylic alcohols with very high enantiomeric excess; only the (R) alcohol is hydrogenated (Scheme 3).[37] Similar catalysts hydrogenate functionalized ketones and α-ketoesters in near-quantitative yields and very high enantiomeric excess. Functional groups capable of chelating to the metal center are essential for the success of the reaction. While 2-bromoacetophenone is converted with high efficiency, acetophenone and 3- and 4-substituted derivatives cannot be hydrogenated satisfactorily.[38] Chelate formation with the substrate may also operate in the double asymmetric hydrogenation of conjugated dienes catalyzed by [Ru$_2$Cl$_4${(R)-BINAP}$_2$]/NEt$_3$. The (S,S)-succinic acid is produced with high stereoselectivity (Scheme 4).[39] Systematic

Scheme 4

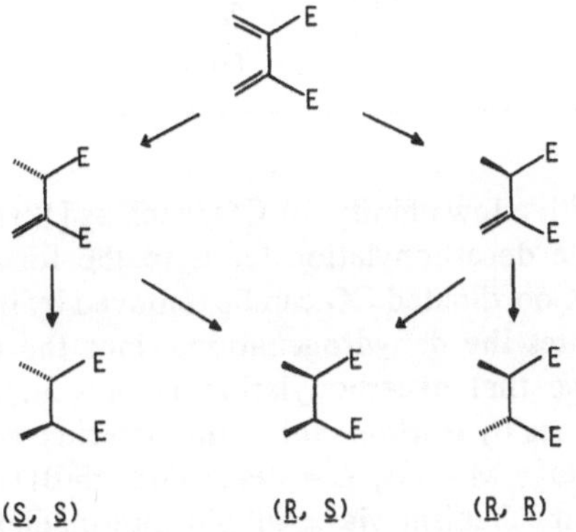

(S, S)   (R, S)   (R, R)

variation of the substituents of the phosphorus atoms in cationic Rh complexes of chelating chiral phosphines suggests that substituents in *axial* position influence the stereoselectivity of cinnamic acid hydrogenation strongly (structure 3), while those in equatorial positions have less influence (structure 4).[40]

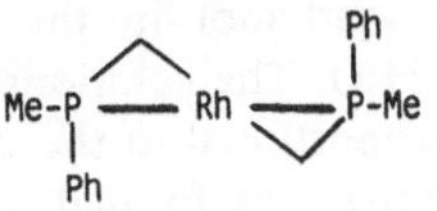

(3)   (4)

high ee   low ee

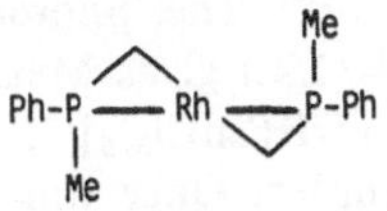

14.3. Hydrogen Transfer and Dehydrogenations

The ruthenium complex $[RuH_2(N_2)(PPh_3)_2]$ catalyzes the dehydrogenation of alcohols under thermal and photochemical conditions and performs much better than $[RuH_2(PPh_3)_4]$ or $[Rh(bipy)_2]Cl$. Irradiation increases the conversion rate. The dihydrogen complex $[RuH_2(H_2)(PPh_3)_3]$ is postulated as an intermediate. The process is particularly efficient for the dehydrogenation of ethylene glycol (Scheme 5).[41] The dehydrogenation of ethanol to give acetaldehyde and H_2 is particularly

Scheme 5

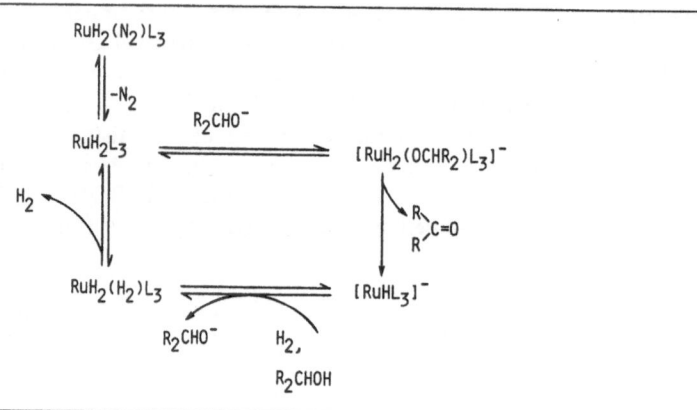

efficient for metals with a low affinity for CO, such as $[PtH(PEt_3)_3]^+$. With Rh and Ru catalysts aldehyde decarbonylation leads to the formation of poorly active carbonyl complexes. Coordinated CO can be removed by irradiation. The presence of base also accelerates the dehydrogenation, since the aldehyde undergoes an aldol condensation so that decarbonylation is prevented.[42] The reduction of α,β-unsaturated ketones by isopropanol in the presence of $[MHCl(CO)L_3]$ (M = Ru, L = PPr_3^i, $PMeBu_2^i$; M = Os, L = PPr_3^i) and $NaBH_4$ is highly selective for the C=C bonds. A mechanism via π-allylic intermediates has been suggested (Scheme 6).[43] The photolytic removal of CO from $[RhX(CO)L_2]$ by irradiation gives catalysts for the generation of H_2 and acetone from isopropanol. The activity increases in the order $PPr_3^i \sim PPh_3 < PEtPh_2 < PEt_2Ph < PEt_3 < PMe_3$ and X = Cl < Br < I. The turnover frequency correlates well with $J(Rh-P)$ and points to an electronic influence of the catalyst activity, although steric factors are clearly also important. The activation energy for the dehydrogenation is small (*ca* 20 kJ mol^{-1}).[44] The photodehydrogenation of methanol in the presence of $IrCl_3/SnCl_2/LiCl$ gives $MeOCH_2OMe$, H_2, and H_2O. The solutions were shown to contain $[IrH(SnCl_3)_5]^{3-}$, and the activity is proportional to the concentration of this complex. Once this relatively stable species was formed, there was no induction period.[45] The mechanism of H transfer from isopropanol to imines catalyzed by $[RhCl(PPh_3)_3]$ in the presence of $NaOPr^i$ has been investigated and

Scheme 6

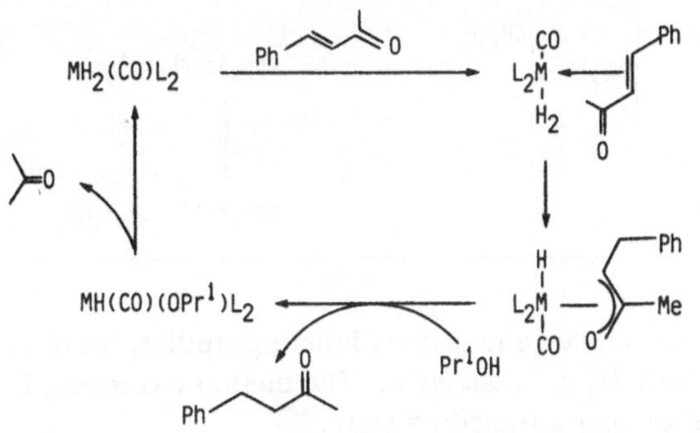

follows in principle the one outlined in Scheme 5. A dimeric hydroxo complex $[L_2Rh(\mu\text{-}OH)_2RhL_2]$ has been isolated and can be used as a catalyst precursor.[46] The dehydro coupling of Et_3SiH with primary amines to give H_2 and Et_3SiNHR is catalyzed by many noble metal complexes, especially by Pd. Ruthenium is also very suitable, especially $[Ru_2(CO)_8(SiEt_3)_2]$ as precursor. A mechanism based on this dinuclear complex has been outlined, although the possibility of side reactions with the amine may introduce complicating features.[47] Oxiranes are stereoselectively hydrogenolyzed by formic acid; zwitterionic palladium allyl complexes are likely intermediates in the reaction (Scheme 7).[48]

Scheme 7

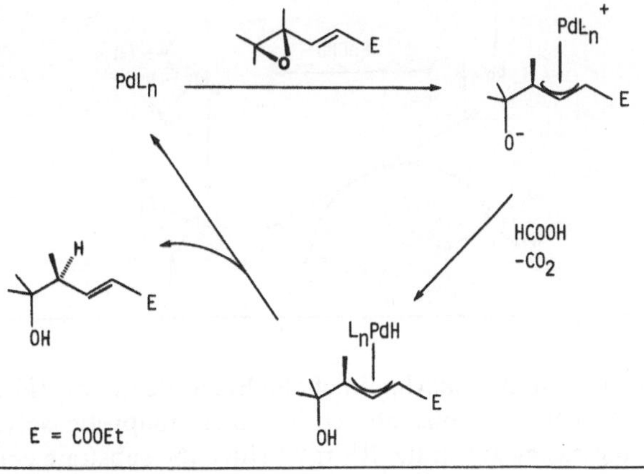

14.4. C—H Activation

The activation of C—H bonds requires as a first step the interaction between an alkane and a metal center. The formation of such a complex has now been

Scheme 8

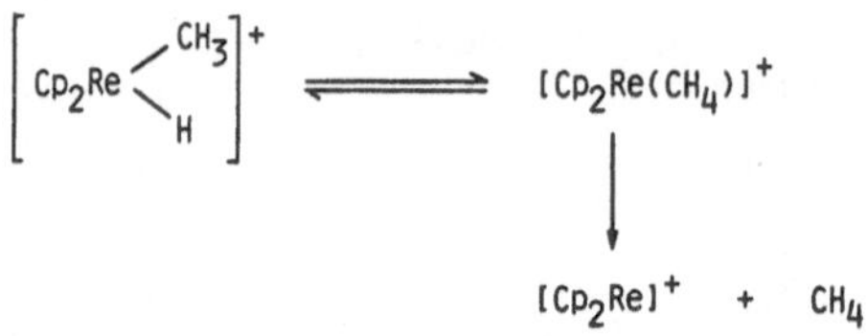

suggested, based on kinetic and isotope labeling studies, for the reductive elimination of $[Cp_2Re(H)(CH_3)]^+$ (Scheme 8). The methane complex is thought to be an intermediate rather than a transition state.[49]

In certain cases, hydrido vinyl complexes show greater stability than alkene complexes, contrary to the general order of stability. One such example is $[LIr(CO)(C_2H_4)]$ which gives $[LIr(H)(C_2H_3)(CO)]$, L = $HB\{C_3N_2(Me)(CF_3)\}_3$.[50] An application of the C—H activation is the hydroacylation of olefins by aldehydes. For example, $[(Ind)Rh(C_2H_4)_2]$ converts ethylene and benzaldehyde into PhC(O)Et. D-labeling studies and *in situ* ^2H NMR under catalytic conditions support the suggested mechanism (Scheme 9); there is no significant aldehyde decarbonylation.[51]

Scheme 9

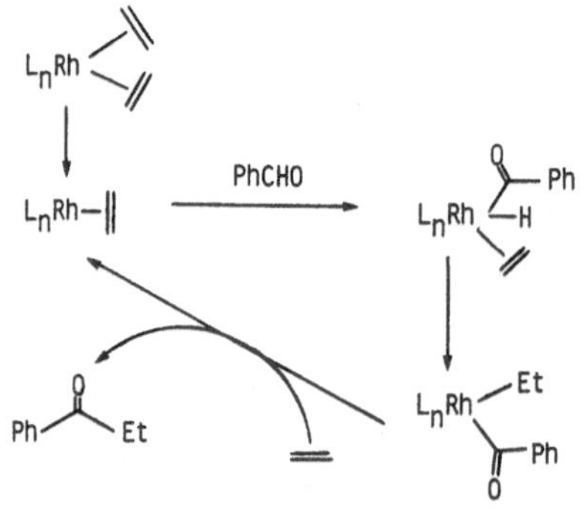

The *intra*molecular hydroacylation of aldehydic olefins is catalyzed by cationic Rh complexes of chelating phosphines in polar nonprotic solvents. The rate decreases with increasing substrate:Rh ratio since the substrate complex with the catalyst inhibits the reaction. However, since this complexation also prevents the decarbonylation of the aldehyde, catalyst deactivation decreases, leading to a higher turnover.[52] The only effective catalytic C—H activations are carried out with $[RhCl(CO)(PMe_3)_2]$ under photolysis which removes CO. Alkanes are converted to alkenes in the order cyclooctane > cyclohexane > *n*-decane ~ *n*-hexane. Up to 200 turnovers cyclooctane per hour are observed. In order to generate terminal alkenes from *n*-alkanes an increase of the PMe_3: Rh ratio to 5–10 is necessary.[53] The dehydrogenation of *n*-alkanes is retarded by the alkene product, which is also isomerized. The evolution of H_2 follows a first-order dependence on the irradiation intensity at 365 nm.[54]

14.5. Hydrosilylation

The complex $[PtCl_2(styrene)_2]$ is a catalyst precursor for the hydrosilylation of terminal olefins and acetylenes, the reduction of carbonyl functions (with pyridine or amines as co-catalysts), and for the formation of R_3SiOR' from R_3SiH and alcohols. During the hydrosilylation the complex is reduced to $[Pt(styrene)_3]$, which could not be isolated but was characterized by NMR.[55] While $[PtCl_2(styrene)_2]$ requires an induction period, $[Pt(styrene)_3]$ is immediately active. Intermediate hydrido complexes have been observed by NMR, possibly of structures (5) or (6). The insertion of styrene into a Pt—H bond appears to be rate

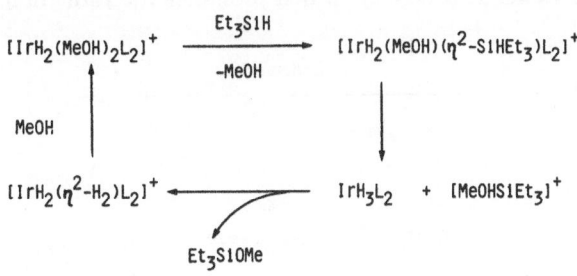

(5)   (6)

determining.[56] Analogous η^2-silane complexes were shown to exist as intermediates in the Ir-catalyzed silyl ether formation (Scheme 10). The fluxional complex $[L_2Ir(H)_2(\eta^2\text{-}Et_3SiH)_2]^+$ was detected by NMR in CD_2Cl_2 at low temperature. The

Scheme 10

$$[IrH_2(MeOH)_2L_2]^+ \xrightarrow[-MeOH]{Et_3SiH} [IrH_2(MeOH)(\eta^2\text{-}SiHEt_3)L_2]^+$$

$$\uparrow MeOH \qquad\qquad \downarrow$$

$$[IrH_2(\eta^2\text{-}H_2)L_2]^+ \xleftarrow{\quad Et_3SiOMe \quad} IrH_3L_2 \;+\; [MeOHSiEt_3]^+$$

η^2-coordination mode does not activate silanes for less nucleophilic substrates than alcohols, and the Ir complex therefore does not catalyze the hydrosilylation of olefins.[57] Another Ir complex, $[Ir(CN)(CO)(dppe)]$, catalyzes the reduction of CO_2 by Me_3SiH at room temperature to give Me_3SiOMe and $(Me_3Si)_2O$. The oxidative addition of Me_3SiH is the initial reaction step.[58] The hydrosilylation-cyclization of α,ω-diynes is catalyzed by Ni(0) complexes, probably following the mechanism outlined in Scheme 11.[59]

Scheme 11

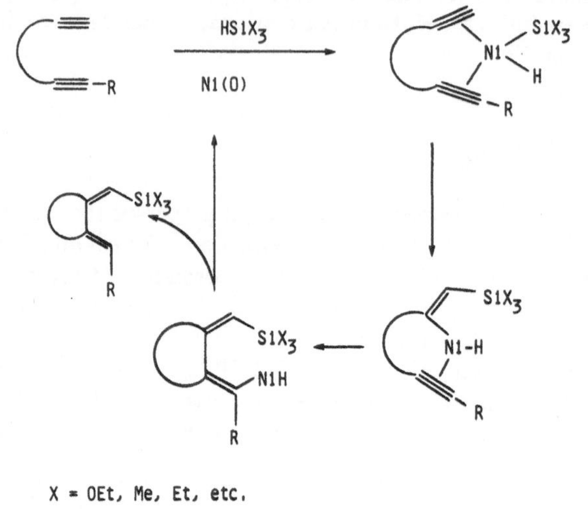

X = OEt, Me, Et, etc.

14.6. Nucleophilic Additions to C=C and C≡C

The amination of norbornene with aniline is catalyzed by [IrCl(C$_2$H$_4$)$_2$L$_2$] (L = PEt$_3$) which is converted into a catalytically active intermediate through loss of C$_2$H$_4$. The mechanism (Scheme 12) is supported by the independent synthesis and isolation of the complexes (**A**) and (**B**). The reaction becomes catalytic in the presence of a Lewis acid (ZnCl$_2$) which promotes the formation of the cationic

Scheme 12

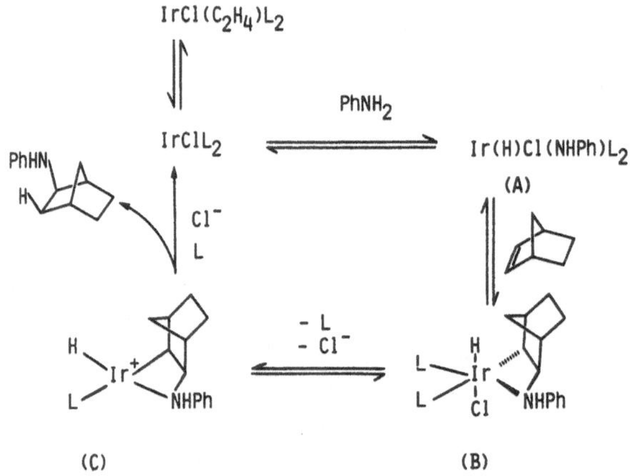

Scheme 13

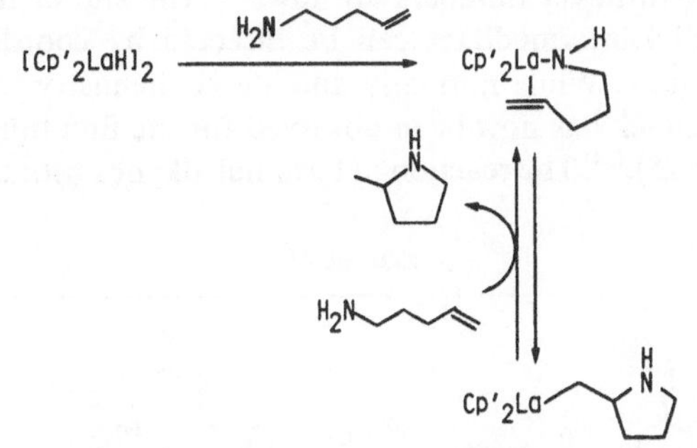

intermediate (**C**). Labeling studies show that aniline is added *cis* to the *exo* face of norbornene. The structure of (**B**) was determined by X-ray crystallography. Traces of water deactivate the catalyst through formation of the structurally characterized inactive hydroxo complex $[\{Ir(H)L_2\}_2(\mu\text{-NHPh})(\mu\text{-OH})(\mu\text{-Cl})]^+Cl^-$.[60] Olefins are hydroaminated by lanthanide hydrides. The mechanism of this reaction was established with the aid of D-labeling studies (Scheme 13). At 60 °C up to 140 turnovers per hour are achieved.[61] The asymmetric hydrocyanation of norbornene is catalyzed by a number of chiral Pd(0) complexes of chelating phosphine ligands at 120 °C. Optical yields are modest because of the high temperatures required. The reaction was followed by NMR. The precursor complex $[(P-P)Pd(C_2H_4)]$ does not exchange readily with most olefins but excess norbornene cleanly displaces C_2H_4 and binds to Pd through the *exo* face.[62] The mechanism of the nickel-catalyzed co-dimerization of vinylcyclohexane with phenylisocyanate is supported by stoichiometric studies (Scheme 14). Small ligands

Scheme 14

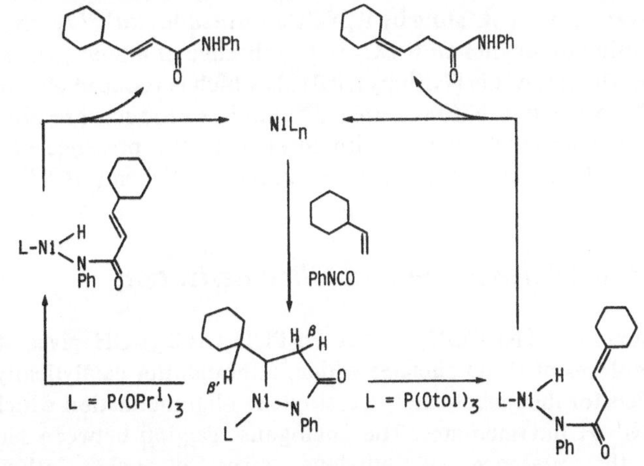

[P(OPr^i)$_3$] favor elimination of β-H, larger ligands [P(O-*o*-tol)$_3$] favor elimination in β' position. The turnover numbers are low.[63] The site of nucleophilic attack on palladium π-allyl intermediates can be directed by coordinating functional groups of the substrate. While normally *anti* stereochemistry would be expected, a *syn* pathway of attack has now been observed for the first time in phosphinated substrates (Scheme 15).[64] The reaction of terminal alkynes with CO$_2$ and secondary

Scheme 15

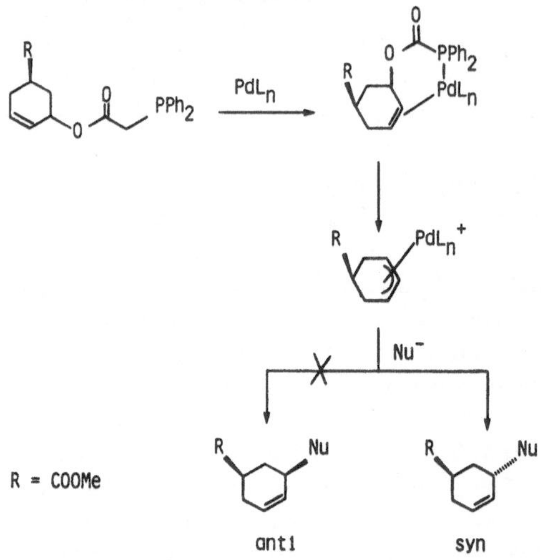

amines in the presence of ruthenium complexes [Ru$_3$(CO)$_{12}$] and [RuCl$_2$(PR$_3$)-(arene)] gives vinyl carbamates. A key step in the reaction sequence appears to be the conversion of a Ru alkyne complex into a vinylide complex Ru=C=CHR, which is attacked at the α-C atom by R$_2$NCO$_2^-$ formed *in situ*.[65] A related reaction, the carboxylation of alkynes with CO$_2$ in the absence of amines, gives substituted acrylic acids. The catalyst is [Ni(bipy)$_3$](BF$_4$)$_2$ which is reduced electrochemically in a single 2e^- wave to a Ni(0) complex.[66] Similar products are obtained in the carboxylation of terminal alkynes with formate in the presence of a Pd/dppb catalyst under CO; a likely mechanism is outlined in Scheme 16.[67]

14.7. Olefin Dimerization and Oligomerization

The reaction of [Ni(COD)$_2$] with Ph$_2$PCH$_2$C(CF$_3$)$_2$OH gives stable nickel hydride complexes of P—O chelates which, although not catalytically active, are seen as models for the nickel-catalyzed ethylene oligomerization which is thought to involve Ni—H intermediates. The analogous reaction between Ni(0) and the ligand . in the presence of ethylene gives a stable ethyl complex,

Scheme 16

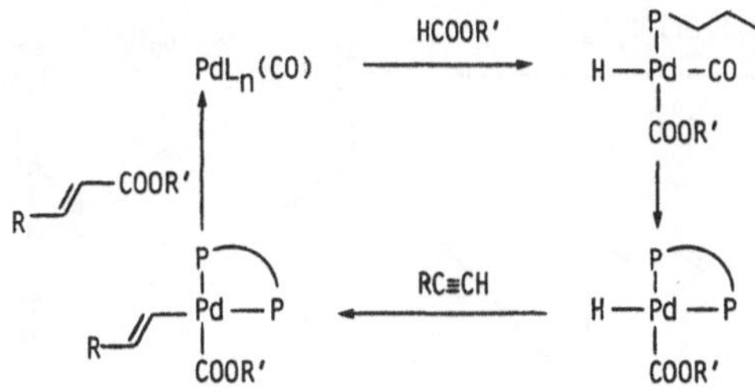

[(P—O)NiEt(PCy$_3$)].[68] The related P—O nickel chelate complex (7) in the presence of Et$_2$AlOEt at 90 °C co-oligomerizes ethylene with higher α-olefins. Analysis of the products reveals that the insertion of olefins gives both primary and secondary

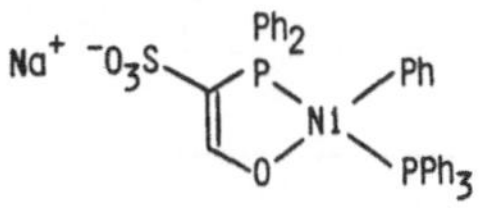

(7)

nickel alkyls; the latter do not permit further insertions and terminate the growth chain by β-H elimination.[69] AlEt$_3$ and Al$_2$Et$_4$O deactivate the catalyst by reduction to Ni(0), while Et$_2$AlCl and EtAlCl$_2$ as additives give a catalyst highly selective for the C$_2$H$_4$ dimerization to butenes. In the presence of Et$_2$AlOR the ratio of chain termination to propagation for ethylene oligomerization is *ca* 0.3:1 over a 50–120 °C temperature range. The number of active centers under these conditions is proportional to the nickel concentration. The activation energy for ethylene oligomerization is estimated as *ca* 27 kJ mol^{-1}.[70] Ethylene is dimerized to 1-butene by (η^6-arene)MR$_2$ (M = Ni $\gg$ Co, R = SiCl$_3$ $\gg$ SiF$_3$ > C$_6$F$_5$). Nickel hydride intermediates are suggested (Scheme 17). The life of the catalyst is greatly prolonged by the addition of EtAlCl$_2$.[71] Zero-valent nickel complexes in the presence of tris(*o*-tolyl)phosphite catalyze intramolecular [4 + 2] dienyne cycloadditions under mild conditions, probably via nickel allyl complexes (Scheme 18).[72] The dehalogenation of substituted dienes R$_2$C=C(X)—C(X)=CR$_2$ with [NiBr$_2$(PPh$_3$)$_2$] in the presence of excess zinc gives a series of four- and six-membered radialenes. The intermediacy of cumulenes, R$_2$C=C=C=CR$_2$, is likely. The reaction is thought to go stepwise through five- and seven-membered nickelacycles from which the products are generated by reductive elimination. The product distribution is strongly dependent on the solvent (C$_6$H$_6$, DMF) and the halide.[73] The intramolecular cyclization of enynes is catalyzed by Pd(0)/P(*o*-tol)$_3$ catalysts in the presence of acid. It is thought that "HPdX" is involved as highly reactive intermediate (Scheme 19).[74]

The strong organometallic Lewis acid [Cp$_2$ZrMe(THF)]$^+$ catalyzes the coupling between propene and α-picoline. Although the reaction is slow (*ca* 1–2

Scheme 17

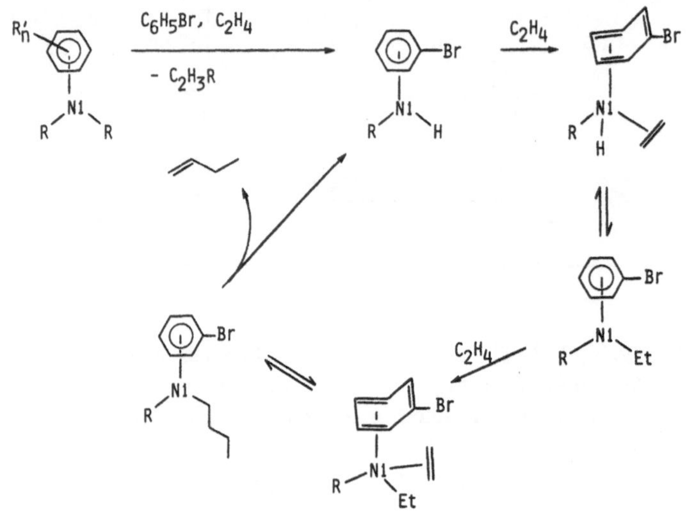

Scheme 18

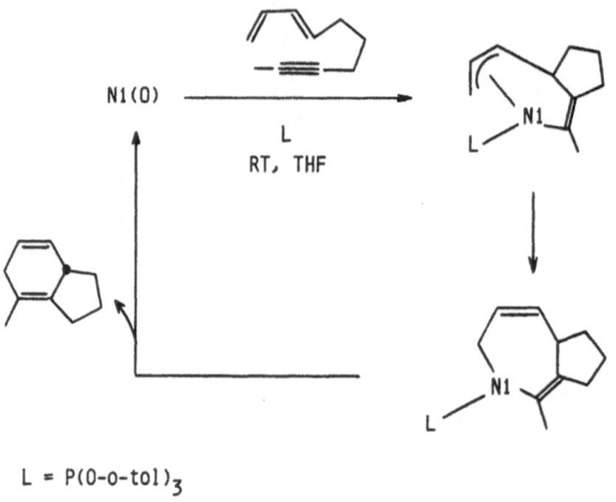

L = P(O-o-tol)$_3$

turnovers per hour) the catalyst has a long life and the reaction proceeds until the substrate is consumed. A mechanism is outlined in Scheme 20.[75]

14.8. Alkyne Oligomerization

A number of models for the cyclooligomerization of alkynes by early transition metal catalysts have been prepared by reducing sterically hindered aryloxide

Scheme 19

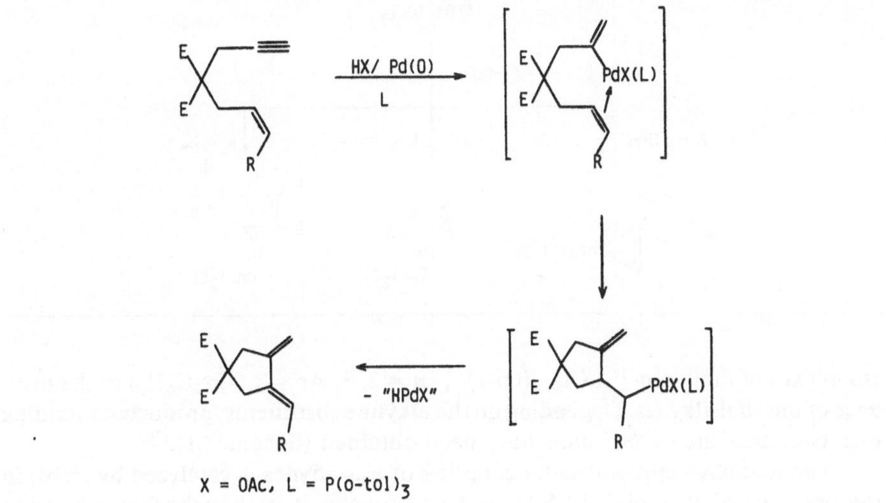

$X = OAc$, $L = P(o\text{-}tol)_3$

Scheme 20

$Zr = [Cp_2Zr]^+$

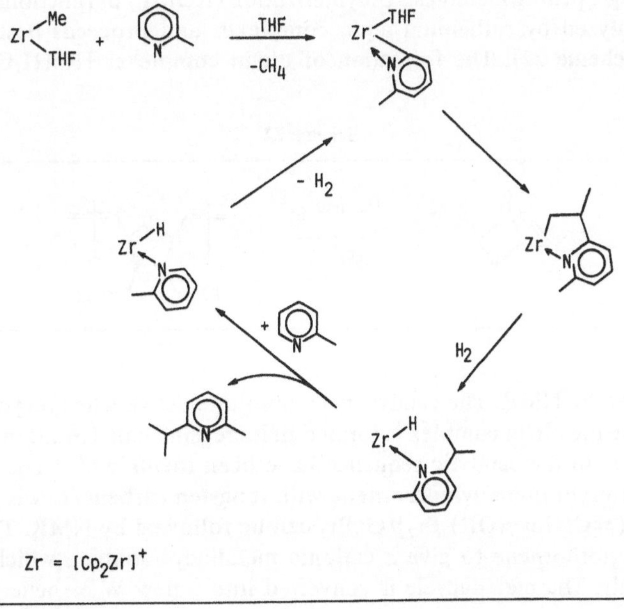

Scheme 21

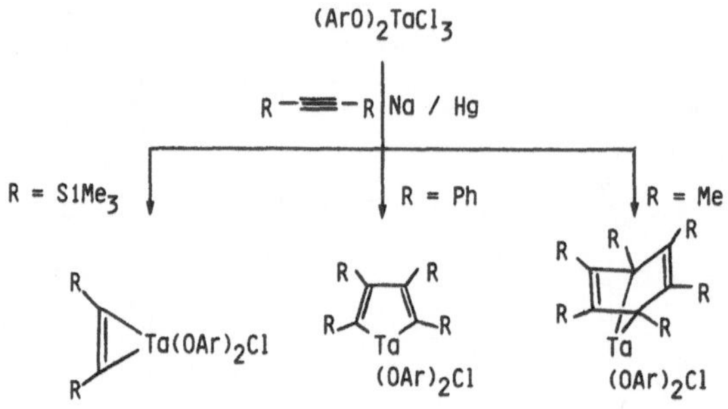

complexes of tantalum, $[TaCl_{5-n}(OAr)_n]$ ($n = 2, 3$; Ar = 2,6-$Pr^i_2C_6H_3$) in the presence of internal alkynes. Depending on the alkyne substituents, products containing one, two, and three C≡C units have been obtained (Scheme 21).[76]

The reductive intramolecular coupling of α,ω-diynes is catalyzed by Pd(0) in the presence of P(*o*-tol)₃, Et₃SiH, and acetic acid. It is thought that a hydrido species, $[Pd(H)(OAc)L_2]$, initiates the reaction. The mechanistic scheme resembles that outlined in Scheme 19. The method is quite general and applicable to a wide range of substrates.[77]

14.9. Metathesis

The ring-opening metathesis polymerization (ROMP) of functionalized olefins can be catalyzed by ruthenium aquo complexes and proceeds best in aqueous solution (Scheme 22). The formation of olefin complexes $[Ru(H_2O)_5(olefin)]^{2+}$

Scheme 22

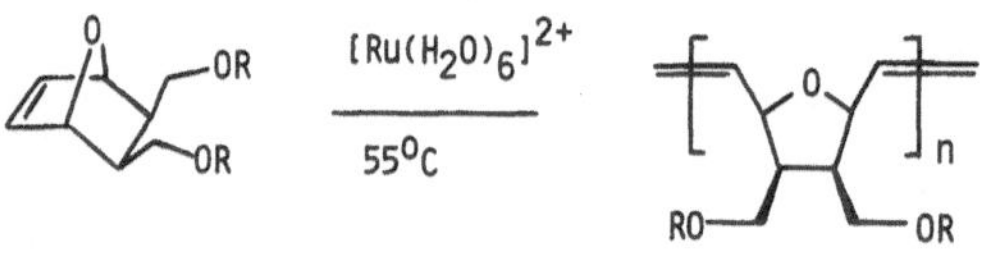

was observed by NMR. The catalyst becomes more active when recycled, presumably because the olefin complex is formed in increasing concentrations. No further intermediates in the catalytic sequence have been identified.[78] The metathetical polymerization of dimethylnorbornene with tungsten carbene/Lewis acid systems such as $[W(=CHBu^t)(OR)_2Br_2]/GaBr_3$ can be followed by NMR. The precursor reacts with norbornene to give a cationic metallacyclobutane which exists as a tight ion pair. The metallacycle is converted into a new W carbene intermediate

Scheme 23

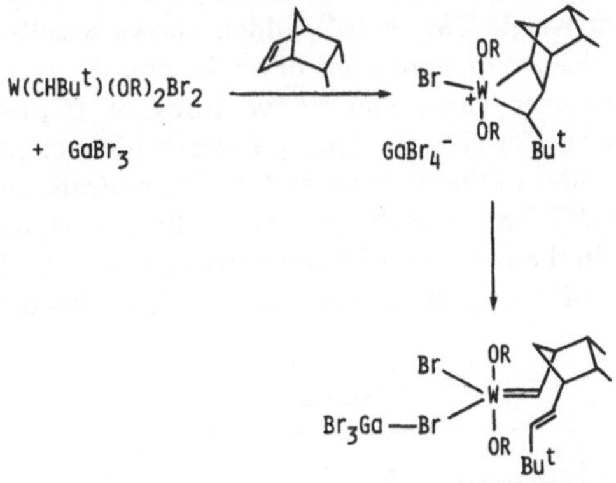

in a first-order reaction, and further reaction with norbornene generates a new metallacycle, thereby propagating the chain (Scheme 23). The relative stabilities of the metallacycle and the carbene complex are strongly influenced by substituents. Introduction of Cl instead of Br increases the half-life of the intermediates, and OPr^i instead of OCH$_2$Bu^t increases the stability of the metallacycle.[79]

14.10. Olefin Polymerization

The mechanism of olefin polymerization has been the subject of theoretical calculations which support the view that ethylene reacts with [Cp$_2$TiMe]$^+$ under insertion to give [Cp$_2$TiC$_3$H$_7$]$^+$.[80] The syndiospecific polymerization of propene is achieved with sterically encumbered hafnium catalysts (**8**) activated by methylaluminoxane. The reaction is carried out in liquid propene at 50–70 °C. Hafnium gives polymers of higher molecular weight but also of higher polydispersity than zirconium. Only 2% of the Hf-produced polymer consisted of isotactic blocks.[81] The Cp rings in the chiral zirconium complex (**9**) show restricted rotation, and a

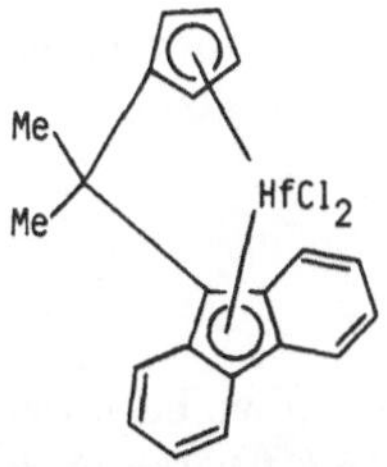

(8) 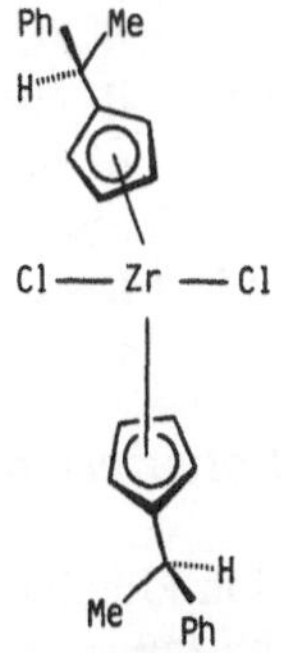   (9)

stereoregular polypropylene can therefore be expected. Activation of the racemic complex with methylaluminoxane at low temperature gives polypropylene of moderate molecular weight ($M_\eta \approx 10^4$) which shows a self-correcting isotactic stereochemistry, indicative of stereocontrol by the chiral metal center, rather than by the growing polymer chain end.[82] Mixtures of [CpMCl₃] and methyl-aluminoxane (M = Ti, Zr) give ethylene polymerization catalysts, accompanied by reduction to hydrides of the trivalent metals. No hydrides are formed if AlMe₃ is used as activator.[83] The isotactic polymerization of styrene is catalyzed by [(allyl)Ni(COD)]⁺ in the presence of one equivalent of PCy₃. The products have relatively low molecular weights. The mechanism is outlined in Scheme 24. The

Scheme 24

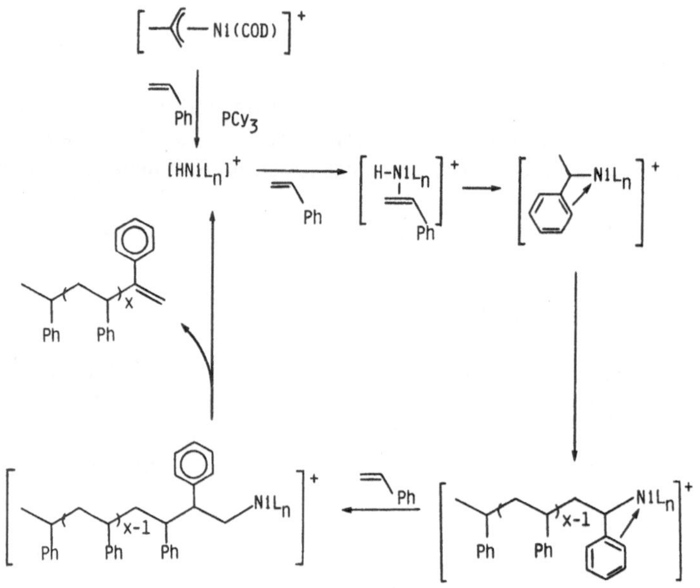

chains are terminated by β-H elimination.[84] Vinylcyclopropanes are polymerized by Pd(0) in the presence of diethylmalonate as initiator under opening of the three-membered ring. Palladium π-allyl complexes are formed as intermediates (Scheme 25). The product has a molecular weight of *ca* 4000.[85]

14.11. Alkyne Polymerizations

The polymerization kinetics of terminal alkynes have been determined. The reaction is initiated by tungsten carbene complexes (**10**); the chelating olefin is displaced by acetylene as the first step in the reaction, followed by formation of

Scheme 25

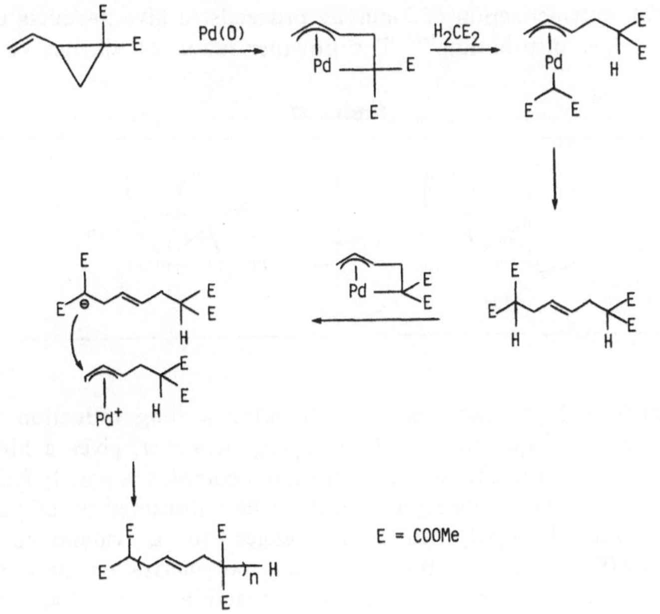

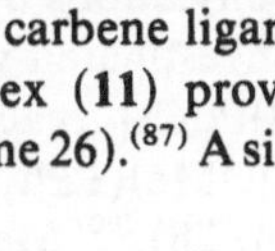

E = COOMe

(10)

a metallacyclobutene which ring-opens to give a new carbene complex.[86] Dinuclear tungsten complexes with bridging carbene ligands also polymerize 1-alkynes; loss of CO from the oligomer complex (11) provides the vacant coordination site required for polymerization (Scheme 26).[87] A similar route is followed by dinuclear

Scheme 26

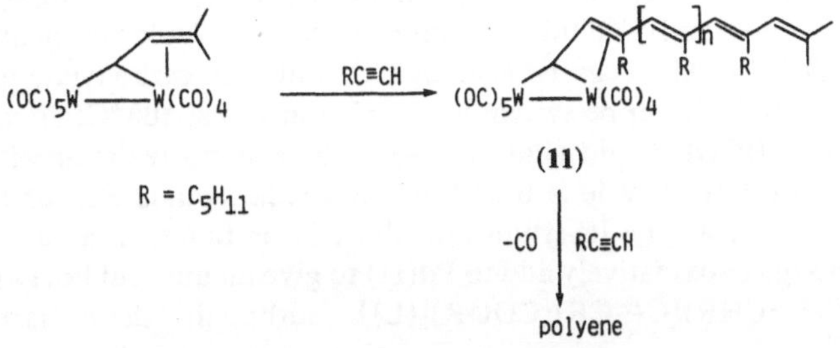

R = C₅H₁₁

$$R = C_5H_{11}$$

(11)

-CO | RC≡CH

polyene

tungsten carbonyls bridged by arylcarbene ligands. The η^3-bonded precursor (12) was crystallographically characterized; it reacts with 2-butyne under insertion (Scheme 27). Polymerization of 2-butyne proceeds to give products of a broad molecular weight distribution.[88] The polymerization of alkynes catalyzed by

Scheme 27

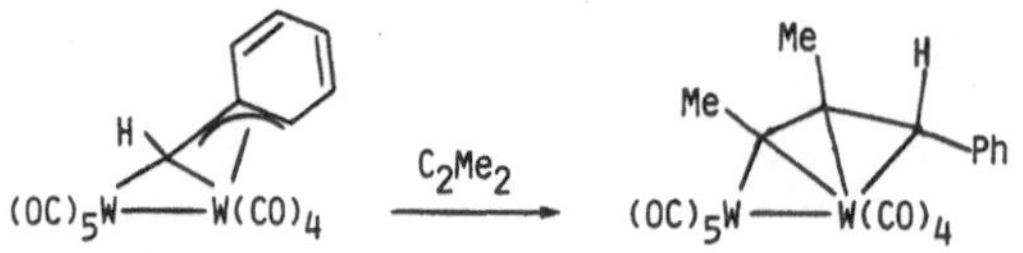

[W(NCMe)$_3$(CO)$_3$] proceeds only slowly after a long induction time. The addition of catalytic quantities of [Cp$_2$Fe]PF$_6$, however, gives a highly active catalyst since ligand substitution in the tungsten complex is greatly facilitated by electron transfer. Two nitrile ligands have to be substituted by alkynes to give [(CO)$_3$W(NCMe)(HC$_2$R$_2$)$_2$], which rearranges to a vinylidene complex, [(CO)$_3$(HC$_2$R)W=C=CHR]; the subsequent polymerization steps follow the pattern expected for carbene complexes: reaction with alkyne to give a metallacyclobutene, followed by ring-opening to give a new tungsten carbene species, etc.[89]

14.12. Carbonylations

14.12.1. Carbonylations of C=C and C≡C

The mechanism of olefin carbonylations by palladium catalysts using methylformate as the source of CO was investigated with model studies. In the presence of triphenylphosphine, [(methallyl)PdCl]$_2$ reacts with methylformate and NaOMe to give isolable [(methallyl)Pd(COOMe)(PPh$_3$)], a model for intermediate (A) in the carbonylation of π-allyl (Scheme 28).[90] Methylformate carbonylates a number of olefins catalyzed, for example, by mono- and polynuclear ruthenium complexes, e.g., [Ru$_3$(CO)$_{12}$] > [H$_4$Ru$_4$(CO)$_{12}$] > [Ru$_3$(CO)$_{12}$]/PCy$_3$ > [Ru(CO)$_3$(PCy$_3$)$_2$] etc. Ruthenium formyl and COOMe complexes are postulated as members of the catalytic cycle.[91] Oxalate can be used as CO source to carbonylate alkynes. The catalyst is a [Pd(dba)$_2$]/dppe system under 82 atm CO at 100 °C. Phenylacetylene gives PhCH=CHCOOR and (PhC≡C—)$_2$. There is no reaction without dppb. The suggested catalytic cycle is based on the oxidative addition of RC≡CH to Pd(0), followed by alkyne insertion into the Pd—H bond. In a subsequent step oxalate is though to oxidatively add to Pd(II) to give an unusual Pd(IV) intermediate, [Pd(CH=CHR)(C≡CR)(COOR)$_2$(L)], although there is no direct evidence.[92]

Scheme 28

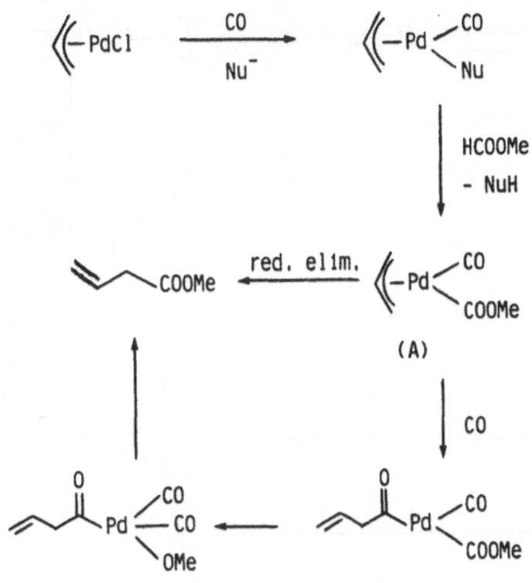

14.12.2 Carbonylation of Organic Halides

Iodobenzene reacts with sodium benzoate in the presence of $Pd(OAc)_2$, PPh_3, and CO to give benzoic anhydride. The mechanism is thought to involve oxidative addition of ArI to Pd(0), followed by exchange of I^- for $ArCO_2^-$ and reductive elimination to the product. The yields depend on the cation of the carboxylate salt.[93] $[Co_2(CO)_8]$ in the presence of CO/H_2 reductively carbonylates benzyl chloride to give $PhCH_2CHO$. Rhodium/cobalt mixed-metal carbonyls give the corresponding alcohol almost exclusively. High-pressure IR showed the presence of $[Co(CO)_4]^-$, $[Rh_2(CO)_4Cl_2]$, and $[Rh_5(CO)_{15}]^-$; mixed-metal species were not detected.[94] The carbonylation of benzyl bromides by $[Fe(CO)_5]$ in the presence of CO under phase-transfer conditions gives mainly ketones and carboxylic acids. IR spectra under 40 atm CO show evidence for $[RC(O)\text{-}Fe(CO)_4]^-$. The formation of the products is explained by two interconnected catalytic cycles (Scheme 29).[95] The mechanism of the double carbonylation of aryl iodides to α-ketoesters has been investigated by studying the reaction of the isolated complexes $[PdPh(I)L_2]$ (I), $[Pd(COPh)(I)L_2]$ (II), and $[Pd(COPh)(ClO_4)L_2]\cdot$acetone (III). Treatment of II with alcohol in the presence of a tertiary amine gives a good selectivity for PhCOCOOR, while III gives only PhCOOR. In contrast to the formation of α-ketoamides, cationic intermediates do not seem to be important in the formation of α-ketoesters.[96] The carbonylation of 1-iodoalkanes containing β-hydrogens is achieved using Pt catalysts in preference to palladium. $[PtCl_2(PPh_3)_2]$ in the presence of K_2CO_3 and 30–70 atm CO in MeOH/dioxane gives alkanoic esters in good yield. A mechanism involving the elimination of HI from RCH_2CH_2I to give $RCH=CH_2$, which is subsequently hydroformylated, is ruled out; added terminal olefins do not react under these conditions. In the presence of H_2 aldehydes are

Scheme 29

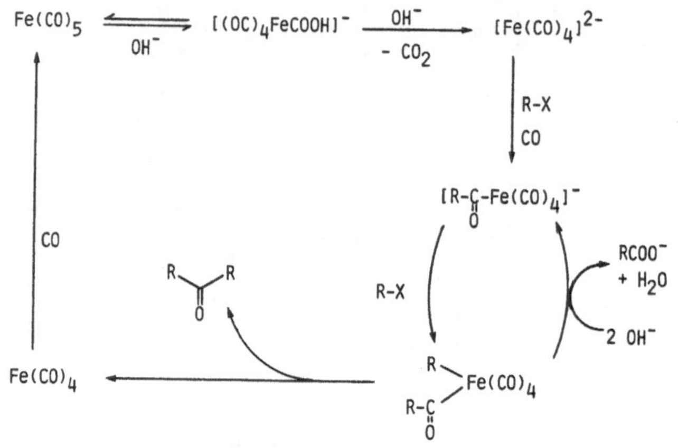

produced.[97] The carbonylation of 1-methoxy-2,7-octadiene by [(methallyl)PdCl]₂ required the addition of Lewis or Brønsted acids in order to generate cationic palladium π-allyl intermediates. It is likely that an (allyl)Pd(COOMe) complex is formed; reductive elimination then transfers the COOMe group to the allyl moiety.[98]

14.12.3. Hydroformylation

A theoretical study of the intermediates of olefin hydroformylation with cobalt complexes based on density functional theory has been carried out. The dissociation of CO from [HCo(CO)₄] gives [HCo(CO)₃], for which a butterfly structure was found to be the most stable. For [RCo(CO)₄] the R group prefers an axial rather than an equatorial position; CO insertion gives [RC(O)Co(CO)₃], which is stabilized if the acyl-O faces the vacant coordination site (η^2-acyl).[99] The course of the hydroformylation of *t*-butylethylene with [Rh₄(CO)₁₂] as the precursor under 50 bar CO and 1 bar H₂ was followed by *in situ* IR at 25 °C. A rhodium acyl complex, (Bu^{t}CH₂CH₂CORh(CO)₄], was identified; it is only stable at room temperature and at least 20 bar CO.[100] Ruthenium(III) EDTA complexes hydroformylate 1-hexene in aqueous media (ethanol:water 8:2) under 50 bar CO/H₂ 1:1 at 130 °C. The precursor K[Ru^{III}(EDTA-H)Cl] is converted into K[Ru^{III}(EDTA)(H₂O)], which reacts with CO under ligand substitution to give K[Ru^{III}(EDTA)(CO)]; this is reduced to H₂ to give the hydride K₂[Ru^{II}(H)(CO)-(EDTA)].[101] The hydroformylation of 1-octene to nonanal and 2-methyloctanal is catalyzed by [Ru₂(μ-O₂CR)₂(CO)₄L₂] [R = Me, Ph, CF₃, CMe₃, L = PPh₃, P(OPh)₃, P(OMe)₃]. The addition of NEt₃ or KOH and H₂O is necessary for the activation of the catalyst. A mechanism based on anionic carboxylate-bridged dinuclear ruthenium complexes has been suggested.[102] The hydroformylation of styrene is catalyzed by [Rh₄(CO)₁₂] to give mainly branched aldehyde. The amount of branched isomer decreases with increasing temperature and decreasing gas pressure, since these conditions facilitate the isomerization of the branched Rh alkyl intermediate $PhCH(Me)$—Rh to the *n*-alkyl $PhCH_2CH_2$—Rh. At room temperature, 98% of the rhodium alkyl is present as the branched isomer. The isomerization process was investigated by deuteroformylation studies. There is no isomerization at room temperature, but at higher temperature β-elimination from the branched alkyl rhodium species takes place more readily than from the linear alkyl intermediate, even at low CO/H_2 pressure.[103] Mixed-metal iron-cobalt clusters hydroformylate 1-pentene; FT-IR, cylindrical internal reflectance measurements, and HPLC analysis of the final reaction solutions suggest catalysis by $[Fe_2Co_2(CO)_{11}(\mu_4\text{-}PPh)_2]$.[104] Another mixed metal system, $[Co_2(CO_8)]/[Ru_3(CO)_{12}]$ at 80 atm CO/H_2 and 90 °C, hydroformylates norbornene to give mainly norbornylmethanol. Ruthenium alone has no hydroformylation activity in this system and is thought to act as a hydrogenation catalyst. The mechanism is thought to involve mainly mononuclear cobalt intermediates. $[Co_2(CO)_8]$ alone gives lactones.[105]

14.12.4. Water-Gas Shift Reactions

The mechanism of the water-gas shift catalyzed by $Pd(OAc)_2/PPh_3$ in 70% aqueous CF_3COOH has been investigated. ^1H, ^{13}C, and ^{31}P NMR suggest that $[PdH(PPh_3)_3]^+$ is an important intermediate. In the presence of ethylene, diethylketone is produced in a hydrocarbonylation reaction. In the absence of CO, C_2H_4 reacts with $[HPdL_3]^+$ to give the unstable complex $[L_3Pd—Et]^+$; both ethyl carbons rapidly exchange at 25 °C. If CO is present, $[L_2Pd—COEt]^+$ is formed; this can react with further ethylene under insertion to $[L_2PdCH_2CH_2COEt]^+$. The suggested mechanism of the water-gas shift is given in Scheme 30.[106] The kinetics of WGS catalyzed by $K[Ru^{II}(EDTA\text{-}H)(CO)]$ in aqueous solution at 20-80 °C under 1-35 bar CO have been determined. The CO consumption depends linearly on the solubility of CO in the solvent medium. The reaction is thought to involve the oxidative addition of H_2O to the $[Ru]^{II}$—CO intermediate to give $[Ru]^{IV}(H)$-$(OH)(CO)$; this is rapidly converted into $[Ru]^{IV}(H)(COOH)$ before decomposing into $[Ru]^{II}(H_2O)$, H_2, and CO_2 ($[Ru] = Ru(EDTA\text{-}H)^-$). Similar intermediates account for the formation of formaldehyde, which is also observed.[107] In a

Scheme 30

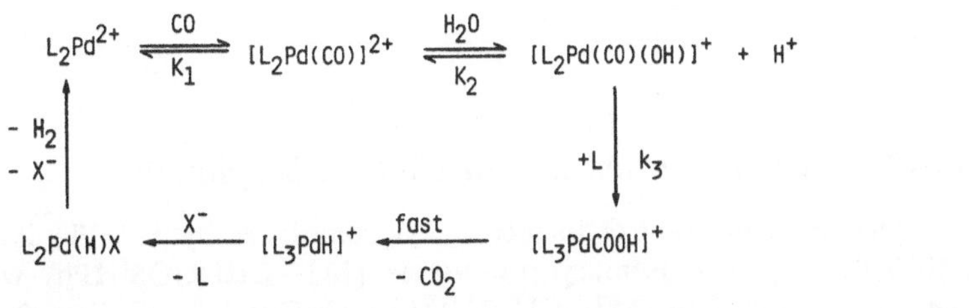

Scheme 31

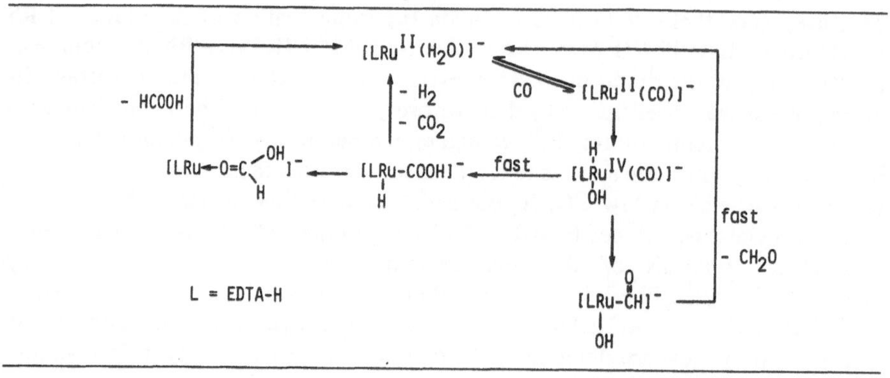

subsequent study formic acid was also detected; the reaction sequences are summarized in Scheme 31.[108] RhCl$_3$ in aqueous pyridine gives stable WGS catalysts which produce about 10^2 moles H$_2$ per Rh and day at 100 °C. Hydrogen formation is first order on [p(CO)] but complex order in [Rh]; *in situ* IR and ^{13}C NMR show the presence of mononuclear catalysts and less active dinuclear species. Of the two suggested mechanisms the one given in Scheme 32 is preferred.[109] The reduction of nitrobenzene to aniline under WGS conditions is catalyzed by ruthenium carbonyl clusters. In the absence of water, the reaction proceeds to PhNHCHO; this was found coordinated to a crystallographically characterized intermediate, [Ru$_3$(CO)$_{10}$(PhNHCHO)]$^-$, and is thought to derive from the reduction of a phenylisocyanate complex.[110]

Scheme 32

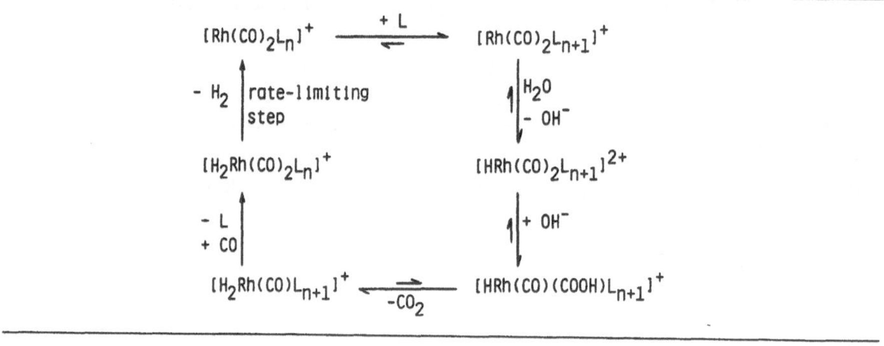

14.12.5. CO Reduction and Alcohol Carbonylation

The reduction of CO in metal acyl complexes by H$_2$SiPh$_2$ is catalyzed by [RhCl(PPh$_3$)$_3$]. The primary product is [M]—CH(R)OSiHPh$_2$ which is subsequently reduced to [M]—CH$_2$R {[M] = CpFe(CO)$_2$, CpFe(CO)(PPh$_3$)}. For

[M] = Mn(CO)$_5$ mixtures of ethylene, propene, 1-butene, and traces of alkanes are formed. The reaction is a model for the Pichler–Schulz mechanism of the CO reduction to hydrocarbons.[111] The hydrocarbonylation of triethylorthoformate is catalyzed by [Rh(CO)$_2$)$_2$I$_2$]$^-$ in the presence of MeI. The products are mainly EtOH and CH$_2$(OEt)$_2$, together with smaller amounts of carbonylation products [EtCOOH, HCOOEt, (EtO)$_2$CO etc.]. [Rh(CO)$_2$Cl]$_2$ in the absence of I$^-$ gives hydrogenation and alcohol hydroformylation products [EtCH(OEt)$_2$], while a halide-free catalyst [Rh$_4$(CO)$_{12}$] is able to hydrogenate but gives no hydroformylation or homologation products.[112] Secondary amines are carbonylated by K[Ru^{III}(EDTA-H)Cl]·2H$_2$O in aqueous media to a mixture of R$_2$NCHO (80%) and R$_2$NCONEt$_2$ (20%). Triethylamine gives exclusively tetraethylurea. The reaction is first order in [Ru], [amine], and dissolved [CO].[113] The carbonylation of BuNH$_2$ to BuNHCHO and (BuNH)$_2$CO is closely related. Reaction of [LRu(CO)]$^-$ with the amine is thought to give [LRu(CO)NH$_2$Bu]$^-$, which is converted in a rate-limiting step into [LRu(H)(CONHBu)]$^-$; in the presence of excess CO this reductively eliminates BuNHCHO and regenerates [LRu(CO)]$^-$ (L = EDTA-H).[114]

14.13. Cross-Coupling Reactions

The mechanism of the homocoupling of bromobenzene catalyzed by [NiCl$_2$(dppe)] under electroreductive conditions has been studied. At low bromobenzene concentration the oxidative addition step to a Ni(0)–dppe intermediate is rate-determining, while at high [PhBr] the reductive elimination from a Ni(III) intermediate, [Ph$_2$NiBr(dppe)], becomes rate-limiting (Scheme 33).[115]

Scheme 33

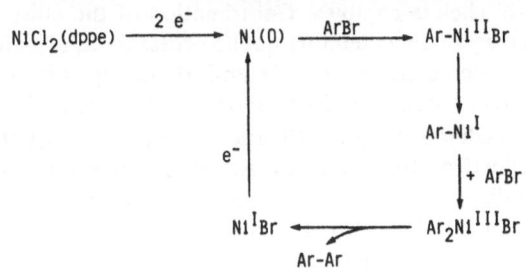

A series of transient intermediates in the cross-coupling of β-bromostyrenes with aryl Grignard reagents catalyzed by palladium have been observed by low-temperature NMR. The styrene first forms an unstable olefin complex (**A**) which undergoes oxidative addition at −60 °C to give [L$_2$Pd(Br)(CH=CHAr)] (**B**) [L$_2$ = (C$_5$H$_4$PPh$_2$)$_2$Fe; Ar = 4-C$_6$H$_4$OMe] (Scheme 34). Subsequent reaction with Ar'MgBr gives a palladium aryl complex (**C**) which is too unstable to be observed

Scheme 34

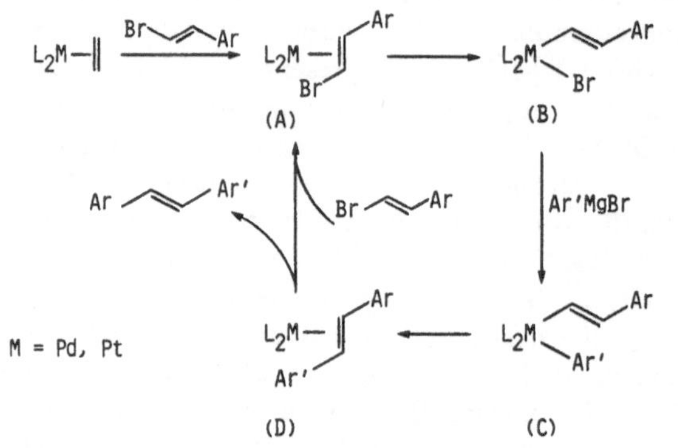

$$L_2M-\| \xrightarrow{Br\diagdown\diagup Ar} \underset{\substack{| \\ Br}}{L_2M-\|}^{Ar} \longrightarrow L_2M\diagdown_{Br}^{Ar}$$

(A)   (B)

$$Ar\diagup\diagdown Ar' \longleftarrow \underset{\substack{| \\ Ar'}}{L_2M-\|}^{Ar} \longleftarrow L_2M\diagdown_{Ar'}^{Ar}$$

M = Pd, Pt

(D)   (C)

and rapidly gives a complex of the coupling product (**D**).[116] This reaction sequence supports the common assumption of the general cross-coupling mechanism and provides information about the relative stabilities of the species involved in the cycle. If platinum is used, some of the intermediates can be isolated. For example, (**A**) (Scheme 34, M = Pt) is stable at room temperature. The transformation of (**A**) into (**B**) takes days in THF but only minutes in CH_2Cl_2, and is apparently catalyzed by electron transfer since it is also catalyzed by traces of $AgSO_3F$. The Pt aryl (**C**) is surprisingly unstable and reductively eliminates under mild conditions. Although platinum is generally not an active catalyst for coupling reactions, over 200 turnovers have been observed at room temperature.[117] Similar NMR investigations of the Pd-catalyzed coupling of $ArCH=CHBr$ with $PhCH_2MgBr$ in the presence of the chiral ferrocenyl phosphine $[Fe(C_5H_4PPh_2)(C_5H_3PPh_2CH(Me)NMe_2)]$ show that intermediates of type (**B**) are present as P—P chelates, but that alkylation to (**C**) gives a P—N chelate complex. Coordination of the chiral amino side-chain to the Grignard reagent may therefore not be necessary to explain the enantioselectivity induced by such chiral ligands. If $PhCH(Me)MgCl$ is used as the coupling partner, the product obtained at 25 °C shows up to 75% ee.[118]

While intermediates of type (**C**) are unstable, $[(allyl)Pd(Ar)PPh_3]$ is identifiable. The reductive elimination of $ArCH_2CH=CH_2$ is enhanced by allylic chloride, which displaces PPh_3 to give a π-complex; the kinetics of this reaction have been determined.[119] The individual steps of coupling reactions involving vinyl triflates have been studied using Pt complexes as models. The triflate forms a π-complex analogous to (**A**) (Scheme 34) in a slow rate-determining step; this rapidly rearranges into $[L_2Pt(OTf)(vinyl)]$. In the presence of free PPh_3 an ionic vinyl complex is formed, $[(vinyl)PtL_3]^+OTf^-$; this was crystallographically characterized.[120] The subsequent alkylation of $[L_2Pt(OTf)(vinyl)]$ with RLi (R = $Bu^tC\equiv C$) gives both *cis*- and *trans*-$[L_2Pt(R)(vinyl)]$. The *cis* isomer reductively eliminates at much lower temperature than the *trans* isomer. The addition of free PPh_3 inhibits the elimination step only in the case of the *trans* complex, which requires an isomerization step to give the coupling product; this isomerization

Scheme 35

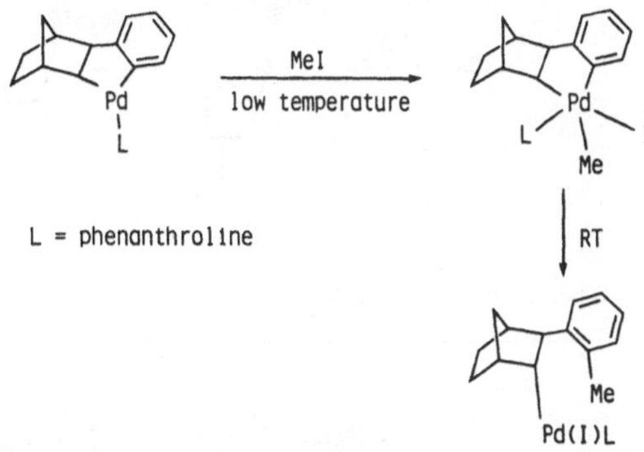

L = phenanthroline

goes via a dissociative pathway.[121] Palladium complexes in oxidation states greater than II have so far not been found in these coupling processes. However, MeI is able to oxidatively add to certain Pd(II) dialkyls to give Pd(IV) products, which are stable between -80 and -20 °C. At room temperature coupling products result (Scheme 35).[122] Benzyl bromides also add oxidatively to Me_2PdL (L = bipy, phen) to give the stable complex $[LPd^{IV}Me_2(CH_2Ar)Br]$, which was characterized by X-ray diffraction. The primary product of the oxidative step is a cationic intermediate $[(ArCH_2)PdMe_2(L)]^+$, which can be characterized by NMR at low temperature and gives the neutral halide complex at 20 °C. Reductive elimination under controlled conditions strongly favors the formation of a Pd(II) benzyl complex and C_2H_6.[123]

The cross-coupling of *n*-alkyl iodides and $NaBPh_4$ under 70 atm CO gives a mixture of products including ketones. Platinum catalysts are superior to Pd because their *n*-alkyl complexes β-eliminate less readily.[124] The reaction of $Cl(CF_2)_6I$ with NEt_3 gives the enamine $Cl(CF_2)_6-CH=CHNEt_2$, $Cl(CF_2)_6 H$, and $[Et_2NH]I$. The catalytic activity of $[M(PPh_3)_4]$ decreases for M = Ni > Pd > Pt. Perfluoroalkyl radicals are likely to be involved in the reaction.[125] Allylic alcohols can be arylated by ArI in the presence of Pd catalysts and NEt_3 as base. The amine also acts as a ligand for Pd. For a series of substituted aryl iodides a linear Hammett relationship was established. The oxidative addition of ArI to Pd(0) is the rate-limiting step. Electron-donating substituents enhance the rate, presumably because they stabilize the transition state by reducing its polarity.[126] Vinyl epoxides react with organostannanes in the presence of palladium catalysts to give mainly 1,4-addition products in high selectivity. The reaction proceeds in polar solvents (DMF, MeCN) via palladium π-allyl intermediates. Cyclic vinyl epoxides give, after hydrolysis, homoallylic alcohols of *trans* stereochemistry.[127] Cyclopentadiene monoepoxide reacts with silyl acetates or phenoxides to give similar 1,4-addition products; again π-allyl Pd intermediates are likely to be involved (Scheme 36).[128]

Substituted α-dienyl-ω-allyl acetates are cyclized by a $[Pd_2(dba)_3]/LiOAc/HOAc$ catalyst in acetonitrile (Scheme 37). The reaction is

Scheme 36

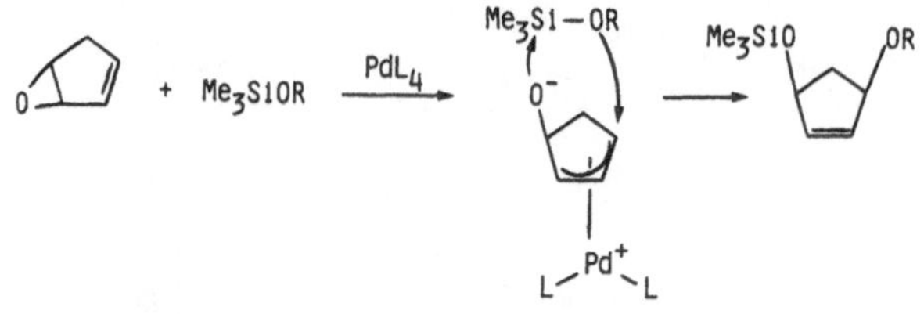

R = COMe, COPh, Ph, p-C$_6$H$_4$NO$_2$

Scheme 37

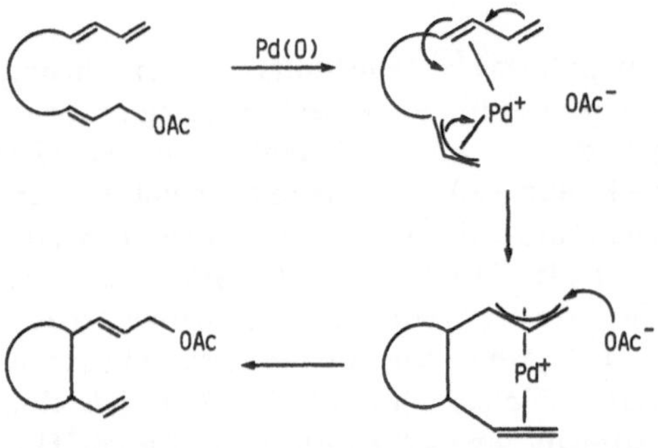

strongly solvent-dependent and is unsuccessful in THF, chloroform, benzene, and DMSO.[129] Other cyclizations utilize the Pd-catalyzed intramolecular aldol reaction of aldehydes with ketone enolates. The reaction is initiated by the oxidative addition of allylic esters to Pd(0) (Scheme 38).[130] Chiral ferrocenyl phosphines with polar

Scheme 38

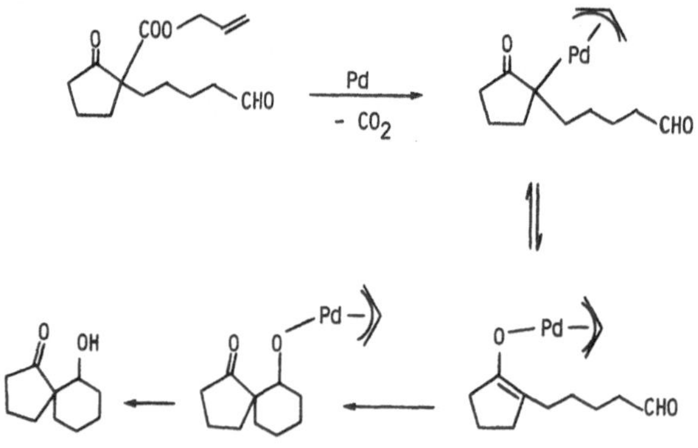

side-chains, e.g., carrying NR_2 and OH functions, direct the attack of amines on palladium π-allyl complexes. The interaction between allyl ligand and an OH substituent in such a complex was confirmed by an X-ray structure. In this way allylic carbonates are converted into allylic amines with very high stereospecificity.[131]

Finally, a different type of coupling reaction is achieved by molybdenum oxo complexes as catalysts. Aldehydes react with diazoacetate in the presence of PPh₃ to give α,β-unsaturated carboxylic esters and Ph₃P=O. Molybdenum carbene intermediates have been suggested (Scheme 39).[132]

Scheme 39

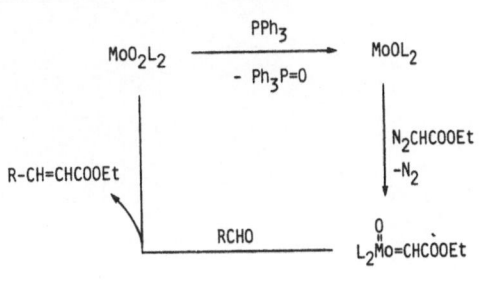

L = S₂C-NEt₂

Part IV

Compilations of Numerical Data

Chapter 15

Volumes of Activation for Inorganic and Organometallic Reactions: A Tabulated Compilation

15.1. Introduction

As in volume 6 of this series[1] this chapter is devoted to a compilation of activation and reaction volume data for inorganic and organometallic reactions: data published during the period July 1988 to December 1989 are included. The application of high-pressure techniques in mechanistic studies has received significant support from various groups. High-pressure equipment has become more readily available and the method is now applied to a wide variety of chemical processes. During the compilation of this contribution we encountered many interesting papers that deal with the application of high-pressure techniques in closely related areas. Reference to these are included in a general way since no direct activation volume data can be included in the tables.

During the past 18 months, a number of review articles dealing with the application of high-pressure techniques appeared.[2-6] In this respect it is important to refer to a number of papers concerning instrumental developments, including a compact, transportable unit and a cell suitable for *in situ* EXAFS spectroscopy.[7-11] Numerous contributions were presented at the AIRAPT/EHPRG meeting held at the University of Paderborn in 1989, for which the proceedings are presently in press as a special issue of a new journal entitled *High Pressure Research.* A number of very interesting papers in the general area of physical chemistry dealing with pressure dependence of self-diffusion,[12] combustion and diffusion flames,[13] heterogeneous catalysis,[14] modeling the effect of pressure on reaction rates,[15] thermodynamics of reactions at high pressure,[16] effect of pressure on stability constants[17] and emission spectra,[18] coordination and speciation changes[19] and corrosion[20] under pressure, and pressure effects on reversible electrode reactions[21] have appeared. In the general area of organometallic chemistry moderate pressures were applied to induce ring-slippage reactions,[22] to study hydroformylation reactions,[23] and to investigate catalytic processes using FITR spectroscopy.[24-26]

An area that has attracted much attention from investigators concerns the general application of pressure tuning spectroscopy. These include pressure-induced solvatochromism[27] studies on transition metal complexes and clusters,[28-40] pressure induced transitions in ferroelectric compounds,[41] and the study of electron-tunneling processes.[42] Another area that has received significant attention deals with bioinorganic and biochemical systems. For this reason a special section has been included in the tables. In addition, high pressure has been applied in structural studies of proteins and related molecules,[43-47] in kinetic studies of biochemical systems,[48,49] in marine microbiology,[50,51] and in enzyme reactions.[52-55] Finally, the tables include some data for electrochemical processes at elevated pressure.

15.2. Data in Tabular Form

The data in the following tables have been arranged as in the past.[1] A similar sequence of reaction types and ordering according to atomic number of the central metal atom have been adopted. In contrast to the previous tabulation, some reaction volume data are included for cases where no ΔV^+ values are available. This is mainly done to illustrate the magnitude of the overall volume changes that occur and that could be expected for a certain type of reaction, assisting the interpretation of ΔV^+ data for related systems. The methods employed to determine $\Delta \bar{V}$ are: (a) from the pressure dependence of the equilibrium constant; (b) from dilatometric or partial molar volume measurements; (c) from kinetic data. Other general remarks are: ΔV^+ data are quoted at ambient pressure—in case of significant curvature in the ln k versus P plots the compressibility coefficient $(\Delta \beta^+)$ is also given; the number of data refers to the different pressures at which kinetic measurements (usually more than one experiment) were performed, with the maximum applied pressure being quoted in the fourth column; concentration is given in mol liter^{-1} (M) or mol kg^{-1} (m).

Abbreviations used in the Table

AcO$^-$	acetate ion
[9]aneS$_3$	1,4,7-trithiacyclononane
ATA	*N*-acetyl-L-tryptophanamide
AT*p*CA	*N*-acetyl-L-tryptophan *p*-chloroanalide
B$_{12a}$	Cob(III)alamin

B_{12r}	Cob(II)alamin
bipy	2,2′-bipyridyl
BTE	*cis*-2,2,7,7-tetramethyl-3,6-dithiaoctane
Bu'	*tert*-butyl radical
CB	chlorobenzene
Cl-phen	2-chloro-1,10-phenanthroline
cp	cyclopentadienyl radical
cyclam	1,4,8,11-tetraazacyclotetradecane
CyDTA	cyclohexyldiaminetetraacetic acid
dab	biacetyl bis(phenylimine)
DCE	1,2-dichloroethane
DCM	dichloromethane
dien	1,5-diamino-3-azapentane
DMF	dimethylformamide
dmp	2,9-dimethyl-1,10-phenanthroline
DMSO	dimethylsulfoxide
dpp	2,9-diphenyl-1,10-phenanthroline
DTH	2,5-dithiahexane
dto	3,6-dithiaoctane
eddaH$_2$	ethylenediamine-N,N′-diacetic acid
edtaH$_4$	ethylenediaminetetraacetic acid
eee	1,8-diamino-3,6-dithiaoctane
ein	ethane-1,2-diimine, HN:CHCH:NH
en	1,2-diaminoethane
Et	ethyl radical
Et$_2$DTC	diethyldithiocarbamate
glyH	glycine
Hdcp^{2-}	2,6-dicarboxy-4-hydroxypyridine
hedtraH$_3$	N-(hydroxyethyl)-ethylenediaminetriacetic acid
HRP	Horseradish peroxidase
hxsb	1,8-bis((2-pyridylmethylene)amino)3,6-diazaoctane
In	Indicator
IS	inner sphere (electron transfer)
LF	ligand field
Ln	a lanthanide ion
M	3,7,11-tribenzyl-3,7,11,17-tetraazabicyclo[11·3·1]heptadeca-1(17),13,15-triene
Mb	myoglobin (Subscripts S, H, D refer to sperm whale, horse, and dog myoglobin, respectively)
MCLT	metal-to-ligand charge transfer
Me	methyl radical

Me$_2$bsb

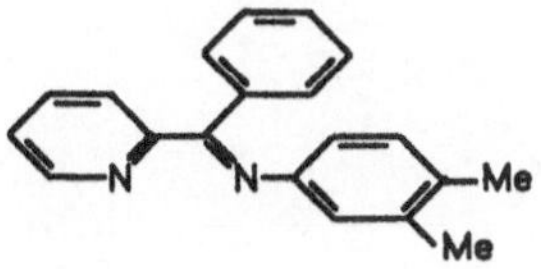

MeIm	1-methylimidazole
3-MP	3-methylpentane
N-eten	*N*-ethylethylenediamine
p-NPT	*p*-nitrophenyl trimethylacetate
OS	outer sphere (electron transfer)
Pada	*trans*-pyridine-2-azo(*p*-dimethylaniline)
Ph	phenyl radical
phen	1,10-phenanthroline
piv	pivalate ion
PMMA	polymethylmethacrylate polymer
Pr^i	*iso*-propyl radical
ptsa	p-toluenesulfonic acid
py	pyridine
Qui	quinoline
RB$_{12}$	alkylcobalamin
S	solvent
SA$_1$*p*NA	succinyl-L-alanine *p*-nitroanilide
SA$_2$*p*NA	succinyl-L-alanyl-L-alanine *p*-nitroanilide
SA$_3$*p*NA	succinyl-L-alanyl-L-alanyl-L-alanine *p*-nitroanilide
sarH	sarcosine
TCE	*sym*-tetrachloroethane
tet	1,9-diamine-3,7-diazanonane
TH	CF_3SO_3H (triflic acid)
tmd	1,3-propanediamine
2,2,4-TMP	2,2,4-trimethylpentane
TMS	tetramethylsilane
TPP	5,10,15,20-tetraphenylporphyrin
trien	1,8-diamino-3,6-diazaoctane

Table 15.1. Volumes of Activation

Reaction	Solvent	T (°C)	P (kbar)	No. of data	$\Delta V^{\ddagger}$ (cm³ mol⁻¹)	$\Delta\beta^{\ddagger}$ (cm³ mol⁻¹ kbar⁻¹)	$\Delta\bar{V}$ (method) (cm³ mol⁻¹)	Ref.	Remarks
Solvent exchange									
$[Mn(HOAc)_6]^{2+}$	HOAc	25	1.75	>12	+0.4 ± 0.7			56	{2.12 × 10⁻² M HClO₄, Diluent d_2-DCM
$[Mn(OAc)_2(HOAc)_4]$	HOAc	25	1.75	>12	+6.7 ± 0.6				0 M HClO₄
$[Co(CN)_5(^{18}OH_2)]^{2-}$	$H_2^{16}O$	17.9	4.0	4	+7.1 ± 0.4	+1.0 ± 0.2		57	pH = 3, μ = 0.35 M (CF₃SO₃H)
$[NiM(DMF)]^{2+}$	DMF	50	2.0	>10	+10.6 ± 0.8			58	¹H NMR
$[NiM(MeCN)]^{2+}$	MeCN	50	2.0	>10	−3.5 ± 0.9			59	¹⁴N NMR
$[Ni(MeCN)_6]^{2+}$	MeCN	59	>2.0	>12	+12.0 ± 0.4	+4.0 ± 0.3			
$[Ni(EtCN)_6]^{2+}$	EtCN	57	>2.0	>12	+13.7 ± 0.5	+2.3 ± 0.2			
$[Ni(PrCN)_6]^{2+}$	PrCN	56	>2.0	>12	+13.1 ± 0.5	+4.1 ± 0.4			
$[Ni(Pr^iCN)_6]^{2+}$	Pr^iCN	55	>2.0	>12	+12.4 ± 0.6	+3.8 ± 0.6			
$[Ni(BuCN)_6]^{2+}$	BuCN	56	>2.0	>12	+14.4 ± 0.4	+5.0 ± 0.4			
$[Ni(PhCN)_6]^{2+}$	PhCN	59	>2.0	>12	+13.1 ± 0.6	+6.3 ± 0.6			
$[Pd(R_3dien)H_2O]^{2+}$ R = H	H₂O	23	2.0	7	−2.8 ± 0.4			60	{¹⁷O NMR, Mn²⁺ as relaxation agent
Me		25			−7.2 ± 0.6			61	{¹⁷O NMR, Mn²⁺ as relaxation agent
Et			2.5	~10	−7.7 ± 1.3			62	{¹⁷O NMR, 2m HClO₄
$[Ln(H_2O)_8]^{3+} + H_2^{*}O \rightarrow$	H₂O								
$[Ln(H_2O)_7(H_2^{*}O)]^{3+} + H_2^{*}O$ (Ln = Tb³⁺)		−4.2	2.5	~10	−5.7 ± 0.5	+0.3 ± 1.6			
(Ln = Dy³⁺)		−5.8			−6.0 ± 0.4	−0.4 ± 1.4			
(Ln = Ho³⁺)		−5.5			−6.6 ± 0.4	−0.6 ± 1.3			
(Ln = Er³⁺)		−5.5			−6.9 ± 0.4	+0.3 ± 1.2			
(Ln = Tm³⁺)		−4.2			−6.0 ± 0.8	+1.2 ± 3.0			
Ligand exchange									
$[Pt(DMSO)_2(DMSO)_2]^{2+} + DMSO$	CD₃NO₂	−8.8	1.8	5	−2.5 ± 0.3			63	O-bonded exchange
		87	2.0		−5 ± 3				S-bonded exchange (estimated)
cis-$[PtR_2L_2] + 2^{*}L \rightleftharpoons$ cis-$[PtR_2^{*}L_2] + 2L$								64	¹H NMR
R = Ph, L = Me₂S	C₆H₆	69,75		~8	+4.7 ± 0.5				
DMSO	CDCl₃	58		9	+5.5 ± 0.8				
R = Me, L = DMSO	C₆H₆	60		11	+4.9 ± 0.5				

(continued)

Table 15.1. (continued)

Reaction	Solvent	T (°C)	P (kbar)	No. of data	$\Delta V^{\ddagger}$ (cm³ mol⁻¹)	$\Delta\beta^{\ddagger}$ (cm³ mol⁻¹ kbar⁻¹)	$\Delta\bar{V}$ (method) (cm³ mol⁻¹)	Ref.	Remarks
Ligand substitution (including substitutions accompanied by redox changes)									
[Cr^{III}(NH₃)₅X]³⁺ + H₂O → [Cr(NH₃)₅(H₂O)]³⁺ + X	H₂O						~10(b)	65	μ = 0.011 M HClO₄
X = DMSO									
HCONH₂		45.0	1.0	3	−3.2 ± 0.1				
OC(NH₂)₂		45.0	1.5	4	−4.8 ± 0.3				
OC(NHMe)₂		50.9	1.38	5	−8.2 ± 0.5				
MeCONMe₂		48.7	1.725	4	−3.8 ± 0.2				
DMF		45.0	1.0	3	−6.2 ± 0.4				
(MeO)₃PO		45.0	1.0	3	−7.4 ± 0.1				
[Cr(NH₃)₅(Cl)]²⁺ + H₂O →	H₂O	45.0	1.0	3	−8.7 ± 0.1			66	[OH⁻] = 0.2 M
[Cr(NH₃)₅(H₂O)]³⁺ + Cl⁻		33.5	1.0	5	+17.0 ± 0.9				μ = 1.0 M (NaClO₄)
[Cr(NH₂CH₃)₅(Cl)]²⁺ + H₂O →	H₂O	25.0	1.0	5	+34.8 ± 1.7				[OH⁻] = 0.5 M
[Cr(NH₂CH₃)₅(H₂O)]³⁺ + Cl⁻									μ = 1.0 M (NaClO₄)
[Cr(NH₃)₅(I)]²⁺ + H₂O →	H₂O	17.2	1.0	5	+22.2 ± 0.6				[OH⁻] = 0.1 M
[Cr(NH₃)₅(H₂O)]³⁺ + I⁻									μ = 1.0 M (NaClO₄)
[Cr(NH₃)₅(OC(CH₃)N(CH₃)₂)]³⁺ + H₂O →	H₂O	29.0	1.0	5	+25.0 ± 0.7				[OH⁻] = 0.1 M
[Cr(NH₃)₅(H₂O)]³⁺ + CH₃CON(CH₃)₂									μ = 1.0 M (NaClO₄)
[Cr^{III}(TPP)(Cl)(L)] $\underset{k_2}{\overset{k_1}{\rightleftharpoons}}$ [Cr^{III}(TPP)(Cl)] + L	toluene	25.0	1.00	5				67	
$\xrightarrow{+\text{MeIm}}_{k_3}$ [Cr^{III}(TPP)(Cl)(MeIm)]									
L = py k_1					+25.7 ± 0.5				
$k_{2/3}$					+1.3 ± 1.1				
Qui k_1					+23.8 ± 0.6				
$k_{2/3}$					−1.9 ± 1.6				
PPh₃ k_1					+19.6 ± 0.2				
$k_{2/3}$					+2.0 ± 0.6				
trans-[Cr(tmd)₂FX]⁺ + H₂O → trans-[Cr(tmd)₂(OH₂)F]²⁺ + X⁻								68	acidic solution
X = F	H₂O	35–50	1.5	4	−3.4 to −2.3				
Cl	H₂O	20–35	1.5	4	−8.8 to −7.7				
[Cr(CO)₄(DTH)] + 2 L → [Cr(CO)₄L₂] + DTH								69	
L = P(OMe)₃	DCE	60.0	1.00	5	+10.1 ± 0.8				
	CB	50.0	1.00	5	+8.8 ± 0.4				
P(OPr^i)₃	DCE	60.0	1.25	5	+10.4 ± 0.4				
P(OPh)₃	DCE	55.5	1.00	5	+9.3 ± 0.5				

Reaction	Solvent	t (°C)	P	n	$\Delta V^{\ddagger}$	$\Delta V^{\ddagger}$(b)	Conditions	Ref.
$[Cr(NH_3)_5(OSO_2CF_3)]^{2+} + S \rightarrow$ $[Cr(NH_3)_5(S)]^{3+} + CF_3OSO_2^-$	CH₃CN	25.2	1.50	4	-8.9 ± 0.5		ionic strength not adjusted	70
	MeOH	20.0			-10.5 ± 0.7			
$[Mn(CO)_5Br] + bipy \rightarrow [Mn(CO)_3(bipy)] + 2\,CO$	MeOH	25.5	1.00	3	$+22$			71
	50% MeOH/H₂O				-20			
$[Mn(CO)_3Cl] + dab \rightarrow [Mn(CO)_3Cl(dab)] + 2\,CO$	toluene	30.0	1.5	4	$+20.6 \pm 0.4$			72
$[Mn(CO)_3Cl] + [9]aneS_3 \rightarrow$ $([9]aneS_3)Mn(CO)_3]Cl + 2\,CO$	acetone	29.5	1.4	5	$+20.6 \pm 2.6$			73
$[Fe^{III}(CN)_5(NO_2)]^{3-} + Solvent \rightarrow$ $[Fe(CN)_5(S)]^{2-} + NO_2^-$	H₂O	25.0	1.0	5	$+2.2 \pm 0.1$ $+1.9 \pm 0.1$		$\mu = 0.5\,M$ (LiClO₄) $[H^+] = 0.2\,M$ (HCl) $[H^+] = 0.5\,M$ (HCl)	74
	DMF	25.0	1.0	5	26.9 ± 1.5		0.5% H₂O	75
	DMSO	25.0	0.75	4	$+25.9 \pm 1.1$		0.5% H₂O	
	MeOH	45.0	1.0	5	$+19.6 \pm 1.8$			
$[Fe^{II}(CN)_5NO_2]^{4-} + H_2O \rightarrow$ $[Fe^{II}(CN)_5(H_2O)]^{3-} + NO_2^-$	H₂O	25.0	1.25	5	$+20.1 \pm 1.0$		$[OH^-] = 0.1\,M$	74
$[Fe(5\text{-}Br\text{-}phen)_3]^{2+} + 2H_2O \rightarrow$ $5\text{-}Br\text{-}phen + [Fe(5\text{-}Br\text{-}phen)_2(H_2O)_2]^{2+}$	H₂O	25–45	1.0	7	$+23.01$	-9.9(b)	$[edtaH_2Na_2] = 0.01\,M$ as Fe²⁺-scavenger; isochoric condition	76
$trans\text{-}[Co(NH_3)_4Cl_2]^+ + H_2O \rightarrow$ $45\%t,55\%c\text{-}[Co(NH_3)_4(H_2O)Cl]^{2+} + Cl^-$	H₂O	16.0	2.0	5	$+1.8$	*(trans)* -12.0(b) *(cis)* -14.0	1 mM HClO₄ for kinetics 7.6 mM HNO₃ for dilatometry	77
$trans\text{-}[Co(en)_2Cl_2]^+ + H_2O \rightarrow$ $74\%t,26\%c\text{-}[Co(en)_2(H_2O)Cl]^{2+} + Cl^-$	H₂O	40.3	2.0	5	-0.3	-9.1(b)	1.1 mM HClO₄ for kinetics 0.6 mM HNO₃ for dilatometry	77
$cis\text{-}[Co(en)_2(H_2O)Cl]^{2+} + Cl^-$	H₂O	30.1	2.0	6	$+1.1$	-18.6(b)	1 mM HClO₄	
$trans\text{-}[Co(trien)Cl_2]^+ + H_2O \rightarrow$	H₂O	10.5			-1.9	-8.8(b)	10 mM HClO₄	
$cis\text{-}\beta\text{-}[Co(trien)(H_2O)Cl]^{2+} + H_2O \rightarrow$	H₂O	31.6			-0.1	-10.0(b)	1 mM HClO₄	
$cis\text{-}\alpha\text{-}[Co(trien)Cl_2]^+ + H_2O \rightarrow$	H₂O	15.0			$+3.2$	$+1.9 \pm 0.8$(b)	9.6 mM HClO₄	
$cis\text{-}\beta\text{-}[Co(trien)Cl_2]^+ + H_2O \rightarrow$	H₂O	30.6			$+0.8$	-7.4(b)	1 mM HClO₄	
$cis\text{-}\alpha\text{-}[Co(trien)(H_2O)Cl]^{2+} + Cl^-$	H₂O	47.1			$+1.0$		10 mM HClO₄	
$cis\text{-}\beta\text{-}[Co(trien)(H_2O)Cl]^{2+} + Cl^-$								
$cis\text{-}\alpha\text{-}[Co(edda)Cl_2]^- + H_2O \rightarrow$								
$cis\text{-}\alpha\text{-}[Co(edda)(H_2O)Cl] + H_2O \rightarrow$								
$cis\text{-}\alpha\text{-}[Co(edda)(H_2O)Cl] + Cl^-$								
$trans\text{-}[Co(en)_2Br_2]^+ + H_2O \rightarrow$ $85\%t,15\%c\text{-}[Co(en)_2(H_2O)Br]^{2+} + Cl^-$	H₂O	25.3	2.0	5				78
$[Co(NH_3)_5X]^{2+} + H_2O \rightarrow$ $[Co(NH_3)_5(H_2O)]^{2+} + X^-$							$[100$ mM HClO₄ $\{\Delta V^* $ at 25°C (Ref.):	78
$X = Cl^-$		65			-6.0		-7.9 cm³ mol⁻¹ (43,44)	
Br^-		54			-6.4		-6.4 cm³ mol⁻¹ (43,44)	
NO_3^-		40			-5.7		-4.9 cm³ mol⁻¹ (43,44)	

(continued)

Table 15.1. (continued)

Reaction	Solvent	T (°C)	P (kbar)	No. of data	$\Delta V^{\ddagger}$ (cm³ mol⁻¹)	$\Delta \beta^{\ddagger}$ (cm³ mol⁻¹ kbar⁻¹)	$\Delta \bar{V}$ (method) (cm³ mol⁻¹)	Ref.	Remarks
[Co(en)₂(NH₃)X]²⁺ + H₂O →			2.0	5				78	
[Co(en)₂(NH₃)(H₂O)]²⁺ + X⁻									
X⁻ = trans-Cl⁻		70			−5.0		−17.3(b)		100 mM HClO₄
trans-Br⁻		60			−3.1		−14.7(b)		1 mM HClO₄
cis-Br⁻		60			−5.6		−16.1(b)		100 mM HClO₄
cis-NO₃⁻		45			−6.1		−13.2(b)		100 mM HClO₄
cis-[Co(en)₂(OH₂)₂]³⁺ + XO₂²⁻/HXO⁻ →	H₂O	55.0	1.38	3				79	pH = 2.0,
H₂O + cis-[Co(en)₂(OH₂)(XO₂)]⁺									μ = 1.0 M (NaClO₄)
X = S					+8.3 ± 0.5				
Se					+8.5 ± 0.4				
cis-[Co(en)₂(OH₂)(XO₄)]⁺ + H₂O →	H₂O	55.0	1.38	3				79	pH = 2.0,
XO₄²⁻ + cis-[Co(en)₂(OH₂)₂]³⁺									μ = 1.0 M (NaClO₄)
X = S					−2.2 ± 0.4				
Se					−3.3 ± 0.7				
Cobalt(III) amine base hydrolysis:	H₂O		1	5				80	
[Co(NH₃)₄OC(Me)N(Me)₂]³⁺ + OH⁻ →		15.4			+43.2 ± 1.7				[OH⁻] = 25 mM
[Co(NH₃)₅OH]²⁺ + MeCONMe₂									μ = 34 mM
[Co(NH₂Me)₅Cl]²⁺ + OH⁻ →		9.5			+32.8 ± 1.7				[OH⁻] = 10 mM
[Co(NH₂Me)₅OH⁻]²⁺ + Cl⁻									μ = 13 mM
[Co(NH₂Et)₅Cl]²⁺ + OH⁻ →		12.2			+31.1 ± 0.5				[OH⁻] = 1 mM
[Co(NH₂Et)₅OH⁻]²⁺ + Cl⁻									μ = 3 mM
cis-[Co(en)₂(NH₃)Cl]²⁺ + OH⁻ →		25			+31.8 ± 0.6				[OH⁻] = 50 mM
78% cis, 22% trans-[Co(en)₂(NH₃)OH]²⁺ + Cl⁻									μ = 56 mM
cis-[Co(en)₂(NH₃)Br]²⁺ + OH⁻ →		25			+30.8 ± 1.0				[OH⁻] = 50 mM
77% cis, 23% trans-[Co(en)₂(NH₃)OH]²⁺ + Br⁻									μ = 56 mM
trans-[Co(en)₂Cl₂]⁺ + OH⁻ →		14.5			+24.3 ± 1.1				[OH⁻] = 15 mM
5% cis, 95% trans-[Co(en)₂(Cl)OH]⁺ + Cl⁻									μ = 16 mM
cis-[Co(en)₂Cl₂]⁺ + OH⁻ →		14.5			+27.9 ± 0.7				[OH⁻] = 50 mM
37% cis, 63% trans-[Co(en)₂(Cl)OH]⁺ + Cl⁻									μ = 51 mM
trans-[Co(en)₂(N₃)Cl]⁺ + OH⁻ →		15.4			+26.7 ± 0.4				[OH⁻] = 50 mM
23% cis, 77% trans-[Co(en)₂(N₃)OH]⁺ + Cl⁻									μ = 50 mM
cis-β-[Co(trien)Cl₂]⁺ + OH⁻ →		3.7			+35.7 ± 1.2				[OH⁻] = 0.005 mM
cis-[Co(trien)(Cl)OH]⁺ + Cl⁻									μ = 100 mM
trans-[Co(RSSR-cyclam)Cl₂]⁺ + OH⁻ →		9.5			+18.7 ± 1.5				[OH⁻] = 1 mM
Cl⁻ + trans-[Co(RSSR-cyclam)(Cl)OH]⁺									μ = 1 mM
cis-[Co(en)₂(NO₂)Cl]⁺ + OH⁻ →		24.4			+20.8 ± 0.3				[OH⁻] = 50 mM
cis-[Co(en)₂(NO₂)OH]⁺ + Cl⁻									μ = 1 mM
cis-[Co(tet)Cl₂]⁺ + OH⁻ →		12.2			+23.1 ± 0.8				[OH⁻] = 1 mM
cis-[Co(tet)(Cl)OH]⁺									μ = 3 mM

Reaction	Solvent	T (°C)	μ	n	$\Delta V^{\ddagger}$	Ref.	Conditions
cis-[Co(tet)(Cl)OH]$^+$ + OH$^-$ →		12.2			+25.9 ± 2.1		[OH$^-$] = 1 mM μ = 3 mM
cis-[Co(tet)(OH)$_2$]$^+$							
trans-RS-[Co(tet)Cl$_2$]$^+$ + OH$^-$ →		12.2			+22.1 ± 0.7	80	[OH$^-$] = 1 mM μ = 3 mM
Cl$^-$ + trans-RS-[Co(tet)(Cl)OH]$^+$		4.7			+20.4 ± 0.7		[OH$^-$] = 2 mM μ = 6 mM
trans-RR(SS)-[Co(tet)Cl$_2$]$^+$ + OH$^-$ →		5.7			+24.4 ± 1.2		[OH$^-$] = 2 mM μ = 6 mM
Cl$^-$ + trans-RS-[Co(tet)(Cl)OH]$^+$							
trans-RS-[Co(tet)(Cl)OH]$^+$ + OH$^-$ →		12.2			+23.8 ± 0.9		[OH$^-$] = 1 mM μ = 3 mM
Cl$^-$ + trans-RS-[Co(tet)(OH)$_2$]$^+$							
[Co(NH$_3$)$_5$(OSO$_2$CF$_3$)]$^{2+}$ + CH$_3$CN →	CH$_3$CN	25.2	1.5	4	−3.1 ± 0.1	70	ionic strength not adjusted
[Co(NH$_3$)$_5$(CH$_3$CN)]$^{3+}$ + CF$_3$OSO$_2^-$	MeOH	17.4			−3.2 ± 0.1		
trans-[Co(N-eten)$_2$Cl$_2$]$^+$ + H$_2$O →	H$_2$O/Bu^{t}OH w/w% ROH:	25	1.5	4		81	0.01 M HCl, μ = 0.5 M (NaCl)
trans-[Co(N-eten)$_2$Cl(H$_2$O)]$^{2+}$ + Cl$^-$	0				+5.83		
	5				+5.61		
	10				+5.61		
	20				+5.00		
	30				+3.95		
	40				+2.55		
s-cis-[Co(eee)X$_2$]$^+$ + H$_2$O →	H$_2$O	40.0	1.5	4		82	0.1 M HClO$_4$
X$^-$ + s-cis-[Co(eee)(H$_2$O)X]$^{2+}$							
X = Cl$^-$					−4.6		
Br$^-$					−4.2		
[Mo(DTH)(CO)$_4$] + 2 L → [Mo(CO)$_4$L$_2$] + DTH						69	
L = P(OMe)$_3$	DCE	40.0	1.0	5	−11.3 ± 0.5		
P(OPr^i)$_3$	DCE	50.0	1.0	5	−10.2 ± 0.8		
P(OPh)$_3$	DCE	40.0	1.0	5	−9.3 ± 0.4		
[Mo(dto)(CO)$_4$] + 2 L → [Mo(CO)$_4$L$_2$] + dto							
L = P(OPr^i)$_3$	DCE	46.0	1.50	4	−9.4 ± 0.2		
[Mo(BTE)(CO)$_4$] + 2 L → [Mo(CO)$_4$L$_2$] + BTE							
L = P(OPr^i)$_3$	DCE	56.0	1.50	4	+3.9 ± 0.3		
[Ru^{III}(edta)(H$_2$O)]$^-$ + L^{n-} →	H$_2$O	25	1.0	6		83	[HOAc/AcO$^-$ buffer] = 0.1 M μ = 0.2 M (Na$_2$SO$_4$)
[Ru^{III}(edta)(L)]$^{(1+n)-}$ + H$_2$O							
L = SC(NH$_2$)$_2$					−6.8 ± 0.6		pH = 5
SC(NHMe)$_2$					−8.8 ± 0.2		pH = 5
SC(NMe$_2$)$_2$					−12.2 ± 0.5		pH = 5
N$_3^-$					−9.0 ± 0.6		pH = 5
					−9.9 ± 0.5		pH = 6
SCN$^-$					−9.6 ± 0.3		pH = 5

(continued)

Table 15.1. (continued)

Reaction	Solvent	T (°C)	P (kbar)	No. of data	$\Delta V^{\ddagger}$ (cm³ mol⁻¹)	$\Delta\beta^{\ddagger}$ (cm³ mol⁻¹ kbar⁻¹)	$\Delta\bar{V}$ (method) (cm³ mol⁻¹)	Ref.	Remarks
$[\text{Ru(hedtra)H}_2\text{O}]^- + \text{L}^{n-} \rightarrow$ $[\text{Ru(hedtra)L}]^{(1+n)-} + \text{H}_2\text{O}$	H_2O	25	100	5				84	$\mu = 0.2\ M$ (Na_2SO_4)
L = SCN⁻					-7.3 ± 0.6				pH = 3 (0.1 M HOAc/OAc⁻ buffer)
					-10.8 ± 0.7				pH = 8.3 (0.025 M HCl/borate buffer)
N_3^-					-14.2 ± 1.2				pH = 8.3 (0.025 M HCl/borate buffer)
					-4.1 ± 0.7				pH = 3 (0.1 M HOAc/OAc⁻ buffer)
$\text{SC(NH}_2)_2$					-7.1 ± 0.5				pH = 8.3 (0.025 M HCl/borate buffer)
SC(NHMe)_2					-6.2 ± 0.5				pH = 3 (0.1 M HOAc/OAc⁻ buffer)
$\text{SC(NMe}_2)_2$					-10.4 ± 1.0				pH = 3 (0.1 M HOAc/OAc⁻ buffer)
$[\text{H}_3\text{Ru}_3(\text{COMe})(\text{CO})_9] + \text{CO} \underset{k_r}{\overset{k_f}{\rightleftharpoons}}$ $[\text{HRu}_3(\text{COMe})(\text{CO})_{10}] + \text{H}_2$	toluene		2.0				$+11$(c)	85	
k_f		70		4	$+20 \pm 2$				
k_r		60		5	$+9.6 \pm 0.6$				
$[\text{Rh(NH}_3)_5(\text{OSO}_2\text{CF}_3)]^{2+} + \text{S} \rightarrow$ $[\text{Rh(NH}_3)_5(\text{S})]^{3+} + \text{CF}_3\text{OSO}_2^-$	CH_3CN	34.6	1.50	4	-7.8 ± 0.5			70	ionic strength not adjusted
	MeOH	25.0			-6.6 ± 0.1				
$[\text{Pd(R}_3\text{dien)H}_2\text{O}]^{2+} + \text{L} \rightarrow [\text{Pd(R}_3\text{dien)L}]^{2+} + \text{H}_2\text{O}$	H_2O							61	$\mu = 0.1\ M$ (NaClO_4) pH ≤ 5
R = Me L = $\text{SC(NH}_2)_2$		15.0	0.95	5	-9.3 ± 0.4				
SC(NHMe)_2		25.0	0.95	5	-9.1 ± 0.6				
$\text{SC(NMe}_2)_2$		25.0	0.97	5	-13.4 ± 0.7				
R = Et L = $\text{SC(NH}_2)_2$		15.0	1.0	6	-8.3 ± 0.3				
SC(NHMe)_2		25.0	1.0	6	-10.2 ± 0.6				
$\text{SC(NMe}_2)_2$		25.0	1.0	6	-12.7 ± 0.6				
$[\text{Pd(R}_3\text{dien)X}]^{(2-n)+} + \text{H}_2\text{O} \underset{k_r}{\overset{k_f}{\rightleftharpoons}}$ $[\text{Pd(R}_3\text{dien)H}_2\text{O}]^{2+} + \text{X}^{n-}$	H_2O	25.0	1.0	5				86	$\begin{cases} \mu = 0.1\ (\text{NaClO}_4) \\ [\text{OH}^-] = 0.01 \end{cases}$
R = Me X = Br⁻ k_f					-11.4 ± 0.7		-4.1 ± 1.1(c)		
I⁻ k_r					-9.7 ± 0.3		$+0.2 \pm 1.5$(c)		
N_3^- k_f					-11.7 ± 0.7		-1.3 ± 1.0(c)		
k_r					-10.1 ± 0.3				
$\text{C}_2\text{O}_4^{2-}$ k_f					-10.4 ± 0.6		-4.5 ± 1.0(c)		(pH ≈ 5)
k_r					-5.9 ± 0.4				(pH ≈ 5)

Reaction		Solvent	T/°C			$\Delta V^{\ddagger}$		Ref	Conditions
R = Et X = Cl⁻	k_r		34			−5.9 ± 0.3	−4.9 ± 0.8(a)		
Br⁻	k_t					−11.6 ± 0.5	−5.2 ± 0.9(c)		
	k_t		34			−5.8 ± 0.4	+0.2 ± 1.5(c)		
I⁻	k_t		34			−6.8 ± 0.9	−7.0 ± 0.6		
N₃⁻	k_r					−11.3 ± 0.4	−0.3 ± 0.8(c)		
	k_t					−11.4 ± 0.4			
C₂O₄²⁻	k_t		25			−7.6 ± 0.8	−3.7 ± 1.0(c)		
	k_r					−3.9 ± 0.2			
$[Pd(Me_4dien)py]^{2+} + S \rightarrow [Pd(Me_4dien)S]^{2+} + py$ (S = solvent)									
		H₂O	25.0	1.5	7	−3.1 ± 0.1		87	pH > 9: [OH⁻] = 0.01 M
		MeOH	25.0	1.5	7	−6 ± 1			0.1 M ptsa
		EtOH	25.0	1.0	5	−4.1			0.1 M ptsa
		DMSO	30.0	0.8	5	≈0			0.1 M ptsa
		DMF	30.0	1.0	5	−3.4 ± 0.3			0.1 M ptsa
		MeCN	30.0	1.0	5	−2.4 ± 0.6			0.1 M ptsa

Isomerization and fluxionality

Reaction	Solvent	T/°C			$\Delta V^{\ddagger}$	Ref	Conditions
$[Cr(NH_3)_5(OCHNH_2)]^{3+} \rightarrow$ $[Cr(NH_3)_5(NHCHO)]^{2+} + H^+$	H₂O	25.0	1.0	5	−7.6 ± 0.8	66	[OH⁻] = 0.1 M {μ = 0.1 M (NaClO₄)
$[Cr(NH_3)_5(OC(NH_2)_2]^{3+} \rightarrow$ $[Cr(NH_3)_5(NHCONH_2)]^{2+} + H^+$	H₂O	23.5	1.0	5	−9.8 ± 0.9		[OH⁻] = 1.0 M {μ = 1.0 M (NaClO₄)
$[Fe(CO)_2(1,3\text{-cyclooctadiene})(PPh_3)]$ fluxionality	toluene	−73	2	9	+0.5 ± 0.2	88	¹H NMR
(MeO/OMe–Ph₃P–Fe structure) fluxionality	CD₂Cl₂	−33	2	9	+5.3 ± 0.3	88	¹H NMR
$trans\text{-}[Co(en)_2(OH_2)(XO_4)]^+ \rightleftharpoons$ $cis\text{-}[Co(en)_2(OH_2)(XO_4)]^+$	H₂O	41.0	1.38	5	+7.2 ± 0.4	79	pH = 1.0
$ttt\text{-}[RuCl_2(CO)_2(Ph_2MeP)_2] \rightarrow$ $ccc\text{-}[RuCl_2(CO)_2(Ph_2MeP)_2]$	CHCl₃	63.2	1.36		+19 ± 2	89	
$ttt\text{-}[RuCl_2(CO)_2(PhMe_2P)_2] \rightarrow$ $ccc\text{-}[RuCl_2(CO)_2(PhMe_2P)_2]$	CB	70.4	1.36		+19 ± 2		
$t\text{-}[RuCl_2(CO)(Ph_2MeP)_3] \rightarrow$ $c\text{-}[RuCl_2(CO)(Ph_2MeP)_3]$	HCCl₃	34.0	1.36		+15 ± 2		
$t\text{-}[RuCl_2(CO)(PhMe_2P)_3] \rightarrow$ $c\text{-}[RuCl_2(CO)(PhMe_2P)_3]$	TCE	64.3	1.36		+16 ± 2		
$[Re_2(piv)X_2]$ X = Cl, Br	crystalline or PMMA matrix		75	>10	>10	90	

(continued)

Table 15.1. (continued)

Reaction	Solvent		T (°C)	P (kbar)	No. of data	$\Delta V^{\ddagger}$ (cm³ mol⁻¹)	$\Delta \beta^{\ddagger}$ (cm³ mol⁻¹ kbar⁻¹)	$\Delta \bar{V}$ (method) (cm³ mol⁻¹)	Ref.	Remarks
Electron transfer										
[Mn(CNC(CH₃)₃)₆]⁺/²⁺										
OS self-exchange	CH₃CN		6.0	2.0	10	−12 ± 1	−4 ± 1		91	{counterion BF₄⁻ by ⁵⁵Mn NMR μ = 0.69 *M*
	MeOH		6.0		6	−20 ± 2	−9 ± 3			μ = 0.17
	C₆H₅CN		12.0		9	−9 ± 2	−8 ± 4			μ = 0.52
	EtOH		7.0		9	−16 ± 2	−6 ± 2			μ = 0.16
	acetone		6.0		10	−20 ± 2	−7 ± 1			μ = 0.16
	(MeO)₃PO		6.0		6	−10 ± 2				μ = 0.16
	Et₂CO		7.0		7	−22 ± 2	−9 ± 4			μ = 0.18
	DCM		0.0		11	−18 ± 2	−7 ± 1			μ = 0.11
mol ratios	CH₃CN : CH₂Cl₂	9:1	1.0	2.0	10	−12 ± 2	−5 ± 2			μ = 0.17
		4:1	0.0		14	−14 ± 2	−4 ± 2			μ = 0.18
		7:3	1.0		23	−15 ± 2	−5 ± 7			μ = 0.16
		3:2	0.0		13	−15 ± 2	−5 ± 2			μ = 0.14
		1:1	1.0		14	−14 ± 2	−4 ± 2			μ = 0.12
		2:3	2.0		19	−19 ± 3	−10 ± 3			μ = 0.12
		3:7	1.0		13	−15 ± 3	−5 ± 3			μ = 0.12
		1:4	2.0		14	−13 ± 3	−4 ± 3			μ = 0.12
	CH₃CN : CH₂Cl₂	1:9	2.0	2.0	8	−15 ± 3	−4 ± 3			μ = 0.11
	CH₃CN : BrC₆H₅	4:1			10	−12 ± 4	−5 ± 3			μ = 0.15
		1:1			18	−15 ± 3	−7 ± 4			μ = 0.15
		1:4			14	−13 ± 3	−10 ± 5			μ = 0.11
	DCM : BrC₆H₅	4:1	1.0		14	−17 ± 2	−7 ± 2			μ = 0.08
		1:1	1.0		19	−21 ± 3	−15 ± 3			μ = 0.11
		3:7	3.0		18	−16 ± 2	−5 ± 2			μ = 0.14
[Mn(CNC₆H₁₁)₆]⁺/²⁺	OS self-exchange									
	CH₃CN		2.0	2.0	11	−17 ± 1	−6 ± 1			{Counterion BF₄⁻ by ⁵⁵Mn NMR μ = 0.20 *M*
	MeOH		2.0		18	−16 ± 2				μ = 0.18 *M*
	acetone		3.0		17	−20 ± 2	−8 ± 2			μ = 0.19 *M*
	DCM		3.0		15	−21 ± 4	−8 ± 4			μ = 0.12 *M*
	BrC₆H₅		2.0		10	−9 ± 2				μ = 0.17 *M*
mol ratios		7:3	3.0		21	−17 ± 2	−3 ± 2			μ = 0.09 *M*
	DCM : BrC₆H₅	1:1			18	−18 ± 2	−3 ± 3			μ = 0.10 M
		3:7			18	−10 ± 4				μ = 0.10 M

Reaction	Solvent						Ref	Notes
$[Fe(phen)_3]^{2+}/[Fe(phen)_3]^{3+}$ self-exchange								¹H NMR
as bisulfate salt	D_2O/D_2SO_4	3.0	2.1	8–10	-2.2 ± 0.1		92	$\mu = 0.3$ mol kg⁻¹
as perchlorate salt	CD_3CN	3.8	2.1	8–10	-5.9 ± 0.5			$\mu = 0.18$ mol kg⁻¹
$[Fe(cp)_2]/[Fe(cp)_2]^+$ self-exchange	CD_3CN	−8.5 to 10	2.0	9	-7 ± 1		93	$\mu = [Fe(cp)_2]$
Peroxodisulfate oxidation of $[Fe^{II}(CN)_4(LL)]^{2-}$		25.2						
LL = etn	H_2O		1.04	3	$+4.6$		94	
phen	H_2O		1.38	3	-2.1			
	60% DMSO/H_2O		1.38	3	-3.6			
Me_6bsb	H_2O		1.04	3	-10.2			
	H_2O		1.04	3	-7.7			
	60% DMSO/H_2O		1.04	3	-8.3			
Peroxodisulfate oxidation of $[Fe^{II}(CN)_2(bipy)_2]$								
$[Co(NH_3)_5(H_2O)]^{3+} + [Fe(CN)_6]^{4-}$	up to 57 wt% glycerol-H_2O	30	1.0	6	$+27.9 \pm 0.1$		95	{data extrapolated to constant viscosity}
$[Co^{II}(CyDTA)]^{2-} + [Fe(CN)_6]^{3-} \rightarrow [Co(CyDTA)]^- + [Fe(CN)_6]^{4-}$	H_2O	25	1.2	4	$+6.52$		96	pH < 10.8
$[Cu(dmp)_2]^{+/2+}$ self-exchange	CD_3CN	38.4	2.02		-3.4 ± 0.6		97	$\mu = 0.5\ M$ (NaClO₄)
	$(CD_3)_2CO$	29.5	2.02		-7.8 ± 0.6			$\mu = 0.094$ mol kg⁻¹ (CF₃SO₃K)
$[Ru^0(cp)_2] + [Ru^{II}(cp)_2Br]$ IS self-exchange	CD_3CN	34–55	2.0	10	-3.0 ± 0.2		98	$\mu = 0.121$ mol kg⁻¹ (CF₃SO₃K); ¹H NMR
Intramolecular								
$[(NH_3)_5Ru^{II}\!\!-\!\!S\diamond\!\diamond\,S\!\!-\!\!Ru^{III}(NH_3)_5]^{5+}$	D_2O	24	1.5	6	-7.5 ± 0.2		42	¹H NMR
Addition, ion-pair formation								
$B(OH)_3 + H_2O \rightleftharpoons H^+ + B(OH)_4^-$	H_2O	25	2	16	-3.1 ± 0.3	-28.9 ± 0.3	99	Indicator method, $\mu = 0.1\ m$ (NaCl)
					-4.8 ± 0.1	-31.8 ± 0.1		$\mu = 1.0\ m$ (NaCl)
$Na^+ + B(OH)_4^- \rightleftharpoons NaB(OH)_4$					$+1.6 \pm 0.1$	$+8.4 \pm 0.4$		extrapolated to $\mu = 0$, $\mu = 0.1\ m$ (NaCl)
						$+5.9 \pm 1.0$		$\mu = 1.0\ m$ (NaCl)
for thymol blue: $HIn^- \rightleftharpoons H^+ + In^{2-}$					$+1.8 \pm 0.1$	$+7.0 \pm 0.2$		$\mu = 1\ M$ (Me₄NCl); independent of μ
					-3 ± 0.8	-17.4 ± 0.7(a)		
Complex formation and dissociation								
$[V(H_2O)_6]^{3+} + SCN^- \underset{k_r}{\overset{k_f}{\rightleftharpoons}} [V(H_2O)_5NCS]^{2+} + H_2O$	H_2O	25	2.0	9	-9.4 ± 0.6		100	[TH] $= 0.4\ M$, $\mu = 0.6\ M$
k_f					-17.9 ± 0.7	$+8.5 \pm 1.2$(a, c)		$0.014 < $ pH $ < 1.7$
k_r								

(continued)

Table 15.1. (continued)

Reaction	Solvent	T (°C)	P (kbar)	No. of data	ΔV^* (cm³ mol⁻¹)	$\Delta\beta^*$ (cm³ mol⁻¹ kbar⁻¹)	$\Delta\bar{V}$ (method) (cm³ mol⁻¹)	Ref.	Remarks
$VO_2^+ + H_2O_2 \rightleftharpoons VO(O_2)^+ + H_2O$ [H⁺] independent path	H₂O	20.1	1.6	8	+2.8 ± 1.0			101	{$\mu = 3.00\ M$ (NaClO₄/HClO₄)
path with 1/[H⁺] dependence					+9.9 ± 1.7				
path with [H⁺] dependence					+14.2 ± 3.2				
$VO(O_2)_2 + H_2O_2 \rightleftharpoons VO(O_2)_2 + H_2O$			2.0	8	0.0 ± 0.2				
$[Fe(hxsb)]^{2+} \xrightarrow{OH^-} Fe^{2+} + hxsb$	H₂O	25	1.34	5	+13.3 ± 1.9			102	[OH⁻] = 0.33 M
	50% MeOH				+14.0				[OH⁻] = 0.33 M
	75% MeOH			4	+6.2 ± 0.6				[OH⁻] = 0.33 M
	75% MeOH				+5.5 ± 1.2				[OH⁻] = 0.10 M
	85% MeOH				+6.8				[OH⁻] = 0.33 M
	35% Pr^iOH		1.01	3	+9				[OH⁻] = 0.10 M
	66.7% Pr^iOH				−2.5				[OH⁻] = 0.10 M
$[Fe(hxsb)]^{2+} \xrightarrow{OH^-} Fe^{2+} + hxsb$	17% Bu^tOH	25	1.34	4	+14.1 ± 1.1			102	[OH⁻] = 0.33 M
	38% Bu^tOH			2	~ +12				[OH⁻] = 0.10 M
	50% Bu^tOH		1.18	3	−4.2				[OH⁻] = 0.10 M
$[Mn(DMF)_6]^{2+} + Et_2DTC^- \rightarrow [Mn(DMF)_4Et_2DTC]^+ + 2\ DMF$	DMF	−45	1.20	7	+9.5 ± 0.5			103	$\mu = 0.10\ M$ (NaClO₄)
$[Fe(DMF)_6]^{2+} + Et_2DTC^- \rightarrow [Fe(DMF)_4(Et_2DTC)]^+ + 2\ DMF$	DMF	−35	1.60	9	+12.3 ± 0.8				$\mu = 0.10\ M$ (NaClO₄)
$[Fe(DMF)_6]^{2+} + Pada \underset{k_r}{\overset{k_f}{\rightleftharpoons}} [Fe(DMF)_4Pada]^{2+} + 2\ DMF$ k_f	DMF	−35	1.60	9	+7.5 ± 1.0		−2.0 ± 0.3(a)		$\mu = 0.10\ M$ (NaClO₄)
k_r					+9.5 ± 1.3				
$[Co(DMF)_6]^{2+} + Et_2DTC^- \rightarrow [Co(DMF)_4Et_2DTC]^+ + 2\ DMF$	DMF	−35	1.60	9	+12.1 ± 0.6				$\mu = 0.10\ M$ (NaClO₄)
$[Ni(H_2O)_6]^{2+} + OAc^- \rightleftharpoons [Ni(H_2O)_5(OAc)]^+ + H_2O$	H₂O	25.0					+8.0 ± 1.5(b)	104	$\mu = 0.10\ M$ (NaClO₄)
$[Ni(H_2O)_6]^{2+} + gly^- \rightleftharpoons [Ni(H_2O)_4(gly)]^+ + 2\ H_2O$	H₂O						+11.2 ± 0.2(b)		
$[Ni(H_2O)_4(gly)]^+ + gly^- \rightleftharpoons [Ni(H_2O)_2(gly)_2] + 2\ H_2O$							+12.0 ± 0.5(b)		
$[Ni(H_2O)_6]^{2+} + sar^- \rightleftharpoons [Ni(H_2O)_4(sar)]^+ + 2\ H_2O$	H₂O						+11.7 ± 0.5(b)		
$[Ni(H_2O)_4(sar)]^+ + sar^- \rightleftharpoons [Ni(H_2O)_2(sar)_2]$	H₂O						+9.9 ± 0.8(b)		
$[Ni(H_2O)_6]^{2+} + en \rightleftharpoons [Ni(H_2O)_4(en)]^{2+} + 2\ H_2O$	H₂O						+5.2 ± 0.5(b)		
$[Ni(H_2O)_4(en)]^{2+} + en \rightleftharpoons [Ni(H_2O)_2(en)_2]^{2+} + 2\ H_2O$	H₂O						+5.6 ± 0.9(b)		

Reaction	Solvent	T (°C)			$\Delta V^\ddagger$	ΔV	Ref	Conditions
$[Ni(H_2O)_6]^{2+} + edda^{2-} \rightleftharpoons [Ni(H_2O)_2(edda)] + 4 H_2O$	H_2O					$+28.6 \pm 0.2$(b)		
$[Cu(H_2O)_6]^{2+} + Cl\text{-phen} \underset{k_r}{\overset{k_f}{\rightleftharpoons}} [Cu(H_2O)_4(Cl\text{-phen})]^{2+} + 2 H_2O$ $\quad k_f$ $\quad k_r$	H_2O	25	2.0	4	$+7.1 \pm 0.5$; $+5.2$ (by difference)	$+1.9 \pm 0.4$(a)	105	$\left\{ \begin{array}{l} \mu = 0.05\ M\ (NaClO_4) \\ 5.7 < pH < 5.9 \end{array} \right.$
$[Zn(H_2O)_6]^{2+} + Cl\text{-phen} \underset{k_r}{\overset{k_f}{\rightleftharpoons}} [Zn(H_2O)_4(Cl\text{-phen})]^{2+} + 2 H_2O$ $\quad k_f$ $\quad k_r$	H_2O	25	2.0	4		$+0.9$(c)		$\left\{ \begin{array}{l} \mu = 0.05\ M\ (NaClO_4) \\ 5.7 < pH < 5.8 \end{array} \right.$
$[Pd(H_2O)_4]^{2+} + MeCN \rightarrow$ $[Pd(H_2O)_3(MeCN)]^{2+} + H_2O$ $[Pd(H_2O)_3(MeCN)]^{2+} + H_2O \rightarrow$ $[Pd(H_2O)_4]^{2+} + MeCN$	H_2O	5	2.0	4–7	$+5.0 \pm 0.4$; $+4.1 \pm 0.4$; -4.0 ± 0.8; -1.5 ± 0.5	-2.5 ± 0.4(a)	106	$1.00\ M\ HClO_4$
$[La(H_2O)_n]^{3+} + Hdcp^{2-} \underset{k_r}{\overset{k_f}{\rightleftharpoons}} [La(H_2O)_{n-3}(Hdcp)]^+ + 3 H_2O$ $\quad k_r$ $\quad k_f$	H_2O	25.0	2.0	8	-4.9 ± 0.3; $+7$ (calculated)	$+12.1 \pm 1.5$(a)	107	^{139}La NMR
$[Ce(H_2O)_9]^{3+} \rightleftharpoons [Ce(H_2O)_8]^{3+} + H_2O$	H_2O	25	2	6		$+10.9 \pm 0.3$	108	
Photochemical								
$[W(CO)_5(4\text{-X-py})] + P(OEt)_3 \rightarrow$ $[W(CO)_5(P(OEt)_3)] + 4\text{-X-py}$ $\quad$ X = H, NC, OAc	toluene	25.0	2.0	9	$+5.7 \pm 0.3$; $+6.3$; $+9.9$		109	$\lambda = 436$ nm (LF)
$[W(CO)_4(phen)] + PEt_3 \rightarrow$ $cis\text{-}[W(CO)_3(phen)PEt_3] + CO$ LF excitation MLCT excitation	toluene	25.0	2.0	9	$+8.1 \pm 0.5$; -12.0 ± 0.7		110	excitation $\lambda = 366$ nm; excitation $\lambda = 546$ nm
Photophysical								
$[Pt_2(P_2O_5H_5)_4]^{4-}$: $^1A_{1g}$ ground state $\rightarrow$ $^3A_{2u}$ excited state $^*Pt_2$ $^1A_{1g} + 0.01\ M\ TlNO_3 \rightarrow$ $^3A_{2u}$ excited state $^*Pt_2Tl^+$	H_2O	8.5–22.0			$+0.5 \pm 0.3$; -10.6 ± 0.9		111	by photoacoustic calorimetry
Emission lifetimes MLCT state of $[Cu(dmp)]^+$ no quencher 0.3 M CH_3CN 0.3 M CH_3OH	DCM	23.1	3.0 2.64 2.64	4 2 2	-3.4 ± 0.2 -4.3 ± 0.2 -4.0 ± 0.2		112	

(continued)

Table 15.1. (continued)

Reaction	Solvent	T (°C)	P (kbar)	No. of data	ΔV^* (cm³ mol⁻¹)	$\Delta\beta^*$ (cm³ mol⁻¹ kbar⁻¹)	$\Delta\bar{V}$ (method) (cm³ mol⁻¹)	Ref.	Remarks
MLCT state of [Cu(dpp)]⁺									
no quencher			3.0	5	−1.6 ± 0.2			113	
0.3 M CH₃CN	4:1 v/v EtOH/MeOH	−113	2.64	2	−1.2 ± 0.2				
[Ru(bipy)₃]²⁺	toluene		10	9	+0.47			109	
[W(CO)₅(4-Ac-O-py)]*			3	2	+0.1 ± 0.3				
[W(CO)₅(4-CN-py)]*			3	2	+1.2 ± 0.3				
Bioinorganic and biochemical									
[B₁₂ₐ—H₂O]⁺ + L"⁻ → [B₁₂ₐ—L]⁽¹⁻ⁿ⁾⁺ + H₂O	H₂O	25	5	5				114	
L"⁻ = [Fe(CN)₆]⁴⁻					+16.2 ± 1.2				$\mu = 0.13\ M$ (NaClO₄), pH = 6.0
[Feᴵᴵ(CN)₅NO]²⁻					+8.9 ± 0.5				$\mu = 0.1\ M$ (NaClO₄), pH ≈ 6
[Feᴵᴵᴵ(CN)₅H₂O]²⁻					+8.2 ± 0.8				$\mu = 0.1\ M$ (NaClO₄), pH ≈ 6
N₃⁻					+6.9 ± 0.2				$\mu = 0.5\ M$ (NaClO₄), pH = 6.4
α-D-glucose mutarotation to β-D-glucose over Al₂O₃ adsorption coefficient	DMSO	40	0.9	4			~0	14	
forward reaction					+22 ± 4				
reverse reaction					+20 ± 3				
(Mb₅···O₂) → Mb₅O₂ (recombination)	H₂O	20	2.0	7	+3.5			115	{nanosecond flash experiment pH = 7.8
(Mb₅···O₂) → Mb₅ + O₂ {O₂ escape from geminate pair}					+16.7				
Mb₅ + O₂ → (Mb₅···O₂) {and association with geminate pair}					+10.4		−6.3(c)		
Mb₅ + O₂ → Mb₅O₂ (overall association)					+4.6				
Mb₅ + CO → (Mb₅···CO)					−9.2				
(Mbₕ···O₂) → MbₕO₂ (recombination)	H₂O	20	2.0	7	−8.4				
(Mbₕ···O₂) → Mbₕ + O₂ {O₂ escape from geminate pair}					+11.2				
Mbₕ + O₂ → (Mbₕ···O₂) {and association with geminate pair}					+12.3		+1.1(c)		
Mbₕ + O₂ → MbₕO₂ (overall association)					+3.8				
Mbₕ + CO → (Mbₕ···CO)					−12.7				
(Mb_D···O₂) → Mb_DO₂ (recombination)	H₂O	20	2.0	7	−17.8				
(Mb_D···O₂) → Mb_D + O₂ {O₂ escape from geminate pair}					−2.1				
Mb_D + O₂ → (Mb_D···O₂) {and association with geminate pair}					+8.6		+10.7(c)		
Mb_D + O₂ → Mb_DO₂ (overall association)					+0.0				
Mb_D + CO → (Mb_D···CO)					−18.8				

System	Solvent					Conditions	Ref
HRP + CO → HRP—CO	H$_2$O	1.2	4	34	−23.6 ± 1		52
				20	−23.7 ± 1		
				4.2	−26.9 ± 1		
	40% ethylene glycol			20	−6.98 ± 0.02		
				0.5	−10.5 ± 0.6		
				−10	−14.6 ± 0.8		
	50% MeOH			20	−9.4 ± 0.2		
				10.3	−6 ± 0.2		
				4.2	−5.5 ± 0.2		
				0	−5.2 ± 1		
				−9.8	−2.3 ± 0.3		
				−20	−1.6 ± 0.4		
Cytochrome c oxidase:							
Reduced oxidase, 445 nm ⇌ 443 nm	H$_2$O	1.5		7.4	≥0	pH 7	116
Aerobic steady state ⇌ partially reduced oxidase					−76	fluorescence quenching	
Porphyrin c–cytochrome b$_3$ complex formation	H$_2$O	2	>10	5	−50	1 mM bis tris (pH 7.0) 1 mM < μ < 10 mM (KCl)	117
Histone (H2A-H2B)$_2$ dissociation	H$_2$O	2.2	>10	20	+90 ± 4	μ = 0.2 M (NaCl)	118
					+75 ± 5	μ = 2.0 M (NaCl)	
tetramer–dimer equilibrium					+108 ± 4	μ = 0.2 M (NaCl)	
octamer–tetramer equilibrium					+168 ± 6	μ = 2.0 M (NaCl)	
octamer–dimer equilibrium					+143 ± 2	μ = 0.2 M (NaCl)	
octamer–dimer equilibrium					+142 ± 2	μ = 2.0 M (NaCl)	
Yeast hexokinase dissociation:	H$_2$O	2.8	>6		+116 to +154	fluorescence methods pH 7.5	119
P2 monomer–dimer equilibrium				0	+119	varies with method	
				0	+167	0.1 M Na$_2$SO$_4$	
				30	+162		
				20	+111	pH 6.0	
				20	+155	pH 9.0	
				20	+141	pH 7.5	
P1 monomer–dimer equilibrium				0	+135	pH 7.5 0.1 M Na$_2$SO$_4$	

α-chymotrypsin Michaelis–Menten kinetics:

$$E + S \underset{k_{-1}}{\overset{k_1}{\rightleftharpoons}} ES \underset{k_{-2}}{\overset{k_2}{\rightleftharpoons}} ET \overset{k_3}{\longrightarrow} ES' + P1 \underset{k_{-3}}{\overset{k_3}{\rightleftharpoons}}$$

$$ET \overset{k_4}{\longrightarrow} E + P2$$

(continued)

Table 15.1 (continued)

Reaction	Solvent	T (°C)	P (kbar)	No. of data	ΔV^* (cm³ mol⁻¹)	$\Delta\beta^*$ (cm³ mol⁻¹ kbar⁻¹)	$\Delta\bar{V}$ (method) (cm³ mol⁻¹)	Ref.	Remarks
Hydrolysis of:									
p-NPT	H_2O	25	2					120	pH 7.8
$k_{1(acyl)}\,K_s$					-24 ± 1		-14 ± 1		0.005 M Tris buffer
ATA $\quad K_3 k_{(deacyl)}\quad K_M = K_s k_3/k_2$					-2 ± 1		$+6\pm1$		
$k_{(acyl)} = k_2 K_1\quad K_M = K_s$					-3 ± 2		-3 ± 2		{0.005 M tris buffer {0.05 M tris buffer 10% (v/v) Me₂SO
ATpCA $\quad k_{2(acyl)}\quad K_M = K_s/K_1$					$+10\pm2$		$+12\pm2$		
SA₁pNA $\quad k_{(acyl)}$					-30 ± 2				0.05 M tris buffer
SA₂pNA $\quad k_{(acyl)}$					-33 ± 2				0.05 M tris buffer
SA₃pNA $\quad k_{2(acyl)}\quad K_M = K_s/K_1$					$+15\pm2$		$+5\pm2$		0.1 M tris buffer
Reactions with free electrons and electrodeposition									
$e^- + CO_2 \underset{k_r}{\overset{k_f}{\rightleftharpoons}} CO_2^-$	3-MP	80	2.5	>6				121	
k_f					-44		-290(c)		
k_r					$+246$				
k_f	2,2,4-TMP	58	2.5	>6	-67		-230(c)		
k_r					$+164$				
k_f	TMS	23	~-2.5	>6	-68		-174(c)		
k_r			~-0.6	6	$+106$				
Electrodeposition									
CO $\quad$ no additive	H_2O	23	1.7		$+12.3\pm0.3$			122	0.1 M K₂SO₄ electrolyte plus
0.1 M–3.5 M KCl					$+6.0\pm0.4$				0.01 M NaClO₄ or KCl
0.01 M KSCN					$+0.0$				
Ni $\quad$ no additive					$+14.0\pm0.5$				0.1 M K₂SO₄ electrolyte plus
up to 0.1 M 2-propene sulfonate					$+14.0\pm0.4$				0.01 M KCl
0.01 M 2-propene sulfonate					$+2.9\pm0.1$				
0.01 M–0.05 M 2-naphthalene sulfonate					$+5.1\pm0.2$				
0.02 M–0.1 M 1,5-naphthalene disulfonate					$+7.2\pm0.3$				
Ag $\quad$ no additive					$+9.5\pm0.4$				0.1 M KNO₃
5 M–10 M NH₃ plus 0–0.1 M KSCN					$+20.6\pm0.5$				
0.5 M KCN					$+0.0$				

References

References for Chapter 1

1. M. A. Fox and M. Chanon (eds.), *Photoinduced Electron Transfer. Part A. Conceptual Basis*, Elsevier, Amsterdam (1988).
2. M. A. Fox and M. Chanon (eds.), *Photoinduced Electron Transfer. Part B. Experimental Techniques and Medium Effects*, Elsevier, Amsterdam (1988).
3. M. A. Fox and M. Chanon (eds.), *Photoinduced Electron Transfer. Part C. Organic Substrates*, Elsevier, Amsterdam (1988).
4. M. A. Fox and M. Chanon (eds.), *Photoinduced Electron Transfer. Part D. Inorganic Substrates and Applications*, Elsevier, Amsterdam (1988).
5. R. F. Khairutdinov, K. I. Zamaraev, and V. P. Zhandov, *Comprehensive Chemical Kinetics*, Vol. 30, *Electron Tunneling in Chemistry. Chemical Reactions over Large Distances* (R. G. Compton, ed.), Elsevier, Amsterdam (1989).
6. K. F. Purcell and B. Blaive, in Ref. 1, Chapter 1.3, p. 123.
7. R. A. Marcus, *Discuss. Faraday Soc.*, **29**, 21 (1960).
8. K. V. Mikkelsen and M. A. Ratner, *Int. J. Quantum Chem., Quantum Chem. Symp.*, **22**, 707 (1988).
9. K. V. Mikkelsen and M. A. Ratner, *J. Phys. Chem.*, **93**, 1759 (1989).
10. K. V. Mikkelsen and M. A. Ratner, *J. Chem. Phys.*, **90**, 4237 (1989).
11. J. K. Reimers and N. S. Hush, *Chem. Phys.*, **134**, 323 (1989).
12. B. S. Brunschwig and N. Sutin, *J. Am. Chem. Soc.*, **111**, 7454 (1989).
13. Y. Hu and S. Mukamel, *J. Chem. Phys.*, **91**, 6973 (1989).
14. M. Bixon, J. Jortner, M. E. Michel-Beyerle, and A. Ogrodnik, *Biochim. Biophys. Acta*, **977**, 273 (1989).
15. K. V. Mikkelsen, J. Ulstrup, and M. G. Zakaraya, *J. Am. Chem. Soc.*, **111**, 1315 (1989).
16. M. Bixon and J. Jortner, *Chem. Phys. Lett.*, **159**, 17 (1989).
17. J. N. Onuchic and P. G. Wolynes, *J. Phys. Chem.*, **92**, 6495 (1988).
18. D. N. Beratan and J. N. Onuchic, *J. Chem. Phys.*, **89**, 6195 (1989).
19. M. E. Michel-Beyerle, M. Bixon, and J. Jortner, *Chem. Phys. Lett.*, **151**, 188 (1988).
20. M. Bixon and J. Jortner, *J. Phys. Chem.*, **92**, 7148 (1988).
21. T. Kakitani and N. Mataga, *J. Phys. Chem.*, **92**, 5059 (1988).
22. R. A. Marcus, *J. Phys. Chem.*, **93**, 3078 (1989).
23. E. K. Knapp and S. F. Fischer, *J. Chem. Phys.*, **90**, 354 (1989).
24. D. F. Calef, in Ref. 1, Chapter 1.9, p. 362.
25. B. Bagchi, *Ann. Rev. Phys. Chem.*, **40**, 115 (1989).
26. R. M. Nielson, G. E. McManis, M. N. Golovin, and M. J. Weaver, *J. Phys. Chem.*, **92**, 3441 (1988).

27. R. M. Nielson, M. N. Golovin, G. E. McManis, and M. J. Weaver, *J. Am. Chem. Soc.*, **110**, 1745 (1988).

28. G. E. McManis, R. M. Neilson, A. Gochev, and M. J. Weaver, *Inorg. Chem.*, **27**, 1827 (1988).

29. G. E. McManis, R. M. Neilson, A. Gochev, and M. J. Weaver, *J. Am. Chem. Soc.*, **111**, 5533 (1989).

30. D. E. Richardson, C. S. Christ, D. Sharpe, and J. E. Eyler, *J. Am. Chem. Soc.*, **109**, 3894 (1987).

31. M. D. Newton, in: *The Challenge of d and f Electrons* (D. R. Salahub and M. C. Zerner, eds.), p. 378, ACS Symp. Ser. 394, American Chemical Society, Washington, D.C. (1989).

32. D. A. Geselowitz, *Inorg. Chim. Acta*, **154**, 225 (1988).

33. D. L. Dexter, *J. Chem. Phys.*, **21**, 836 (1953).

34. J. F. Endicott, *Acc. Chem. Res.*, **21**, 59 (1988).

35. S. Mukamel and Y. J. Yan, *Acc. Chem. Res.*, **22**, 301 (1989).

36. J. M. Lawson, D. C. Craig, M. N. Padden-Row, J. Kroon, and J. W. Verhoeven, *Chem. Phys. Lett.*, **164**, 120 (1989).

37. H. Oevering, J. W. Verhoeven, M. N. Padden-Row, and J. M. Warman, *Tetrahedron*, **45**, 4751 (1989).

38. M. N. Padden-Row, A. M. Oliver, M. C. R. Symons, E. Costaris, S. S. Wong, and J. W. Verhoeven, in: *Photoconversion Processes for Energy and Chemicals* (D. O. Hall and G. Grassi, eds.), p. 79, Elsevier, New York (1989).

39. J. W. Verhoeven, J. Kroon, M. N. Padden-Row, and A. M. Oliver, in: *Photoconversion Processes for Energy and Chemicals* (D. O. Hall and G. Grassi, eds.), p. 100, Elsevier, New York (1989).

40. R. D. Cannon, *Electron Transfer Reactions*, Butterworths, London (1980).

41. R. A. Marcus and N. Sutin, *Biochim. Biophys. Acta*, **811**, 265 (1985).

42. B. S. Brunschwig, S. Ehrenson, and N. Sutin, *J. Am. Chem. Soc.*, **106**, 6858 (1984).

43. A. Weller, *Z. Phys. Chem. N.F.*, **133**, 93 (1982).

44. H. B. Gray and B. G. Malmström, *Biochemistry*, **28**, 7499 (1989).

45. M. A. Cusanovic, J. T. Hazzard, T. E. Meyer, and G. Tollin, *J. Macromol. Sci., Chem.*, **A26**, 433 (1989).

46. L. P. Pan, B. Durham, J. Wolinseka, and F. S. Millett, *Biochemistry*, **27**, 7180 (1988).

47. B. Durham, L. P. Pan, J. E. Long, and F. Millett, *Biochemistry*, **28**, 8659 (1989).

48. T. J. Meade, H. B. Gray, and J. R. Winkler, *J. Am. Chem. Soc.*, **111**, 4353 (1989).

49. M. D. Johnson, J. R. Miller, N. S. Green, and G. L. Closs, *J. Phys. Chem.*, **93**, 1173 (1989).

50. G. L. Closs, L. T. Calcaterra, N. J. Green, K. W. Penfield, and J. R. Miller, *J. Phys. Chem.*, **90**, 3673 (1986).

51. G. L. Closs, M. D. Johnson, J. R. Miller, and P. Piotrowiak, *J. Am. Chem. Soc.*, **111**, 3751 (1989).

52. A. B. P. Lever, *Inorganic Electronic Spectroscopy* (2nd ed.), Elsevier, Amsterdam (1984).

53. R. C. Weask and M. J. Astle (ed.), *CRC Handbook of Chemistry and Physics*, CRC Press, Boca Raton, Florida (1980).

54. A. M. Oliver, D. C. Craig, M. N. Padden-Row, J. Kroon, and J. W. Verhoeven, *Chem. Phys. Lett.*, **250**, 366 (1988).

55. H. Oevering, M. N. Padden-Row, M. Heppner, A. M. Oliver, E. Costaris, J. W. Verhoeven, and N. S. Hush, *J. Am. Chem. Soc.*, **109**, 3258 (1987).

56. R. A. Marcus, *Chem. Phys. Lett.*, **146**, 13 (1988).

57. R. A. Marcus, in: *The Photosynthetic Bacterial Reaction Center* (J. Breton and A. Vermeglio, eds.), p. 389, Plenum Press, New York (1988).

58. M. Bixon, J. Jortner, M. Plato, and M. E. Michel-Beyerle, in: *The Photosynthetic Reaction Center* (J. Breton and A. Vermeglio, eds.), p. 399, Plenum Press, New York (1988).

59. M. E. Michel-Beyerle, M. Bixon, and J. Jortner, *Chem. Phys. Lett.*, **151**, 188 (1988).

60. M. Bixon and J. Jortner, *J. Phys. Chem.*, **92**, 7148 (1988).

61. S. Boxer, R. A. Goldstein, D. J. Lockhart, T. R. Middenhorf, and L. Takiff, *J. Phys. Chem.*, **93**, 8280 (1989).

62. D. J. Lockhart, R. F. Goldstein, and S. G. Boxer, *J. Chem. Phys.*, **89**, 1408 (1988).
63. S. F. Fischer, in: *The Photosynthetic Reaction Center* (J. Breton and A. Vermeglio, eds.), p. 425, Plenum Press, New York (1988).
64. M. Lösche, G. Feher, and M. Y. Okamura, in: *The Photosynthetic Reaction Center* (J. Breton and A. Vermeglio, eds.), p. 151, Plenum Press, New York (1988).
65. M. Bixon, J. Jortner, M. E. Michel-Beyerle, and A. Ogrodnik, *Biochim. Biophys. Acta*, **977**, 273 (1989).
66. J. Deisenhofer and H. Michel, *Science*, **245**, 1463 (1989).
67. D. Gust and T. A. Moore, *Science*, **244**, 35 (1989).
68. Y. Hatano, M. Saito, T. Kakitani, and N. Mataga, *J. Phys. Chem.*, **92**, 1008 (1988).
69. A. Yoshimori, T. Kakitani, Y. Enbomoto, and M. Mataga, *J. Phys. Chem.*, **93**, 8316 (1989).
70. T. Kakitani and N. Mataga, *J. Phys. Chem.*, **91**, 6277 (1987).
71. E. A. Carter and J. T. Hynes, *J. Phys. Chem.*, **93**, 2184 (1989).
72. M. Tachiya, *Chem. Phys., Lett.*, **159**, 505 (1989).
73. M. Tachiya, *J. Phys. Chem.*, **93**, 7050 (1989).
74. D. Y. Yang and R. I. Cukier, *J. Chem. Phys.*, **91**, 281 (1989).
75. I. R. Gould, J. E. Moser, B. Armitage, and S. Farid, *J. Am. Chem. Soc.*, **111**, 1917 (1989).
76. I. R. Gould, R. Moody, and S. Farid, *J. Am. Chem. Soc.*, **110**, 7242 (1988).
77. I. R. Gould and S. Farid, *J. Am. Chem. Soc.*, **110**, 7883 (1988).
78. N. R. Kestner, J. Logan, and J. Jortner, *J. Phys. Chem.*, **78**, 2148 (1974).
79. P. P. Levin, P. F. Pluzhnikov, and V. A. Kuzmin, *Chem. Phys.*, **137**, 331 (1989).
80. L. F. Cooley, C. E. L. Headford, C. M. Elliott, and D. F. Kelley, *J. Am. Chem. Soc.*, **110**, 6673 (1988).
81. P. Chen, R. Duesing, G. Tapolsky, and T. J. Meyer, *J. Am. Chem. Soc.*, **111**, 8305 (1989).
82. R. Engelman and J. Jortner, *Mol. Phys.*, **18**, 145 (1970).
83. M. Maroncelli, J. MacInnis, and G. R. Flemming, *Science*, **243**, 1674 (1989).
84. P. G. Wolynes, *J. Chem. Phys.*, **86**, 5133 (1987).
85. I. Rips, J. Klafter, and J. Jortner, *J. Chem. Phys.*, **88**, 3246 (1988).
86. I. Rips, J. Klafter, and J. Jortner, *J. Chem. Phys.*, **88**, 4288 (1988).
87. G. E. McManis and M. J. Weaver, *J. Chem. Phys.*, **90**, 1720 (1989).
88. T. Fonesca, *J. Chem. Phys.*, **91**, 2869 (1989).
89. M. Morillo and R. I. Cukier, *J. Chem. Phys.*, **89**, 6736 (1988).
90. M. Morillo, D. Y. Yang, and R. I. Cukier, *J. Chem. Phys.*, **90**, 5711 (1989).
91. A. B. Helman, *Chem. Phys.*, **133**, 271 (1989).
92. D. Huppert, V. Ittah, and E. M. Kosower, *Chem. Phys. Lett.*, **144**, 15 (1988).
93. P. Finch, H. Heitele, and Michel-Beyerle, *Chem. Phys.*, **138**, 1 (1988).
94. H. Lu, F. H. Long, R. M. Bowman, and K. B. Eisenthal, *J. Phys. Chem.*, **93**, 27 (1989).
95. Y. Gaudel, S. Pommeret, A. Migus, and A. Antonetti, *J. Phys. Chem.*, **93**, 3880 (1989).
96. H. Miyasaka, S. Ojima, and N. Mataga, *J. Phys. Chem.*, **93**, 3380 (1989).
97. Y. Hirata and N. Mataga, *J. Phys. Chem.*, **93**, 7539 (1989).
98. N. Mataga, H. Yao, H. Okada, Y. Kanda, and A. Harriman, *Chem. Phys.*, **131**, 473 (1989); A. Osuka, K. Maruyama, N. Mataga, T. Asahi, I. Yamazaki, and N. Tamai, *J. Am. Chem. Soc.*, **112**, 4958 (1990).
99. S. B. Piepho, *J. Am. Chem. Soc.*, **110**, 6319 (1988).
100. L.-T. Zhang, J. Ko, and H. J. Ondrechen, *J. Phys. Chem.*, **93**, 3030 (1989).
101. J. Katriel and M. A. Ratner, *J. Phys. Chem.*, **93**, 5065 (1989).
102. J. T. Hupp and J. Weydert, *Inorg. Chem.*, **26**, 2657 (1987).
103. N. A. Lewis and Y. S. Obeng, *J. Am. Chem. Soc.*, **111**, 7624 (1989).

References for Chapter 2

1. A. C. Dema and R. N. Bose, *Inorg. Chem.*, **28**, 2711 (1989).
2. M. Zhu, A. O. Oyetunjii, K. Lu, and J. E. Earley, *Polyhedron*, **8**, 577 (1989).

3. P. Huston, J. H. Espenson, and A. Bakac, *Inorg. Chem.*, **28**, 3671 (1989).
4. A. Neves and C. V. Franco, *Inorg. Chim. Acta*, **160**, 223 (1989).
5. R. J. Balahura and M. D. Johnson, *Inorg. Chem.*, **28**, 4548 (1989).
6. R. J. Balahura, M. D. Johnson, and T. Black, *Inorg. Chem.*, **28**, 3933 (1989).
7. M. J. Sisley and R. B. Jordon, *Inorg. Chem.*, **28**, 2714 (1989).
8. W. L. Purcell, *Inorg. Chem.*, **28**, 2312 (1989).
9. N. A. Lewis, W. L. Purcell, and D. V. Taveras, *Inorg. Chem.*, **28**, 133 (1989).
10. A. Neves, W. Walz, K. Wieghardt, B. Nuber, and J. Weiss, *Inorg. Chem.*, **27**, 2484 (1988).
11. T. Katsuyama, A. Bakac, and J. H. Espenson, *Inorg. Chem.*, **28**, 339 (1989).
12. M. A. Malati and A. Sear, *Polyhedron*, **8**, 1874 (1989).
13. M. E. Brynildson, A. Bakac, and J. H. Espenson, *Inorg. Chem.*, **27**, 2592 (1988).
14. S. K. Ghosh, K. Laali, and E. S. Gould, *Inorg. Chem.*, **27**, 4224 (1988).
15. C. Minero, E. Pramauro, and E. Pelizzetti, *J. Phys. Chem.*, **92**, 4670 (1988).
16. C. Millan and H. Diebler, *J. Chem. Soc., Dalton Trans.*, 2397 (1988).
17. C. Sharp and A. G. Sykes, *J. Chem. Soc., Dalton Trans.*, 2579 (1988).
18. B.-L. Ooi, C. Sharp, and A. G. Sykes, *J. Am. Chem. Soc.*, **111**, 125 (1989).
19. P. W. Dimmock, J. McGinnis, B.-L. Ooi, and A. G. Sykes, *Inorg. Chim. Acta*, **165**, 3 (1989).
20. P. W. Dimmock, B.-L. Ooi, and A. G. Sykes, *Polyhedron*, **8**, 2791 (1989).
21. B.-L. Ooi, A. L. Petrou, and A. G. Sykes, *Inorg. Chem.*, **27**, 3626 (1988).
22. L. M. Schwane and R. C. Thompson, *Inorg. Chem.*, **28**, 3938 (1989).
23. K. A. Anderson and S. Wherland, *Inorg. Chem.*, **28**, 601 (1989).
24. K. Zahir, J. H. Espenson, and A. Bakac, *Inorg. Chem.*, **27**, 3144 (1988).
25. S. Lee, A. Bakac, and J. H. Espenson, *Inorg. Chem.*, **28**, 1367 (1989).
26. C. A. Simmons, A. Bakac, and J. H. Espenson, *Inorg. Chem.*, **28**, 581 (1989).
27. M. Stebler, R. M. Nielson, W. F. Siems, J. P. Hunt, H. W. Dodgen, and S. Wherland, *Inorg. Chem.*, **27**, 2893 (1988).
28. D. H. Macartney and D. W. Thompson, *Inorg. Chem.*, **28**, 2195 (1989).
29. D. J. Kuchynka and J. K. Kochi, *Inorg. Chem.*, **27**, 2574 (1988).
30. D. J. Kuchynka and J. K. Kochi, *Inorg. Chem.*, **28**, 855 (1989).
31. L. K. Woo and J. G. Goll, *J. Am. Chem. Soc.*, **111**, 3755 (1989).
32. H. Doine and T. W. Swaddle, *Can. J. Chem.*, **66**, 2763 (1988).
33. K. Kirchner, S.-Q. Dang, M. Stebler, H. W. Dodgen, S. Wherland, and J. P. Hunt, *Inorg. Chem.*, **28**, 3604 (1989).
34. K. Kirchner, H. W. Dodgen, S. Wherland, and J. P. Hunt, *Inorg. Chem.*, **28**, 604 (1989).
35. G. E. McManis, R. M. Nielson, A. Gochev, and M. J. Weaver, *J. Am. Chem. Soc.*, **111**, 5533 (1989).
36. R. Schmid, K. Kirchner, and V. N. Sapunov, *Inorg. Chem.*, **28**, 4167 (1989).
37. I. Krack and R. van Eldik, *Inorg. Chem.*, **28**, 851 (1989).
38. G.-H. Lee, L. Della Ciana, and A. Haim, *J. Am. Chem. Soc.*, **111**, 2535 (1989).
39. D. W. Dixon, S. Woehler, X. Hong, and A. M. Stolzenberg, *Inorg. Chem.*, **27**, 3682 (1988).
40. S. Fukuzumi, S. Mochizuki, and T. Tanada, *Inorg. Chem.*, **28**, 2459 (1989).
41. M. D. Purugganan, C. V. Kumar, N. J. Turro, and J. K. Barton, *Science*, **241**, 1645 (1988).
42. P. Berhard and A. M. Sargeson, *Inorg. Chem.*, **27**, 2582 (1988).
43. H.-J. Kuppers, K. Wieghardt, S. Steenken, B. Nuber, and J. Weiss, *Z. Anorg. Allg. Chem.*, **573**, 43 (1989).
44. P. Bernhard, A. M. Sargeson, and F. C. Anson, *Inorg. Chem.*, **27**, 2754 (1988).
45. J. C. Dobson and T. J. Meyer, *Inorg. Chem.*, **27**, 3283 (1988).
46. T. E. Mallouk, J. S. Krueger, J. E. Mayer, and C. M. G. Dymond, *Inorg. Chem.*, **28**, 3507 (1989).
47. K. Ohkubo, T. Hamada, T. Inaoka, and H. Ishida, *Inorg. Chem.*, **28**, 2021 (1989).
48. K. Ohkubo, H. Ishida, T. Hamada, and T. Inaoka, *Chem. Lett.*, 1545 (1989).
49. G. W. Bushnell, D. G. Fortier, and A. McAuley, *Inorg. Chem.*, **27**, 2626 (1988).
50. R. J. Geue, A. J. Hendry, and A. M. Sargeson, *J. Chem. Soc., Chem. Commun.*, 1646 (1989).
51. J. Gribble and S. Wherland, *Inorg. Chem.*, **28**, 2859 (1989).

52. E. K. Byrne and K. H. Theopold, *J. Am. Chem. Soc.*, 111, 3887 (1989).
53. M. A. S. Aquino, D. A. Foucher, and D. H. Macartney, *Inorg. Chem.*, 28, 3357 (1989).
54. K. R. Howes, A. Bakac, and J. H. Espenson, *Inorg. Chem.*, 28, 579 (1989).
55. S. Ronco, B. van Vlierberg, and G. Ferraudi, *Inorg. Chem.*, 27, 3453 (1988).
56. N. Sadler, S. L. Scott, A. Bakac, J. H. Espenson, and M. S. Ram, *Inorg. Chem.*, 28, 3951 (1989).
57. S. L. Anliker, M. W. Beach, H. D. Lee, and D. W. Margerum, *Inorg. Chem.*, 27, 3809 (1988).
58. J.-F. Wang, K. Kumar, and D. W. Margerum, *Inorg. Chem.*, 28, 3481 (1989).
59. A. McAuley, C. J. Macdonald, L. Spencer, and P. R. West, *J. Chem. Soc., Daltons Trans.*, 2279 (1988).
60. A. McAuley and T. W. Whitcombe, *Inorg. Chem.*, 27, 3090 (1988).
61. A. Peloso, *Gazz. Chim. Ital.*, 118, 787 (1988).
62. C. M. Groeneveld, J. van Rijn, J. Reedijk, and G. W. Canters, *J. Am. Chem. Soc.*, 110, 4893 (1988).
63. J. A. Goodwin, L. J. Wilson, D. M. Stanbury, and R. A. Scott, *Inorg. Chem.*, 28, 42 (1989).
64. H. Doine, Y. Yano, and T. W. Swaddle, *Inorg. Chem.*, 28, 2319 (1989).
65. M. M. Bernardo, P. V. Robandt, R. R. Schroeder, and D. B. Rorabacher, *J. Am. Chem. Soc.*, 111, 1224 (1989).
66. Y. Sulfab, *Polyhedron*, 8, 2409 (1989).
67. A. Hussein, Y. Sulfab, and M. Nasreldin, *Inorg. Chem.*, 28, 157 (1989).
68. K. Libson, M. N. Doyle, R. W. Thomas, T. Nelesnik, M. Woods, J. C. Sullivan, R. C. Elder, and E. Deutsch, *Inorg. Chem.*, 27, 3614 (1988).
69. M. S. Corraine and J. D. Atwood, *Inorg. Chem.*, 28, 3781 (1989).
70. D. C. Eisenberg, S. A. Kinsley, and A. Steitwieser, *J. Am. Chem. Soc.*, 111, 5769 (1989).
71. H. B. Gray and B. G. Malmstrom, *Biochemistry*, 28, 7499 (1989).
72. O. Farver and I. Pecht, *Coord. Chem. Rev.*, 94, 17 (1989).
73. *J. Inorg. Biochem.*, 36, 204 (1989).
74. J. D. Rush, F. Levine, and W. H. Koppenol, *Biochemistry*, 27, 5876 (1988).
75. M. P. Jackman, J. McGinnis, R. Powls, G. A. Salmon, and A. G. Sykes, *J. Am. Chem. Soc.*, 110, 5880 (1988).
76. L. M. Peerey and N. M. Kostic, *Biochemistry*, 28, 1861 (1989).
77. O. Farver and I. Pecht, *FEBS Lett.*, 244, 379 (1989).
78. S. Ozaki, J. Hirose, and Y. Kidani, *Inorg. Chem.*, 27, 3746 (1988).
79. J. A. Cowan, R. K. Upmacis, D. N. Beratan, J. N. Onuchic, and H. B. Gray, *Ann. N.Y. Acad. Sci.*, 550, 68 (1988).
80. J. A. Cowan and H. B. Gray, *Inorg. Chem.*, 28, 2074 (1989).
81. M. J. Natan and B. M. Hoffman, *J. Am. Chem. Soc.*, 111, 6468 (1989).
82. E. Magner and G. McLendon, *Biochem. Biophys. Res. Commun.*, 159, 472 (1989).
83. K. Tsukahara, *J. Am. Chem. Soc.*, 111, 2040 (1989).
84. P. Osvath, G. A. Salmon, and A. G. Sykes, *J. Am. Chem. Soc.*, 110, 7114 (1988).
85. B. E. Bowler, T. J. Meade, S. L. Mayo, J. H. Richards, and H. B. Gray, *J. Am. Chem. Soc.*, 111, 8757 (1989).
86. T. J. Meade, H. B. Gray, and J. R. Winkler, *J. Am. Chem. Soc.*, 111, 4353 (1989).
87. P. L. Drake, R. T. Hartshorn, J. McGinnis, and A. G. Sykes, *Inorg. Chem.*, 28, 1361 (1989).
88. D. W. Conrad and R. A. Scott, *J. Am. Chem. Soc.*, 111, 3461 (1989).
89. L. P. Pan, B. Durham, J. Wolinska, and F. Millett, *Biochemistry*, 27, 7180 (1988).
90. B. Durham, L. P. Pan, J. E. Long, and F. Millett, *Biochemistry*, 28, 8659 (1989).
91. J. D. Rush and W. H. Koppenol, *Biochim. Biophys. Acta*, 936, 187 (1988).
92. G. Cheddar, T. E. Meyer, M. A. Cusanovich, C. D. Stout, and G. Tollin, *Biochemistry*, 28, 6318 (1989).
93. K. C. Cho, W. F. Chu, C. L. Choy, and C. M. Che, *Biochim. Biophys. Acta*, 934, 161 (1988).
94. K. C. Cho, W. F. Chu, C. L. Choy, and C. M. Che, *Biochim. Biophys. Acta*, 973, 53 (1989).
95. S. Sakaki, Y. Nishijima, H. Koga, and K. Ohkubo, *Inorg. Chem.*, 28, 4061 (1989).
96. K. Kobayashi, H. Une, and K. Hayashi, *J. Biol. Chem.*, 264, 7976 (1989).

97. J. E. Morgan, P. M. Li, D.-J. Jang, M. A. El-Sayed, and S. I. Chan, *Biochemistry*, **28**, 6975 (1989).

98. B. Michel and H. R. Bosshard, *Biochemistry*, **28**, 244 (1989).

99. B. Michel, A. G. Mauk, and M. R. Bosshard, *FEBS Lett.*, **243**, 149 (1989).

100. F. E. Summers and J. E. Erman, *J. Biol. Chem.*, **263**, 14267 (1988).

101. M. A. Miller, J. T. Hazzard, J. M. Mauro, S. L. Edwards, P. C. Simons, G. Tollin, and J. Kraut, *Biochemistry*, **27**, 9081 (1988).

102. V. Niviere, E. C. Hatchikian, P. Bianco, and J. Haladjian, *Biochim. Biophys. Acta*, **935**, 34 (1988).

103. C. H. Kim, T. E. King, and C. Balny, *Biochim. Biophys. Res. Commun.*, **163**, 276 (1989).

104. M. Sola, J. A. Cowan, and H. B. Gray, *J. Am. Chem. Soc.*, **111**, 6627 (1989).

105. R. T. Hartshorn, M.-C. Lim, and A. G. Sykes, *Inorg. Chem.*, **27**, 4603 (1988).

References for Chapter 3

1. P. Neta, R. E. Huie, A. B. Ross, and B. Albert, *J. Phys. Chem. Ref. Data*, **17**, 1027 (1988).

2. C. Marsh and J. O. Edwards, *Prog. React. Kinet.*, **15**, 35 (1989).

3. A. E. Martell, A. K. Basak, and G. J. Raleigh, *Pure Appl. Chem.*, **60**, 1325 (1988).

4. B. H. J. Bielski, *Basic Life Sci.*, **49**, (Oxygen Radical Biol. Med.), 123 (1988).

5. F. P. Pruchnik, *Pure Appl. Chem.*, **61**, 795 (1989).

6. J. R. Pladziewicz, P. Slattum, and M. R. V. Sahyun, *J. Imaging Sci.*, **33**, 119 (1989).

7. M. D. Roundhill, H. B. Gray, and C.-M. Che, *Acc. Chem. Res.*, **22**, 55 (1989).

8. K. M. Kadish, Redox Chem. Interfacial Behav. Biol. Mod. (Proc. Int. Symp. Redox Mech. Interfacial Prop. Mol. Biol. Importance), **3rd**, 27 (1987).

9. D. H. Macartney, *Rev. Inorg. Chem.*, **9**, 101 (1988).

10. S. S. Eaton and G. R. Eaton, *Electron Spin Reson.*, **11(B)**, 258 (1989).

11. W. C. Trogler, *Int. J. Chem. Kinet.*, **19**, 1025 (1987).

12. D. R. Tyler, *Prog. Inorg. Chem.*, **36**, 125 (1988).

13. W. L. Purcell, *Polyhedron*, **8**, 563 (1989).

14. C. Varotsis, W. H. Woodruff, and G. T. Babcock, *J. Am. Chem. Soc.*, **111**, 6439 (1989).

15. T. J. Lomis, M. G. Elliott, S. Siddiqui, M. Moyer, R. R. Koepsel, and R. E. Shepherd, *Inorg. Chem.*, **28**, 2369 (1989).

16. H. Kurosaki, H. Anan, and E. Kimura, *Nippon Kagaku Kaishi*, 691 (1988); *Chem. Abstr.*, **109**: 103534a.

17. T. Funabiki, T. Konishi, S. Tada, and S. Yoshida, *Stud. Org. Chem.* (*Amsterdam*), **33** (Role Oxygen Chem. Biochem.), 495 (1988).

18. S. Fukuzumi, S. Mochizuki, and T. Tanaka, *Inorg. Chem.*, **28**, 2459 (1989).

19. B.-L. Ooi, C. Sharp, and A. G. Sykes, *J. Am. Chem. Soc.*, **111**, 125 (1989).

20. K. Weighardt, U. Bossek, A. Neves, B. Nuber, and J. Weiss, *Inorg. Chem.*, **28**, 432 (1989).

21. H. Arzoumanian, J.-F. Petrignani, M. Pierrot, F. Ridouane, and J. Sanchez, *Inorg. Chem.*, **27**, 3377 (1988).

22. L. O. Spreer, A. Leone, A. C. Maliyackel, J. W. Otvos, and M. Calvin, *Inorg. Chem.*, **27**, 2401 (1988).

23. K. J. Brewer, M. Calvin, R. S. Lumpkin, J. W. Otvos, and L. O. Spreer, *Inorg. Chem.*, **28**, 4446 (1989).

24. M. M. T. Khan, C. Sreelatha, S. A. Mirza, G. Ramachandraiah, and S. H. R. Abdi, *Inorg. Chim. Acta*, **154**, 103 (1988).

25. M. M. T. Khan, S. A. Mirza, C. Sreelatha, S. H. R. Abdi, and Z. A. Shaikh, *Stud. Org. Chem.* (*Amsterdam*), **33** (Role Oxygen Chem. Biochem.), 211 (1988).

26. S. Davis and D. R. Drago, *Inorg. Chem.*, **27**, 4759 (1988).

27. P. Bernhard, A. M. Sargeson, and F. C. Anson, *Inorg. Chem.*, **27**, 2754 (1988).

28. R. J. Motekaitis and A. E. Martell, *J. Chem. Soc., Chem. Commun.*, 1020 (1988).

29. M. G. Basallote, D. Chen, and A. E. Martell, *Inorg. Chem.*, **28**, 3494 (1989).
30. K. Sakata and M. Hashimoto, *Inorg. Chim. Acta*, **146**, 145 (1988).
31. K. D. Karlin, P. Ghosh, R. W. Cruse, A. Farooq, Y. Gultneh, R. R. Jacobson, N. J. Blackburn, R. W. Strange, and J. Zubieta, *J. Am. Chem. Soc.*, **110**, 6769 (1988).
32. R. W. Cruse, S. Kaderli, K. D. Karlin, and A. D. Zuberbühler, *J. Am. Chem. Soc.*, **110**, 6882 (1988).
33. K. D. Karlin, B. T. Cohen, A. Farooq, S. Liu, and J. Zubieta, *Inorg. Chim. Acta*, **153**, 9 (1988).
34. J. A. Goodwin, G. A. Bodager, L. J. Wilson, D. M. Stanbury, and W. R. Scheidt, *Inorg. Chem.*, **28**, 35 (1989).
35. M. A. El-Sayed, A. Abu-Raqabah, G. Davies, and A. El-Toukhy, *Inorg. Chem.*, **28**, 1909 (1989).
36. M. F. Cabral, J. Cabral, J. Trocha-Grimshaw, K. P. McKillop, S. M. Nelson, and J. Nelson, *J. Chem. Soc., Dalton Trans.*, 1351 (1989).
37. O. Nakamura, M. Adachi, and I. Ogino, *Osaka Kogyo Gijutsu Shikensho Kiho*, **38**, 202 (1987); *Chem. Abstr.*, **108**; 227521t.
38. K. Zahir, J. H. Espenson, and A. Bakac, *J. Am. Chem. Soc.*, **110**, 5059 (1988).
39. J. L. Bear, C.-L. Yao, F. J. Capdevielle, and K. M. Kadish, *Inorg. Chem.*, **27**, 3782 (1988).
40. K. R. Howes, A. Bakac, and J. H. Espenson, *Inorg. Chem.*, **27**, 3147 (1988).
41. M. Kumar, G. J. Colpas, R. O. Day, and M. J. Maroney, *J. Am. Chem. Soc.*, **111**, 8323 (1989).
42. M. G. Alonso-Amigo and S. Schlick, *J. Phys. Chem.*, **93**, 7526 (1989).
43. P. Bonchev, M. Miteva, and G. Gencheva, *Pure Appl. Chem.*, **61**, 897 (1989).
44. P. Schreiber, K. Wieghardt, U. Froerke, and H. J. Haupt, *Inorg. Chem.*, **27**, 2111 (1988).
45. J. D. Rush and W. H. Koppenol, *J. Am. Chem. Soc.*, **110**, 4957 (1988).
46. E. N. Rizkalla, M. A. E. Mansour, and S. S. Anis, *Transition Metal Chem.* (*London*), **14**, 131 (1989).
47. S. Rahhal and H. W. Richter, *Radiat. Phys. Chem.*, **32**, 129 (1988).
48. V. G. Isak, T. N. Nguyen, and A. Ya. Sychev, *Zh. Fiz. Khim.*, **62**, 1527 (1988); *Chem. Abstr.*, **109**: 80594d.
49. L. A. Grigor'eva, E. I. Dodin, and V. M. Kazazis, *Zh. Anal. Khim.*, **42**, 2197 (1987); *Chem. Abstr.*, **109**: 221484g.
50. S. K. Ghosh, K. Laali, and E. S. Gould, *Inorg. Chem.*, **27**, 4224 (1988).
51. G. Rabai, K. Kustin, and I. R. Epstein, *J. Am. Chem. Soc.*, **111**, 3870 (1989).
52. G. Rabai, K. Kustin, and I. R. Epstein, *J. Am. Chem. Soc.*, **111**, 8271 (1989).
53. H. J. Arnikar, A. H. Kapadi, A. S. Gohad, and S. B. Bhosale, *J. Chim. Phys. Phys.-Chim. Biol.*, **85**, 707 (1988).
54. Y. V. Geletti, V. V. Laurushko, and G. V. Lubimova, *J. Chem. Soc., Chem. Commun.*, 936 (1988).
55. M. Y. El-Sheikh, A. F. M. Habib, F. M. Ashmawy, A. H. Gemeay, and A. B. Zaki, *Transition Metal Chem.* (*London*), **13**, 96 (1988).
56. Y. Zhong and P. K. Lim, *J. Am. Chem. Soc.*, **111**, 8398 (1989).
57. P. K. Lim and Y. Zhong, *J. Am. Chem. Soc.*, **111**, 8404 (1989).
58. Y. Luo, M. Orban, K. Kustin, and I. R. Epstein, *J. Am. Chem. Soc.*, **111**, 4541 (1989).
59. Y. Luo, K. Kustin, and I. R. Epstein, *Inorg. Chem.*, **27**, 2489 (1988).
60. R. W. Cruse, S. Kaderli, C. J. Meyer, A. D. Zuberbühler, and K. D. Karlin, *J. Am. Chem. Soc.*, **110**, 5020 (1988).
61. M. Masarwa, H. Cohen, D. Meyerstein, D. L. Hickman, A. Bakac, and J. H. Espenson, *J. Am. Chem. Soc.*, **110**, 4293 (1988).
62. C. K. Ranganathan, T. Ramasami, D. Ramaswamy, and M. Santappa, *Inorg. Chem.*, **28**, 1306 (1989).
63. S. O. Travin, J. Fiala, and A. Vlcek, *Khim. Fiz.*, **6**, 660 (1987); *Chem. Abstr.*, **107**: 65473d.
64. N. Sadler, S. L. Scott, A. Bakac, J. H. Espenson, and M. S. Ran, *Inorg. Chem.*, **28**, 3951 (1989).
65. A. F. Ghiron and R. C. Thompson, *Inorg. Chem.*, **27**, 4766 (1988).
66. L. M. Schwane and R. C. Thompson, *Inorg. Chem.*, **28**, 3938 (1989).
67. R. Banerjee, A. Das, and S. Dasgupta, *J. Chem. Soc., Dalton Trans.*, 1645 (1989).

68. M. M. T. Khan, H. C. Bajaj, and M. R. H. Siddiqui, *J. Mol. Catal.*, **44**, 279 (1988).
69. L. J. Csanyi, Z. M. Galbacs, L. Nagy, and I. Horvath, *Acta Chim. Hung.*, **123**, 123 (1986).
70. T. Tomohiro, T. Laitalainen, T. Shimura, and Y. Okuno, *Stud. Org. Chem.* (*Amsterdam*), **33** (Role Oxygen Chem. Biochem.), 557 (1988).
71. A. Zanardo, R. A. Michelin, F. Pinna, and G. Strukul, *Inorg. Chem.*, **28**, 1648 (1989).
72. P. Battioni, J. P. Renaud, J. F. Bartolli, M. Reina-Artiles, M. Fort, and D. Mansuy, *J. Am. Chem. Soc.*, **110**, 8462 (1988).
73. I. G. Tananaev and V. I. Dzyubenko, *Radiokhimiya*, **30**, 842 (1988); *Chem. Abstr.*, **110**: 142256b.
74. G. A. Milovanovic, M. A. Sekheta, and A. Vassilios, *Microchem. J.*, **37**, 263 (1988).
75. D. T. Sawyer, M. S. McDowell, L. Spencer, and P. K. S. Tsang, *Inorg. Chem.*, **28**, 1166 (1989).
76. M. A. Ansari, J. Chandrasekaran, and S. Sarkar, *Polyhedron*, **7**, 471 (1988).
77. V. F. Toropov, Yu. G. Shtyrlin, and A. R. Garifzyanov, *React. Kinet. Catal. Lett.*, **38**, 381 (1989).
78. J. R. Lindsay Smith, P. N. Balasubramanian, and T. C. Bruice, *J. Am. Chem. Soc.*, **110**, 7411 (1988).
79. P. N. Balasubramanian, J. R. Lindsay Smith, M. J. Davies, T. W. Kaaret, and T. C. Bruice, *J. Am. Chem. Soc.*, **111**, 1477 (1989).
80. M. E. Brynildson, *Energy Res. Abstr.*, **13**, Abstr. No. 29293 (1988).
81. S.-I. Murahashi, T. Naota, and K. Yonemura, *J. Am. Chem. Soc.*, **110**, 8256 (1988).
82. T.-C. Lau, C.-M. Che, W.-O. Lee, and C.-K. Poon, *J. Chem. Soc., Chem. Commun.*, 1406 (1988).
83. G. Strukul, R. Sinigalia, A. Zanardo, F. Pinna, and R. A. Michelin, *Inorg. Chem.* **28**, 554 (1989).
84. P. S. Dixit and K. Srininvasan, *Inorg. Chem.*, **27**, 4507 (1988).
85. D. T. Sawyer, H. Sugimoto, and H.-C. Tung, *Stud. Org. Chem.* (*Amsterdam*), **33**, 407 (1988).
86. D. L. Hickman, A. Nanthakumar, and H. M. Goff, *J. Am. Chem. Soc.*, **110**, 6384 (1988).
87. J. R. Kincaid, A. J. Schneider, and K.-J. Paeng, *J. Am. Chem. Soc.*, **111**, 735 (1989).
88. K. R. Rodgers and H. M. Goff, *J. Am. Chem. Soc.*, **110**, 7049 (1988).
89. J. T. Groves and M. K. Stern, *J. Am. Chem. Soc.*, **110**, 8628 (1988).
90. W. H. Leung and C. M. Che, *J. Am. Chem. Soc.*, **111**, 8812 (1989).
91. C.-M. Che, W.-C. Chung, and T.-F. Lai, *Inorg. Chem.*, **27**, 2801 (1988).
92. J. T. Groves and Y. Watanabe, *J. Am. Chem. Soc.*, **110**, 8443 (1988).
93. H. Tomiyasu and G. Gordon, *Ozone: Sci. Eng.*, **11**, 59 (1989).
94. Y. Watanabe and Y. Ishimura, *J. Am. Chem. Soc.*, **111**, 410 (1989).
95. A. J. Castellino and T. C. Bruice, *J. Am. Chem. Soc.*, **110**, 7512 (1988).
96. D. Ostović and T. Bruice, *J. Am. Chem. Soc.*, **110**, 6906 (1988).
97. W. H. Leung and C. M. Che, *Inorg. Chem.*, **28**, 4619 (1989).
98. M. Bressan and A. Morvillo, *Inorg. Chem.*, **28**, 950 (1989).
99. J. F. Kinneary, J. S. Albert, and C. J. Burrows, *J. Am. Chem. Soc.*, **110**, 6124 (1988).
100. M. J. Nappa and R. J. McKinney, *Inorg. Chem.*, **27**, 3740 (1988).
101. M. W. Rophael, N. S. Petro, and L. B. Khalil, *J. Power Sources*, **22**, 133 (1988).
102. T. G. Costner and N. Ganapathisubramanian, *Inorg. Chem.*, **28**, 3620 (1989).
103. P. A. Lay and H. Taube, *Inorg. Chem.*, **28**, 3561 (1989).
104. M. P. Heyward and C. F. Wells, *J. Chem. Soc., Dalton Trans.*, 1331 (1988).
105. B. K. Park, K. J. Kim, and J. S. Lim, *Taehan Hwahakhoe Chi*, **33**, 309 (1989); *Chem. Abstr.*, **111**: 202805n.
106. S. K. Ghosh and E. S. Gould, *Inorg. Chem.*, **27**, 4228 (1988).
107. A. R. Nadig, H. S. Yathirajan, and Rangaswamy, *J. Indian Chem. Soc.*, **65**, 164 (1988).
108. D. Esjornson, D. R. Derringer, P. E. Fansick, and R. A. Walton, *Inorg. Chem.*, **28**, 2821 (1989).
109. D. R. Root and R. A. Walton, *Inorg. Chem.*, **28**, 2503 (1989).
110. J. Iqbal, D. W. A. Sharp, and J. M. Winfield, *J. Chem. Soc., Dalton Trans.*, 461 (1989).
111. L. Li, R. E. Perrier, D. R. Eaton, and M. J. McGlinchey, *Can. J. Chem.*, **67**, 1868 (1989).
112. M. R. Goyal, R. K. Mittal, and Y. K. Gupta, *Indian J. Chem., Sect. A*, **27A**, 487 (1988).
113. M. R. Goyal, P. Bhatnagar, R. K. Mittal, and Y. K. Gupta, *Indian J. Chem., Sect. A*, **28A**, 382 (1989).

114. M. R. Goyal, P. Bhatnagar, R. K. Mittal, and Y. K. Gupta, *Indian J. Chem., Sect. A*, **28A**, 280 (1989).
115. G. Bazsa, I. Lengyel, and W. Linert, *J. Chem. Soc., Faraday Trans. 1*, **85**, 3273 (1989).
116. V. Zang, M. Kotowski, and R. van Eldik, *Inorg. Chem.*, **27**, 3279 (1988).
117. M. R. Rhodes and T. J. Meyer, *Inorg. Chem.*, **27**, 4772 (1988).
118. D. M. Stanbury, R. Martinez, E. Tseng, and C. E. Miller, *Inorg. Chem.*, **27**, 4277 (1988).
119. J. A. Craig and R. H. Holm, *J. Am. Chem. Soc.*, **111**, 2111 (1989).
120. Y. Yamamoto and K. Hidetaka, *Bull. Chem. Soc. Jpn.*, **60**, 1299 (1987).
121. W. Micklitz, G. Muller, B. Huber, J. Riede, F. Rashwan, J. Heinze, and B. Lippert, *J. Am. Chem. Soc.*, **110**, 7084 (1988).
122. I. Willner, N. Lapidot, and A. Riklin, *J. Am. Chem. Soc.*, **111**, 1883 (1989).
123. C. Brewer, *Inorg. Chim. Acta*, **150**, 189 (1988).
124. M. T. Stershic, L. K. Keefer, B. P. Sullivan, and T. J. Meyer, *J. Am. Chem. Soc.*, **110**, 6884 (1988).
125. F. Monacelli and G. O. Morpurgo, *Inorg. Chim. Acta*, **149**, 139 (1988).
126. F. Wang and L. M. Sayre, *Inorg. Chem.*, **28**, 169 (1989).
127. K. Hegetschweiler and P. Saltman, *Inorg. Chem.*, **25**, 107 (1986).
128. H. A. Garrera, J. J. Cosa, and C. M. Previtali, *J. Photochem. Photobiol. A*, **47**, 143 (1989).
129. N. Kitamura, R. Obata, H. B. Kim, and S. Tazuke, *J. Phys. Chem.*, **93**, 5764 (1989).
130. C. B. Castellani, A. Buttafava, O. Carugo, and A. Poggi, *J. Chem. Soc., Dalton Trans.*, 1497 (1988).
131. J. S. Miller, D. T. Glatzhofer, D. M. O'Hare, W. M. Reiff, A. Chakraborty, and A. J. Epstein, *Inorg. Chem.*, **28**, 2930 (1989).
132. C. L. Hill, D. A. Bouchard, M. Kadkhodayan, M. M. Williamson, J. A. Schmidt, and E. F. Hilinski, *J. Am. Chem. Soc.*, **110**, 5471 (1988).
133. K. A. Kumar, P. Sivaswaroop, K. J. Rao, and P. V. K. Rao, *Transition Metal Chem. (London)*, **12**, 441 (1987).
134. F. Sanchez, M. J. Nasarre, M. M. Graciani, R. Jimenez, M. L. Moya, J. Burgess, and M. J. Blandamer, *Transition Metal Chem. (London)*, **13**, 150 (1988).
135. K. E. Linder, J. C. Dewan, and A. Davison, *Inorg. Chem.*, **28**, 3820 (1989).
136. S. Zhang and R. E. Shepherd, *Inorg. Chem.*, **27**, 4712 (1988).
137. J. A. Oblabe and A. Haim, *Inorg. Chem.*, **28**, 3278 (1989).
138. S. Nishida and M. Kimura, *J. Chem. Soc., Dalton Trans.*, 357 (1989).
139. M. Orbán and I. R. Epstein, *J. Am. Chem. Soc.*, **111**, 2891 (1989).
140. M. D. Johnson and R. J. Balahura, *Inorg. Chem.*, **27**, 3104 (1988).
141. C. A. Simmons, A. Bakac, and J. H. Espenson, *Inorg. Chem.*, **28**, 581 (1989).
142. M. N. Bhattacharjee, M. K. Chaudhuri, and N. S. Islam, *Inorg. Chem.*, **28**, 2420 (1989).
143. C. King, D. D. Heinrich, G. Garzon, J. C. Wang, and J. P. Fackler, Jr., *J. Am. Chem. Soc.*, **111**, 2300 (1989).
144. W. R. Scheidt, Y. J. Lee, and M. G. Finnegan, *Inorg. Chem.*, **27**, 4725 (1988).
145. K. K. Sen Gupta, S. Das, and S. Sen Gupta, *Transition Metal Chem. (London)*, **12**, 417 (1987).
146. M. H. Conklin and M. R. Hoffmann, *Environ. Sci. Technol.*, **22**, 891 (1988).
147. M. H. Conklin and M. R. Hoffmann, *Environ. Sci. Technol.*, **22**, 899 (1988).
148. L. J. Kirschenbaum, I. Kouadio, and E. Mentasti, *Polyhedron*, **8**, 1299 (1989).
149. K. K. Sen Gupta, S. Das, and S. Sen Gupta, *J. Chem. Res. (S)*, 112 (1989).
150. V. M. Bobba, G. Giraudi, and E. Mentasti, *Transition Metal Chem. (London)*, **13**, 256 (1988).
151. H. Li, *Wuji Huaxue*, **4**, 114 (1988); *Chem. Abstr.*, **110**: 82957u.
152. K. D. Fogelman, D. M. Walker, and D. W. Margerum, *Inorg. Chem.*, **28**, 986 (1989).
153. E. A. Betterton and M. R. Hoffmann, *J. Phys. Chem.*, **92**, 5962 (1988).
154. J. Kraft and R. Van Eldik, *Inorg. Chem.*, **28**, 2306 (1989).
155. K. K. Sen Gupta, S. Das, and S. Sen Gupta, *Transition Metal Chem. (London)*, **13**, 155 (1988).
156. M. Kimura and M. Ishibashi, *Inorg. Chim. Acta*, **129**, 69 (1987).
157. D. Dolphin, A. Matsumoto, and C. Shortman, *J. Am. Chem. Soc.*, **111**, 411 (1989).
158. R. L. Cook, M. Woods, J. C. Sullivan, and E. H. Appelman, *Inorg. Chem.*, **28**, 3352 (1989).

159. R. L. Cook, M. Woods, J. C. Sullivan, and E. H. Appelman, *Inorg. Chem.*, **28**, 3349 (1989).
160. Yu. V. Polenov, V. V. Budanov, and G. D. Kublashuili, *Izv. Vyssh. Uchebn. Zaved., Khim. Khim. Tekhnol.*, **31**, 58 (1988).
161. A. Hammershoei, R. M. Harkhorn, and A. M. Sargeson, *J. Chem. Soc., Chem. Commun.*, 1267 (1988).
162. S. B. Rabin, Ph.D. Dissertation, *Diss. Abstr. Int. B*, **49**, Pt. 1, 5300 (1989).
163. W. P. Lu and D. P. Kelley, *J. Gen. Microbiol.*, **134**, 865 (1988).
164. A. R. Butler, A. M. Calsy-Harrison, C. Glidewell, and E. Sorenson, *Polyhedron*, 7, 1197 (1988).
165. S. Colonna, N. Goggero, A. Manfredi, L. Casella, and M. Gulotti, *J. Chem. Soc., Chem. Commun.*, 1451 (1988).
166. M. Kumar, R. O. Day, G. J. Colpas, and M. J. Maroney, *J. Am. Chem. Soc.*, 111, 5974 (1989).
167. B. R. James, A. Pacheco, S. J. Rettig, and J. A. Ivers, *Inorg. Chem.*, 27, 2414 (1988).
168. M. Aresta, E. Quaranta, and I. Tommasi, *J. Chem. Soc., Chem. Commun.*, 450 (1988).
169. H. J. Krueger and R. H. Holm, *Inorg. Chem.*, **28**, 1148 (1989).
170. S. J. Brown, S. E. Hudson, D. W. Stephan, and P. K. Mascherak, *Inorg. Chem.*, **28**, 468 (1989).
171. P. S. K. Leung, E. A. Betterton, and M. R. Hoffmann, *J. Phys. Chem.*, **93**, 430 (1989).
172. P. O'Brien and Z. Ozolins, *Inorg. Chim. Acta*, **161**, 261 (1989).
173. S. Kitagawa, H. Seki, F. Kametani, and H. Sakurai, *Inorg. Chim. Acta*, **152**, 251 (1988).
174. M. M. Giurgiu, *Stud. Univ. Babes-Bolyai Chem.*, **32**, 15 (1987).
175. F. J. Andres Ordax, A. Arrizabalaga, F. J. Bravo, and M. Y. Fernandez de Aranguiz, *An. Quim. Ser. A*, **85**, 224 (1989); *Chem. Abstr.*, **111**: 181788e.
176. L. J. Kirschenbaum and R. K. Panda, *Polyhedron*, 7, 2753 (1988).
177. R. J. Balahura, M. Johnson, and T. Black, *Inorg. Chem.*, **28**, 3933 (1989).
178. R. A. Holwerda, *J. Inorg. Biochem.*, **33**, 131 (1988).
179. S. K. Ghosh and E. S. Gould, *Inorg. Chem.*, **28**, 3651 (1989).
180. R. Trismitro and H. N. Po, *J. Coord. Chem.*, **17**, 1 (1988).
181. Yu. A. Maletin, N. G. Strizhakova, T. V. Verkhoulyuk, and I. A. Sheka, *Teor. Eksp. Khim.*, **24**, 450 (1988); *Chem. Abstr.*, **109**: 177510m.
182. V. Singh, *J. Inst. Chem. (India)*, **59**, 248 (1987).
183. M. G. Kanatzidis and S.-P. Huang, *Inorg. Chem.*, **28**, 4667 (1989).
184. R. R. Babu, P. Vani, and L. S. A. Dikshitulu, *Indian J. Chem., Sect. A*, **26A**, 1027 (1987).
185. M. Hojo and D. T. Sawyer, *Inorg. Chem.*, **28**, 1201 (1989).
186. K. Ishikawa, S. Fukuzumi, and T. Tanaka, *Inorg. Chem.*, **28**, 1661 (1989).
187. P. J. Toscano, E. Barren, and A. L. Seligson, *Organometallics*, **2**, 2085 (1989).
188. We J. Wang and T. M. Yu, *Synth. Met.*, **29**, F145 (1988).
189. M. Kumar, R. O. Day, G. J. Colpas, and M. J. Maroney, *J. Am. Chem. Soc.*, 111, 5974 (1989).
190. E. M. Rubtsov, *Radiokhimiya*, **30**, 699 (1988); *Chem. Abstr.*, **110**: 32965z.
191. D. Fox and C. F. Wells, *J. Chem. Soc., Dalton Trans.*, 151 (1989).
192. A. Mills and A. Cook, *J. Chem. Soc., Faraday Trans. 1*, **84**, 1691 (1988).
193. M. P. Hayward and C. F. Wells, *Inorg. Chim. Acta*, **148**, 241 (1988).
194. S. Truffer-Caron, E. Ianoc, and P. Lerch, *Inorg. Chem. Acta*, **149**, 119 (1988).
195. P. Uma, P. K. Rao, and M. N. Sashry, *React. Kinet. Catal. Lett.*, **39**, 255 (1989).
196. S. E. Castillo-Blum, D. T. Richens, and A. G. Sykes, *Inorg. Chem.*, **28**, 954 (1989).
197. A. Kamkar and J. P. Day, *Iran. J. Chem. Eng.*, **10–11**, Pt. A, 3 (1988).
198. Yu. A. Maletin, K. E. Gulyanitskii, V. V. Orda, and I. I. Maletina, *Ukr. Khim. Zh. (Russ. Ed.)*, **54**, 816 (1988); *Chem. Abstr.*, **110**: 102565v.
199. C. S. Reddy and E. V. Sundaram, *React. Kinet. Catal. Lett.*, **36**, 131 (1988).
200. A. J. Elliott, S. Geersten, and G. V. Buxton, *J. Chem. Soc., Faraday Trans. 1*, 1101 (1988).
201. W. M. Song, K. Kustin, and I. R. Epstein, *J. Phys. Chem.*, **93**, 4698 (1989).
202. J. C. Nagy, K. Kumar, and D. W. Margerum, *Inorg. Chem.*, 27, 2773 (1988).
203. A. Peloso, *J. Chem. Soc., Dalton Trans.*, 1577 (1988).
204. N. A. Lewis, W. L. Purcell, and D. V. Taueras, *Inorg. Chem.*, **28**, 133 (1989).
205. G. Rabai and I. R. Epstein, *Inorg. Chem.*, **28**, 732 (1989).

206. E. C. Edblom, Y. Luo, M. Orban, K. Kustin, and I. R. Epstein, *J. Phys. Chem.*, **93**, 2722 (1989).

207. J. J. Jwo and E. F. Chang, *J. Chin. Chem. Soc. (Taipei)*, **36**, 115 (1989).

208. L. P. Tikhonova, F. S. Moshkovich, and A. S. Kovalenko, *Ukr. Khim. Zh. (Russ. Ed.)*, **54**, 1045 (1988); *Chem. Abstr.*, **110**: 102576x.

209. A. A. Abdel-Khalek and M. M. Elsemongy, *Bull. Chem. Soc. Jpn.*, **61**, 4407 (1988).

210. A. A. Abdel-Khalek and M. M. Elsemongy, *Transition Metal Chem. (London)*, **14**, 206 (1989).

211. D. E. Linn, Jr., S. K. Ghosh, and E. S. Gould, *Inorg. Chem.*, **28**, 3225 (1989).

212. R. E. Rodriguez and H. C. Kelly, *Inorg. Chem.*, **28**, 589 (1989).

213. R. W. Lee, P. C. Nakagaki, and T. C. Bruice, *J. Am. Chem. Soc.*, **111**, 1368 (1989).

214. Z. Zhang, X. Zhang, J. Lu, and C. Pan, *Yejin Fenxi*, **7**, 6 (1987); *Chem. Abstr.*, **109**: 103793j.

215. H. Zhao, H. Li, and Z. Qiu, *Fenxi Huaxue*, **16**, 480 (1988); *Chem. Abstr.*, **109**: 103809c.

216. P. Ruoff and R. M. Noyes, *J. Phys. Chem.*, **93**, 7394 (1989).

217. *J. Phys. Chem.*, **93**(7), (1989).

218. *J. Chem. Ed.*, **66**(3), (1989).

219. R. N. Mehrotra and L. J. Kirschenbaum, *Inorg. Chem.*, **28**, 4327 (1989).

220. D. E. Linn, Jr. and E. S. Gould, *Inorg. Chem.*, **27**, 3140 (1988).

221. K. K. S. Gupta and J. Karak, *J. Chem. Res. (S)*, 258 (1989).

222. G. Sarala, P. J. Rao, B. Sethuram, and T. N. Rao, *Transition Metal Chem. (London)*, **13**, 113 (1988).

223. P. Sivaswaroop, K. A. Kumar, and P. V. K. Rao, *React. Kinet. Catal. Lett.*, **39**, 33 (1989).

224. P. Hendry and A. Sargeson, *Inorg. Chem.*, **29**, 92, 97 (1990).

225. Ya. A. Dorfman, G. S. Polimbetova, I. V. Kelman, D. M. Doroshkevich, T. V. Petrova, B. A. Mansurov, A. K. Bikmukhametova, and I. M. Yukht, *Koord. Khim.*, **15**, 77 (1989); *Chem. Abstr.*, **110**: 180124q.

226. Chi Ming Che and Kwok Yin Wong, *J. Chem. Soc., Dalton Trans.*, 2065 (1989).

227. W. Kläui and Choong-Eui Sung, *Inorg. Chem.*, **28**, 3845 (1989).

228. A. Okumaura and Y. Daido, *Inorg. Chim. Acta*, **144**, 63 (1988).

229. P. Bernhard and F. C. Anson, *Inorg. Chem.*, **27**, 4574 (1988).

230. M. Perrée-Fauvet, A. Gaudemer, J. Bonvoisin, J. J. Girerd, C. Boucly-Goester, and P. Boucly, *Inorg. Chem.*, **28**, 3533 (1989).

231. Q. Lao, M. Shen, Q. Peng, and Z. Zhang, *Gaodeng Xuexiao Huaxue Suebao*, **10**, 335 (1989); *Chem. Abstr.*, **112**: 29783z.

232. A. S. Dubrovina, A. I. Malkova, L. M. Artem'eva, and V. I. Tupikov, *Zh. Fiz. Khim.*, **62**, 1904 (1988); *Chem. Abstr.*, **109**: 157244z.

233. P. V. Bernhardt, G. A. Lawrance, and D. F. Sangster, *Inorg. Chem.*, **27**, 4055 (1988).

234. C. Lierse, K. H. Schmidt, and J. C. Sullivan, *Radiochim. Acta*, **44–45** (Pt. 1), 71 (1988).

235. A. V. Gogolev, V. P. Shilov, and A. K. Pikaer, *Izv. Akad. Nauk SSR, Ser. Khim.*, 1001 (1988); *Chem. Abstr.*, **109**: 44800j.

236. E. Bothe and R. K. Broszkiewicz, *Inorg. Chem.*, **28**, 2988 (1989).

237. W. L. Waltz, J. Lilie, A. Goursot, and H. Chermette, *Inorg. Chem.*, **28**, 2247 (1989).

238. U. K. Kläning, B. H. J. Bielski, and K. Sehested, *Inorg. Chem.*, **28**, 2717 (1989).

239. M. Faraggi and M. H. Klapper, *J. Am. Chem. Soc.*, **110**, 5753 (1988).

240. P. Osvath, G. A. Salmon, and A. G. Sykes, *J. Am. Chem. Soc.*, **110**, 7114 (1988).

241. S. Goldstein, G. Czapski, H. Cohen, and D. Meyerstein, *J. Am. Chem. Soc.*, **110**, 3903 (1988).

242. K. Tsukahara and R. G. Wilkins, *Inorg. Chem.*, **28**, 1605 (1989).

243. T. V. Khmelinskii, V. F. Plyusnin, and V. P. Grivin, *Izv. Akad. Nauk SSSR, Ser. Khim.*, 1247 (1988); *Chem. Abstr.*, **109**: 157204m.

244. M. R. Faraggi and M. H. Klapper, *Radiat. Phys. Chem.*, **32**, 293 (1988).

245. K. R. Howes, A. Bakac, and J. H. Espenson, *Inorg. Chem.*, **28**, 579 (1989).

246. W. A. Nugent and T. V. RajanBabu, *J. Am. Chem. Soc.*, **110**, 8561 (1988).

247. M. J. Hynes and D. F. Kelly, *J. Chem. Soc., Chem. Commun.*, 849 (1988).

248. P. Martinez and D. Uribe, *J. Chim. Phys.*, **78**, 47 (1981).

249. M. B. Davies, R. J. Mortimer, and T. R. Vine, *Inorg. Chim. Acta*, **146**, 59 (1988).

250. A. Yu. Nazarenko, N. A. Lipkovskaya, and I. A. Poluyanovich, *Ukr. Khim. Zh.* (*Russ. Ed.*), **54**, 578 (1988); *Chem. Abstr.*, 109: 177470y.

251. S. K. Ghosh and E. S. Gould, *Inorg. Chem.*, **28**, 1538 (1989).

252. S. K. Ghosh and E. S. Gould, *Inorg. Chem.*, **28**, 1948 (1989).

253. M. M. Taqui Khan, R. S. Shukla, and A. P. Rao, *Inorg. Chem.*, **28**, 452 (1989).

254. O. A. Travina and S. O. Travin, *Khim. Fiz.*, **6**, 350 (1987); *Chem. Abstr.*, **107**; 65425q.

255. S. Banwart, S. Davies, and W. Stumm, *Colloids Surf.*, **39**, 303 (1989).

256. J. J. Ruiz, J. M. Rodriguez-Mellado, M. Dominguez, and A. Aldaz, *J. Chem. Soc., Faraday Trans. 1*, **85**, 1567 (1989).

257. G. R. Buettner, *J. Biochem. Biophys. Methods*, **16**, 27 (1988).

258. A. McAuley, C. J. Macdonald, L. Spencer, and P. R. West, *J. Chem. Soc., Dalton Trans.*, 2279 (1988).

259. C. Benelli, A. Dei, D. Gatteschi, and L. Pardi, *J. Am. Chem. Soc.*, **110**, 6897 (1988).

260. G. A. Abakumov, V. A. Garnov, V. I. Neudchikov, and V. K. Cherkasov, *Dokl. Akad. Nauk SSSR*, **304**, 107 (1989); *Chem. Abstr.*, **110**: 241558h.

261. S. Ernest, P. Haenel, J. Jordanov, W. Kaim, V. Kasack, and E. Roth, *J. Am. Chem. Soc.*, **111**, 1733 (1989).

262. L. A. deLearie and C. G. Pierpont, *Inorg. Chem.*, **27**, 3842 (1988).

263. D. K. Smith, L. M. Tender, G. A. Lane, S. Licht, and M. S. Wrighton, *J. Am. Chem. Soc.*, **111**, 1099 (1989).

264. M. R. Malachowski, M. G. Davidson, and J. N. Hoffman, *Inorg. Chim. Acta*, **157**, 91 (1989).

265. R. B. Jordan and Jinhuang Xu, *Pure Appl. Chem.*, **60**, 1205 (1988).

266. Jinhuang Xu and R. B. Jordan, *Inorg. Chem.*, **27**, 4563 (1988).

267. M. Tachibana and M. Iwaizumi, *Nippon Kagaku Kaishi*, 701 (1988); *Chem. Abstr.*, **109**: 84996j.

268. M. R. Wasielewski, D. G. Johnson, W. A. Svec, K. M. Kersey, D. E. Cragg, and D. W. Minsek, Photochem. Energy Convers., Proc. Int. Conf. Photochem. Convers. Storage Solar Energy, **7th**, 135 (1988).

269. W. H. Kim and G. R. Choppin, *Inorg. Chem.*, **27**, 2771 (1988).

270. J. D. Druliner and E. Wasserman, *J. Am. Chem. Soc.*, **110**, 5270 (1988).

271. R. G. Cunninghame, L. R. Hanton, S. D. Jensen, B. H. Robinson, and J. Simpson, *Organometallics*, **6**, 1470 (1987).

272. N. Zhang, C. M. Mann, and P. A. Shapley, *J. Am. Chem. Soc.*, **110**, 6591 (1988).

273. R. A. Leising and K. J. Takeuchi, *J. Am. Chem. Soc.*, **110**, 4079 (1988).

274. W. K. Seok and T. J. Meyer, *J. Am. Chem. Soc.*, **110**, 7358 (1988).

275. B. Divisia-Blohorn and M. Paltrier, *Inorg. Chim. Acta*, **131**, 65 (1987).

276. T. Katsuyama, A. Bakac, and J. H. Espenson, *Inorg. Chem.*, **28**, 339 (1989).

277. H. C. Sutton, *J. Chem. Soc., Faraday Trans. 1*, **85**, 883 (1989).

278. R. Ruppert, S. Herrmann and E. Steckhan, *J. Chem. Soc., Chem. Commun.*, 1150 (1988).

279. I. I. Creaser, M. J. Lancaster, G. A. Lawrance, A. M. Sargeson, and J. C. Sullivan, *Aust. J. Chem.*, **41**, 1275 (1988).

280. R. Banerjee, K. Das, A. Das, and S. Dasgupta, *Inorg. Chem.*, **28**, 585 (1989).

281. R. P. Farrell, R. J. Judd, P. A. Lay, R. Bramley, and J. Y. Ji, *Inorg. Chem.*, **28**, 3401 (1989).

282. M. Mitewa and R. Bontcher, *Coord. Chem.*, **61**, 214 (1985).

283. C. R. Steffan, A. Bakac, and J. H. Espenson, *Inorg. Chem.*, **28**, 2992 (1989).

284. S. Nishida, Y. Harima, and K. Yamashita, *Inorg. Chem.*, **28**, 4073 (1989).

285. K. Subramani and V. S. Srinivasan, *J. Chem. Soc., Dalton Trans.*, 1413 (1988).

286. R. K. Panda and L. J. Kirschenbaum, *J. Chem. Soc., Dalton Trans.*, 217 (1989).

287. E. Fujita, D. J. Szalda, C. Creutz, and N. Sutin, *J. Am. Chem. Soc.*, **110**, 4870 (1988).

288. C. M. Bolinger, N. Story, B. P. Sullivan, and T. J. Meyer, *Inorg. Chem.*, **27**, 4582 (1988).

289. K. Bouzouita, J. Desmaison, and M. Billy, *J. Less-Common Met.*, **138**, 349 (1988).

290. E. L. Kuksenko, V. A. Golodov, and M. T. Abilou, *React. Kinet. Catal. Lett.*, **38**, 215 (1989).

291. H. J. Lawson and J. D. Atwood, *J. Am. Chem. Soc.*, **110**, 3680 (1988).

292. K. S. Sun and C. H. Cheng, *J. Am. Chem. Soc.*, **110**, 6744 (1988).

293. Y. Zhen, W. G. Feighery, C. K. Lai, and J. D. Atwood, *J. Am. Chem. Soc.*, **111**, 7832 (1989).
294. A. M. Stolzenberg and M. J. Stershic, *J. Am. Chem. Soc.*, **110**, 5397 (1988).
295. M. S. Ram, A. Bakac, and J. H. Espenson, *Inorg. Chem.*, **27**, 2011 (1988).
296. M. S. Ram, A. Bakac, and J. H. Espenson, *Inorg. Chem.*, **27**, 4231 (1988).
297. G. B. Maiya, B. C. Han, and K. M. Kadish, *Langmuir*, **5**, 645 (1989).
298. R. H. Hill, *J. Chem. Soc.*, *Chem. Commun.*, 293 (1989).
299. E. Kita and R. B. Jordan, *Inorg. Chem.*, **28**, 1549 (1989).
300. C. P. Andrieux, I. Gallardo, and J.-M. Savéant, *J. Am. Chem. Soc.*, **111**, 1620 (1989).
301. D. Lexa, J.-M. Savéant, K.-B. Su, and D.-L. Wang, *J. Am. Chem. Soc.*, **110**, 7617 (1988).
302. Y. Sorek, H. Cohen, and D. Meyerstein, *J. Chem. Soc.*, *Faraday Trans. 1*, **85**, 1169 (1989).
303. A. Sauer, H. Cohen, and D. Meyerstein, *Inorg. Chem.*, **28**, 2511 (1989).
304. A. Bakac, V. Butkovic, J. H. Espenson, J. Lourić, and M. Orhanović, *Inorg. Chem.*, **28**, 4323 (1989).
305. K. Kusaba, H. Ogino, A. Bakac, and J. H. Espenson, *Inorg. Chem.*, **28**, 970 (1989).
306. A. Bakac and J. H. Espenson, *Inorg. Chem.*, **28**, 3901 (1989).
307. A. Bakac and J. H. Espenson, *Inorg. Chem.*, **28**, 4319 (1989).
308. A. Bakac and J. H. Espenson, *J. Am. Chem. Soc.*, **110**, 3453 (1988).
309. M. Tada, T. Nakamura, and M. Matsumoto, *J. Am. Chem. Soc.*, **110**, 4647 (1988).
310. A. Sauer, H. Cohen, and D. Meyerstein, *Inorg. Chem.*, **27**, 4578 (1988).
311. J. D. Rush and B. H. J. Bielski, *Inorg. Chem.*, **28**, 3947 (1989).
312. D. E. Cabelli, J. D. Rush, M. J. Thomas, and B. H. J. Bielski, *J. Phys. Chem.*, **93**, 3579 (1989).
313. S. Goldstein, G. Czapski, H. Cohen, and D. Meyerstein, *Inorg. Chem.*, **27**, 4130 (1988).
314. M. A. Brusa, L. J. Perissinotti, and A. J. Colussi, *Inorg. Chem.*, **27**, 4474 (1988).
315. L. Brammer, N. G. Connelly, J. Edwin, W. E. Geiger, A. G. Orpen, and J. B. Sheridan, *Organometallics*, **7**, 1259 (1988).

References for Chapter 4

1. J. F. Stanton, W. N. Lipscomb, and R. J. Bartlett, *J. Am. Chem. Soc.*, **111**, 5165 (1989).
2. J. F. Stanton, W. N. Lipscomb, and R. J. Bartlett, *J. Am. Chem. Soc.*, **111**, 5173 (1989).
3. R. Greatrex, N. N. Greenwood, and S. D. Waterworth, *J. Chem. Soc.*, *Chem. Commun.*, 925 (1988).
4. J. J. Ott, C. A. Brown, and B. M. Gimarc, *Inorg. Chem.*, **28**, 4269 (1989).
5. S. Wu and M. Jones, Jr., *J. Am. Chem. Soc.*, **111**, 5373 (1989).
6. M. L. McKee, *J. Phys. Chem.*, **93**, 3426 (1989).
7. G. A. Olah, R. Aniszfelo, G. K. Surya Prakash, R. E. Williams, K. Lammertsma, and O. F. Güner, *J. Am. Chem. Soc.*, **110**, 7885 (1988).
8. J. H. Davis, Jr. and R. N. Grimes, *Inorg. Chem.*, **27**, 4213 (1988).
9. M. Yalpani, J. Serwatowski, and R. Köster, *Chem. Ber.*, **122**, 3 (1989).
10. A. Brandl, P. Kölle, and H. Nöth, *Chem. Ber.*, **122**, 419 (1989).
11. L. A. Jackson and C. W. Allen, *J. Chem. Soc.*, *Dalton Trans.*, 2423 (1989).
12. D. A. Saulys, J. Castillo, and J. A. Morrison, *Inorg. Chem.*, **28**, 1619 (1989).
13. J. N. Kirwan and B. P. Roberts, *J. Chem. Soc.*, *Perkin Trans. 2*, 539 (1989).
14. V. Paul and B. P. Roberts, *J. Chem. Soc.*, *Perkin Trans. 2*, 1895 (1988).
15. V. Paul and B. P. Roberts, *J. Chem. Soc.*, *Perkin Trans. 2*, 1183 (1988).
16. S. Padmaja, V. Ramakrishnan, J. Rajaran, and J. C. Kuriacose, *J. Chem. Soc.*, *Faraday Trans. 1*, **85**, 2249 (1989).
17. J. E. Crooks and J. P. Donnellan, *J. Chem. Soc.*, *Perkin Trans. 2*, 331 (1989).
18. N. P. Botting and B. C. Challis, *J. Chem. Soc.*, *Chem. Commun.*, 1585 (1989).
19. R. D. Gillard, N. Garif, and J. Pedrosa de Jesus, *Polyhedron*, **8**, 775 (1989).
20. Yu. Yu. Kharitonov, M. A. Ismail, and M. A. Sarukhamov, *Russ. J. Inorg. Chem.*, **33**, 1117 (1988).

21. J. R. Jones, K. T. Walkin, J. P. Davey, and E. Buncel, *J. Phys. Chem.*, **93**, 1362 (1989).
22. G. L. Larson, *J. Organomet. Chem.*, **374**, 1 (1989).
23. H.-G. Woo and T. D. Tilley, *J. Am. Chem. Soc.*, **111**, 8043 (1989).
24. L. S. Chang and J. Y. Corey, *Organometallics*, **8**, 1885 (1989).
25. M. Fujino and H. Isaka, *J. Chem. Soc., Chem. Commun.*, 466 (1989).
26. B. Becker and W. Wojnowski, *J. Organomet. Chem.*, **346**, 287 (1988).
27. B. Becker, R. J. P. Corriu, C. Guérin, and B. J. L. Henner, *J. Organomet. Chem.*, **369**, 147 (1989).
28. J. L. Brefort, R. J. P. Corriu, C. Guérin, and B. J. L. Henner, *J. Organomet. Chem.*, **370**, 9 (1989).
29. C. Chuit, R. J. P. Corriu, and C. Reyé, *J. Organomet. Chem.*, **358**, 57 (1988).
30. R. Damrauer, L. W. Burggraf, L. P. Davis, and M. S. Gordon, *J. Am. Chem. Soc.*, **110**, 6601 (1988).
31. J. Boyer, C. Brelière, F. Carré, R. J. P. Corriu, A. Kpoton, M. Poirier, G. Royo, and J. C. Young, *J. Chem. Soc., Dalton Trans.*, 43 (1989).
32. D. A. Dixon, W. R. Hertler, B. Chase, W. B. Farnham, and F. Davidson, *Inorg. Chem.*, **27**, 4012 (1988).
33. R. Köster, G. Seidel, B. Wrackmeyer, K. Horchler, and D. Schlosser, *Angew. Chem., Int. Ed. Engl.*, **28**, 918 (1989).
34. B. Wrackmeyer and H. Zhou, *J. Organomet. Chem.*, **375**, 1 (1989).
35. M. Kira, K. Sato, C. Kabuto, and H. Sakurai, *J. Am. Chem. Soc.*, **111**, 3747 (1989).
36. S. Gronert, R. Glaser, and A. Streitwieser, *J. Am. Chem. Soc.*, **111**, 3111 (1989).
37. S. E. Johnson, J. S. Payne, R. O. Day, J. M. Holmes, and R. R. Holmes, *Inorg. Chem.*, **28**, 3190 (1989).
38. S. E. Johnson, R. O. Day, and R. R. Holmes, *Inorg. Chem.*, **28**, 3182 (1989).
39. J. A. Deiters, R. R. Holmes, and J. M. Holmes, *J. Am. Chem. Soc.*, **110**, 7672 (1988).
40. S. E. Johnson, J. A. Deiters, R. O. Day, and R. R. Holmes, *J. Am. Chem. Soc.*, **111**, 3250 (1989).
41. R. Damrauer, B. O'Connell, S. E. Danoahey, and R. Simon, *Organometallics*, **8**, 1167 (1989).
42. C. Brelière, R. J. P. Corriu, G. Royo, and J. Zwecker, *Organometallics*, **8**, 1834 (1989).
43. M. Kira, K. Sato, and H. Sakurai, *J. Am. Chem. Soc.*, **110**, 4599 (1988).
44. K. M. Johnson and B. P. Roberts, *J. Chem. Res. (S)*, 352 (1989).
45. I. G. Green, K. M. Johnson, and B. P. Roberts, *J. Chem. Soc., Perkin Trans. 2*, 1963 (1989).
46. K. M. Johnson and B. P. Roberts, *J. Chem. Soc., Perkin Trans. 2*, 1111 (1989).
47. A. J. McKinley, T. Karatsu, G. M. Wallraff, R. D. Miller, R. Sooriyakumaran, and J. Michl, *Organometallics*, **7**, 2567 (1988).
48. P. Boudjouk, U. Samaraweera, R. Sooriyakumaran, J. Chrusciel, and K. R. Anderson, *Angew. Chem., Int. Ed. Engl.*, **27**, 1355 (1988).
49. K. M. Welsh, J. Michl, and R. West, *J. Am. Chem. Soc.*, **110**, 6689 (1988).
50. H. Shizuka, K. Murata, Y. Arai, K. Tonokura, H. Tanaka, H. Matsumoto, Y. Nagai, G. Gillette, and R. West, *J. Chem. Soc., Faraday Trans. 1*, **85**, 2369 (1989).
51. I. N. Jung, D. H. Pae, B. R. Yoo, M. E. Lee, and P. R. Jones, *Organometallics*, **8**, 2017 (1989).
52. M. G. Steinmetz, B. S. Udayakumae, and M. S. Gordon, *Organometallics*, **8**, 530 (1989).
53. N. Wiberg, K. Schurz, G. Müller, and J. Rieve, *Angew. Chem., Int. Ed. Engl.*, **27**, 935 (1988).
54. J. B. Lambert and W. Schile, *J. Am. Chem. Soc.*, **110**, 6364 (1988).
55. S. K. Shin and J. L. Beauchamp, *J. Am. Chem. Soc.*, **111**, 900 (1989).
56. J. J. Eisch and C. S. Chiu, *J. Organomet. Chem.*, **358**, C1 (1988).
57. J. C. Ellington, Jr. and E. M. Arnett, *J. Am. Chem. Soc.*, **110**, 7778 (1988).
58. S. Rubinsztajn, M. Cypryk, and J. Chojnowski, *J. Organomet. Chem.*, **367**, 27 (1989).
59. S. L. Jones and C. J. M. Stirling, *J. Chem. Soc., Chem. Commun.*, 1153 (1988).
60. S. D. Kinrade and T. W. Swaddle, *Inorg. Chem.*, **27**, 4259 (1988); S. D. Kinrade and T. W. Swaddle, *Inorg. Chem.*, **27**, 4253 (1988).
61. C. T. G. Knight, A. R. Thompson, A. C. Kunwar, H. S. Gutowsky, E. Oldfield, and R. J. Kirkpatrick, *J. Chem. Soc., Dalton Trans.*, 275 (1989).
62. A. V. McCormick, A. T. Bell, and C. J. Radke, *J. Phys. Chem.*, **93**, 1737 (1989).
63. J. Chmielecka, J. Chojnowski, W. A. Stańczyk, and C. A. Eaborn, *J. Chem. Soc., Perkin Trans. 2*, 865 (1989).

64. M. Cypryk, S. Rubinsztajn, and J. Chojnowski, *J. Organomet. Chem.*, **377**, 197 (1989).
65. K. Ito and T. Ibaraki, *Int. J. Chem. Kinet.*, **21**, 757 (1989).
66. I. Hasegawa and S. Sakka, *Bull. Chem. Soc. Jpn.*, **61**, 4087 (1988).
67. U. Scheim, R. Lehnert, A. Porzel, and K. Rühlmann, *J. Organomet. Chem.*, **356**, 141 (1988).
68. U. Scheim, A. Porzel, and K. Rühlmann, *J. Organomet. Chem.*, **354**, 31 (1988).
69. E. Suzuki, M. Okamoto, and Y. Ono, *Chem. Lett.*, 1487 (1989).
70. R. J. Klingler, T. R. Krause, and J. W. Rathke, *J. Organomet. Chem.*, **352**, 81 (1988).
71. J. Kowalski and J. Chojnowski, *J. Organomet. Chem.*, **356**, 285 (1988).
72. R. J. Perry, *Organometallics*, **8**, 906 (1989).
73. S. Tomoda, M. Shimoda, Y. Takeuchi, Y. Kajii, K. Obi, I. Tanaka, and K. Honda, *J. Chem. Soc., Chem. Commun.*, 910 (1988).
74. A. J. Shusterman, B. E. Landrum, and R. L. Miller, *Organometallics*, **8**, 1851 (1989).
75. K. Mochida, M. Wakasa, Y. Nakadaira, Y. Sakaguchi, and H. Hayashi, *Organometallics*, **7**, 1869 (1988).
76. W. Ando and T. Tsumaraya, *Organometallics*, **7**, 1882 (1988).
77. T. Tsumaraya, S. Sato, and W. Ando, *Organometallics*, **7**, 2015 (1988).
78. M. P. Egorov, S. P. Kolesnikov, O. M. Nefedov, and A. Krebs, *J. Organomet. Chem.*, **375**, C5 (1989).
79. G. Billeb, H. Brauer, S. Maslov, and W. P. Neumann, *J. Organomet. Chem.*, **373**, 11 (1989).
80. J. Barrau, G. Rima, M. El-Amine, and J. Satgé, *J. Organomet. Chem.*, **345**, 39 (1988).
81. J. B. Lambert and W. Schilf, *Organometallics*, **7**, 1659 (1988).
82. N. S. Hosmane, M. S. Islam, B. S. Pinkston, U. Siriwardane, J. J. Banewicz, and J. A. Maguire, *Organometallics*, **7**, 2340 (1988).
83. J.-Y. Park and Y.-N. Lee, *J. Phys. Chem.*, **92**, 6294 (1988).
84. D. M. Stanbury, M. M. de Maine, and G. Goodloe, *J. Am. Chem. Soc.*, **111**, 5496 (1989).
85. A. Castro, M. Mosquera, M. F. Rodriguez Prieto, J. A. Santaballa, and J. V. Tato, *J. Chem. Soc., Perkin Trans. 2*, 1963 (1988).
86. V. Zang, M. Kotowski, and R. van Eldik, *Inorg. Chem.*, **27**, 3279 (1988).
87. M. J. Crookes and D. L. H. Williams, *J. Chem. Soc., Perkin Trans. 2*, 1319 (1989).
88. M. J. Crookes and D. L. H. Williams, *J. Chem. Soc., Perkin Trans. 2*, 1339 (1989).
89. M. J. Crookes, P. Roy, and D. L. H. Williams, *J. Chem. Soc., Perkin Trans. 2*, 1015 (1989).
90. M. J. Crookes and D. L. H. Williams, *J. Chem. Soc., Perkin Trans. 2*, 759 (1989).
91. H. M. S. Patel and D. L. H. Williams, *J. Chem. Soc., Perkin Trans. 2*, 339 (1989).
92. A. Coello, F. Meijide, and J. V. Tato, *J. Chem. Soc., Perkin Trans. 2*, 1677 (1989).
93. L. Abia, A. Castro, E. Iglesias, J. R. Leis, and M. E. Pena, *J. Chem. Res. (S)*, 106 (1989).
94. F. Meijide and G. Stedman, *J. Chem. Res. (S)*, 232 (1989).
95. A. Castro, M. González, F. Meijide, and M. Mosquera, *J. Chem. Soc., Perkin Trans. 2*, 2021 (1988).
96. E. Iglesias and D. L. H. Williams, *J. Chem. Soc., Perkin Trans. 2*, 343 (1989).
97. L. K. Keefer, J. A. Hrabie, B. D. Hilton, and D. Wilbur, *J. Am. Chem. Soc.*, **110**, 7459 (1988).
98. F. Radner, *Acta. Chem. Scand.*, **43**, 903 (1989).
99. S. M. N. Y. F. Oh and D. L. H. Williams, *J. Chem. Soc., Perkin Trans. 2*, 755 (1989).
100. A. Castro, J. R. Leis, and M. E. Pena, *J. Chem. Soc., Perkin Trans. 2*, 1861 (1989).
101. M. R. Crampton, J. K. Scranage, and P. Golding, *J. Chem. Res. (S)*, 72 (1989).
102. S. M. N. Y. F. Oh and D. L. H. Williams, *J. Chem. Res. (S)*, 264 (1989).
103. H. Zollinger, *Helv. Chim. Acta*, **71**, 1661 (1988).
104. G. C. M. Bourke, Ph.D. thesis, University of Wales (1985).
105. F. T. Bonner, C. E. Donald, and M. N. Hughes, *J. Chem. Soc., Dalton Trans.*, 527 (1989).
106. M. N. Hughes, P. E. Wimbledon, and G. Stedman, *J. Chem. Soc., Dalton Trans.*, 533 (1989).
107. E. Weeg-Aerssens, J. M. Tiedje, and B. A. Averill, *J. Am. Chem. Soc.*, **110**, 6851 (1988).
108. M. Horváth, I. Lengyel, and G. Bazsa, *Int. J. Chem. Kinet.*, **20**, 687 (1988).
109. I. Lengyel, I. Nagy, and G. Bazsa, *J. Phys. Chem.*, **93**, 2801 (1989).
110. D. M. Stanbury, R. Martinez, E. Tseng, and C. E. Miller, *Inorg. Chem.*, **27**, 4277 (1988).
111. D. Gaude, G. Gellon, R. Le Goaller, and J.-L. Pierre, *Can. J. Chem.*, **67**, 104 (1989).

112. A. Boughriet, A. Coumare, J. C. Fischer, and M. Wartez, *Int. J. Chem. Kinet.*, **20**, 775 (1988).
113. F. Cacace, M. Attinà, and G. de Petris, *J. Am. Chem. Soc.*, **111**, 5481 (1989).
114. J. F. Johnston, J. H. Ridd, and J. P. B. Sandall, *J. Chem. Soc., Chem. Commun.*, 244 (1989).
115. D. J. Belson and A. N. Strachan, *J. Chem. Soc., Perkin Trans. 2*, 15 (1989).
116. M. W. Melhuish, R. B. Moodie, M. A. Payne, and K. Schofield, *J. Chem. Soc., Perkin Trans. 2*, 1637 (1988).
117. E. Carvalho, J. Iley, F. Norberto, and E. Rosa, *J. Chem. Res. (S)*, 260 (1989).
118. K. Tanaka, T. Matsui, and T. Tanaka, *Chem. Lett.*, 1827 (1989).
119. K. Tanaka, R. Wakita, and T. Tanaka, *J. Am. Chem. Soc.*, **111**, 2428 (1989).
120. M. R. Rhodes and T. J. Meyer, *Inorg. Chem.*, **27**, 4772 (1988).
121. J. E. Toth and F. C. Anson, *J. Am. Chem. Soc.*, **111**, 2444 (1989).
122. J. A. Craig and R. H. Holm, *J. Am. Chem. Soc.*, **111**, 2111 (1989).
123. R. Tian, V. Balaji and J. Michl, *J. Am. Chem. Soc.*, **110**, 7225 (1988).
124. R. P. Saxon and M. Yoshimine, *J. Phys. Chem.*, **93**, 3130 (1989).
125. B. T. Gowda and P. Ramachandra, *J. Chem. Soc., Perkin Trans. 2*, 1067 (1989).
126. E. Athanasiou-Malaki and M. A. Koupparis, *Talanta*, **36**, 431 (1989).
127. R. G. Harfmann and S. R. Crouch, *Talanta*, **36**, 261 (1989).
128. D. Littlejohn, A. R. Wizansky, and S. G. Chang, *Can. J. Chem.*, **67**, 1596 (1989).
129. J. Passmore and M. J. Schriver, *Inorg. Chem.*, **27**, 2749 (1988).
130. R. G. Keese and A. W. Castleman, Jr., *J. Am. Chem. Soc.*, **111**, 9015 (1989).
131. U. Schülke, R. Kayser, and P. Neumann, *Z. Anorg. Allg. Chem.*, **576**, 273 (1989).
132. D. Herschlag and W. P. Jencks, *J. Am. Chem. Soc.*, **111**, 7579 (1989); D. Herschlag and W. P. Jencks, *J. Am. Chem. Soc.*, **111**, 7587 (1989).
133. L. Z. Avila and J. W. Frost, *J. Am. Chem. Soc.*, **110**, 7904 (1988).
134. J. Burgess, N. Blundell, P. M. Cullis, C. D. Hubbard, and R. Mistra, *J. Am. Chem. Soc.*, **110**, 7900 (1988).
135. P. R. Norman, A. Tate, and P. Rich, *Inorg. Chim. Acta*, **145**, 211 (1988).
136. J. R. Morrow and W. C. Trogler, *Inorg. Chem.*, **28**, 2330 (1989); J. R. Morrow and W. C. Troggler, *Inorg. Chem.*, **27**, 3387 (1988).
137. T. H. Fife and M. P. Pujari, *J. Am. Chem. Soc.*, **110**, 7790 (1988).
138. L. D. Quin, N. N. Sadanani, and X.-P. Wu, *J. Am. Chem. Soc.*, **111**, 6852 (1989).
139. P. S. Hammond, J. S. Forster, C. N. Lieske, and H. D. Durst, *J. Am. Chem. Soc.*, **111**, 7860 (1989).
140. R. A. Moss, B. Wilk, K. Krogh-Jespersen, J. T. Blair, and W. D. Westbrook, *J. Am. Chem. Soc.*, **111**, 250 (1989).
141. K. E. DeBruin, C. W. Tang, D. M. Johnson, and R. L. Wilde, *J. Am. Chem. Soc.*, **111**, 5871 (1989).
142. A. E. H. de Keijzer, L. H. Koole, and H. M. Buck, *J. Am. Chem. Soc.*, **110**, 5995 (1988).
143. N. J. Noble and B. V. L. Potter, *J. Chem. Soc., Chem. Commun.*, 1194 (1989).
144. R. N. Mehrota and L. J. Kirschenbaum, *Inorg. Chem.*, **28**, 4327 (1989).
145. K. K. Sen Gupta, J. Karak, and A. Mahapatra, *J. Chem. Res. (S)*, 366 (1989).
146. L. A. Jackson and P. J. Harris, *Inorg. Chem.*, **27**, 4338 (1988).
147. W. F. Deutsch and R. A. Shaw, *J. Chem. Soc., Dalton Trans.*, 1757 (1988).
148. O. Johnson, M. Murray, and G. Woodward, *J. Chem. Soc., Dalton Trans.*, 821 (1989).
149. R. H. Neilson and P. Wisian-Neilson, *Chem. Rev.*, **88**, 541 (1988).
150. A. Tarassou, M. L. Thompson, R. C. Haltiwanger, T. G. Hill, and A. D. Norman, *Inorg. Chem.*, **27**, 3382 (1988).
151. J. M. Barendt, E. G. Bent, R. C. Haltiwanger, and A. D. Norman, *Inorg. Chem.*, **28**, 2334 (1989).
152. T. G. Hill, R. C. Haltiwanger, T. R. Prout, and A. D. Norman, *Inorg. Chem.*, **28**, 3461 (1989).
153. R. A. Shaw and D. A. Watkins, *J. Chem. Soc., Dalton Trans.*, 2591 (1988).
154. L. R. Avens, R. A. Wolcott, L. V. Cribbs, and J. L. Mills, *Inorg. Chem.*, **28**, 200 (1989).
155. L. R. Avens, L. V. Cribbs, and J. L. Mills, *Inorg. Chem.*, **28**, 205 (1989).
156. L. R. Avens, L. V. Cribbs, and J. L. Mills, *Inorg. Chem.*, **28**, 211 (1989).

157. A. H. Cowley, P. C. Knüppel, and C. M. Nunn, *Organometallics*, **8**, 2490 (1989).
158. M. Gouygou, C. Tachon, M. Koenig, and G. Etemad-Moghadam, *New J. Chem.*, **13**, 315 (1989).
159. M. C. Démarcq, *J. Chem. Soc., Dalton Trans.*, 2221 (1988).
160. B. E. Maryannoff and A. B. Reitz, *Chem. Rev.*, **89**, 863 (1989).
161. E. Vedejs and T. J. Fleck, *J. Am. Chem. Soc.*, **111**, 5861 (1989).
162. A. Igau, A. Baceiredo, H. Grützmacher, H. Pritzkov, and G. Bertrand, *J. Am. Chem. Soc.*, **111**, 6853 (1989).
163. J. A. van Doorn and N. Meijboom, *J. Chem. Soc., Perkins Trans. 2*, 1309 (1989).
164. B. M. Gimarc and J. J. Ott, *Inorg. Chem.*, **28**, 2560 (1989).
165. R. Ali and K. B. Dillon, *J. Chem. Soc., Dalton Trans*, 2077 (1988).
166. R. Minkwitz and A. Liedtke, *Inorg. Chem.*, **28**, 4238 (1989).
167. J. A. Deiters, R. R. Holmes, and J. M. Holmes, *J. Am. Chem. Soc.*, **110**, 7672 (1988).
168. M. R. Banks and R. F. Hudson, *J. Chem. Soc., Perkin Trans. 2*, 463 (1989).
169. L. Andrews and Z. Mielke, *Inorg. Chem.*, **28**, 4001 (1989).
170. L. Andrews, R. Withnall, and B. W. Moores, *J. Phys. Chem.*, **93**, 1279 (1989).
171. H. Grützmacher and H. Pritzkow, *Chem. Ber.*, **122**, 1417 (1989).
172. L. Refkik, P. Loiseau, Y. Madaule, P. Tisnes, and J.-G. Wolf, *J. Chem. Res. (S)*, 20 (1989).
173. G. López-Cueto and C. Ubide, *Can. J. Chem.*, **66**, 2855 (1988).
174. S. Nishida and M. Kimura, *J. Chem. Soc., Dalton Trans.*, 357 (1989).
175. U. K. Kläning, B. H. J. Bielski, and K. Sehested, *Inorg. Chem.*, **28**, 2717 (1989).
176. K. Zahir, J. H. Espenson, and A. Bakac, *J. Am. Chem. Soc.*, **110**, 5059 (1988).
177. G. Braathen, P.-T. Chou, and H. Frei, *J. Phys. Chem.*, **92**, 6610 (1988).
178. K. Nahm and C. S. Foote, *J. Am. Chem. Soc.*, **111**, 1909 (1989).
179. V. V. Shereshovets, N. M. Korotaeva, R. K. Yanbaev, V. D. Komissarov, and G. A. Tolstikov, *React. Kinet. Catal. Lett.*, **38**, 243 (1989).
180. Y. Luo, K. Kustin, and I. R. Epstein, *Inorg. Chem.*, **27**, 2489 (1988).
181. M. Hojo and D. T. Sawyer, *Inorg. Chem.*, **28**, 1201 (1989).
182. R. C. Graham, *Talanta*, **36**, 585 (1989).
183. P. Dubois, J. P. Lelieur, and G. Lepoutre, *Inorg. Chem.*, **28**, 195 (1989).
184. P. Dubois, J. P. Lelieur, and G. Lepoutre, *Inorg. Chem.*, **28**, 2489 (1989).
185. K. T. Bestari, R. T. Boeré, and R. T. Oakley, *J. Am. Chem. Soc.*, **111**, 1579 (1989).
186. D. P. N. Satchell and R. S. Satchell, *J. Chem. Res. (S)*, 102 (1989).
187. D. P. N. Satchell and R. S. Satchell, *J. Chem. Res. (S)*, 262 (1988).
188. J. H. Ramsden, R. S. Drago, and R. Riley, *J. Am. Chem. Soc.*, **111**, 3958 (1989).
189. K. Laali, H. Y. Chen, and R. J. Gerzina, *J. Organomet. Chem.*, **348**, 199 (1988).
190. F. Belleisia, F. Ghelfi, U. M. Pagnoni, and A. Pinetti, *J. Chem. Res. (S)*, 182 (1989).
191. V. A. Golodov and L. V. Kashnikova, *Russ. Chem. Rev.*, **57**, 1029 (1988).
192. D. Sülzle, M. Verhoeven, J. K. Terlouw, and H. Schwarz, *Angew. Chem., Int. Ed. Engl.*, **27**, 1533 (1988).
193. L. Carlsen and H. Egsgaard, *J. Chem. Res. (S)*, 180 (1989).
194. E. A. Betterton and M. R. Hoffman, *J. Phys. Chem.*, **92**, 5962 (1988).
195. J. Kraft and R. van Eldik, *J. Chem. Soc., Chem. Commun.*, 790 (1989); J. Kraft and R. van Eldik, *Inorg. Chem.*, **28**, 2297 (1989); J. Kraft and R. van Eldik, *Inorg. Chem.*, **28**, 2306 (1989).
196. K. K. Sen Gupta, S. Das, and S. S. Sen Gupta, *J. Chem. Res. (S)*, 112 (1989).
197. B. C. Faust, M. R. Hoffmann, and D. W. Bahnemann, *J. Phys. Chem.*, **93**, 6371 (1989).
198. A. B. Burg, *Inorg. Chem.*, **28**, 1295 (1989).
199. P. Sanecki and E. Rokaszewski, *Can. J. Chem.*, **66**, 3056 (1988).
200. A. Arcoria, F. P. Ballistreri, E. Spina, G. A. Tomaselli, and A. Maccarone, *J. Chem. Soc., Perkin Trans. 2*, 1793 (1988).
201. G. Asensio, E. G. Núñez, M. J. Rodrigo, and T. Varea, *Chem. Ber.*, **122**, 1799 (1989).
202. J. E. Bennett, G. Brunton, B. C. Gilbert, and P. E. Whittall, *J. Chem. Soc., Perkin Trans. 2*, 1359 (1988).

203. F. Calderazzo, A. Morvillo, G. Pelizzi, R. Poli, and F. Ungari, *Inorg. Chem.*, **27**, 3730 (1988).
204. J. C. Pleasants, W. Guo, and D. L. Rabenstein, *J. Am. Chem. Soc.*, **111**, 6553 (1989).
205. B. Bildstein, K. Giselbrecht, and F. Sladky, *Chem. Ber.*, **122**, 2279 (1989).
206. K. E. Grung, C. Rømming, and J. Songstad, *Acta Chem. Scand.*, **43**, 518 (1989).
207. R. Grand and P. Vivarelli, *J. Chem. Res. (S)*, 186 (1989).
208. R. S. Laitinen, P. Pekonen, Y. Hiltunen, and T. A. Pakkanen, *Acta Chem. Scand.*, **43**, 436 (1989).
209. J. D. Miller and T. A. Tahir, *J. Chem. Soc., Dalton Trans.*, 457 (1989).
210. D. Dakternieks, R. Di Giacomo, R. W. Gable, and B. F. Hoskins, *J. Am. Chem. Soc.*, **110**, 6762 (1988).
211. D. Dakternieks, R. Di Giacomo, R. W. Gable, and B. F. Hoskins, *J. Am. Chem. Soc.*, **110**, 6541 (1988).
212. A. M. Bond, D. Dakternieks, R. Di Giacomo, and A. F. Hollenkamp, *Inorg. Chem.*, **28**, 1510 (1989).
213. A. F. Janzen, K. Alam, M. Jang, B. J. Blackburn, and A. S. Secco, *Can. J. Chem.*, **66**, 1308 (1988).
214. A. F. Janzen and M. Jang, *Can. J. Chem.*, **67**, 71 (1989).
215. S. Rozen, *Acc. Chem. Res.*, **21**, 307 (1988).
216. M. Murakami, K. Takahashi, and Y. Kondo, *J. Labelled Compd. Radiopharm.*, **25**, 572 (1988).
217. R. J. Gillespie and J. Liang, *J. Am. Chem. Soc.*, **110**, 6053 (1988).
218. R. L. Cook, M. Woods, E. H. Appelman, and J. C. Sullivan, *Inorg. Chem.*, **28**, 3352 (1989).
219. M. W. Melhuish and R. B. Moodie, *J. Chem. Soc., Perkin Trans. 2*, 667 (1989).
220. D. Matte, B. Solastiouk, A. Merlin, and X. Deglise, *Can. J. Chem.*, **67**, 786 (1989).
221. M. Ferriol, J. Gazet, R. Rizk-Ouaini, and M.-T.S.-C. Adad (Saugier-Cohen Adad), *Inorg. Chem.*, **28**, 3808 (1989).
222. F. Minisci, E. Vismara, F. Fontana, E. Platone, and G. Faraci, *J. Chem. Soc., Perkin Trans. 2*, 123 (1989).
223. J. R. Lindsay Smith, L. C. McKeer, and J. M. Taylor, *J. Chem. Soc., Perkin Trans. 2*, 1529 (1989).
224. J. R. Lindsay Smith, L. C. McKeer, and J. M. Taylor, *J. Chem. Soc., Perkin Trans. 2*, 1537 (1989).
225. B. T. Gowda and J. I. Bhat, *J. Ind. Chem Soc.*, **65**, 512 (1988).
226. B. T. Gowda and J. I. Bhat, *J. Ind. Chem. Soc.*, **65**, 159 (1988).
227. B. T. Gowda and B. S. Sherigara, *Int. J. Chem. Kinet.*, **21**, 31 (1989).
228. A. R. Nadig, H. S. Yathirajan, and D. Rangaswamy, *J. Ind. Chem. Soc.*, **65**, 164 (1988); A. R. Nadig, H. S. Yathirajan, and D. Rangaswamy, *J. Ind. Chem. Soc.*, **65**, 136 (1988).
229. J. C. Nagy, K. Kumar, and D. W. Margerum, *Inorg. Chem.*, **27**, 2773 (1988).
230. K. D. Fogelman, D. M. Walker, and D. W. Margerum, *Inorg. Chem.*, **28**, 986 (1989).
231. W. M. Song, K. Kustin, and I. R. Epstein, *J. Phys. Chem.*, **93**, 4698 (1989).
232. A. Mills and A. Cook, *J. Chem. Soc., Faraday Trans. 1*, **84**, 1691 (1988).
233. B. C. Gilbert, J. K. Stell, W. J. Peet, and K. J. Radford, *J. Chem. Soc., Faraday Trans. 1*, **84**, 3319 (1988).
234. A. J. Colussi, R. W. Redmond, and J. C. Scaiano, *J. Phys. Chem.*, **93**, 4783 (1989).
235. G. Bellucci, R. Bianchini, C. Chiappe, and R. Ambrosetti, *J. Am. Chem. Soc.*, **111**, 199 (1989).
236. G. Kshirsagar and R. J. Field, *J. Phys. Chem.*, **92**, 7074 (1988).
237. G. Rábai and I. R. Epstein, *Inorg. Chem.*, **28**, 732 (1989).
238. G. K. Muthakia and S. B. Jonnalagadda, *J. Phys. Chem.*, **93**, 4751 (1989).
239. F. Radner, *Acta Chem. Scand.*, **43**, 481 (1989).
240. Y. L. Wang, J. C. Nagy, and D. W. Margerum, *J. Am. Chem. Soc.*, **111**, 7838 (1989).
241. B. T. Gowda and J. I. Bhat, *Int. J. Chem. Kinet.*, **21**, 621 (1989).
242. R. H. Simoyi, I. R. Epstein, and K. Kustin, *J. Phys. Chem.*, **93**, 2792 (1989).
243. M. Varga and P. Ruoff, *J. Phys. Chem.*, **93**, 261 (1989).
244. G. J. Schrobilgen, *J. Chem. Soc., Chem. Commun.*, 863 (1988).

245. H. J. Frohn and S. Jakobs, *J. Chem. Soc., Chem. Commun.*, 625 (1989).
246. T. B. Patrick and S. Nadji, *J. Fluorine Chem.*, **39**, 415 (1988).
247. R. L. Cook, M. Woods, and J. C. Sullivan, *Inorg. Chem.*, **28**, 3349 (1989).
248. U. K. Kläning and E. H. Appelman, *Inorg. Chem.*, **27**, 3760 (1988).
249. *J. Chem. Ed.*, **66**, 187 ff (1989).
250. *J. Phys. Chem.*, **93**(7), 2689 ff (1989).
251. G. Nicolis and P. Gray (eds.), Conference Proceedings "Spatial Inhomogeneities and Transient Behaviour in Chemical Systems," Brussels 1987, Manchester Press (1988). .
252. Preprints of Lectures "Dynamics of Exotic Phenomena in Chemistry," Hajduszoboszlo, Hungarian Academy of Sciences and Hungarian Chemical Society (1989).
253. H. D. Försterling and Z. Noszticzius, *J. Phys. Chem.*, **93**, 2740 (1989).
254. E. W. Hansen and P. Ruoff, *J. Phys. Chem.*, **93**, 2696 (1989); E. W. Hansen and P. Ruoff, *J. Phys. Chem.*, **93**, 264 (1989).
255. H. Degn and F. R. Lauritsen, *J. Phys. Chem.*, **93**, 2781 (1989).
256. J. Krempaský and M. Smrčinová, *Coll. Czech. Chem. Commun.*, **54**, 1232 (1989).
257. M. Dolnik, J. Finkeová, I. Schreiber, and M. Marek, *J. Phys. Chem.*, **93**, 2764 (1989).
258. J. Weiner, F. W. Schneider, and K. Bar-Eli, *J. Phys. Chem.*, **93**, 2704 (1989).
259. M. F. Crowley and I. R. Epstein, *J. Phys. Chem.*, **93**, 2496 (1989).
260. M. Melicherčík and L. Treindl, *J. Phys. Chem.*, **93**, 7652 (1989).
261. P. Ruoff and R. M. Noyes, *J. Chem. Phys.*, **89**, 6247 (1988).
262. R. M. Noyes, R. J. Field, H. D. Försterling, E. Körös, and P. Ruoff, *J. Phys. Chem.*, **93**, 270 (1989).
263. P. Ruoff and R. M. Noyes, *J. Phys. Chem.*, **93**, 7394 (1989).
264. P. P. Jha, B. N. Prased, and R. K. Prasad, *J. Ind. Chem. Soc.*, **65**, 177 (1988).
265. D. Balasubramanian and G. A. Rodley, *J. Phys. Chem.*, **92**, 5995 (1988).
266. T. Vajda, A. Rockenbauer, and M. Györ, *Int. J. Chem. Kinet.*, **20**, 661 (1988).
267. P. G. Sørensen and F. Hynne, *J. Phys. Chem.*, **93**, 5467 (1989).
268. R. P. Rastogi, I. Das, and A. Sharma, *J. Chem. Soc., Faraday Trans. 1*, **85**, 2011 (1989).
269. P. Ruoff, M. Varga, and E. Körös, *Acc. Chem. Res.*, **21**, 326 (1988).
270. M. Handlířová, *React. Kinet. Catal. Lett.*, **37**, 145 (1988).
271. J.-J. Jwo and E.-F. Chang, *J. Phys. Chem.*, **93**, 2388 (1989).
272. E. C. Edblom, Y. Luo, M. Orbān, K. Kustin, and I. R. Epstein, *J. Phys. Chem.*, **93**, 2722 (1989).
273. W. Šcifgosz and S. Pokrzywnicki, *Acta Chem. Scand.*, **43**, 926 (1989).
274. L. Adamčiková and P. Ševčik, *Z. Phys. Chem. (Frankfurt)*, **162**, 21 (1989).
275. Y. Sasaki, *Bull. Chem. Soc. Jpn.*, **61**, 4071 (1988).
276. Y. Sasaki, *Bull. Chem. Soc. Jpn.*, **61**, 1479 (1988).
277. Y. Luo and I. R. Epstein, *J. Phys. Chem.*, **93**, 1398 (1989).
278. G. Rábai and M. T. Beck, *J. Phys. Chem.*, **92**, 4831 (1988).
279. R. H. Simoyi, I. R. Epstein, and K. Kustin, *J. Phys. Chem.*, **93**, 1689 (1989).
280. S. Anić and L. Z. Kolar-Anić, *J. Chem. Soc., Faraday Trans. 1*, **84**, 3413 (1988).
281. J. P. Laplante, *J. Phys. Chem.*, **93**, 3882 (1989).
282. G. Rábai, M. T. Beck, K. Kustin, and I. R. Epstein, *J. Phys. Chem.*, **93**, 2853 (1989).
283. G. Rábai and I. R. Epstein, *J. Phys. Chem.*, **93**, 7556 (1989).
284. G. Rábai, K. Kustin, and I. R. Epstein, *J. Am. Chem. Soc.*, **111**, 8271 (1989).
285. G. Rábai, K. Kustin, and I. R. Epstein, *J. Am. Chem. Soc.*, **111**, 3870 (1989).
285. G. Rábai, K. Kustin, and I. R. Epstein, *J. Am. Chem. Soc.*, **111**, 3870 (1989).
286. S. D. Furrow, *J. Phys. Chem.*, **93**, 2817 (1989).
287. A. Nagy and L. Treindl, *J. Phys. Chem.*, **93**, 2807 (1989).
288. M. Orbán and I. R. Epstein, *J. Am. Chem. Soc.*, **111**, 8543 (1989).
289. M. Morita, K. Iwamoto, and M. Seno, *Bull. Chem. Soc. Jpn.*, **61**, 3467 (1988).
290. R. T. Powell, T. Oskin, and N. Ganapathisubramanian, *J. Phys. Chem.*, **93**, 2718 (1989).
291. T. G. Costner and N. Ganapathisubramanian, *Inorg. Chem.*, **28**, 3621 (1989).
292. J. Amrehn, P. Rezch, and F. W. Schneider, *J. Phys. Chem.*, **92**, 3318 (1988).

293. M. Orbán and I. R. Epstein, *J. Am. Chem. Soc.*, **111**, 2891 (1989).
294. Y. Luo, M. Orbán, K. Kustin, and I. R. Epstein, **111**, 4541 (1989).
295. P. Resch, R. J. Field, and F. W. Schneider, *J. Phys. Chem.*, **93**, 2783 (1989).
296. P. Resch, R. J. Field, F. W. Schneider, and M. Burger, *J. Phys. Chem.*, **93**, 8181 (1989).
297. J. D. Druliner and E. Wasserman, *J. Am. Chem. Soc.*, **110**, 5270 (1988).
298. M. Schell, F. N. Albahadily, J. Safar, and Y. Xu, *J. Phys. Chem.*, **93**, 4806 (1989).
299. M. R. Bassett and J. L. Hudson, *J. Phys. Chem.*, **93**, 2731 (1989).
300. O. Lev, A. Wolffberg, L. M. Pismen, and M. Sheintuch, *J. Phys. Chem.*, **93**, 1661 (1989).
301. D. G. Leaist, *Aust. J. Chem.*, **41**, 469 (1988).
302. I. Nagypal and I. R. Epstein, *J. Chem. Phys.*, **89**, 6925 (1988).
303. F. Sagués and J. M. Sancho, *J. Chem. Phys.*, **89**, 3793 (1988).
304. L. Györgyi and R. J. Field, *J. Phys. Chem.*, **92**, 7079 (1988).
305. M. T. Beck and I. P. Nagy, *J. Phys. Chem.*, **93**, 7755 (1989).
306. M. Sheintuch and D. Luss, *J. Phys. Chem.*, **93**, 5727 (1989).
307. R. Larter, C. G. Steinhetz, and B. D. Aguda, *J. Chem. Phys.*, **89**, 6506 (1988).
308. L. Szirovicza, I. Nagypál, and E. Boga, *J. Am. Chem. Soc.*, **111**, 2842 (1989).
309. G. Póta, I. Lengyel, and G. Bazsa, *J. Chem. Soc.*, *Faraday Trans. 1*, **85** 3871 (1989).
310. G. Bazsa, I. Lengyel, and W. Linert, *J. Chem. Soc.*, *Faraday Trans. 1*, **85**, 3273 (1989).
311. C. Vidal and A. Pagola, *J. Phys. Chem.*, **93**, 2711 (1989).
312. Z. Nagy-Ungvarai, S. C. Muller, J. J. Tyson, and B. Hess, *J. Phys. Chem.*, **93**, 2760 (1989).
313. Z. Nagy-Ungvarai, J. J. Tyson, and B. Hess, *J. Phys. Chem.*, **93**, 707 (1989).
314. J. Maselko, J. S. Reckley, and K. Showalter, *J. Phys. Chem.*, **93**, 2774 (1989).
315. W. Jahnke, W. E. Skaggs, and A. T. Winfree, *J. Phys. Chem.*, **93**, 740 (1989).
316. A. T. Winfree and W. Jahnke, *J. Phys. Chem.*, **93**, 2823 (1989).
317. M. Frankowicz and A. L. Kawcyński, *J. Phys. Chem.*, **93**, 2755 (1989).
318. Y. Brechet and J. S. Kirkaldy, *J. Chem. Phys.*, **90**, 1499 (1989).

References for Chapter 5

1. R. van Eldik, T. Asano, and W. J. Le Noble, *Chem. Rev.*, **89**, 549 (1989).
2. M. Kotowski and R. van Eldik, *Coord. Chem. Rev.*, **93**, 19 (1989).
3. M. J. Blandamer and J. Burgess, *Trans. Metal Chem.*, **13**, 1 (1988).
4. J. Burgess and A. Pelizzetti, *Gazz. Chim. Ital.*, **118**, 803 (1988).
5. J. F. Stoddart and R. Zarzycki, *Recl. Trav. Chim. Pays-Bas*, **107**, 515 (1988); D. R. Alston, A. M. Z. Slawin, J. F. Stoddart, D. J. Williams, and R. Zarzycki, *Angew. Chem., Int. Ed. Engl.*, **27**, 1184 (1988).
6. R. Ballardini, M. T. Gandolfi, L. Prodi, M. Ciano, V. Balzani, F. H. Kohnke, H. Shahriari-Zavareh, N. Spencer, and J. F. Stoddart, *J. Am. Chem. Soc.*, **111**, 7072 (1989).
7. P. Umapathy, *Coord. Chem. Rev.*, **95**, 129 (1989).
8. A. Laovi, J. Kozelka, and J.-C. Chottard, *Inorg. Chem.*, **27**, 2751 (1988).
9. N. Goswami, L. L. Bennett-Slavin, and R. N. Bose, *J. Chem. Soc., Chem. Commun.*, 432 (1989).
10. R. E. Norman and P. J. Sadler, *Inorg. Chem.*, **27**, 3583 (1988).
11. T. G. Appleton, J. W. Connor, J. R. Hall, and P. D. Prenzler, *Inorg. Chem.*, **28**, 2030 (1989).
12. A. R. Hermes, R. J. Morris, and G. S. Girolami, *Organometallics*, **7**, 2373 (1988); J. J. H. Edema, S. Gambarotta, F. van Bolhuis, W. J. J. Smeets, and A. L. Spec, *Inorg. Chem.*, **28**, 1407 (1989).
13. K. Kojima, *Bull. Chem. Soc. Jpn.*, **61**, 385 (1988).
14. L. C. A. de Carvalho, Y. Pérès, M. Dartiguenave, Y. Dartiguenave, and A. L. Beauchamp, *Acta Crystallogr., Sect. C*, **C45**, 159 (1989).
15. N. Habadie, M. Dartiguenave, Y. Dartiguenave, J. F. Britten, and A. L. Beauchamp, *Organometallics*, **8**, 2564 (1989).
16. E. G. Lundquist, W. E. Streib, and K. G. Caulton, *Inorg. Chim. Acta*, **159**, 23 (1989).

17. J. A. Rahn, D. J. O'Donnell, A. J. Palmer, and J. H. Nelson, *Inorg. Chem.*, **28**, 2631 (1989).
18. T. S. Cameron, H. C. Clark, A. Linden, and M. J. Hampden-Smith, *Inorg. Chim. Acta*, **162**, 9 (1989).
19. A. J. Paviglianiti, D. J. Minn, W. C. Fultz, and J. L. Burmeister, *Inorg. Chim. Acta*, **159**, 65 (1989).
20. A. J. Paviglianiti, W. C. Fultz, and J. L. Burmeister, *Synth. React. Inorg. Met.-Org. Chem.*, **19**, 469 (1989).
21. G. Annibale, L. Cattalini, V. Bertolasi, V. Ferretti, G. Gilli, and M. L. Tobe, *J. Chem. Soc., Dalton Trans.*, 1265 (1989).
22. Y. Ducommun, L. Helm, A. E. Merbach, B. Hellquist, and L. I. Elding, *Inorg. Chem.*, **29**, 377 (1989).
23. O. Gröning and L. I. Elding, *Inorg. Chem.*, **28**, 3366 (1989).
24. L. Cattalini, M. Bonivento, G. Michelon, M. L. Tobe, and A. T. Treadgold, *Gazz. Chim. Ital.*, **118**, 725 (1988).
25. L. Cattalini, G. Chesa, G. Marangoni, B. Pitteri, and E. Celon, *Inorg. Chem.*, **28**, 1944 (1989).
26. L. Canovese, L. Cattalini, G. Chesa, and M. L. Tobe, *J. Chem. Soc., Dalton Trans.*, 2135 (1988).
27. C. Ammann, P. S. Pregosin, and A. Scrivanti, *Inorg. Chim. Acta*, **155**, 217 (1989).
28. S. Toyota, Y. Yamada, M. Kaneyoshi, and M. Oki, *Bull. Chem. Soc. Jpn.*, **62**, 1509 (1989).
29. S. Yamazaki and T. Ama, *Bull. Chem. Soc. Jpn.*, **62**, 931 (1989).
30. T. Kemmitt, W. Levason, and M. Webster, *Inorg. Chem.*, **28**, 692 (1989).
31. J. J. Pienaar, E. L. Breet, and R. van Eldik, *Inorg. Chim. Acta*, **155**, 249 (1989).
32. J. J. Pienaar, M. Kotowski, and R. van Eldik, *Inorg. Chem.*, **28**, 373 (1989).
33. J. Berger, M. Kotowski, R. van Eldik, U. Frey, L. Helm, and A. E. Merbach, *Inorg. Chem.*, **28**, 3759 (1989).
34. M. Kotowski, S. Begum, J. G. Leipoldt, and R. van Eldik, *Inorg. Chem.*, **27**, 4472 (1988).
35. F. L. Wimmer and S. Wimmer, *Inorg. Chim. Acta*, **149**, 1 (1988).
36. G. Alibrandi, M. Cusumano, A. Gianetto, and D. Minniti, *J. Chem. Soc., Dalton Trans.*, 375 (1989).
37. N. Gupta, R. M. Naik, and P. C. Nigam, *Inorg. Chim. Acta*, **160**, 103 (1989).
38. A. K. Das, S. Gangopadhyay, and D. Banerjea, *Trans. Metal Chem.*, **14**, 73 (1989).
39. B. Hellquist, L. I. Elding, and Y. Ducommun, *Inorg. Chem.*, **27**, 3620 (1988).
40. U. Frey, L. Helm, A. Merbach, and R. Romeo, *J. Am. Chem. Soc.*, **111**, 8161 (1989).
41. G. Alibrandi, D. Minniti, L. M. Scolaro, and R. Romeo, *Inorg. Chem.*, **28**, 1939 (1989).
42. R. L. Brainard, W. R. Nutt, T. R. Lee, and G. M. Whitesides, *Organometallics*, **7**, 2379 (1988).
43. S. Toyota and M. Oki, *Bull. Chem. Soc. Jpn.*, **61**, 699 (1988).
44. G. Strukul, R. Sinigalia, A. Zanardo, F. Pinna, and R. A. Michelin, *Inorg. Chem.*, **28**, 554 (1989).
45. T. D. Newbound, M. R. Colsman, M. M. Miller, G. P. Wulfsberg, O. P. Anderson, and S. H. Strauss, *J. Am. Chem. Soc.*, **111**, 3762 (1989).
46. M. Martinez and G. Muller, *J. Chem. Soc., Dalton Trans.*, 1669 (1989).
47. Y. Li, Q. Zhang, R. Chen, Z. Zhang, and X. Wang, *J. Mol. Sci., Int. Ed.*, **4**, 135 (1986).
48. B. Büsing, H. Elias, I. Eslick, and K. J. Wannowius, *Inorg. Chim. Acta*, **150**, 223 (1988).
49. P. Segla and H. Elias, *Inorg. Chim. Acta*, **149**, 259 (1988).
50. T. R. Griffiths and N. J. Phillips, *J. Chem. Soc., Dalton Trans.*, 325 (1989).
51. F. P. Emmenegger, *Inorg. Chem.*, **28**, 2210 (1989).
52. N. F. Curtis, *Aust. J. Chem.*, **41**, 1665 (1988).
53. N. Saddler, S. L. Scott, A. Bakac, J. H. Espenson, and M. S. Ram, *Inorg. Chem.*, **28**, 3951 (1989).
54. J. H. Potgeiter, *J. Organomet. Chem.*, **366**, 369 (1989).
55. W. Duczmal, B. Marciniec, and E. Sliwinska, *Trans. Metal Chem.*, **14**, 105 (1989).
56. H. El-Amouri, A. A. Bahsoun, and J. A. Osborn, *Polyhedron*, **7**, 2035 (1988).
57. F. Araghizadeh, D. M. Branan, N. W. Hoffman, J. H. Jones, E. A. McElroy, N. C. Miller, D. L. Ramage, A. B. Salazar, and S. H. Young, *Inorg. Chem.*, **27**, 3752 (1988).

58. S. Elmroth, L. H. Skibsted, and L. I. Elding, *Inorg. Chem.*, **28**, 2703 (1989).
59. K. J. S. Gupta, S. Das, and S. S. Gupta, *Trans. Metal Chem.*, **13**, 261 (1988).
60. J. R. Roper, H. Paulus, and H. Elias, *Inorg. Chem.*, **28**, 2323 (1989).
61. T. P. E. Auf der Heyde and H.-B. Bürgi, *Inorg. Chem.*, **28**, 3960, 3970, 3982 (1989).
62. N. Koga, S. Q. Jin, and K. Morokuma, *J. Am. Chem. Soc.*, **110**, 3417 (1988).
63. S. P. Ermer, R. S. Shinomoto, M. A. Deming, and T. C. Flood, *Organometallics*, **8**, 1377 (1989).
64. J.-K. Shen, Y.-C. Gao, Q.-Z. Shi, and F. Basolo, *Inorg. Chem.*, **28**, 4304 (1989).
65. J. Iqbal, D. W. A. Sharp, and J. M. Winfield, *J. Chem. Soc., Dalton Trans.*, 461 (1989).
66. T. Sakai, Z. Taira, S. Yamazaki, and T. Ama, *Polyhedron*, **8**, 1989 (1989).
67. P. Brüggeller, T. Hübner, and A. Gieren, *Z. Naturforsch. B*, **44B**, 800 (1989).
68. J. L. Gambaro, W. H. Hohman, and D. W. Meek, *Inorg. Chem.*, **28**, 4154 (1989).
69. A. de Renzi, M. Funicello, G. Morelli, A. Panunzi, and C. Pellecchia, *J. Chem. Soc., Perkin Trans. 2*, 427 (1987); V. G. Albano, F. Demartin, A. de Renzi, and G. Morelli, *Inorg. Chim. Acta*, **149**, 253 (1988); V. G. Albano, C. Castellari, G. Morelli, and A. Vitagliano, *Gazz. Chim. Ital.*, **119**, 235 (1989).
70. M. E. Cucciolito, V. de Felice, A. Panunzi, and A. Vitagliano, *Organometallics*, **8**, 1180 (1989).
71. F. P. Fanizzi, L. Maresca, G. Natile, M. Lanfranchi, A. M. Manotti-Lanfredi, and A. Tiripichio, *Inorg. Chem.*, **27**, 2422 (1988).
72. G. Hunter, A. McAuley, and T. W. Whitcombe, *Inorg. Chem.*, **27**, 2634 (1988); A. McAuley and T. W. Whitcombe, *Inorg. Chem.*, **27**, 3090 (1988).
73. G. Reid, A. J. Blake, T. I. Hyde, and M. Schröder, *J. Chem. Soc., Chem. Commun.*, 1397 (1988).
74. A. A. H. van der Zeijden, G. van Koten, J. A. M. Wouters, W. F. A. Wijsmuller, D. M. Grove, W. J. J. Smeets, and A. L. Spek, *J. Am. Chem. Soc.*, **110**, 5354 (1988).
75. R. J. Kulawiec and R. H. Crabtree, *Organometallics*, **7**, 1891 (1988).
76. G. J. Arsenault, C. M. Anderson, and R. J. Puddephatt, *Organometallics*, **7**, 2094 (1988).
77. R. Usón, J. Forniés, M. Tomás, I. Ara, and J. M. Casas, *Inorg. Chem.*, **28**, 2388 (1989); R. Usón, J. Forniés, M. Tomás, and J. M. Casas, *Angew. Chem. Int. Ed. Engl.*, **28**, 748 (1989).
78. R. G. Raptis, J. P. Fackler, H. H. Murray, and L. C. Porte, *Inorg. Chem.*, **28**, 4057 (1989).
79. J. K. Nagle, A. L. Balch, and M. M. Olmstead, *J. Am. Chem. Soc.*, **110**, 319 (1988).
80. J. H. Welch, R. D. Bereman, and P. Singh, *Inorg. Chim. Acta*, **163**, 93 (1989).
81. X.-Y. Zhou and N. M. Kostić, *Inorg. Chem.*, **27**, 4402 (1988).
82. P. P. Deutsch and R. Eisenberg, *Chem. Rev.*, **88**, 1147 (1988).
83. M. J. Burk, M. P. McGrath, R. Wheeler, and R. H. Crabtree, *J. Am. Chem. Soc.*, **110**, 5034 (1988).
84. F. Maseras, M. Duran, A. Lledós, and J. Bertrán, *Inorg. Chem.*, **28**, 2984 (1989).
85. M. Y. Darensbourg, M. Ludwig, and C. G. Riordan, *Inorg. Chem.*, **28**, 1630 (1989).
86. R. A. Michelin and R. Ros, *J. Chem. Soc., Dalton Trans.*, 1149 (1989).
87. I. A. Krol, V. Yu. Kukushkin, Z. A. Starikova, V. M. Tkachuk, and B. V. Zhadanov, *Zh. Obshch. Khim.*, **58**, 2625 (1988); *J. Gen. Chem. USSR* (*Engl. Transl.*), **58**, 2336 (1988).
88. M. L. Tobe, A. T. Treadgold, and L. Cattalini, *J. Chem. Soc., Dalton Trans.*, 2347 (1988).
89. S. Komiya, I. Endo, S. Ozaki, and Y. Ishizaki, *Chem. Lett.*, 63 (1988).
90. G. Alibrandi, M. Cusumano, D. Minniti, and L. M. Scolaro, *Inorg. Chem.*, **28**, 342 (1989).
91. A. R. Siedle, R. A. Newmark, and R. D. Howells, *Inorg. Chem.*, **27**, 2473 (1988).
92. T. V. O'Halloran and S. J. Lippard, *Inorg. Chem.*, **28**, 1289 (1989).

References for Chapter 6

1. D. A. House, in: *Mechanisms of Inorganic and Organometallic Reactions* (M. V. Twigg, ed.), Vol. 6, Plenum Press, New York (1989).
2. J. E. Newbery, *Annu. Rep. Prog. Chem., Sect. A*, **84**, 173 (1987).
3. R. Colton, *Coord. Chem. Rev.*, **90**, 1 (1988).

4. T. C. Bruice, *Aldrichimica Acta*, **21**, 87 (1988).

5. J. M. Garrison and T. C. Bruice, *J. Am. Chem. Soc.*, **111**, 191 (1989).

6. F. A. Luzzio and F. S. Guziec, *Org. Prep. Proced. Int.*, **20**, 533 (1988); *Chem. Abstr.*, **110**, 172325 (1989).

7. M. K. Chandhuri, *J. Mol. Catal.*, **44**, 129 (1988).

8. L. Mønsted and O. Mønsted, *Coord. Chem. Rev.*, **94**, 109 (1989).

9. J. Springborg, *Adv. Inorg. Chem.*, **32**, 55 (1988).

10. P. H. Andersen, *Coord. Chem. Rev.*, **94**, 47 (1989).

11. R. Banerjee, *Coord. Chem. Rev.*, **68**, 145 (1985).

12. R. van Eldik, T. Asano, and W. J. Le Noble, *Chem. Rev.*, **89**, 549 (1989).

13. P. A. Lay, Abstracts of papers 27th ICCC Queensland, Australia, #T89 (1989).

14. N. J. Curtis, G. A. Lawrance, and R. van Eldik, *Inorg. Chem.*, **28**, 333 (1989).

15. N. J. Curtis, S. Suvachittanont, and R. van Eldik, Abstracts of papers, 27th ICCC Queensland, Australia, #T90 (1989).

16. S. Suvachittanont and R. van Eldik, *Inorg. Chem.*, **28**, 3660 (1989).

17. A. S. Babaqi, M. T. Mohamed, and S. S. Massound, *Transition Metal Chem.*, **13**, 201 (1988).

18. A. Sauer, H. Cohen, and D. Meyerstein, *Inorg. Chim. Acta*, **155**, 101 (1989).

19. P. Kita and R. B. Jordan, *Inorg. Chem.*, **28**, 3489 (1989).

20. J. Chatlas and P. Kita, *Pol. J. Chem.*, **60**, 741 (1986); *Chem. Abstr.*, **107**, 142241 (1987).

21. J. Chatlas, E. Kita, and P. Kita, *Pol. J. Chem.*, **62**, 653 (1988).

22. D. A. House, to be published.

23. P. Riccieri, E. Zinato, and A. Damiani, *Inorg. Chem.*, **27**, 3755 (1988).

24. J. J. Chung, H. T. Kim, and S. O. Bek, *Taehan Hwahak Hoechi*, **33**, 164 (1989); *Chem. Abstr.*, **111**, 45922 (1989).

25. M. J. Sisley and R. B. Jordan, *Inorg. Chem.*, **28**, 2714 (1989).

26. N. R. Anipindi and V. A. Raman, *Chim. Acta Turc.*, **16**, 233 (1988); *Chem. Abstr.*, **111**, 13099 (1989).

27. A. D. Kirk and A. M. Ibrahim, *Inorg. Chem.*, **27**, 4567 (1988).

28. D. A. House, *Inorg. Chem.*, **27**, 2587 (1988).

29. P. Kita, *Pol. J. Chem.*, **56**, 913 (1982).

30. E. Kita, *Pol. J. Chem.*, **61**, 361 (1987); *Chem. Abstr.*, **109**, 62016 (1988).

31. K. R. Ashley and J. Kuo, *Inorg. Chem.*, **27**, 3556 (1988).

32. K. R. Ashley and I. Trent, *Inorg. Chim. Acta*, **163**, 159 (1989).

33. M. Inamo, T. Sumi, N. Nakagawa, S. Funahashi, and M. Tanaka, *Inorg. Chem.*, **28**, 2688 (1989).

34. H. Stunzi, L. Spiccia, F. P. Rotzinger, and W. Marty, *Inorg. Chem.*, **28**, 66 (1989).

35. L. Spiccia, W. Marty, and R. Giovanoli, *Inorg. Chem.*, **27**, 2660 (1988).

36. Kabir-ud-Din, I. A. Khan, and M. Shahid, *Transition Met. Chem.*, **12**, 557 (1987).

37. I. A. Khan, M. Shahid, and Kabir-ud-Din, *Transition Met. Chem.*, **12**, 393 (1987).

38. H. G. M. Mustofy and G. S. De, *Transition Metal Chem.*, **13**, 196 (1988).

39. D. L. Huizenga and H. M. Patterson, *Anal. Chim. Acta*, **206**, 263 (1988).

40. E. Kita, *Pol. J. Chem.*, **62**, 359 (1988); *Chem. Abstr.*, **111**, 45921 (1989).

41. A. K. Gangopadhyay and G. S. De, *Indian J. Chem., Sect. A*, **27A**, 778 (1988); *Chem. Abstr.*, **111**, 13076 (1989).

42. K. Govindaraju, T. Ramasami, and D. Ramaswamy, *J. Inorg. Biochem.*, **35**, 127 (1989).

43. K. Govindaraju, T. Ramasami, and D. Ramaswamy, *J. Inorg. Biochem.*, **35**, 137 (1989).

44. Y. Abe and H. Ogino, *Bull. Chem. Soc. Jpn.*, **62**, 56 (1989).

45. M. J. Sisley and R. B. Jordan, *Inorg. Chem.*, **27**, 4483 (1988).

46. P. Kita, *Pol. J. Chem.*, **62**, 653 (1988).

47. G. H. Searle, G. S. Bull, and D. A. House, *J. Chem. Educ.*, **66**, 605 (1989).

48. P. Kita and A. Swinarski, *Z. Anorg. Allg. Chem.*, **464**, 195 (1980).

49. D. A. House and R. van Eldik, to be published.

50. P. Guardado, G. A. Lawrance, and R. van Eldik, *Inorg. Chem.*, **28**, 976 (1989).

51. B. G. Gafford and R. A. Holwerda, *Inorg. Chem.*, **28**, 60 (1989).

52. M. Wawrzenczyk, *Z. Phys. Chem.* (*Leipzig*), **269**, 993 (1988).

53. A. A. Abdel-Khalek and M. M. Elsemongy, *Transition Met. Chem.*, **13**, 340 (1988).

54. A. A. Abdel-Khalek and M. M. Elsemongy, *Transition Met. Chem.*, **13**, 35 (1988).

55. A. A. Abdel-Khalek and M. M. Elsemongy, *Bull. Chem. Soc. Jpn.*, **61**, 4407 (1988).

56. A. A. Abdel-Khalek and M. M. Elsemongy, *Transition Met. Chem.*, **14**, 206 (1989).

57. M. M. Elsemongy, *Transition Metal Chem.*, **14**, 206 (1989).

58. M. E. Brynildson, A. Bakac, and J. H. Espenson, *Inorg. Chem.*, **27**, 2592 (1988).

59. T. Katsuyama, A. Bakac, and J. H. Espenson, *Inorg. Chem.*, **28**, 339 (1989).

60. K. R. Howes, C. G. Pippin, J. C. Sullivan, D. Meisel, J. H. Espenson, and A. Bakac, *Inorg. Chem.*, **27**, 2932 (1988).

61. S. Ronco, B. Van Vlierberge, and G. Ferrandi, *Inorg. Chem.*, **27**, 3453 (1988).

62. H. Ogino, T. Nagata, and K. Ogino, *Inorg. Chem.*, **28**, 3656 (1989).

63. R. V. Kasza and Y. Koga, *Inorg. Chim. Acta*, **162**, L15 (1989).

64. P. Biscarni and R. Kuroda, *Inorg. Chim. Acta*, **154**, 209 (1988).

65. K. Zahir, J. H. Espenson, and A. Bakac, *J. Am. Chem. Soc.*, **110**, 5059 (1988).

66. A. M. Ghaith, L. S. Forster, and J. V. Rund, *J. Phys. Chem.*, **92**, 6197 (1988).

67. P. Huston, J. H. Espenson, and A. Bakac, *Inorg. Chem.*, **28**, 3671 (1989).

68. P. S. Cartwright and R. D. Gillard, *Polyhedron*, **8**, 1453 (1989).

69. S. Lee, A. Bakac, and J. H. Espenson, *Inorg. Chem.*, **28**, 1367 (1989).

70. C. A. Simmons, A. Bakac, and J. H. Espenson, *Inorg. Chem.*, **28**, 581 (1989).

71. C. R. Steffan, A. Bakac, and J. H. Espenson, *Inorg. Chem.*, **28**, 2992 (1989).

72. Reference 1, Section 6.8.1, No. 24.

73. K.-W. Lee and P. E. Hoggard, *Polyhedron*, **8**, 1557 (1989).

74. K. V. Goodwin, W. T. Pennington, and J. D. Petersen, *Inorg. Chem.*, **28**, 2016 (1989).

75. A. D. Kirk and C. Namasivayam, *J. Phys. Chem.*, **93**, 5488 (1989).

76. R. R. Ruminski, M. H. Healy, and W. F. Coleman, *Inorg. Chem.*, **28**, 1666 (1989).

77. C. K. Ryu, R. B. Lessard, D. Lynch, and J. F. Endicott, *J. Phys. Chem.*, **93**, 1752 (1989).

78. R. B. Lessard, J. F. Endicott, M. W. Perkovic, and L. A. Ochrymowycz, *Inorg. Chem.*, **28**, 2574 (1989).

79. C. A. Bignozzi, M. T. Indelli, and F. Scandola, *J. Am. Chem. Soc.*, **111**, 5192 (1989).

80. A. Damiani, P. Riccieri, and E. Zinato, *Photochem. Photophys. Coord. Compd. Proc.*, 7th Int. Symp. (H. Yersin, ed.), p. 35, Springer, Berlin (1987); *Chem. Abstr.*, **109**, 201110 (1988).

81. A. D. Kirk, C. Namasivayam, W. Riske, and D. Ristic-Petrovic, *Inorg. Chem.*, **28**, 972 (1989).

82. D. A. House, in: *Mechanisms of Inorganic and Organometallic Reactions* (M. V. Twigg, ed.), Vol. 4, Plenum Press, New York (1986).

83. L. Golic and N. Bule, *Acta Crystallogr.*, **44C**, 2065 (1988).

84. A. Hazell, *Acta Crystallogr.*, **44C**, 1362 (1988).

85. W. Zhang, Y. Fan, S. Zhan, X. Wang, and W. Han, *Jiegou Huaxne*, **6**, 274 (1987); *Chem. Abstr.*, **109**, 203679 (1988).

86. F. Lloret, M. Julve, M. Mollar, I. Castro, J. Lactorre, J. Faus, X. Solans, and I. Morgenstern-Badaru, *J. Chem. Soc., Dalton Trans.*, 729 (1989).

87. D. J. Radanovic, S. F. Trifunovic, S. B. Grujic, C. Maricondi, M. Parvez, and B. E. Douglas, *Inorg. Chim. Acta*, **157**, 33 (1989).

88. S. Kaizaki, M. Hayashi, K. Umakoshi, and S. Ooi, *Bull. Chem. Soc. Jpn.*, **61**, 3519 (1988).

89. K. Swaminathan, U. C. Sinha, C. Chatterjee, V. S. Yadava, and V. M. Padmanabhan, *Acta Crystallogr.*, **45C**, 21 (1989).

90. G. Svensson and J. Albertsson, *Acta Crystallogr.*, **44C**, 1386 (1988).

91. W. Zhang, Y. Fan, Q. Kong, and H. Xu, *Huaxue Xuebao*, **47**, 163 (1989); *Chem. Abstr.*, **111**, 87836 (1989).

92. Y. Sakabe, H. Sakaguchi, H. Takayanagi, H. Ogura, and Y. Iitaka, Abstracts of papers 27th ICCC Queensland, Australia, #T70 (1989).

93. D. A. House, V. McKee, and W. T. Robinson, *Inorg. Chim. Acta*, **157**, 15 (1989).

94. V. I. Bondar, K. A. Potekhim, T. F. Rau, S. P. Rozman, and T. Iu, *Dokl. Akad. Nauk SSSR*, **300**, 1372 (1988); *Chem. Abstr.*, **109**, 241216 (1988).
95. E. Bang, J. Eriksen, J. Glerup, L. Mønsted, and O. Mønsted, Abstracts 14th Int. Symp. Macrocyclic Chem., Queensland, Australia P45/T (1989).
96. B. G. Gafford, C. O'Rear, J. H. Zhang, C. J. O'Connor, and R. A. Holwerda, *Inorg. Chem.*, **28**, 1720 (1989).
97. K. M. Corbin, D. J. Hodgson, M. H. Lynn, K. Michelsen, K. M. Nielsen, and E. Pedesen, *Inorg. Chim. Acta*, **159**, L129 (1989).
98. R. J. Judd, T. W. Hambley, and P. A. Lay, *J. Chem. Soc., Dalton Trans.*, 2205 (1989).
99. J. Shamir, *Inorg. Chim. Acta*, **156**, 163 (1989).
100. S. Kaizaki and N. Sakagami, Abstracts of papers, 27th ICCC Queensland, Australia, #T64 (1989).
101. S. Kaizaki and M. Hayashi, *J. Chem. Soc., Chem. Commun.*, 613 (1988).
102. D. J. Radanovic, M. I. Djuvan, M. M. Djorovic, and B. E. Douglas, *Inorg. Chim. Acta*, **146**, 199 (1988).
103. P. Andersen, A. Døssing, J. Glerup, and M. Rude, *Acta Chem. Scand.*, to appear.
104. F. Palacio, M. C. Moron, J. Pons, J. Casbo, K. E. Merabet, and R. L. Carlin, *Phys. Lett.*, **135**, 231 (1989).
105. V. Subramaniam, and P. E. Hoggard, *Inorg. Chim. Acta*, **155**, 161 (1989).
106. M. Johda, M. Suzuki, and A. Uehara, *Bull. Chem. Soc. Jpn.*, **62**, 738 (1989).
107. C. Gldewell and J. S. McKechnie, *J. Chem. Educ.*, **65**, 1015 (1988).
108. E. A. Sato, S. Kida, and I. Murase, *Inorg. Chem.*, **28**, 800 (1989).
109. G. B. Kauffman, N. Sugisaka, and I. K. Reid, *Inorg. Synth.*, **25**, 139 (1989).
110. C. G. Young, *J. Chem. Educ.*, **65**, 918 (1988).
111. W. L. Purcell, *Polyhedron*, **8**, 563 (1989).
112. N. A. Lewis, W. L. Purcell, and D. V. Traveras, *Inorg. Chem.*, **28**, 133 (1989).
113. F. Ferranti and A. Indelli, *J. Chem. Soc. Faraday Trans. 1*, **85**, 2241 (1989).
114. A. Sauer, H. Cohen, and D. Meyerstein, *Inorg. Chem.*, **27**, 4578 (1988).
115. R. J. Balahura, M. Johnson, and T. Black, *Inorg. Chem.*, **28**, 3933 (1989).
116. W. L. Purcell, *Inorg. Chem.*, **28**, 2312 (1989).
117. A. Neves and C. V. Franco, *Inorg. Chim. Acta*, **160**, 223 (1989).
118. B.-L. Ooi, C. Sharp, and A. G. Sykes, *J. Am. Chem. Soc.*, **111**, 125 (1989).
119. K. Zahir, J. H. Espenson, and A. Bakac, *Inorg. Chem.*, **27**, 3144 (1988).
120. A. Bakac and J. H. Espenson, *Inorg. Chem.*, **28**, 3901 (1989).
121. S. K. Ghosh, K. Laali, and E. S. Gould, *Inorg. Chem.*, **27**, 4224 (1988).
122. S. K. Ghosh and E. S. Gould, *Inorg. Chem.*, **27**, 4228 (1988).
123. S. K. Ghosh and E. S. Gould, *Inorg. Chem.*, **28**, 1948 (1989).
124. S. K. Ghosh and E. S. Gould, *Inorg. Chem.*, **28**, 1538 (1989).
125. S. K. Ghosh and E. S. Gould, *Inorg. Chem.*, **28**, 3651 (1989).
126. C. K. Ranganathan, T. Ramasami, D. Ramaswamy, and M. Santappa, *Inorg. Chem.*, **28**, 1306 (1989).
127. R. P. Farrell, R. J. Judd, P. A. Lay, R. Bramley, and J.-Y. Ji, *Inorg. Chem.*, **28**, 3401 (1989).
128. X. Shi and N. S. Dalal, *Biochem. Biophys. Res. Commun.*, **156**, 137 (1988); *Chem. Abstr.*, **110**, 2346 (1989).
129. M. Branca, G. Micera, and A. Dessi, *Inorg. Chim. Acta*, **153**, 61 (1988).
130. M. P. Alvarez Macho, *Rev. Roum. Chim.*, **33**, 857 (1988).
131. T. W. Hambley, R. J. Judd, P. Lay, R. Bramley, and J.-Y. Ji, Abstracts of papers, 27th ICCC Queensland, Australia, #W46 (1989).
132. R. P. Farrell, R. J. Judd, P. A. Lay, N. E. Dixon, R. S. U. Baker, and A. M. Bonin, *Chem. Res. Toxicol* (in press).
133. P. O'Brien and G. Wang, *Inorg. Chim. Acta*, **162**, L27 (1989).
134. Y. Sulfab and M. Nasreldin, Abstracts of papers, 27th ICCC Queensland, Australia, #W57 (1989).
135. K. K. Banerji, *J. Chem. Soc., Perkin Trans. 2*, 547 (1988).

136. K. K. Banerji, *J. Chem. Soc., Perkin Trans. 2*, 639 (1978).

137. K. K. Banerji, *J. Chem. Soc., Perkin Trans. 2*, 2065 (1988).

138. G. V. Rao and P. K. Saiprakash, *Oxid. Commun.*, **11**, 33 (1988); *Chem. Abstr.*, **110**, 135622 (1989).

139. D. A. Dixon, N. P. Sadler, and T. P. Dasgupta, Abstracts of papers, 198th ACS National Meeting, Florida, #94 (1989).

140. K. K. Sen Gupta, S. Sen Gupta, S. K. Mandal, and S. N. Basu, *Indian J. Chem.*, **27A**, 23 (1988); *Chem. Abstr.*, **109**, 110757 (1988).

141. K. Rajasekavan, T. Baskavan, and C. Gnansekavan, *Indian J. Chem.*, **26A**, 956 (1987); *Chem. Abstr.*, **109**, 229916 (1988).

142. G. S. Gokavi and J. R. Raju, *Indian J. Chem.*, **26A**, 494 (1988); *Chem. Abstr.*, **109**, 216835 (1988).

143. M. P. Alvarez Macho, *An. Quim, Ser. A*, **84**, 14 (1988).

144. K. K. Sen Gupta, M. Adhikari, and S. Sen Gupta, *React. Kinet. Catal. Lett.*, **38**, 313 (1989); *Chem. Abstr.*, **111**, 56814 (1989).

145. P. S. Kalidos and S. Vangalur, *Indian J. Chem.*, **26A**, 1039 (1987).

146. S. S. Anis, M. A. Mohammed, and S. S. Labib, *Microchem. J.*, **37**, 345 (1988); *Chem. Abstr.*, **110**, 114057 (1989).

147. P. O. I. Virtanen and R. Lindroos-Heinanen, *Acta Chem. Scand.*, **42B**, 411 (1988).

148. G. Lopez-Cueto and M. Duarte, *Can. J. Chem.*, **67**, 279 (1989).

149. R. Banerjee, A. Bhattacharya, P. K. Das, and A. K. Chakraburtty, *J. Chem. Soc., Dalton Trans.*, 1557 (1988).

150. G. Lunn and E. B. Sansone, *J. Chem. Educ.*, **66**, 443 (1989).

151. W. Mazurek, P. J. Nichols, and B. O. West, Abstracts of papers 27th ICCC Queensland, Australia, #T87 (1989).

152. K. Meenal and P. Roopakalyani, *J. Indian Chem. Soc.*, **65**, 624 (1988).

153. L. V. S. P. Rao and S. B. Rao, *J. Indian Chem. Soc.*, **65**, 404 (1988); *Chem. Abstr.*, **110**, 74624 (1989).

154. L. E. Early and D. Rai, *Am. J. Sci.*, **289**, 180 (1989).

155. G. S. Gokavi and J. R. Raju, *Oxid. Commun.*, **11**, 205 (1988); *Chem. Abstr.*, **111**, 133479 (1989).

156. P. R. Norman, A. Tate, and P. Rich, *Inorg. Chim. Acta*, **145**, 211 (1988).

157. T. Sekine, K. Inaba, T. Morimoto, and H. Aikawa, *Bull. Chem. Soc. Jpn.*, **61**, 1131 (1988).

References for Chapter 7

1. M. C. Ghosh, P. Bhattacharya, and P. Banerjee, *Coord. Chem. Rev.*, **91**, 1 (1988).

2. M. J. Blandamer and J. Burgess, *Transition Met. Chem.*, **13**, 1 (1988).

3. S. S. Massoud and R. M. Milburn, *Polyhedron*, **8**, 275 (1989).

4. N. J. Curtis, G. A. Lawrance, and R. van Eldik, *Inorg. Chem.*, **28**, 329 (1989).

5. Y. Kitamura, S. Taneda, and K. Kuroda, *Inorg. Chim. Acta*, **156**, 31 (1989).

6. W. G. Jackson, A. G. Kuzmission, J. N. Cooper, and J. C. Hendry, *Inorg. Chem.*, **28**, 1553 (1989).

7. M. E. Kastner, D. A. Smith, A. G. Kuzmission, J. N. Cooper, T. Tyree, and M. Yearik, *Inorg. Chim. Acta*, **158**, 185 (1989).

8. A. S. Babaqi, M. T. Mohamed, S. S. Massoud, and M. F. Amira, *Transition Met. Chem.*, **13**, 201 (1988).

9. A. Panda and N. C. Naik, *Transition Met. Chem.*, **14**, 439 (1989).

10. D. A. Buckingham, I. I. Olsen, and A. M. Sargeson, *Aust. J. Chem.*, **20**, 597 (1967).

11. N. E. Brasch, D. A. Buckingham, C. R. Clark, and K. S. Finnie, *Inorg. Chem.*, **28**, 3386 (1989).

12. D. A. Buckingham, C. R. Clark, and M. J. Gaudin, *Inorg. Chem.*, **27**, 293 (1988).

13. W. G. Jackson and B. H. Dutton, *Inorg. Chem.*, **28**, 525 (1989).

14. E. Ahmed, C. Chatterjee, C. J. Cooksey, M. L. Tobe, G. Williams, and M. Humanes, *J. Chem. Soc., Dalton Trans.*, 645 (1989).
15. D. A. House and M. L. Tobe, *J. Chem. Soc., Dalton Trans.*, 853 (1989).
16. G. C. Pradhan, R. K. Nanda, and A. C. Dash, *Transition Met. Chem.*, **14**, 309 (1989).
17. N. Das, R. K. Nanda, and A. C. Dash, *Transition Met. Chem.*, **14**, 261 (1989).
18. Y. Kitamura, G. A. Lawrance, and R. van Eldik, *Inorg. Chem.*, **28**, 333 (1989).
19. R. W. Hay and F. M. McLaren, *Transition Met. Chem.*, **14**, 361 (1989).
20. D. Chatterjee and G. S. De, *Transition Met. Chem.*, **14**, 277 (1989).
21. A. K. Gangopadhyay and G. S. De, *Transition Met. Chem.*, **14**, 241 (1989).
22. W. G. Jackson, *Inorg. Chim. Acta*, **149**, 101 (1988).
23. D. A. Gooden, H. A. Ewart, and K. E. Hyde, *Inorg. Chim. Acta*, **166**, 263 (1989).
24. S. S. Massoud and R. M. Milburn, *Polyhedron*, **8**, 415 (1989).
25. I. M. Sidamed and A. M. Ismail, *Transition Met. Chem.*, **13**, 193 (1988).
26. I. M. Sidahmed and A. M. Ismail, *Transition Met. Chem.*, **12**, 449 (1987).
27. Y. Kitamura, T. Takamoto, and K. Yoshitani, *Inorg. Chem.*, **27**, 1382 (1988).
28. T. Li, K. W. Pfahler, and J. H. Worrell, *J. Coord. Chem.*, **19**, 61 (1988).
29. G. G. Sadler and T. P. Dasgupta, *Inorg. Chim. Acta*, **167**, 233 (1989).
30. T. Taura, *Inorg. Chim. Acta*, **158**, 33 (1989).
31. P. G. Yohannes, N. Bresciani-Pahor, L. Randaccio, E. Zangrando, and L. G. Marzilli, *Inorg. Chem.*, **27**, 4738 (1988).
32. M. J. Bjerrum, M. Gajhede, E. Larsen, and J. Springborg, *Inorg. Chem.*, **27**, 3960 (1988).
33. P. J. Toscano and E. Barren, *J. Chem. Soc., Chem. Commun.*, 1159 (1989).
34. G. Stochel, R. van Eldik, H. Kunkely, and A. Vogler, *Inorg. Chem.*, **28**, 4314 (1989).
35. N. Bresciani-Pahor, L. Randaccio, E. Zangrando, and P. A. Marzilli, *J. Chem. Soc., Dalton Trans.*, 1941 (1989).
36. D. Datata and G. Tomba Sharma, *J. Chem. Soc., Dalton Trans.*, 115 (1989).
37. H. M. Marques, T. J. Egan, J. H. Marsh, J. R. Mellor, and O. Q. Munro, *Inorg. Chim. Acta*, **166**, 249 (1989).

References for Chapter 8

1. M. F. Summers, J. van Rijn, J. Reedijk, and L. G. Marzilli, *J. Am. Chem. Soc.*, **108**, 4254 (1986).
2. R. W. Hay, M. P. Pujari, and R. Bembi, *Transition Met. Chem.*, **11**, 261 (1986); A. Evers and R. D. Hancock, *Inorg. Chim. Acta*, **160**, 245 (1989).
3. P. V. Bernhardt, L. S. Curtis, N. F. Curtis, G. A. Lawrance, B. W. Skelton, and A. H. White, *Aust. J. Chem.*, **42**, 797 (1989).
4. W. Collier, L. S. Curtis, N. F. Curtis, and I. J. Pomer, *Aust. J. Chem.*, **42**, 1611 (1989).
5. B. E. Schwederski and D. W. Margerum, *Inorg. Chem.*, **28**, 3472 (1989); B. E. Schwerderski, F. Basile-D'Alessandro, P. N. Dickson, H. D. Lee, J. M. T. Raycheba, and D. W. Margerum, *Inorg. Chem.*, **28**, 3477 (1989).
6. M. K. Moi, C. F. Meares, and S. J. DeNardo, *J. Am. Chem. Soc.*, **110**, 6266 (1988).
7. J. P. L. Cox, K. J. Jankowski, R. Kataky, D. Parker, N. R. A. Beeley, B. A. Boyce, M. A. W. Eaton, K. Millar, A. T. Millican, A. Harrison, and C. Walker, *J. Chem. Soc., Chem. Commun.*, 797 (1989).
8. W. Radecka-Paryzek and E. Luks, *Polyhedron*, **9**, 475 (1990).
9. G. Bombieri, F. Benetollo, W. T. Hawkins, A. Polo, and L. M. Vallarino, *Polyhedron*, **8**, 1923 (1989); G. Bombieri, F. Benetollo, A. Polo, L. de Cola, W. T. Hawkins, and L. M. Vallarino, *Polyhedron*, **8**, 2157 (1989).
10. J. L. Sessler, T. Murai, and G. Hemmi, *Inorg. Chem.*, **28**, 3390 (1989).
11. S. V. Deshpande, R. Subramanian, M. J. McCall, S. J. DeNardo, G. L. DeNardo, and C. F. Meares, *J. Nucl. Med.*, **31**, 218 (1990).
12. H. R. Mäcke, A. Riesen, and W. Ritter, *J. Nucl. Med.*, **30**, 1235 (1989).

13. A. Riesen, T. A. Kaden, W. Ritter, and H. R. Mäcke, *J. Chem. Soc., Chem. Commun.*, 460 (1989).

14. A. S. Craig, I. M. Helps, D. Parker, H. Adams, N. A. Bailey, M. G. Williams, J. M. A. Smith, and G. Ferguson, *Polyhedron*, **8**, 2481 (1989).

15. A. S. Craig, I. M. Helps, K. J. Jankowski, D. Parker, N. R. A. Beeley, B. A. Boyce, M. A. W. Eaton, A. T. Millican, K. Millar, A. Phipps, S. K. Rhind, A. Harrison, and C. Walker, *J. Chem. Soc., Chem. Commun.*, 794 (1989).

16. E. Brücher, G. Laurenczy, and Z. Makra, *Inorg. Chim. Acta*, **139**, 141 (1987).

17. E. R. Souaya, *Z. Phys. Chem. (Leipzig)*, **269**, 1017 (1988).

18. A. C. Muscatello, G. R. Choppin, and W. D'Olieslager, *Inorg. Chem.*, **28**, 993 (1989).

19. Y. Ducommun, L. Helm, G. Laurenczy, and A. E. Merbach, *Inorg. Chim. Acta*, **158**, 3 (1989).

20. V. Ferreira, R. A. Krause, and S. Larsen, *Inorg. Chim. Acta*, **145**, 29 (1988).

21. P. M. Treichel and G. J. Essenmacher, *Inorg. Chem.*, **15**, 146 (1976).

22. M. R. Weaver and J. M. Harris, *Anal. Chem.*, **61**, 1001 (1989).

23. R. B. Martin, J. Savory, S. Brown, R. L. Bertholf, and M. R. Wills, *Clin. Chem.*, **33**, 405 (1987).

24. J. C. Dobson and H. Taube, *Inorg. Chem.*, **28**, 1310 (1989).

25. R. A. Henderson, G. J. Leigh, and C. J. Pickett, *J. Chem. Soc., Dalton Trans.*, 425 (1989).

26. Bee-Lean Ooi, C. Sharp, and A. G. Sykes, *J. Am. Chem. Soc.*, **111**, 125 (1989).

27. Bee-Lean Ooi and A. G. Sykes, *Inorg. Chem.*, **28**, 3799 (1989).

28. Bee-Lean Ooi, M. Martinez, T. Shibahara, and A. G. Sykes, *J. Chem. Soc., Dalton Trans.*, 2239, 2683 (1988).

29. Bee-Lean Ooi, M. Martinez, and A. G. Sykes, *J. Chem. Soc., Chem. Commun.*, 1324 (1988).

30. I. M. Potgieter, S. S. Basson, A. Roodt, and J. G. Leipoldt, *Transition Met. Chem.*, **13**, 209 (1988).

31. S. A. Roberts, C. G. Young, W. E. Cleland, R. B. Ortega, and J. H. Enemark, *Inorg. Chem.*, **27**, 3044 (1988).

32. A. Roodt, J. G. Leipoldt, S. S. Basson, and I. M. Potgieter, *Transition Met. Chem.*, **13**, 336 (1988).

33. P. Schneiber, K. Wieghardt, U. Flörke, and H.-J. Haupt, *Inorg. Chem.*, **27**, 2111 (1988).

34. P. Mishra, R. M. Naik, and P. C. Nigam, *Transition Met. Chem.*, **13**, 267 (1988).

35. R. Alberto, P. Bläuenstein, G. Anderegg, and A. Albinati, *J. Lab. Cmpd. Radiopharm.*, **26**, 258 (1989).

36. T. Konno, M. J. Heeg, and E. Deutsch, *Inorg. Chem.*, **28**, 1694 (1989).

37. H. Kido and Y. Hatakeyama, *Inorg. Chem.*, **27**, 3623 (1988).

38. W. Hirth, S. Jurisson, K. Linder, T. Feld, and A. D. Nunn, *J. Lab. Cmpd. Radiopharm.*, **26**, 48 (1989).

39. E. N. Treher, L. C. Francesconi, J. Z. Gougoutas, M. F. Malley, and A. D. Nunn, *Inorg. Chem.*, **28**, 3411 (1989).

40. R. Alberto, P. Bläuenstein, P. A. Schubinger, and G. Anderegg, *J. Lab. Cmpd. Radiopharm.*, **26**, 47 (1989).

41. J. L. Martin, J. Yuan, C. E. Lunte, R. C. Elder, W. R. Heineman, and E. Deutsch, *Inorg. Chem.*, **28**, 2899 (1989).

42. H. F. Kung, Bo-Li Liu, and S. Pan, *Appl. Radiat. Isot.*, **40**, 677 (1989).

43. J. C. Hung, M. Corlija, W. A. Volkert, and R. A. Holmes, *J. Nucl. Med.*, **29**, 1568 (1988).

44. Jun Lu and M. J. Clarke, *Inorg. Chem.*, **27**, 4761 (1988).

45. Jun Lu and M. J. Clarke, *Inorg. Chem.*, **28**, 2315 (1989).

46. J. Baldas, J. Bonnyman, and G. A. Williams, *Inorg. Chem.*, **25**, 150 (1986).

47. J. Baldas, J. F. Boas, and J. Bonnyman, *Aust. J. Chem.*, **42**, 639 (1989).

48. A. Alagui, M. Apparu, and A. du Moulinet d'Hardemare, F. Riche, and M. Vidal, *Appl. Radiat. Isot.*, **40**, 813 (1989).

49. J. R. Ballinger, K. Y. Gulenchyn, and M. N. Hassan, *Appl. Radiat. Isot.*, **40**, 547 (1989).

50. C. M. Kennedy, M. V. Mikelsons, B. L. Lawson, and T. C. Pinkerton, *Appl. Radiat. Isot.*, **39**, 213 (1988).

51. A. Cusanelli and D. Sutton, *J. Chem. Soc., Chem. Commun.*, 1719 (1989).
52. Xiao-Liang Luo and R. H. Crabtree, *Inorg. Chem.*, **28**, 3775 (1989).
53. W. Purcell, A. Roodt, S. S. Basson, and J. G. Leipoldt, *Transition Met. Chem.*, **14**, 224 (1989).
54. J. Ferry, P. McArdle, and M. J. Hynes, *J. Chem. Soc., Dalton Trans.*, 767 (1989).
55. D. H. Macartney and L. J. Warrack, *Can. J. Chem.*, **67**, 1774 (1989).
56. Gyu-Hwan Lee, L. della Ciane, and A. Haim, *J. Am. Chem. Soc.*, **111**, 2535 (1989).
57. D. H. Macartney, *Rev. Inorg. Chem.*, **9**, 101 (1988).
58. See p. 195 of Volume 5 and p. 201 of Volume 6 of this series.
59. S. M. Bradley, H. Doine, H. R. Krouse, M. J. Sisley, and T. W. Swaddle, *Aust. J. Chem.*, **41**, 1323 (1988).
60. S. Wieland, R. van Eldik, D. R. Crane, and P. C. Ford, *Inorg. Chem.*, **28**, 3663 (1989).
61. A. Kathó and M. T. Beck, *Inorg. Chim. Acta*, **154**, 99 (1988).
62. Yu. G. Gol'tsev and V. V. Zhilinskaya, *Russ. J. Inorg. Chem.*, **34**, 221 (1989).
63. G. Stochel, R. van Eldik, E. Hejmo, and Z. Stasicka, *Inorg. Chem.*, **27**, 2767 (1988).
64. G. Stochel and R. van Eldik, *Inorg. Chim. Acta*, **155**, 95 (1989).
65. H. E. Toma and M. S. Takasugi, *Polyhedron*, **8**, 941 (1989).
66. N. Y. M. Iha, H. E. Toma, and J. F. de Lima, *Polyhedron*, **7**, 1687 (1988).
67. Y. Kuroda, N. Tanaka, M. Goto, and T. Sakai, *Inorg. Chem.*, **28**, 997 (1989).
68. Y. Kuroda, N. Tanaka, M. Goto, and T. Sakai, *Inorg. Chem.*, **28**, 2163 (1989).
69. J. K. Beattie and K. J. McMahon, *Aust. J. Chem.*, **41**, 1315 (1988).
70. M. J. Blandamer, J. Burgess, H. J. Cowles, I. M. Horn, J. B. F. N. Engberts, S. A. Galema, and C. D. Hubbard, *J. Chem. Soc., Faraday Trans. 1*, **85**, 3733 (1989).
71. M. J. Blandamer, J. Burgess, H. J. Cowles, A. J. De Young, J. B. F. N. Engberts, S. A. Galema, S. J. Hill, and I. M. Horn, *J. Chem. Soc., Chem. Commun.*, 1141 (1988).
72. C. C. Deb, D. K. Hazra, and S. C. Lahiri, *Z. Phys. Chem. (Leipzig)*, **270**, 777 (1989).
73. F. Ibáñez, J. G. Santos, M. A. Francois, and S. Gallardo, *J. Chem. Soc., Dalton Trans.*, 1105 (1989).
74. M. J. Blandamer, J. Burgess, P. Guardado, and C. D. Hubbard, *J. Chem. Soc., Faraday Trans. 1*, **85**, 735 (1989).
75. M. J. Blandamer, B. Briggs, J. Burgess, D. Elvidge, P. Guardado, A. W. Hakin, S. Radulović, and C. D. Hubbard, *J. Chem. Soc., Faraday Trans. 1*, **84**, 2703 (1988).
76. M. J. Blandamer and J. Burgess, *Transition Met. Chem.*, **13**, 1 (1988).
77. M. J. Blandamer, J. Burgess, H. J. Cowles, I. M. Horn, N. J. Blundell, and J. B. F. N. Engberts, *J. Chem. Soc., Chem. Commun.*, 1233 (1989).
78. J. Burgess and C. D. Hubbard, *Inorg. Chem.*, **27**, 2548 (1988).
79. L. L. Martin, K. S. Hagen, A. Hauser, R. L. Martin, and A. M. Sargeson, *J. Chem. Soc., Chem. Commun.*, 1313 (1988).
80. J. Burgess and E. Pelizzetti, *Gazz. Chim. Ital*, **118**, 803 (1988); *Prog. React. Kinet.*, in press.
81. F. Ortega and E. Rodenas, *Can. J. Chem.*, **67**, 305 (1989).
82. M. Tubino, E. J. S. Vichi, and I. K. Lauff, *Chem. Scr.*, **29**, 201 (1989).
83. J. Faus, M. Julve, J. M. Amigo, and T. Debaerdemaeker, *J. Chem. Soc., Dalton Trans.*, 1681 (1989).
84. A. R. Butler, A. M. Calsy, and C. Glidewell, *Polyhedron*, **8**, 175 (1989).
85. J. Springborg, P. C. Wilkins, and R. G. Wilkins, *Acta Chem. Scand.*, **43**, 967 (1989).
86. M. Momenteau, B. Loock, C. Tetreau, D. Lavalette, A. Croisy, C. Schaeffer, C. Huel, and J.-M. Lhoste, *J. Chem. Soc., Perkin Trans. 2*, 249 (1987).
87. A. Desbois, M. Momenteau, and M. Lutz, *Inorg. Chem.*, **28**, 825 (1989).
88. Hee Cheon Lee and E. Oldfield, *J. Am. Chem. Soc.*, **111**, 1584 (1989).
89. A. Maldotti, G. Varni, R. Amadelli, C. Bartocci, and A. Ferri, *New J. Chem.*, **12**, 819 (1988).
90. D. V. Stynes, *Pure Appl. Chem.*, **60**, 561 (1988).
91. P. Basu, S. Pal, and A. Chakravorty, *J. Chem. Soc., Chem. Commun.*, 977 (1989).
92. P. K. Das, S. G. Bhattacharya, R. Banerjee, and D. Banerjea, *J. Coord. Chem.*, **19**, 311 (1989); N. M. Shuaib, M. S. El-Ezaby, and O. Al-Hussaini, *Polyhedron*, **8**, 1477 (1989).

93. N. Gupta and P. C. Nigam, *Transition Met. Chem.*, **13**, 367 (1988).
94. D. F. Evans and D. A. Jakubovic, *J. Chem. Soc., Dalton Trans.*, 2927 (1988).
95. M. Nakamura, *Inorg. Chim. Acta*, **161**, 73 (1989).
96. M. Inamo, T. Sumi, N. Nakagawa, S. Funahashi, and M. Tanaka, *Inorg. Chem.*, **28**, 2688 (1989).
97. J. W. Buchler, C. Dreher, and G. Herget, *Ann. Chem.*, 43 (1988).
98. G. M. Miskelly, W. S. Webley, C. R. Clark, and D. A. Buckingham, *Inorg. Chem.*, **27**, 3773 (1988).
99. I. Bertini, J. Hirose, C. Luchinat, L. Messori, M. Piccioli, and A. Scozzafava, *Inorg. Chem.*, **27**, 2405 (1988).
100. W. R. Harris and P. K. Bali, *Inorg. Chem.*, **27**, 2687 (1988).
101. P. K. Bali and W. R. Harris, *J. Am. Chem. Soc.*, **111**, 4457 (1989).
102. J. B. Porter, E. R. Huehns, and R. C. Hider, *Baillière's Clin. Haematol.*, **2**, 257, 260, 261, 279 (1989).
103. W. E. Jones, R. A. Smith, M. T. Abramo, M. D. Williams, and J. Van Houten, *Inorg. Chem.*, **28**, 2281 (1989).
104. S. Tachiyashiki, N. Nagao, and K. Mizumachi, *Chem. Lett.*, 1001 (1988).
105. S. Tachiyashiki and K. Mizumachi, *Chem. Lett.*, 1153 (1989).
106. S. G. Farina, W. Yuey, C. Ambrose, and P. E. Hoggard, *Inorg. Chim. Acta*, **148**, 97 (1988).
107. B. E. Buchanan, E. McGovern, P. Harkin, and J. G. Vos, *Inorg. Chim. Acta*, **154**, 1 (1988).
108. H.-F. Suen, S. W. Wilson, M. Pomerantz, and J. L. Walsh, *Inorg. Chem.*, **28**, 786 (1989).
109. H. B. Ross, M. Boldagi, D. P. Rillema, C. B. Blanton, and R. P. White, *Inorg. Chem.*, **28**, 1013 (1989).
110. R. R. Ruminski, T. Cockroft, and M. Shoup, *Inorg. Chem.*, **27**, 4026 (1988).
111. L. De Cola, F. Barigelletti, V. Balzani, P. Belser, A. von Zelewsky, F. Vögtle, F. Ebmeyer, and S. Grammenudi, *J. Am. Chem. Soc.*, **110**, 7210 (1988); F. Barigelletti, L. De Cola, V. Balzani, P. Belser, A. von Zelewsky, F. Vögtle, F. Ebmeyer, and S. Grammenudi, *J. Am. Chem. Soc.*, **111**, 4662 (1989).
112. J. A. Arce Sagues, R. D. Gillard, and P. A. Williams, *Transition Met. Chem.*, **14**, 110 (1989).
113. R. A. Leising, J. S. Ohman, and K. J. Takeuchi, *Inorg. Chem.*, **27**, 3804 (1988).
114. J. L. Walsh and C. C. Yancey, *Polyhedron*, **8**, 1223 (1989).
115. W. Śliwa, *Transition Met. Chem.*, **14**, 321 (1989).
116. E. C. Constable and T. A. Leese, *Inorg. Chim. Acta*, **146**, 55 (1988).
117. M. A. Khalifa, *Polyhedron*, **8**, 851 (1989).
118. P. Hambright, *Inorg. Chim. Acta*, **157**, 95 (1989).
119. R. L. Sernaglia and D. W. Franco, *Inorg. Chem.*, **28**, 3485 (1989).
120. J. C. Nascimento Filho, J. M. de Rezende, B. D. S. L. Neto, and D. W. Franco, *Inorg. Chim. Acta*, **145**, 111 (1988).
121. A. E. Almaraz, L. A. Gentil, and J. A. Olabe, *J. Chem. Soc., Dalton Trans.*, 1973 (1989).
122. A. L. Tokman, L. A. Gentil, and J. A. Olabe, *Polyhedron*, **8**, 2091 (1989).
123. A. R. Butler, A. M. Calsy-Harrison, C. Glidewell, and I. L. Johnson, *Inorg. Chim. Acta*, **146**, 187 (1988).
124. H. Doine, H. Fukutomi, and R. D. Cannon, *J. Chem. Soc., Dalton Trans.*, 2007 (1988).
125. P. K. L. Chan, B. R. James, D. C. Frost, P. K. H. Chan, and Hong-Liang Hu, *Can. J. Chem.*, **67**, 508 (1989).
126. E. Alensio, G. Mestroni, G. Nardin, W. M. Attia, M. Calligaris, G. Sava, and S. Zorzet, *Inorg. Chem.*, **27**, 4099 (1988).
127. D. W. Krassowski, J. H. Nelson, K. R. Brower, D. Hauenstein, and R. A. Jacobson, *Inorg. Chem.*, **27**, 4294 (1988); D. W. Krassowski, K. Reimer, H. E. LeMay, and J. H. Nelson, *Inorg. Chem.*, **27**, 4307 (1988).
128. H. Nagao, H. Nishimura, H. Funato, Y. Ichikawa, F. S. Howell, M. Mukaida, and H. Kakihana, *Inorg. Chem.*, **28**, 3955 (1989).
129. A. Basu, T. G. Kasar, and N. Y. Sapre, *Inorg. Chem.*, **27**, 4539 (1988).

130. Y. Hoshino, R. Takahashi, K. Shimizu, G. P. Satô, and K. Aoki, *Bull. Chem. Soc. Jpn.*, **62**, 993 (1989).
131. Y. Ilan and A. Kfir, *Inorg. Chim. Acta*, **156**, 221 (1989).
132. M. M. Taqui Khan, G. Ramachandraiah, and R. S. Shukla, *Inorg. Chem.*, **27**, 3274 (1988).
133. U. C. Sarma, K. P. Sarma, and R. K. Poddar, *Polyhedron*, **7**, 1727 (1988).
134. U. C. Sarma and R. K. Poddar, *Polyhedron*, **7**, 1737 (1988).
135. H. C. Bajaj and R. van Eldik, *Inorg. Chem.*, **27**, 4052 (1988).
136. H. C. Bajaj and R. van Eldik, *Inorg. Chem.*, **28**, 1980 (1989).
137. M. M. Taqui Khan, A. Hussain, M. A. Moiz, and R. M. Naik, *Polyhedron*, **8**, 2199 (1989).
138. N. G. Maksimov, O. K. Shpigunova, T. T. Bbanakova, N. M. Sinitsyn, E. D. Korniets, V. E. Volkov, and I. D. Danilov, *Russ. J. Inorg. Chem.*, **34**, 234 (1989).
139. J. C. Dobson and T. J. Meyer, *Inorg. Chem.*, **27**, 3274 (1988).
140. W. D. Harman and H. Taube, *J. Am. Chem. Soc.*, **110**, 5403 (1988).
141. M. Sekine, W. D. Harman, and H. Taube, *Inorg. Chem.*, **27**, 3604 (1988).
142. R. Cordone, W. D. Harman, and H. Taube, *J. Am. Chem. Soc.*, **111**, 5969 (1989).
143. R. Cordone, W. D. Harman, and H. Taube, *J. Am. Chem. Soc.*, **111**, 2896 (1989).
144. W. D. Harman, J. F. Wishart, and H. Taube, *Inorg. Chem.*, **28**, 2411 (1989).
145. W. D. Harman, J. C. Dobson, and H. Taube, *J. Am. Chem. Soc.*, **111**, 3061 (1989).
146. W. D. Harman and H. Taube, *J. Am. Chem. Soc.*, **110**, 7906 (1988).
147. P. A. Lay, R. H. Magnuson, and H. Taube, *Inorg. Chem.*, **28**, 3001 (1989).
148. J. C. Dobson and T. J. Meyer, *Inorg. Chem.*, **28**, 2013 (1989).
149. E. M. Kober, J. V. Kaspar, B. P. Sullivan, and T. J. Meyer, *Inorg. Chem.*, **27**, 4587 (1988).
150. A. J. Deeming, M. N. Meah, N. P. Randle, and K. I. Hardcastle, *J. Chem. Soc., Dalton Trans.*, 2211 (1989).
151. W. D. Harman and H. Taube, *Inorg. Chem.*, **27**, 3261 (1988).
152. W. D. Harman, M. Sekine, and H. Taube, *J. Am. Chem. Soc.*, **110**, 5725 (1988).
153. S. Suvachittanont and R. van Eldik, *Inorg. Chem.*, **28**, 3660 (1989).
154. R. D. Gillard and J. D. Pedrosa de Jesus, *Polyhedron*, **8**, 1163 (1989).
155. M. M. Muir and L. M. Torres, *Inorg. Chim. Acta*, **164**, 33 (1989).
156. P. D. Clark, J. H. Machin, J. F. Richardson, N. I. Dowling, and J. B. Hyne, *Inorg. Chem.*, **27**, 3526 (1988).
157. G. Hunter and N. Kilcullen, *J. Chem. Soc., Dalton Trans.*, 2115 (1989).
158. E. C. Constable, *Polyhedron*, **8**, 83 (1989).
159. L. Mønsted, O. Mønsted, and L. H. Skibsted, *Acta Chem. Scand.*, **43**, 128 (1989).
160. C. J. Arpey, S. E. Jacobs, C. Terzakis, A. Williams, and P. S. Sheridan, *Inorg. Chem.*, **27**, 3447 (1988).
161. D. Magde, G. E. Rojas, and L. H. Skibsted, *Inorg. Chem.*, **27**, 2900 (1988).
162. B. B. Wayland, V. L. Coffin, and M. D. Farnos, *Inorg. Chem.*, **27**, 2745 (1988).
163. M. A. S. Aquino and D. H. Macartney, *Inorg. Chem.*, **27**, 2868 (1988).
164. S. E. Castillo-Blum, D. T. Richens, and A. G. Sykes, *Inorg. Chem.*, **28**, 954 (1989).
165. P. S. Cartwright, R. D. Gillard, E. R. J. Sillanpaa, and J. Valkonen, *Polyhedron*, **7**, 2143 (1988).
166. J. D. Scott and R. J. Puddephat, *Organometallics*, **5**, 2522 (1986).
167. P. D. Akrivos and L. P. Aldridge, *Inorg. Chim. Acta*, **145**, 63 (1988).
168. P. F. Kelly and J. D. Woollins, *Polyhedron*, **8**, 2907 (1989).
169. E. W. Abel, T. P. J. Coston, K. G. Orrell, and V. Sik, *J. Chem. Soc., Dalton Trans.*, 711 (1989).
170. E. W. Abel, T. P. J. Coston, K. M. Higgins, K. G. Orrell, V. Šik, and T. S. Cameron, *J. Chem. Soc., Dalton Trans.*, 701 (1989).
171. E. W. Abel, D. G. Evans, J. R. Koe, V. Šik, P. A. Bates, and M. B. Hursthouse, *J. Chem. Soc., Dalton Trans.*, 985 (1989).
172. S. Toyota, Y. Yamada, M. Kaneyoshi, and M. Oki, *Bull. Chem. Soc. Jpn.*, **62**, 1509 (1989); S. Yamazaki and T. Anna, *Bull. Chem. Soc. Jpn.*, **62**, 931 (1989).
173. E. W. Abel, *Chem. Brit.*, 148 (1990).

References for Chapter 9

1. M. Kotowski and R. van Eldik, *Coord. Chem. Rev.*, **93**, 19 (1989).
2. R. van Eldik, T. Asano, and W. J. Le Noble, *Chem. Rev.*, **89**, 549 (1989).
3. H.-D. Projahn, C. Dreher, and R. van Eldik, *J. Am. Chem. Soc.*, **112**, 17 (1990).
4. B. B. Hasinoff, *Biochemistry*, **13**, 311 (1974).
5. H. P. Graves and C. Detellier, *J. Am. Chem. Soc.*, **110**, 6019 (1988).
6. H. D. H. Stover and C. Detellier, *J. Phys. Chem.*, **93**, 3174 (1989).
7. M. Bisnaire, C. Detellier, and D. Nadon, *Can. J. Chem.*, **60**, 3071 (1982).
8. H. D. H. Stöver, Ph.D Thesis, University of Ottawa, 1986.
9. D. L. Hughes, *J. Chem. Soc., Dalton Trans.*, 2374 (1975).
10. Y. Hou and A. I. Popov, *Spectrosc. Lett.*, **22**, 87 (1989).
11. D. P. Cobranchi, G. R. Phillips, D. E. Johnson, R. M. Barton, D. J. Rose, L. J. Rodriguez, and S. Petrucci, *J. Phys. Chem.*, **93**, 1396 (1989).
12. J. C. Lockhart, M. B. McDonnell, W. Clegg, M. N. Stuart Hill, and M. Todd, *J. Chem. Soc., Dalton Trans.*, 203 (1989).
13. J. J. Dechter, *Prog. Inorg. Chem.*, **29**, 285 (1982).
14. B. G. Cox, J. Stroka, and H. Schneider, *Inorg. Chim. Acta*, **147**, 9 (1988).
15. B. G. Cox, J. Stroka, I. Schneider, and H. Schneider, *J. Chem. Soc., Faraday Trans. I*, **85**, 187 (1989).
16. B. G. Cox, P. Firman, I. Schneider, and H. Schneider, *Inorg. Chem.*, **27**, 4018 (1989).
17. A. M. Albrecht-Gary, S. Blanc-Parasote, D. W. Boyd, G. Dauphin, G. Jeminet, J. Juillard, M. Prudhomme, and C. Tissier, *J. Am. Chem. Soc.*, **111**, 8598 (1989).
18. R. M. Snyder, C. K. Mirabelli, and S. T. Crook, *Biochem. Pharmacol.*, **35**, 923 (1986).
19. P. N. Dickson, A. Wehrli, and G. Geier, *Inorg. Chem.*, **27**, 2921 (1988).
20. J. A. Goodwin, L. J. Wilson, D. M. Stanbury, and R. A. Scott, *Inorg. Chem.*, **28**, 42 (1989).
21. C. Cossy, L. Helm, and A. E. Merbach, *Helv. Chim. Acta*, **70**, 1516 (1987).
22. A. E. Merbach, *Pure Appl. Chem.*, **54**, 1479 (1982); **59**, 161 (1987).
23. Y. Ducommun, P. J. Nichols, and A. E. Merbach, *Inorg. Chem.*, **28**, 2643 (1989).
24. A. K. Pondit, A. K. Das, and D. Banerjea, *Transition Met. Chem.*, **13**, 459 (1988).
25. L. Fielding and P. Moore, *J. Chem. Soc., Dalton Trans.*, 873 (1989).
26. M. Ishii, S. Funahashi, and M. Tanaka, *Inorg. Chem.*, **27**, 3192 (1988).
27. G. Laurenczy, Y. Ducommon, and A. E. Merbach, *Inorg. Chem.*, **28**, 3024 (1989).
28. K. Tamura, *J. Phys. Chem.*, **93**, 7358 (1989).
29. C. M. Madeyski, J. P. Michael, and R. D. Hancock, *Inorg. Chem.*, **23**, 1487 (1984).
30. S. P. Kasparzyk and R. G. Wilkins, *Inorg. Chem.*, **21**, 3349 (1982).
31. R. W. Hay, M. P. Pujari, W. T. Moodie, S. Craig, D. T. Richens, A. Perotti, and L. Ungaretti, *J. Chem. Soc., Dalton Trans.*, 2605 (1987).
32. F. McLaren, P. Moore, and A. M. Wynn, *J. Chem. Soc., Chem. Commun.*, 798 (1989).
33. M.-S. Chao and C.-S. Chung, *Inorg. Chem.*, **28**, 686 (1989).
34. L.-H. Chen and C.-S. Chung, *Inorg. Chem.*, **28**, 1402 (1989).
35. B. E. Schwederski, F. Basile-D'Allessandro, P. N. Dickson, H. D. Lee, J. M. T. Raycheba, and D. W. Margerum, *Inorg. Chem.*, **28**, 3477 (1989).
36. B. E. Schwerderski and D. W. Margerum, *Inorg. Chem.*, **28**, 3472 (1989).
37. M. Hauröder, M. Schütz, K. J. Wannowius, and Horst Elias, *Inorg. Chem.*, **28**, 736 (1989).
38. J. R. Röper, H. Paulus, and H. Elias, *Inorg. Chem.*, **28**, 2323 (1989).
39. P. Segla and H. Elias, *Inorg. Chim. Acta*, **149**, 259 (1988).
40. A. Odani, S. Deguchi, and O. Yamauchi, *Inorg. Chem.*, **25**, 62 (1986).
41. B. Busing, H. Elias, I. Eslick, and K. J. Wannowius, *Inorg. Chim. Acta*, **150**, 223 (1988).
42. V. Zang, M. Kotowski, and R. van Eldik, *Inorg. Chem.*, **27**, 3279 (1988).
43. Y. Ducommun, L. Helm, A. E. Merbach, B. Hellquist, and L. I. Elding, *Inorg. Chem.*, **28**, 337 (1989).
44. D. Minniti, G. Alibrandi, M. L. Tobe, and R. Romeo, *Inorg. Chem.*, **26**, 3956 (1987).

45. G. Alibrandi, D. Minniti, L. Monsu Scolaro, and R. Romeo, *Inorg. Chem.*, **28**, 1939 (1989).
46. B. Hellquist, L. I. Elding, and Y. Ducommun, *Inorg. Chem.*, **27**, 3620 (1988).
47. Y. Ducommun, A. E. Merbach, B. Hellquist, and L. I. Elding, *Inorg. Chem.*, **26**, 1759 (1987).
48. J. Berger, M. Kotowski, R. van Eldik, U. Frey, L. Helm, and A. E. Merbach, *Inorg. Chem.*, **28**, 3759 (1989).
49. J. J. Pienaar, M. Kotowski, and R. van Eldik, *Inorg. Chem.*, **28**, 373 (1989).
50. M. Kotowski, S. Begun, J. G. Leipoldt, and R. van Eldik, *Inorg. Chem.*, **27**, 4472 (1988).
51. J. J. Pienaar, E. L. Breet, and R. van Eldik, *Inorg. Chim. Acta*, **155**, 249 (1989).
52. A. Kumar, S. Gangopadhyay, and D. Banerjea, *Transition Met. Chem.*, **14**, 73 (1989).
53. N. Gupta, R. M. Naik, and P. C. Nigam, *Inorg. Chim. Acta*, **160**, 103 (1989).
54. D. C. Cheesman, A. P. Arnold, and D. L. Rabenstein, *J. Am. Chem. Soc.*, **110**, 6359 (1988).
55. G. Geier and H. Gross, *Inorg. Chim. Acta*, **156**, 91 (1989).
56. B. J. Plankey and H. H. Patterson, *Inorg. Chem.*, **28**, 4331 (1989).
57. D. Hugi-Cleary, L. Helm, and A. E. Merbach, *Helv. Chim. Acta*, **68**, 545 (1985).
58. B. J. Plankey, H. H. Patterson, and C. S. Cronan, *Environ. Sci. Technol.*, **20**, 160 (1986).
59. B. J. Plankey and H H. Patterson, *Environ. Sci. Technol.*, **22**, 1454 (1988).
60. I. Banyai and J. Glaser, *J. Am. Chem. Soc.*, **111**, 3186 (1989).
61. M. J. Sisley and R. B. Jordan, *Inorg. Chem.*, **27**, 4483 (1988).
62. K. R. Ashley and J. Kuo, *Inorg. Chem.*, **27**, 3556 (1988).
63. M. Inamo, T. Sumi, N. Nakagawa, S. Funahashi, and M. Tanaka, *Inorg. Chem.*, **28**, 2688 (1989).
64. G. Stochel and R. van Edik, *Inorg. Chim. Acta*, **155**, 95 (1989).
65. G. Stochel, R. van Eldik, E. Hejmo, and Z. Stasicka, *Inorg. Chem.*, **27**, 2767 (1988).
66. H. C. Bajaj and R. van Eldik, *Inorg. Chem.*, **27**, 4052 (1988).
67. M. M. Taqui Khan and R. M. Naik, *Polyhedron*, **7**, 463 (1988).
68. H. C. Bajaj and R. van Eldik, *Inorg. Chem.*, **28**, 1980 (1989).
69. S. Elmroth, L. H. Skibsted, and L. I. Elding, *Inorg. Chem.*, **28**, 2703 (1989).
70. L. H. Skibsted, *Adv. Inorg. Bioinorg. Mech.*, **4**, 137 1(986).
71. C. Cossy, A. C. Barnes, and J. E. Enderby, *J. Chem. Phys.*, **90**, 3254 (1989).
72. C. Cossy, L. Helm, and A. E. Merbach, *Inorg. Chem.*, **28**, 2699 (1989).
73. C. Cossy, and A. E. Merbach, *Pure Appl. Chem.*, **60**, 1785 (1988).
74. D. P. Fay, D. Litchinsky, and N. Purdie, *J. Phys. Chem.*, **73**, 544 (1969).
75. Y. Ducommun, L. Helm, G. Laurenczy, and A. E. Merbach, *Inorg. Chim. Acta*, **158**, 3 (1989).
76. Y. Ducommun, L. Helm, G. Laurenczy, and A. E. Merbach, *Magn. Reson. Chem.*, **26**, 1023 (1988).
77. R. J. Clarke, J. H. Coates, and S. F. Lincoln, *Inorg. Chim. Acta*, **153**, 21 (1988).
78. C. A. Chang, P. H.-L. Chang, V. K. Manchanda, and S. P. Kasprzyk, *Inorg. Chem.*, **27**, 3786 (1988).
79. S. Funahashi, K. Ishihara, M. Inamo, and M. Tanaka, *Inorg. Chim. Acta*, **157**, 65 (1989).
80. M. A. Islam and R. C. Thompson, *Inorg. Chem.*, **28**, 4419 (1989).
81. M. A. S. Aquino and D. H. Macartney, *Inorg. Chem.*, **27**, 2868 (1988).
82. H. J. Callot, A.-M. Albrecht-Gary, M. Al Joubbeh, B. Metz, and F. Metz, *Inorg. Chem.*, **28**, 3633 (1989).
83. T. Yamaguchi, T. Sasaki, A. Nagasawa, T. Ito, N. Koga, and K. Morokuma, *Inorg. Chem.*, **28**, 4311 (1989).

References for Chapter 10

1. (a) N. M. J. Brodie, L. Chen, and A. J. Poë, *Int. J. Chem. Kinet.*, **20**, 467 (1988); (b) A. J. Poë, *Pure Appl. Chem.*, **60**, 1209 (1988).
2. M. M. Rahman, H.-Y. Liu, K. Eriks, A. Prock, and W. P. Giering, *Organometallics*, **8**, 1 (1989).
3. L. N. Zakharov, Y. N. Saf'Yanov, and G. A. Domrachev, *Inorg. Chim. Acta*, **160**, 77 (1989).

4. K. Ericks, W. P. Giering, H.-Y. Liu, and A. Prock, *Inorg. Chem.*, **28**, 1759 (1989).

5. G. M. Bodner, M. P. May, and L. E. McKinney, *Inorg. Chem.*, **19**, 1951 (1980).

6. S. Y. Krueger, R. Poli, A. L. Rheingold, and D. L. Staley, *Inorg. Chem.*, **28**, 4599 (1989).

7. H.-R. Jaw and J. I. Zink, *Inorg. Chem.*, **27**, 3421 (1988).

8. (a) L. Wang, X.-M. Zhu, and K. G. Spears, *J. Am. Chem. Soc.*, **110**, 8695 (1988); L. Wang, X.-M. Zhu, and K. G. Spears, *J. Phys. Chem.*, **93**, 2 (1989); (b) J. D. Simon and X.-L. Xie, *J. Phys. Chem.*, **93**, 291 (1989); (c) M. Lee and C. B. Harris, *J. Am. Chem. Soc.*, **111**, 8963 (1989); (d) A. G. Joly and K. A. Nelson, *J. Phys. Chem.*, **93**, 2876 (1989); (e) C. H. Langford, C. Moralejo, and D. K. Sharma, *Inorg. Chim. Acta*, **126**, L11 (1987).

9. J. M. Morse, Jr., G. H. Parker, and T. J. Burkey, *Organometallics*, **8**, 2471 (1989).

10. R. D. Davy and M. B. Hall, *Inorg. Chem.*, **28**, 3524 (1989).

11. T. L. Brown, *Inorg. Chem.*, **28**, 3229 (1989).

12. T. J. Marks (ed.), *Polyhedron*, **7**, Nos. 16–17, 1409 (1988).

13. G. K. Yang, V. Vaida, and K. S. Peters, *Polyhedron*, **7**, 1619 (1988).

14. G. P. Smith, *Polyhedron*, **7**, 1605 (1988).

15. R. A. Jackson and A. J. Poë, *Inorg. Chem.*, **17**, 997 (1978).

16. (a) H. E. Bryndza, P. J. Domaille, W. Tam, L. K. Fong, R. A. Paciello, and J. E. Bercaw, *Polyhedron*, **7**, 1441 (1988); (b) L. E. Schock, A. M. Seyam, M. Sabat, and T. J. Marks, *Polyhedron*, **7**, 1517 (1988); (c) J. Halpern, *Polyhedron*, **7**, 1483 (1988); (d) A. R. Bulls, J. E. Bercaw, J. M. Manriquez, and M. E. Thompson, *Polyhedron*, **7**, 1409 (1988).

17. T. W. Koenig, B. P. Hay, and R. G. Finke, *Polyhedron*, **7**, 1499 (1988).

18. A. A. Gonzalez, K. Zhang, S. P. Nolan, R. L. de la Vega, S. L. Mukerjee, and C. D. Hoff, *Organometallics*, **7**, 2429 (1988).

19. G. Schmidt, H. Paulus, R. van Eldik, and H. Elias, *Inorg. Chem.*, **27**, 3211 (1988).

20. J.-K. Shen, Y.-C. Gao, Q.-Z. Shi, and F. Basolo, *Inorg. Chem.*, **28**, 4304 (1989).

21. L. Chen and A. J. Poë, *Inorg. Chem.*, **28**, 3641 (1989).

22. Y. Sulfab, F. Basolo, and A. L. Rheingold, *Organometallics*, **8**, 2139 (1989).

23. D. J. Darensbourg, K. M. Sanchez, and J. Reibenspies, *Inorg. Chem.*, **27**, 3636 (1988).

24. G. A. Luinstra, J. H. Teuben, and H.-H. Brintzinger, *J. Organomet. Chem.*, **375**, 183 (1989).

25. D. W. Krassowski, J. H. Nelson, K. R. Brower, D. Hauenstein, and R. A. Jacobson, *Inorg. Chem.*, **27**, 4294 (1988).

26. S. P. Ermer, R. S. Shinomoto, M. A. Deming, and T. C. Flood, *Organometallics*, **8**, 1377 (1989).

27. H. E. Bryndza, P. J. Domaille, R. A. Paciello, and J. E. Bercaw, *Organometallics*, **8**, 379 (1989).

28. D. W. Krassowski and J. H. Nelson, *J. Organomet. Chem.*, **356**, 93 (1988).

29. D. W. Krassowski, K. Reimer, H. E. LeMay, Jr., and J. H. Nelson, *Inorg. Chem.*, **27**, 4307 (1988).

30. J. D. Meinhart, E. V. Anslyn, and R. H. Grubbs, *Organometallics*, **8**, 583 (1989).

31. K. J. Asali, G. J. van Zyl, and G. R. Dobson, *Inorg. Chem.*, **27**, 3314 (1988).

32. A. A. Gonzalez, K. Zhang, and C. D. Hoff, *Inorg. Chem.*, **28**, 4285 (1989).

33. B. S. Creaven, F.-W. Grevels, and C. Long, *Inorg. Chem.*, **28**, 2231 (1989).

34. N. B. Pahor, L. Randaccio, E. Zangrando, and P. A. Marzilli, *J. Chem. Soc., Dalton Trans.*, 1941 (1989).

35. S. Roy, R. J. Puddephatt, and J. D. Scott, *J. Chem. Soc., Dalton Trans.*, 2121 (1989).

36. (a) W. D. Harman, M. Sekine, and H. Taube, *J. Am. Chem. Soc.*, **110**, 5725 (1988); (b) W. D. Harman and H. Taube, *J. Am. Chem. Soc.*, **110**, 7906 (1988).

37. P. A. Chaloner and G. T. L. Broadwood-Strong, *J. Organomet. Chem.*, **362**, C21 (1989).

38. R. M. Bullock, *J. Chem. Soc., Chem. Commun.*, 165 (1989).

39. A. J. L. Pombeiro, S. S. P. R. Almeida, M. F. C. G. Silva, J. C. Jeffrey, and R. L. Richards, *J. Chem. Soc., Dalton Trans.*, 2381 (1989).

40. K. Zhang, A. A. Gonzalez, and C. D. Hoff, *J. Am. Chem. Soc.*, **111**, 3627 (1989).

41. G. Pacchioni, *J. Organomet. Chem.*, **377**, C13 (1989).

42. R. bin Ali, J. Burgess, and A. T. Casey, *J. Organomet. Chem.*, **362**, 305 (1989).

43. G. R. Dobson and J. E. Cortés, *Inorg. Chem.*, **27**, 3308 (1988); G. R. Dobson and J. E. Cortés, *Inorg. Chem.*, **28**, 539 (1989).

44. D. P. Drolet, L. Chan, and A. J. Lees, *Organometallics,* **7,** 2502 (1988).
45. J. H. Potgieter, *J. Organomet. Chem.,* **366,** 369 (1989).
46. D. J. Wink, N.-F. Wang, and B. T. Creagan, *Organometallics,* **8,** 561 (1989).
47. C. Moralejo, C. H. Langford, and D. K. Sharma, *Inorg. Chem.,* **28,** 2205 (1989).
48. S. Zhang and G. R. Dobson, *Inorg. Chim. Acta,* **165,** 11 (1989).
49. P. H. Wermer and G. R. Dobson, *J. Coord. Chem.,* **20,** 125 (1989).
50. K. J. Asali, S. S. Basson, J. S. Tucker, B. C. Hester, J. E. Cortés, H. H. Awad, and G. R. Dobson, *J. Am. Chem. Soc.,* **109,** 5386 (1987).
51. S. Zhang and G. R. Dobson, *Inorg. Chem.,* **28,** 324 (1989).
52. E. P. Wasserman, R. G. Bergman, and C. B. Moore, *J. Am. Chem. Soc.,* **110,** 6076 (1988).
53. R. B. Hitam and A. J. Rest, *Organometallics,* **8,** 1598 (1989).
54. H. Angermund, A. K. Bandyopadhyay, F.-W. Grevels, and F. Mark, *J. Am. Chem. Soc.,* **111,** 4656 (1989).
55. T. Majima, T. Ishii, Y. Matsumoto, and M. Takami, *J. Am. Chem. Soc.,* **111,** 2417 (1989).
56. M. Kucharska-Zon and A. J. Poë, in: *Photochemistry and Photophysics of Coordination Compounds* (H. Yersin and A. Vogler, eds.), pp. 231–234, Springer-Verlag, Berlin (1987).
57. S. Wieland, R. van Eldik, D. R. Crane, and P. C. Ford, *Inorg. Chem.,* **28,** 3663 (1989).
58. S. Wieland and R. van Eldik, *J. Chem. Soc., Chem. Commun.,* 367 (1989).
59. L. L. Costanzo, S. Giuffrida, G. De Guidi and G. Condorelli, *J. Organomet. Chem.,* **349,** 235 (1988).
60. Y.-M. Wuu and M. S. Wrighton, *Organometallics,* **7,** 1839 (1988).
61. P. E. Bloyce, A. J. Rest, I. Whitwell, W. A. G. Graham, and R. Holmes-Smith, *J. Chem. Soc., Chem. Commun.,* 846 (1988).
62. S. M. Howdle and M. Poliakoff, *J. Chem. Soc., Chem. Commun.,* 1099 (1989).
63. G. Teixeira, T. Avilés, A. R. Dias, and F. Pina, *J. Organomet. Chem.,* **353,** 83 (1988).
64. R. H. Hill and B. J. Palmer, *Organometallics,* **8,** 1651 (1989).
65. H. K. van Dijk, D. J. Stufkens, and A. Oskam, *J. Am. Chem. Soc.,* **111,** 541 (1989).
66. H. K. van Dijk, J. J. Kok, D. J. Stufkens, and A. Oskam, *J. Organomet. Chem.,* **362,** 163 (1989).
67. P. C. Servaas, D. J. Stufkens, and A. Oskam, *Inorg. Chem.,* **28,** 1780 (1989).
68. R. R. Andréa, W. G. J. de Lange, D. J. Stufkens, and A. Oskam, *Inorg. Chem.,* **28,** 318 (1989).
69. H. K. van Dijk, J. van der Haar, D. J. Stufkens, and A. Oskam, *Inorg. Chem.,* **28,** 75 (1989).
70. N. S. Crossley, J. C. Green, A. Nagy, and G. Stringer, *J. Chem. Soc., Dalton Trans.,* 2139 (1989).
71. E. U. van Raaij and H.-H. Brintzinger, *J. Organomet. Chem.,* **356,** 315 (1988).
72. V. S. Leong and N. J. Cooper, *Organometallics,* **7,** 2058 (1988).
73. G.-H. Lee, S.-M. Peng, I.-C. Tsung, D. Mu, and R.-S. Liu, *Organometallics,* **8,** 2248 (1989).
74. R. S. Herrick, A. B. Frederick, and R. R. Duff, Jr., *Organometallics,* **8,** 1120 (1989).
75. H. Ahmed, D. A. Brown, N. J. Fitzpatrick, and W. K. Glass, *Inorg. Chim. Acta,* **164,** 5 (1989).
76. M. Cheong and F. Basolo, *Organometallics,* **7,** 2041 (1988).
77. A. K. Kakkar, N. J. Taylor, and T. B. Marder, *Organometallics,* **8,** 1765 (1989).
78. R. M. Nielson and M. J. Weaver, *Organometallics,* **8,** 1636 (1989).
79. A. Habib, R. S. Tanke, E. M. Holt, and R. H. Crabtree, *Organometallics,* **8,** 1225 (1989).
80. M. L. H. Green and L.-L. Wong, *J. Chem. Soc., Dalton Trans.,* 2133 (1989).
81. J. Arnold, G. Wilkinson, B. Hussain, and M. B. Hursthouse, *Organometallics,* **8,** 1362 (1989).
82. H. Kurosawa, K. Ishii, Y. Kawasaki, and S. Murai, *Organometallics,* **8,** 1756 (1989).
83. G. E. Herberich, U. Büschges, B. A. Dunne, B. Hessner, N. Klaff, D. P. J. Köffer, and K. Peters, *J. Organomet. Chem.,* **372,** 53 (1989).
84. J. P. Fawcett, R. A. Jackson, and A. J. Poë, *J. Chem. Soc., Chem. Commun.,* 733 (1975).
85. M. J. Therien and W. C. Trogler, *J. Am. Chem. Soc.,* **110,** 4942 (1988).
86. (a) R. N. McDonald and P. L. Schell, *Organometallics,* **7,** 1806 (1988); (b) R. N. McDonald and P. L. Schell, *Organometallics,* **7,** 1820 (1988).
87. H.-Y. Liu, M. N. Golovin, D. A. Fertal, A. A. Tracey, K. Eriks, W. P. Giering, and A. Prock, *Organometallics,* **8,** 1454 (1989).
88. G. W. Dillow, G. Nicol, and P. Kebarle, *J. Am. Chem. Soc.,* **111,** 5465 (1989).
89. D. J. Kuchynka and J. K. Kochi, *Inorg. Chem.,* **28,** 855 (1989).

90. J. Ruiz and D. Astruc, *J. Chem. Soc., Chem. Commun.*, 815 (1989).
91. W. C. Watkins, D. H. Macartney, and M. C. Baird, *J. Organomet. Chem.*, **377**, C52 (1989).
92. P. R. Drake and M. C. Baird, *J. Organomet. Chem.*, **363**, 131 (1989).
93. K. Y. Lee and J. K. Kochi, *Inorg. Chem.*, **28**, 567 (1989).
94. F. Mao, D. R. Tyler, and D. Keszler, *J. Am. Chem. Soc.*, **111**, 130 (1989).
95. Y. H. Pan and D. P. Ridge, *J. Am. Chem. Soc.*, **111**, 1150 (1989)
96. C.-P. Cheng, H.-S. Chen, and S.-R. Wang, *J. Organomet. Chem.*, **359**, 71 (1989).
97. E. B. Milosavljević, L. Solujić, D. W. Krassowski, and J. H. Nelson, *J. Organomet. Chem.*, **352**, 177 (1988).
98. N. C. Hallinan, G. Morelli, and F. Basolo, *J. Am. Chem. Soc.*, **110**, 6585 (1988).
99. M. Chanon, M. Julliard, and J. C. Poite, *Paramagnetic Organometallic Species in Activation/Selectivity, Catalysis*, Kluwer Academic Publishers, Dordrecht (1989).
100. F. Rourke and J. A. Crayston, *J. Chem. Soc., Chem. Commun.*, 1264 (1988).
101. S. Mönkeberg, E. U. van Raaij, H. Kiesele, and H.-H. Brintzinger, *J. Organomet. Chem.*, **365**, 285 (1989).
102. B. Olbrich-Deussner and W. Kaim, *J. Organomet. Chem.*, **361**, 335 (1989).
103. A. M. Bond, R. Colton, and T. F. Mann, *Organometallics*, **7**, 2224 (1988).
104. R. L. Keiter, E. A. Keiter, K. H. Hecker, and C. A. Boecker, *Organometallics*, **7**, 2466 (1988).
105. M. S. Loonat, L. Carlton, J. C. A. Boeyens, and N. J. Coville, *J. Chem. Soc., Dalton Trans.*, 2407 (1989).
106. J.-K. Shen, Y.-C. Gao, Q.-Z. Shi, and F. Basolo, *Organometallics*, **8**, 2144 (1989).
107. C.-K. Lai, W. G. Feighery, Y.-Q. Zhen, and J. D. Atwood, *Inorg. Chem.*, **28**, 3929 (1989).
108. M. S. Corraine and J. D. Atwood, *Inorg. Chem.*, **28**, 3781 (1989).
109. (a) Y.-Q. Zhen, W. G. Feighery, C.-K. Lai, and J. D. Atwood, *J. Am. Chem. Soc.;* **111**, 7832 (1989); (b) Y.-Q. Zhen, and J. D. Atwood, *J. Am. Chem. Soc.*, **111**, 1506 (1989).
110. T. J. Oyer and M. S. Wrighton, *Inorg. Chem.*, **27**, 3689 (1988).
111. N. J. Coville and P. Johnston, *J. Organomet. Chem.*, **363**, 343 (1989).
112. P. Johnston, G. J. Hutchings, and N. J. Coville, *J. Am. Chem. Soc.*, **111**, 1902 (1989).
113. M. P. Castellani and D. R. Tyler, *Organometallics*, **8**, 2113 (1989).
114. A. Darchen, H. Mousser, and H. Patin, *J. Chem. Soc., Chem. Commun.*, 968 (1988).
115. A. Darchen, E. K. Lhadi, and H. Patin, *J. Organomet. Chem.*, **363**, 137 (1989).
116. H. Patin, A. Darchen, and E. K. Lhadi, *J. Organomet. Chem.*, **375**, 91 (1989).
117. D. Osella, M. Botta, R. Gobetto, R. Amadelli, and V. Carassiti, *J. Chem. Soc., Dalton Trans.*, 2519 (1988).
118. R. T. Baker, J. C. Calabrese, P. J. Krusic, M. J. Therien, and W. C. Trogler, *J. Am. Chem. Soc.*, **110**, 8392 (1988).
119. P. J. Krusic, R. T. Baker, J. C. Calabrese, J. R. Morton, K. F. Preston, and Y. L. Page, *J. Am. Chem. Soc.*, **111**, 1262 (1989).
120. V. Riera, M. A. Ruiz, F. Villafañe, C. Bois, and Y. Jeannin, *J. Organomet. Chem.*, **375**, C23 (1989).
121. M. J. Chetcuti and K. A. Green, *Organometallics*, **7**, 2450 (1988).
122. A. M. Joshi and B. R. James, *J. Chem. Soc., Chem. Commun.*, 1785 (1989).
123. S. S. Belt, F.-W. Grevels, W. E. Klotzbücher, A. McCamley, and R. N. Perutz, *J. Am. Chem. Soc.*, **111**, 8373 (1989).
124. N. M. J. Brodie and A. J. Poë, *Inorg. Chem.*, **27**, 3156 (1988).
125. N. M. J. Brodie, L. Chen, A. J. Poë, and J. F. Sawyer, *Acta Crystallogr.*, **C45**, 1314 (1989).
126. J. J. Venter and M. A. Vannice, *Inorg. Chem.*, **28**, 1634 (1989); J. J. Venter and M. A. Vannice, *J. Am. Chem. Soc.*, **111**, 2377 (1989).
127. T. Chin-Choy, W. T. A. Harrison, G. D. Stucky, N. Keder, and P. C. Ford, *Inorg. Chem.*, **28**, 2028 (1989).
128. S.-H. Han, G. L. Geoffroy, B. D. Dombek, and A. L. Rheingold, *Inorg. Chem.*, **27**, 4355 (1988).
129. (a) J. A. Partin and M. G. Richmond, *J. Organomet. Chem.*, **353**, C13 (1988); (b) D. S. Dumond, S. Hwang, and M. G. Richmond, *Inorg. Chim. Acta*, **160**, 135 (1989).

130. N. M. J. Brodie, R. Huq, J. Malito, S. Markiewicz, A. J. Poë, and V. C. Sekhar, *J. Chem. Soc., Dalton Trans.*, 1933 (1989).

131. J. A. DiBenedetto, D. W. Ryba, and P. C. Ford, *Inorg. Chem.*, **28**, 3503 (1989).

132. A. E. Friedman and P. C. Ford, *J. Am. Chem. Soc.*, **111**, 551 (1989).

133. B. Ambwani, S. K. Chawla, and A. J. Poë, *Polyhedron*, **7**, 1939 (1988).

134. L. Chen and A. J. Poë, *Can. J. Chem.*, **67**, 1924 (1989).

135. J.-K. Shen, Y.-C. Gao, Q.-Z. Shi, and F. Basolo, *Inorg. Chem.*, **27**, 4236 (1988).

136. N. M. J. Brodie, Ph.D. Thesis, University of Toronto (1989).

137. R. Shojaie and J. D. Atwood, *Inorg. Chem.*, **27**, 2558 (1988).

138. R. H. E. Hudson, A. J. Poë, C. N. Sampson, and A. Siegel, *J. Chem. Soc., Dalton Trans.*, 2235 (1989).

139. K. Hinkelmann, J. Heinze, H.-T. Schacht, J. S. Field, and H. Vahrenkamp, *J. Am. Chem. Soc.*, **111**, 5078 (1989).

140. (a) S. Ching and D. F. Shriver, *J. Am. Chem. Soc.*, **111**, 3238 (1989); (b) S. Ching and D. F. Shriver, *J. Am. Chem. Soc.*, **111**, 3243 (1989); (c) S. Ching, M. Sabat, and D. F. Shriver, *Organometallics*, **8**, 1047 (1989).

141. J. B. Keister and C. C. O. Onyeso, *Organometallics*, **7**, 2364 (1988).

142. J. Anhaus, H. C. Bajaj, R. van Eldik, L. R. Nevinger, and J. B. Keister, *Organometallics*, **8**, 2903 (1989).

143. J. W. Ziller, D. K. Bower, D. M. Dalton, J. B. Keister, and M. R. Churchill, *Organometallics*, **8**, 492 (1989).

144. A. M. Bradford, M. C. Jennings, and R. J. Puddephatt, *Organometallics*, **8**, 2367 (1989).

145. T. A. Pakkanen, J. Pursiainen, T. T. Venäläinen, and T. Pakkanen, *J. Organomet. Chem.*, **372**, 129 (1989).

146. I. T. Horváth, M. Garland, G. Bor, and P. Pino, *J. Organomet. Chem.*, **358**, C17 (1988).

147. J. R. Kennedy, P. Selz, A. L. Rheingold, W. C. Trogler, and F. Basolo, *J. Am. Chem. Soc.*, **111**, 3615 (1989).

148. (a) J. T. Jaeger and H. Vahrenkamp, *Organometallics*, **7**, 1746 (1988); (b) J. T. Jaeger, J. S. Field, D. Collison, G. P. Speck, B. M. Peake, J. Hähnle, and H. Vahrenkamp, *Organometallics*, **7**, 1753 (1988).

149. E. P. Kyba, M. C. Kerby, R. P. Kashyap, J. A. Mountzouris, and R. E. Davis, *Organometallics*, **8**, 852 (1989).

150. A. K. Bandyopadhyay, R. Khattar, and T. P. Fehlner, *Inorg. Chem.*, **28**, 4434 (1989).

151. R. D. Pergola, L. Garlaschelli, S. Martinengo, F. Demartin, M. Manassero, and N. Masciocchi, *J. Chem. Soc., Dalton Trans.*, 2307 (1988).

152. A. Fumagalli, S. Martinengo, V. G. Albano, D. Braga, and F. Grepioni, *J. Chem. Soc., Dalton Trans.*, 2343 (1989).

153. S. R. Drake, B. F. G. Johnson, and J. Lewis, *J. Chem. Soc., Dalton Trans.*, 505 (1989).

154. S. R. Drake, B. F. G. Johnson, and J. Lewis, *J. Chem. Soc., Chem. Commun.*, 1033 (1988).

155. S. P. Nolan, R. L. de la Vega, S. L. Mukerjee, A. A. Gonzalez, K. Zhang, and C. D. Hoff, *Polyhedron*, **7**, 1491 (1988).

156. K. R. Lane and R. R. Squires, *Polyhedron*, **7**, 1609 (1988).

157. L. Verslúis, T. Ziegler, E. J. Baerends, and W. Ravenek, *J. Am. Chem. Soc.*, **111**, 2018 (1989).

158. M. Bassetti, G. J. Sunley, and P. M. Maitlis, *J. Chem. Soc., Chem. Commun.*, 1012 (1988).

159. G. Cardaci, G. Reichenbach, G. Bellachioma, B. Wassink, and M. C. Baird, *Organometallics*, **7**, 2475 (1988).

160. L. J. Johnston and M. C. Baird, *J. Organomet. Chem.*, **358**, 405 (1988).

161. T. C. Forschner and A. R. Cutler, *J. Organomet. Chem.*, **361**, C41 (1989).

162. J. R. Moss and L. G. Scott, *J. Organomet. Chem.*, **363**, 351 (1989).

163. M. Moran, C. Pascual, I. Cuadrado, J. R. Masaguer, and J. Losada, *J. Organomet. Chem.*, **363**, 157 (1989).

164. C. Bianchini, D. Masi, A. Meli, M. Peruzzini, and F. Zanobini, *J. Am. Chem. Soc.*, **110**, 6411 (1988).

165. F. Ozawa, T. Son, K. Osakada, and A. Yamamoto, *J. Chem. Soc., Chem. Commun.*, 1067 (1989).

166. J. A. Kampmeier and T.-Z. Liu, *Inorg. Chem.*, **28**, 2228 (1989).

167. J. Arnold, T. D. Tilley, A. L. Rheingold, S. J. Geib, and A. M. Arif, *J. Am. Chem. Soc.*, **111**, 149 (1989).

168. V. L. Coffin, W. Brennen, and B. B. Wayland, *J. Am. Chem. Soc.*, **110**, 6063 (1988).

169. T. A. Shackleton and M. C. Baird. *Organometallics*. **8**. 2225 (1989).

170. C. K. Ghosh and W. A. G. Graham, *J. Am. Chem. Soc.*, **111**, 375 (1989).

171. D. L. Lichtenberger, G. P. Darsey, G. E. Kellogg, R. D. Sanner, V. G. Young, Jr., and J. R. Clark, *J. Am. Chem. Soc.*, **111**, 5019 (1989).

172. J. López, A. Romero, A. Santos, A. Vegas, A. M. Echavarren, and P. Noheda, *J. Organomet. Chem.*, **373**, 249 (1989).

173. M. Etienne and J. E. Guerchais, *J. Chem. Soc., Dalton Trans.*, 2187 (1989).

174. P. DeShong, G. A. Slough, D. R. Sidler, P. J. Rybczynski, W. von Philipsborn, R. W. Kunz, B. E. Bursten, and T. W. Clayton, Jr., *Organometallics*, **8**, 1381 (1989).

175. A. K. Kakkar, N. J. Taylor, J. C. Calabrese, W. A. Nugent, D. C. Roe, E. A. Connaway, and T. B. Marder, *J. Chem. Soc., Chem. Commun.*, 990 (1989).

176. (a) J. R. Crook, B. Chamberlain, and R. J. Mawby, *J. Chem. Soc., Dalton Trans.*, 465 (1989); (b) J. M. Bray and R. J. Mawby, *J. Chem. Soc., Dalton Trans.*, 589 (1989).

177. J. S. Merola and R. T. Kacmarcik, *Organometallics*, **8**, 778 (1989).

178. S. Sakaki and K. Ohkubo, *Inorg. Chem.*, **28**, 2583 (1989); S. Sakaki and K. Ohkubo, *Organometallics*, **8**, 2970 (1989).

179. C. Bo and A. Dedieu, *Inorg. Chem.*, **28**, 304 (1989).

180. D. J. Darensbourg, K. M. Sanchez, J. Reibenspies, and A. J. Rheingold, *J. Am. Chem. Soc.*, **111**, 7094 (1989).

181. A. Shaver, B. S. Lum, P. Bird, and K. Arnold, *Inorg. Chem.*, **28**, 1900 (1989).

182. M. Herberhold and A. F. Hill, *J. Organomet. Chem.*, **353**, 243 (1988).

183. M. I. Bruce, M. J. Liddell, M. R. Snow, and E. R. T. Tiekink, *J. Organomet. Chem.*, **352**, 199 (1988).

184. D. J. Darensbourg, C. G. Bauch, J. Reibenspies, and A. J. Rheingold, *Inorg. Chem.*, **27**, 4203 (1988).

185. D. S. Glueck, F. J. Hollander, and R. G. Bergman, *J. Am. Chem. Soc.*, **111**, 2719 (1989).

186. P. Legzdins, G. B. Richter-Addo, B. Wassink, F. W. B. Einstein, R. H. Jones, and A. C. Willis, *J. Am. Chem. Soc.*, **111**, 2097 (1989).

References for Chapter 11

1. C. L. Hill (ed.), *Activation and Functionalization of Alkanes*, Wiley, New York (1989).

2. W. D. Jones and F. J. Feher, *Acc. Chem. Res.*, **22**, 91 (1989).

3. V. D. Makhaev and A. P. Borisov, *Russ. Chem. Rev.*, **57**, 1162 (1988).

4. J. M. Brown and N. A. Cooley, *Chem. Rev.*, **88**, 1031 (1988).

5. A. G. Ginzburg, *Russ. Chem. Rev.*, **57**, 1175 (1988).

6. P. P. Deutsch and R. Eisenberg, *Chem. Rev.*, **88**, 1147 (1988).

7. H. Schwarz, *Acc. Chem. Res.*, **22**, 282 (1989).

8. P. B. Armentrout and J. L. Beauchamp, *Acc. Chem. Res.*, **22**, 315 (1989).

9. D. W. Evans, G. R. Baker, and G. R. Newkome, *Coord. Chem. Rev.*, **93**, 155 (1989).

10. T. M. Gomes Carneiro, D. Matt, and P. Braunstein, *Coord. Chem. Rev.*, **96**, 49 (1989).

11. M. Tanaka, *ChemTech*, 59 (1989).

12. T. Sakakura, T. Sodeyama, and M. Tanaka, *New J. Chem.*, **13**, 737 (1989).

13. D. M. Heinekey, N. G. Payne, and G. K. Schulte, *J. Am. Chem. Soc.*, **110**, 2303 (1988).

14. B. Chaudret, G. Commenges, F. Jalon, and A. Otero, *J. Chem. Soc., Chem. Commun.*, 210 (1989).

15. A. Antinolo, B. Chaudret, G. Commenges, M. Fajardo, F. Jalon, R. H. Morris, A. Otero, and C. T. Schweltzer, *J. Chem. Soc., Chem. Commun.*, 1210 (1988).
16. D. H. Jones, J. A. Labinger, and D. P. Weitekamp, *J. Am. Chem. Soc.*, 111, 3087 (1989).
17. K. W. Zilm, D. M. Heinekey, J. M. Millar, N. G. Payne, and P. Demou, *J. Am. Chem. Soc.*, 111, 3088 (1989).
18. C. Bianchini, M. Peruzzini, and F. Zanobini, *J. Organomet. Chem.*, 354, C19 (1988).
19. C. Bianchini, C. Mealli, A. Meli, M. Peruzzini, and F. Zanobini, *J. Am. Chem. Soc.*, 110, 8725 (1988).
20. F. Maseras, M. Duran, A. Lledos, and J. Bertran, *Inorg. Chem.*, 28, 2984 (1989).
21. M. T. Bautista, K. A. Earl, P. A. Maltby, R. H. Morris, C. T. Schweitzer, and A. Sella, *J. Am. Chem. Soc.*, 110, 7031 (1988).
22. D. G. Hamilton and R. H. Crabtree, *J. Am. Chem. Soc.*, 110, 4126 (1988).
23. J. S. Ricci, T. F. Koetzle, M. T. Bautista, T. M. Hofstede, R. H. Morris, and J. F. Sawyer, *J. Am. Chem. Soc.*, 111, 8823 (1989).
24. S. Antoniutti, G. Albertin, P. Amendola, and E. Bordignon, *J. Chem. Soc., Chem. Commun.*, 229 (1989).
25. J. A. K. Howard, S. A. Mason, O. Johnson, I. C. Diamond, S. Crennell, P. A. Keller, and J. L. Spencer, *J. Chem. Soc., Chem. Commun.*, 1502 (1988).
26. X.-L. Luo and R. H. Crabtree, *Inorg. Chem.*, 28, 3775 (1989).
27. D. G. Hamilton, X.-L. Luo, and R. H. Crabtree, *Inorg. Chem.*, 28, 3198 (1989).
28. F. A. Cotton and R. L. Luck, *Inorg. Chem.*, 28, 6 (1989).
29. M. T. Costello and R. A. Walton, *Inorg. Chem.*, 27, 2563 (1988).
30. F. A. Cotton and R. L. Luck, *Inorg. Chem.*, 28, 2181 (1989).
31. F. A. Cotton and R. L. Luck, *J. Am. Chem. Soc.*, 111, 5757 (1989).
32. D. M. Lunder, M. A. Green, W. E. Streib, and K. G. Caulton, *Inorg. Chem.*, 28, 4527 (1989).
33. F. A. Cotton and R. L. Luck, *Inorg. Chem.*, 28, 4522 (1989).
34. F. A. Cotton and R. L. Luck, *J. Chem. Soc., Chem. Commun.*, 1277 (1988).
35. R. H. Morris, K. A. Earl, R. L. Luck, N. J. Lazarowych, and A. Sella, *Inorg. Chem.*, 26, 2674 (1987).
36. A. A. Gonzalez, K. Zhang, S. P. Nolan, R. Lopez de la Vega, S. L. Mukerjee, and C. D. Hoff, *Organometallics*, 7, 2429 (1988).
37. A. A. Gonzalez and C. D. Hoff, *Inorg. Chem.*, 28, 4295 (1989).
38. K. Zhang, A. A. Gonzalez, and C. D. Hoff, *J. Am. Chem. Soc.*, 111, 3627 (1989).
39. P. Mura, A. Segre, and S. Sostero, *Inorg. Chem.*, 28, 2853 (1989).
40. P. Bergamini, S. Sostero, O. Traverso, P. Mura, and A. Segre, *J. Chem. Soc., Dalton Trans.*, 2367 (1989).
41. M. S. Chinn, D. M. Heinekey, N. G. Payne, and C. D. Sofield, *Organometallics*, 8, 1824 (1989).
42. R. L. Sweany, M. A. Polito, and A. Moroz, *Organometallics*, 8, 2305 (1989).
43. R. L. Sweany and A. Moroz, *J. Am. Chem. Soc.*, 111, 3577 (1989).
44. S. M. Howdle and M. Poliakoff, *J. Chem. Soc., Chem. Commun.*, 1099 (1989).
45. A. M. Joshi and B. R. James, *J. Chem. Soc., Chem. Commun.*, 1785 (1989).
46. T. Tsukahara, H. Kawano, Y. Ishii, T. Takahashi, M. Saburi, Y. Uchida, and S. Akutagawa, *Chem. Lett.*, 2055 (1988).
47. A. Andriollo, M. A. Esteruelas, U. Meyer, L. A. Oro, R. A. Sanchez-Delgado, E. Sola, C. Valero, and H. Werner, *J. Am. Chem. Soc.*, 111, 7431 (1989).
48. G. Alberin, S. Antoniutti, and E. Bordignon, *J. Am. Chem. Soc.*, 111, 2072 (1989).
49. G. Jia and D. W. Meek, *J. Am. Chem. Soc.*, 111, 757 (1989).
50. C. Hampton, W. R. Cullen, B. R. James, and J.-P. Charland, *J. Am. Chem. Soc.*, 110, 6918 (1988).
51. E. G. Lundquist, J. C. Huffman, K. Folting, and K. G. Caulton, *Angew. Chem., Int. Ed. Engl.*, 27, 1165 (1988).
52. G. Marinelli, I. El-Idrissi Rachidi, W. E. Streib, O. Eisenstein, and K. G. Caulton, *J. Am. Chem. Soc.*, 111, 2346 (1989).

53. T. J. Johnson, J. C. Huffman, K. G. Caulton, S. A. Jackson, and O. Eisenstein, *Organometallics*, **8**, 2073 (1989).
54. R. U. Kirss, T. C. Eisenschmid, and R. Eisenberg, *J. Am. Chem. Soc.*, **110**, 8564 (1988).
55. D. Morton and D. J. Cole-Hamilton, *J. Chem. Soc., Chem. Commun.*, 1154 (1988).
56. D. Dawson, R. A. Henderson, A. Hills, and D. L. Hughes, *Polyhedron*, **8**, 1870 (1989).
57. T. Arligaie, B. Chaudret, and R. H. Morris, *Polyhedron*, 7, 2031 (1988).
58. M. R. A. Blomberg, J. Schule, and P. E. M. Siegbahn, *J. Am. Chem. Soc.*, **111**, 6156 (1989).
59. T. Ziegler, W. Cheng, E. J. Baerends, and W. Ravenek, *Inorg. Chem.*, **27**, 3458 (1988).
60. P. B. Armentrout and R. Georgiadis, *Polyhedron*, **7**, 1573 (1988).
61. R. Georgiadis, E. R. Fisher, and P. B. Armentrout, *J. Am. Chem. Soc.*, **111**, 4251 (1989).
62. A. R. Bulls, J. E. Bercaw, J. M. Manriquez, and M. E. Thompson, *Polyhedron*, **7**, 1409 (1988).
63. L. E. Schock and T. J. Marks, *J. Am. Chem. Soc.*, **110**, 7701 (1988).
64. T. J. Marks, M. R. Gagne, S. P. Nolan, L. E. Schock, A. M. Seyam, and D. Stern, *Pure Appl. Chem.*, **61**, 1665 (1989).
65. S. P. Nolan, D. Stern, and T. J. Marks, *J. Am. Chem. Soc.*, **111**, 7844 (1989).
66. M. Tilset and V. D. Parker, *J. Am. Chem. Soc.*, **111**, 6711 (1989).
67. B. A. Vaartstra, K. N. O'Brien, R. Eisenberg, and M. Cowie, *Inorg. Chem.*, **27**, 3668 (1988).
68. B. A. Vaartstra and M. Cowie, *Inorg. Chem.*, **28**, 3138 (1989).
69. A. L. Balch, B. J. Davis, F. Neve, and M. M. Olmstead, *Organometallics*, **8**, 1000 (1989).
70. M. J. Burk, M. P. McGrath, R. Wheeler, and R. H. Crabtree, *J. Am. Chem. Soc.*, **110**, 5034 (1988).
71. P. J. Desrosiers, R. S. Shinomoto, M. A. Deming, and T. C. Flood, *Organometallics*, **8**, 2861 (1989).
72. R. T. Edidin, K. M. Hennessy, A. E. Moody, S. J. Okrasinski, and J. R. Norton, *New J. Chem.*, **12**, 475 (1988).
73. R. H. E. Hudson, A. J. Poe, C. N. Sampson, and A. Siegel, *J. Chem. Soc., Dalton Trans.*, 2235 (1989).
74. J. Anhaus, H. C. Bajaj, R. van Eldik, L. R. Nevinger, and J. B. Keister, *Organometallics*, **8**, 2903 (1989).
75. A. M. Arif, T. A. Bright, R. A. Jones, and C. M. Nunn, *J. Am. Chem. Soc.*, **110**, 6894 (1988).
76. R. U. Kirss and R. Eisenberg, *J. Organomet. Chem.*, **359**, C22 (1989).
77. S.-C. Chang, R. H. Hauge, Z. H. Kafafi, J. L. Margrave, and W. E. Billups, *J. Am. Chem. Soc.*, **110**, 7975 (1988).
78. T. Ziegler, V. Tschinke, L. Fan, and A. D. Becke, *J. Am. Chem. Soc.*, **111**, 9177 (1989).
79. P. O. Stoutland, R. G. Bergman, S. P. Nolan, and C. D. Hoff, *Polyhedron*, **7**, 1429 (1988).
80. H. E. Bryndza, P. J. Domaille, W. Tam, L. K. Fong, R. A. Paciello, and J. E. Bercaw, *Polyhedron*, **7**, 1441 (1988).
81. P. J. Toscano, A. L. Seligson, M. T. Curran, A. T. Skrobutt, and D. C. Sonnenberger, *Inorg. Chem.*, **28**, 166 (1989).
82. D. L. Lichtenberger, G. P. Darsey, G. E. Kellogg, R. D. Sanner, V. G. Young, Jr., and J. R. Clark, *J. Am. Chem. Soc.*, **111**, 5019 (1989).
83. E. Matarasso-Tchiroukhine and G. Jaouen, *Can. J. Chem.*, **66**, 2157 (1988).
84. X.-L. Luo and R. H. Crabtree, *J. Am. Chem. Soc.*, **111**, 2527 (1989).
85. D. L. Lichtenberger and A. Rai-Chaudhuri, *J. Am. Chem. Soc.*, **111**, 3583 (1989).
86. U. Schubert, E. Kunz, B. Harkers, J. Willnecker, and J. Meyer, *J. Am. Chem. Soc.*, **111**, 2572 (1989).
87. G. Parkin and J. E. Bercaw, *Organometallics*, **8**, 1172 (1989).
88. R. M. Bullock, C. E. L. Headford, K. M. Hennessy, S. E. Kegley, and J. R. Norton, *J. Am. Chem. Soc.*, **111**, 3897 (1989).
89. G. L. Gould and D. M. Heinekey, *J. Am. Chem. Soc.*, **111**, 5502 (1989).
90. E. P. Wasserman, R. G. Bergman, and C. B. Moore, *J. Am. Chem. Soc.*, **110**, 6076 (1988).
91. S. T. Belt, F.-W. Grevels, W. E. Klotzbucher, A. McCamley, and R. N. Perutz, *J. Am. Chem. Soc.*, **111**, 8373 (1989).

92. S. B. Duckett, D. M. Haddleton, S. A. Jackson, R. N. Perutz, M. Poliakoff, and R. K. Upmacis, *Organometallics*, 7, 1526 (1988).

93. M. B. Sponsler, B. H. Weiller, P. O. Stoutland, and R. G. Bergman, *J. Am. Chem. Soc.*, 111, 6841 (1989).

94. B. H. Weiller, E. P. Wasserman, R. G. Bergman, C. B. Moore, and G. C. Pimentel, *J. Am. Chem. Soc.*, 111, 8288 (1989).

95. C. T. Spillett and P. C. Ford, *J. Am. Chem. Soc.*, 111, 1932 (1989).

96. P. E. Bloyce, A. J. Rest, I. Whitwell, W. A. G. Graham, and R. Holmes-Smith, *J. Chem. Soc., Chem. Commun.*, 846 (1988).

97. W. L. Evans, L. R. Chamberlain, T. A. Ulibarri, and J. W. Ziller, *J. Am. Chem. Soc.*, 110, 6423 (1988).

98. T. G. P. Harper, R. S. Shinomoto, M. A. Deming, and T. C. Flood, *J. Am. Chem. Soc.*, 110, 7915 (1988).

99. C. C. Cummins, S. M. Baxter, and P. T. Wolczanski, *J. Am. Chem. Soc.*, 110, 8731 (1988).

100. A. R. Siedle and R. A. Newmark, *J. Am. Chem. Soc.*, 111, 2058 (1989).

101. A. R. Siedle, R. A. Newmark, K. A. Brown-Wensley, R. P. Skarjune, L. C. Haddad, K. O. Hodgson, and A. L. Roe, *Organometallics*, 7, 2078 (1988).

102. A. R. Siedle, *New J. Chem.*, 13, 719 (1989).

103. Y. Fujiwara, T. Jintoku, and Y. Uchida, *New J. Chem.*, 13, 649 (1989).

104. S. W. Buckner, T. J. MacMahon, G. D. Byrd, and B. S. Freiser, *Inorg. Chem.*, 28, 3511 (1989).

105. L. S. Sunderlin and P. B. Armentrout, *J. Am. Chem. Soc.*, 111, 3845 (1989).

106. K. K. Irikura and J. L. Beauchamp, *J. Am. Chem. Soc.*, 111, 75 (1989).

107. Y. Huang and B. S. Freiser, *J. Am. Chem. Soc.*, 111, 2387 (1989).

108. S. T. Belt, S. B. Duckett, M. Helliwell, and R. N. Perutz, *J. Chem. Soc., Chem. Commun.*, 928 (1989).

109. W. D. Jones and L. Dong, *J. Am. Chem. Soc.*, 111, 8722 (1989).

110. R. Cordone, W. D. Harmon, and H. Taube, *J. Am. Chem. Soc.*, 111, 2896 (1989).

111. M. Lavin, E. M. Holt, and R. H. Crabtree, *Organometallics*, 8, 99 (1989).

112. C. Bianchini, D. Masi, A. Meli, M. Peruzzini, and F. Zanobini, *J. Am. Chem. Soc.*, 110, 6411 (1988).

113. J. S. Merola, *Organometallics*, 8, 2975 (1989).

114. R. L. Brainard, W. R. Nutt, T. R. Lee, and G. M. Whitesides, *Organometallics*, 7, 2379 (1988).

115. D. H. Berry and Q. Jiang, *J. Am. Chem. Soc.*, 111, 8049 (1989).

116. R. T. Price, R. A. Andersen, and E. L. Muetterties, *J. Organomet. Chem.*, 376, 407 (1989).

117. W. D. Jones, R. P. Duttweiler, Jr., F. J. Feher, and E. T. Hessell, *New J. Chem.*, 13, 725 (1989).

118. C. K. Ghosh, D. P. S. Rodgers, and W. A. G. Graham, *J. Chem. Soc., Chem. Commun.*, 1511 (1988).

119. C. K. Ghosh and W. A. G. Graham, *J. Am. Chem. Soc.*, 111, 375 (1989).

120. P. J. Walsh, F. J. Hollander, and R. G. Bergman, *J. Am. Chem. Soc.*, 110, 8729 (1988).

121. J. F. Hartwig, R. A. Andersen, and R. G. Bergman, *J. Am. Chem. Soc.*, 111, 2717 (1989).

122. P. O. Stoutland and R. G. Bergman, *J. Am. Chem. Soc.*, 110, 5732 (1988).

123. C. K. Ghosh, J. K. Hoyano, R. Krentz, and W. A. G. Graham, *J. Am. Chem. Soc.*, 111, 5480 (1989).

124. R. S. Tanke and R. H. Crabtree, *Inorg. Chem.*, 28, 3444 (1989).

125. M. J. Fernandez, M. J. Rodriquez, L. A. Oro, and F. J. Lahoz, *J. Chem. Soc., Dalton Trans.*, 2073 (1989).

126. H. Werner, T. Dirnberger, and M. Schulz, *Angew. Chem., Int. Ed. Engl.*, 27, 948 (1988).

127. H. Suzuki, H. Omori, and Y. Moro-Oka, *Organometallics*, 7, 2579 (1988).

128. S. T. Belt, S. B. Duckett, D. M. Haddleton, and R. N. Perutz, *Organometallics*, 8, 748 (1989).

129. J. W. Faller and C. J. Smart, *Organometallics*, 8, 602 (1989).

130. A. R. Siedle, R. A. Newmark, M. R. V. Sahyun, P. A. Lyon, S. L. Hunt, and R. P. Skarjune, *J. Am. Chem. Soc.*, 111, 8346 (1989).

131. J. Wu and R. G. Bergman, *J. Am. Chem. Soc.*, 111, 7628 (1989).

132. P. Chow, D. Zargarian, N. J. Taylor, and T. B. Marder, *J. Chem. Soc., Chem. Commun.*, 1545 (1989).
133. C. Bianchini, A. Meli, M. Peruzzini, J. A. Ramirez, A. Vacca, F. Vizza, and F. Zanobini, *Organometallics*, **8**, 337 (1989).
134. M. Helliwell, K. M. Stell, and R. J. Mawby, *J. Organomet. Chem.*, **356**, C32 (1988).
135. R. C. Hemond, R. P. Hughes, D. J. Robinson, and A. L. Rheingold, *Organometallics*, **7**, 2239 (1988).
136. R. P. Hughes, M. E. King, D. J. Robinson, and J. M. Spotts, *J. Am. Chem. Soc.*, **111**, 8919 (1989).
137. L. M. Lech and B. S. Freiser, *Organometallics*, **7**, 1948 (1988).
138. R. L. Hettich and B. S. Freiser, *Organometallics*, **8**, 2447 (1989).
139. L. M. Lech, J. R. Gord, and B. S. Freiser, *J. Am. Chem. Soc.*, **111**, 8588 (1989).
140. R. Stepnowski and J. Allison, *J. Am. Chem. Soc.*, **111**, 449 (1989).
141. M. Brookhart, D. M. Lincoln, A. F. Volpe, Jr., and G. F. Schmidt, *Organometallics*, **8**, 1212 (1989).
142. A. E. Derome, M. L. H. Green, and L.-L. Wong, *New J. Chem.*, **13**, 747 (1989).
143. D. M. Heinekey, S. T. Michel, and G. K. Schulte, *Organometallics*, **8**, 1241 (1989).
144. S. S. Kristjansdottir, A. E. Moody, R. T. Weberg, and J. R. Norton, *Organometallics*, **7**, 1983 (1988).
145. G. Erker, R. Zwettler, and C. Kruger, *Chem. Ber.*, **122**, 1377 (1989).
146. P. Knappe and N. Rosch, *J. Organomet. Chem.*, **359**, C5 (1989).
147. R.-M. Catala, D. Cruz-Garritz, P. Sosa, P. Terreros, H. Torrens, A. Hills, D. L. Hughes, and R. L. Richards, *J. Organomet. Chem.*, **359**, 219 (1989).
148. T. Arliguie, B. Chaudret, F. Jalon, and F. Lahoz, *J. Chem. Soc., Chem. Commun.*, 998 (1988).
149. F. A. Cotton and R. L. Luck, *Inorg. Chem.*, **28**, 3210 (1989).
150. N. Carr, B. J. Dunne, A. G. Orpen, and J. L. Spencer, *J. Chem. Soc., Chem. Commun.*, 926 (1988).
151. F. Ozawa, J. W. Park, P. B. Mackenzie, W. P. Schaefer, L. M. Henling, and R. H. Grubbs, *J. Am. Chem. Soc.*, **111**, 1319 (1989).
152. H. van der Heijden, C. J. Schaverien, and A. G. Orpen, *Organometallics*, **8**, 255 (1989).
153. J. Albert, J. Granell, and J. Sales, *Synth. React. Inorg. Met.-Org. Chem.*, **19**, 1009 (1989).
154. R. M. Ceder, M. Gomez, and J. Sales, *J. Organomet. Chem.*, **361**, 391 (1989).
155. A. A. H. van der Zeijden, G. van Koten, R. Luijk, and D. M. Grove, *Organometallics*, **7**, 1556 (1988).
156. A. A. H. van der Zeijden, G. van Koten, R. Luijk, R. A. Nordemann, and A. L. Spek, *Organometallics*, **7**, 1549 (1988).
157. S. Park, M. P. Johnson, and D. M. Roundhill, *Organometallics*, **8**, 1700 (1989).
158. D. C. Griffiths and G. B. Young, *Organometallics*, **8**, 875 (1989).
159. K. R. Ballard, I. M. Gardiner, and D. E. Wigley, *J. Am. Chem. Soc.*, **111**, 2159 (1989).
160. S. T. Carter, W. Clegg, V. C. Gibson, T. P. Kee, and R. D. Sanner, *Organometallics*, **8**, 253 (1989).
161. M. A. Gallop, B. F. G. Johnson, J. Lewis, A. McCamley, and R. N. Perutz, *J. Chem. Soc., Chem. Commun.*, 1071 (1988).
162. V. I. Pavlovski and A. L. Poznyak, *Z. Chem.*, **29**, 6 (1989).
163. J. L. Kerschner, P. E. Fanwick, I. P. Rothwell, and J. C. Huffman, *Organometallics*, **8**, 1431 (1989).
164. J. L. Kerschner, J. S. Yu, P. E. Fanwick, I. P. Rothwell, and J. C. Huffman, *Organometallics*, **8**, 1414 (1989).
165. J. L. Kerschner, I. P. Rothwell, J. C. Huffman, and W. E. Streib, *Organometallics*, **7**, 1871 (1988).
166. G. D. Smith, P. E. Fanwick, and I. P. Rothwell, *J. Am. Chem. Soc.*, **111**, 750 (1989).
167. A. J. Blake, C. O. Dietrich-Buchecker, T. I. Hyde, J.-P. Sauvage, and M. Schroder, *J. Chem. Soc., Chem. Commun.*, 1663 (1989).

168. J. Dupont, N. Beydoun, and M. Pfeffer, *J. Chem. Soc., Dalton Trans.*, 1715 (1989).
169. E. C. Alyea, G. Ferguson, J. Malito, and B. L. Ruhl, *Can. J. Chem.*, **66**, 3162 (1988).
170. M. Nonoyama, *Inorg. Chim. Acta*, **157**, 9 (1989).
171. A. Guarnieri, A. W. Parkins, *J. Organomet. Chem.*, **361**, 399 (1989).
172. O. Juanes, J. de Mendoza, and J. C. Rodriquez-Ubis, *J. Organomet. Chem.*, **363**, 393 (1989).
173. M. T. Periera, M. Pfeffer, and M. A. Rotteveel, *J. Organomet. Chem.*, **375**, 139 (1989).
174. M. T. Periera, J. M. Vila, A. Suarez, E. Gayoso, and M. Gayoso, *Gaz. Chim. Ital.*, **118**, 783 (1988).
175. H.-J. Kneuper and J. R. Shapley, *New J. Chem.*, **12**, 479 (1988).
176. R. Zoet, G. van Koten, K. Vrieze, J. Jansen, K. Goubitz, and C. H. Stam, *Organometallics*, **7**, 1565 (1988).
177. S. B. Colbran, P. T. Irele, B. F. G. Johnson, F. J. Lahoz, J. Lewis, and P. R. Raithby, *J. Chem. Soc., Dalton Trans.*, 2023 (1989).
178. R. Zoet, G. van Koten, K. Vrieze, A. J. M. Duisenberg, and A. L. Spek, *Inorg. Chim. Acta*, **148**, 71 (1988).
179. R. D. Adams and J. E. Babin, *New J. Chem.*, **12**, 641 (1988).
180. K. I. Hardcastle, H. Minassian, A. J. Arce, Y. De Sanctis, and A. J. Deeming, *J. Organomet. Chem.*, **368**, 119 (1989).
181. J. M. Ressner, P. C. Wernett, C. S. Kraihanzel, and A. L. Rheingold, *Organometallics*, **7**, 1661 (1988).
182. E. Farnetti, G. Nardin, and M. Graziani, *J. Chem. Soc., Chem. Commun.*, 1264 (1988).
183. J. Arnold, G. Wilkinson, B. Hussain, and M. B. Hursthouse, *Organometallics*, **8**, 415 (1989).
184. E. C. Morrison and D. A. Tocher, *Inorg. Chim. Acta*, **157**, 139 (1989).
185. K. Roder and H. Werner, *J. Organomet. Chem.*, **362**, 321 (1989).
186. G. J. Sunley, P. del C. Menanteau, H. Adams, N. A. Bailey, and P. M. Maitlis, *J. Chem. Soc., Dalton Trans.*, 2415 (1989).
187. O. J. Scherer, M. Florchinger, K. Gobel, J. Kaub, and W. S. Sheldrick, *Chem. Ber.*, **121**, 1265 (1988).
188. J. M. Cooney, L. H. P. Gommans, L. Main, and B. K. Nicholson, *J. Organomet. Chem.*, **349**, 197 (1988).
189. N. P. Robinson, L. Main, and B. K. Nicholson, *J. Organomet. Chem.*, **349**, 209 (1988).
190. P. Lahuerta, J. Paya, E. Peris, M. A. Pellinghelli, and A. Tiripicchio, *J. Organomet. Chem.*, **373**, C5 (1989).
191. C. Schulze and H. Schwarz, *Chimia*, **42**, 297 (1988).
192. S. Karrass, K. Eller, C. Schulze, and H. Schwarz, *Angew. Chem., Int. Ed. Engl.*, **28**, 607 (1989).
193. S. Karrass and H. Schwarz, *Helv. Chim. Acta*, **72**, 633 (1989).
194. S. Karrass, T. Prusse, K. Eller, and H. Schwarz, *J. Am. Chem. Soc.*, **111**, 9018 (1989).
195. T. Prusse and H. Schwarz, *Organometallics*, **8**, 2856 (1989).
196. D. Schroder and H. Schwarz, *Chimia*, **43**, 317 (1989).
197. G. Czekay, K. Eller, D. Schroder, and H. Schwarz, *Angew. Chem., Int. Ed. Engl.*, **28**, 1277 (1989).
198. G. Czekay, T. Drewello, K. Eller, W. Zummack, and H. Schwarz, *Organometallics*, **8**, 2439 (1989).
199. G. Czekay, T. Drewello, and H. Schwarz, *J. Am. Chem. Soc.*, **111**, 4561 (1989).
200. R. M. Stepnowski and J. Allison, *Organometallics*, **7**, 2097 (1988).
201. K. Eller, T. Drewello, W. Zummack, T. Allspach, U. Annen, M. Regitz, and H. Schwarz, *J. Am. Chem. Soc.*, **111**, 4228 (1989).
202. K. Eller and H. Schwarz, *Organometallics*, **8**, 1820 (1989).
203. N. Steinruck and H. Schwarz, *Organometallics*, **8**, 759 (1989).
204. P. A. M. van Koppen, D. B. Jacobson, A. Illies, M. T. Bowers, M. Hanratty, and J. L. Beauchamp, *J. Am. Chem. Soc.*, **111**, 1991 (1989).
205. W. de Graaf, J. Boersma, W. J. J. Smeets, A. L. Spek, and G. van Koten, *Organometallics*, **8**, 2907 (1989).

206. S. Roy, R. Puddephatt, and J. D. Scott, *J. Chem. Soc., Dalton Trans.*, 2121 (1989).
207. S. E. Himmel and G. B. Young, *Organometallics*, **7**, 2440 (1988).
208. P. J. Stang and M. H. Kowalski, *J. Am. Chem. Soc.*, **111**, 3356 (1989).
209. S. K. Thomson and G. B. Young, *Organometallics*, **8**, 2068 (1989).
210. F. Ozawa, K. Kurihara, M. Fujimori, T. Hidaka, T. Toyoshima, and A. Yamamoto, *Organometallics*, **8**, 180 (1989).
211. H. Kurosawa, H. Ohnishi, M. Emoto, Y. Kawasaki, and S. Murai, *J. Am. Chem. Soc.*, **110**, 6272 (1988).
212. Y. Hayashi, K. Matsumoto, Y. Nakamura, and K. Isobe, *J. Chem. Soc., Dalton Trans.*, 1519 (1989).
213. H.-A. Brune, R. Klotzbucher, and G. Schmidtberg, *J. Organomet. Chem.*, **365**, 389 (1989).
214. H.-A. Brune, R. Klotzbucher, K. Berhalter, and T. Debaerdemaeker, *J. Organomet. Chem.*, **369**, 321 (1989).
215. A. Becalska and R. H. Hill, *J. Am. Chem. Soc.*, **111**, 4346 (1989).
216. C. H. Winter and J. A. Gladysz, *J. Organomet. Chem.*, **354**, C33 (1988).
217. M. J. Poss, A. M. Arif, and T. G. Richmond, *Organometallics*, **7**, 1669 (1988).
218. C. M. Anderson, R. J. Puddephatt, G. Furguson, and A. L. Lough, *J. Chem. Soc., Chem. Commun.*, 1297 (1989).
219. P. J. Stang, M. H. Kowalski, M. D. Schiavelli, and D. Longford, *J. Am. Chem. Soc.*, **111**, 3347 (1989).
220. A. J. Canty, A. A. Watson, B. W. Skelton, and A. H. White, *J. Organomet. Chem.*, **367**, C25 (1989).
221. N. Hadj-Bagheri and R. J. Puddephatt, *Polyhedron*, **7**, 2695 (1988).
222. W. D. McGhee, F. J. Hollander, and R. G. Bergman, *J. Am. Chem. Soc.*, **110**, 8428 (1988).
223. M. H. Chisholm, B. W. Eichhorn, and J. C. Huffman, *Organometallics*, **8**, 67 (1989).
224. S. M. Baxter, G. S. Ferguson, and P. T. Wolczanski, *J. Am. Chem. Soc.*, **110**, 4231 (1988).
225. C. Blandy, S. A. Locke, S. J. Young, and N. E. Schore, *J. Am. Chem. Soc.*, **110**, 7540 (1988).
226. K. I. Goldberg and R. G. Bergman, *J. Am. Chem. Soc.*, **110**, 4853 (1988).
227. H. Omori, H. Suzuki, Y. Take, and Y. Moro-Oka, *Organometallics*, **8**, 2270 (1989).
228. S. J. Young, B. Kellenberger, J. H. Reibenspies, S. E. Himmel, M. Manning, O. P. Anderson, and J. K. Stille, *J. Am. Chem. Soc.*, **110**, 5744 (1988).
229. S. Komiya and I. Endo, *Chem. Lett.*, 1709 (1988).
230. R. D. Brost, D. O. K. Fjeldsted, and S. R. Stobart, *J. Chem. Soc., Chem. Commun.*, 488 (1989).

References for Chapter 12

1. N. Mensi and S. S. Isied, *J. Am. Chem. Soc.*, **109**, 7882 (1987).
2. See, for example, S. S. Isied, J. Lyon, A. Vassilian, and G. Worosila, *J. Liq. Chromatogr.*, **5**, 537 (1982); S. S. Isied, J. Lyon, and A. Vassilian, *J. Am. Chem. Soc.*, **104**, 3910 (1982).
3. G. C. Baumann, J. A. Potenza, and S. S. Isied, *Inorg. Chim. Acta*, **156**, 85 (1989).
4. R. M. Hartshorn, A. C. Willis, and A. M. Sargeson, *J. Chem. Soc., Chem. Commun.*, 1269 (1988).
5. A. Hammershoi, R. M. Hartshorn, and A. M. Sargeson, *J. Chem. Soc., Chem. Commun.*, 1226 (1988).
6. A. Hammershoi, R. M. Hartshorn, and A. M. Sargeson, *J. Chem. Soc., Chem. Commun.*, 1267 (1988).
7. N. J. Curtis, A. Hammershoi, L. M. Nicolas, A. M. Sargeson, and K. J. Watson, *Acta Chem. Scand.*, **A41**, 36 (1987).
8. J. T. Groves and L. A. Baron, *J. Am. Chem. Soc.*, **111**, 5442 (1989).
9. J. Chin and M. Banaszczyk, *J. Am. Chem. Soc.*, **111**, 2724 (1989).
10. P. Hendry and A. M. Sargeson, *J. Am. Chem. Soc.* **111**, 2521 (1989).
11. G. H. Rawji and R. M. Milburn, *Inorg. Chim. Acta*, **150**, 227 (1988).

12. D. P. Fairlie, W. G. Jackson, and G. M. McLaughlin, *Inorg. Chem.*, **28**, 1983 (1989).
13. L. M. Engelhardt, A. R. Gainsford, G. J. Gainsford, B. T. Golding, J. McB. Harrowfield, A. J. Herlt, A. M. Sargeson, and A. H. White, *Inorg. Chem.*, **27**, 4551 (1988).
14. L. Roecker, J. D. Lydon, A. C. Willis, A. M. Sargeson, and E. Deutsch, *Aust. J. Chem.*, **42**, 339 (1989).
15. P. S. Braterman (ed.), *Reactions of Coordinated Ligands*, vol. 2, Plenum Press, New York (1989).
16. R. K. Ray, M. K. Bandyopadhyay, and G. B. Kauffman, *Polyhedron*, **8**, 757 (1989).
17. V. Y. Kukushkin, I. G. Zenkevich, V. K. Belsky, V. E. Konovalov, A. I. Moiseev, and E. O. Sidorov, *Inorg. Chim. Acta*, **166**, 79 (1989).
18. A. M. Shallaby, M. S. Soliman, R. M. El-Shazely, and M. M. Mostafa, *Synth. React. Inorg. Met.-Org. Chem.*, **18**, 807 (1988).
19. Y. N. Belokon, V. I. Tararov, T. F. Savel'eva, S. V. Vitt, E. A. Paskonova, S. Ch. Dotdayev, Y.A. Borisov, Y. T. Struchkov, A. S. Batasanov, and V. M. Velikov, *Inorg. Chem.*, **27**, 4046 (1988).
20. F. P. Fanizzi, F. P. Intini, and G. Natile, *J. Chem. Soc., Dalton Trans.*, 947 (1989).
21. J. Barker, M. Kilner, M. M. Mahmoud, and S. C. Wallwork, *J. Chem. Soc., Dalton Trans.*, 947 (1989).
22. G. Süss-Fink, G. Hermann, and G. F. Schmidt, *Polyhedron*, **7**, 2341 (1988).
23. E. C. Constable and J. M. Holmes, *Polyhedron*, **7**, 2531 (1988).
24. A. Sahajpal and P. Thornton, *Polyhedron*, **7**, 2715 (1988).
25. Z. Szeverényi and L. I. Simándi, *J. Mol. Catal.*, **51**, 155 (1989).
26. M. M. Shoukry, E. M. Khairy, and A. Saeed, *Transition Metal Chem. (London)*, **13**, 146 (1988).
27. M. M. Shoukry, E. M. Khairy, and A. Saeed, *Transition Metal Chem. (London)*, **13**, 379 (1988).
28. R. W. Hay, A. K. Basak, M. P. Pujari, and A. Perotti, *J. Chem. Soc., Dalton Trans.*, 197 (1989).
29. D. Tschudin, A. Riesen, and T. A. Kaden, *Helv. Chim. Acta*, **72**, 131 (1989).
30. M. P. Suh. C.-H. Kwak, and J. Suh, *Inorg. Chem.*, **28**, 50 (1989).
31. F. Ahmad, S. Sabir, and S. Kumar, *Transition Metal Chem. (London)*, **13**, 388 (1988).
32. G. S. Reddy, G. V. Reddy, and G. G. Smith, *Inorg. Chim. Acta*, **166**, 55 (1989).
33. Y. N. Belokon, A. N. Popov, N. I. Chernoglazova, M. B. Saporovskaya, V. I. Bakhmutov, and V. M. Belikov, *J. Chem. Soc., Chem. Commun.*, 1337 (1988).
34. D. N. Marjit, U. S. Sharma, and D. K. Jaiswal, *Curr. Sci.*, **57**, 542 (1988).
35. J. R. Morrow and W. C. Trogler, *Inorg. Chem.*, **28**, 2330 (1989).
36. P. Hendry and A. M. Sargeson, *J. Am. Chem. Soc.*, **111**, 2521 (1989).
37. F. M. Menger and T. Tsuno, *J. Am. Chem. Soc.*, **111**, 4903 (1989).
38. D. J. Irvine, D. J. Cole-Hamilton, J. Barnes, and P. K. G. Hodgson, *Polyhedron*, **8**, 1575 (1989).
39. P. Bergamini, S. Sostero, O. Traverso, T. J. Kemp, and P. G. Pringle, *J. Chem. Soc., Dalton Trans.*, 2017 (1989).
40. J. A. Rahn, A. Delian, and J. H. Nelson, *Inorg. Chem.*, **28**, 215 (1989).
41. J. A. Rahn, M. S. Holt, G. A. Gray, N. W. Alcock, and J. H. Nelson, *Inorg. Chem.*, **28**, 217 (1989).
42. L. Sokiji'c, E. B. Milosavljevic, J. H. Nelson, N. W. Alcock, and J. Fischer, *Inorg. Chem.*, **28**, 3453 (1989).
43. M. L. Brader, E. W. Ainscough, E. N. Baker, and A. M. Brodie, *Polyhedron*, **8**, 2219 (1989).
44. E. W. Ainscough, E. N. Baker, A. G. Bingham, and A. M. Brodie, *J. Chem. Soc., Dalton Trans.*, 39 (1989).
45. M. Kumar, G. J. Colpas, R. O. Day, and M. J. Moroney, *J. Am. Chem. Soc.*, **111**, 8323 (1989).
46. L. Ehrenberg, M. Harms-Ringdahl, I. Fedorcsák, and F. Granath, *Acta Chem. Scand.*, **43**, 177 (1989).
47. R. F. Jamieson, W. Linert, and A. Tschinkowitz, *J. Chem. Soc., Dalton Trans.*, 2109 (1988).
48. P. J. Nichols and M. W. Grant, *Aust. J. Chem.*, **42**, 1085 (1989).
49. M. F. Iskander, L. El-Sayed, N. J. Lees-Gayed, and G. R. M. Tawfik, *Inorg. Chim. Acta*, **155**, 243 (1989).

50. S. Massoud, L. El-Sayed, and M. F. Iskander, *Inorg. Chim. Acta*, **160**, 209 (1989).
51. P. S. Cartwright and R. D. Gillard, *Polyhedron*, **8**, 1453 (1989).
52. R. D. Gillard and D. P. J. Hall, *J. Chem. Soc., Chem. Commun.*, 1163 (1988).
53. E. C. Constable, *Polyhedron*, **8**, 83 (1989).
54. E. C. Constable and C. E. Housecroft, *Transition Metal Chem.* (*London*), **13**, 19 (1988).
55. L. M. Kohriger, A. K. McMullen. P. E. Fanwick. and I. P. Rothwell, *Polyhedron*, **8**, 77 (1989).
56. P. B. Hitchcock, K. R. Seddon, J. E. Turp, Y. Z. Yousif, J. A. Zora, E. C. Constable, and O. Wernberg, *J. Chem. Soc., Dalton Trans.*, 1837 (1988).
57. A. Cantarero, J. M. Amigó, J. Faus, M. Julve, and T. Debaerdemaeker, *J. Chem. Soc., Dalton Trans.*, 2033 (1988).
58. I. Castro, J. Faus, M. Julve, M. Mollar, A. Monge, and E. Gutierrez-Puebla, *Inorg. Chim. Acta*, **161**, 97 (1989).
59. J. Faus, M. Julve, J. M. Amigó, and T. Debaerdemaeker, *J. Chem. Soc., Dalton Trans.*, 1681 (1989).
60. M. G. Elliott and R. E. Shepherd, *Transition Metal Chem.* (*London*), **14**, 251 (1989).
61. J. A. A. Sagues, R. D. Gillard, and P. A. Williams, *Transition Metal Chem.* (*London*), **14**, 110 (1989).
62. E. C. Constable and T. A. Leese, *Inorg. Chim. Acta*, **146**, 55 (1988).
63. J. Albert, J. Granell, and J. Sales, *Polyhedron*, **8**, 2725 (1989).
64. D. W. Allen, P. E. Cropper, and I. W. Nowell, *Polyhedron*, **8**, 1039 (1989).
65. C. Pomp, K. Wieghardt, B. Nuber, and J. Weiss, *Inorg. Chem.*, **27**, 3789 (1988).
66. L. Huang, F. Ozawa, K. Osakada, and A. Yamamoto, *Organometallics*, **8**, 2065 (1989).
67. J. Powell, C. Coutoure, and M. R. Gregg, *J. Chem. Soc., Chem. Commun.*, 1208 (1988).
68. W. Keim, *J. Organomet. Chem.*, **372**, 15 (1989).
69. M. E. Dry, *J. Organomet. Chem.*, **372**, 117 (1989).
70. B. D. Dombek, *J. Organomet. Chem.*, **372**, 151 (1989).
71. D. B. Jacobson, J. R. Gord, and B. S. Freiser, *Organometallics*, **8**, 2957 (1989).
72. P. J. Desrosiers, R. S. Shinomoto, M. A. Deming, and T. C. Flood, *Organometallics*, **8**, 2861 (1989).
73. R. M. Waymouth and R. H. Grubbs, *Organometallics*, **7**, 1631 (1988).
74. T. J. Mazenec, *J. Catal.*, **98**, 115 (1986).
75. S. Georgiou and J. Gladysz, *Tetrahedron*, **42**, 1109 (1986) and references cited therein.
76. M. Brookhart and R. C. Buck, *J. Am. Chem. Soc.*, **111**, 559 (1989).
77. J. I. Seeman and S. G. Davies, *J. Am. Chem. Soc.*, **107**, 6533 (1985) and references cited therein.
78. M. Brookhart and R. C. Buck, *J. Organomet. Chem.*, **370**, 111 (1989).
79. D. R. Senn, A. Wong, A. T. Patton, M. Marsi, C. E. Strouse, and J. A. Gladysz, *J. Am. Chem. Soc.*, **110**, 6096 (1988).
80. W. K. Wan, K. Zaw, and P. M. Henry, *Organometallics*, **7**, 1677 (1988).
81. N. Gregor, K. Zaw, and P. M. Henry, *Organometallics*, **3**, 1251 (1984) and references cited therein.
82. M. Hodgson, D. Parker, R. J. Taylor, and G. Ferguson, *Organometallics*, **7**, 1761, (1988).
83. M. Brookhart and D. M. Lincoln, *J. Am. Chem. Soc.*, **110**, 8719 (1988).
84. H. Kurosawa, K. Ishii, Y. Kawasaki, and S. Murai, *Organometallics*, **8**, 1756 (1989).
85. E. Cesarotti, M. Grassi, and L. Prati, *J. Chem. Soc., Dalton Trans.*, 161 (1989).
86. S. G. Davies, M. L. H. Green, and D. M. P. Mingos, *Tetrahedron*, **34**, 3047 (1978).
87. D. Mandon and D. Astruc, *Organometallics*, **8**, 2372 (1989).
88. S. D. Ittel, J. F. Whitney, Y. K. Chung, P. G. Williard, and D. A. Sweigart, *Organometallics*, **7**, 1323 (1988) and references cited therein.
89. R. D. Pike, T. J. Alavosus, C. A. Camaioni-Neto, J. C. Williams, and D. A. Sweigart, *Organometallics*, **8**, 2631 (1989).
90. R. D. Pike, W. J. Ryan, G. B. Carpenter, and D. A. Sweigart, *J. Am. Chem. Soc.*, **111**, 8535 (1989).
91. A. J. Pearson and M. P. Burello, *J. Chem. Soc., Chem. Commun.*, 1332 (1989).

92. J. A. Heppert, M. A. Morgenstern, D. M. Scherubel, F. Takusagawa, and M. R. Shaker, *Organometallics*, **7**, 1715 (1988).
93. S. Ostrowski and M. Makosza, *J. Organomet. Chem.*, **367**, 95 (1989).
94. T. S. Cameron, M. D. Clerk, A. Linden, K. C. Sturge, and M. J. Zaworotko, *Organometallics*, **7**, 2571 (1988).
95. S. C. Huckett and R. J. Angelici, *Organometallics*, **7**, 1491 (1988).
96. I. Ojima and H. B. Kwon, *J. Am. Chem. Soc.*, **110**, 5617 (1988).
97. S. G. Davies, J. I. Seeman, and I. H. Williams, *Tetrahedron Lett.*, **27**, 619 (1986) and references cited therein.
98. L. S. Leibeskind and M. E. Welker, *Organometallics*, **2**, 194 (1983) and references cited therein.
99. M. F. N. N. Carvalho, R. A. Henderson, A. J. L. Pombeiro, and R. L. Richards, *J. Chem. Soc., Chem. Commun.*, 1796 (1989).
100. D. R. Senn, A. Wong, A. T. Patton, M. Marsi, C. E. Strouse, and J. A. Gladysz, *J. Am. Chem. Soc.*, **110**, 6096 (1988).
101. M. Brookhart, J. Yoon, and S. K. Noh, *J. Am. Chem. Soc.*, **111**, 4117 (1989).
102. J. W. Faller and D. L. Linebarrier, *J. Am. Chem. Soc.*, **111**, 1937 (1989).
103. M. Brookhart, S. K. Noh, F. J. Timmers, and Y. H. Hong, *Organometallics*, **7**, 2458 (1988).
104. J. F. Hoover and J. M. Stryker, *Organometallics*, **8**, 2973 (1989).

References for Chapter 13

1. K. G. Orrell, in: *Mechanisms of Inorganic and Organometallic Reactions* (M. V. Twigg, ed.), Vol. 6, Chapter 13, Plenum Press, New York (1989).
2. B. E. Mann, *Adv. Organomet. Chem.*, **28**, 397 (1988).
3. K. G. Orrell, *Coord. Chem. Rev.*, **96**, 1 (1989).
4. G. Fraenkel and W. R. Winchester, *J. Am. Chem. Soc.*, **111**, 3794 (1989).
5. G. Fraenkel and W. R. Winchester, *J. Am. Chem. Soc.*, **110**, 8720 (1988).
6. D. A. Femec, M. E. Silver, and R. C. Fay, *Inorg. Chem.*, **28**, 2789 (1989).
7. G. Erker, R. Schlund, M. Albrecht, and C. Sarter, *J. Organomet. Chem.*, **353**, C27 (1988).
8. K. Tatsumi, I. Matsubara, Y. Inoue, A. Nakamura, K. Miki, and N. Kasai, *J. Am. Chem. Soc.*, **111**, 7766 (1989).
9. A. A. Ismail, F. Sauriol, and I. S. Butler, *Inorg. Chem.*, **28**, 1007 (1989).
10. R. A. Brown, S. Endud, J. Friend, J. M. Hill, and M. W. Whiteley, *J. Organomet. Chem.*, **339**, 283 (1988).
11. C. G. Kreiter, G. Michael, and J. Kaub, *J. Organomet. Chem.*, **355**, 149 (1988).
12. C. Bessenbacher and W. Kaim, *J. Organomet. Chem.*, **369**, 83 (1989).
13. B. Davis, F. W. B. Einstein, P. G. Glavina, T. Jones, R. K. Pomeroy, and P. Rushman, *Organometallics*, **8**, 1030 (1989).
14. R. V. Honeychuck and W. H. Hersh, *J. Am. Chem. Soc.*, **111**, 6056 (1989).
15. M. Brookhart, J. Yoon, and S. K. Noh, *J. Am. Chem. Soc.*, **111**, 4117 (1989).
16. R. S. Bly, R. K. Bly, M. M. Hossain, L. Lebioda, and M. Raja, *J. Am. Chem. Soc.*, **110**, 7723 (1988).
17. J. Loset, L. Helm, A. Merbach, R. Roulet, F. Grepioni, and D. Braga, *Helv. Chim. Acta*, **71**, 1458 (1988).
18. J. M. Brown, P. A. Chaloner, and G. A. Morris, *J. Chem. Soc., Perkin Trans. 2*, 1583 (1987).
19. J. M. Brown, P. L. Evans, and A. R. Lucy, *J. Chem. Soc., Perkin Trans. 2*, 1589 (1987).
20. J. M. Brown and A. G. Kent, *J. Chem. Soc., Perkin Trans. 2*, 1597 (1987).
21. A. R. Siedle, R. A. Newmark, and R. D. Howells, *Inorg. Chem.*, **27**, 2473 (1988).
22. C. Bianchini, A. Meli, F. Laschi, F. Vizza, and P. Zanello, *Inorg. Chem.*, **28**, 227 (1989).
23. R. L. Geerts, J. C. Huffman, D. E. Westerberg, K. Folting, and K. G. Caulton, *New J. Chem.*, **12**, 455 (1988).
24. F. Maassarani, M. Pfeffer, and G. Van Koten, *Organometallics*, **8**, 871 (1989).

25. J. Fawcett, W. Henderson, M. D. Jones, R. D. W. Kemmitt, D. R. Russell, B. Lam, S. K. Kang, and T. A. Albright, *Organometallics*, **8**, 1991 (1989).
26. R. Baumgaertner and H. A. Brune, *J. Organomet. Chem.*, **350**, 115 (1988).
27. N. W. Alcock, A. W. G. Platt, and P. G. Pringle, *J. Chem. Soc., Dalton Trans.*, 139 (1989).
28. A. G. Avent, P. D. Lickiss, and A. Pidcock, *J. Organomet. Chem.*, **341**, 281 (1988).
29. G. Klebe, *J. Organomet. Chem.*, **332**, 35 (1987).
30. D. Baudry, P. Dorion, and M. Ephritikhine, *J. Organomet. Chem.*, **356**, 165 (1988).
31. R. E. Cramer, S. Roth, F. Edelmann, M. A. Bruck, K. C. Cohn, and J. W. Gilje, *Organometallics*, **8**, 1192 (1989).
32. S. Denholm, G. Hunter, and T. J. R. Weakley, *J. Chem. Soc., Dalton Trans.*, 2789 (1987).
33. G. Hunter, T. J. R. Weakley and W. Weissensteiner, *J. Chem. Soc., Perkin Trans. 2*, 1633 (1987).
34. F. W. Grevels, J. Jacke, C. Krueger, and Y. H. Tsay, *Organometallics*, **8**, 293 (1989).
35. P. R. Brown, M. L. H. Green, P. M. Hare, and J. A. Bandy, *Polyhedron*, **7**, 1819 (1988).
36. J. P. McNally and N. J. Cooper, *Organometallics*, **7**, 1704 (1988).
37. M. Arthurs, H. K. Al-Daffaee, J. Haslop, G. Kubal, M. D. Pearson, P. Thatcher, and E. Curzon, *J. Chem. Soc., Dalton Trans.*, 2615 (1987).
38. L. Carlton and J. L. Davidson, *J. Chem. Soc., Dalton Trans.*, 2071 (1988).
39. J. L. Davidson and G. Vasapollo, *J. Chem. Soc., Dalton Trans.*, 2855 (1988).
40. E. M. Armstrong, P. K. Baker, M. E. Harman, and M. B. Hursthouse, *J. Chem. Soc., Dalton Trans.*, 295 (1989).
41. H. G. Alt, H. E. Engelhardt, A. Razavi, M. D. Rausch, and R. D. Rogers, *Z. Naturforsch., B*, **43**, 438 (1988).
42. G. K. Anderson, R. J. Cross, L. Manojlovic-Muir, K. W. Muir, and M. Rocamora, *Organometallics*, **7**, 1520 (1988).
43. P. Courtot, V. Labed, R. Pichon, and J. Y. Salaun, *J. Organomet. Chem.*, **359**, C9 (1989).
44. J. L. Davidson, *J. Chem. Soc., Dalton Trans.*, 2715 (1987).
45. J. R. Bleeke and D. J. Rauscher, *Organometallics*, **7**, 2328 (1988).
46. H. Sitzmann, *J. Organomet. Chem.*, **354**, 203 (1988).
47. J. Okuda, *J. Organomet. Chem.*, **356**, C43 (1988).
48. J. Okuda and E. Herdtweck, *Chem. Ber.*, **121**, 1899 (1988).
49. F. Carre, E. Colomer, R. J. P. Corriu, and M. Lheureux, *J. Organomet. Chem.*, **331**, 29 (1987).
50. G. E. Herberich and M. Negele, *J. Organomet. Chem.*, **350**, 81 (1988).
51. H. Kurosawa, H. Ohnishi, M. Emoto, Y. Kawasaki, and S. Murai, *J. Am. Chem. Soc.*, **110**, 6272 (1988).
52. G. H. Lee, S. M. Peng, I. C. Tsung, D. Mu, and R. S. Liu, *Organometallics*, **8**, 2248 (1989).
53. K. W. Nugent, J. K. Beattie, and L. D. Field, *J. Phys. Chem.*, **93**, 5371 (1989).
54. S. T. Abu-Orabi and P. Jutzi, *J. Organomet. Chem.*, **329**, 169 (1987).
55. S. T. Aru-Orabi and P. Jutzi, *J. Organomet. Chem.*, **347**, 307 (1988).
56. G. Parkin and J. E. Bercaw, *Polyhedron*, **7**, 2053 (1988).
57. M. Brookhart, R. C. Buck, and E. Danielson III, *J. Am. Chem. Soc.*, **111**, 567 (1989).
58. C. M. Lukehart and W. R. True, *Organometallics*, **7**, 2387 (1988).
59. J. Merkert, R. M. Nielson, M. J. Weaver, and W. E. Geiger, *J. Am. Chem. Soc.*, **111**, 7084 (1989).
60. R. M. Nielson and M. J. Weaver, *Organometallics*, **8**, 1636 (1989).
61. R. Benn, R. Mynott, I. Topalovic, and F. Scott, *Organometallics*, **8**, 2299 (1989).
62. K. Muellen, N. T. Allison, J. Lex, H. Schmickler, and E. Vogel, *Tetrahedron*, **43**, 3225 (1987).
63. H. El-Amouri, A. A. Bahsoun, and J. A. Osborn, *Polyhedron*, **7**, 2035 (1988).
64. M. Grassi, B. E. Mann, B. T. Pickup, and C. M. Spencer, *J. Chem. Soc., Dalton Trans.*, 2649 (1987).
65. E. W. Abel, K. G. Orrell, K. B. Qureshi, V. Sik, and D. Stephenson, *J. Organomet. Chem.*, **353**, 337 (1988).
66. R. C. Hemond, R. P. Hughes, and A. L. Rheingold, *Organometallics*, **8**, 1261 (1989).
67. M. J. Bermejo, J. I. Ruiz, X. Solans, and J. Vinaixa, *Inorg. Chem.*, **27**, 4385 (1988).
68. G. Hunter, A. McAuley, and T. W. Whitcombe, *Inorg. Chem.*, **27**, 2634 (1988).

69. M. Brookhart, D. M. Lincoln, A. F. Volpe, Jr., and G. F. Schmidt, *Organometallics*, **8**, 1212 (1989).
70. F. Bouachir, B. Chaudret, F. Dahan, and I. Tkatchenko, *New J. Chem.*, **11**, 527 (1987).
71. B. Buchmann, U. Piantini, W. Von Philipsborn, and A. Salzer, *Helv. Chim. Acta*, **70**, 1487 (1987).
72. R. M. Bullock, C. E. L. Headford, K. M. Hennessy, S. E. Kegley, and J. R. Norton, *J. Am. Chem. Soc.*, **111**, 3897 (1989).
73. G. L. Crocco, and J. A. Gladysz, *J. Am. Chem. Soc.*, **110**, 6110 (1988).
74. C. Bianchini, D. Masi, A. Meli, M. Peruzzini, J. A. Ramirez, A. Vacca, and F. Zanobini, *Organometallics*, **8**, 2179 (1989).
75. T. Schaeffer, G. H. Penner, C. S. Takeuchi, and C. Beaulieu, *Can. J. Chem.*, **67**, 1283 (1989).
76. D. Stalke, U. Klingebiel, and G. M. Sheldrick, *J. Organomet. Chem.*, **341**, 119 (1988).
77. R. D. Thomas and K. E. Rowland, *Magn. Reson. Chem.*, **26**, 111 (1988).
78. M. S. Samples and C. H. Yoder, *J. Organomet. Chem.*, **332**, 69 (1987).
79. E. P. Cullen, J. Doherty, A. R. Manning, P. McArdle, and D. Cunningham, *J. Organomet. Chem.*, **348**, 109 (1988).
80. A. L. Nivorozhkin, T. G. Takhirov, E. V. Sukholenko, N. I. Borisenko, V. Minkin, and O. A. D'yachenko, *Izv. Akad. Nauk SSSR, Ser. Khim.*, 1198 (1988).
81. A. L. Nivorozhkin, E. V. Sukholenko, L. E. Nivorozhkin, N. I. Borisenko, V. I. Minkin, Yu. K. Grishin, O. A. D'yachenko, T. G. Takhirov, and D. B. Tagiev, *Polyhedron*, **8**, 569 (1989).
82. G. T. Crisp, G. Salem, S. B. Wild, and F. S. Stephens, *Organometallics*, **8**, 2360 (1989).
83. S. Yamazaki and T. Ama, *Bull. Chem. Soc. Jpn.*, **62**, 931 (1989).
84. S. Toyota, Y. Yamada, M. Kaneyoshi, and M. Oki, *Bull. Chem. Soc. Jpn.*, **62**, 1509 (1989).
85. E. W. Abel, M. A. Beckett, D. Ellis, K. G. Orrell, V. Sik, and D. Stephenson, *J. Chem. Res. (S)*, 232 (1988).
86. J. Heidrich and W. Beck, *J. Organomet. Chem.*, **354**, 91 (1988).
87. E. W. Abel, D. E. Budgen, I. Moss, K. G. Orrell, and V. Sik, *J. Organomet. Chem.*, **362**, 105 (1989).
88. E. W. Abel, N. J. Long, K. G. Orrell, A. G. Osborne, V. Sik, P. A. Bates, and M. B. Hursthouse, *J. Organomet. Chem.*, **367**, 275 (1989).
89. E. W. Abel, N. J. Long, K. G. Orrell, A. G. Osborne, and V. Sik, *J. Organomet. Chem.*, **378**, 473 (1989).
90. E. W. Abel, T. P. J. Coston, K. M. Higgins, K. G. Orrell, V. Sik, and T. S. Cameron, *J. Chem. Soc., Dalton Trans.*, 701 (1989).
91. E. W. Abel, T. P. J. Coston, K. G. Orrell, and V. Sik, *J. Chem. Soc. Dalton Trans.*, 711 (1989).
92. P. K. Byers, A. J. Canty, R. T. Honeyman, and A. W. Watson, *J. Organomet. Chem.*, **363**, C22 (1989).
93. T. Klapoetke, H. Koepf, and P. Gowik, *Polyhedron*, **6**, 1923 (1987).
94. E. Liepins, G. Ancens, N. P. Erchak, and E. Lukevics, *Zh. Obshch. Khim.*, **58**, 384 (1988).
95. A. E. Aliev, A. A. Fomichev, A. V. Varlamov, and N. S. Prostakov, *Khim. Geterotsikl, Soedin.*, 1577 (1987).
96. C. A. Mike, T. Nelson, J. Graham, A. W. Cordes, and N. T. Allison, *Organometallics*, **7**, 2573 (1988).
97. N. Albrecht and E. Weiss, *J. Organomet. Chem.*, **355**, 89 (1988).
98. C. P. Casey, M. S. Konings, S. R. Marder, and Y. Takezawa, *J. Organomet. Chem.*, **358**, 347 (1988).
99. G. Erker, R. Nolte, G. Tainturier, and A. Rheingold, *Organometallics*, **8**, 454 (1989).
100. K. A. Azam, A. J. Deeming, M. S. B. Felix, P. A. Bates, and M. B. Hursthouse, *Polyhedron*, **7**, 1793 (1988).
101. J. R. Lockemeyer, T. B. Rauchfuss, A. L. Rheingold, and S. R. Wilson, *J. Am. Chem. Soc.*, **111**, 8828 (1989).
102. S. A. Benyunes and P. A. Chaloner, *J. Organomet. Chem.*, **341**, C50 (1988).
103. C. G. Kreiter, G. Wendt, and J. Kaub, *Chem. Ber.*, **122**, 215 (1989).

104. A. Ricalton and M. W. Whiteley, *J. Organomet. Chem.*, **361**, 101 (1989).

105. Yu. A. Ustynuk, A. K. Shestakova, V. A. Chertkov, N. N. Zemlyanskii, I. V. Borisova, A. I. Gusev, E. B. Chuklanova, and E. A. Chernyshev, *J. Organomet. Chem.*, **335**, 43 (1987).

106. J. Edwin, W. E. Geiger, and C. H. Bushweller, *Organometallics*, **7**, 1486 (1988).

107. M. A. Kulzick, R. T. Price, R. A. Andersen, and E. L. Muetterties, *J. Organomet. Chem.*, **333**, 105 (1987).

108. M. D. Fryzuk, W. E. Piers, F. W. B. Einstein, and T. Jones, *Can. J. Chem.*, **67**, 883 (1989).

109. D. Obendorf, M. Probst, P. Peringer, H. Falk, and N. Mueller, *J. Chem. Soc., Dalton Trans.*, 1709 (1988).

110. R. Benn, E. Janssen, H. Lehmkuhl, and A. Rufinska, *J. Organomet. Chem.*, **333**, 169 (1987).

111. S. M. Baxter, G. S. Ferguson, and P. T. Wolczanski, *J. Am. Chem. Soc.*, **110**, 4231 (1988).

112. D. H. Berry and J. E. Bercaw, *Polyhedron*, **7**, 759 (1988).

113. G. Erker, F. Sosna, R. Zwettler, and C. Krueger, *Organometallics*, **8**, 450 (1989).

114. J. Heck and G. Rist, *J. Organomet. Chem.*, **342**, 45 (1988).

115. J. Heck, K. A. Kriebisch, and H. Mellinghoff, *Chem. Ber.*, **121**, 1753 (1988).

116. M. Ferrer, O. Rossell, M. Seco, and P. Braunstein, *J. Chem. Soc., Dalton Trans.*, 379 (1989).

117. S. Toefke and U. Behrens, *J. Organomet. Chem.*, **338**, 29 (1988).

118. M. H. Chisholm, B. W. Eichhorn, K. Folting, and J. C. Huffman, *Organometallics*, **8**, 49 (1989).

119. M. H. Chisholm, B. W. Eichhorn, and J. C. Huffman, *Organometallics*, **8**, 67 (1989).

120. M. H. Chisholm, B. W. Eichhorn, and J. C. Huffman, *Organometallics*, **8**, 80 (1989).

121. M. H. Chisholm, J. A. Heppert, J. C. Huffman, and C. D. Ontiveros, *Organometallics*, **8**, 976 (1989).

122. N. C. Schroeder, R. Funchess, R. A. Jacobson, and R. J. Angelici, *Organometallics*, **8**, 521 (1989).

123. A. Rubello, P. Vogel, and G. Chapuis, *Helv. Chim. Acta*, **70**, 1638 (1987).

124. D. J. Elliot, H. A. Mirza, R. J. Puddephatt, D. G. Holah, A. N. Hughes, R. H. Hill, and W. Xia, *Inorg. Chem.*, **28**, 3282 (1989).

125. C. P. Morley, *Organometallics*, **8**, 800 (1989).

126. Y. Fuchita, M. Nakashima, K. Hiraki, M. Kawatani, and K. Ohnuma, *J. Chem. Soc., Dalton Trans.*, 785 (1988).

127. B. T. Heaton, *Pure Appl. Chem.*, **60**, 1757 (1988).

128. D. T. Clark, K. A. Sutin, R. E. Perrier, and M. J. McGlinchey, *Polyhedron*, **7**, 2297 (1988).

129. M. F. D'Agostino and M. J. McGlinchey, *Polyhedron*, **7**, 807 (1988).

130. M. H. Chisholm, D. L. Clark, and M. J. Hampden-Smith, *J. Am. Chem. Soc.*, **111**, 574 (1989).

131. S. Aime, M. Botta, O. Gambino, R. Gobetto, and D. Osella, *J. Chem. Soc., Dalton Trans.*, 1277 (1989).

132. S. S. D. Brown, I. D. Salter, V. Sik, I. J. Colquhoun, W. McFarlane, P. A. Bates, and M. B. Hursthouse, *J. Chem. Soc., Dalton Trans.*, 2177 (1988).

133. S. S. D. Brown, P. J. McCarthy, I. D. Salter, P. A. Bates, M. B. Hursthouse, I. J. Colquhoun, W. McFarlane, and M. Murray, *J. Chem. Soc., Dalton Trans.*, 2787 (1988).

134. S. S. D. Brown, I. D. Salter, A. J. Dent, G. F. M. Kitchen, A. G. Orpen, P. A. Bates, and M. B. Hursthouse, *J. Chem. Soc., Dalton Trans.*, 1227 (1989).

135. P. Ewing and L. J. Farrugia, *Organometallics*, **8**, 1246 (1989).

136. L. J. Farrugia, *Organometallics*, **8**, 2410 (1989).

137. S. L. Bassner, G. L. Geoffroy, and A. L. Rheingold, *Polyhedron*, **7**, 791 (1988).

138. S. Hajela, B. M. Novak, and E. Rosenberg, *Organometallics*, **8**, 468 (1989).

139. M. N. Ackermann, D. E. Adams, J. Pranata, and C. F. Yamauchi, *J. Organomet. Chem.*, **369**, 55 (1989).

140. J. L. Zuffa, S. J. Kivi, and W. L. Gladfelter, *Inorg. Chem.*, **28**, 1888 (1989).

141. V. J. Johnston, F. W. B. Einstein, and R. K. Pomeroy, *Organometallics*, **7**, 1867 (1988).

142. B. E. Mann, B. T. Pickup, and A. K. Smith, *J. Chem. Soc., Dalton Trans.*, 889 (1989).

143. A. Strawczynski, R. Ros, and R. Roulet, *Helv. Chim. Acta*, **71**, 867 (1988).

144. A. Strawczynski, R. Ros, R. Roulet, F. Grepioni, and D. Braga, *Helv. Chim. Acta*, **71**, 1885 (1988).

145. W. E. Lindsell, N. M. Walker, and A. S. F. Boyd, *J. Chem. Soc., Dalton Trans.*, 675 (1988).
146. A. Fumagalli, R. Della Pergola, F. Bonacina, L. Garlaschelli, M. Moret, and A. Sironi, *J. Am. Chem. Soc.*, 111, 165 (1989).
147. S. Bordini, B. T. Heaton, C. Seregni, L. Strona, R. J. Goodfellow, M. B. Hursthouse, M. Thornton-Pett, and S. Martinengo, *J. Chem. Soc., Dalton Trans.*, 2103 (1988).
148. W. R. Cullen, S. T. Chacon, M. I. Bruce, F. W. B. Einstein, and R. H. Jones, *Organometallics*, 7, 2273 (1988).
149. M. A. Gallop, B. F. G. Johnson, J. Lewis, and P. R. Raithby, *J. Chem. Soc., Chem. Commun.*, 1809 (1987).
150. E. Boyar, A. J. Deeming, M. S. B. Felix, S. E. Kabir, T. Adatia, R. Bhusate, M. McPartlin, and H. R. Powell, *J. Chem. Soc., Dalton Trans.*, 5 (1989).
151. H. B. Davis, F. W. B. Einstein, V. J. Johnston, and R. K. Pomeroy, *J. Am. Chem. Soc.*, 110, 4451 (1988).
152. E. Wehman, G. Van Koten, C. J. M. Erkamp, D. M. Knotter, J. T. B. H. Jastrzebski, and C. H. Stam, *Organometallics*, 8, 94 (1989).
153. G. Douglas, L. Manojlovic-Muir, K. W. Muir, M. C. Jennings, B. R. Lloyd, M. Rashidi, and R. J. Puddephatt, *J. Chem. Soc., Chem. Commun.*, 149 (1988).
154. E. W. Abel, K. G. Orrell, and D. Stephenson, *J. Organomet. Chem.*, 373, 401 (1989).
155. P. Sundberg, B. F. G. Johnson, J. Lewis, and P. R. Raithby, *J. Organomet. Chem.*, 354, 131 (1988).
156. G. Predieri, A. Tiripicchio, C. Vignali, and E. Sappa, *J. Organomet. Chem.*, 342, C33 (1988).
157. A. A. Koridze, O. A. Kizas, P. V. Petrovskii, N. E. Kolobova, Yu. T. Struchkov, and A. I. Yanovskii, *J. Organomet. Chem.*, 338, 81 (1988).
158. H. J. Kneuper and J. R. Shapley, *New J. Chem.*, 12, 479 (1988).
159. T. Beringhelli, G. D'Alfonso, G. Ciani, A. Sironi, and H. Molinari, *J. Chem. Soc., Dalton Trans.*, 1281 (1988).
160. S. R. Drake, B. F. G. Johnson, and J. Lewis, *J. Chem. Soc., Dalton Trans.*, 1517 (1988).
161. P. A. W. Dean, J. J. Vittal, and M. H. Trattner, *Inorg. Chem.*, 26, 4245 (1987).
162. P. A. W. Dean, V. Manivannan, and J. J. Vittal, *Inorg. Chem.*, 28, 2360 (1989).
163. M. Bown, T. Jelinek, B. Stibr, S. Hermanek, X. L. R. Fontaine, N. N. Greenwood, J. D. Kennedy, and M. Thornton-Pett, *J. Chem. Soc., Chem. Commun.*, 974 (1988).
164. G. Ferguson, J. D. Kennedy, X. L. R. Fontaine, Faridoon, and T. R. Spalding, *J. Chem. Soc., Dalton Trans.*, 2555 (1988).
165. D. M. Schubert, C. B. Knobler, P. A. Wegner, and M. F. Hawthorne, *J. Am. Chem. Soc.*, 110, 5219 (1988).

References for Chapter 14

1. A. E. Martell and D. T. Sawyer (eds.), *Oxygen Complexes and Oxygen Activation by Transition Metals*, Plenum Press, New York (1988).
2. *Pure Appl. Chem.*, 60, Part 1 (1988).
3. B. Walther, *Z. Chem.*, 29, 117, (1989).
4. I. Ojima, *Chem. Rev.*, 88, 1011 (1988).
5. B. C. Gates and H. H. Lamb, *J. Mol. Catal.*, 52, 1 (1989).
6. W. Keim, *J. Mol. Catal.*, 52, 19 (1989).
7. P. M. Maitlis, I. M. Saez, N. J. Meanwell, K. Isobe, A. Nutton, A. Vaguez de Miguel, D. W. Bruce, S. Okeya, P. M. Bailey, D. G. Andrews, P. R. Ashton, and I. R. Johnstone, *New J. Chem.*, 13, 419 (1989).
8. D. Milstein, *Acc. Chem. Res.*, 21, 428 (1988).
9. R. J. Angelici, *Acc. Chem. Res.*, 21, 387 (1988).
10. A. Sen, *Acc. Chem. Res.*, 21, 421 (1988).
11. W. Oppolzer, *Angew. Chem.*, 101, 39 (1989).

12. J. J. Brunet, D. Neilbecker, and F. Niedercorn, *J. Mol. Catal.*, **49**, 235 (1989).
13. W. J. Scott and J. E. McMurry, *Acc. Chem. Res.*, **21**, 47 (1988).
14. L. S. Hegedus, *Angew. Chem.*, **100**, 1147 (1988).
15. L. Markó, *J. Organomet. Chem.*, **357**, 481 (1988).
16. M. Bochmann, in: F. G. A. Stone and E. W. Abel (eds.), *Specialist Periodical Reports: Organometallic Chemistry*, Vol. 17, p. 362 ff, The Royal Society of Chemistry (1989).
17. M. Bochmann in: F. G. A. Stone and E. W. Abel (eds.), *Specialist Periodical Reports: Organometallic Chemistry*, Vol. 18, p. 349 ff, The Royal Society of Chemistry (1989).
18. *J. Organomet. Chem.*, **370** (1989).
19. R. Noyori, *Chem. Soc. Rev.*, **18**, 187 (1989).
20. G. Consiglio and R. M. Waymouth, *Chem. Rev.*, **89**, 257 (1989).
21. K. A. Jørgensen, *Chem. Rev.*, **89**, 431 (1989).
22. W. D. Harman and H. Taube, *J. Am. Chem. Soc.*, **110**, 7906 (1988).
23. M. A. Esteruelas, E. Sola, L. A. Oro, U. Meyer, and H. Werner, *Angew. Chem.*, **100**, 1621 (1988).
24. C. Bianchini, C. Mealli, A. Meli, M. Peruzzini, and F. Zanobini, *J. Am. Chem. Soc.*, **110**, 8725 (1988).
25. C. Bianchini, A. Meli, F. Laschi, J. A. Ramirez, P. Zanello, and A. Vacca, *Inorg. Chem.*, **27**, 4429 (1988).
26. E. G. Lundquist, J. C. Huffman, K. Folting, and K. G. Caulton, *Angew. Chem.*, **100**, 1236 (1988).
27. R. U. Kirss, T. C. Eisenschmid, and R. Eisenberg, *J. Am. Chem. Soc.*, **110**, 8564 (1988).
28. R. U. Kirss and R. Eisenberg, *J. Organomet. Chem.*, **359**, C22 (1989).
29. M. M. Taqui Khan, S. A. Samad, Z. Shirin, and M. R. H. Siddiqui, *J. Mol. Catal.*, **54**, 81 (1989).
30. M. M. Taqui Khan, S. A. Samad, and M. R. H. Siddiqui, *J. Mol. Catal.*, **53**, 23 (1989).
31. M. M. Taqui Khan, B. Taqui Khan, and P. J. Reddy, *J. Mol. Catal.*, **54**, 171 (1989).
32. H. Moriyama, A. Yabe, and F. Matsui, *J. Mol. Catal.*, **50**, 195 (1989).
33. A. Bose and C. R. Saha, *J. Mol. Catal.*, **49**, 271 (1989).
34. R. H. Fish, H. S. Kim, J. E. Babin, and R. D. Adams, *Organometallics*, **7**, 2250 (1988).
35. C. Bergounhou, P. Fompeyrine, G. Commenges, and J. J. Bonnet, *J. Mol. Catal.*, **48**, 285 (1988).
36. T. Tsukahara, H. Kawano, Y. Ishii, T. Takahashi, M. Saburi, Y. Uchida, and S. Akutagawa, *Chem. Lett.*, 2055 (1988).
37. M. Kitamura, I. Kasahara, K. Manabe, R. Noyori, and H. Takaya, *J. Org. Chem.*, **53**, 710 (1988).
38. M. Kitamura, T. Okhuma, S. Inoue, N. Sayo, H. Kumobayashi, S. Akutagawa, T. Ohta, H. Takaya, and R. Noyori, *J. Am. Chem. Soc.*, **110**, 629 (1988).
39. H. Muramatsu, H. Kawano, Y. Ishii, M. Saburi, and Y. Uchida, *J. Chem. Soc., Chem. Commun.*, 769 (1989).
40. U. Nagel and B. Rieger, *Organometallics*, **8**, 1534 (1989).
41. D. Morton and D. J. Cole-Hamilton, *J. Chem. Soc., Chem. Commun.*, 1154 (1988).
42. D. Morton, D. J. Cole-Hamilton, I. D. Utuk, M. Paneque-Sosa, and M. Lopez-Poveda, *J. Chem. Soc. Dalton Trans.*, 489 (1989).
43. M. A. Esteruelas, E. Sola, L. A. Oro, H. Werner, and U. Meyer, *J. Mol. Catal.*, **53**, 43 (1989).
44. K. Nomura, Y. Saito, and S. Shinoda, *J. Mol. Catal.*, **52**, 99 (1989).
45. K. Nomura, Y. Saito, and S. Shinoda, *J. Mol. Catal.*, **50**, 303 (1989).
46. H. A. Brune, J. Unsin, R. Hemmer, and M. Reichhardt, *J. Organomet. Chem.*, **369**, 335 (1989).
47. C. Biran, Y. D. Blum, R. Glaser, D. S. Tse, K. A. Youngdahl, and R. M. Laine, *J. Mol. Catal.*, **48**, 183 (1988).
48. M. Oshima, H. Yamazaki, I. Shimizu, M. Nisar, and J. Tsuji, *J. Am. Chem. Soc.*, **111**, 6280 (1989).
49. G. L. Gould and D. M. Heinekey, *J. Am. Chem. Soc.*, **111**, 5502 (1989).
50. C. K. Ghosh, J. K. Hoyano, R. Krentz, and W. A. G. Graham, *J. Am. Chem. Soc.*, **111**, 5480 (1989).

51. T. B. Marder, D. C. Roe, and D. Milstein, *Organometallics*, 7, 1451 (1988).
52. D. P. Fairlie and B. Bosnich, *Organometallics*, 7, 946 (1988).
53. T. Sakakura, T. Sodeyama, and M. Tanaka, *New J. Chem.*, 13, 737 (1989).
54. K. Nomura and Y. Saito, *J. Mol. Catal.*, 370, 57 (1989).
55. W. Caseri and P. S. Pregosin, *Organometallics*, 7, 1373 (1988).
56. W. Caseri and P. S. Pregosin, *J. Organomet. Chem.*, 356, 259 (1988).
57. X. L. Luo and R. H. Crabtree, *J. Am. Chem. Soc.*, 111, 2527 (1989).
58. T. C. Eisenschmid and R. Eisenberg, *Organometallics*, 8, 1822 (1989).
59. K. Tamao, K. Kobayashi, and Y. Ito, *J. Am. Chem. Soc.*, 111, 6478 (1989).
60. A. L. Casalnuovo, J. C. Calabrese, and D. Milstein, *J. Am. Chem. Soc.*, 110, 6738 (1988).
61. M. R. Gagné and T. J. Marks, *J. Am. Chem. Soc.*, 111, 4108 (1989).
62. M. Hodgson, D. Parker, R. J. Taylor, and G. Ferguson, *Organometallics*, 7, 1761 (1988).
63. H. Hoberg and D. Guhl, *Angew. Chem.*, 101, 1091 (1989).
64. I. Starý and P. Kočouský, *J. Am. Chem. Soc.*, 111, 4981 (1989).
65. R. Mahé, Y. Sasaki, C. Bruneau, and P. H. Dixneuf, *J. Org. Chem.*, 54, 1518 (1989).
66. E. Dunach and J. Périchon, *J. Organomet. Chem.*, 352, 239 (1988).
67. H. Alper, M. Saldana-Maldonado, and I. J. B. Lin, *J. Mol. Catal.*, 49, L27 (1988).
68. U. Müller, W. Keim, C. Krüger, and P. Betz, *Angew. Chem.*, 101, 1066 (1989).
69. Y. V. Kissin, *J. Polym. Sci: Part A: Polym. Chem.*, 27, 605 (1989).
70. Y. V. Kissin and D. L. Beach, *J. Polym. Sci. Part A: Polym. Chem.*, 27, 147 (1989).
71. S. B. Choe, H. Kanai, and K. J. Klabunde, *J. Am. Chem. Soc.*, 111, 2875 (1989).
72. P. A. Wender and T. E. Jenkins, *J. Am. Chem. Soc.*, 111, 6432 (1989).
73. M. Iyoda, S. Tanaka, H. Otani, M. Nose, and M. Oda, *J. Am. Chem. Soc.*, 110, 8494 (1988).
74. B. M. Trost, D. C. Lee, and F. Rise, *Tetrahedron Lett.*, 30, 651 (1989).
75. R. F. Jordan and D. F. Taylor, *J. Am. Chem. Soc.*, 111, 778 (1989).
76. J. R. Strickler, P. A. Wexler, and D. E. Wigley, *Organometallics*, 7, 2067 (1988).
77. B. M. Trost and D. C. Lee, *J. Am. Chem. Soc.*, 110, 7255 (1988).
78. B. M. Novak and R. H. Grubbs, *J. Am. Chem. Soc.*, 110, 7542 (1988).
79. J. Kress, J. A. Osborn, V. Amir-Ebrahimi, K. J. Ivin, and J. J. Rooney, *J. Chem. Soc., Chem. Commun.*, 1164 (1988).
80. C. A. Jolly and D. S. Marynick, *J. Am. Chem. Soc.*, 111, 7968 (1989)
81. J. A. Ewen, R. L. Jones, A. Razavi, and J. D. Ferrara, *J. Am. Chem. Soc.*, 110, 6255 (1988).
82. G. Erker, R. Nolte, Y. H. Tsay, and C. Krüger, *Angew. Chem.*, 101, 642 (1989).
83. U. Bueschges and J. C. W. Chien, *J. Polym. Sci., Part A: Polym. Chem.*, 27, 1525 (1989).
84. J. R. Ascenso, A. R. Dias, P. T. Gomes, C. C. Romao, Q. T. Pham, D. Neibecker, and I. Tkatchenko, *Macromolecules*, 22, 998 (1989).
85. M. Suzuki, S. Sawada, and T. Saegusa, *Macromolecules*, 22, 1505 (1989).
86. A. Soum and M. Fontanille In: M. Fontanille and A. Guyot (eds.), *Recent Advances in Mechanistic and Synthetic Aspects of Polymerization*, p. 375, Reidel, Dordrecht (1987).
87. D. Meziane, A. Soum, and M. Fontanielle, *Makromol. Chem.*, 189, 1407 (1988).
88. H. Fischer, J. Schmid, and J. Riede, *J. Organomet. Chem.*, 355, 219 (1988).
89. M. H. Desbois and D. Astruc, *New J. Chem.*, 13, 595 (1989).
90. W. Keim, J. Becker, P. Kraneburg, and R. Greven, *J. Mol. Catal.*, 54, 37 (1989).
91. W. Keim and J. Becker, *J. Mol. Catal.*, 54, 95 (1989).
92. H. Alper and M. Saldana-Maldonado, *Organometallics*, 8, 1124 (1989).
93. I. Pri-Bar and H. Alper, *J. Org. Chem.*, 54, 36 (1989).
94. M. Ishino and T. Deguchi, *J. Mol. Catal.*, 52, L17 (1989).
95. H. des Abbayes, J. C. Clément, P. Laurent, G. Tanguy, and N. Thilmont, *Organometallics*, 7, 2293 (1988).
96. H. Yamashita, T. Sakakura, T. Kobayashi, and M. Tanaka, *J. Mol. Catal.*, 48, 69 (1988).
97. R. Takeuchi, Y. Tsuji, M. Fujita, T. Kondo, and Y. Watanabe, *J. Org. Chem.*, 54, 1831 (1989).
98. M. C. Bonnet, J. Coombes, B. Manzano, D. Neibecker, and I. Tkatchenko, *J. Mol. Catal.*, 52, 263 (1989).
99. L. Versluis, T. Ziegler, E. J. Baerends, and W. Ravenek, *J. Am. Chem. Soc.*, 111, 2018 (1989).

100. M. Garland and G. Bor, *Inorg. Chem.*, **28**, 410 (1989).
101. M. M. Taqui Khan, S. B. Halligudi, and S. H. R. Abdi, *J. Mol. Catal.*, **48**, 313 (1988).
102. J. Jenck, P. Kalck, E. Pinelli, M. Siani, and A. Thorez, *J. Chem. Soc., Chem. Commun.*, 1428 (1988).
103. R. Lazzaroni, A. Raffaelli, R. Settambolo, S. Bertozzi, and G. Vitulli, *J. Mol. Catal.*, **50**, 1 (1989).
104. M. G. Richmond, *J. Mol. Catal.*, **54**, 199 (1989).
105. Y. Ishii, M. Sato, H. Matsuzaka, and M. Hidai, *J. Mol. Catal.*, **54**, L13 (1989).
106. V. N. Zudin, V. D. Chinakov, V. M. Nekipelov, V. A. Rogov, V. A. Likholobov, and Y. I. Yermakov, *J. Mol. Catal.*, **52**, 27 (1989).
107. M. M. Taqui Khan, S. B. Halligudi, and S. Shukla, *Angew. Chem.*, **100**, 1803 (1988).
108. M. M. Taqui Khan, S. B. Halligudi, N. Nageswara Rao, and S. Shukla, *J. Mol. Catal.*, **51**, 161 (1989).
109. A. J. Pardey and P. C. Ford, *J. Mol. Catal.*, **53**, 247 (1989).
110. S. Bhaduri, H. Khwaja, K. Sharma, and P. G. Jones, *J. Chem. Soc., Chem. Commun.*, 515 (1989).
111. M. Akita, O. Matani, and Y. Moro-oka, *J. Chem. Soc., Chem. Commun.*, 527 (1989).
112. A. M. Raspolli Galetti, G. Brace, and G. Sbrana, *J. Organomet. Chem.*, **356**, 221 (1988).
113. M. M. Taqui Khan, S. B. Halligudi, and S. H. R. Abdi, *J. Mol. Catal.*, **48**, 325 (1988).
114. M. M. Taqui Khan, S. B. Halligudi, S. Shukla, and S. H. R. Abdi, *J. Mol. Catal.*, **51**, 129 (1989).
115. C. Amatore and A. Jutand, *Organometallics*, **7**, 2203 (1988).
116. J. M. Brown and N. A. Cooley, *J. Chem. Soc., Chem. Commun.*, 1345 (1988).
117. J. M. Brown, N. A. Cooley, and D. W. Price, *J. Chem. Soc., Chem. Commun.*, 458 (1989).
118. K. V. Baker, J. M. Brown, N. A. Cooley, G. D. Hughes, and R. J. Taylor, *J. Organomet. Chem.*, **370**, 397 (1989).
119. H. Kurosawa, M. Emoto, and Y. Kawasaki, *J. Organomet. Chem.*, **346**, 137 (1988).
120. P. J. Stang, M. H. Kowalski, M. D. Schiavelli, and D. Longford, *J. Am. Chem. Soc.*, **111**, 3347 (1989).
121. P. J. Stang, and M. H. Kowalski, *J. Am. Chem. Soc.*, **111**, 3356 (1989).
122. M. Catellani and G. P. Chiusoli, *J. Organomet. Chem.*, **346**, C27 (1988).
123. A. J. Canty, A. A. Watson, B. W. Skelton, and A. H. White, *J. Organomet. Chem.*, **367**, C25 (1989).
124. T. Kondo, Y. Tsuji, and Y. Watanabe, *J. Organomet. Chem.*, **345**, 397 (1988).
125. Y. Z. Huang and Q. L. Zhou, *J. Org. Chem.*, **52**, 3552 (1987).
126. R. Benhaddou, S. Czernecki, G. Ville, and A. Zegar, *Organometallics*, **7**, 2435 (1988).
127. A. M. Echavarren, D. R. Teuting, and J. K. Stille, *J. Am. Chem. Soc.*, **110**, 4039 (1988).
128. D. R. Deardorff, S. Shambayati, R. G. Linde, and M. M. Dunn, *J. Org. Chem.*, **53**, 189 (1988).
129. B. M. Trost and J. I. Luengo, *J. Am. Chem. Soc.*, **110**, 8239 (1988).
130. J. Nokami, T. Mandai, H. Watanabe, H. Ohyama, and J. Tsuji, *J. Am. Chem. Soc.*, **111**, 4126 (1989).
131. T. Hayashi, A. Yamamoto, Y. Ito, E. Nishioka, H. Miura, and K. Yanagi, *J. Am. Chem. Soc.*, **111**, 6301 (1989).
132. X. Lu, H. Fang, and Z. Ni, *J. Organomet. Chem.*, **373**, 77 (1989).

References for Chapter 15

1. R. van Eldik and K. J. Schneider, in: *Mechanisms of Inorganic and Organometallic Reactions* (M. V. Twigg, ed.), Vol. 6, p. 437, Plenum Press, New York (1989). .
2. F.-G. Klärner, *Chem. Unserer Zeit*, **23**, 53 (1989).
3. C. Cossy and A. E. Merbach, *Pure Appl. Chem.*, **60**, 1785 (1988).
4. M. Kotowski and R. van Eldik, *Coord. Chem. Rev.*, 93, 19 (1989).
5. R. van Eldik, T. Asano, and W. J. le Noble, *Chem. Rev.*, **89**, 549 (1989).

6. G. Demazeau, *Chem. Scr.*, 28, 21 (1988).
7. K. Hara and I. Morishima, *Rev. Sci. Instrum.*, 59, 2397 (1988).
8. M. Spitzer, F. Gartig, and R. van Eldik, *Rev. Sci. Instrum.*, 59, 2092 (1988).
9. S. Wieland and R. van Eldik, *Rev. Sci. Instrum.*, 60, 955 (1989).
10. K. Krtschil and F. Strohbusch, *Ber. Bunsenges. Phys. Chem.*, 93, 1437 (1989).
11. T. L. Neils and J. M. Burlitch, *J. Catal.*, 118, 79 (1989).
12. F. X. Prielmeier, E. W. Lang, R. J. Speedy, and H.-D. Lüdemann, *Ber. Bunsenges. Phys. Chem.*, 92, 1111 (1988).
13. W. Schilling and E. U. Franck, *Ber. Bunsenges. Phys. Chem.*, 92, 631 (1988).
14. M. Sato, J. Nakatani, S. Ozawa, and Y. Ogina, *Nippon Kagaku Kaishi*, 11, 1794 (1988).
15. B. Gavish, *J. Chem. Soc., Faraday Trans. 1*, 85, 1199 (1989).
16. S. D. Hamann, *Bull. Chem. Soc. Jpn.*, 62, 2126 (1989).
17. H. L. Lee, Z. U. Bae, Y. C. Park, and J. H. Yun, *J. Korean Chem. Soc.*, 32, 30 (1988).
18. T. Hiraga, N. Kitamura, H.-B. Kim, and S. Tazuke, *J. Phys. Chem.*, 93, 2940 (1989).
19. X. Xue, J. F. Stebbins, M. Kanzaki, and R. G. Tronnes, *Science*, 245, 962 (1989).
20. A. M. Beccaria, P. Fiordiponti, and G. Mattogno, *Corros. Sci.*, 29, 403 (1989).
21. W. M. Flarsheim, A. J. Bard, and K. P. Johnston, *J. Phys. Chem.*, 93, 4234 (1989).
22. S. Mönkeberg, E. van Raaij, H. Kiesele, and H.-H. Brintzinger, *J. Organomet. Chem.*, 365, 285 (1989).
23. I. T. Horváth, R. V. Kastrup, A. A. Oswald, and E. J. Mozeleski, *Catal. Lett.*, 2, 85 (1989).
24. A. J. Kunin, M. D. Noirot, and W. L. Gladfelter, *J. Am. Chem. Soc.*, 111, 2739 (1989).
25. F. M. Hoffmann, *J. Chem. Phys.*, 90, 2816 (1989).
26. H. Arakawa, K. Takeuchi, T. Matsuzaki, and Y. Sugi, *J. Jpn. Petroleum Inst.*, 31, 335 (1988).
27. W. S. Hammack, D. N. Hendrickson, and H. G. Drickamer, *J. Phys. Chem.*, 93, 3483 (1989).
28. W. S. Hammack, H. G. Drickamer, and D. N. Hendrickson, *Chem. Phys. Lett.*, 151, 469 (1988).
29. R. T. Roginski, J. R. Shapley, and H. G. Drickamer, *J. Phys. Chem.*, 92, 4316 (1988).
30. R. T. Roginski, A. Moroz, D. N. Hendrickson, and H. G. Drickamer, *J. Phys. Chem.*, 92, 4319 (1988).
31. J. K. McCusker, M. Zvagulis, H. G. Drickamer, and D. N. Hendrickson, *Inorg. Chem.*, 28, 1380 (1989).
32. W. S. Hammack, A. J. Conti, D. N. Hendrickson, and H. G. Drickamer, *J. Am. Chem. Soc.*, 111, 1738 (1989).
33. M. A. Stroud, H. G. Drickamer, M. H. Zietlow, H. B. Gray, and B. I. Swanson, *J. Am. Chem. Soc.*, 111, 66 (1989).
34. T. Hiraga, T. Uchido, N. Kitamura, H.-B. Kim, S. Tazuke, and T. Yagi, *J. Am. Chem. Soc.*, 111, 7466 (1989).
35. K. L. Bray and H. G. Drickamer, *J. Phys. Chem.*, 93, 7604 (1989).
36. K. L. Bray, H. G. Drickamer, E. A. Schmitt, and D. N. Hendrickson, *J. Am. Chem. Soc.*, 111, 2849 (1989).
37. H. G. Drickamer and K. L. Bray, *Int. Rev. Phys. Chem.*, 8, 41 (1989).
38. A. Lechner and G. Gliemann, *J. Am. Chem. Soc.*, 111, 7469 (1989).
39. R. bin Ali, P. Banerjee, J. Burgess, and A. Smith, *Trans. Met. Chem.*, 13, 107 (1988).
40. J. L. Coffer, J. R. Shapley, and H. G. Drickamer, *Polyhedron*, 8, 801 (1989).
41. S. Endo, T. Chino, S. Tsuboi, and K. Koto, *Nature*, 340, 452 (1989).
42. N. A. Lewis, Y. S. Obeng, D. V. Traveras, and R. van Eldik, *J. Am. Chem. Soc.*, 111, 924 (1989).
43. L. Heremans and K. Heremans, *J. Mol. Struct.*, 214, 305 (1989).
44. M. Iqbal and R. E. Verrall, *J. Biol. Chem.*, 263, 4159 (1988).
45. R. G. Alden, J. D. Satterlee, J. Mintorovitch, I. Constantinidis, M. R. Ondrias, and B. I. Swanson, *J. Biol. Chem.*, 264, 1933 (1989).
46. P. T. T. Wong and K. Heremans, *Biochim. Biophys. Acta*, 956, 1 (1988).
47. D.-G. Park, S. S. Nam, K. Kim, and H. Kim, *Biochim. Biophys. Acta*, 973, 19 (1989).
48. R. F. Tilton, Jr. and G. A. Petsko, *Biochem.*, 27, 6574 (1988).
49. C. Balny and A. B. Hooper, *Eur. J. Biochem.*, 176, 273 (1988).

50. J. F. Miller, C. M. Nelson, J. M. Ludlow, N. N. Shah, and D. S. Clark, *Biotechnol. Bioeng.*, **34**, 1015 (1989).

51. J. F. Miller, N. N. Shah, C. M. Nelson, J. M. Ludlow, and D. S. Clark, *Appl. Environ. Microbiol.*, **54**, 3039 (1988).

52. C. Balny and F. Travers, *Biophys. Chem.*, **33**, 237 (1989).

53. S. Kunugi, K. Tanabe, K. Yamashita, Y. Morikawa, T. Ito, T. Kondoh, K. Hirata, and A. Nomura, *Bull. Chem. Soc. Jpn.*, **62**, 514 (1989).

54. S. Makimoto and Y. Tanaguchi, *Tanpakushitsu Kakusan Koso*, **34**, 105 (1989).

55. S. Kunugi, *Tanpakushitsu Kakusan Koso*, **34**, 113 (1989).

56. M. Ishii, S. Funahashi, and M. Tanaka, *Inorg. Chem.*, **27**, 3192 (1988).

57. S. M. Bradley, H. Doine, H. R. Krouse, M. J. Sisley, and T. W. Swaddle, *Aust. J. Chem.*, **41**, 1323 (1988).

58. L. Fielding and P. Moore, *J. Chem. Soc., Dalton Trans.*, 873 (1989).

59. M. Ishii, S. Funahashi, K. Ishihara, and M. Tanaka, *Bull. Chem. Soc. Jpn.*, **62**, 1852 (1989).

60. L. Helm, A. E. Merbach, M. Kotowski, and R. van Eldik, *High Pressure Research*, **2**, 49 (1989).

61. J. Berger, M. Kotowski, R. van Eldik, U. Frey, L. Helm, and A. E. Merbach, *Inorg. Chem.*, **28**, 3759 (1989).

62. C. Cossy, L. Helm, and A. E. Merbach, *Inorg. Chem.*, **28**, 2699 (1989).

63. Y. Ducommun, L. Helm, A. E. Merbach, B. Hellquist, and L. I. Elding, *Inorg. Chem.*, **28**, 377 (1989).

64. U. Frey, L. Helm, A. E. Merbach, and R. Romeo, *J. Am. Chem. Soc.*, **111**, 8161 (1989).

65. N. J. Curtis, G. A. Lawrance, and R. van Eldik, *Inorg. Chem.*, **28**, 329 (1989).

66. P. Guardado, G. A. Lawrance, and R. van Eldik, *Inorg. Chem.*, **28**, 976 (1989).

67. M. Inamo, T. Sumi, N. Nakagawa, S. Funahashi, and M. Tanaka, *Inorg. Chem.*, **28**, 2688 (1989).

68. J. J. Chung, H. T. Kim, and S. O. Bek, *J. Korean Chem. Soc.*, **33**, 164 (1989).

69. H. H. Awad, C. B. Dobson, G. R. Dobson, J. G. Leipoldt, K. Schneider, R. van Eldik, and H. E. Wood, *Inorg. Chem.*, **28**, 1654 (1989).

70. S. Suvachittanont and R. van Eldik, *Inorg. Chem.*, **28**, 3660 (1989).

71. R. bin Ali, J. Burgess, and P. Guardado, *Trans. Met. Chem.*, **13**, 126 (1989).

72. G. Schmidt, H. Paulus, R. van Eldik, and H. Elias, *Inorg. Chem.*, **27**, 3211 (1988).

73. H. Elias, G. Schmidt, H.-J. Küppers, M. Saher, K. Wieghardt, B. Nuber, and J. Weiss, *Inorg. Chem.*, **28**, 3021 (1989).

74. G. Stochel, R. van Eldik, E. Hejmo, and Z. Stasicka, *Inorg. Chem.*, **27**, 2767 (1988).

75. G. Stochel and R. van Eldik, *Inorg. Chim. Acta*, **155**, 95 (1989).

76. M. J. Blandamer, J. Burgess, H. J. Cowles, I. M. Horn, J. B. F. N. Engberts, S. A. Galema, and C. Hubbard, *J. Chem. Soc., Faraday Trans. 1*, **85**, 3733 (1989).

77. Y. Kitamura, T. Takamoto, and K. Kuroda, *Inorg. Chim. Acta*, **159**, 181 (1989).

78. Y. Kitamura, S. Taneda, and K. Kuroda, *Inorg. Chim. Acta*, **156**, 31 (1989).

79. A. D. Fowless, G. A. Lawrance, D. R. Stranks, T. R. Sullivan, and N. Vanderhoek, *Aust. J. Chem.*, **41**, 1263 (1988).

80. Y. Kitamura, G. A. Lawrence, and R. van Eldik, *Inorg. Chem.*, **28**, 333 (1989).

81. Y. C. Park and Y. J. Cho, *Bull. Korean Chem. Soc.*, **9**, 1 (1988).

82. J. J. Chung and S. O. Bek, *J. Korean Chem. Soc.*, **32**, 318 (1988).

83. H. C. Bajaj and R. van Eldik, *Inorg. Chem.*, **27**, 4052 (1988).

84. H. C. Bajaj and R. van Eldik, *Inorg. Chem.*, **28**, 1980 (1989).

85. J. Anhaus, H. C. Bajaj, R. van Eldik, L. R. Nevinger, and J. B. Keister, *Organometallics*, **8**, 2903 (1989).

86. J. J. Pienaar, M. Kotowski, and R. van Eldik, *Inorg. Chem.*, **28**, 373 (1989).

87. M. Kotowski, S. Begum, J. G. Leipoldt, and R. van Eldik, *Inorg. Chem.*, **27**, 4472 (1988).

88. J. Ioset, L. Helm, A. Merbach, R. Roulet, F. Grepioni, and D. Braga, *Helv. Chim. Acta*, **71**, 1458 (1989).

89. D. W. Krassowski, J. H. Nelson, K. R. Brower, D. Hauenstein, and R. A. Jacobson, *Inorg. Chem.*, **27**, 4294 (1988).

90. R. T. Roginski, T. L. Carroll, A. Moroz, B. R. Whittlesey, J. R. Shapley, and H. G. Drickamer, *Inorg. Chem.*, **27**, 3701 (1988).
91. M. Stebler, R. M. Nielson, W. F. Siems, J. P. Hunt, H. W. Dodgen, and S. Wherland, *Inorg. Chem.*, **27**, 2893 (1988).
92. H. Doine and T. W. Swaddle, *Can. J. Chem.*, **66**, 2763 (1988).
93. K. Kirchner, S.-Q. Dang, M. Stebler, H. W. Dodgen, S. Wherland, and J. P. Hunt, *Inorg. Chem.*, **28**, 3604 (1989).
94. N. J. Blundell, J. Burgess, and C. D. Hubbard, *Inorg. Chim. Acta*, **155**, 165 (1989).
95. I. Krack and R. van Eldik, *Inorg. Chem.*, **28**, 851 (1989).
96. Y. C. Park and S. S. Kim, *J. Korean Chem. Soc.*, **33**, 273 (1989).
97. H. Doine, Y. Yano, and T. W. Swaddle, *Inorg. Chem.*, **28**, 2319 (1989).
98. K. Kirchner, H. W. Dodgen, S. Wherland and J. P. Hunt, *Inorg. Chem.*, **28**, 604 (1989).
98. L. M. Rowe, L. B. Tran, and G. Atkinson, *J. Solution Chem.*, **18**, 675 (1989).
100. P.-Y. Sauvageat, Y. Ducommun, and A. E. Merbach, *Helv. Chim. Acta*, **72**, 1801 (1989).
101. S. Funahashi, K. Ishihara, M. Inamo, and M. Tanaka, *Inorg. Chim. Acta*, **157**, 65 (1989).
102. J. Burgess and C. Hubbard, *Inorg. Chem.*, **27**, 2548 (1988).
103. Y. Ducommun, P. J. Nichols, and A. E. Merbach, *Inorg. Chem.*, **28**, 2643 (1989).
104. T. Amari, S. Funahashi, and M. Tanaka, *Inorg. Chem.*, **27**, 3368 (1988).
105. G. Laurenczy, Y. Ducommun, and A. E. Merbach, *Inorg. Chem.*, **28**, 3024 (1989).
106. B. Hellquist, L. I. Elding, and Y. Ducommun, *Inorg. Chem.*, **27**, 3620 (1988).
107. Y. Ducommun, L. Helm, G. Laurenczy, and A. E. Merbach, *Inorg. Chim. Acta*, **158**, 3 (1989).
108. G. Laurenczy and A. E. Merbach, *Helv. Chim. Acta*, **71**, 1971 (1988).
109. S. Wieland, R. van Eldik, D. R. Crane, and P. C. Ford, *Inorg. Chem.*, **28**, 3663 (1989).
110. S. Wieland and R. van Eldik, *J. Chem. Soc., Chem. Commun.*, 367 (1989).
111. M. S. Herman and J. L. Goodman, *J. Am. Chem. Soc.*, **111**, 9105 (1989).
112. D. R. Crane, J. DiBenedetto, C. E. A. Palmer, D. R. McMillin, and P. C. Ford, *Inorg. Chem.*, **27**, 3698 (1988).
113. T. Hiraga, N. Kitamura, H.-B. Kim, S. Tazuke, and N. Mori, *J. Phys. Chem.*, **93**, 2940 (1989).
114. G. Stochel, R. van Eldik, H. Kunkely, and A. Vogler, *Inorg. Chem.*, **28**, 4314 (1989).
115. S. Adachi and I. Morishima, *J. Biol. Chem.*, **264**, 18896 (1989).
116. J. A. Kornblatt, G. H. bon Hoa, and K. Heremans, *Biochemistry*, **27**, 5122 (1988).
117. J. A. Kornblatt, G. H. bon Hoa, L. Eltis, and A. G. Mauk, *J. Am. Chem. Soc.*, **110**, 5909 (1988).
118. S. F. Scarlata, T. Ropp, and C. A. Royer, *Biochemistry*, **28**, 6637 (1989).
119. K. Ruan and G. Weber, *Biochemistry*, **27**, 3295 (1988).
120. Y. Taniguchi and S. Makimoto, *J. Mol. Catal.*, **47**, 323 (1988).
121. M. Nishikawa, K. Itoh, and R. Holroyd, *J. Phys. Chem.*, **92**, 5262 (1988).
122. T. C. Franklin and S. A. Mathew, *J. Electrochem. Soc.*, **136**, 3627 (1989).

Index